ANDERSONIAN LIBRARY
WITHDRAWN
FROM
LIBRARY
STOCK
UNIVERSITY OF STRATHCLYDE

Surface Phenomena in Enhanced Oil Recovery

Surface Phenomena in Enhanced Oil Recovery

Edited by

Dinesh O. Shah

University of Florida
Gainesville, Florida

PLENUM PRESS · NEW YORK AND LONDON

Library of Congress Cataloging in Publication Data

Main entry under title:

Surface phenomena in enhanced oil recovery.

"Papers presented at the Symposium on Surface Phenomena in Enhanced Oil Recovery held at Stockholm, Sweden, during August 20-25, 1979" – Pref.

Bibliography: p.

Includes index.

1. Secondary recovery of oil – Congresses. 2. Surface active agents – Congresses. 3. Surface chemistry – Congresses. 4. Oil field flooding – Congresses. I. Shah, D. O. (Dinesh Ochhavlal), 1938– . II. Symposium on Surface Phenomena in Enhanced Oil Recovery (1979: Stockholm, Sweden)

TN871.S7678 622'.3382 81-8704

ISBN 0-306-40757-4 AACR2

Proceedings of a symposium on Surface Phenomena in Enhanced Oil Recovery, organized as part of the Third International Conference on Surface and Colloid Science, and held August 20–25, 1979, in Stockholm, Sweden

A Division of Plenum Publishing Corporation
233 Spring Street, New York, N.Y. 10013

Printed in the United States of America

PREFACE

It is with great pleasure and satisfaction that I present to the international scientific community this collection of papers presented at the symposium on Surface Phenomena in Enhanced Oil Recovery held at Stockholm, Sweden, during August 20-25, 1979. It has been an exciting and exhausting experience to edit the papers included in this volume.

The proceedings cover six major areas of research related to chemical flooding processes for enhanced oil recovery, namely, 1) Fundamental aspects of the oil displacement process, 2) Microstructure of surfactant systems, 3) Emulsion rheology and oil displacement mechanisms, 4) Wettability and oil displacement mechanisms, 5) Adsorption, clays and chemical loss mechanisms, and 6) Polymer rheology and surfactant-polymer interactions. This book also includes two invited review papers, namely, "Research on Enhanced Oil Recovery: Past, Present and Future," and "Formation and Properties of Micelles and Microemulsions" by Professor J. J. Taber and Professor H. F. Eicke respectively.

This symposium volume reflects the current state-of-art and our understanding of various surface phenomena in enhanced oil recovery processes. The participation by researchers from various countries in this symposium reflects the global interest in this area of research and the international effort to develop the science and technology of enhanced oil recovery processes.

It is difficult to summarize all the phenomena discussed in this volume. However, major topics include ultralow interfacial tension, phase behavior, microstructure of surfactant systems, optimal salinity concept, middle-phase microemulsions, interfacial rheology, flow of emulsions in porous media, wettability of rocks, rock-fluid interactions, surfactant loss mechanisms, precipitation and redissolution of surfactants, coalescence of drops in emulsions and in porous media, surfactant mass transfer across interfaces, equilibrium vs. dynamic properties of surfactant/oil/brine systems, mechanisms of oil displacement in porous media, ion-exchange in clays, polymer rheology and surfactant-polymer interactions.

The volume includes discussions of various phenomena at the molecular, microscopic and macroscopic levels. I hope that this book will serve its intended objective of reflecting our current understanding of surface phenomena in enhanced oil recovery processes. It is also my earnest hope that this volume will be useful to researchers in this area as a valuable reference source.

Dinesh O. Shah
Professor of Chemical Engineering,
Anesthesiology and Biophysics
University of Florida, Gainesville,
Florida 32611, U.S.A.

ACKNOWLEDGMENTS

The Symposium on Surface Phenomena in Enhanced Oil Recovery was held at Stockholm, Sweden, during August 20-25, 1979. The Symposium was organized within the framework of Third International Conference on Surface and Colloid Science. Plenary and main lectures presented at the Conference have been published in the May 1980 issue of the official IUPAC journal Pure and Applied Chemistry, copies of which can be obtained from Pergamon Press, Oxford, UK. Republication of the lectures of J. J. Taber and H. F. Eicke in this volume has been carried out with the permission of IUPAC.

I wish to convey my sincere thanks and appreciation to the chairmen of the Conference Scientific Committee, Professor Stig Friberg and Professor Per Stenius, for their support of the symposium and Professors J. J. Taber and H. F. Eicke for their permission to republish their invited lectures in this symposium volume. I also wish to thank the following session chairmen for their assistance; Professor T. Fort, Jr., Carnegie-Mellon University, Professor D. T. Wasan, Illinois Institute of Technology, Professor S. S. Marsden, Stanford University, Mr. L. W. Holm, Union Oil Company, and Professor C. L. McCormick, University of Southern Mississippi. Special thanks are extended to Plenum Publishing Company for the prompt publication of the symposium proceedings.

I am grateful to my colleagues, post-doctoral associates and students for their assistance throughout this project. Specifically, I wish to thank Dr. Sunder Ram and Mrs. Anita S. Ram for their editorial assistance and Mrs. Jeanne Ojeda for the excellent typing of the manuscripts. I also wish to convey my sincere thanks and appreciation to all the authors and co-authors of the papers, without whose assistance and cooperation this symposium proceedings would not have been completed.

I wish to convey my sincere thanks and appreciation to my colleagues in the Chemical Engineering and Petroleum Engineering Departments at Stanford University for their cooperation and assistance in this endeavor, where I spent the entire year of 1979 as a visiting professor of Chemical Engineering, Petroleum Engineering and the Institute for Energy Studies. I am very grateful to Professor John C. Biery, chairman of the Chemical Engineering Department and Professor J. H. Modell, chairman of the Anesthesiology Department at the University of Florida, for their kind assistance and encouragement. Finally, I wish to convey my sincere thanks and appreciation to my secretary, Mrs. Derbra Scott, for her assistance in coordinating my activities while I was at Stanford.

Dinesh O. Shah

CONTENTS

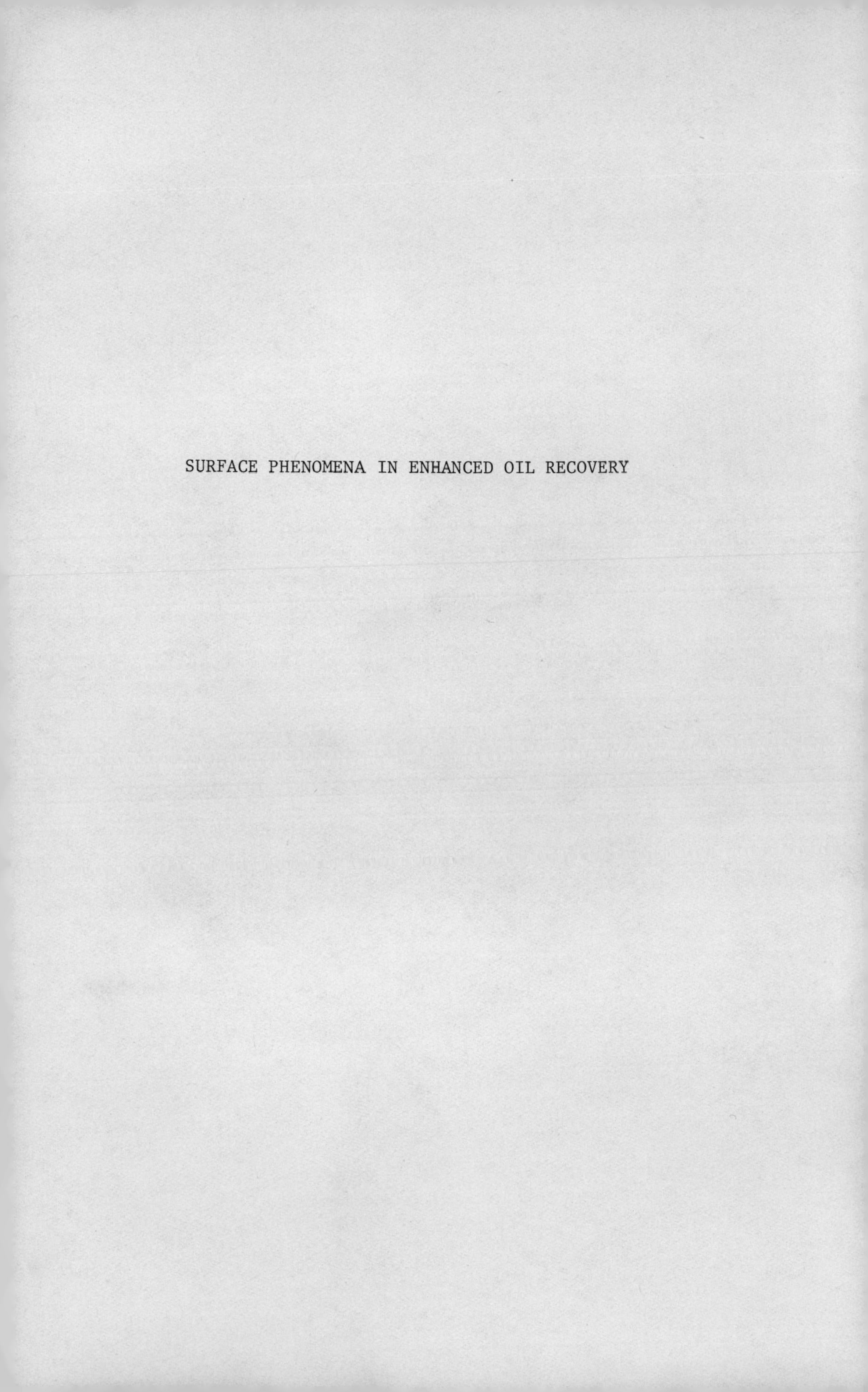

SURFACE PHENOMENA IN ENHANCED OIL RECOVERY

INTRODUCTION

D. O. Shah

Departments of Chemical Engineering and Anesthesiology
University of Florida
Gainesville, Florida 32611, U.S.A.

In view of the world-wide shortage of petroleum and the fact that almost 65% to 70% of original oil-in-place is left in the reservoirs at the end of waterflooding, the importance of tertiary oil recovery methods to produce additional oil can hardly be overemphasized. The enhanced oil recovery methods can be divided in two major groups, namely, thermal processes and chemical flooding processes. The in situ combustion, steam injection, wet combustion, etc., fall in the first category whereas the caustic flooding, surfactant flooding, micellar-polymer flooding, CO_2 flooding, etc., fall in the second category of processes.

At the end of the waterflooding, the residual oil is believed to be in the form of discontinuous oil ganglia trapped in the pores of rocks in the reservoir. The two major forces acting on an oil ganglion are the viscous forces and the capillary forces. In an oil recovery process, the microscopic displacement efficiency is determined by the ratio of these forces. Melrose and Brandner (1) suggested a parameter, the capillary number which is the ratio of the viscous to capillary forces (Figure 1). It is generally recognized that at the end of the waterflooding, the capillary number is around 10^{-6}. In order to recover additional oil by a chemical flooding process, the capillary number has to be increased to around 10^{-3} to 10^{-2}. Under practical reservoir conditions, this change in capillary number of 3 to 4 orders of magnitude can be achieved by decreasing interfacial tension at the oil/brine interface. Generally, the crude oil/brine interfacial tension can be in the range of 20 to 30 dynes/cm. Using an appropriate surfactant system, this interfacial tension can be reduced to 10^{-3} or 10^{-4} dynes/cm under given conditions. In

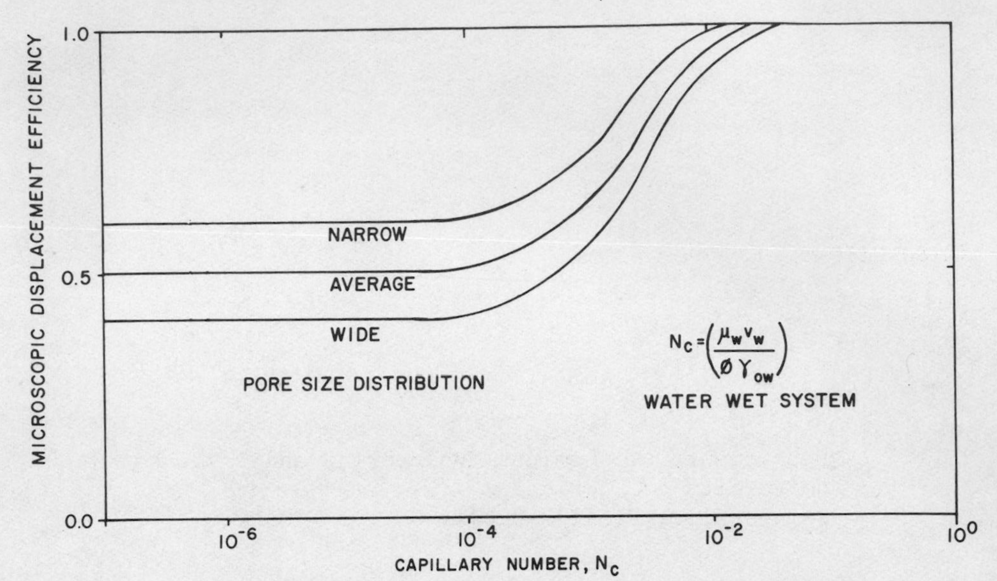

Fig. 1. The effect of capillary number, N_c, on the microscopic oil displacement efficiency in porous media of various size distribution (Ref. 1).

other words, if one considers interfacial tension at the oil ganglia/brine interface as the major factor resisting the deformation of the ganglion during the passage through narrow channels between sand particles (or pore necks), the reduction in interfacial tension decreases this resistance to deformation.

Figure 2 shows a schematic three-dimensional view of a petroleum reservoir. For example, one can use an inverse five-spot pattern to inject a surfactant solution in the reservoir. As shown at the bottom of Figure 2, the reduction in interfacial tension leads to mobilization of oil ganglia which produces an oil bank. This oil bank is propagated to the production wells by the use of a polymer buffer solution followed by the drive water. Since surfactant and polymer slugs are the major components of a surfactant/polymer flooding process, a brief introduction is given regarding the structural aspects of these solutions.

Figure 3 shows the molecular aggregation process in an aqueous surfactant solution (2,3). At very low surfactant concentrations, the surfactant remains as monomers in the solution in equilibrium with adsorbed film of surfactant molecules (Figure 3B). As the surfactant concentration is increased, the surfactant molecules begin to form aggregates or micelles in a very narrow range of concentration called the critical micelle concentration (cmc) (Figure 3A). Further increase in surfactant concentration results in the formation of more micelles with monomer concentration remaining constant. The interior of micelles provides a nonpolar

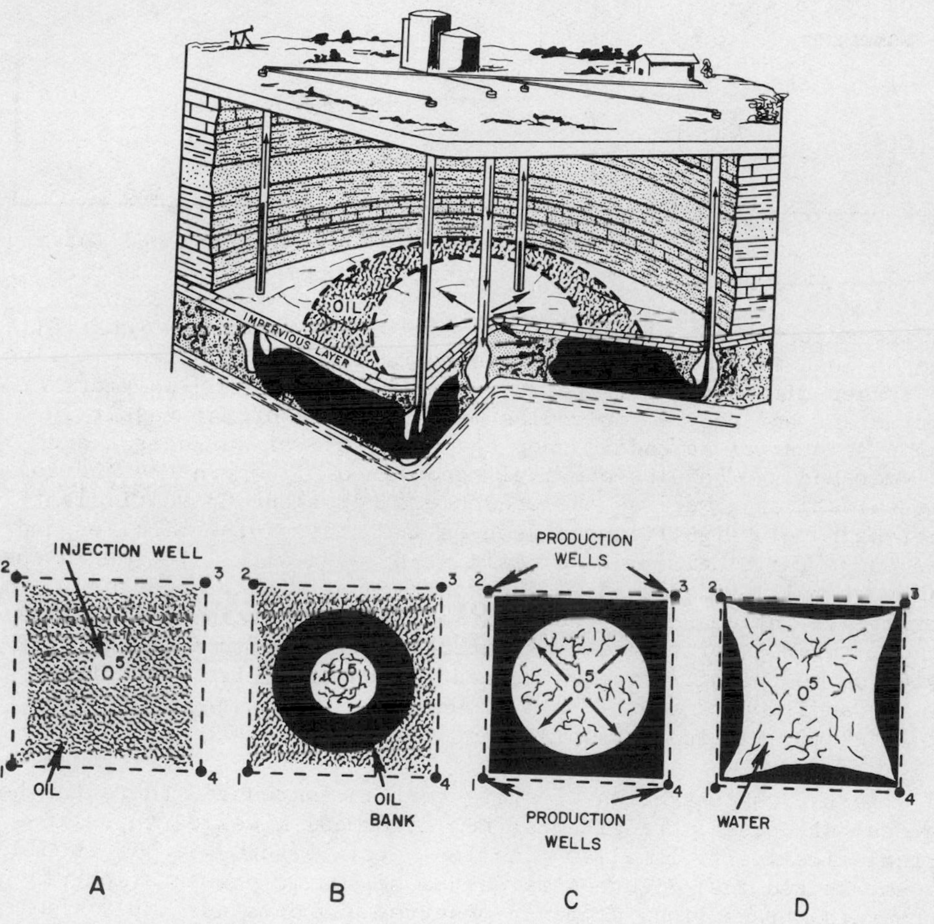

Fig. 2. Displacement of oil in petroleum reservoirs by water or chemical flooding (five-spot pattern).

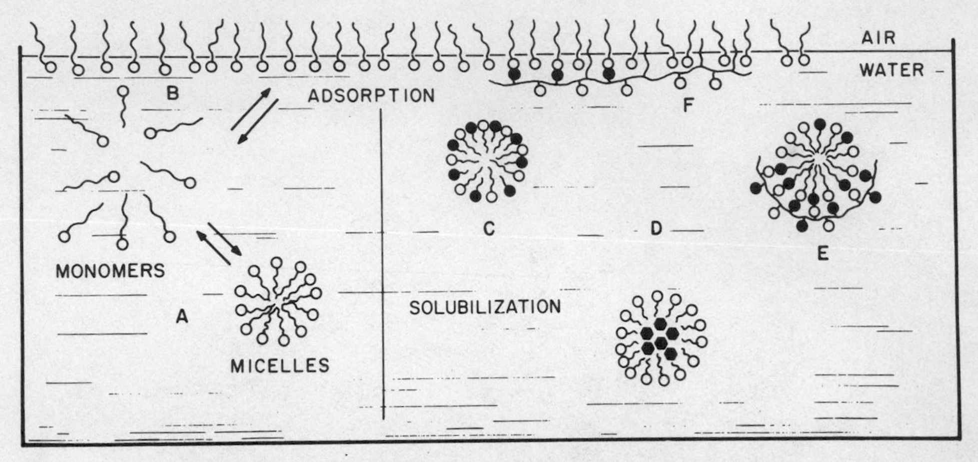

Fig. 3. Adsorption, micelle formation, solubilization and interactions at the micelle surface.

environment for solubilization of oil soluble species. Thus, oil can be solubilized within micelles resulting into swollen micelles of larger dimensions (Figure 3D). In the presence of polymers, surfactant and polymer molecules may associate or may result in phase separation depending upon the structure of surfactant and polymer and the physico-chemical conditions. Temperature and salt concentration as well as the presence of divalent or multivalent cations have a significant effect on the size of the micelles and the cmc. The micelles are dynamic structures with lifetime in the range of milli-seconds (4). The kinetics of micelles can be strikingly influenced by the addition of short chain alcohols (5). Hence, the rate of transfer of surfactant molecules from the bulk solution to the oil/brine interface can be significantly influenced by the addition of short chain alcohols (6), presumably due to kinetics of the micellar solutions.

As the concentration of surfactant is increased, there can be several structural transitions from spherical micelles to cylindrical micelles to lamellar micelles, etc. (Figure 4). It should be emphasized that Figure 4 is only a schematic presentation of structural transitions commonly observed and does not imply that every surfactant undergoes these structural transitions. However, in general, many surfactants have shown similar structural transitions depending upon the surfactant structure, concentration and physico-chemical conditions such as temperature, pH and the presence of electrolytes. It should be emphasized that the monomer concentration and the driving force for surfactant molecules to adsorb at the interface will change with these structural transitions. Therefore, the microstructure of surfactant solution is important in relation to the magnitude of interfacial tension, the extent of solubilization, rheological properties, adsorption and

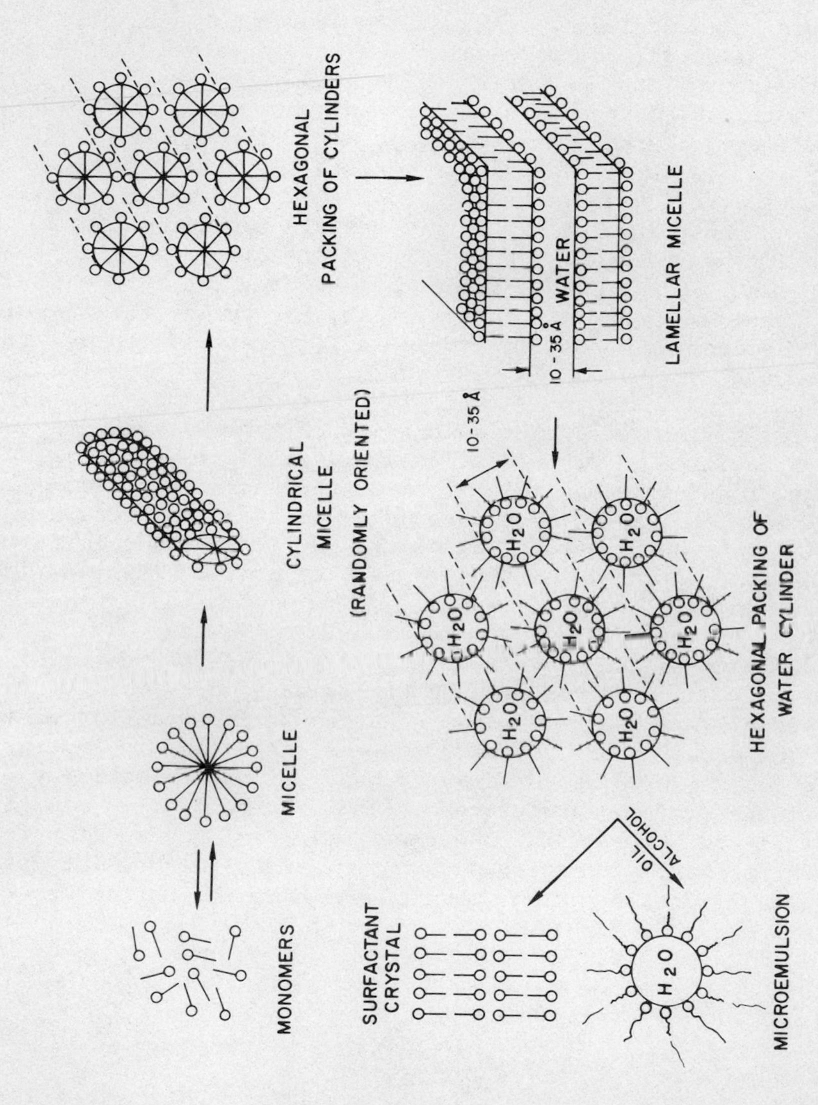

Fig. 4. Structure formation in surfactant solution.

other chemical loss mechanisms. The stability of these structures is also important in relation to the rate of mass transfer from bulk solution to the oil/brine interface.

The Mechanism of Displacement of Oil Ganglia in Porous Media

Figure 5 schematically illustrates the role of ultralow interfacial tension for the movement of an entrapped oil ganglion through the narrow neck of pore. As mentioned previously, the ultralow interfacial tension minimizes the work necessary to deform the interface of the oil ganglion as it moves through the narrow neck of pores. With the use of appropriate surfactant systems, a large number of oil ganglia can be mobilized. The subsequent coalescence of these oil ganglia is a necessary condition to form an oil bank. It has been known that high interfacial viscosity prevents whereas low interfacial viscosity promotes the coalescence of oil drops. As shown in Figure 6, thus, a very low interfacial viscosity is desirable for coalescence of oil ganglia to form an oil/brine bank.

Figure 7 illustrates the propagation of the oil bank and its subsequent coalescence with additional oil ganglia. If ultralow interfacial tension is not maintained at the surfactant slug/oil bank interface, considerable oil would be lost due to entrapment process (6). Hence, an efficient oil recovery process requires the presence of ultralow interfacial tension at the trailing edge of the oil bank.

In addition to the factors mentioned earlier (e.g., ultralow interfacial tension and coalescence of oil ganglia), mobility control of the oil bank and surfactant slug is an important requirement for a successful oil recovery process. As shown in Figure 8, polymer solutions are used as drive fluids for proper mobility control of the oil recovery process. The dispersion of surfactant, polymer, oil and brine during the flow could lead to emulsion formation and/or phase-separation due to surfactant-polymer incompatibility (7). Efforts should be made to minimize the formation of

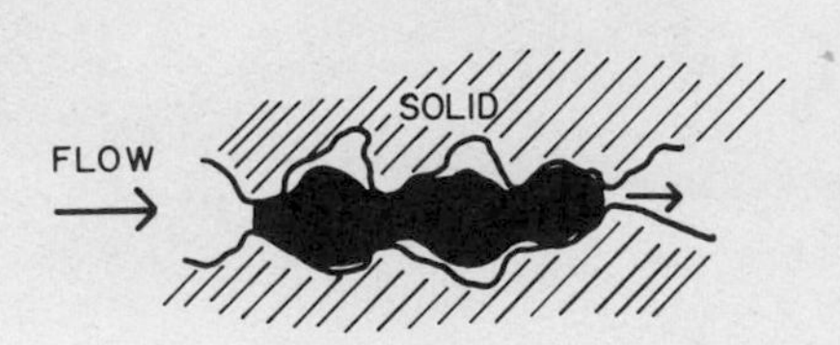

Fig. 5. The effect of interfacial tension on the movement of oil ganglia through narrow neck of pores. For the movement of the oil ganglia a very low oil/water interfacial tension is desirable ≈ 0.001 dynes/cm.

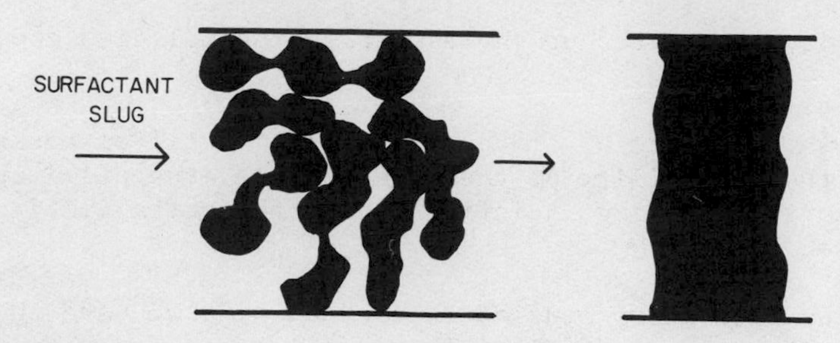

Fig. 6. Displaced oil ganglia must coalesce to form a continuous oil bank.* For this a very low interfacial viscosity is desirable.

Fig. 7. An efficient propagation of an oil bank* through porous media requires ultralow interfacial tension at the trailing edge of the oil bank.

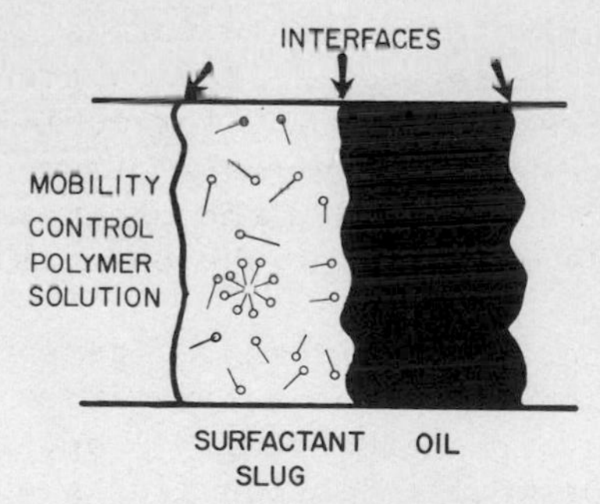

Fig. 8. The flow of oil bank,* surfactant slug and the polymer solution through porous media results in the formation of emulsions and mixed surfactant plus polymer region. The formation of stable emulsions and high viscosity surfactant plus polymer structures should be minimized for efficient oil recovery.

---

*It should be emphasized that the oil bank does include some brine with it.

high viscosity structures or phases during the oil displacement process.

As shown in Figure 9, the contact angle of oil on rock surface depends on the wettability of the rock. Hence, the oil displacement by a surfactant slug is influenced by the wettability of the rock surface.

From the laboratory studies on a crude oil, we have shown (8) that the surface charge density of an oil ganglion can strikingly influence its displacement efficiency. High surface charge density leads to low interfacial tension, low interfacial viscosity and strong electrical repulsion between oil droplets and sand particles (Figure 10).

An oil/brine/surfactant/alcohol system often forms a middle phase microemulsion in an appropriate salinity range. The salinity at which the middle phase microemulsion contains an equal volume of oil and brine is defined as the optimal salinity (9). At the optimal salinity, the interfacial tension is in the millidynes/cm range at both oil/microemulsion and microemulsion/brine interfaces, and the oil recovery is maximum (6,9). Moreover, we have shown (10) that at optimal salinity, the coalescence time or phase-separation time is minimum for oil/brine/surfactant/alcohol systems. When these systems are pumped through porous media, a minimum pressure drop or apparent viscosity is observed at the optimal salinity (10). All these phenomena occurring at optimal salinity are summarized in Figure 11. In a recent study, we have also found that the surfactant loss in porous media is minimum at the optimal salinity. Therefore, besides ultralow interfacial tension, a favorable coalescence process for mobilized oil ganglia and the minimum apparent viscosity (or minimum ΔP) of the oil bank and the minimum surfactant loss are the other factors contributing towards the maximum oil recovery at the optimal salinity.

Figure 12 shows three major areas of research for enhanced oil recovery, namely, chemistry of reservoir components and injection fluids, and reservoir engineering. A better understanding of complex interactions between injection fluids and reservoir components (e.g., oil, brine, rocks, clays and minerals) is highly desirable for a successful application of various oil recovery processes. The economic feasibility and optimization are the two important criteria which must be considered for implementation of various oil recovery processes.

This symposium includes two overview invited lectures by Professor J. J. Taber on "Enhanced Oil Recovery: Past, Present and Future" and Professor H. F. Eicke on "Aggregation in Surfactant Solutions: Formation and Properties of Micelles and Micro-

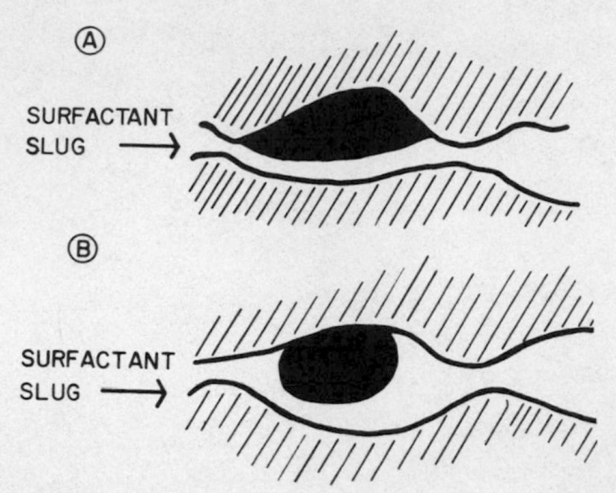

Fig. 9. The effect of wettability of rock surface on the contact angle of an oil drop in porous media. (A) represents oil-wet rock surface whereas (B) represents water-wet rock surface.

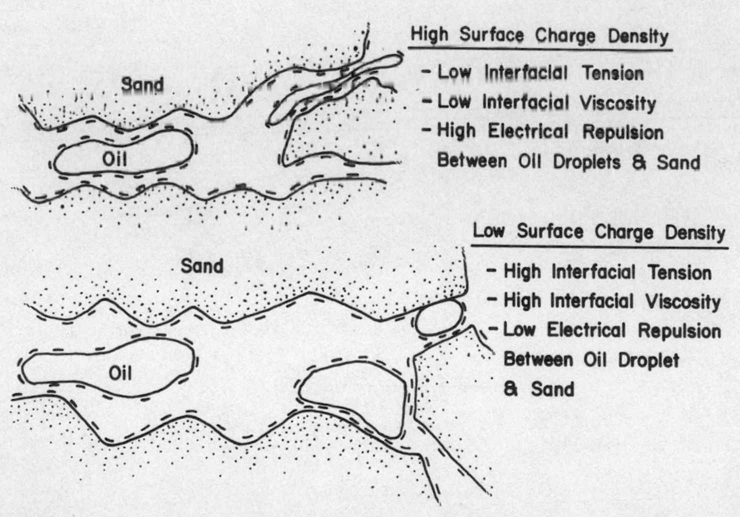

Fig. 10. The effect of surface charge density of oil/brine interface on the oil displacement process (Ref. 8).

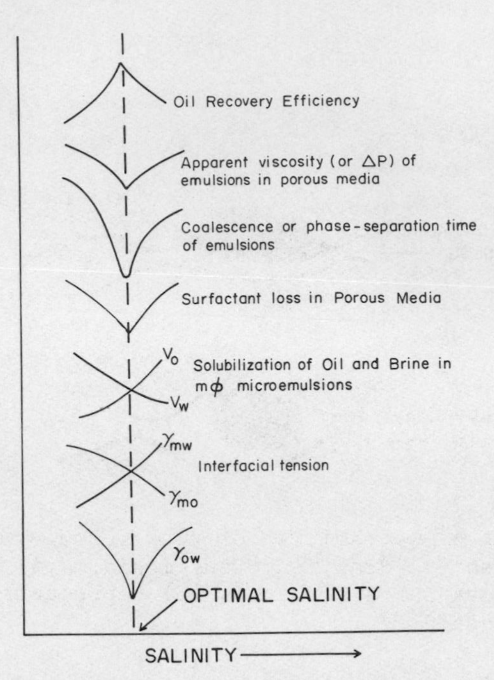

Fig. 11. Various phenomena occurring at the optimal salinity in oil/brine/surfactant/alcohol systems in relation to oil displacement efficiency.

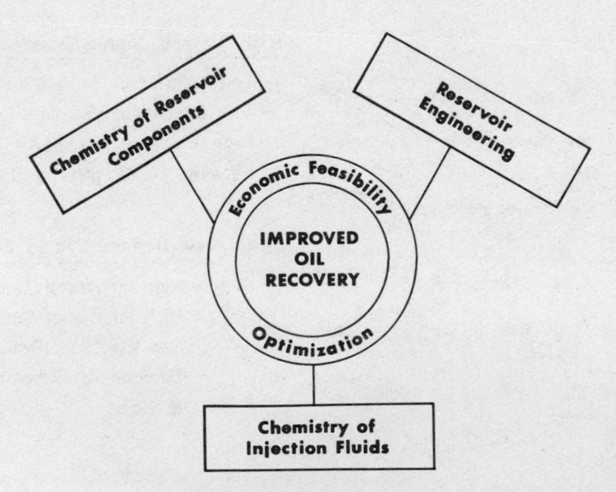

Fig. 12. Three major areas of research for enhanced oil recovery.

emulsions." The six sessions of this symposium focus on the following aspects of oil displacement: 1) Fundamental aspects of oil displacement process, 2) Microstructure of surfactant systems, 3) Emulsion rheology and oil displacement mechanisms, 4) Wettability and oil displacement mechanisms, 5) Adsorption, clays, and chemical loss mechanisms, and 6) Polymer rheology and surfactant-polymer interactions. It is hoped that the papers of this symposium will reflect the current state of art in oil recovery processes and will highlight the problems still remaining to be solved.

ACKNOWLEDGEMENTS

The author wishes to express his sincere thanks and appreciation to the National Science Foundation-RANN, ERDA and the Department of Energy (Grant No. DE-AC1979BC10075) and the consortium of the following Industrial Associates for their generous support of the University of Florida Enhanced Oil Recovery Research Program during the past seven years: 1) Alberta Research Council, Canada, 2) American Cyanamid Co., 3) Amoco Production Co., 4) Atlantic Richfield Co., 5) BASF-Wyandotte Co., 6) British Petroleum Co., England, 7) Calgon Corp., 8) Cities Service Oil Co., 9) Continental Oil Co., 10) Ethyl Corp., 11) Exxon Production Research Co., 12) Getty Oil Co., 13) Gulf Research and Development Co., 14) Marathon Oil Co., 15) Mobil Research and Development Co., 16) Nalco Chemical Co., 17) Phillips Petroleum Co., 18) Shell Development Co., 19) Standard Oil of Ohio, 20) Stepan Chemical Co., 21) Sun Oil Chemical Co., 22) Texaco Inc., 23) Union Carbide Corp., 24) Union Oil Co., 25) Westvaco Inc., and 26) Witco Chemical Co.

REFERENCES

1. J. C. Melrose and C. F. Brandner, J. Canadian Petr. Tech., 58, Oct.-Dec., 1974.
2. D. O. Shah, Chem. Eng. Edn., 11, 14 (1977).
3. V. K. Bansal and D. O. Shah, in "Micellization, Solubilization, and Microemulsions," K. L. Mittal, ed., Vol. 1, Plenum Press, New York, p. 87, 1977.
4. H. Hoffmann, H. Nüsslein, and W. Ulbricht, in "Micellization, Solubilization, and Microemulsions," K. L. Mittal, ed., Vol. 1, Plenum Press, New York, p. 263, 1977.
5. R. Zana, in "Surface Phenomena in Enhanced Oil Recovery," D. O. Shah, ed., Plenum Press, New York (in press).
6. M. Y. Chiang and D. O. Shah, SPE 8988 presented at the Fifth International Symposium on Oilfield and Geothermal Chemistry held in Stanford, California, May 28-30, 1980.
7. S. P. Trushenski, in "Improved Oil Recovery by Surfactant and Polymer Flooding," D. O. Shah and R. S. Schechter, eds., Academic Press, Inc., New York, p. 555, 1977.
8. M. Y. Chiang, K. S. Chan, and D. O. Shah, J. Canadian Petr. Tech., 17, 1, Oct.-Dec., 1978.
9. R. L. Reed and R. N. Healy, in "Improved Oil Recovery by Surfactant and Polymer Flooding," D. O. Shah and R. S. Schechter, eds., Academic Press, Inc., New York, p. 383, 1977.

10. S. Vijayan, C. Ramachandran, H. Doshi, and D. O. Shah, in "Surface Phenomena in Enhanced Oil Recovery," D. O. Shah, ed., Plenum Press, New York (in press).

I: FUNDAMENTAL ASPECTS OF THE OIL DISPLACEMENT PROCESS

RESEARCH ON ENHANCED OIL RECOVERY: PAST, PRESENT AND FUTURE

J. J. Taber

New Mexico Petroleum Recovery Research Center
New Mexico Institute of Mining and Technology
Socorro, New Mexico 87801, U.S.A.

Past and present research on Enhanced Oil Recovery is reviewed with emphasis on the surface phenomena involved. The nature of capillary pressure phenomena in porous media has been understood for some time, and much research has been devoted towards the alteration of the surface forces which prevent the efficient displacement of oil by water. Early work often treated surface active agents as wetting agents designed to remove the oil from the solid surface by classical detergent action. More recent work has recognized the strong influence of oil-water interfacial tension on the displacement of discontinuous oil blobs or ganglia. Therefore, surfactant systems are now being developed to produce the lowest possible oil-water interfacial tensions by adjusting the various components and thus, the phase behavior in the total system. In addition to interfacial tension, the phase behavior itself can strongly influence the oil displacement. The surfactant work, current work in blob mechanics, current research in CO_2 flooding, and past results in alcohol flooding all indicate that an expanding oil phase is very important for effective oil displacement. Therefore, much current research is directed toward methods which utilize materials (including gases such as CO_2 and mobility control agents) to dislodge oil blobs, or to prevent their entrapment by maintaining a continuous oil phase and improving sweep efficiency during displacement. The general direction of future research on enhanced oil recovery is predicted.

INTRODUCTION

According to legend, the first successful attempts to force oil from underground rocks by the injection of another fluid occurred by accident. In the early oil production days in Pennsylvania, wells were often abandoned improperly, and surface water was allowed to enter the productive sand zone (1,2). Actually, the first deliberate waterflood for oil recovery may have occurred in Sweden prior to 1740 when "running water was used to produce crude oil from galleries cut into the rocks bearing strata of 'tar and sand'" (3). Early operators in America feared that water entering the productive zone would "drown" the oil wells and for years, many states had laws which prohibited the injection of water into oil-producing sands. However, the increased production from waterflooding was observed consistently and by 1940, it was considered to be "unquestionably the most efficient method ever devised for increasing oil recovery" (1).

In spite of the enormous effectiveness of waterflooding, the early engineers soon realized that waterflooding still bypassed some oil. A patent was granted in 1917 for the addition of alkali to the flooding water (4), and by 1925 (5), engineers were describing how the surface forces which were responsible for holding the oil in the rock might be altered for better oil recovery. This paper will examine some of these surface chemistry methods and comment on the "state of the art" in enhanced recovery today. Because this is a conference on surface chemistry, emphasis will be placed on methods that alter or eliminate the surface forces which exist between oil and water or between any of the fluids or the fluids and solids which are found in petroleum reservoirs. Recovery methods which depend primarily on heat will not be reviewed.

EARLY WORK ON DISTRIBUTION OF OIL, WATER AND GAS IN THE POROUS ROCKS

From the earliest times, man has recognized that the material called petroleum comes from rocks (3). Although the mechanics of the flow of petroleum, water and gas from these rocks are still being studied, the early petroleum engineers soon recognized that the oil must flow through very small passages between the sand grains, and that capillary forces must be involved wherever interfaces occur between two fluids. The earliest published work on the large resistance to flow caused by a series of bubbles in a capillary was that by Jamin (6) in 1859.

The early oil production experts were well aware of this effect. Indeed, Herold (7), in a comprehensive book dealing with mechanisms of oil production, bases his entire description of Paleozoic production (half of the book) on the fact that oil and gas are distributed uniformly as tiny droplets of oil and little bubbles of gas (he ignores water) which occur in sequence through all pore space within the rock. Just as Jamin (6) found that a series of gas bubbles in a perfectly smooth capillary tube could build up large resistances to flow, Herold (7) pointed out that one could visualize thousands of gas bubbles interspersed between the oil phase in either smooth or rough capillaries, and that large pressures could be built up even though the pressure across an individual gas bubble was very small. Although Jamin's original effect was observed only in smooth capillary tubes, writers now apply the name to all situations which involve resistance to flow whether it be from the hysteresis of the contact angle in smooth capillary tubes, the change in the radius of the capillary in a pore neck, or a change in interfacial tension at the forward and trailing edge of the bubble (8).

The distribution of the oil, gas and water in the porous medium was better understood when Botset and Wyckoff (9) carried out the first experiments on relative permeability. They showed that either oil or gas would flow only if a specific minimum saturation of the phase in question existed in the flow region of the porous material. Some of the early workers also recognized that either the oil or gas droplets could be discontinuous, and in this condition, would be hard to displace by flowing water because of the Jamin effect. In 1927, Uren and Fahmy (10) investigated a number of "factors" which affect the recovery of petroleum from unconsolidated sands by waterflooding. Table 1 lists these factors and the general results observed by Uren and Fahmy. With one exception (rate), the results observed by Uren and Fahmy are similar to generalizations which most experts in this field claim today after work of more than 50 years.

From Table 1, it is noted that waterflooding is more effective in coarse sand packs (therefore high permeability), in sands of high porosity, in sands with the most uniform textures, in sands which seem to have an intermediate wettability, with oils of lower rather than high viscosities, with low oil-water interfacial tensions, at high temperatures of the waterflood, and with flood waters which contain alkaline additives, especially those agents which react with sand to produce silicates. Most of Uren and Fahmy's conclusions have been borne out by the hundreds of comprehensive studies which have been conducted since that time. As one who has worked at different periods of his life in this field, it is almost embarrassing to examine Uren and Fahmy's work of 1927 and realize that we still have not answered some of the questions which Uren explored with his rather simple experiments. In any event, it is clear that Uren and other early workers recognized that the capillary forces are responsible for holding the oil in the rock because the capillaries are so small that the surface forces far exceed the viscous forces which are available from waterflooding gradients in the reservoir. Therefore, it was logical that they should turn to materials which could modify the oil-water interfacial tension or the oil-water-rock contact angle to reduce the magnitude of the surface forces which hold the oil back during a displacement process. These attempts to change the surface forces and permit water to release or displace oil more

Table 1. Factors influencing recovery of petroleum (after Uren and Fahmy, 1927, Ref. 10)

Factor (Condition)	Condition for maximum recovery	Actual Value of "Variable" giving best recovery	Maximum recovery of oil at "Best Condition" % O.I.P.
1. Grain size	Largest grain size	20 to 40 mesh	41.3
2. Porosity	Highest porosity	41.1%	37.5
3. Variable grain size or texture	Most uniform	(all 40-48 mesh)	54.8
4. Coating on grains (wettability)	Smooth carbon	Intermediate wettability	84
5. Oil viscosity	Lower oil viscosity	≃ 1 cp (gasoline)	67
6. Temperature	Highest temperature	125°F	70.5
7. Rate	Slow rates (with additives)	< 1 ft/day	59
8. Interfacial tension	Lowest IFT	8 dynes/cm	71
9. Salts in solution (acid or alkaline)	Highly alkaline	1.0 normal sodium carbonate solution (also lowest IFT)	71

effectively may be classed into three broad and often overlapping areas, i.e., attempts to: (1) change wettability, (2) change oil-water interfacial tension, or (3) remove the interface entirely (miscible flooding). Each of these areas will be treated separately although the interactions and overlap between them may be very strong.

WETTABILITY

Wettability must seem like the weather to most students of enhanced recovery research: until recently everyone talked about it but no one did anything about it. Now, however, professional rainmakers (using results of good surface and colloid research) do attempt to change the weather even though controversy surrounds their efforts. So it is with wettability, much has been learned in the last 50 years, but it is difficult to turn the knowledge directly into increased oil recovery.

Wettability effects in capillary action were recognized long ago. The classic Jamin effect depends, of course, on the difference between advancing and receding contact angles, and without some preferential wetting, there can be no capillary forces. [In this paper, it can be assumed that contact angles are measured through the aqueous phase unless stated otherwise. The basic concepts and equations of capillarity, including contact angles are not reviewed here, but the reader is referred to a good analysis and review of the role of capillary effects in oil recovery in the work of Melrose and Brandner (11).] The influence of wettability on oil recovery, however, has been difficult to study systematically because of the nature of the surface forces involved. Contact angles can be measured on smooth silica plates with known liquids, and various additives can be studied in these laboratory-controlled systems. However, it is extremely difficult to study contact angles inside the porous medium, particularly if it is a rock made up of scores of different minerals of many shapes and sizes.

Most early authors seemed to agree that wettability has an important effect on the recovery of oil during waterflooding. For example, Uren and Fahmy (10) commented on wettability in 1927. As Table 1 indicates, the best oil recovery (84%) was observed by them when they treated their sand to provide a smooth carbon-like surface on the sand grains. However, they were working with large grain sizes and were flooding from the bottom up, so gravity aided the oil production.

A detailed description of the wettability effects was presented by Benner, Riches and Bartell in 1938 (12) when they described not only wettability, but other aspects of the influence of surface chemistry on oil recovery. As director of API Project

27, Bartell and co-workers (13) emphasized that there should be no controversy about whether high energy surfaces such as silica are inherently water-wet or oil-wet. From a thermodynamic point of view, silica must always be wetted by water in preference to any hydrocarbon or mixtures of hydrocarbons. According to Bartell, this is true even if the hydrocarbons are crude oils with their many different substances, some of which are known to be surface active. He examined the competition between water and crude oils for silica surfaces, and he found no crude oils (even with their many nonhydrocarbon constituents) which had greater adhesion tensions for silica than did pure water. He did point out, however, that there were large differences between the wetting properties of crude oils, and that some might be found eventually which would wet silica better than water.

Benner, Riches and Bartell (12) may also be responsible for one of the first detailed descriptions of the trapping of oil in capillaries during waterflooding, when capillary forces are much more important than the viscous forces. Using the classical pore-doublet model, they described the reasons for the advance of water in smaller capillaries (in water-wet systems), and the resultant trapping of segments of oil in larger capillaries. They carried out experiments to show conclusively that countercurrent flow of both oil and water is often observed in porous media wherein water is displacing oil under capillary control.

Bartell and co-workers (14) also presented an early analysis of the importance of advancing and receding contact angles with respect to oil recovery. Although they apparently did not recognize the saturation versus pressure differences which are always observed when fluids are allowed to imbibe into or are drained from a porous medium (i.e., capillary pressure hysteresis), they did study advancing and receding contact angles for several crude oils in packed silica powders. They found that the advancing and receding water-oil contact angles varied over a very wide range for the different crude oils. To their surprise, they found many crude oils with advancing contact angles (measured through the aqueous phase, of course) which were considerably greater than 90°. Therefore, Bartell was forced to conclude that not all rocks behave in a water-wet fashion for every crude oil studied. However, he also found that the <u>receding</u> contact angle was less than 90° in every case. Thus, the hysteresis in a great many systems was often so great that advancing contact angles might indicate an oil-wet system and receding contact angles a water-wet system. With a system of this type, it is clear that neither the water nor oil can advance into a rock by capillarity alone. He concluded that spontaneous displacement of oil by water should occur only in those cases where both advancing and receding contact angles are less than 90°, and that no spontaneous movement should occur where the two angles are on opposite sides of 90°. Since that time, this non-imbibition of either oil or water has been observed by several workers (15-17).

The importance of the hysteresis of the contact angle on the oil displacement mechanism has been recognized and commented on by many other workers in the field of oil recovery. Melrose (18) has studied the problem in some detail and points out that the complexity of the geometrical shapes in the individual interconnected pores can have a very large influence on this contact angle hysteresis. His model predicts that the "pore structure in typical reservoir rock types of porous solids can lead to hysteresis effects in capillary pressures even if a zero value of the contact angle is maintained" (18). Morrow (17) studied the effect of surface roughness on hysteresis and found that the difference between advancing and receding angles could be related to the "intrinsic" angle (equilibrium angle) which is measured in the absence of roughness. In the intermediate wetting range, the difference between advancing and receding angles can exceed 100° which means that there is a very wide range over which the porous medium will not imbibe either water or oil.

Morrow (17) conducted his experiments on polytetrafluoroethylene (PTFE) which is a low energy surface with well-studied surface characteristics. The use of this material has permitted contact angle and other wettability studies with less ambiguity than with crude oils on the high energy surfaces of reservoir rocks. With PTFE it was possible to use pure fluid pairs with reproducible contact angles and known wettabilities inside the porous medium. Using these systems, Morrow and McCaffery (19) studied immiscible displacements over a wide range of wetting conditions and found displacement behavior to be systematically related to the contact angle. They found the greatest differences between drainage and imbibition relative permeability curves in the intermediate wetting conditions. They found that the relative permeability to the wetting phase increased markedly as the advancing contact angle increased toward and through 90°. Thus, the implication is that water breakthrough occurs earlier in going from water-wet to oil-wet systems.

In the course of measuring imbibition capillary pressures, Morrow (20) also determined residual non-wetting phase saturations as a function of the intrinsic contact angle. For systems which spontaneously imbibe, he found that the residual oil values increased as the intrinsic contact angle was increased from 0° to 62°, the limit at which spontaneous imbibition occurs. Therefore, for systems which imbibe, the best recovery should be obtained from strongly water-wet systems.

The work of Morrow (20) or Morrow and McCaffery (19) helps to explain some of the observations of others on real reservoir fluid and rock systems, although their conclusions may appear to differ from those of earlier workers. Moore and Slobod (15) noted that oil recovery at water breakthrough in water-wet systems is much higher than in oil-wet systems. Moore and Slobod (15) also presented convincing laboratory evidence to show that the capillary forces in strongly water-wet systems were responsible for trapping oil in the larger pore spaces. Almost none of this "residual" oil could be produced after water breakthrough at any reasonable flooding rates. This "permanently trapped" residual oil varied from 37 to 45% in four different water-wet reservoir sandstones. With oil-wet systems, the oil recovery at water breakthrough was always less, the breakthrough residuals ranging from 40 to 54%. Moore and Slobod (15) observed their best oil recoveries at intermediate wettabilities (by their definition, advancing contact angles between 30 and 90° in the water phase) where residual oil values ranging from 24 to only 29% were observed in the same sandstones. They were convinced that the best recoveries should be obtained with contact angles of exactly 90°, since they assumed that capillary forces at the oil-water interface must vanish when the cos θ approaches zero. However, they seemed to be unaware of the increased effort required to mobilize the trapped oil because of the large hysteresis between advancing and receding contact angles in these intermediate regions. The detailed work of Morrow (21) and Melrose and Brandner (11) is in agreement that the mobilization of trapped oil should be the most difficult in this intermediate wettability region. However, Morrow also points out that the <u>prevention of entrapment</u> should be the easiest here, i.e., at an advancing contact angle of a little less than 90° (81° in Morrow's system) when imbibition forces do fall to zero (21). The sparse information on intermediate wetting in real oil-rock systems seems to show that lower residual oil values are indeed found in this intermediate region if sufficient pore volumes are injected (15).

Methods for Improving Recovery by Changing Wettability

The early researchers in oil recovery seemed to believe that the main function of waterflooding additives was to remove oil from the solid surface. In 1917, Squires (4) obtained his United States patent which involved the addition of alkaline materials to floodwaters and from 1925 through 1928, Nutting (5,22-24) described his "soda process" for petroleum recovery. He showed that dissolved salts such as sodium carbonate and silicates were more effective than water for removing oil from rock surfaces. In 1956, Reisberg and Doscher (25) carried out experiments with sodium hydroxide and with strong detergents plus sodium hydroxide to show conclusively that the adhering crude oils could be removed very effectively from glass or silica by these high pH solutions. Although there continued to be some uncertainty about the quantitative aspect of the wettability of most of the world's oil reservoirs, laboratory work, especially that of Wagner and Leach (26) in 1959, showed that drastic changes in the pH of the water would change the wettability of rocks from predominantly oil-wet to predominantly water-wet, and that increased oil recovery could be obtained thereby. Wagner and Leach first emphasized the use of acids to reverse this wettability, but they found that the reservoir rocks consumed so much acid that it appeared to be impractical. In 1962, Leach et al. (27) demonstrated that sodium hydroxide is effective for changing rock surfaces from oil-wet to water-wet. They reported on an early field trial using this material and demonstrated improved recovery, although the economics at that time were uncertain because of the low cost of crude oil. Emery et al. (28) also reported on a field trial which used a strong solution of sodium hydroxide (2%), but they obtained an oil recovery of only 2.34% of the pore volume and also felt that the economics were uncertain.

By tradition, most of the efforts to change wettability with either strong acids or alkali have been directed toward making substances more water-wet. However, Cooke, Williams and Kolodzie (29) described a field trial wherein a water-wet reservoir was actually changed to a system which was more oil-wet with a small increase in recovery resulting from the injection of an alkaline material. In order to effect this unusual wettability reversal from water-wet to oil-wet, it was necessary to achieve the right conditions of temperature, pH and, especially, salt concentration. In this particular field trial (water-wet to oil-wet), an increased oil cut from 9 to 17% was obtained but no economic evaluation was made.

Perhaps the most successful field trial utilizing inexpensive sodium hydroxide is that reported by Graue and Johnson in 1974 (30). Although it is not considered to be a wettability reversal project, it is included here because of the use of sodium hydroxide. Graue and Johnson (30) reported an increased oil recovery of approximately 400,000 barrels of crude oil in the Whittier Field in California by what they describe as an emulsification and entrapment mechanism. [A good review of this process, along with a classification of various alkali flooding mechanisms, is given by Johnson (31).] This emulsification and entrapment process is designed for very viscous oils which respond poorly to waterflooding because of the poor sweep efficiency. By converting much of the oil to a fine emulsion within the porous medium, the flow of water is partially blocked in some areas and the overall sweep efficiency of the water in the reservoir is improved. Therefore, the improved recovery comes from a more complete waterflood of the reservoir rather than from a better displacement of oil from the capillaries by sodium hydroxide.

Throughout the long history of laboratory and field experiments dealing with wettability of rock-oil-water surfaces, it has been difficult to obtain a clear consensus about the true wetting characteristics of the crude oil-water systems in contact with reservoir rock. One of the more definitive studies on this subject was the work by Treiber et al. (32) in 1972 which reported on the laboratory evaluation of the wettability of 55 different oil producing reservoirs from different parts of the world. They divided these reservoirs, which included sandstones and carbonates, into three arbitrary divisions; namely, water-wet from zero to 75°, oil-wet from 105 to 180°, and intermediate-wet from 75 to 105°, all contact angles being measured through the aqueous phase. From their careful laboratory studies, they concluded that 66% of the reservoirs were oil-wet, 27% were water-wet, and only 7% were in the intermediate range where neither oil nor water wetted the surface very strongly. From this study, it would appear that the search for agents to change the wettability from oil-wet to water-wet might be a fruitful avenue for research to improve oil recovery. Morrow (20) has pointed out that it would be better to classify intermediate wettability as systems which do not spontaneously imbibe either phase. This increases the range from 62° (advancing) to 133° (receding) according to his work on well-characterized surfaces. Based on the contact angle measurements of Treiber et al. (32), this would put 47% of the reservoirs into the intermediate category with 26% classified as water-wet and only 27% as oil-wet.

One of the significant aspects of wettability information related to oil recovery is the fact that nature may be providing the best possible wettability conditions for optimum recovery in some of the reservoirs. For example, Richardson et al. (33) studied a large number of "fresh" and aged cores taken from the Woodbine Reservoir of the East Texas field. If they cleaned these Woodbine cores and resaturated them with oil to carry out a normal waterflood for residual oil determination, they observed the usual laboratory residual oil values of about 30% of pore volume. However, if they conducted their waterfloods on fresh cores, they observed much lower residuals of only 15-18% of pore volume. In some parts of the reservoir, the residual oil values were reduced to less than 10% of the pore volume. These residual oil values are lower than those obtainable by most recovery methods even with good laboratory formulations unless very high combinations of flooding rate and low interfacial tensions are used (see the next section).

In an attempt to understand how nature is able to achieve such low residual oil saturations without help of engineers or laboratory specialists, Salathiel in 1973 (34) described a method for changing the wettability of rock surfaces in an unusual way. By treating cores which were saturated with typical values of oil and water, he was able to generate a mixed-wettability condition in the laboratory wherein surfaces in the larger pores were primarily oil-wet and rock surfaces in the smaller pores remained water-wet. With this special mixed wettability condition, it appears that the oil relative permeability can remain finite to very low saturations, with the oil apparently draining or flowing in very small rivulets until extremely low values of residual oil saturation can be obtained. With this laboratory-generated system, he feels he has reproduced the situation which nature may have providently provided in the East Texas field.

This is not to be taken as a concluding note that all should be left to nature and that those of us engaged in enhanced oil recovery should cease to search for methods where we can alter wettability to improve oil recovery, but it does suggest that our knowledge is far from complete and that our task remains a difficult one. The author of this paper is reminded of his first meeting many years ago with his new research director, an experienced oil recovery expert in one of the world's major oil corporations. He was a powerful man, and he shook the room with the demand that "we need to be struck by lightning with wettability." It is the opinion of this author that the lightning bolt has yet to strike.

OIL-WATER INTERFACIAL TENSION AND SURFACTANTS FOR OIL RECOVERY

Early Work with Surface Active Agents

In the early days of chemical flooding, surface active agents were thought of primarily as detergents and the term "wetting-agent" was often used by the companies or salesmen attempting to promote the use of these materials in waterflooding. In much of the early work, these materials were assumed to be effective primarily for releasing oil from the solid surface of the rock (25). The laboratory experimenters were alert to the need for interfacial tension reduction and some correlations were attempted in this area, but the laboratory results were not consistent. Some of the early workers even reasoned that since water displaces oil largely by capillary action, better displacement would occur if the interfacial tension could be increased to increase the capillary driving force. Some of the early experimenters also reported that very little improvement in oil recovery could be expected with many of the surface active materials (35). One detailed study indicated that oil recovery actually decreased slightly when the oil-water interfacial tension was decreased in the experiments (36).

Although there was no consensus on the effectiveness of detergents for oil displacement, it was clear in the early work that a very serious problem was the adsorption of the materials on the large surface area of the rock. Not only was a large amount of the detergent lost to the rock, but this adsorption prevented the advance of detergent through the formation along with the bank of water in the waterflood. In 1968, Taber (37) pointed out that very high concentrations of detergent (about 10%) would permit an advance of the detergent front at a rate which was only a little less than that of the advancing flood water. These high concentrations also appeared to recover oil more effectively than the low concentrations, but the actual recovery varied with conditions of the laboratory test. It appeared to Taber that the most sensitive variable in detergent flooding, at least in laboratory experiments, was the rate of flooding. As long as effective detergents with or without sodium hydroxide were used, additional oil could be produced whenever the rate of flooding was increased. Therefore, in his early work citing possibilities of the use of detergents in waterfloods, he restricted recovery predictions to situations where the laboratory flooding rates were close to field rates with the result that predicted oil recovery by surfactants was rather low (37).

Taber also observed that the widely different responses for oil recovery with various detergents and at various rates could be understood if the experiments were carried out at equivalent values of the ratio of viscous to interfacial forces. For the displacement of presumed, non-wetting residual oil, he observed that no oil could be displaced until a critical value of $\Delta P/L\sigma$ had been reached, where ΔP is the pressure drop over the distance L and σ is the oil-water interfacial tension (38). This value in English engineering units for Berea sandstone was about 2-5 (psi/ft)/(dyne/cm), or approximately 45-113 kPa/mN (kiloPascal/milliNewton) in SI units. Taber observed this ratio with several fluids and under a number of different conditions, varying interfacial tension, length and pressure gradient independently. In every case, after the critical gradient had been exceeded, additional oil could be recovered as long as the value of $\Delta P/L\sigma$ was raised even further. Although it was difficult to accomplish, essentially all of the oil could be displaced from the Berea sandstone if the value of $\Delta P/L\sigma$ was made large enough. However, the rates and the resulting pressure gradients were more than 1,000 times those experienced in the field.

The Capillary Number Concept

A number of authors have studied the ratio of viscous to capillary forces and the effect of the ratio or dimensionless group on oil recovery. Table 2 lists several of these groups which have been described by various authors (11,15,21,38-51). Many are dimensionless and are often referred to collectively as either the "capillary number" or the "critical displacement ratio" by workers dealing with surface phenomena and oil recovery. Although the experimental data are still rather limited as far as capillary number results are concerned for different types of rocks, the consensus by the various workers is good. Different authors have examined the pore dimensions and geometry in both synthetic and real systems to calculate the critical value of the

Table 2. Capillary number or "displacement ratio" correlating groups

Reference	Author(s)	Year Published	Porous Media	Correlating Group
39	Fairbrother and Stubbs	1935	Capillary	$\sqrt{\frac{v\mu}{\sigma}}$
40	Leverett	1939	Sandstone	$\frac{LP_c}{D\Delta P}$
41	Brownell and Katz	1947	Sandstone	$\frac{K}{g\cos\theta}\frac{\Delta P}{L\sigma}$
42	Ojeda, Preston and Calhoun	1953	Sandstone	$\frac{\sigma}{\Delta P}$
15	Moore and Slobod	1956	Sandstone	$\frac{v\mu}{\sigma\cos\theta}$
43	Saffman and Taylor	1958	Hele-Shaw Cell	$\frac{v\mu}{\sigma}$
38	Taber	1969	(Berea) Sandstone	$\frac{\Delta P}{L\sigma}$
44	Foster	1973	(Berea) Sandstone	$\frac{v\mu}{\phi\sigma}$
45	Lefebvre duPrey	1973	Teflon, Steel, and Aluminum	$\frac{\sigma}{v\mu}$

Table 2 (Continued)

Reference	Author(s)	Year Published	Porous Media	Correlating Group
11	Melrose and Brandner	1974	Unconsolidated Glass Beads	$\frac{Kk_{rw}}{\sigma}\frac{\Delta P}{L\sigma}$ or $\frac{v\mu}{\phi\sigma}$
46	Ehrlich, Hasiba and Raimondi	1974	Sandstone	$\frac{K\Delta P}{\phi L\sigma}$
47	Abrams	1975	Sandstone, Limestone	$\frac{v\mu_w}{\sigma}\left(\frac{\mu_w}{\mu_o}\right)^{0.4}$
48	MacDonald and Dullien	1976	Sandstone	$\frac{\bar{\ell} \quad \Delta P}{(1/\bar{D}_e - 1/\bar{D})L\sigma}$
49	Reed and Healy	1977	Various	$\frac{K\Delta P}{\cos\theta \; L\sigma}$
50	Stegemeier	1977	Analysis of Various	$\frac{\phi N_{Le}^2 k_{rw} \Psi^2}{2f}$
21	Morrow	1978	Teflon	$\frac{v\mu}{\sigma Z_{imb}\cos\theta_A}$
51	Oh and Slattery	1979	Model Pore	$\frac{r_n \Delta P}{2\sigma}$

capillary number at which either residual oil will be displaced or at which better oil displacement will occur during waterflooding. For example, Melrose and Brandner (11) have calculated these values and estimated the results for three different types of porous media, generalizing them as those with large, intermediate, and small differences in the capillary constrictions. The displacement efficiencies predicted for each of these media are given by Figure 1. MacDonald and Dullien (48), Oh and Slattery (51), and Stegemeier (50) have made calculations and have compared them with experimental data of their own or from the literature. In general, the calculations agree with experimental data. For example, Figure 2 shows that Stegemeier's calculated results for the remaining oil saturations at various capillary number values agree rather well with the data presented by Taber (38) as long as the advancing contact angle of the displacing phase is taken as 60°.

The results of several other studies to illustrate the desaturation of either the residual non-wetting (usually oil) phase or the wetting phase at various capillary numbers were compared by Stegemeier and are reproduced here as Figure 3. In general, the results show that the initial desaturation of the non-wetting phase occurs over several orders of magnitudes of the capillary number and at lower viscous to capillary ratios than reported by Melrose and Brandner (Figure 1). However, the results of the displacements of discontinuous non-wetting fluid [oil, for Taber (38) and Foster (44), gas for Wagner and Leach (52)], from similar sandstones are fairly close with significant removal of the fluid starting at capillary numbers from 10^{-6} to 10^{-5}. The work of Wagner and Leach is noteworthy because they achieved very low interfacial tensions with no additives in their gas-liquid systems near the critical point. In addition, they carried out their studies at slow rates which were close to flooding rates in the field, and their work is often cited as direct evidence for the need for low interfacial tensions for effective displacement at field rates. For the desaturation of continuous or at least partially connected oil, the work of Abrams (47), Lefebvre duPrey (45), as well as Moore and Slobod (15), shows that desaturation of the continuous oil starts at a much lower ratio of the viscous to capillary forces, i.e., a capillary number of about 10^{-7}. However, after the initial oil displacement in their experiments, the oil must become discontinuous at the lower saturations and Figure 3 shows that their values are then closer to those of Foster or Taber.

The work of Dombrowski and Brownell (53) shows that it is much harder to displace the residual wetting phase from a porous medium (in this case, small glass beads). Figure 3 indicates that the capillary number required for their displacement of the wetting phase is two orders of magnitude higher than that required for the displacement of most non-wetting phases. Jenks et al. (54)

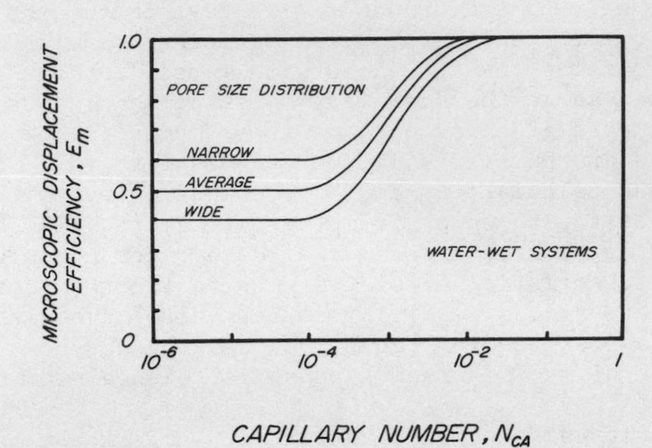

Fig. 1. Correlation of microscopic displacement with capillary number. (After Melrose & Brandner, Ref. 11.)

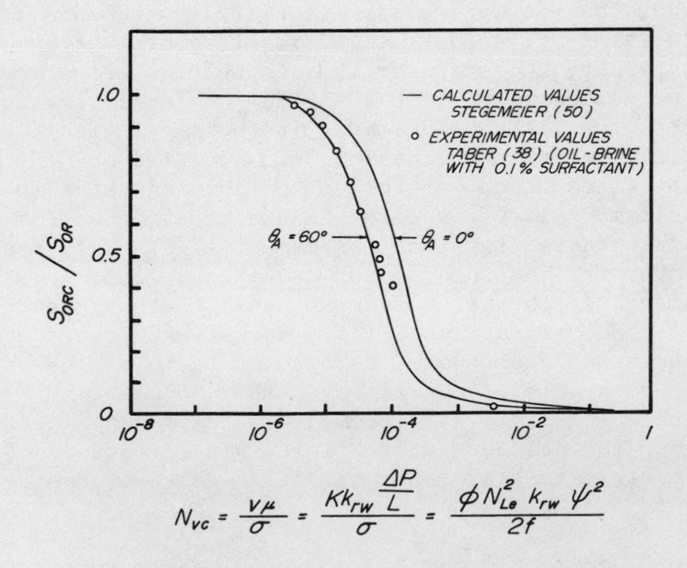

Fig. 2. Calculated and experimental residual oil. (After Stegemeier, Ref. 50.)

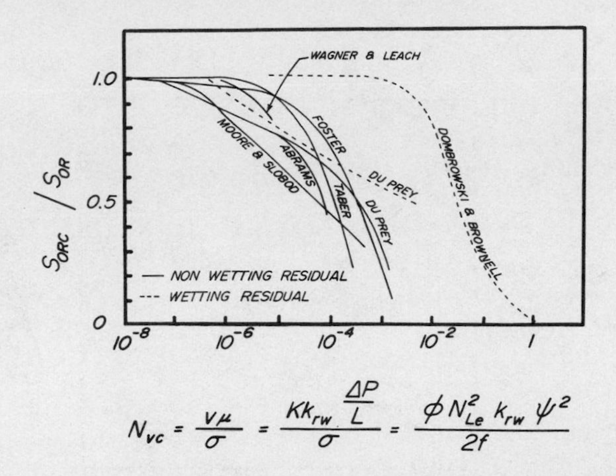

Fig. 3. Average experimental recoveries of residual phases. (After Stegemeier, Ref. 50.)

also presented a clear picture of the increased ratio of viscous to capillary forces which are required for the displacement of the wetting (water) phase compared to the oil phase in Berea sandstone. At a pressure gradient of 2,000 psi/ft, they observed that only 4% of the residual water was displaced by oil, but about 40% of the residual oil was produced by a waterflood on a similar Berea core at the same conditions.

Other than the information summarized and compared in Figure 3, the experimental data on displacement of residual oil at supercritical values of the capillary number are not extensive. Taber, Kirby and Schroeder (55) carried out experiments to determine the critical values of $\Delta P/L\sigma$ for several sandstones which had permeabilities ranging from 40.8 to 2,190 millidarcies. Prior to the critical displacement test, all rock samples were waterflooded at slow (subcritical) rates until each sample contained its normal residual oil value. The flooding rate was then increased until measurable quantities of residual oil were produced, i.e., until the critical value of $\Delta P/L\sigma$ was attained.

The results which are given in Figure 4 indicate a definite trend towards higher $\Delta P/L\sigma$ critical values at lower permeabilities. Since the experiments were done on different rocks from several formations (55), a precise correlation between permeability and the critical displacement ratio should not be expected. However, where data are available in the literature, the values do appear to fit the general trend in Figure 4. For example, Jenks et al. (54,56) observed that a critical $\Delta P/L\sigma$ value of 8.9 (psi/ft)/(dyne/cm) (201 kPa/mN) was required for the displacement of residual oil from their Berea sandstone. Although this is higher than the critical values observed for Berea by Taber et al., the value

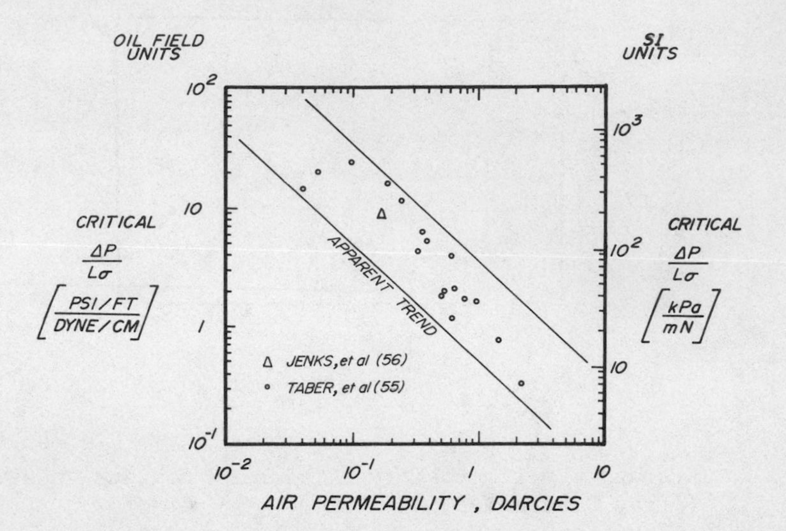

Fig. 4. Relationship between permeability and the critical value of $\Delta P/L\sigma$. (After Taber et al., Ref. 55.)

obtained by Jenks et al. fits the trend because their Berea had a permeability of only 150 to 180 md compared to Berea permeabilities of 300 to 900 md for the other Berea data points in Figure 4. (Sandstones other than Berea are also plotted on Figure 4.)

As pointed out above, most authors have preferred to use a dimensionless form of the capillary number. However, it is often more convenient in laboratory and field calculations to use the simple critical displacement ratio $\Delta P/L\sigma$ which has the units of L^{-2}. For example, by using methods outlined by Stosur and Taber (57), it is possible to calculate the critical displacement ratio at points close to the wellbore or deeper in the reservoir. Their work indicates that this may be important when certain logging methods are used to estimate the in situ residual oil saturation remaining after a waterflood. From Figure 4, it appears that the critical displacement ratio is a property of the rock, and it is hoped that more values will be determined for future work in enhanced oil recovery.

Some laboratory experiments may be easier to interpret in terms of quantitative $\Delta P/L\sigma$ values instead of a dimensionless capillary number. For example, when experiments with surfactants are carried out at constant rates (and thus a constant $v\mu/\sigma$) on rocks of different permeabilities, it appears that the oil recovery actually increases for those rocks with lower permeabilities. Clearly, the better recovery is not a "reverse permeability effect" but it comes about because the $\Delta P/L\sigma$ is much higher in tighter rocks when all experiments are carried out at a constant rate. In general, almost all of the experimental and theoretical work indicates that the displacement of oil increases with higher values of the ratio of viscous to interfacial forces. An exception is the recent work by Sayyouh et al. (58) which indicates that at very slow rates (less than 1 ft/day), the recovery appears to increase slightly with a further decrease in rate. Therefore, the complex nature of the oil displacement process still warrants the continuing research emphasis which it is receiving. In addition to the work cited above, current progress is being made by the investigation of gravity forces along with the viscous forces for the mobilization of trapped oil. Morrow (21) reports that gravity alone may dislodge oil when interfacial tensions are low enough to bring the system into the critical capillary number region at reservoir rates. In addition, the current work by Larson and Scriven (59), who are applying certain facets of percolation theory to the mechanics of oil displacement, is bringing new insight into the whole process.

Micellar-Polymer Fluids

The evidence that low oil-water interfacial tensions will enhance oil recovery has led to an enormous effort by many experts to prepare surfactant formulations which will be effective in underground reservoirs. In contrast to the early work with low concentrations, between 1957 and 1968 several workers proposed the use of somewhat higher concentrations in special formulations to overcome adsorption problems and to provide the optimum interfacial tension conditions (37,60-64). These high surfactant concentrations could be dissolved in either oil or water and normally contain other additives to improve the stability or overall effectiveness. In 1968 Gogarty and Tosch (65), and Davis and Jones (66) published their first papers dealing with "Oil Recovery and Displacement Mechanisms with Micellar Solutions." They described their surfactant fluids as microemulsions which are effective with either oil or water as the external phase. In addition to the three main components (surfactant, hydrocarbon and water), a "cosurfactant" (usually an alcohol) and an inorganic salt are normally added to provide the optimum formulation for oil displacement. Similar oil external fluids called "soluble oils" which were developed by Holm et al. (60,67,68) also have proved to be effective oil recovery formulations. These micellar fluids, or microemulsions as defined by Schulman et al. (69), are not only excellent recovery agents, but they appear to reduce the adsorption loss to the reservoir under certain conditions. However, the adsorption problem is far from solved, though good work by a number of investigators is continuing (70-74). A review of the surfactant adsorption literature is beyond the scope of this paper.

Good recovery with microemulsions also requires good mobility control, i.e., to prevent viscous fingering of the drive water into the surfactant bank or of the surfactant into the reservoir oil-.

water bank, the viscosities of the injected fluids must be increased in most cases (75,76). The microemulsion viscosity can be controlled by adjusting the components in the formulation, and the drive water or "mobility buffer" is normally thickened by adding a natural or synthetic, water-soluble polymer (77). However, the polymer mobility control agents present additional problems for the surface chemists. Several authors have shown that the large polymer molecules will not penetrate the smallest pores of typical rocks and, thus, move faster in the larger pores than the average advance of the waterfront (78-81). Because of this "inaccessible pore volume," the polymer solution invades the microemulsion fluid to bring about a phase separation of the surfactant-rich phase, according to Trushenski (82). If the interfacial tension is high enough, the surfactant-rich phase can be trapped in the pores much like residual oil. Trushenski describes various methods for reducing this adverse surfactant-polymer interaction including the use of more surfactant solubilizers, i.e., cosolvents such as alcohols (82).

The never-ending quest for ultra low oil-water interfacial tensions in these oil recovery systems was given a big boost by the pioneering development of the spinning drop apparatus by Cayias, Schechter and Wade (83) at the University of Texas for interfacial tension measurement [first suggested by Vonnegut (84)]. When measuring the interfacial tension of pure hydrocarbons against an aqueous surfactant solution, Cayias et al. (85) showed that the minimum interfacial tension was found against only one of the hydrocarbons in a homologous series. This minimum interfacial tension was very specific, not only for normal alkanes but for branched, cyclic and aromatic hydrocarbons as well. From these correlations, and by using a series of surfactants, Cash et al. (86,87) were able to assign an "equivalent alkane carbon number" (EACN) to any hydrocarbon or mixture of hydrocarbons including crude oils. Their work indicates that most crude oils appear to have an EACN ranging from 6.2 to 8.6, i.e., from the point of view of interfacial tension, crude oils behave as if they were comprised primarily of hydrocarbons in the range of hexane to nonane. This information has permitted many surfactant investigators to work with pure hydrocarbons such as octane (or specific hydrocarbon mixtures) with more confidence that their laboratory hydrocarbon is not an unrealistic model for crude oils. To determine the surfactant molecular weight or other property most effective for interfacial tension lowering, the University of Texas group also developed the concept of n_{min}, i.e., the specific EACN which produces the lowest interfacial tension for a given surfactant or mixture of surfactants (88,89). This n_{min} concept proved to be useful for studying the variables, such as temperature, age of system, and the concentrations of added electrolyte or cosurfactant. For example, they noted that the n_{min} increases with electrolyte concentration; also, n_{min} decreases with increasing temperature for anionic surfactants and increases for nonionic surfactants. Temperature effects have also been reported by Burkowsky and Marx (90) who observed consistent minima in interfacial tension at the relatively high temperatures of 60 and 76°C for surfactant-crude oil systems for two German oil fields. The best oil recoveries were also obtained when displacement experiments were conducted at the temperatures of lowest interfacial tensions.

The urgent need for the preparation of the very best formulations and, therefore, for understanding microemulsions has led to an upsurge of basic and applied work in the area of micellar fluids. Since this subject will be reviewed thoroughly by others, only applied aspects related to oil recovery can be touched on here. Although the oil-water-surfactant systems are very complex, especially after cosolvents and electrolytes have been added, a number of workers have been able to describe the systems and adjust the required properties to provide the best oil recovery. The work of Reed, Healy (49,91-94) and their co-workers is noteworthy for providing a clearer picture of the physical nature of the system, especially the gross phase behavior under the influence of many variables. Their studies on the addition of electrolytes to the surfactant systems has brought a sense of order into what had been a rather confusing picture. Ever since Harkins (95) showed that extremely low interfacial tensions can be obtained by, in effect, salting out the surfactant so that it goes to the oil-water interface, others (38,96-98) have utilized this added salt concentration to lower the oil-water interfacial tension. Reed, Healy and their co-workers developed a system of "optimal salinity" for maximizing the desired property by the judicious addition of the electrolyte. In general, optimal salinity is usually thought of as that specific salt concentration in an oil-water-surfactant system which produces the lowest interfacial tension, and Reed et al. showed that this optimal salinity affects several of the interrelated properties of the microemulsion system. It can be measured in at least four different ways, namely, the salt concentration at which the lowest interfacial tension is observed, the salt concentration which provides the most favorable phase behavior for oil recovery, the salt concentration which provides equal solubility parameters (as defined by Healy and Reed), and the salt concentration at which contact angles between each liquid phase and the solid reach similar equilibrium values. The optimal salinity measured by any one of the four methods on a particular system agrees reasonably well with that obtained with the other methods. Perhaps, of even more importance, the work of Reed, Healy et al. (49,91-94) demonstrated that oil recovery was maximized near the optimal salinity value (as measured by any of the methods), and that oil recovery correlated well with the capillary number.

The concept of optimal salinity has aided many other workers as they have pursued a better understanding of these microemulsions through various types of fundamental studies. Miller et al. (99) presented a model for the phase separation which occurs as the salinity is increased until the middle or surfactant phase is observed. Miller's model indicates that the repulsive forces shift sufficiently with the addition of salt until the aggregates come close enough together to force an actual phase separation at the optimal salinity. His studies indicate that this middle or surfactant phase is water continuous (99). With careful studies using the polarizing microscope, Benton, Fort and Miller (100) showed that there are two distinct structures, i.e., one, at low salinities, and one at the high salinity values. In the intermediate region near optimal salinity, both structures were observed. Hwan et al. (101) observed the effect of salinity on the actual drop sizes in the microemulsions. By studies with an ultracentrifuge they found that drop sizes increased in a regular way with salinity, reaching a maximum in the region of the so-called optimal salinity and then decreasing at the higher salinity values. Their studies also confirmed the existence of oil in water microemulsions at low salinities and water in oil microemulsions at high salinities. Between these two regions the ultracentrifuge work indicates that the drops are characteristic of both oil-continuous behavior and water-continuous behavior, thus adding additional evidence to the dual nature of these micellar fluids in the intermediate salinity ranges. As early as 1976, Scriven (102) had also presented an interesting model for an equilibrium bi-continuous structure of the microemulsion systems, i.e., his analysis showed that both oil and water may be continuous in these intermediate salinity regions of greatest interest.

Chan and Shah (103) also studied relationships between microemulsions and optimal salinity by observing the effects of additives and temperature changes which can bring about the phase transition from lower to middle to upper phase in these systems. In general, their work indicates that the surfactant partition coefficient between the oil phase and the excess brine phase is unity at the optimal parameter value. Their work indicates that there is a strong similarity between the interfacial tension behavior of low concentration systems and those of high concentration systems. Bansal and Shah (104) also showed that the salt tolerance of surfactant systems can be extended to rather high salt concentrations by mixing ethoxylated sulfonates with the usual petroleum sulfonate materials. An optimal salinity as high as 32% sodium chloride was observed in one of the mixed systems which was also characterized by very low oil-water interfacial tensions.

This section started with a discussion of the need for low interfacial tensions between oil and water for good oil recovery. It is clear that this sense of urgency in finding low interfacial tensions has contributed to a much clearer understanding of complex systems containing surface-active materials plus oil, water and other additives. It is also clear that systems which achieve low interfacial tensions may exhibit very complicated phase behavior characteristics, and that the combination of interfacial tension and the phase behavior of the fluids should be important for maximum oil recovery. The connection between phase behavior and oil recovery will be discussed more fully in the next section.

MISCIBLE FLOODING AND EFFECT OF PHASE BEHAVIOR ON OIL RECOVERY

Introduction to Miscible Processes

The allure of miscible displacement for complete oil recovery is obvious; it is the only method which permits 100% recovery, even in theory. As long as two phases exist in small capillaries, some trapping of the oil phase will normally occur even under the best conditions. However, if the two fluids are truly miscible, and no other fluid phase is present (seldom true in oil reservoirs), there can be no capillary forces to trap off segments of oil. If the "miscible slug" is large enough, all of the oil can be displaced to achieve a theoretical recovery of 100%. However, to date no practical method has been devised which can provide complete miscibility in real reservoirs (where both oil and water are present) so 100% recovery remains a distant goal.

The common miscible processes can be generalized very loosely by the three-component phase diagram in Figure 5. The fluid which is miscible with both oil and water (or oil and gas) at the prescribed conditions is shown at the apex of the ternary diagram. As long as the percentage of this "miscibilizing fluid" (or amphipathic solvent) is high enough, the three-component mixture will be single phase and no interfaces should exist between the displaced and displacing fluid. Below this concentration, two phases exist and capillarity will return to interfere with the efficient displacement of the oil. It will be shown that the normal presence of both oil and water flowing in advance of a miscible driving fluid apparently prevents the attainment of true miscibility in large segments of the rock because of the capillary trapping of oil.

Figure 5 illustrates the conventional diagram normally used for oil-water-alcohol (or surfactant) processes. For CO_2 or hydrocarbon-miscible processes, authors conventionally rotate the ternary diagram 120° clockwise, but the illustrated principle is the same. The binodal curve is normally higher for alcohol and lower for surfactants than the curve shown in Figure 5. Although no real reservoir process can be completely miscible in the exact

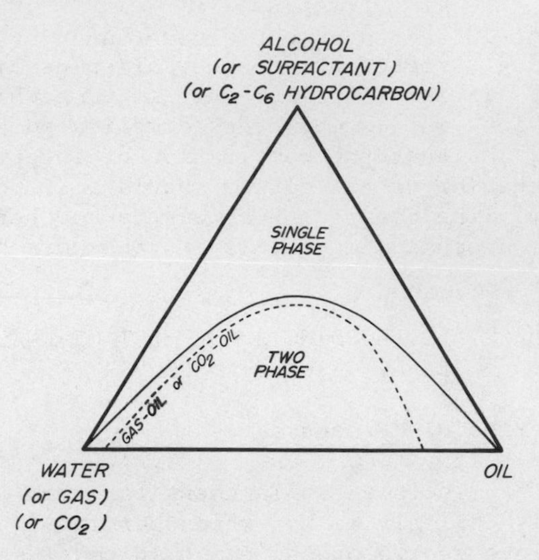

Fig. 5. General phase behavior to illustrate miscible type processes for enhanced oil recovery.

sense, the following are often referred to as miscible: (the names refer to the displacing phase)

- Hydrocarbon miscible processes
- CO_2 processes
- Miscible slug or alcohol processes
- Micellar processes (certain segments of some processes)

These will be discussed briefly with some comments on the surface chemistry involved. This is an apparent contradiction since in true miscible systems there are no interfaces.

Hydrocarbon Miscible Processes

The original efforts to conduct miscible displacements made use of gaseous or liquid hydrocarbons in the low molecular range. At one time, methane and liquefied petroleum gas (LPG) were readily available and inexpensive compared to crude oil. In the 50's, several published papers presented research results which indicated that these low molecular weight hydrocarbons could be miscible with the crude oil and a driving gas under the proper conditions in the reservoir (105-109). The techniques may be classified broadly as High Pressure Gas Drive (105-107), Enriched Gas Drive (110,111) and Miscible (Hydrocarbon) Slug Flooding (108,109). The miscible hydrocarbon slug process is the easiest to visualize: a quantity of propane or LPG is injected into the reservoir and followed by dry gas. At most common reservoir temperatures and pressures, the propane will be miscible with both the reservoir oil and the driving gas. The enriched process develops miscibility by adding some C_2-C_6 hydrocarbons to the injected dry gas (methane) to bring the three-component system closer to the single phase region in Figure 5. The high pressure gas drive achieves miscibility by vaporizing enough of the C_2-C_6 (or higher intermediates) to bring the two phase system up into the single phase region. This "vaporizing gas drive" mechanism is explained by Hutchinson and Braun (112).

All of these hydrocarbon-miscible processes can recover oil, but there are many difficulties. Mobility control is a very serious problem, although techniques such as alternate water and gas injection (WAG) can be used to reduce the mobility of the lighter, driving fluids (113). In the United States, the intermediate hydrocarbons as well as methane are in short supply and are valuable as fuel or chemical feedstocks. Therefore, much attention in the United States is now focused on materials such as CO_2 or other inert gases which have no fuel value but which can provide some miscible displacement under certain conditions. In parts of the world where light hydrocarbons and methane are available, the hydrocarbon-miscible process and various modes of gas injection should increase oil recovery for many years.

Carbon Dioxide Processes

Carbon dioxide (CO_2) processes are usually listed with miscible processes even though much of the research which is now underway is directed toward finding an answer to the question: What are the conditions required for miscibility (if it can be attained), and what is the nature of the phase behavior of the CO_2-crude oil system? Normally, CO_2 is not completely miscible with crude oils, but it is very soluble in most oils at reservoir conditions.

Early work with carbon dioxide for enhanced recovery was divided between carbonated waterflooding and "CO_2 slug" flooding, the latter often being called "CO_2 miscible flooding" or by now just CO_2 flooding. In 1963, Holm (114) showed that a CO_2 slug, injected at reservoir pressures was much more effective than carbonated waterflooding if equivalent quantities of CO_2 were used in the displacements. Therefore, the emphasis has swung from carbonated water to the injection of very large quantities of CO_2 at reservoir pressures to attempt a miscible or "partially miscible" displacement.

The CO_2 flooding process has been described by several authors (115-119). A good review of the process (including laboratory and field work) with estimates of the future recovery potential has been given by Stalkup (120). While not normally miscible at reservoir conditions, the CO_2 has a high solubility in the oil, and it does perform an effective job of displacing oil, at least in laboratory experiments. Investigators have suggested several factors which aid in this improved recovery:

1. Swelling of oil: For tertiary recovery, this may be one of the most important effects because oil cannot flow until it is continuous and has finite relative permeability (118,119).

2. Viscosity reduction: The viscosity of the "swollen oil" will be much less than that of the original crude oil at reservoir temperatures if large quantities of CO_2 are dissolved in the oil.

3. Vaporization of crude oil: The lighter hydrocarbons are vaporized by the high pressure CO_2 in a process analogous to a high pressure vaporizing gas drive. In this way, miscibility can be generated in the reservoir to provide an effective displacement of the crude oil.

4. Blowdown recovery: At the end of high pressure laboratory experiments, additional oil is recovered when the pressure is lowered, and a similar effect should be noted toward the end of a field displacement, but this production will, of course, only be realized in the final stages of the process.

5. Interfacial effects: Although early published reports clained that CO_2 can form detergents or alter the wettability of the rock-oil system (121), there is no strong evidence that this was a major effect in the carbonated water case for which it was clained. However, interfacial effects may be very important when the system is close to miscibility which is a goal of the process design for most fields (122).

In general, CO_2 flooding is a very complicated process. The phase behavior is difficult to pin down; at times, 3 phases appear in the CO_2-hydrocarbon system, and the actual flow mechanisms in these complicated, near-miscible regions are not well understood. Laboratory studies have indicated that full miscibility is not necessarily required in displacements because good recovery can be obtained when several phases are flowing (116,117,122,123).

It can be presumed that interfacial effects must be very important in these near miscible regions, and at least one study has shown that oil recovery from CO_2 injection decreased markedly in lower permeability sandpacks, presumably because the critical capillary number requirements for displacement are higher in low permeability media (124).

In conclusion, it appears that CO_2 flooding requires the same types of conditions which are required for alcohol or micellar processes, i.e., a high oil saturation to provide good oil permeability and less trapping of the oil phase, along with near miscibility and low interfacial tensions between the displaced and displacing phases. The complete experimental evidence for these "requirements" is still lacking for CO_2 flooding. The good evidence for the controlling factors in alcohol-miscible and micellar-polymer flooding will be examined in the next sections.

Alcohol-Water-Oil Miscible Processes

Alcohol flooding is discussed here in some detail because it provides a basis for understanding the relationship between phase behavior, interfacial tension, and oil recovery. It is recognized that no successful field trial of the alcohol process has ever been carried out.

The use of alcohol as a miscible driving fluid for the displacement of oil has a number of advantages over other materials. Certain low molecular weight alcohols are miscible in all proportions with both the driving water and with some reservoir oils. If not miscible with oils of high molecular weights, miscible displacement of the oil can be still achieved by the injection of an intermediate hydrocarbon such as a gasoline fraction. By choosing the proper alcohol, and if necessary, an advance slug of hydrocarbon, it is possible to be in the single phase region in Figure 5 for essentially any combination of water, oil, and alcohol. Another big advantage of alcohol is that it can be driven through the reservoir with water and the water can be thickened with a polymer. Therefore, it is possible to choose a system which has a much more favorable mobility ratio than the previous systems which used methane, light hydrocarbon slugs, or carbon dioxide.

Since the use of alcohol provides the possibility of a true miscible system, 100% oil recovery is theoretically possible and the early work in this area indicated that some of these theoretical promises would be borne out (125-127). The early interest led to a field trial which failed because the required hydrocarbons of low to intermediate molecular weight were not injected in advance of the alcohol slug and the required phase behavior characteristics were not met (128,129).

In 1960, Gatlin and Slobod (130), by utilizing concepts developed by Wilson et al. (131), proposed their "piston-like" displacement theory which explained that the oil and water flowing simultaneously in front of the miscible alcohol piston would adjust their saturations to form stabilized banks whose saturations could be calculated from the relative permeability and viscosity of the fluids. They carried out a number of experiments which initially appeared to agree quite well with this straightforward theory.

Taber, Kamath and Reed (132) attempted to confirm the piston-like theory on relatively long, single pieces of consolidated sandstone (Torpedo and Berea), and in addition, they attempted to determine if the slug deterioration did follow the miscible mixing theories, i.e., did it deteriorate (lose miscibility) in proportion to the square root of the distance travelled by the slug. They found that the displacement was not at all piston-like, and it appeared that oil but not water was leaking into the alcohol piston. The reason for this inefficient and non-piston-like displacement is found in the ternary phase diagram for isopropyl alcohol, oil and water used in the initial system (Figure 6-A). Because of the slope of the tie lines, the oil phase diminishes as the alcohol concentration increases at the leading edge of the alcohol slug. Therefore, the oil saturation in the bank falls below the normal residual oil saturation and much of this oil becomes trapped in a fashion similar to regular waterflooding. As alcohol of higher concentration continues to flow past these trapped ganglia of oil, the oil is extracted so that 100% recovery is still possible eventually, but the displacement is no longer piston-like because the average forward velocity of the oil mass is now much less than that of the water and alcohol. To overcome the poor recovery caused by shrinking of the oil phase in the alcohol transition zone, Taber et al. (132) chose an alcohol system such as the one using tertiary butyl alcohol illustrated in Figure 6-B which permits an expanding oil phase in the region of increasing alcohol concentration as the system approaches the region of miscibility. By convention, these two different phase diagrams are referred to as plait point-right and plait point-left, or, rather loosely, as a shrinking oil phase and an expanding oil phase during the displacement process.

The recovery with this expanding oil phase was much better; the oil recovery per unit of alcohol injected increased by a factor of seven by moving the plait point of the ternary phase diagram from the right to the left side to permit an expanding oil phase. This improvement in displacement efficiency with an expanding oil phase (which keeps the oil phase continuous) was confirmed quickly by several others (133-135), and, more recently, by researchers working with surfactants (see below).

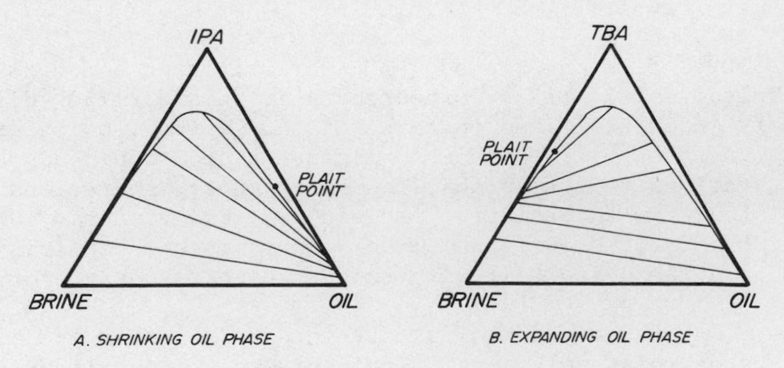

Fig. 6. Typical phase behavior of alcohol (or surfactant)-oil-water systems. (After Taber, Kamath and Reed, Ref. 132.)

Gatlin and Slobod's (130) displacement appeared to be piston-like because they had a very low residual oil saturation to ordinary waterflooding, and they used high flooding rates. Contrary to the assumption by some experts that there should be no interfacial tension or capillary number effects in alcohol displacements because the system is miscible, experiments show that rate effects are extremely important. Taber and Meyer (136) showed that the deviation from piston-like displacement is at a maximum at rates from about 1 to 12 ft/day. When the rate of displacement was increased beyond 12 ft/day, the recovery improved steadily with ever increasing rates until at the limit of their experiment at 70 ft/day, their displacements were almost piston-like. Gatlin and Slobod (130) carried out their experiments at even higher rates of 120 to 144 ft/day in their unconsolidated sandpack inside a steel pipe. At these conditions, the value of $\Delta P/L\sigma$ at most alcohol concentrations, would be well above the critical value. Therefore, in Gatlin and Slobod's experiments, there was no oil trapping (or oil leaking into the alcohol piston) and the piston-like theory appeared to be confirmed by experiment. Additional situations wherein oil trapping occurs even with an expanding oil phase will be discussed along with the relationship between phase behavior and surfactants in the next section.

Phase Behavior and Micellar-Polymer Fluids

The importance of phase behavior on oil recovery was mentioned in the section on micellar polymer fluids. Healy, Reed and Stenmark (93) showed that most common types of phase behavior for these complicated surfactant-oil-water systems can be classified as lower, middle and upper phase microemulsions. Their lower phase microemulsion corresponds to the alcohol-oil-water phase behavior in Figure 6-A while the upper phase corresponds to the slope of the tie lines and the phase behavior in Figure 6-B. The phase diagram for a surfactant system which has a middle phase microemulsion as described by Healy, Reed and Stenmark is not illustrated here. Nelson and Pope (137) have also investigated the relationship between phase behavior and oil recovery. They prefer to call the phase behavior exhibited in Figure 6-A as Type II- (the slope of the tie lines is negative) and Figure 6-B as Type II+ (slope of the tie lines is positive for expanding oil phase). For microemulsions with a prominent middle phase, they used the same designation as Winsor (138) and Healy et al. (93) [i.e., Type III to illustrate the three phases].

In general, Nelson and Pope (137) agree with the observations of Reed and Healy (49) that the best oil recovery should be obtained with microemulsions exhibiting the Type III behavior because this is normally near the optimal salinity with interfacial tensions at a minimum between both the oil and water phases in contact with the middle microemulsion phase. However, all of the data reported by Nelson and Pope indicate that the recovery from the Type II+ surfactant system (expanding oil phase) was equally good. They observed an oil recovery of 89% of the residual oil in both the Type III and the Type II+ displacement experiments.

Larson (139) also carried out a detailed analysis of the influence of phase behavior on surfactant flooding. He constructed mathematical models to account for phase behavior, low interfacial tension, dispersion and other mechanisms to determine conditions under which good oil recovery can be obtained. He also found that the best recovery should come from Type II+, that is an expanding oil phase with the plait point on the left. Larson goes even further in his analysis to claim that good recovery can be achieved from phase behavior alone without the requirement for low interfacial tensions. However, without the aid of low interfacial tension, the good recovery is delayed according to Larson (139).

Therefore, it appears that this section could be concluded with the summary that the best oil recovery with fluids which bring about low interfacial tension can come either from fluids which are at their optimal salinity so that a fluid of very low interfacial tension is interposed between the oil phase and the brine phase, or, in the event that the surfactant-oil-water system is a two-phase mixture, it is essential that the displacement process occur with the Type II+ mechanism as described by Nelson or Larson, i.e., an expanding oil phase should be part of the displacement process in the region where surfactant or alcohol concentration is increasing.

Unfortunately, when it comes to multi-phase flow in porous media, nature does not permit us to come to such straightforward conclusions. Taber and Meyer (136) observed what might be termed an "inverse viscosity effect." During a long series of experiments in which they were able to provide an expanding oil phase throughout the displacement process (simply by adding oil to the injected alcohol), they noted that oil recovery actually decreased in some cases when their target oil in the Berea sandstone had a low viscosity. This adverse displacement of a low viscosity oil with a very favorable mobility ratio, in a presumed miscible displacement, was contrary to all information on miscible displacement at that time. They noted that even though phase behavior was much more favorable, as, for example with pentane, the recovery was actually somewhat less than the recovery of Soltrol (a C_{10}-C_{12} hydrocarbon) because the heavier and more viscous hydrocarbon was trapped less in the oil-water bank which was being displaced by the alcohol piston. This increased trapping of less viscous oil is observed only when oil and water banks are flowing simultaneously as they must when they are being displaced with a fluid miscible with each of them. According to the stabilized bank theory, an oil of a high viscosity will automatically flow with a much higher saturation in the stabilized bank than an oil of lower viscosity. Raimondi and Torcaso (140) have shown that the trapping of the oil phase when both oil and water are flowing simultaneously is a function of the saturation of that oil phase in the bank. Therefore, oils with high viscosities will have high oil saturations in the stabilized bank ahead of the alcohol piston and relatively little capillary trapping of this viscous oil will take place. On the other hand, oils with low viscosities, such as iso-octane or pentane will be flowing at oil saturations much closer to the normal residual saturation. Therefore, the amount of oil trapped is somewhat higher for the low-viscosity oils. This means that the alcohol is unable to produce all of the oil in this stabilized bank except by extraction even for systems with an expanding oil phase. The data of Taber and Meyer (136) indicate that the optimum oil recovery for normal hydrocarbons should come with an oil having a viscosity of about 2 or 3 centipoises. For oils which are more viscous than this, the phase diagram becomes less favorable, and with oils having viscosities somewhat less than 3 centipoises, the "Raimondi and Torcaso" type trapping in simultaneously flowing banks takes over and displacement becomes inefficient regardless of the favorable phase diagram employed. Therefore, it appears that any displacement, wherein oil and water are both pushed ahead of a miscible solvent, will be aided by low interfacial tensions because of the difficulty of avoiding oil trapping at some point in the displacement process.

In conclusion, it appears that a miscible displacement can approach the theoretical value (in which the slug deteriorates in proportion to the square root of the distance travelled) only in those cases where there is no water (or 2nd phase) in advance of the miscible slug.

FUTURE DIRECTIONS FOR ENHANCED OIL RECOVERY

As hydrocarbons become more valuable and also less available in the United States and Canada, it is clear that the increased use of inert gases such as carbon dioxide, nitrogen and waste flue gases will be emphasized. Laboratory research will continue to try to understand the mechanisms and researchers will try to devise better techniques for using these inexpensive low-viscosity fluids more effectively. Research on efforts to use that almost incompressible fluid, water, to drive the special gases through the reservoir will undoubtedly receive more emphasis.

For those tertiary oil recovery processes where thickened water must serve as the efficient driving fluid, the direction of the future research emerges from the discussions in the previous sections. The engineers and surface chemists will continue to try to find formulation which not only provide very low interfacial tensions between the oil and water but which will function to keep oil phase expanding or flowing at a high saturation during the displacement process. Recent work with microwave techniques indicate that the transitory high oil saturations at the leading edge of the oil-water bank during micellar flooding come from hysteresis in the fractional flow curve (141). This work may provide insight into possible new methods for optimizing this oil bank saturation at the bank-micellar (or solvent) interface to minimize deleterious capillary trapping in near-miscible systems.

New experimental work on oil-water interfacial viscosities and the coalescence of oil droplets will also be carried out because some authors have shown that these phenomena can affect oil recovery, especially in some oil-water-surfactant systems (142-145). However, to date it has been difficult to separate the net contributions of coalescence or interfacial viscosity from the well-documented influence of interfacial tension on oil recovery.

There already is a tendency to use higher quantities of certain alcohols as cosurfactants in these formulations, and it is anticipated that the use of these materials will increase. Although alcohols are expensive, the potential advantages for oil recovery are so great that future research may continue to examine the possibility of using alcohols as the main slug material for some processes.

There will also be efforts to improve the effectiveness of recovery processes by changing the geometry of the standard injection well-production well system. With these recovery fluids becoming more expensive, it is anticipated that there will be efforts to drill long lateral drain holes to put the fluids more precisely where they are needed in the reservoir. Drilling technology is improving and strides are being made which permit horizontal drilling with a shorter turning radius at the bottom of the well. Some of this emphasis has come with the increased need for in situ processes for coal gasification, but there are good reasons to use this technology to convert the inefficient radial, 5-spot pattern into a flood which may approach a linear drive. In addition to the marked improvement in efficiency for placing the fluids, horizontal drain holes can increase the flooding rate and the resultant $\Delta P/L\sigma$ ratio by a factor of 2 to 4 compared to the ordinary 5-spot patterns (38,96).

New attempts to recover the valuable fluids which are injected in recovery processes can be expected. Most field operations with micellar-polymer fluids are designed with the understanding that all of the surfactant and other chemicals are sacrificed to the reservoir. The cost of the lost chemicals must be made up by a very large increase in oil production. As hydrocarbons become more valuable, it appears that a system of alcohols may become practical if a large portion of these alcohols can be recovered by continuing the waterflooding beyond the point at which most of the oil has been produced. Economic recovery of the injected solvents will be difficult because there will be large quantities of single phase fluids composed of oil plus alcohol dissolved in large amounts of water and complete separation of these fluids (by distillation, etc.) might take more energy than that contained in the solvents and hydrocarbons recovered. Therefore, it has been suggested that solar energy (which is too expensive today) might provide some non-hydrocarbon energy for at least partial separation of the alcohol and water produced. Not all of the alcohol, oil and water mixture would need to be separated; some of it could be reinjected as a field process is expanded.

In conclusion, much has been learned since Uren, Fahmy and others started investigating the possibility of improving oil recovery by using materials to alter the surface properties of the oil-water-solid system. An ideal displacement process for displacing oil should include good mobility control and a method for keeping the oil phase continuous while low oil-water interfacial tension is maintained.

Present formulations and methods are designed now to provide just these conditions economically, but the difficulty of propelling the recovery fluids through the formation while maintaining these optimum conditions will keep surface chemists occupied for many years to come.

SYMBOLS

D average pore diameter [L] (Leverett)
$\bar{D}$ median pore diameter [L] (MacDonald and Dullien)
$\bar{D}_e$ mean pore entry diameter [L] (MacDonald and Dullien)
E_m microscopic displacement efficiency [Dimensionless]
f multiple oil filament length to radius ratio $\equiv \frac{\Delta L}{a}$ [Dimensionless], where ΔL = length of multipore oil mass [L] and a = average radius of trapped oil mass [L] (Stegemeier)
g gravitational constant $\equiv$ 32.2 (ft/sec)/sec [LT^{-2}]
K total (absolute) permeability [L^2]
k_{rw} relative permeability to displacing wetting phase (usually water) [Dimensionless]
L length over which pressure drop occurs [L]
ℓ mean length of an oil ganglion [L] (MacDonald and Dullien)
N_{ca} capillary number, $\equiv \frac{v\mu}{\phi\sigma} = \frac{Kk_{rw}\Delta P}{\phi \quad L\sigma}$ [Dimensionless] (Foster; Melrose and Brandner)
N_{Le} Leverett number $\equiv J_d K/\phi$ [Dimensionless], where J_d = drainage curvature, reciprocal of sum of interface radii [L^{-1}] (Stegemeier)

N_{vc} viscous/capillary number $\equiv \frac{v\mu}{\sigma} = \frac{Kk_{rw}\Delta P}{L\sigma}$ [Dimensionless] (Stegemeier)

P_c capillary pressure ("displacement" pressure as used by Leverett), normally $P_c = \frac{2\sigma\cos\theta}{r}$ $[ML^{-1}T^{-2}]$, where r = pore radius

ΔP pressure drop $[ML^{-1}T^{-2}]$

r_n pore neck radius [L] (Oh and Slattery)

v Darcy velocity of displacing wetting phase (usually water) $[LT^{-1}]$

Z_{imb} curvature correction factor for imbibition [Dimensionless] (Morrow)

Greek

θ contact angle at phase boundary between wetting and non-wetting phase (usually water and oil) respectively, normally measured through wetting phase [Dimensionless]

μ effective viscosity of displacing wetting phase (usually water) $\equiv \mu_w$ $[ML^{-1}T^{-1}]$

μ_w same as μ $[ML^{-1}T^{-1}]$

μ_o effective viscosity of displaced non-wetting phase (usually oil) $[ML^{-1}T^{-1}]$

σ interfacial tension between wetting and non-wetting phase (usually water and oil) $[MT^{-2}]$

ϕ porosity of medium [Dimensionless]

ψ geometrical factor for curvature of front and back of trapped oil mass [Dimensionless]

REFERENCES

1. E. DeGolyer, in "Elements of the Petroleum Industry," E. DeGolyer, ed., AIME, New York, 289 (1940).
2. R. C. Smith, "Mechanics of Secondary Oil Recovery," R. E. Kreiger, Huntington, New York (1975).
3. R. J. Forbes, "Studies in Early Petroleum History," E. J. Brill, ed., Leiden, Netherlands, 52 (1958).
4. F. Squires, U.S. Patent No. 1,238,355 (1917).
5. P. G. Nutting, Ind. Eng. Chem., 17, 1035 (1925).
6. J. M. Jamin, Compt. Rend., 50, 172, 311, 385 (1860).
7. S. C. Herold, "Oil Well Drainage," Stanford University Press, Stanford, California (1941).
8. J. C. Calhoun, Jr., "Fundamentals of Reservoir Engineering," University of Oklahoma Press, Norman, Oklahoma, 123 (1953).
9. R. D. Wyckoff and G. H. Botset, Physics, 7, 325 (1936).
10. L. D. Uren and E. H. Fahmy, Trans. AIME, 77, 318 (1927).
11. J. C. Melrose and C. F. Brandner, J. Can. Pet. Tech., 13, 54 (1974).

12. F. C. Benner, W. W. Riches, and F. E. Bartell, "API Drilling and Production Practice," 442 (1938).
13. F. C. Benner and F. E. Bartell, "API Drilling and Production Practice," 341 (1941).
14. F. C. Benner, C. G. Dodd, and F. E. Bartell, "API Drilling and Production Practice," 169 (1942).
15. T. F. Moore and R. L. Slobod, Prod. Monthly, 20, 20 (Aug. 1956).
16. E. Ammott, Trans. AIME, 216, 156 (1959).
17. N. R. Morrow, J. Can. Pet. Tech., 14, 42 (1975).
18. J. C. Melrose, Soc. Pet. Eng. J., 5, 259 (1965).
19. N. R. Morrow and F. G. McCaffery, in "Wetting, Spreading and Adhesion," J. J. Padday, ed., Academic Press, Inc., London, 289 (1978).
20. N. R. Morrow, J. Can. Pet. Tech., 15, 49 (1976).
21. N. R. Morrow, Paper No. 78-29-24, presented at the 29th meeting of the Pet. Soc. of CIM, Calgary, June 13 (1978).
22. P. G. Nutting, Oil & Gas J., 25, 45, 76, 150 (1927).
23. P. G. Nutting, Oil & Gas J., 25, 50, 32, 106 (1927).
24. P. G. Nutting, Oil & Gas J., 27, 22, 146, 238 (1928).
25. J. Reisberg and T. M. Doscher, Prod. Monthly, 21, 43 (Nov. 1956).
26. O. R. Wagner and R. O. Leach, Trans. AIME, 216, 65 (1959).
27. R. O. Leach, O. R. Wagner, H. W. Wood, and C. F. Harpke, Trans. AIME, 225, 206 (1962).
28. L. W. Emery, N. Mungan, and R. W. Nicholson, J. Pet. Tech., 1569 (1970).
29. C. E. Cooke, Jr., R. E. Williams, and P. A. Kolodzie, J. Pet. Tech., 1365 (1974).
30. D. J. Graue and C. E. Johnson, Jr., J. Pet. Tech., 1353 (1974).
31. C. E. Johnson, Jr., J. Pet. Tech., 85 (1976).
32. L. E. Treiber, D. L. Archer, and W. W. Owens, Soc. Pet. Eng. J., 12, 531 (1972).
33. J. G. Richardson, F. M. Perkins, and J. S. Osoba, Trans. AIME, 204, 89 (1955).
34. R. A. Salathiel, J. Pet. Tech., 1216 (1973).
35. R. A. Guereca and H. S. Butler, Prod. Monthly, 19, 21 (Jan. 1955).
36. H. T. Kennedy and G. T. Guerrero, Trans. AIME, 201, 124 (1954).
37. J. J. Taber, Trans. AIME, 213, 186 (1959).
38. J. J. Taber, Soc. Pet. Eng. J., 9, 3 (1969).
39. F. Fairbrother and A. E. Stubbs, J. Chem. Soc., 527 (1935).
40. M. C. Leverett, Trans. AIME, 132, 149 (1939).
41. L. E. Brownell and D. L. Katz, Chem. Eng. Progr., 42, 601 (1947).
42. E. Ojeda, F. Preston, and J. C. Calhoun, Prod. Monthly, 18, 20 (Dec. 1953).
43. P. G. Saffman and G. Taylor, Proc. Royal Soc. London, Ser. A, 245, 312 (1958).

44. W. R. Foster, J. Pet. Tech., 205 (1973).
45. E. G. Lefebvre duPrey, Soc. Pet. Eng. J., 13, 39 (1973).
46. R. Ehrlich, H. H. Hasiba, and P. Raimondi, J. Pet. Tech., 1335 (1974).
47. A. Abrams, Soc. Pet. Eng. J., 15, 437 (1975).
48. I. F. MacDonald and F.A.L. Dullien, Soc. Pet. Eng. J., 7 (1976).
49. R. L. Reed and R. N. Healy, in "Improved Oil Recovery by Surfactant and Polymer Flooding," D. O. Shah and R. S. Schechter, eds., Academic Press, Inc., New York, 383 (1977).
50. G. L. Stegemeier, in "Improved Oil Recovery by Surfactant and Polymer Flooding," D. O. Shah and R. S. Schechter, eds., Academic Press, Inc., New York, New York, 55 (1977).
51. S. G. Oh and J. C. Slattery, Soc. Pet. Eng. J., 19, 83 (1979).
52. O. R. Wagner and R. O. Leach, Soc. Pet. Eng. J., 6, 335 (1966).
53. H. S. Dombrowski and L. E. Brownell, Ind. Eng. Chem., 46, 1207 (1954).
54. L. H. Jenks, J. D. Huppler, N. R. Morrow, and R. A. Salathiel, J. Pet. Tech., 932 (1969).
55. J. J. Taber, J. C. Kirby, and F. U. Schroeder, AIChE Symp. Ser. No. 127, 53 (1973).
56. L. H. Jenks, J. D. Huppler, N. R. Morrow, and R. A. Salathiel, J. Can. Pet. Tech., 7, 172 (1968).
57. J. J. Stosur and J. J. Taber, J. Pet. Tech., 865 (1976).
58. M.H.M. Sayyouh, S. M. Farouq Ali, and C. D. Stahl, paper SPE 7639, presented at the SPE-AIME Eastern Regional Meeting, Washington, D.C., Nov. 1-3 (1978).
59. R. G. Larson, L. E. Scriven, and H. T. Davis, Nature, 268, 409 (1977).
60. L. W. Holm and G. G. Bernard, U.S. Patent No. 3,006,411 (1958).
61. A. K. Csaszar, U.S. Patent No. 2,356,205 (1942).
62. W. B. Gogarty and R. W. Olson, U.S. Patent No. 3,254,714 (1962).
63. S. C. Jones, U.S. Patent No. 3,497,006 (1967).
64. S. C. Jones, U.S. Patent No. 3,506,070 (1967).
65. W. B. Gogarty and W. C. Tosch, J. Pet. Tech., 1407 (1968); Trans. AIME, 243, 1407 (1968).
66. J. A. Davis, Jr. and S. C. Jones, J. Pet. Tech., 1415 (1968); Trans. AIME, 243, 1415 (1968).
67. L. W. Holm, U.S. Patent No. 3,537,520 (1970).
68. L. W. Holm, J. Pet. Tech., 1475 (1971); Trans. AIME, 251, 1475 (1971).
69. J. H. Schulman, W. Stoeckenius, and L. M. Prince, J. Phys. Chem., 63, 1677 (1959).
70. P. Somasundaran and H. S. Hanna, in "Improved Oil Recovery by Surfactant and Polymer Flooding," D. O. Shah and R. S. Schechter, eds., Academic Press, Inc., New York, New York, 205 (1977).

71. H. S. Hanna and P. Somasundaran, in "Improved Oil Recovery by Surfactant and Polymer Flooding," D. O. Shah and R. S. Schechter, eds., Academic Press, Inc., New York, New York, 253 (1977).
72. E. W. Malmberg and L. Smith, in "Improved Oil Recovery by Surfactant and Polymer Flooding," D. O. Shah and R. S. Schechter, eds., Academic Press, Inc., New York, New York, 275 (1977).
73. W. W. Gale and E. I. Sandvik, Soc. Pet. Eng. J., 13, 191 (1973).
74. F. J. Trogus, R. S. Schechter, G. A. Pope, and W. H. Wade, J. Pet. Tech., 769 (1979).
75. W. B. Gogarty, H. P. Meabon, and H. W. Milton, Jr., J. Pet. Tech., 141 (1970).
76. W. B. Gogarty and J. A. Davis, Jr., paper SPE 3806, presented at the SPE-AIME Improved Oil Recovery Symposium, Tulsa, Oklahoma, April 16-19, 1972.
77. B. B. Sandiford, in "Improved Oil Recovery by Surfactant and Polymer Flooding," D. O. Shah and R. S. Schechter, eds., Academic Press, Inc., New York, New York, 487 (1977).
78. R. Dawson and R. B. Lantz, Soc. Pet. Eng. J., 12, 448 (1972).
79. S. P. Trushenski, D. C. Dauben, and D. R. Parrish, Soc. Pet. Eng. J., 14, 633 (1974).
80. S. Vela, D. W. Peaceman, and E. I. Sandvik, paper SPE 5102, presented at the SPE-AIME 49th Annual Fall Meeting, Houston, Texas (1974).
81. J. G. Dominguez and G. P. Willhite, paper SPE 5835, presented at the SPE-AIME Improved Oil Recovery Symposium, Tulsa, Oklahoma, March 22-24 (1976).
82. S. P. Trushenski, in "Improved Oil Recovery by Surfactant and Polymer Flooding," D. O. Shah and R. S. Schechter, eds., Academic Press, Inc., New York, New York, 555 (1977).
83. J. L. Cayias, R. S. Schechter, and W. H. Wade, Am. Chem. Soc. Symp. Ser. No. 8, 235 (1975).
84. B. Vonnegut, Rev. Sci. Instrum., 13, 6 (1942).
85. J. L. Cayias, R. S. Schechter, and W. H. Wade, J. Coll. Interface Sci., 59, 31 (1977).
86. R. L. Cash, J. L. Cayias, G. Fournier, D. J. MacAllister, T. Schares, R. S. Schechter, and W. H. Wade, J. Coll. Interface Sci., 59, 39 (1977).
87. R. L. Cash, J. L. Cayias, G. Fournier, J. K. Jacobson, T. Schares, R. S. Schechter, and W. H. Wade, paper SPE 5813, presented at the SPE-AIME Improved Oil Recovery Symposium, Tulsa, Oklahoma, March 22-24 (1976).
88. J. K. Jacobson, J. C. Morgan, R. S. Schechter, and W. H. Wade, paper SPE 6002, presented at the SPE-AIME 51st Annual Meeting, New Orleans, Louisiana, Oct. 3-6 (1976).
89. R. L. Cash, J. L. Cayias, G. Fournier, K. J. Jacobson, C. A. LeGear, T. Schares, R. S. Schechter, and W. H. Wade, Oil Chemists Soc., Short Course at Hershey, Pennsylvania, June (1975).

90. M. Burkowsky and C. Marx, Erdoel-Erdgas Zeitschrift (Int'l Edition, 93, 33 (1977).
91. R. N. Healy and R. L. Reed, Soc. Pet. Eng. J., 14, 491 (1974); Trans. AIME, 257.
92. R. N. Healy, R. L. Reed, and C. W. Carpenter, Soc. Pet. Eng. J., 15, 87 (1975); Trans. AIME, 259, 87 (1975).
93. R. N. Healy, R. L. Reed, and D. G. Stenmark, Soc. Pet. Eng. J., 16, 147 (1976); Trans. AIME, 261, 147 (1976).
94. R. N. Healy and R. L. Reed, paper SPE 5817, presented at SPE-AIME Improved Oil Recovery Symposium, Tulsa, Oklahoma, Mar. 22-24 (1976).
95. W. D. Harkins and H. Zollman, J. Am. Chem. Soc., 48, 69 (1926).
96. J. H. Henderson and J. J. Taber, U.S. Patent No. 3,199,586 (1965).
97. P. M. Wilson, L. C. Murphy, and W. R. Foster, paper SPE 5812, presented at the SPE-AIME Improved Oil Recovery Symposium, Tulsa, Oklahoma, Mar. 22-24 (1976).
98. M. C. Puerto and W. W. Gale, paper SPE 5814, presented at the SPE-AIME Improved Oil Recovery Symposium, Tulsa, Oklahoma, Mar. 22-24 (1976).
99. C. A. Miller, R. Hwan, W. J. Benton, and T. Fort, Jr., J. Coll. Interface Sci., 61, 554 (1977).
100. W. J. Benton, T. Fort, Jr., and C. A. Miller, paper SPE 7579, presented at the SPE-AIME 53rd Annual Fall Meeting, Houston, Texas, Oct. 1-3 (1978).
101. R. N. Hwan, C. A. Miller, and T. Fort, Jr., J. Coll. Interface Sci., 68, 221 (1979).
102. L. E. Scriven, Nature, 263, 123 (1976).
103. K. S. Chan and D. O. Shah, paper SPE 7896, presented at the 1979 SPE-AIME Int'l Symp. on Oilfield and Geothermal Chemistry, Houston, Texas, Jan. 22-24 (1979).
104. V. K. Bansal and D. O. Shah, Soc. Pet. Eng. J., 18, 167 (1978).
105. R. L. Slobod and H. A. Koch, Jr., Oil Gas J., 51, 48, 85, 115 (1953).
106. E. R. Brownscombe, Oil & Gas J., 53, 6 133 (1954).
107. L. R. Kern, O. K. Kimbler, and R. Wilson, J. Pet. Tech., 16, (May 1958).
108. H. T. Kennedy, Oil & Gas J., 51, 8, 58, 69 (1952).
109. H. A. Koch, Jr. and R. L. Slobod, Trans. AIME, 210, 40 (1957).
110. D. M. Kehn, G. T. Pyndus, and M. G. Gaskell, J. Pet. Tech., 45 (June 1958).
111. H. J. Welge, E. J. Johnson, S. P. Dwing, Jr., and F. H. Brinkman, paper SPE 1525-G, presented at SPE-AIME Annual Fall Meeting, Denver, Colorado, Oct. (1960).
112. C. A. Hutchinson, Jr. and P. H. Braun, Trans. AIME, 219, 229 (1960).
113. B. H. Caudle and A. B. Dyes, Trans. AIME, 213, 281 (1958).
114. L. W. Holm, Prod. Monthly, 27, 6 (Sept. 1963).
115. D. M. Beeson and H. D. Ortloff, Trans. AIME, 216, 388 (1959).
116. L. W. Holm, Trans. AIME, 216, 225 (1959).

117. J. J. Rathmell, F. I. Stalkup, and R. C. Hassinger, paper SPE 3483, SPE-AIME, Dallas, Texas (1971).
118. L. W. Holm and V. A. Josendal, J. Pet. Tech., 1427 (1974).
119. E.T.S. Huang and J. H. Tracht, paper SPE 4735, presented at the SPE-AIME Improved Oil Recovery Symposium, Tulsa, Oklahoma, April (1974).
120. F. I. Stalkup, paper SPE 7042, presented at the SPE-AIME Improved Oil Recovery Symposium, Tulsa, Oklahoma, April 16-19 (1972).
121. J. W. Martin, Prod. Monthly, 15, 13 (May 1951).
122. A. Rosman and E. Zana, paper SPE 6723, presented at the 52nd Annual Fall Meeting, Denver, Colorado, Oct. 9-12 (1977).
123. R. S. Metcalfe and L. Yarborough, paper SPE 7061, presented at the SPE-AIME Improved Oil Recovery Symposium, Tulsa, Oklahoma, April 16-19 (1972).
124. H. Sarhosh, M.S. Thesis, Texas A & M University (1977).
125. R. A. Morse, German Patent No. 849,534, July (1952).
126. J. A. Sievert, J. M. Dew, and F. R. Conley, Trans. AIME, 213, 228 (1958).
127. R. L. Slobod, Prod. Monthly, 22, 24 (Feb. 1958).
128. N. Breston, Prod. Monthly, 24, 22 (Sept. 1960).
129. R. L. Slobod, Prod. Monthly, 26, No. 3, 2 (1962).
130. C. Gatlin and R. L. Slobod, Trans. AIME, 219, 46 (1960).
131. L. A. Wilson, R. J. Wygal, D. W. Reed, R. L. Gergins, and J. H. Henderson, Trans. AIME, 213, 146 (1958).
132. J. J. Taber, I.S.K. Kamath, and R. L. Reed, Soc. Pet. Eng. J., 1, 195 (1961).
133. L. W. Holm and A. K. Csaszar, Soc. Pet. Eng. J., 2, 129 (1962).
134. S. M. Farouq Ali and C. D. Stahl, Prod. Monthly, 27, No. 1, 2 (Jan. 1963).
135. E. J. Burcik, Prod. Monthly, 26, No. 1, 2 (March 1962).
136. J. J. Taber and W. K. Meyer, Soc. Pet. Eng. J., 4, 37 (1964).
137. R. C. Nelson and G. A. Pope, Soc. Pet. Eng. J., 18, 325 (1978).
138. P. A. Winsor, "Solvent Properties of Amphiphilic Compounds," Butterworth's Scientific Publications, London (1954).
139. R. G. Larson, paper SPE 6744, presented at the SPE-AIME 52nd Annual Fall Meeting, Denver, Colorado, Oct. 9-12 (1977).
140. P. Raimondi and M. A. Torcaso, Trans. AIME, 231, 49 (1964).
141. R. E. Gladfelter and S. P. Gupta, paper SPE 7577, presented at the SPE-AIME 53rd Annual Fall Meeting, Houston, Texas, Oct. 1-3 (1978).
142. R. J. Mannheimer and R. S. Schechter, J. Coll. Interface Sci., 32, 195 (1970).
143. J. C. Slattery, AIChE J., 20, 1145 (1974).
144. D. T. Wasan and V. Mohan, in "Improved Oil Recovery by Surfactant and Polymer Flooding," D. O. Shah and R. S. Schechter, eds., Academic Press, Inc., New York, New York, 161 (1977).

145. R. W. Flumerfelt and A. C. Payatakes, in "Enhanced Oil Recovery and Improved Drilling Technology," Progress Review No. 19, USDOE, BETC - 79/3, 59 (Sept. 1979).

THE PHYSICO-CHEMICAL CONDITIONS NECESSARY TO PRODUCE ULTRALOW INTERFACIAL TENSION AT THE OIL/BRINE INTERFACE

K. S. Chan* and D. O. Shah

Departments of Chemical Engineering and Anesthesiology
University of Florida
Gainesville, Florida 32611, U.S.A.

In a petroleum sulfonate/isobutanol/dodecane/brine system, there are two regions of ultralow interfacial tension (IFT), one at low surfactant concentrations (0.1-0.2%) and the other at higher surfactant concentrations (4 to 10%). In the low concentration range, the oil/brine/surfactant/alcohol system is a two-phase system, whereas at high surfactant concentrations, it becomes a three-phase system in which a middle phase microemulsion is in equilibrium with excess brine and oil. For low surfactant concentration systems, we have shown that the ultralow IFT minimum corresponds to the onset of micellization and partition coefficient of surfactant near unity. This correlation was observed for the effect of surfactant concentration, salt concentration and oil chain length on the interfacial tension. The minimum in interfacial tension corresponds to a maximum electrophoretic mobility of oil droplets. This correlation was also observed for the effect of caustic on several crude oils.

In high surfactant concentration systems, a middle phase microemulsion forms in equilibrium with excess oil and brine in a given salinity range. The middle phase microemulsion contains equal volumes of oil and brine and practically all of the surfactant at a specific salinity defined as the optimal salinity of the given system. The interfacial tension of the two interfaces, middle phase/brine and middle phase/oil, depends on the extent of solubilization of oil and brine in the middle phase microemulsion. The higher the solubilization of oil and brine in the middle phase, the lower is the interfacial tension at both these interfaces. We have

*Present Address: Dr. K. S. Chan, SOHIO Laboratories, 3092 Broadway Ave., Cleveland, Ohio 44115, U.S.A.

shown that the optimal salinity increases with the oil chain length for a given surfactant system. The optimal salinity decreases with an increase in the alcohol chain length. As the oil chain length increases the solubilization of oil and brine in the middle phase microemulsion at the optimal salinity decreases. The formation of middle phase microemulsions appears to be similar to coacervation phenomenon and quite different from the solubilization of brine in oil external microemulsions. The $\ell \rightarrow m \rightarrow u$ transition can be achieved by changing one of the following variables: Salinity, Temperature, Alcohol Chain Length, Oil Chain Length, Oil/Brine Ratio, Surfactant Concentration, Surfactant Solution/Oil Ratio, and the Molecular Weight of Surfactant.

INTRODUCTION

It is well recognized that the two major forces acting on an entrapped oil ganglion in the reservoir pore structure are viscous forces and capillary forces. Melrose and Brandner (1) have shown that the capillary number must be increased from 10^{-6} to 10^{-2} for mobilization of entrapped oil ganglia in order to improve the microscopic displacement efficiency. Under practical reservoir conditions, this change can be brought about by decreasing the interfacial tension from about 20 dynes to 10^{-3} dynes/cm. It has been shown by several investigators (2-7) that using a suitable surfactant system, an ultralow interfacial tension can be achieved at the oil/brine interface. At low surfactant concentrations, the oil/brine/surfactant/alcohol systems form two phases whereas often at high surfactant concentrations, middle phase microemulsions form in equilibrium with excess oil and brine (8,9). The interfacial tension of the middle phase microemulsion with excess oil or excess brine depends upon the magnitude of solubilization of oil and brine in the middle phase. The higher the solubilization of oil and brine in the middle phase, the lower is the interfacial tension. The salinity at which the middle phase microemulsion contains equal volumes of oil and brine is defined as the optimal salinity. The oil recovery is found to be maximum at or near the optimal salinity (8,10). At optimal salinity, the phase separation time or coalescence time of emulsions and the apparent viscosity of these emulsions in porous media are found to be minimum (11,12). Therefore, it appears that upon increasing the salinity, the surfactant migrates from the lower phase to middle phase to upper phase in an oil/brine/surfactant/alcohol system. The $\ell \rightarrow m \rightarrow u$ transition can be achieved by also changing any of the following variables: Temperature, Alcohol Chain Length, Oil/Brine Ratio, Surfactant Solution/Oil Ratio, Surfactant Concentration and Molecular Weight of Surfactant. The present paper summarizes our extensive studies on the low and high surfactant concentration systems and related phenomena necessary to achieve ultralow interfacial tension in oil/brine/surfactant/alcohol systems.

EXPERIMENTAL

The petroleum sulfonates TRS 10-80 (≈ 80% active) and TRS 10-410 (≈ 60% active) were supplied by Witco Chemicals Company. The average equivalent weight of these surfactants was 420. Alkyl monophosphate ester, Klearfac AA-270 (≈ 90% active) was supplied by BASF-Wyandotte Corporation. Paraffinic oils and short chain alcohols of > 99% purity were purchased from Chemical Samples Company. All chemicals were used as received. Aqueous solutions of surfactants were prepared on weight basis.

Low surfactant concentration systems: The systems were equilibrated by taking two-thirds aqueous solution and one-third oil by volume in 250 ml separatory funnels maintained at 25°C. After shaking vigorously for about 30 minutes using a mechanical shaker, the solutions were allowed to stand for about 3 to 6 weeks until clear interfaces were obtained. The equilibrated oil and aqueous phases were then separated from the top and bottom of the funnel respectively for various physico-chemical measurements.

For mixed surfactant systems, the effect of Klearfac AA-270 on the interfacial tension was studied by gradually replacing TRS 10-410 with Klearfac AA-270 and keeping the total surfactant concentration at 0.2%.

Interfacial tensions between the equilibrated oil and aqueous phases were measured by a spinning drop tensiometer using the method described by Schechter and co-workers (2,3,5). Each drop of oil was spun until a constant interfacial tension was reached. The temperature was maintained at 25 ± 1°C.

High surfactant concentration systems: The petroleum sulfonate and alcohol of given quantities were dissolved in equal volumes of oil and brine. The system was shaken vigorously and was rotated on a slow rotor for several days and was then allowed to stand until the phase volumes did not show any change with time. The surfactant concentration was determined by two-phase dye-titration or UV-absorbance measurements.

RESULTS AND DISCUSSION

The effect of surfactant concentration on interfacial tension in the TRS 10-410/IBA/Dodecane/Brine system is shown in Figure 1. It is evident that there are two regions in which ultralow interfacial tension is observed, i.e., in the low concentration region around 0.1% and the other around 4% surfactant concentration. In the low concentration region, the system forms two phases, namely, oil and brine whereas in the high concentration region the middle phase microemulsion is in equilibrium with excess oil and brine

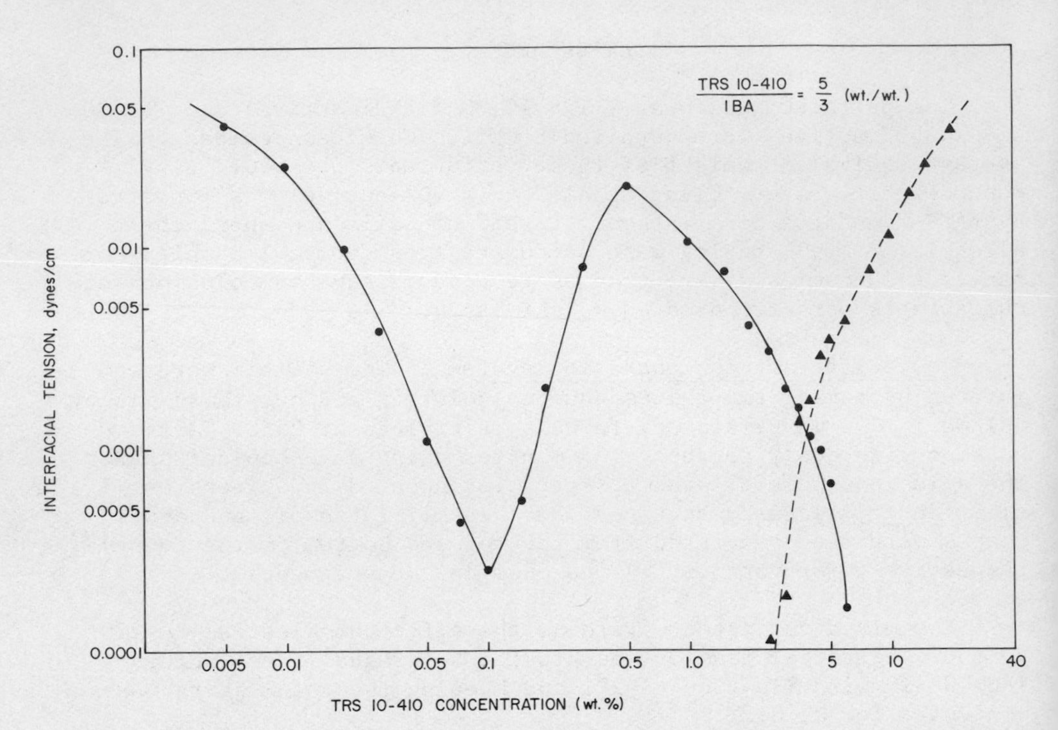

Fig. 1. Effect of surfactant concentration on interfacial tension of TRS 10-410 + IBA in 1.5% NaCl solution with dodecane.

(13). First, we will discuss the molecular mechanism responsible for the ultralow interfacial tension in the low concentration region and, subsequently, we shall discuss the phenomena related to the three-phase system.

Figure 2 summarizes the effect of NaCl concentration, oil chain length and surfactant concentration on the interfacial tension in this system. It is evident that ultralow interfacial tension minimum occurs at a specific salinity or specific oil chain length or specific surfactant concentration. Using light scattering, osmotic pressure, surface tension, dye solubilization and various other techniques, we have shown that the ultralow interfacial tension minimum in this system correlates with the onset of micellization in the aqueous phase and the partition coefficient of surfactant near unity (14). Baviere (15) has also shown that the partition coefficient is near unity at the salinity where minimum interfacial tension occurs.

Figure 3 schematically illustrates the effect of NaCl concentration on the partitioning of surfactant molecules in an oil/brine system. At low salinities, most of the surfactant molecules are found in the brine phase whereas at very high salinities, most of the surfactant molecules are found in the oil phase. Hence, as the

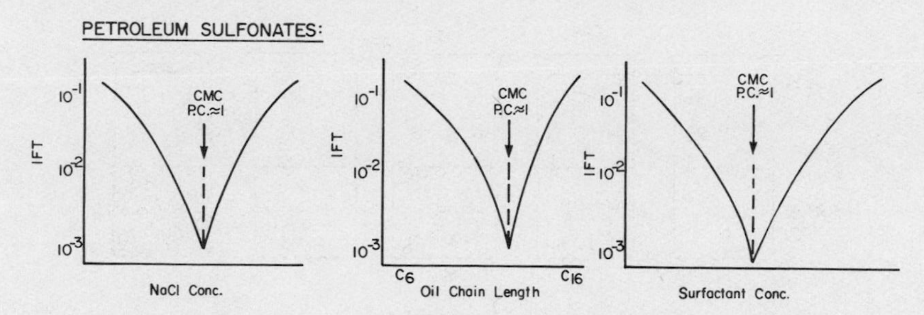

Fig. 2. The effect of salt concentration, oil chain length and surfactant concentration on interfacial tension.

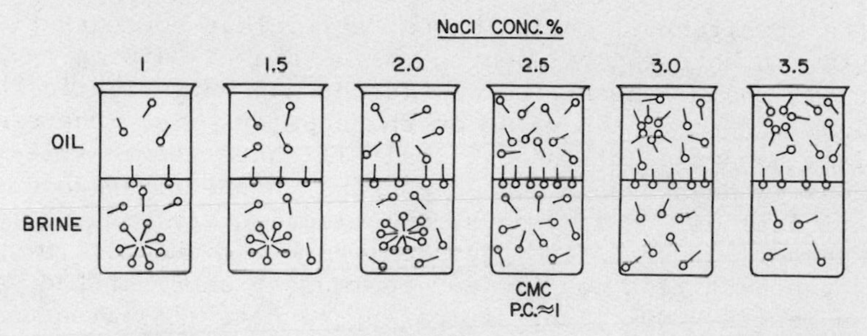

Fig. 3. The effect of salt concentration on the partitioning of surfactant and micelle formation.

salinity is increased, the surfactant molecules gradually migrate from the aqueous to the oil phase. At a specific salinity, the surfactant concentration in the oil and aqueous phase becomes equal leading to the partition coefficient of unity. Also there will be a specific salinity at which the surfactant concentration in the aqueous phase reaches the value of the 'apparent cmc'. We are using the words apparent cmc because the micellization in this case occurs in the presence of dissolved oil and salt and hence it is likely to be different from the cmc of the same surfactant in pure water without oil. In the system studied in this paper, both the apparent cmc and the partition coefficient near unity occurred at the same salinity. However, we have found (16) that in other systems, this may occur at different salinities. Thus, it appears that the minimum in ultralow IFT corresponds to the onset of micellization rather than to the partition coefficient of unity.

The effect of oil chain length on the partition coefficient of surfactant in the oil/brine system is explained in Figure 4. We have shown (14) that as the oil chain length is increased, the partitioning of the surfactant in the oil phase decreases and consequently the surfactant concentration in brine increases. There-

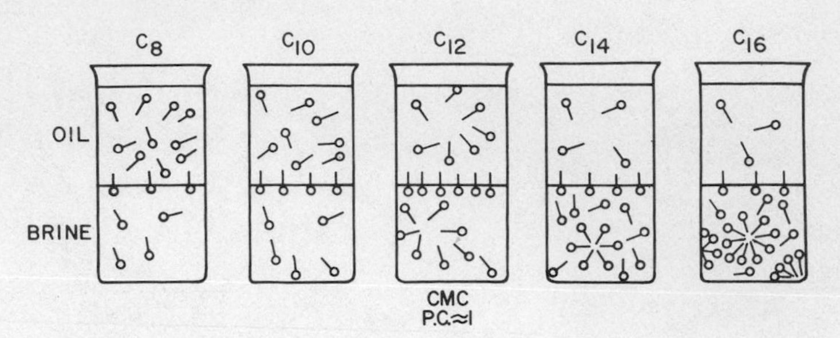

Fig. 4. The effect of oil chain length on the partitioning of surfactant and micelle formation.

fore, at a specific oil chain length, surfactant concentration becomes equal in both the oil and the brine phase. Similarly, at a specific oil chain length, the surfactant concentration in the aqueous phase reaches the value of the apparent cmc. Therefore, on the left hand side of the critical oil chain length (Figure 4), the aqueous phase has no micelles whereas on the right hand side of the critical oil chain length, the micelles are present in the aqueous phase. In the present system, both the onset of micellization and the partition coefficient near unity occur at the same oil chain length. Our light scattering results (14) also confirm that the micelles begin to form at the same chain length where the minimum IFT is observed.

Figure 5 schematically illustrates the effect of surfactant concentration on partitioning of the surfactant and micellization phenomena. One can consider that the petroleum sulfonate is having both oil soluble and water soluble species. At very low surfactant concentrations (e.g., C_1), water soluble species will remain in brine and the oil soluble species will partition into oil. At the oil/brine interface both these species can be adsorbed. However, as the surfactant concentration is increased, the concentration of oil soluble species in oil and water soluble species in brine increases. At a specific concentration (e.g., C_3), the water soluble species can reach to its cmc and begin to form micelles. Above this critical concentration, the oil soluble species can remain in the micelles. In other words, once the water soluble species begin to form micelles in the aqueous phase, the oil soluble species can either solubilize in micelles or partition into oil. The solubilization of the oil soluble species in the micelles decreases the apparent cmc and hence the surfactant monomer concentration. The reduction in monomer concentration results in the decrease of surface concentration of the water soluble species and hence the interfacial tension increases. The more oil soluble species dissolve into the micelles, the greater is the decrease in the apparent cmc and hence the corresponding decrease in the monomer concentration resulting in the increase in interfacial tension. The explanation as shown in Figure 5 explains the

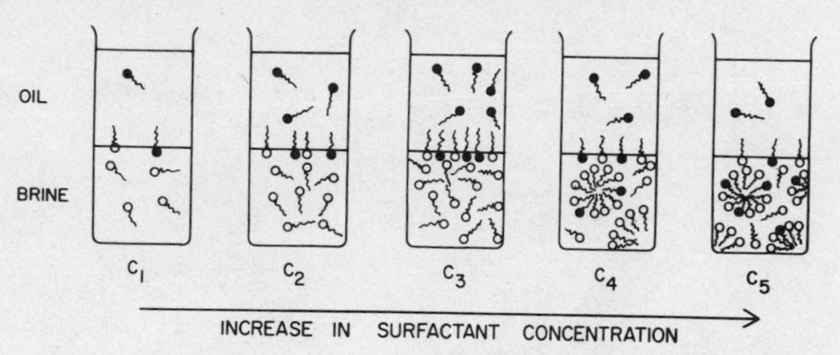

Fig. 5. The molecular mechanism for the effect of surfactant concentration on interfacial and surface tensions.

effect of surfactant concentration on interfacial tension in the systems studied in this paper.

Figure 6 shows the effect of surfactant concentration on interfacial tension and electrophoretic mobility of oil droplets (14). It is evident that the minimum in interfacial tension corresponds to a maximum in electrophoretic mobility and hence in zeta potential at the oil/brine interface. Similar to the electrocapillary effect observed in mercury/water systems, we believe that the high surface charge density at the oil/brine interface also contributes to lowering of the interfacial tension. This correlation was also observed for the effect of caustic concentration on the interfacial tension of several crude oils (Figure 7). Here also, the minimum interfacial tension and the maximum electrophoretic mobility occurred in the same range of caustic concentration (17). Similar correlation for the effect of salt concentration on the interfacial tension and electrophoretic mobility of a crude oil was also observed (18). Thus, we believe that surface charge density at the oil/brine interface is an important component of the ultralow interfacial tension.

An overview of several variables that influence the interfacial tension and the important processes that occur in oil/brine/surfactant systems, namely, the partitioning of the surfactant and the effective cmc or monomer concentration of the surfactant are shown in Figure 8. These two processes will determine the three major components of the interfacial tension, namely, surface concentration of the surfactant, surface charge density and solubilization of oil or brine in each other. Hence, as shown in Figure 9, the magnitude of interfacial tension can be adjusted by changing any of these three variables. The surface concentration is the major variable, then surface charge density and then solubilization of oil or brine. By adjusting these variables, one can push the system to ultralow interfacial tension. Using this concept, we were able to decrease the interfacial tension by increasing the surface charge density upon the addition of alkyl monophosphate esters in this system.

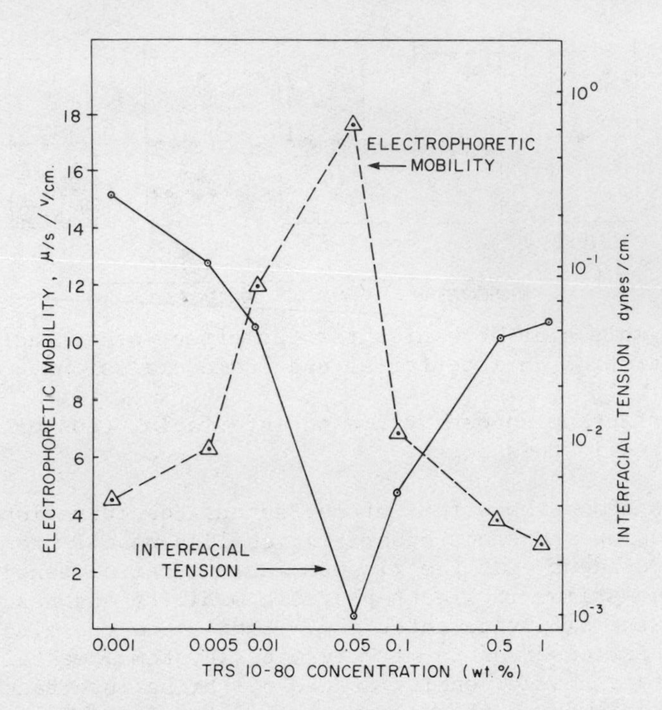

Fig. 6. The effect of TRS 10-80 concentration on interfacial tension and electrophoretic mobility of n-octane droplets in surfactant solutions, Temp. ≈ 28°C.

Figure 10 shows the interfacial tension of an oil/brine system by a mixture of petroleum sulfonate (TRS 10-410) and alkyl monophosphate ester (Klearfac AA-270). As the proportion of phosphate ester increases, the magnitude of interfacial tension decreases and there is a considerable broadening of the interfacial tension minimum. Moreover, the minimum shifts to a higher salinity. We explained this on the basis of the number of anionic atoms per polar group. In the petroleum sulfonate, there is one negatively charged oxygen per sulfonate group whereas in the phosphate group, there are two negatively charged oxygens per polar group. Therefore, as shown in Figure 11, the mixed micelles of the petroleum sulfonate and the alkyl monophosphate esters will have a greater surface charge density at the micelle surface as well as at the oil/brine interface. We believe that these results support our conclusion that the surface charge density is an important component of the ultralow interfacial tension.

The Formation of Middle Phase Microemulsions in High Surfactant Concentration Systems

The effect of salinity on the surfactant rich phase in the oil/brine/surfactant/alcohol systems is shown in Figure 12. It is

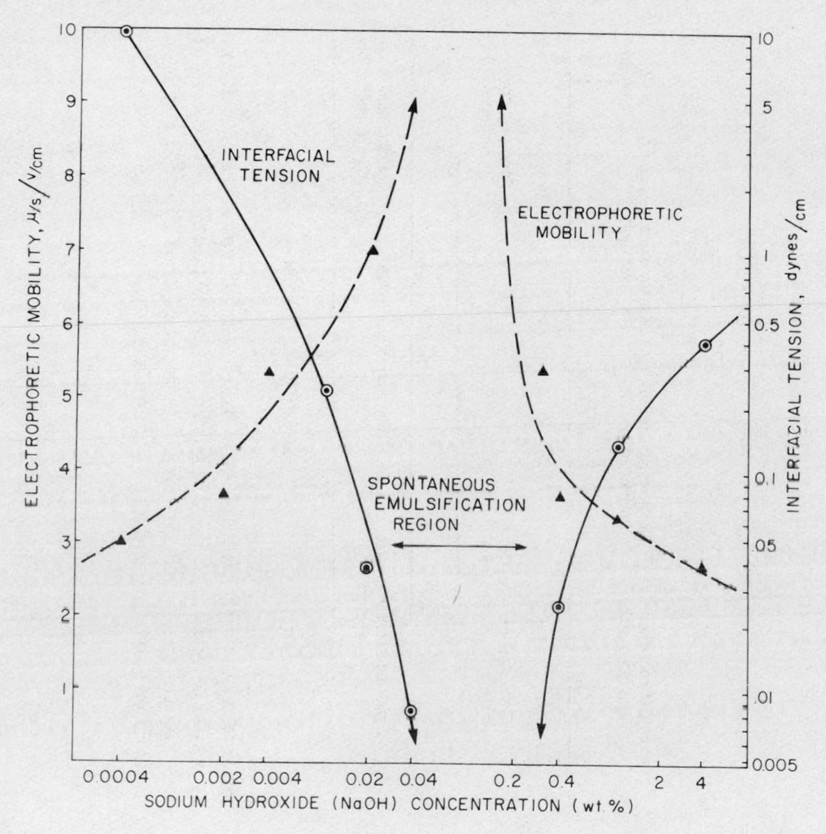

Fig. 7. Electrophoretic mobility and interfacial tension for crude oil-NaOH solutions.

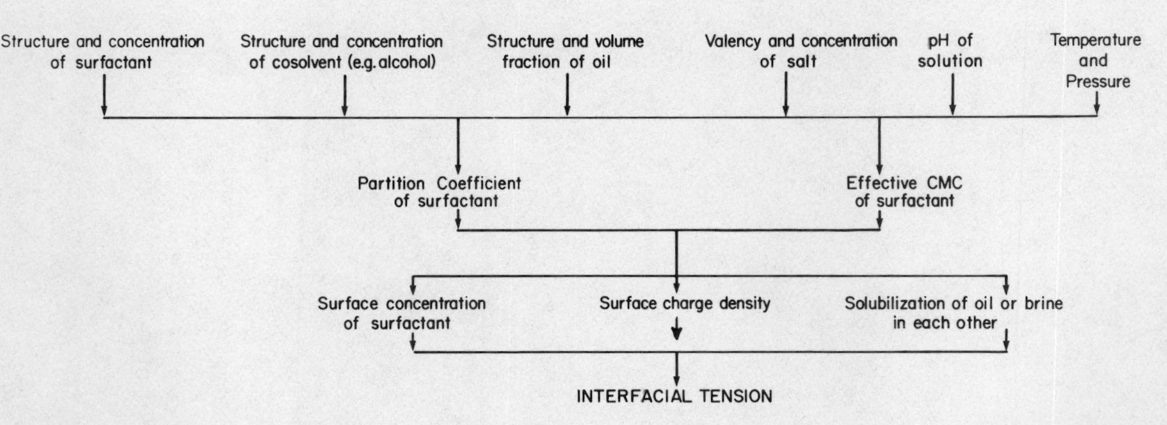

Fig. 8. An overview of the effect of several variables on processes occurring in oil/brine/surfactant/alcohol systems in relation to the three parameters influencing the magnitude of interfacial tension.

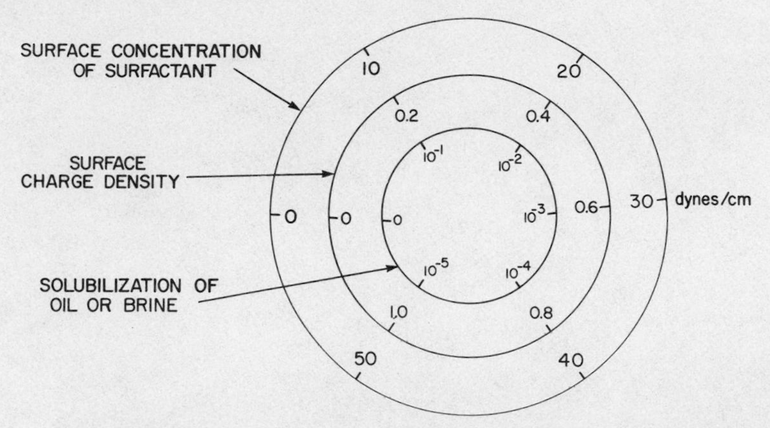

Fig. 9. A schematic presentation of three components of interfacial tension in the oil/brine/surfactant systems.

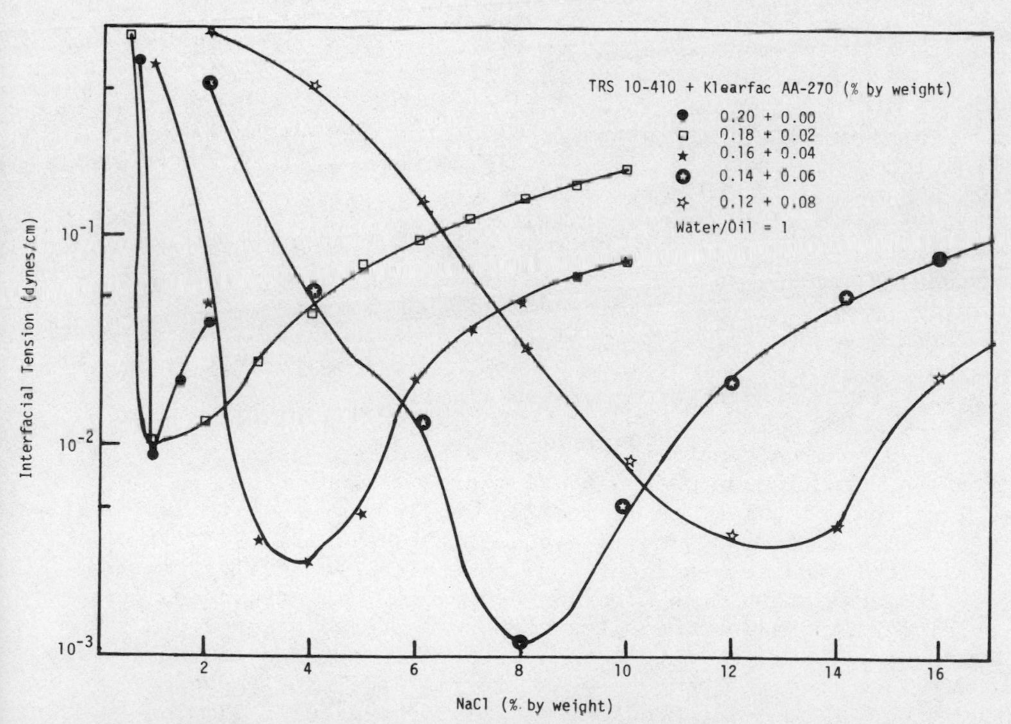

Fig. 10. Effect of salinity on the magnitude and broadening of the interfacial tension by the addition of Klearfac AA-270 in 0.2% surfactants + 0.12% IBA/n-octane system.

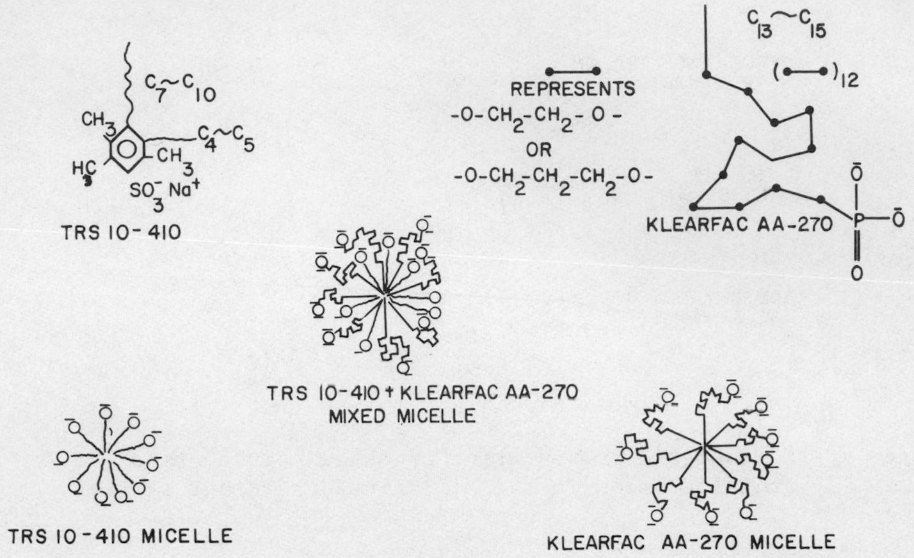

Fig. 11. The schematic presentation of the increase in the surface charge density at the micellar surface and oil/water interface due to mixing of Klearfac AA-270 with petroleum sulfonate TRS 10-410.

evident that as the salinity increases the surfactant migrates from the lower phase to middle phase to upper phase microemulsions. The salinity at which equal volumes of oil and brine are solubilized in the middle phase microemulsion is defined as the optimal salinity (8). It has been further shown that the higher the magnitude of solubilization, the greater is the reduction in interfacial tension at the oil/microemulsion or brine/microemulsion interface (8,19-21). The optimal salinity can be shifted to a higher value by incorporation of ethoxylated sulfonates or alcohols in the surfactant formulation (19-21).

Figure 13 schematically illustrates our proposed mechanism for the formation of middle phase microemulsions (13). At low salinities, micelles are formed in the aqueous phase in equilibrium with oil. As the salinity increases, the solubilization of oil within the micelles increases and the thickness of electrical double layer around the micelles decreases. The reduction in repulsive forces allows micelles to approach each other closely and subsequently a micelle-rich phase separates out due to the density difference from the aqueous phase forming the middle phase microemulsion. Hence, the middle phase microemulsion is similar to coacervation process in micellar solution where a micelle-rich phase separates out upon addition of salts. The presence of oil only contributes towards the solubilization of oil within the micelles. Ultimately, at higher salinities, surfactant preferen-

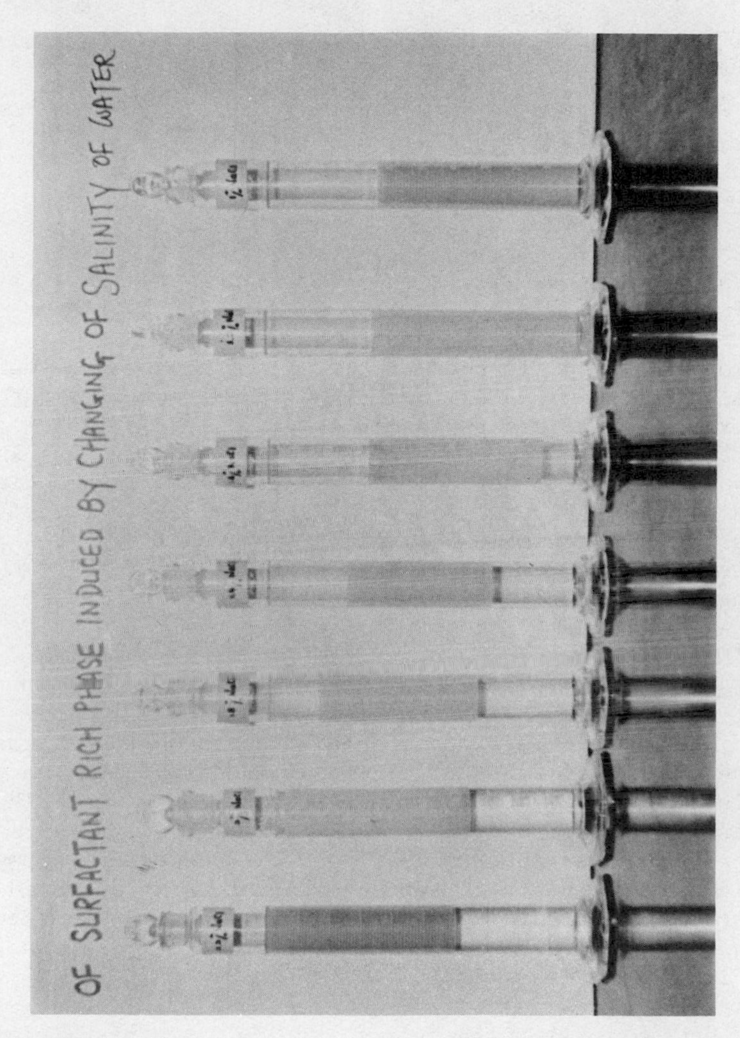

Fig. 12. The formation of lower phase, middle phase and upper phase microemulsions due to increase in salinity of oil/brine/surfactant/alcohol systems.

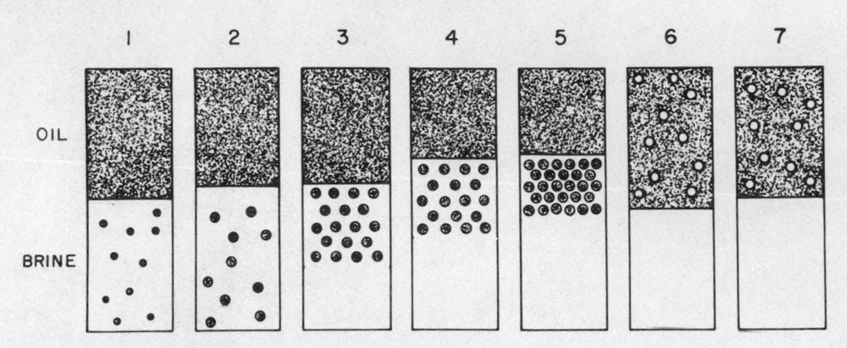

Fig. 13. A schematic presentation of the mechanism of formation of the middle phase microemulsions upon increasing salinity.

tially partitions into the oil phase forming reverse micelles or water-in-oil microemulsions. Figure 14 shows the freeze-fracture electronmicrograph of a middle phase microemulsion at optimal salinity. It clearly shows the swollen microemulsion droplets in a continuous aqueous phase. Our additional studies (22) on sodium/calcium equivalent ratio strongly support the conclusion that at the optimal salinity, the middle phase microemulsion is water-external. However, it is well recognized that the microemulsion structure changes in the salinity range over which middle phase microemulsion exists because the solubilization of oil and brine in the middle phase microemulsion changes depending upon the salinity.

Figure 15 shows the effect of oil chain length on the optimal salinity in the present system (23). It is evident that as the oil chain length increases, the optimal salinity also increases. The interfacial tension at the corresponding optimal salinity also increases with the chain length of oil. Figure 16 shows the volume of the middle phase and the surfactant concentration in the middle phase as a function of the oil chain length. It is clear that the solubilization of oil and brine in the middle phase decreases as the oil chain length increases (23).

Figure 17 schematically illustrates the $\ell \rightarrow m \rightarrow u$ transition in the oil/brine/surfactant/alcohol systems in relation to several variables. Therefore, it appears that the surfactant migration is a general phenomenon that can be brought about by changing of several variables. Figure 18 schematically illustrates the comparison of low surfactant concentration and high surfactant concentration systems. It is interesting that in low surfactant concentration systems (two phase systems) the ultralow interfacial

Fig. 14. Freeze-fracture electronmicrograph of a middle phase micro-emulsion in the Exxon system [4% emulsifier (MEAC 120 XS/TAA::63/37) + 48% oil (ISOPAR M/HAN::90/10) + 48% brine (1.2% NaCl) in D_2O.] The bar represents 0.5 micron.

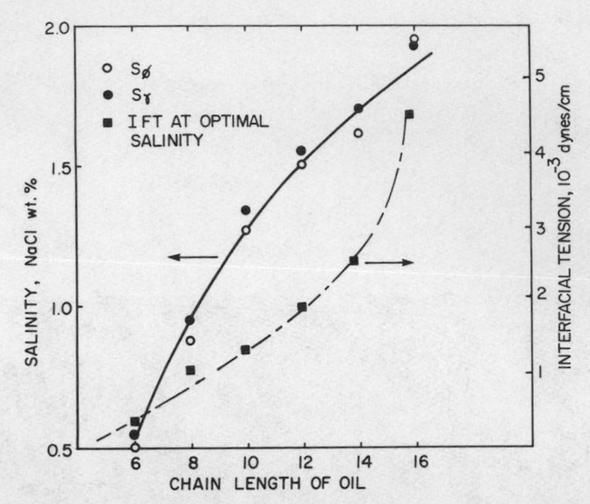

Fig. 15. The effect of oil chain length on the optimal salinity and the corresponding interfacial tension of 5% TRS 10-410 + 3% IBA system.

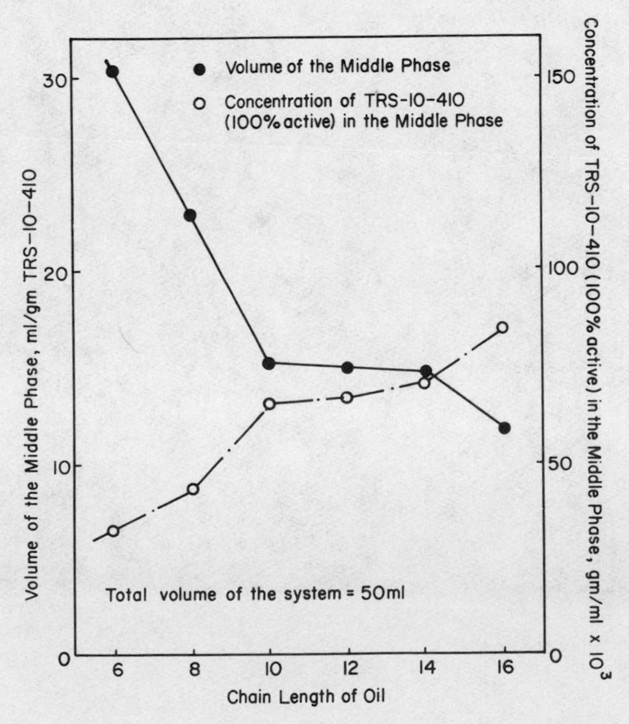

Fig. 16. Volume of the middle phase and concentration of TRS 10-410 (100% active) in the middle phase at optimal salinity for different chain length of oils.

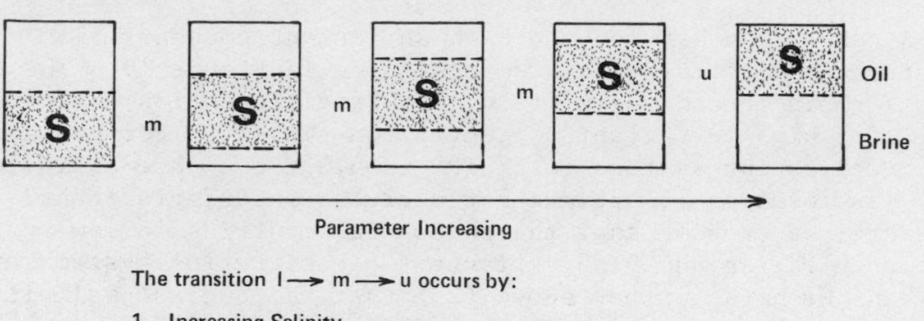

Fig. 17. Factors influencing the formation of ℓ-, m-, u- microemulsions for oil/brine/surfactant/alcohol systems.

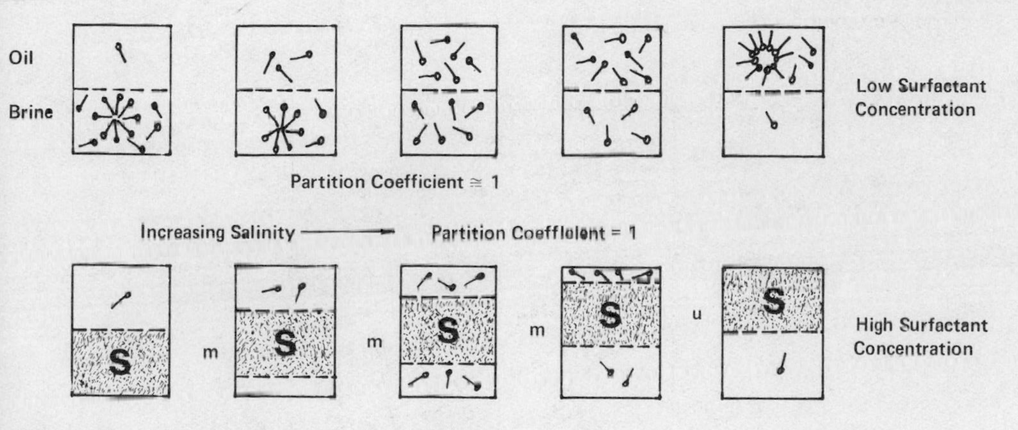

Fig. 18. A comparison of phenomena occurring in low and high surfactant concentration systems.

tension is observed near the partition coefficient unity (14). In the high surfactant concentration systems, at optimal salinity the partition coefficient of surfactant in excess oil and excess brine is also near unity (13). In other words, the middle phase microemulsion seems to be a reservoir of excess surfactant, whereas the excess oil and brine phase resemble that of low surfactant concentration systems. We have also found that at optimal salinity the interfacial tension between excess oil and excess brine is minimum due to partition coefficient near unity (13).

A comparison between the high surfactant concentration and low surfactant concentration systems is shown in Figure 19. The optimal oil chain length for ultralow interfacial tension is found to be C_{12} for high surfactant concentration systems. For low surfactant concentration systems (0.1% TRS 10-410), the same chain length (C_{12}) was found to be optimum for ultralow interfacial tension. Therefore, we propose that the phenomena occurring in low surfactant concentration and high surfactant concentration systems are similar. We have further shown (23) that the optimal salinity decreases with alcohol chain length (23). In other words, the higher the alcohol chain length, the lower is the optimal salinity of a given surfactant/alcohol/oil/brine system. The effect of alcohol on the optimal salinity is related to the brine solubility of the alcohol. The higher the solubility of the alcohol in brine, the higher is the optimal salinity (24).

In summary, the formation of middle phase microemulsion at the optimal salinity is an important phenomenon with respect to ultralow interfacial tension, solubilization, rate of coalescence and oil displacement efficiency in porous media (10,11,25). Also, the optimal salinity can be shifted to a desired value by adjusting several variables.

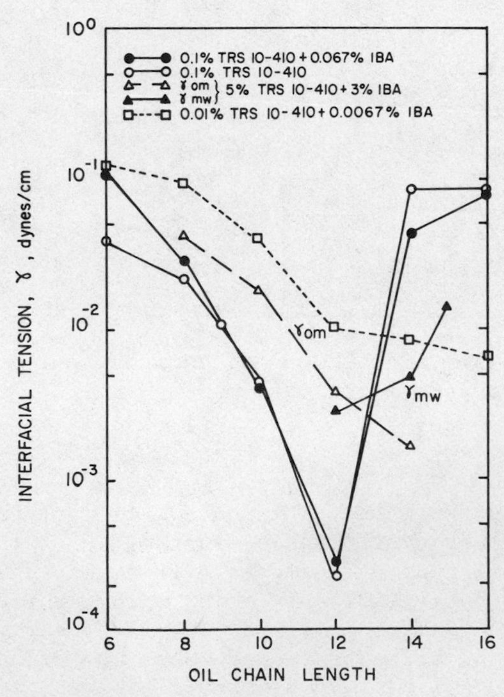

Fig. 19. Effect of oil chain length on the interfacial tension in high and low surfactant concentration systems.

ACKNOWLEDGMENTS

We wish to express our sincere appreciation and thanks to Mr. S. N. Rashid for obtaining results on mixed surfactant systems containing alkyl monophosphate ester (Klearfac AA-270) and to the National Science Foundation-RANN, ERDA and the Department of Energy (Grant No. DE-AC1979BC10075) and the consortium of the following Industrial Associates for their generous support of the University of Florida Enhanced Oil Recovery Research Program: 1) Alberta Research Council, Canada, 2) American Cyanamid Co., 3) Amoco Production Co., 4) Atlantic Richfield Co., 5) BASF-Wyandotte Co., 6) British Petroleum Co., England, 7) Calgon Corp., 8) Cities Service Oil Co., 9) Continental Oil Co., 10) Ethyl Corp., 11) Exxon Production Research Co., 12) Getty Oil Co., 13) Gulf Research and Development Co., 14) Marathon Oil Co., 15) Mobil Research and Development Co., 16) Nalco Chemical Co., 17) Phillips Petroleum Co., 18) Shell Development Co., 19) Standard Oil of Ohio, 20) Stepan Chemical Co., 21) Sun Oil Chemical Co., 22) Texaco Inc., 23) Union Carbide Corp., 24) Union Oil Co., 25) Westvaco Inc., and 26) Witco Chemical Co.

REFERENCES

1. J. C. Melrose and C. F. Brandner, J. Can. Pet. Tech., 54 (Oct.-Dec., 1974).
2. I. H. Doe, W. H. Wade, and R. S. Schechter, J. Colloid Interface Sci., 59, 525 (1977).
3. J. L. Cayias, R. S. Schechter, and W. H. Wade, J. Colloid Interface Sci., 59, 31 (1977).
4. P. M. Wilson and C. F. Brandner, paper presented at the 165th National Meeting of the American Chemical Society, Dallas, Texas, April 8-13, 1973.
5. J. C. Morgan, R. S. Schechter, and W. H. Wade, in "Improved Oil Recovery by Surfactant and Polymer Flooding," D. O. Shah and R. S. Schechter, eds., Academic Press, New York, p. 101, 1977.
6. K. Shinoda and S. Friberg, Adv. Colloid Interface Sci., 4, 281 (1975).
7. D. R. Anderson, M. S. Bidner, H. T. Davis, C. D. Manning, and L. E. Scriven, SPE 5811 presented at the Improved Oil Recovery Symposium held in Tulsa, Oklahoma, March 22-24, 1976.
8. R. L. Reed and R. N. Healy, in "Improved Oil Recovery by Surfactant and Polymer Flooding," D. O. Shah and R. S. Schechter, eds., Academic Press, New York, p. 383, 1977.
9. M. Y. Chiang, Ph.D. Thesis, University of Florida, 1979.

10. M. Y. Chiang and D. O. Shah, SPE 8988 presented at the 5th International Symposium on Oilfield and Geothermal Chemistry held in Stanford, California, May 28-30, 1980.
11. S. Vijayan, C. Ramachandran, H. Doshi, and D. O. Shah, in "Surface Phenomena in Enhanced Oil Recovery, D. O. Shah, ed., Plenum Press, New York (in press).
12. J. E. Vinatieri, Soc. Pet. Eng. J., 20, 402 (1980).
13. K. S. Chan and D. O. Shah, SPE 7869 presented at the International Symposium on Oilfield and Geothermal Chemistry held in Houston, Texas, January 22-24, 1979.
14. K. S. Chan and D. O. Shah, J. Disp. Sci. Tech., 1, 55 (1980).
15. M. Baviere, SPE 6000 presented at the 51st Annual Fall Technical Conference and Exhibition of the Society of Petroleum Engineers of AIME, New Orleans, Louisiana, October 3-6, 1976.
16. J. C. Noronha, Ph.D. Thesis, University of Florida, 1980.
17. V. K. Bansal, K. S. Chan, R. McCullough, and D. O. Shah, J. Canad. Pet. Tech., 17(1), Jan.-March, 1978.
18. M. Y. Chiang, K. S. Chan, and D. O. Shah, J. Canad. Pet. Tech., 17(4), Oct.-Dec., 1978.
19. V. K. Bansal and D. O. Shah, J. Colloid Interface Sci., 65, 451 (1978).
20. V. K. Bansal and D. O. Shah, Soc. Pet. Eng. J., 167 (June, 1978).
21. V. K. Bansal and D. O. Shah, J. Amer. Oil Chem. Soc., 55, 367 (1978).
22. S. I. Chou and D. O. Shah, J. Colloid Interface Sci. (in press).
23. W. C. Hsieh and D. O. Shah, SPE 6594 presented at the International Symposium on Oilfield and Geothermal Chemistry held in La Jolla, California, June 27-28, 1977.
24. W. C. Hsieh, Ph.D. Thesis, University of Florida, 1977.
25. D. O. Shah, in "Surface Phenomena in Enhanced Oil Recovery," D. O. Shah, ed., Plenum Press, New York (in press).

FURTHER STUDIES ON PHASE RELATIONSHIPS IN CHEMICAL FLOODING

R. C. Nelson

Shell Development Company
P. O. Box 481
Houston, Texas 77001, U.S.A.

Additional experimental evidence continues to support the idea that local phase equilibrium is approached closely in an oil reservoir under chemical flood. Since phases formed inside the pore space of the rock generally are not of equal mobility, components of the chemical slug may separate during the flood if they partition differently among the phases. Under certain conditions excess brine phases, generated in the mixing zone between the chemical slug and the oil bank, can sweep ahead of surfactant-rich, microemulsion phases creating, thereby, an in situ preflood.

During the early stages of a chemical flood, before the mixing zone between the drive and the chemical slug begins to overlap the mixing zone between the chemical slug and the oil bank, Phase Volume Diagrams provide useful information regarding the number, volume fractions, compositions and properties of phases in those mixing zones. After those mixing zones overlap, similar pertinent information is provided by Salinity Requirement Diagrams.

For most surfactant systems used in chemical flooding, optimal salinity for oil displacement in the presence of multivalent cations decreases as surfactant concentration decreases. One reason for conducting a chemical flood in a salinity gradient is to keep the surfactants at optimal salinity as their concentration is reduced by adsorption and dispersion during the flood.

Multivalent cations affect phase behavior, hence, optimal salinity, more than the effect of an equal molar quantity of monovalent cations; and the multivalent to monovalent cation effectiveness ratio increases with decreasing surfactant concentration. Consequently, ion exchange during a chemical flood can influence the oil-displacing effectiveness of the flood, especially late in the flood when surfactant concentration is low. Another reason for conducting a chemical flood in a salinity gradient is to establish a condition of "favorable" ion exchange.

The effectiveness of a chemical flood depends upon phase compositions established in the mixing zones--particularly the mixing zone between the drive and chemical slug--so the oil-displacing effectiveness of a chemical flooding system cannot be predicted by screening tests which involve only the full-strength chemical slug and the oil it is intended to displace.

INTRODUCTION

In a previous paper (1), we presented evidence from laboratory chemical floods that the same equilibrium phases which form in laboratory sample tubes form also inside the rock during a chemical flood. We concluded that a reservoir under chemical flood can be treated as a series of connected mixing cells in each cell of which phase equilibrium is attained. Thus, in addition to helping us understand, predict and represent the phases we observe in laboratory sample tubes, phase diagrams also help us understand the mechanisms of the chemical flooding process.

Figure 1 is reproduced from Reference (1). It is typical of pseudo phase diagrams for chemical flooding systems. The surfactants, brine and oil are considered as being single, pure components. In this case, the isobutyl alcohol in the system has been included with the surfactant. The figure depicts a "Type III" phase environment; one-phase, two-phase and three-phase regions appear. The reader is referred to Reference (1) for more detailed discussion of the figure and for complete discussion of phase environment types and their significance in chemical flooding.

The inadequacies of representing multicomponent chemical flooding systems by three pseudo components and techniques for choosing better pseudo components have been discussed recently in the literature (2,3). To the extent that we can ignore or compensate for those inadequacies, we could in principle follow or predict the entire compositional course of a chemical flood if we had phase diagrams for all combinations of surfactant, brine and oil which were to be present at any point in the pore space of the rock at any time during the flood and if we knew the fractional flow characteristics of all those phases in the presence of the other phases. The compositional simulators address that task (4,5). In this paper we show how a relatively few observations of phase behavior, with brine salinity and surfactant concentration as variables, can be of considerable value in formulating chemical flooding systems and in understanding performance observed in laboratory chemical floods.

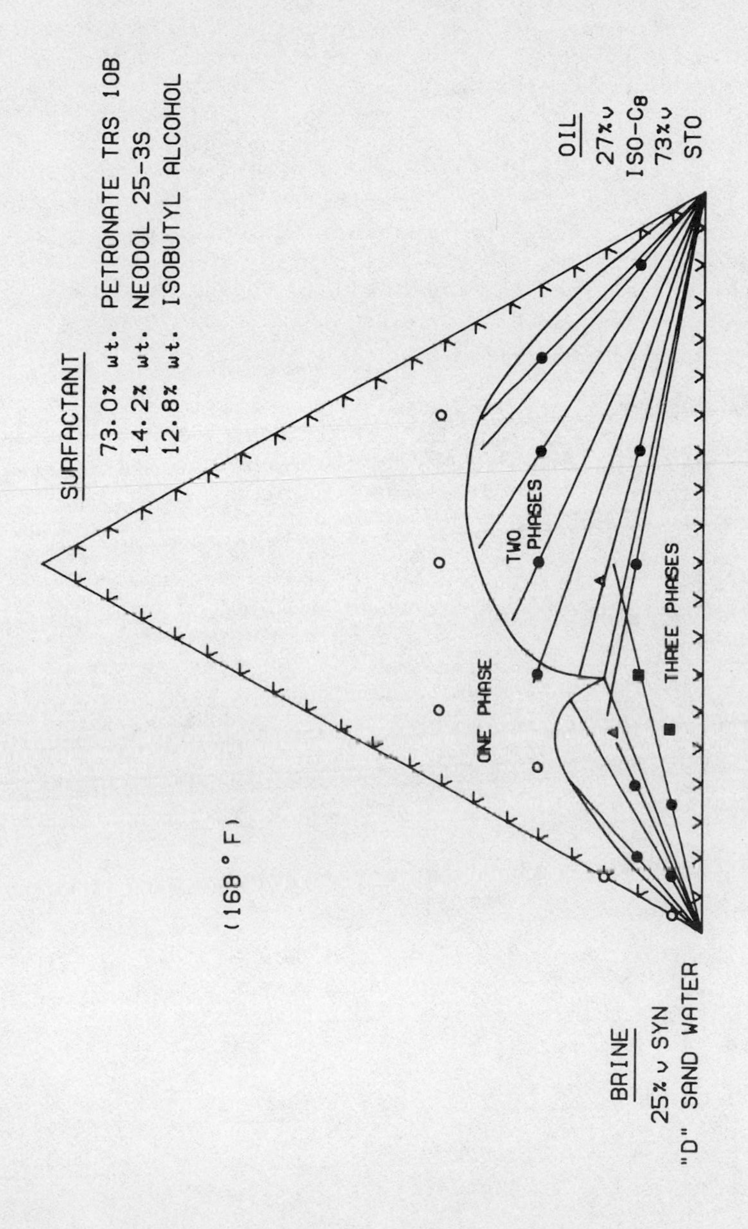

Fig. 1. A typical pseudo phase diagram

We illustrate and emphasize that optimal salinity for most chemical flooding surfactant systems is a function of surfactant concentration, particularly in the presence of multivalent cations. Since most reservoir brines contain multivalent cations, failure to recognize this point can lead to serious errors in formulating the chemical slug and drive. As surfactant concentration decreases, due to adsorption and dispersion, during the course of a chemical flood with a finite chemical slug, optimal salinity usually decreases. A convenient way to maintain optimal oil-displacing activity as the surfactant concentration decreases, is to conduct the chemical flood in a salinity gradient. That is, the salinity of the chemical slug should be less than the salinity of the formation or preflood brine, and the salinity of the drive should be less than the salinity of the chemical slug. The advantages of conducting a chemical flood in a salinity gradient have been pointed out in the literature (1,4-6).

In this paper we utilize "Phase Volume Diagrams" in discussing the results of floods with continuous chemical floods. Such diagrams depict equilibrium volumes, compositions and properties for varying combinations of chemical slug, brine and oil. Phase Volume Diagrams are used by others too (7,8).

We introduce in this paper the "Salinity Requirement Diagram", a representation of the phase behavior and optimal salinity of a given surfactant system in the presence of a given oil as a function of surfactant concentration and the salinity of the brine to which the surfactant is exposed. We illustrate the utility of Salinity Requirement Diagrams in formulating chemical flooding systems and in understanding the results of laboratory chemical floods. Chemical flooding research by others appears to be moving in this direction (9).

EXPERIMENTAL

All laboratory chemical floods were run at flow rates of one foot per day at 168°F in two-inch-square Berea sandstone cores 22 inches long. Permeability of the cores to brine was near 600 millidarcies, and the porosity of the cores was about 20 percent.

The oil was a blend of 27 volume percent isooctane and 73 volume percent stock tank oil. That blend is designed to give an oil of the same viscosity and of approximately the same solvency characteristics as the live crude in a particular oil reservoir of interest.

The formation brine and the source of salinity for the chemical slugs and the drives was "Synthetic 'D' Sand Water" (SDSW). "D" Sand Water is the waterflood brine in the oil reservoir of interest. It contains about 120,000 ppm total dissolved salts, about 1700 ppm of calcium ion and about 1300 ppm of magnesium ion.

The primary surfactant in all chemical slugs and surfactant blends is Stepan Chemical Company's Petrostep 450--a 60 percent active petroleum sulfonate of approximately 450 average equivalent weight. The cosurfactant is Shell Chemical Company's NEODOL® 25-3S--a 60 percent active alcoholethoxysulfate of about 440 average equivalent weight. The amounts of primary surfactant and cosurfactant in chemical slugs and surfactant blends are expressed on an "as supplied" basis.

Drive brines were thickened with Kelco Company's Kelzan MF--a polysaccharide biopolymer. From the results of steady state flow experiments we calculate that the maximum reciprocal mobility of the clean oil bank in these chemical floods is about 8 cp. To provide mobility control between the chemical slug and the oil bank and between the polymer drive and the chemical slug, we formulate the chemical slugs to a viscosity between 12 and 15 cp and the polymer drives to a viscosity between 20 and 30 cp at 168°F and 7.3 sec^{-1}. However, as we illustrate later, new phases are formed as the chemical slug mixes with oil, formation brine and drive during the flood, and polymer added to the chemical slug to provide mobility control often does not stay in the same phase as the surfactants. Consequently, mobility control during the flood can become complex.

RESULTS AND DISCUSSION

Continuous Chemical Slug--Phase Volume Diagrams

Consider two chemical floods. Both are continuous with the same chemical slug: 6.32 weight percent Petrostep 450 and 1.58 weight percent NEODOL 25-3S in a brine of 36 volume percent SDSW in distilled water. The first ten percent pore volume of chemical slug in each flood is tagged with tritiated water. This radioactive tagging allows us to compare the rate at which the leading edge of the surfactant bank transports through the core with the rate at which the brine injected with the leading edge of surfactant transports through the core. The two floods were run as described in the Experimental section. The only deliberate difference between the two floods is that one starts conventionally with formation brine and oil at waterflood residual in the core while the other starts with only formation brine in the core. Results are shown in Figure 2.

The upper graph of the figure relates to the flood with only formation brine in the core. Tritium-tagged chemical slug was

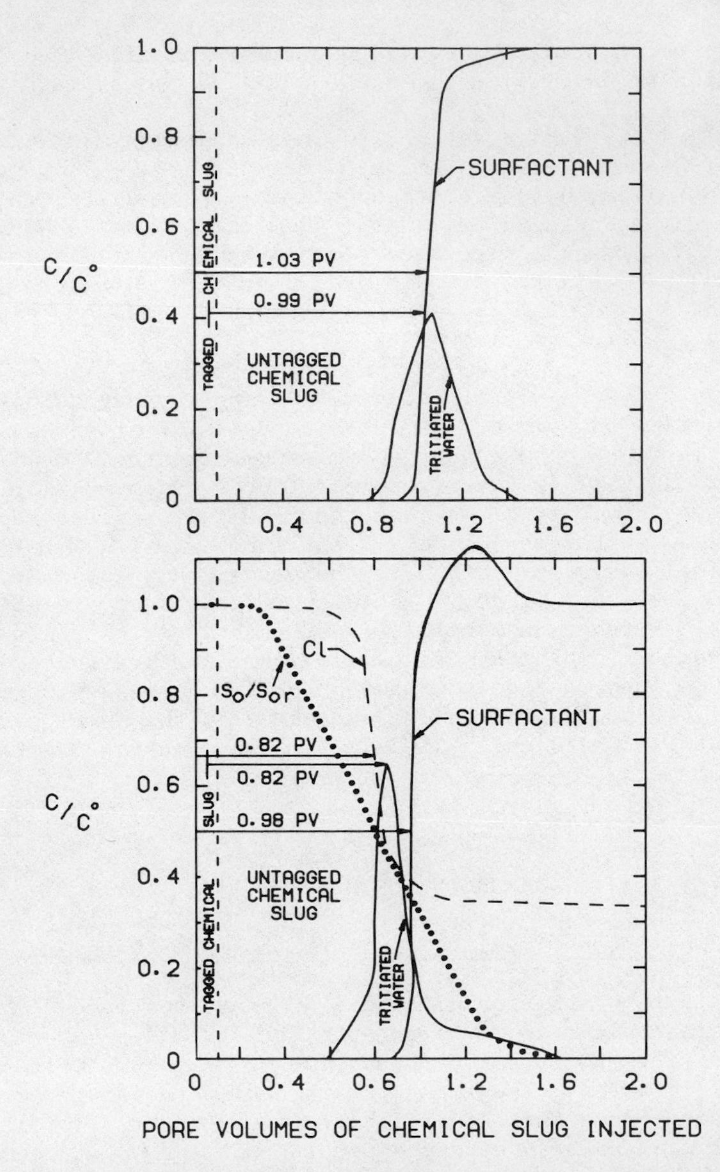

Fig. 2. Retardation of surfactant compared to brine in the presence and absence of oil.

injected for the first 0.10 pore volumes. The tritiated slug was followed immediately by non-tagged chemical slug of the same composition. The concentration of tritiated water in the effluent liquids peaked at 0.99 pore volumes of total chemical slug injected. (Measurement is from the 0.05 pore volumes point since that is where the median of the tagged chemical slug was when the switch to the non-tagged chemical slug was made.) Thus, the tritium peak was only one percent off the pore volume determined by saturating the core with formation brine.

Surfactant concentration in the effluent liquids reached half of its injected concentration at 1.03 pore volumes of total chemical slug injected. The 0.04 pore volumes lag of the surfactant front behind the tritium front for this chemical slug of 0.0936 milliequivalents (meq) of surfactant per milliliter translates to an adsorption of surfactant by the core of 0.038 meq per 100 grams of rock.

The lower graph of Figure 2 relates to the flood with the same chemical slug and formation brine, but with oil at waterflood residual saturation in the core. This time, the tritium concentration in the effluent liquids peaked at 0.82 pore volumes of total chemical slug injected--0.17 pore volumes earlier than when there was no oil in the core. The tritiated water peaked earlier when there was oil in the core due to the pore volume occupied by the oil. According to the oil production curve, 16 percent of the pore volume of the core was occupied by oil when the tritium peaked.

If all of the surfactant were traveling in the same phase as the tritiated water but were being retarded by adsorption on the rock, we would expect the surfactant bank to arrive at the effluent end of the core later than the tritium by a fractional pore volume corresponding to the amount of surfactant adsorbed plus the fractional pore volume of oil produced during the retardation period. When the retardation period is expressed as a fractional pore volume, the fractional pore volume of oil produced during that period is equal to the rate of change of residual oil saturation with pore volumes of fluids injected times the retardation fractional pore volume. But the rate of change of residual oil saturation with pore volumes of fluids injected is just the oil cut of the clean oil bank, i.e., the oil cut just before surfactant breakthrough. Thus,

$$PV_{Surf.} = PV_{Tracer} + (1 + V_o/V_T)\, PV_{Adsp.}$$

where

$PV_{Surf.}$ = pore volume at which surfactant bank arrives

PV_{Tracer} = pore volume at which tracer arrives. Either the tritiated water or chloride can be used as the tracer in this experiment.

V_o/V_T = oil cut in clean oil bank, i.e., volume of oil being produced divided by total volume of oil and brine being produced.

$PV_{Adsp.}$ = amount of surfactant adsorbed on the rock expressed as fractional pore volume of chemical slug.

If we assume that in these chemical floods in water-wet cores the surfactant sees as much rock surface when oil is present at or below waterflood residual as it sees when there is no oil present at all, we can take $PV_{Adsp.} = 0.04$ from the flow experiment run under the same conditions in the absence of oil. The oil cut in the clean oil bank in this flood was 0.34, and $PV_{Tracer} = 0.82$ by both tritiated water and chloride.

With these figures we calculate that the surfactant bank should have arrived at the effluent end of the core at 0.87 pore volumes of fluids injected provided that adsorption of surfactant on the rock was the only mechanism of retardation. The surfactant bank actually arrived, as measured by $C/C° = 0.5$, at 0.98 pore volumes of chemical slug injected. To attribute that additional 0.11 pore volumes of surfactant retardation to increased adsorption requires a mechanism by which the presence of oil in the rock, at waterflood residual saturation and less, more than doubles the amount of surfactant absorbed by the rock. Furthermore, since the concentration of surfactant in the effluent liquids rises above the concentration of surfactant in the chemical slug, any mechanism based on adsorption would have to include a condition under which previously adsorbed surfactant could be produced in the presence of full-strength chemical slug. We believe, rather, that the greater retention of surfactant in the presence of oil is due to the formation inside the core of microemulsion phases which are richer in surfactant and often are more viscous than the chemical slug.

Such microemulsion phases can be seen in the effluent liquids. Table 1 gives composition and viscosity data for the effluent liquids in collection tubes which were half full at 1.09, 1.17 and 1.24 pore volumes of total chemical slug injected.

Considering that the chemical slug used in this continuous chemical flood was formulated to lie below midpoint salinity in the Type III phase region, where the concentration of brine is greater than the concentration of oil in the microemulsion phase, the composition and viscosity of the produced microemulsion are what would be expected. The concentration of surfactant in the produced microemulsion was half again as much as the concentration of surfactant in the chemical slug, and the microemulsion phase occupied 56 to 66 percent of the total volume of produced liquids

Table 1. Composition and Properties of Effluent Liquids

Pore Volumes of Chemical Slug Injected	Volume Percent in Collection Tube			Volume Percent in Microemulsion			Viscosity of Microemulsion (Cp at 7.3 sec^{-1}, 168°F)	C/C°
	Brine	Micro-emulsion	Oil	Surfactant	Brine	Oil		
1.09	35	56	9	12	68	20	11	1.02
1.17	30	64	6	12	68	20	12	1.07
1.24	28	66	6	12	67	21	13	1.15

in the three collection tubes. Consequently, even though oil and brine were being produced simultaneously with the microemulsion, the overall concentration of surfactant in those sample tubes, not just the concentration of surfactant in the produced microemulsion, exceeded the concentration of surfactant in the chemical slug ($C/C° > 1$).

Phase Volume Diagrams

Essentially the same phases produced in this chemical flood in the presence of oil can be observed in laboratory sample tubes by constructing what we call a "Phase Volume Diagram". In one sample tube we equilibrate, at reservoir temperature, 70 volume percent of chemical slug and 30 volume percent of oil. (The chemical slug to oil ratio usually is not critical. For some diagrams we prefer to use an 80/20 ratio.) In other sample tubes we equilibrate oil, chemical slug and formation brine keeping the volume percent of oil constant while decreasing the volume ratio of chemical slug to formation brine. A suitable set of oil/chemical slug/formation brine volume percents is 30/70/0, 30/65/5, 30/60/10, 30/55/15, 30/50/20, 30/40/30, 30/30/40, 30/20/50 and 30/10/60.

Volume fractions of the equilibrium phases are recorded, and the composition of the surfactant-rich (microemulsion) phase, when there is just one such phase, is calculated by material balance. In making that calculation, we assume that essentially all of the surfactant is in one phase. That assumption becomes less valid, and sometimes has to be rejected, for the last one or two (most dilute in surfactant) sample tubes. Some surfactant must always be present in excess oil and excess brine phases in equilibrium with a surfactant-rich microemulsion phase. At higher total surfactant concentrations, say the first seven of the tubes in the above set, the amount of surfactant in the excess phases is small relative to the total. At lower surfactant concentrations the amount of surfactant in the excess phases becomes a more significant fraction of the total.

Occasionally, two microemulsion phases form at the higher surfactant concentrations. When that happens, only the average composition of those two phases can be calculated by material balance. A refinement of the pseudo ternary phase diagram representation of these systems, involving a two-phase region below the three-phase region of a Type III diagram, helps to explain some of these conditions which are anomalous to the more simple ternary representation. However, for the construction of Phase Volume Diagrams and Salinity Requirement Diagrams, discussed later, the more simple ternary representation is satisfactory.

Phase Volume Diagrams, constructed from equilibrium volumes and properties measured in a set of sample tubes of overall composition such as the set described above, relate to the mixing zone in which the chemical slug is mixing with oil and formation brine. As we move from the sample tube in which the oil/chemical slug/formation brine ratio is 30/70/0 to the sample tube in which that ratio is 30/10/60, we are moving forward through an approximation of that mixing zone.

Figure 3 presents the Phase Volume Diagram for the oil/chemical slug/formation brine combination used in the flow experiments discussed above. In the upper part of the diagram we plot phase volume fractions observed at equilibrium as a function of the amount of chemical slug replaced by formation brine in each sample tube. The points plotted at X = 0,5,10,15,20,30,40,50 and 60 show phase volume fractions observed for the oil/chemical slug/formation brine set of overall compositions: 30/70/0, 30/65/5, 30/60/10, etc.

The diagram shows that this chemical slug gives three phases at equilibrium in all of the sample tubes except the last one (60). The volume fraction of the surfactant-rich, middle (microemulsion) phase decreases as chemical slug is replaced by formation brine. That decrease in volume fraction is due partly to the fact that the total amount of surfactant in the sample tube is decreasing, as chemical slug is replaced by formation brine, and partly to the fact that the brine in the tube is becoming more saline. (The chemical slug is made up in 36 percent SDSW; the formation brine is 100 percent SDSW.) The dotted lines on the right of the diagram indicate that in that part of the diagram the combination of low overall surfactant concentration and high brine salinity is causing a significant fraction of the surfactant to be in the "oil" phase even though a middle phase is still present. Finally, with only 10 volume percent of chemical slug contributing to the contents of the tube (60) no middle phase is apparent, i.e., most of the surfactant is in an "upper phase" microemulsion.

Compositions of the microemulsion phases are given across the bottom of the diagram. Volume percent surfactant (% Surf.), volume percent brine (% Brine) and volume percent oil (% Oil) in the microemulsion phase is calculated by material balance assuming that all of the surfactant is in that phase. As would be expected from phase diagrams, the percent oil in the middle phase increases and the percent brine in that phase decreases as the overall surfactant concentration decreases and the overall salinity of the brine to which that surfactant is exposed increases (left to right across the diagram). The point at which the volume of oil equals the volume of brine in the middle phase is indicated on the diagram as being the "midpoint salinity"--a condition, discussed later in this report, which always is close to "optimal salinity" for the system.

Notice that the concentration of surfactant in the middle phase increases even though the overall concentration of surfactant

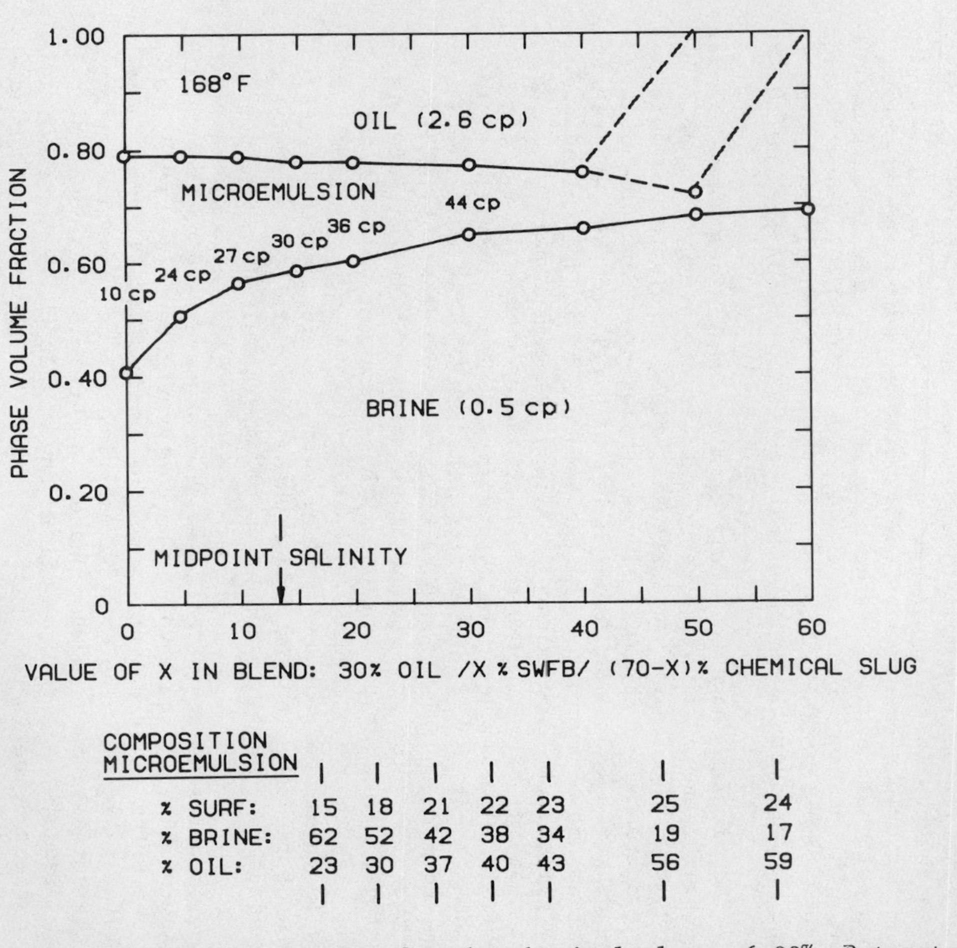

Fig. 3. Phase Volume Diagram for the chemical slug: 6.32%w Petrostep 450, 1.58%w NEODOL 25-3S, 92.10%w (36.2%v SDSW) brine.

in the sample tube decreases as we move from left to right across the diagram. This increase in concentration of surfactant and oil in the middle phase, as chemical slug is replaced by formation brine, causes the viscosity of that phase to increase. Viscosities, measured with a cone and plate viscometer at 168°F and 7.3 sec^{-1} shear rate, are shown at appropriate points on the diagram. Due to experimental difficulties in working with small samples at high temperature and the sensitivity of microemulsion viscosity to shear, the values reported are subject to considerable error. Nevertheless, the trend toward higher viscosity of the surfactant-rich phase, as we move forward through this representation of the mixing zone of the particular, continuous chemical flood under discussion, is clear. Furthermore, such an increase in viscosity is a general trend, although magnitudes vary, for most chemical slugs mixing with oil and formation brines of high salinity. With this information from the Phase Volume Diagram we are in a position to discuss the results of the chemical flood described in the previous section. Before doing that, however, we comment on the effect of polymer in the chemical slug.

The viscosity of the oil used in these experiments is 2.6 cp at 168°F; the viscosity of the brine is 0.5 cp at that temperature. Thus the viscosities of the excess oil (top) and excess brine (bottom) phases in Figure 3 are considerably less than the viscosity of even the least viscous microemulsion (middle) phase. In contrast, examine the Phase Volume Diagram, Figure 4, for the chemical slug: 10.0 weight percent of Petrostep 450 in a brine which is 3.0 volume percent of SDSW. Kelzan MF biopolymer has been added at 1,300 ppm to this chemical slug to raise its viscosity from a few centipoise to about 18 centipoise. When equilibrated with oil, this chemical slug gives two phases--excess oil in equilibrium with a microemulsion which is 9.3 volume percent surfactant, 83.4 volume percent brine and 7.3 volume percent oil. The viscosity of the equilibrium microemulsion is similar to the viscosity of the chemical slug. Apparently, when in contact only with excess oil, the polymer remains in the microemulsion.

Figure 4 shows that when as little as 5/70 of the chemical slug is replaced by SDSW three equilibrium phases form. Remarkably, the viscosity of the microemulsion (middle) phase is quite low while the viscosity of the excess brine (bottom) phase is so high that it does not flow when the sample tube is inverted. Even when 10/70 of the chemical slug is replaced by SDSW, we find that the viscosity of the microemulsion still is less than the viscosity of the original chemical slug; and the excess brine, now a larger fraction of the total, is a very viscous 385 centipoise. The same situation has occurred in every chemical slug we have examined--essentially all of the polymer stays in the most aqueous phase regardless of where the surfactant is!

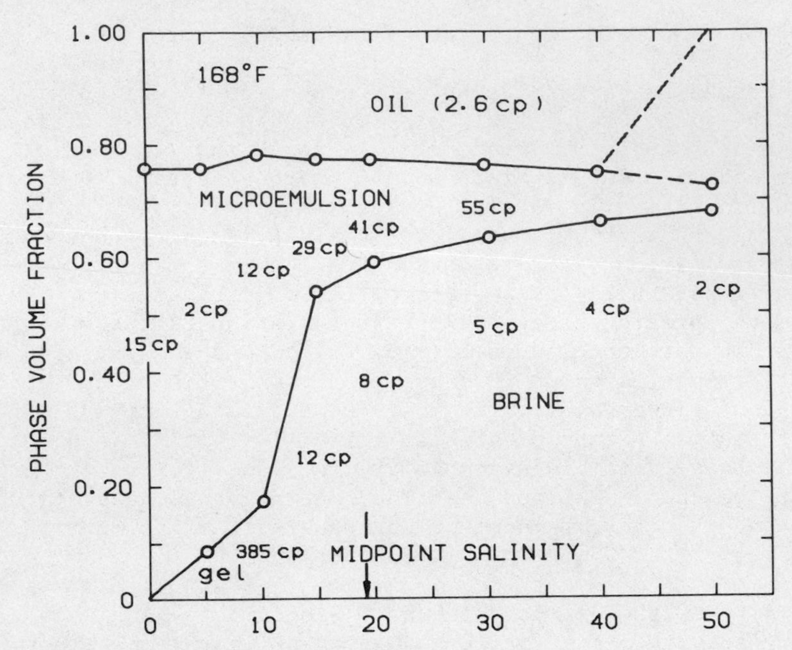

VALUE OF X IN BLEND: 30% OIL /X% SDSW/ (70-X)% CHEMICAL SLUG

COMPOSITION OF MICROEMULSION							
% SURF:	9	10	10	23	27	31	34
% BRINE:	84	82	76	47	35	22	17
% OIL:	7	8	14	30	38	47	49

Fig. 4. Phase Volume Diagram for the chemical slug: 10.0%w Petrostep 450, 90.0%w (3.0%v SDSW) brine, 1300 ppm Kelzan MF biopolymer.

Returning to Figure 3, which relates to the mixing zone of the chemical flood with oil in the core, described in the previous section and in Figure 2, we see that at the leading edge of the mixing zone (the right side of the Phase Volume Diagram) the surfactant-rich phases are considerably more viscous than the excess brine and excess oil phases with which they are in equilibrium. Consequently, the excess brine and excess oil phases tend to flow through the rock faster than the surfactant-rich phases.

As a lagging surfactant-rich phase is overtaken by less saline brine, oil and less viscous surfactant-rich phases coming up from the rear, it re-equilibrates to form a larger, less viscous, surfactant-rich phase. But even that phase is more viscous than the excess brine and excess oil with which it is in equilibrium, so the entire process of surfactant lag and re-equilibration is repeated. In actuality, of course, this process of equilibrium and separation due to mobility differences among equilibrium phases takes place continuously throughout the mixing zone rather than in steps as described here for illustrative purposes. The overall result is a sharpening of the surfactant front--the flow of surfactant in the leading edge of the front mixing zone is retarded more than flow of surfactant further back in that mixing zone when the formation brine is more saline than the chemical slug.

Retardation of surfactant flow in the mixing zone shows up in four ways in the chemical flood under discussion. Three of the four are evident in Figure 2. First, there is the 11 percent pore volume lag of the surfactant behind the brine which was injected concurrently when only a five percent pore volume lag would be anticipated from surfactant adsorption in the absence of oil. Second, is the steepness with which surfactant concentration rises once surfactant finally breaks through, i.e., a very sharp surfactant front. Third, is the fact that average surfactant concentration in the total effluent liquids rises above surfactant concentration in the chemical slug shortly after surfactant breakthrough. Enough of the most retarded surfactant is coming out of the core at the same time as surfactant which has been retarded less to raise $C/C°$ above unity. Fourth, is the evidence presented in Table 1. Notice how closely the composition and viscosity of the surfactant-rich phases produced at 1.09, 1.17 and 1.24 pore volumes of chemical slug injected match those properties of the surfactant-rich, middle phase formed when the chemical slug is equilibrated with oil (X=0 in Figure 3). Apparently, the surfactant-rich, microemulsion phases, which form temporarily in what would be the leading edge of the mixing zone, lag and re-equilibrate to such an extent that little surfactant leaves the core accompanied by brine of salinity much higher than the salinity of the chemical slug.

In Situ Preflood

An interesting consequence of surfactant moving through the core more slowly than the brine with which it is injected, when the salinity of the formation brine is higher than the salinity of the chemical slug, is that the brine of the chemical slug in effect conducts a small in situ preflood. When that condition exists, the salinity of the formation brine has little effect on the amount of oil recovered by the chemical flood.

For example, we conducted two laboratory chemical floods with 12 percent pore volume chemical slugs of the same composition as the continuous chemical slug used in the floods discussed above. Both chemical slugs were made up in 33 percent SDSW. Both drives were 10 percent SDSW thickened with Kelzan biopolymer to 31 centipoises at 168°F and 7.3 sec^{-1}. All other conditions were the same except that the formation brine was 100 percent SDSW in one flood and 20 percent SDSW in the other flood. In spite of this large salinity difference in the formation brines, residual oil saturation after chemical flooding was seven percent pore volume in both floods. Thus, a five-fold difference in salinity of the formation brine made no difference in the effectiveness of the chemical flood!

Consistent with the phase equilibrium concept discussed in this section, there was a considerable difference in the time at which surfactant concentration peaked in the effluent liquids from those two chemical floods. Surfactant lag, brought about by the high salinity formation brine in the one flood, caused surfactant concentration in the effluent liquids to peak about 15 percent pore volume later in the flood with 100 percent SDSW formation brine than in the flood with 20 percent SDSW formation brine.

Finite Chemical Slug--Salinity Requirement Diagrams

While some understanding of the mechanisms of chemical flooding can be obtained from floods with continuous chemical slugs, real chemical floods must be conducted for economic reasons with chemical slugs of finite volume. Usually, that volume is on the order of 10 percent pore volume. Early in the life of a flood with a finite chemical slug, the important regions can be described as front mixing zone, full-strength chemical slug, and rear mixing zone. At that early stage, useful Phase Volume Diagrams can be constructed for the front and rear mixing zones independently. However, the front and rear mixing zones of these small chemical slugs overlap quickly. After that time, the chemical slug, as injected or even as equilibrated with oil, exists nowhere in the core. We find that another expression of equilibrium phase volumes, which we call a "Salinity Requirement Diagram", is a powerful tool for correlating, understanding and predicting the performance of chemical flooding systems.

Salinity Requirement Diagrams

The concept of local phase equilibrium within a reservoir under chemical flood with fractional flow of each phase governed by its fractional mobility and saturation, is well supported by evidence from laboratory chemical floods. Some of that evidence is presented in this paper and in Reference (1). To the extent that this concept represents actuality, we could predict the course of a chemical flood from starting compositions of the rock (for ion exchange and mineral dissolution characteristics), oil, formation brine, chemical slug and drive--if we had phase volume and mobility data on all phases which would form during the flood.

The amount of laboratory work required to collect such data is prohibitive; however, the data-collecting effort can be reduced by representing the multicomponent chemical flooding systems by three pseudo components. The attendant loss of precision depends upon the particular chemical flooding system. Still less laboratory data is required, and still less precision is attained, from what we call a "Salinity Requirement Diagram". As described below, Salinity Requirement Diagrams are constructed from "single point" pseudo phase diagrams. In spite of their approximate nature, Salinity Requirement Diagrams still relate to actual phases formed during a chemical flood; and, because of that, they are a particularly useful tool for understanding and predicting the performance of chemical flooding systems.

Figure 5 is reproduced from Reference (1). It shows schematic pseudo phase diagrams for Type II(-), Type III and Type II(+) phase environments and the phase volume relationships which would be observed in laboratory sample tubes for each type of phase environment at the same overall surfactant-brine-oil content represented by the black point. The significance of these phase environment types in chemical flooding is discussed and illustrated in detail in Reference (1). For our present purpose it is sufficient to know that one of our objectives in designing a chemical flooding system is to keep most of the surfactant in the Type III phase environment for as long as possible during the flood, and to recognize that phase volumes and compositions obtained from a single, well-chosen, overall surfactant-brine-oil composition usually will reveal unambiguously the phase environment type. The black point in Figure 5 illustrates the latter point. An overall composition of 10 percent surfactant, 60 percent brine and 30 percent oil equilibrates to give microemulsion with excess oil for Type II(-), microemulsion with excess oil and excess brine for Type III, and microemulsion with excess brine for Type II(+). Since all phase environments are single-phase at high surfactant concentrations, the overall composition point for determining phase environment type from a single point should lie fairly close to the brine-oil base of the ternary diagram. Furthermore, since we rarely use more than 10 percent of surfactant in a chemical slug, phase behavior at low surfactant concentrations relates better to actual chemical flooding systems.

Regarding the best brine-oil ratio to use, i.e., the horizontal location of the overall composition point, we have two logical choices. Setting the overall composition point in the middle of the diagram (brine volume equals oil volume) gives phase volumes which are easiest to interpret. With that overall composition point the changes are greatest that a Type III system will equilibrate as three phases. Setting the brine-oil ratio to either side of unity increases the changes of a Type III system falling within one of the two-phase nodes. While we usually can tell from the higher

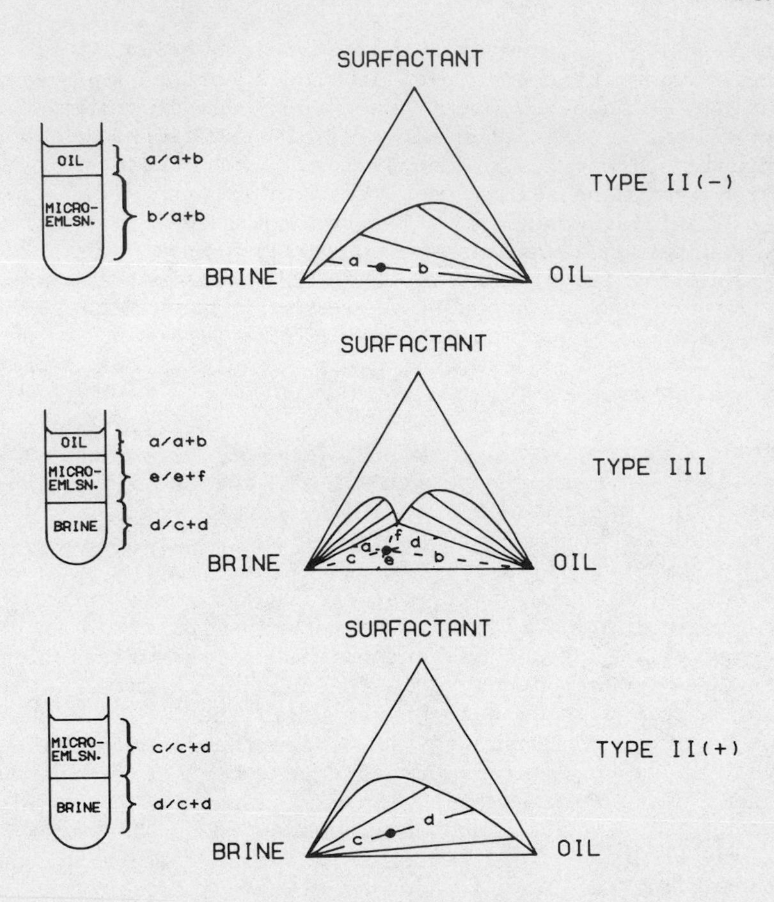

Fig. 5. Ternary representation of phase relationships

degree of phase swelling whether a system is in a two-phase node of a Type III phase environment or in the corresponding Type II phase environment, there is less uncertainty when three phases appear.

On the other hand, a brine-oil ratio of unity does not correspond well to conditions in the region behind the clean oil bank where the surfactants of a chemical flood travel. There the brine-oil ratio is considerably greater than unity. To work at an overall composition point closer to actual surfactant-brine-oil compositions which occur in a reservoir under chemical flood, we have fixed the brine content of the overall composition point of our single-point phase diagrams at 80 volume percent.

As discussed in Reference (1), any change in a surfactant-brine-oil system which increases solubility of the surfactant in the oil relative to the brine shifts the phase environment in the direction: II(-) to III to II(+). If all other variables which affect surfactant solubility are held constant, increasing the salinity of the brine shifts the phase environment in that same direction. For example, the system, 2.0 volume percent of the 80/20, Petrostep 450/NEODOL 25-3S surfactant blend, 80 volume percent brine and 18 volume percent oil, at 168°F is of type II(-) when the brine contains less than about 4.3 weight percent of sodium chloride in distilled water, Type III when the brine contains from about 4.3 to about 16 percent of sodium chloride and Type II(+) when the brine contains more than about 16 percent of sodium chloride.

Healy and Reed (10) found (and others (11) have confirmed) that when the salinity of the brine is such that a Type III phase environment is formed in which the microemulsion (middle) phase contains equal volumes of brine and oil, the microemulsion/excess oil and microemulsion/excess brine interfacial tensions are nearly equal, the sum of those interfacial tensions is at or close to minimum and oil recovery efficiency is at or close to maximum. That "midpoint salinity" for the chemical flooding system under discussion occurs at 7.4 weight percent sodium chloride in distilled water when there is 2.0 volume percent of the surfactant blend in the system.

If the three pseudo components (surfactant, brine and oil) truly behaved as single components, changing the concentration of surfactant in the system would not change the midpoint salinity. (Again, the midpoint salinity is that level of brine salinity which causes the "invariant point" of the Type III phase diagram to be on the line which bisects the ternary diagram and passes through the surfactant apex, i.e., the line on which oil volume equals brine volume.) In reality, however, changing the surfactant concentration does change the midpoint salinity to some extent. With five percent surfactant in the system, midpoint salinity is at 8.6 percent of sodium chloride in the brine; with 0.8 percent surfactant in the system, midpoint salinity is at 6.8 percent sodium chloride in the brine.

This is the type of information that is presented in a Salinity Requirement Diagram. Figure 6 is the Salinity Requirement Diagram for the system under discussion. The vertical bars show, as a function of overall surfactant concentration, the range of brine salinity over which the system is in a Type III phase environment (although not necessarily three phases). The position of the circle on the bar indicates midpoint salinity at that overall surfactant concentration. Optimal salinity for oil-displacement efficiency should be close to that level of salinity. The number within the circle is the volume fraction of surfactant in the "invariant" phase at midpoint salinity. Healy and Reed (12) found lower microemulsion/excess brine and microemulsion/excess oil interfacial tensions for systems in which the volume fraction of surfactant in

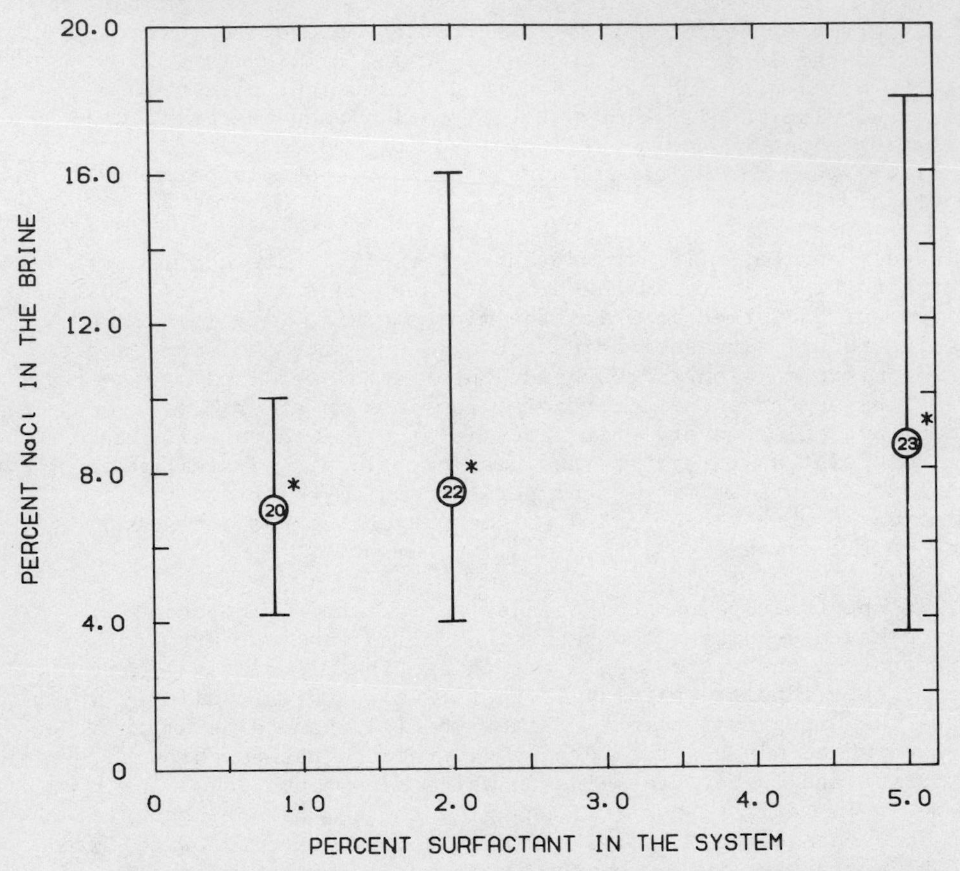

Fig. 6. Salinity Requirement Diagram for 80/20, Petrostep 450/ NEODOL 25-3S, surfactant blend with sodium chloride brines.

the "invariant", microemulsion phase at midpoint salinity was low. Intuitively, it is reasonable that the more brine and oil the microemulsion contains, the lower should be the interfacial tensions between that microemulsion and excess brine and excess oil.

If the overall composition point lies in the single-phase region or in one of the two-phase regions (nodes) of the phase diagram at midpoint salinity, the position of the "invariant" point has to be estimated. The higher the overall surfactant concentration and the better the surfactant system with regard to producing low interfacial tensions at midpoint salinity (i.e., the closer the "invariant" point to the Brine-Oil base of the ternary diagram), the less likely it is that we will see a three-phase system at midpoint salinity. In fact, any time the "invariant" point lies below either a line drawn from the Brine apex through the overall composition point or a line drawn from the Oil apex through the overall composition point only one phase or two equilibrium phases will appear ideally in a Type III phase environment. It sometimes is a point of confusion that in a chemical flood we want to maximize the amount of time the surfactants spend in a Type III phase environment, but we seek surfactant-brine-oil systems the phase diagrams of which exhibit minimum three-phase regions!

The surfactant-brine-oil, pseudo three-component, phase diagram, as presented in References (1) and (10), allows only one microemulsion phase to be present at a time. Occasionally, in preparing Salinity Requirement Diagrams we see systems in which two microemulsions apparently are in equilibrium. Invariably, this occurs near the II(-)/III or III/II(+) transition regions, that is, when the "invariant" point is far to the left or far to the right on the phase diagram. Consequently, such deviations from simple pseudo ternary representation do not interfere with our ability to determine the optimal (midpoint) salinity ("invariant" point in the center of the ternary diagram). Although a more complex ternary representation appears unnecessary when constructing Salinity Requirement Diagrams for screening chemical flooding formulations, such a representation may improve our ability to model chemical floods mathematically.

Relative Effectiveness of Multivalent Cations and Monovalent Cations in Changing Phase Behavior

Figure 6 shows a 21 percent decrease in optimal salinity (as measured by midpoint salinity) when surfactant concentration is lowered from 5.0 to 0.8 percent. A decrease of that magnitude is typical for systems in which the brine contains only monovalent cations. When the brine contains multivalent cations, such as SDSW and most real reservoir brines, optimal salinity is considerably more sensitive to surfactant concentration. Figure 7 is the Salinity Requirement Diagram for the same 80/20, Petrostep 450/NEODOL 25-3S, system we have been discussing, but with SDSW in distilled water rather than sodium chloride in distilled water brines. The figure shows that with SDSW as the source of salinity optimal salinity decreases by 64 percent, as compared to 21 percent, when surfactant concentration is lowered from 5.0 to 0.8 percent.

Comparing midpoint salinities in Figures 6 and 7, we see that at 168°F with 80 volume percent brine, 15 volume percent of the

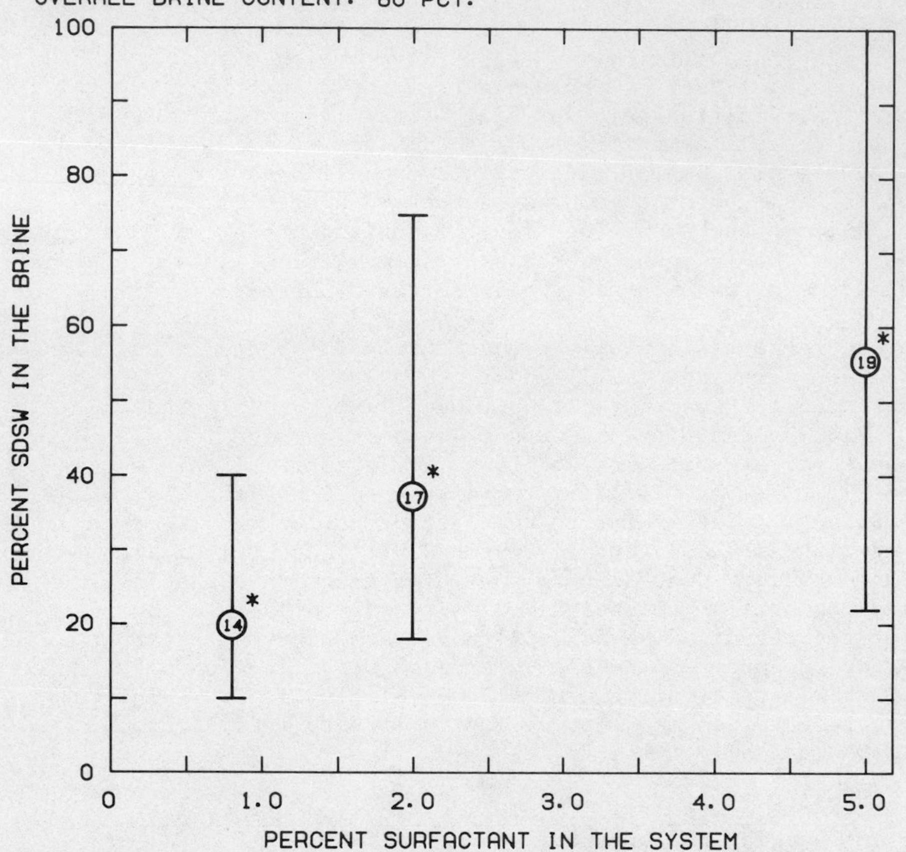

Fig. 7. Salinity Requirement Diagram for 80/20, Petrostep 450/ NEODOL 25-3S, surfactant blend with SDSW brines.

particular oil and 5 volume percent of the particular surfactant blend in the system, an 8.6 percent sodium chloride brine is equivalent to a 55 percent SDSW brine. Now, an 8.6 percent sodium chloride brine contains 1.47 moles/kg of sodium ions; a 55 percent brine contains 1.03 moles/kg of sodium ions and 0.052 moles/kg of multivalent cations. Thus, with 5.0 percent surfactant in the system and 1.03 moles/kg of sodium in the brine, 0.052 moles/kg of calcium, magnesium, barium and strontium ions, as present in SDSW, affects phase behavior to the same extent as 0.44 moles/kg of additional sodium ion.

By the same procedure we find that with 2.0 percent surfactant in the system and 0.69 moles/kg of sodium in the brine, 0.035 moles/kg of SDSW multivalent cations affects phase behavior to the same extent as 0.57 moles/kg of additional sodium ion. And with 0.8 percent surfactant in the system and 0.37 moles/kg of sodium in the brine, we find that 0.019 moles/kg of SDSW multivalent cations affects phase behavior to the same extent as 0.79 moles/kg of additional sodium ion. From these numbers we present in Table 2 relative effectiveness of the divalent ions in SDSW to sodium ions in changing phase behavior of the surfactant-brine-oil system under discussion as a function of overall surfactant concentration in the system.

Table 2. Relative Effectiveness of M^{++} and Na^+ in Changing Phase Behavior.

Percent Surfactant in the System	Moles of M^{++} to Moles of Na^+ Effectiveness Ratio
5.0	8.5
2.0	16
0.8	42

It is apparent from the table that the phase behavior of the system under discussion is much more sensitive to the multivalent cation concentration at low surfactant concentrations than at high surfactant concentrations. This means that the exact ionic composition of the brine in the surfactant bank is more critical near the end of a chemical flood than it is in the beginning. It means also that the effect of ion exchange on the phase behavior and, hence, on the oil displacing activity of the surfactant-brine-oil system becomes more pronounced as the chemical flood proceeds.

Absolute values of M^{++}/Na^+ effectiveness and the dependency of those values on surfactant concentration depend upon the particular surfactant or surfactant blend. For example, for Petrostep 450 alone in the same system as the 80/20, Petrostep 450/NEODOL 25-3S, blend we have been discussing, the M^{++}/Na^+ effectiveness ratio rises from 21 to 67 as the surfactant concentration is lowered from 5.0 to 2.0 percent. On the other hand, judging from how "flat" its Salinity Requirement Diagram is, the M^{++}/Na^+ effectiveness ratio for NEODOL 25-3S by itself does not change much with surfactant concentration. (The midpoint salinity of NEODOL 25-3S drops only 11 percent, from 185 percent to 165 percent SDSW, as its concentration in the subject system is decreased from 5.0 to 0.8 percent.)

The "steepness" of a Salinity Requirement Diagram indicates how rapidly the M^{++}/Na^{+} effectiveness ratio changes with surfactant concentration. If that ratio did not change at all with surfactant concentration, Salinity Requirement Diagrams made with multivalent cations in the brine would parallel the "flat" diagrams, such as Figure 6, found when the only cations in the brine are sodium.

The greater effectiveness of multivalent cations as compared to monovalent cations in changing phase behavior has been recognized recently in the literature. Fleming, et al. (7), report a calcium to sodium effectiveness ratio of 14 and a magnesium to sodium effectiveness ratio of 16 for North Burbank crude oil and a surfactant system composed of 5.0 percent Petronate TRS 10B and 3.0 percent isobutyl alcohol. That paper does not indicate whether the ratios were determined at other surfactant concentrations.

Salinity Requirement Diagrams for Various Organic Sulfonates

Considerable differences exist in Salinity Requirement Diagrams of various organic sulfonates. Table 3 shows midpoint salinities and the percent surfactant in the "invariant" microemulsion phase at midpoint salinity for three Stepan petroleum sulfonates.

Witco's petroleum sulfonate, Petronate TRS 12B, is similar to Petrostep 465, and Witco's Petronate TRS 10B falls between Petrostep 465 and Petrostep 450. The Salinity Requirement Diagram of Amoco's (polybutene) Sulfonate 151 lies even lower than the diagrams for Petrostep 465 and Petronate TRS 12B. Midpoint salinity for that sulfonate at 5.0 percent surfactant in the system is about five percent SDSW.

Table 3 illustrates how surfactants of quite different midpoint salinities at high surfactant concentration can exhibit similar midpoint salinities at low surfactant concentration. The increase in slope of the Salinity Requirement Diagram in the order Petrostep 465 < Petrostep 450 < Petrostep 420, correlates with decreasing average molecular weight of the petroleum sulfonate; however, we believe that the amount of disulfonate in those products increases in that same order.

Looking at Table 3 again, one may conclude that Petrostep 420 has a higher "salinity tolerance" than Petrostep 450 or Petrostep 465. The expression "salinity tolerance" should be used with caution since Petrostep 420 also has a higher "salinity requirement" than the other two petroleum sulfonates. For example, with 5.0 percent petroleum sulfonate in the system and, say, 10 percent of SDSW in the brine, Petrostep 465 would be more active than Petrostep 420, because 10 percent SDSW is much closer to the midpoint (optimal) salinity of Petrostep 465. Furthermore, at the same

Table 3. Data from the Salinity Requirement Diagrams for Three Petroleum Sulfonates

Percent Surfactant in System:	Optimal Salinity, e.g., Midpoint Salinity (Percent SDSW)			% Surfactant in Middle Phase at Midpoint Salinity		
	0.8	2.0	5.0	0.8	2.0	5.0
Petrostep 465	0.5	2.5	8.0	3.0	6.0	12.0
Petrostep 450	1.0	9.0	26.0	-	20.0	26.0
Petrostep 420	1.5	17.0	57.0	10.0	30.0	40.0

concentration of surfactant in the system, Petrostep 465 probably is more active at its optimal salinity than Petrostep 420 is at its optimal salinity. We reach that conclusion from the percent surfactant in the middle phase at midpoint salinity. As mentioned earlier, interfacial tensions tend to be higher as the concentration of surfactant in the microemulsion, "invariant" middle phase is higher at midpoint salinity. In general, the higher the salinity requirement of a surfactant system the higher is the concentration of surfactant in the microemulsion phase at midpoint salinity. One criterion in seeking better surfactant systems for chemical flooding is high salinity requirement concurrent with low surfactant concentration in the microemulsion phase at midpoint salinity.

The Utility of Salinity Requirement Diagrams

As an example of the utility of Salinity Requirement Diagrams in understanding and explaining the results of chemical floods, we choose a set of four laboratory floods in which only the salinity of the polymer drive was varied. Experimental conditions for the set were as described in the Experimental section except that the chemical slug was 5.45 weight percent Petrostep 450, 1.56 weight percent NEODOL 25-3S and 0.78 weight percent isobutyl alcohol in a brine composed of 36.2 volume percent of SDSW in distilled water. Five hundred parts per million of Kelzan MF biopolymer were added to provide mobility control.

The Salinity Requirement Diagram for this 70/20/10, Petrostep 450/NEODOL 25-3S/IBA, system is almost identical to the diagram for the 80/20, Petrostep 450/NEODOL 25-3S, system we have been discussing. (The two systems perform equally well under the same conditions.) In drawing the Salinity Requirement Diagram for the 70/20/10 system, Figure 8, we have used midpoint salinity and Type III range determined at 0.8, 2.0 and 5.0 percent surfactant concentration as before. Within the shaded region the phase

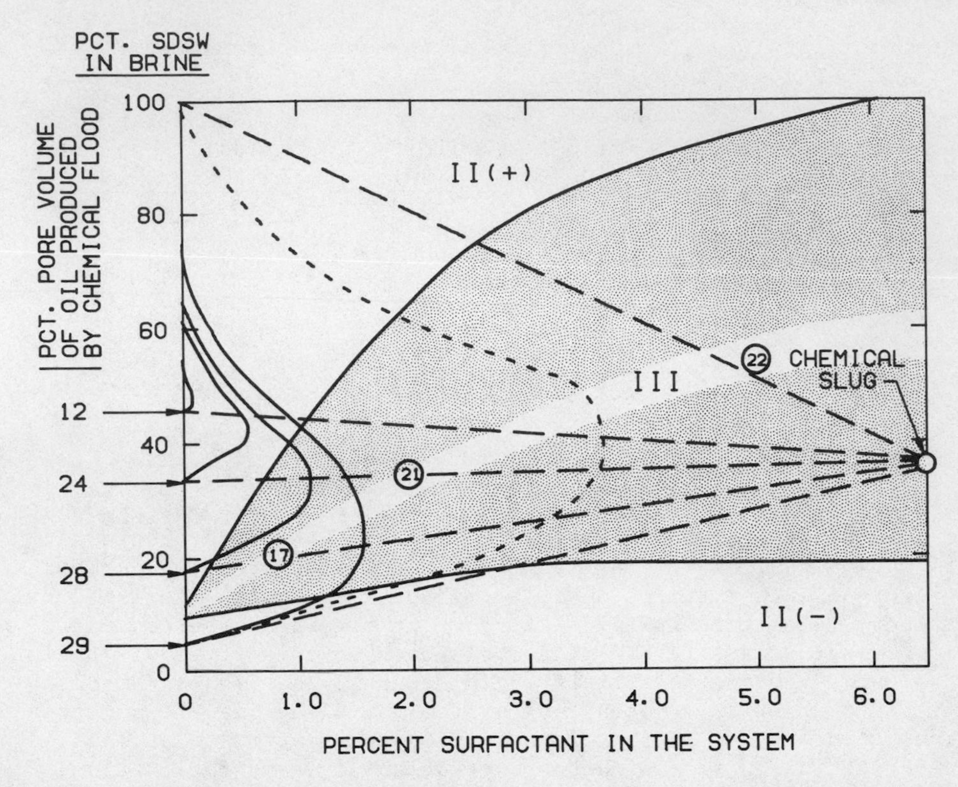

Fig. 8. Surfactant concentration versus salinity superimposed upon Salinity Requirement Diagram for 70/20/10, Petrostep 450/ NEODOL 25-3S/isobutyl alcohol, blend.

environment of the system is Type III. Above the Type III region the phase environment is Type II(+). Below the Type III region the phase environment is Type II(-). The light band near the center of the Type III region indicates midpoint salinity near which, we repeat, oil-displacing activity is optimal. Numbers within the light band indicate volume percent of surfactant in the middle phase of the Type III phase environment at midpoint salinity whether or not middle phases actually were observed in the sample tubes used to generate the phase volume data from which the Salinity Requirement Diagram was constructed. As mentioned previously, middle phase/excess brine and middle phase/excess oil interfacial tensions tend to be lower the lower the percent of surfactant in the middle phase at midpoint salinity.

Superimposed on the Salinity Requirement Diagram of Figure 8 are data relevant to the set of four chemical floods. In each flood the formation brine was 100 percent SDSW; no preflood was used. The 12 percent pore volume chemical slug used in each flood was of the composition shown in the figure after, in effect, mixing the chemical slug with enough oil to comply with the requirement that all points on the Salinity Requirement Diagram represent overall surfactant-brine-oil combinations in which the brine component is 80 percent of the total. (The chemical slug is 7.5 volume percent surfactant and 92.5 volume percent brine. Adding 15.6 volume percent of oil brings the overall composition to 6.5 percent surfactant, 80.0 percent brine and 13.5 percent oil.)

The only deliberate difference among the four chemical floods was the salinity of the biopolymer drives. As indicated by the arrows in Figure 8, the drive brines contained 5, 16, 32 and 45 volume percent SDSW.

Oil recovery efficiency for each chemical flood, expressed as percent pore volume of oil produced by the flood, is indicated by the number of the left of each arrow. That is, the percent pore volumes of oil produced by the chemical floods were 29, 28, 24, and 12 for the floods with drives containing 5, 16, 32 and 45 percent SDSW respectively. The average waterflood residual oil saturation, before the chemical flood, in these experiments was 31.9 percent pore volume.

The dashed lines in Figure 8 represent the approximate relationship between surfactant concentration and salinity early in each chemical flood. Before the front mixing zone (chemical slug mixing with oil and formation brine) and the rear mixing zone (chemical slug mixing with oil and drive) begin to overlap, surfactant concentration as a function of salinity in the front mixing zone was the same for the four chemical floods. That relationship is approximated by the dashed line between 100 percent SDSW and the composition point for the chemical slug. Similarly, before the front mixing zone and the rear mixing zone begin to overlap, surfactant concentration as a function of salinity in each rear mixing zone is approximated by the dashed line between the composition point for the chemical slug and the point on the ordinate corresponding to the percent SDSW in the drive for each flood.

Once the front and rear mixing zones begin to overlap, chemical slug, as such, no longer exists anywhere in the core. The surfactant concentration versus salinity curve rises from zero near 100 percent formation brine to a maximum to the left of the composition point for the chemical slug, then falls back to zero at the salinity of the drive. The dotted curve in Figure 8 illustrates the general shape of such a surfactant concentration versus salinity curve some time after overlap of the mixing zones in the chemical flood with five percent SDSW in the drive. The exact surfactant concentration versus salinity curve cannot be determined experimentally without disturbing the flood. Such curves can be estimated, however, by mathematical representations of the model (4,5).

Surfactant concentration versus salinity can be measured in the effluent liquids without disturbing the chemical flood. The two types of curves--one, observing at many points along the core at a single point in time; the other, observing at a single point along the core (the outflow end) as time passes--are similar in shape. The solid curves in Figure 8 depict surfactant concentration versus salinity relationships measured in the effluent liquids of the four chemical floods.

Two features of these effluent liquid curves are apparent. One is that the salinity at which the surfactant concentration is maximum is higher than the salinity of the drive. The other is that the total amount of surfactant produced in the effluent liquids, which is related to the area under the surfactant concentration versus salinity curve, decreases as the salinity of the drive is increased. The percent of injected surfactant retained by the core was 46, 69, 82 and 98 for the floods with drive salinities of 5, 16, 32, and 45 percent SDSW respectively.

While some of this increase in retention of surfactant with increasing salinity of the drive may be due to increased adsorption of surfactant on the rock, we believe that most of it is caused by increased trapping of surfactant-rich phases. Figure 8 shows that the higher the salinity of the drive in the four floods the greater was the fraction of surfactant exposed to a Type II(+) phase environment. In the two-phase region of the Type II(+) phase environment, the surfactant-rich phase is in equilibrium with an "excess brine" phase. As discussed earlier in this paper, the less mobile surfactant-rich (microemulsion) phase usually moves more slowly through the core than the excess brine phase. Furthermore, interfacial tensions, except near the plait point, usually are not as low in the Type II(+) phase environment as in the Type III phase environment. Consequently, surfactant-rich, microemulsion phases can become trapped in the rock during a chemical flood much like oil is trapped in a rock during waterflood. Surfactant-rich phase trapping is most pronounced when Type II(+) conditions exist in the rear mixing zone; and the more surfactant exposed to Type II(+) conditions in the rear mixing zone, the more surfactant is retained by the core by trapping. Surfactant-rich phase trapping is one of the subjects discussed by Gupta and Trushenski (6).

Since the effectiveness of a chemical flood can depend so much on mixing between the chemical slug, oil and formation brine and particularly on mixing between the chemical slug, oil and drive brine, results from floods with large pore volume or continuous chemical slugs must be used cautiously in designing chemical flooding systems. Optimal salinity for a small pore volume chemical slug is not the same as optimal salinity for a continuous chemical slug, especially when the brines involved in the system contain multivalent cations.

It is apparent from results such as these that chemical flooding effectiveness cannot be predicted by screening methods which examine the isolated chemical slug or even the chemical slug in contact with the oil it is intended to recover. Conditions to which the surfactants will be exposed inside the rock, as they mix with oil, formation brine and drive, as their concentration is reduced by adsorption and dispersion, and as they are affected by ion exchange, must be considered. Properties of the full-strength chemical slug or properties of that slug in contact with oil reveal, at best, a potential for oil-displacement activity. That potential may or may not be realized depending upon the brine the chemical slug must face in the formation and particularly upon the drive that will follow the chemical slug into the formation.

Ion Exchange

Figure 8 shows that all of the surfactant produced in the flood with 32 percent SDSW in the drive was in brine which, according to the Salinity Requirement Diagram, was saline enough to promote Type II(+) behavior. Nevertheless, oil desaturation by the chemical flood remained moderately high at 24 percent pore volume. One reason for the moderately good performance of that chemical flood is that, although the surfactant was above optimal salinity at the end of the flood, considerable surfactant did travel near optimal salinity for much of the flood. (Visualize the surfactant concentration versus salinity curve as it changed during the flood from the two dashed lines representing conditions early in the flood to the measured conditions in the effluent liquids.) Another reason is that all of these floods were run under conditions of "favorable" ion exchange.

The possible effects on chemical flooding performance of cation exchange between clays equilibrated against a formation brine and brines introduced into the reservoir as prefloods, chemical slugs and drives have been researched and reported by others (13-15). We define here as "favorable" those conditions which cause reservoir clays to replace multivalent cations with monovalent cations in the region in which the surfactant is traveling, that is, conditions which cause the brine in the surfactant bank to be "softened" partially by ion exchange with the reservoir clays.

Conducting a chemical flood with a salinity gradient which decreases from formation brine to chemical slug to drive provides favorable ion exchange if the only source of salinity in the chemical slug and drive is formation brine. In most of the chemical floods discussed in this paper the formation brine was 100 percent SDSW and smaller amounts of that same brine were used to provide salinity to the chemical slug and drive. In that situation, ion exchange is favorable in all mixing zones. Recalling, from the previous section that the relative effectiveness of multivalent cations and monovalent cations in changing phase behavior increases with decreasing surfactant concentration, it is apparent that the effect of favorable ion exchange will be more pronounced at low surfactant concentrations than at high surfactant concentrations. In other words, favorable ion exchange will tend to flatten the salinity requirement versus surfactant concentration curve by raising that curve on the low surfactant concentration side of the Salinity Requirement Diagram. This effect of favorable ion exchange--a slight raising at low surfactant concentration of the active region in a Salinity Requirement Diagram--may be responsible in part for the moderately good oil recovery indicated in Figure 8 for the chemical flood in which the drive contained 32 percent SDSW.

CONCLUSIONS

Additional experimental evidence continues to support the idea that local phase equilibrium is approached closely in a sandstone core under chemical flood at flow rates comparable to flow rates in the field. Since phases formed inside the core are not generally of equal mobility, components of the chemical slug may separate during a chemical flood if they partition differently among the phases.

When flooding with a continuous chemical slug or early in a flood with a small, finite volume chemical slug, Phase Volume Diagrams provide useful information regarding the mixing zones. Phase Volume Diagrams are generated by equilibrating chemical slug, brine and oil at different chemical slug to brine ratios. For the front mixing zone, the brine is formation or waterflood brine; for the rear mixing zone, the brine is drive.

After the front and rear mixing zones overlap, Salinity Requirement Diagrams provide the pertinent information. Salinity Requirement Diagrams, constructed from single-point phase diagrams, approximate the dependence on surfactant concentration of phase environment type and optimal salinity for oil displacement. When the brines contain multivalent cations, optimal salinity for most surfactant systems is a direct function of surfactant concentration. That is, the optimal salinity for oil displacement is lower at low surfactant concentrations than it is at high surfactant concentrations. Because of that, optimal salinity for most chemical flooding surfactant systems in the presence of actual reservoir brines decreases during the course of the chemical flood. One reason for conducting a chemical flood in a salinity gradient is to keep the surfactants at optimal salinity as their concentration is reduced by adsorption and dispersion during the flood.

Another reason for conducting a chemical flood in a salinity gradient is to establish a condition of "favorable" ion exchange, that is, a condition under which the reservoir clays partially soften the brine in the surfactant bank. Actual optimal salinity at low surfactant concentrations will be a little higher than that read from a Salinity Requirement Diagram when the reservoir clays are replacing multivalent cations in the brine with monovalent cations in that region of the Salinity Requirement Diagram.

Since optimal salinity in the presence of multivalent cations decreases as surfactant concentration decreases for most chemical flooding systems, laboratory chemical flooding results can be misleading if the actual reservoir brine is simulated by a brine containing only monovalent cations. For the same reason, equally misleading results can be obtained from laboratory core floods using continuous or large pore volume chemical slugs.

Another consequence of optimal salinity being higher for a full-strength chemical slug than for the same slug at lower surfactant concentrations is that a chemical flood designed with the salinity of the chemical slug and drive optimal for the full-strength chemical slug is likely to go over-optimal in use. One characteristic of an over-optimal chemical flood is high retention of surfactant by the core.

The effectiveness of a chemical flood depends on compositions established in the mixing zones--particularly the rear mixing zone between the chemical slug and the drive--so the oil-displacing effectiveness of a chemical flooding system cannot be predicted by a screening test which involves only the chemical slug and the oil it is intended to displace.

ACKNOWLEDGMENTS

The experimental work on which this paper is based was conducted by J. J. Evans, A. F. Roscoe and J. T. Wortham. We thank W. M. Sawyer for his review and the management of Shell Oil Company for permission to publish this paper.

REFERENCES

1. R. C. Nelson and G. A. Pope, Soc. Pet. Eng. J., 18, 325 (1978).
2. S. J. Salter, SPE 7056, presented at SPE Symposium on Improved Oil Recovery, Tulsa, April 16-19, 1978.
3. J. E. Vinatieri and P. D. Fleming III, Soc. Pet. Eng. J., 19, 289 (1979).

4. G. A. Pope and R. C. Nelson, Soc. Pet. Eng. J., 18, 339 (1978).
5. G. A. Pope, B. Wang and K. Tsaur, SPE 7079, presented at SPE Symposium on Improved Oil Recovery, Tulsa, April 16-19, 1978.
6. S. P. Gupta and S. P. Trushenski, Soc. Pet. Eng. J., 19, 116 (1979).
7. P. D. Fleming, D. M. Sitton, J. E. Hessert, J. E. Vinatieri and D. F. Boneau, SPE 7576, presented at 53rd Annual Fall SPE Meeting, Houston, October 1-3, 1978.
8. K. S. Chan and D. O. Shah, SPE 7869, presented at SPE of AIME International Symposium on Oilfield and Geothermal Chemistry, Houston, January 22-24, 1979.
9. C. J. Glover, M. C. Puerto, J. M. Maerker and E. I. Sandvik, Soc. Pet. Eng. J., 19, 183 (1979).
10. R. N. Healy, R. L. Reed and D. G. Stenmark, Soc. Pet. Eng. J., 16, 147 (1976).
11. S. J. Salter, SPE 6843, presented at 52nd Annual Fall SPE Meeting, Denver, October 9-12, 1977.
12. R. N. Healy and R. L. Reed, Soc. Pet. Eng. J., 17, 129 (1977).
13. G. A. Pope, L. W. Lake and F. G. Helfferich, Soc. Pet. Eng. J., 18, 418 (1978).
14. L. W. Lake and F. G. Helfferich, Soc. Pet. Eng. J., 18, 435 (1978).
15. H. J. Hill and L. W. Lake, Soc. Pet. Eng. J., 18, 445 (1978).

SELECTION OF COMPONENTS FOR AN OPTIMAL MICELLAR SYSTEM IN RESERVOIR CONDITIONS

A. Eisenzimmer and J.-P. Desmarquest

Institut Francais du Petrole
1 et 4 Avenue de Bois-Préau
92506 Rueil Malmaison, France

A method is proposed for selecting components and their relative amounts in order to formulate an optimal micellar system between a given oil and a fixed salinity water. The method involves two separate steps: 1) selection of the cosolvent from partitioning between oil and water without any surfactant and, 2) selection of the surfactant from ternary diagrams wherein the amount of the previously selected cosolvent is constant. This study supports the assumption of separate effects for the surfactant and the cosolvent in micellar systems.

INTRODUCTION

Since selection of components for application to actual reservoir conditions is concerned in this study, we have to consider that:

1. Oil composition, brine salinity and composition, and temperature are fixed.

2. Several criteria usually conflicting one with the other have to be taken into account to get an optimal micellar system suitable for injection and flooding.

The first step in selecting an optimal micellar system for any field application is to define a complete one-phase microemulsion without any oil or water phase separation, whose viscosity is in the range of the equivalent viscosity of the oil-bank, and which can be prepared from a minimum amount of surfactant, corresponding to the smallest multiphase area in phase diagrams or to the highest solubilization parameters. Such a system involves a surfactant and a cosurfactant or more exactly a cosolvent which is usually an alcohol.

The influence of alcohols on phase behavior is known as mainly effective inside phases (1-3), for instance, by changing the optimal salinity for a given oil-surfactant system (4-6) or the surfactant partitioning between the oil and brine (7-9). Therefore, it becomes convenient to look at the problem in two separate steps:

1) selection of the cosolvent by observing partitioning of the cosolvent between oil and brine at the reservoir temperature without any added surfactant.

2) selection of the surfactant by determining the solubilization parameters of the total system, oil, brine, the selected cosolvent, and various surfactants.

EXPERIMENTAL

The oil-brine-surfactant system is composed of C_{10}-C_{14} paraffinic oil-cut, 0.5% NaCl solution and TRS 10-80 (average M.W. = 405) alone or in mixture with a heavier petroleum sulfonate, Gerland S85 (average M.W. = 495) in order to adjust the average M.W. to a definite value. The cosolvents considered here are alcohols of low molecular weight (C_1 to C_6) from the normal and iso-series, whose partitioning between oil and water has been determined by gas chromatography, after equilibration at room temperature.

RESULTS

Alcohol partitioning: When the alcohol concentration in each phase is plotted versus the total alcohol content of the system (WOR = 1), three types of partitioning behavior were observed as follows:

1) the water soluble alcohols preferentially partition in the water-phase, irrespective of the total amount of alcohol within the system (Figure 1). In our studies the alcohols exhibiting such behavior were methanol, ethanol and propanol.

2) the oil-soluble alcohols preferentially partition in the oil-phase with a saturating concentration in the water-phase (Figure 2). The alcohols exhibiting such behavior were pentanol and heavier alcohols.

3) the intermediate alcohols which favor either the oil or the water-phase depending upon the total amount in the system

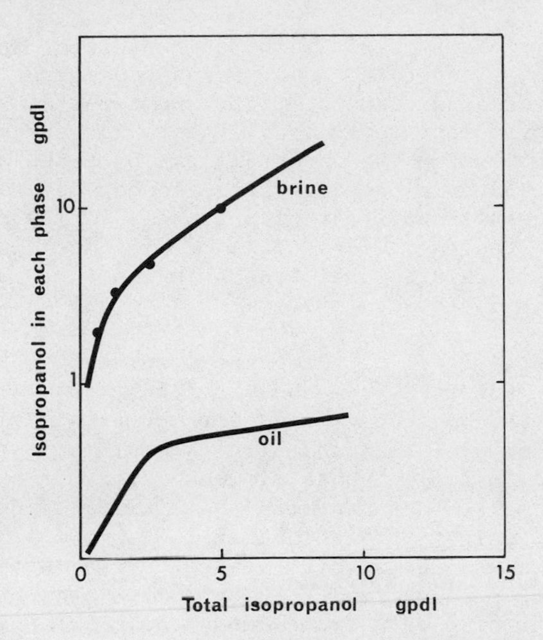

Fig. 1. Isopropanol partitioning between oil and brine as a function of concentration.

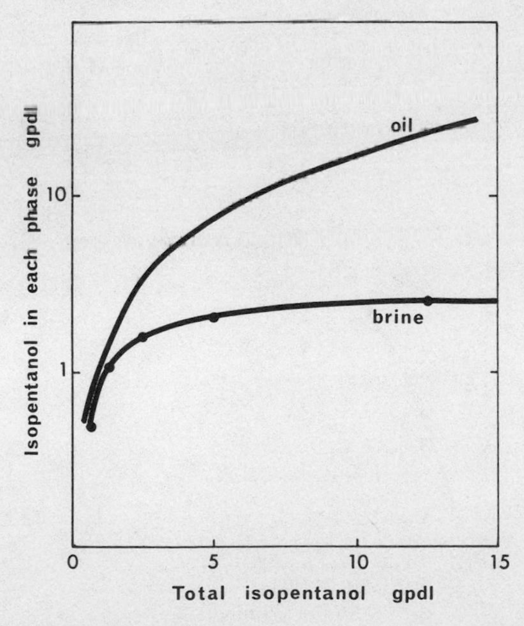

Fig. 2. Isopentanol partitioning between oil and brine as a function of concentration.

(Figure 3). With this type of alcohols a specific concentration can be determined, close to the saturation concentration of the water-phase for the considered salinity and temperature, for which the partition coefficient is one. At this point the alcohol partition does not depend on the water/oil ratio of the system (the corresponding tie-line in the oil-brine-alcohol diagram is horizontal) and the resulting effect on optimal salinity should be minimal. In our studies, this point is 6.25% butanol in oil and brine and we assume this composition is the most efficient one for practical optimization of the system.

Surfactant optimization: Once the cosolvent type and concentration have been selected from the partition curves, the surfactant optimization can be accomplished for the total system. With 6.25% n-butanol, complete oil and brine solubilization at WOR = 1 (i.e. formation of a single phase microemulsion) is obtained for mixtures of the two surfactants with the minimum amount for the average molecular weight of 447 (53% TRS 10-80 and 47% Gerland S85). With this blend the required amount of surfactant to get a single phase is 6% pure sulfonate (Figure 4). The corresponding ternary phase diagram with a fixed amount of n-butanol exhibits minimum height of the binodal curve and the multiphase region of the diagram includes a range of middle phase compositions (Figure 5).

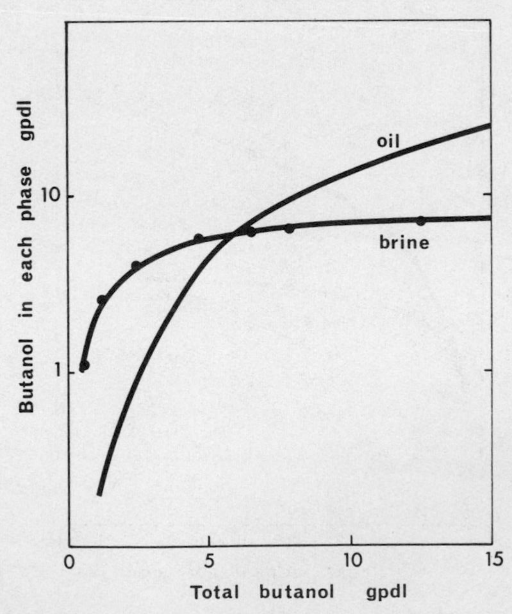

Fig. 3. Butanol partitioning between oil and brine as a function of concentration.

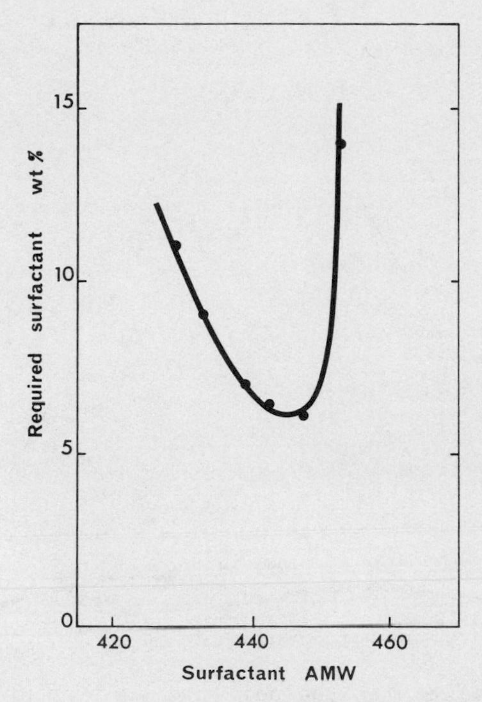

Fig. 4. Required amount of surfactant for a single phase from oil-brine (WOR = 1) and 6.25 gpdl butanol as a function of the surfactant average molecular weight.

DISCUSSION

At this point of the study, the question is to establish whether the system defined this way is actually the optimal system and, furthermore, if it is the only feasible one.

1) As far as pure alcohols (and not blends) are used, this system can be considered as the most suitable for further oil-recovery investigations. The following relationships between the butanol concentration and some important features of the system emphasize its practical optimization:

1a) The required amount of surfactant to completely solubilize oil and brine (WOR = 1) is minimum in the range just below 6.25% butanol but increases beyond this point and also increases for very low alcohol content (Figure 6). This concentration very close to the saturation concentration of the water phase appears as an upper limit for the alcohol content of the oil-brine system.

1b) The regions of various phase behavior are plotted on the same figure showing separate effects of the alcohol and surfactant. In this plane every vertical line lies in a ternary diagram with a

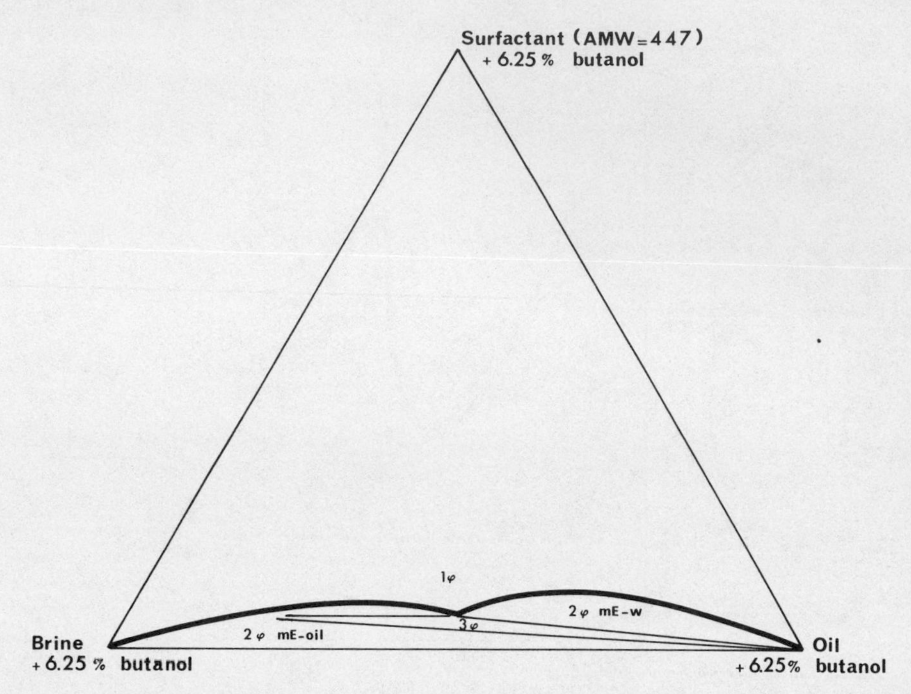

Fig. 5. Phase diagram with the selected surfactant and a constant amount of alcohol.

fixed alcohol content and every inclined one in a pseudo-ternary diagram with constant surfactant/cosolvent ratio. This representation provides elucidation of complex behavior usually encountered in such pseudo-ternary diagrams wherein both alcohol and surfactant are varied simultaneously. Small monophasic areas between two multiphase zones or three-phase lenses inside a large multiphase region (Figure 7) are observed in pseudo-ternary diagrams of constant surfactant/alcohol ratios. On the other hand, it is to be noticed that the system we have selected here could not be represented by a similar pseudo-ternary diagram.

1c) The plot of interfacial tension between oil and brine (WOR = 1) with a constant amount of 5% pure sulfonate as a function of the alcohol concentration (Figure 8) is very similar to the curve of the required amount of surfactant to form single phase microemulsions (Figure 6).

1d) Viscosity of the corresponding microemulsions along the curve of the required amount of surfactant goes through a minimum (Figure 9) at a specific alcohol concentration (~ 6.25%) which is relevant to the mobility of the oil bank (10).

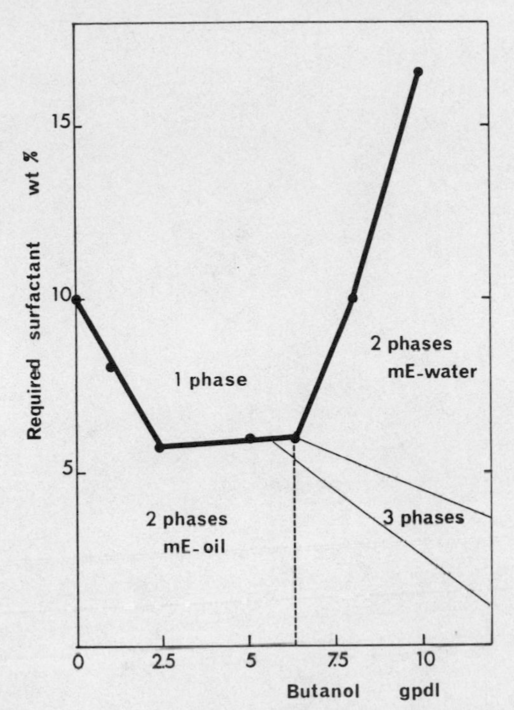

Fig. 6. The effect of butanol concentration on the required amount of surfactant (AMW = 447) for a single phase at WOR = 1.

2) However, from relationship of alcohol partition and saturation under given conditions, many combinations of surfactant-cosolvent may lead to complete solubilization of oil and brine at a given WOR with the same amount of surfactant (Figure 10).

2a) When used with small amounts of either the oil soluble alcohol or the water-soluble one, the surfactant blend can be adjusted to an optimal average molecular weight. For instance, at WOR = 1, average M.W. = 430 in the presence of 1% isopentanol and average M.W. = 485 in the presence of 1% isopropanol. But these combinations are only valid for a given WOR of formulations, because of strong effects of such cosolvents on optimal salinity (9), and they are usually viscous because of their low alcohol content.

2b) When the surfactant is selected without any alcohol, the optimal average molecular weight is 455, very nearly equal to the one determined with the selected 6.25% n-butanol (Figures 4 and 10). The resulting composition is highly viscous without alcohol.

From this series of curves, a cosolvent blend can be formulated for which surfactant optimization is exactly the same as without any alcohol. This cosolvent blend, close to butanol with

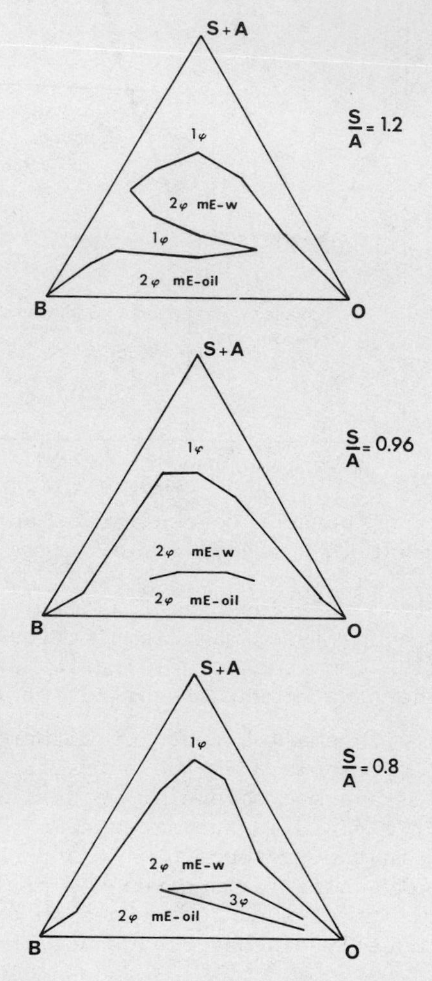

Fig. 7. Pseudo-ternary phase diagrams with various surfactant/alcohol ratios.

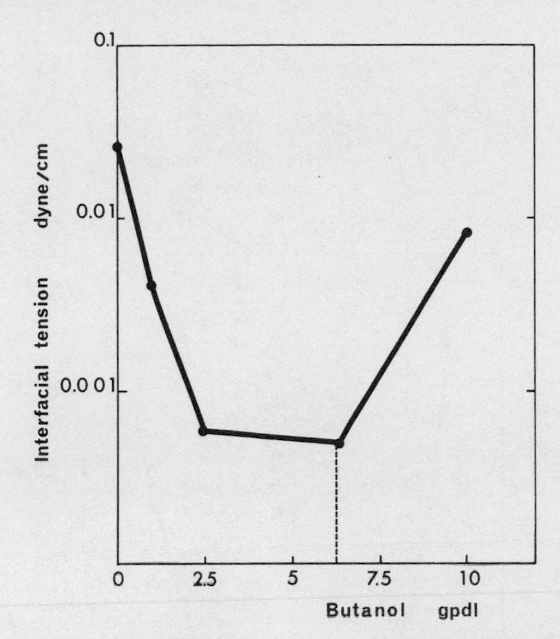

Fig. 8. The effect of butanol concentration on interfacial tension between oil and brine in the presence of 5% pure sulfonate (AMW = 447).

a small fraction of isopropanol, might have been selected from partition curves when investigating alcohol mixtures. It is likely that the resulting system would be optimal due to equal partitioning of both surfactant and cosolvent.

CONCLUSIONS

From many possible surfactant-alcohol combinations, selection of the preferred components of a micellar system for any field application can be conducted in two successive steps. The method leads to a micellar system for which the required amount of surfactant, the interfacial tension for a given composition and viscosity are simultaneously minimized.

Due to the two-step selection path, both cosolvent and surfactant should be equally partitioned between oil and brine: the alcohol type and concentration are directly determined from partitioning behavior. The optimal alcohol concentration is the one for which the partition coefficient is unity, and then the surfactant mixture is adjusted to the total system of oil-brine-cosolvent.

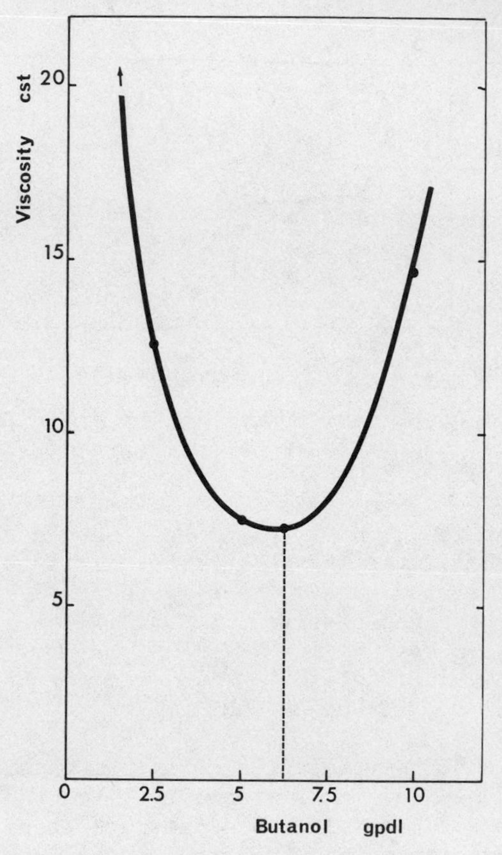

Fig. 9. The effect of butanol concentration on viscosity of the microemulsion with oil-brine (WOR = 1) and the required amount of surfactant.

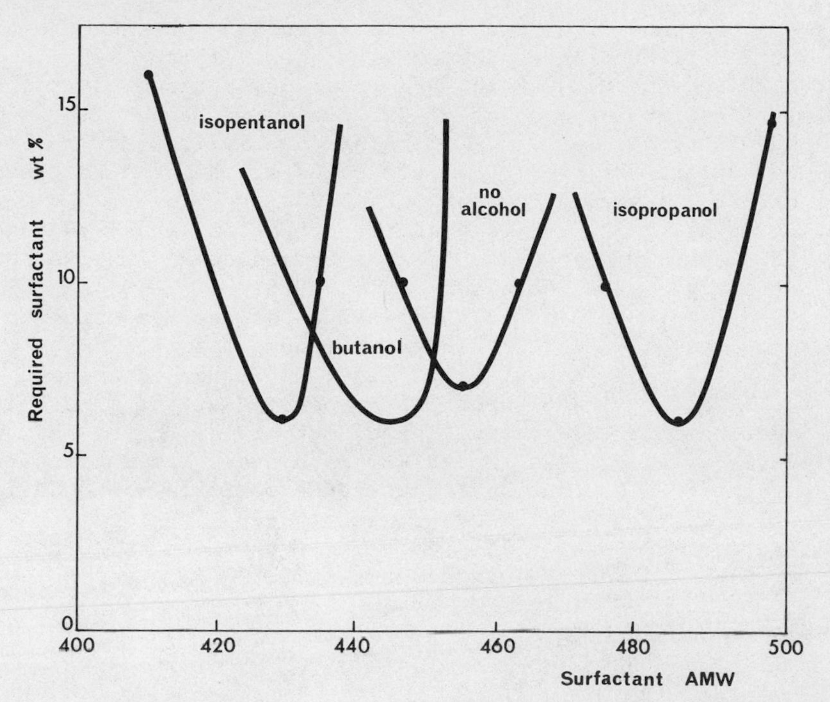

Fig. 10. Required amount of surfactant for a single phase from oil-brine (WOR = 1) and various cosolvents as a function of the surfactant average molecular weight.

This study supports the assumption that the alcohol influences the phase behavior of oil-brine-surfactant system and its effect is related to the saturation concentration in the water-phase. On the other hand, the ternary phase diagram with a constant amount of the cosolvent seems to be more valuable for selecting microemulsion formulations than the pseudo-ternary representation in which both alcohol and surfactant are varied simultaneously.

ACKNOWLEDGMENT

The authors wish to thank T. Auger for his contribution to this study.

REFERENCES

1. W. C. Tosch, S. C. Jones, and A. W. Adamson, J. Colloid Interface Sci., 31, 297 (1969).
2. M. Baviere, SPE Paper 6000 presented at the 51st Fall Meeting of SPE, New Orleans, Louisiana, Oct. 1976.

3. S. C. Jones and K. D. Dreher, SPE Paper 5566 presented at the 50th Fall Meeting of SPE, Dallas, Texas, Oct. 1975.
4. W. C. Hsieh and D. O. Shah, SPE Paper 6594 presented at the International Symposium on Oil Field and Geothermal Chemistry, La Jolla, California, June 1977.
5. S. J. Salter, SPE Paper 6843 presented at the 52nd Fall Meeting of SPE, Denver, Colorado, Oct. 1977.
6. J. L. Salager, J. C. Morgan, R. S. Schechter, and W. H. Wade, SPE Paper 7054 presented at the Symposium on Improved Oil Recovery, Tulsa, Oklahoma, April 1978.
7. S. J. Salter, SPE Paper 7056 presented at the Symposium on Improved Oil Recovery, Tulsa, Oklahoma, April 1978.
8. M. Bourrel, A. M. Lipow, W. H. Wade, R. S. Schechter, and J. L. Salager, SPE Paper 7450 presented at the 53rd Fall Meeting of SPE, Houston, Texas, Oct. 1978.
9. M. Baviere, W. H. Wade, and R. S. Schechter, Paper presented at the 3rd International Conference on Surface and Colloid Science, Stockholm, Sweden, August 1979.
10. J. Labrid, SPE Paper 8325 presented at the 54th Fall Meeting of SPE, Las Vegas, Nevada, Oct. 1979.

THE EFFECT OF SALT, ALCOHOL AND SURFACTANT

ON OPTIMUM MIDDLE PHASE COMPOSITION

M. Baviere,[+] W. H. Wade,[*] and R. S. Schechter[++]

[+]Institut Francais du Petrole B.P. 311, 92506 - Rueil-Malmaison (France). Departments of Chemistry[*] and Petroleum Engineering,[++] The University of Texas at Austin, Austin, Texas 78712, U.S.A.

In this work, salt, alcohol and surfactant effects are investigated, at different water-to-oil ratios to determine their influence on phase behavior. A linear relationship is found between the logarithm of water-oil ratio in the middle phase and the salinity. Moreover, in the middle of the range of salt or alcohol concentration giving a three-phase system, the water-oil ratio in the middle phase has a value close to unity (value defined as an optimum), whatever the overall water-oil ratio may be.

The alcohol concentration in the middle phase is higher than the value calculated according to the solubilization parameters and the alcohol concentration in the excess phases. This implies that the alcohol is involved in the interfacial structures of the middle phase, and is further verified by studies involving changing the surfactant/alcohol ratio.

INTRODUCTION

The formulation of a chemical slug for a micellar-polymer oil recovery process requires an understanding of the surfactant-brine-crude oil phase behavior. It has been established that the most efficient formulations are those which solubilize the greatest volume of oil and water per unit quantity of surfactant and exhibit low interfacial tensions with excess oil and water. The phase diagram of such formulations generally includes an area containing three-phase systems, systems of type III according to Winsor's notation (1).

The rules for selecting components to obtain this phase behavior has been described in several publications (2,3). But there is no extensive study of the exact evolution of the phase composition in a multiphase system containing brine, hydrocarbon, surfactant and alcohol when the parameters are varied.

In this investigation we determined the composition of each phase during the I→III→II transitions obtained by salinity, alcohol, surfactant and water-to-oil volume ratio (WOR) scans. In this way, the "quality" of the middle phase was characterized with regard to solubilization parameter, water-to-oil and surfactant-to-alcohol ratios.

In addition, it will be shown that a composition change in a given system may greatly change the phase compositions, namely that of the middle phase. So, to obtain information on interfacial and structural effects, it is necessary to compare different compositions under such conditions that the type of the system is always the same. Here we chose the symmetrical type III as the reference state which means that the middle phase exhibits a water-to-oil ratio (WOR_{MP}) equal to one. This kind of system, at least when sulfonates are used, corresponds from an interfacial and phase behavior point of view to the optimum conditions for the oil recovery process, i.e. lowest interfacial tension and high solubilization parameters for water and oil.

Changes in the composition of the micellar slug occur in porous media, especially because of dilution by both brine and oil and the adsorption of surfactant on the rock and for this reason evolution of the optimum state is of practical importance.

EXPERIMENTAL

Materials: Alcohols and normal octane were supplied by Fisher Scientific Company and Phillips Petroleum Company, respectively, and were 99+ mole % purity. One surfactant used was a synthetic sodium dodecyl orthoxylene sulfonate provided by Exxon Chemical Company. After deoiling, the product is a light beige crystal. In addition a deoiled petroleum sulfonate, TRS 10-80, supplied by Witco Chemical Company and a monoisomeric species, sodium 4 phenyldodecyl sulfonate (99.5 mole % purity) synthesized in our laboratory were studied.

Methods: The approach was to study multiphase systems containing brine (water and NaCl), hydrocarbon, alcohol and sulfonate, in which one parameter was varied in order to produce the I→III→II transitions. All the mixtures had a volume of 10 ml and were gently shaken several times and allowed to equilibrate, at 29°C for at least 12 hours until stable phase volumes and clear phases were obtained. Then the volumes were measured (the error in assuming volume additivity was less than 0.5%) and each phase separated and analyzed. Water, n-octane and alcohols were measured by gas-solid chromatography (Varian 3700 Gas Chromatograph) using a Porapak S 100-120 mesh.

Because of relatively low salinity level, the salt distribution which occurs between the aqueous phase and water in sulfonate rich micellar phase (4) was not taken into consideration. The sulfonate concentration in the sulfonate rich micellar phase was determined by phase volume measurement while being certain that all the sulfonate was in the phase. In fact, ultra-violet spectrophotometry results show that the sulfonate concentration in the excess phases is at least 100 times less than in the sulfonate rich micellar phase (5).

RESULTS AND DISCUSSION

The effect of salinity: Table 1 shows how the phase compositions vary when the salinity is increased in the following system: brine, octane (WOR = 1.1), sulfonate (1 wt %) and isopropanol (3 wt %). Below 27 g/l NaCl, the system is of type I, i.e. a sulfonate rich micellar solution in equilibrium with excess oil. By increasing salinity, type III systems are formed and finally, above 38 g/l NaCl, type II systems appear. When there is only one excess phase, Table 1 regards the middle phase as the sulfonate-rich micellar phase. Thus, at low salinities the composition of the micellar phase in equilibrium with an excess oil phase is tabulated for convenience as a middle phase in equilibrium with an upper phase. There is practically no variation in the compositions of the excess phases attending the I→III→II transitions resulting from increasing the electrolyte composition. Table 1 shows that the isopropanol concentration in the excess water and excess hydrocarbon phases is essentially constant and equal, on the average to 7.1 and 0.15 vol %, respectively. It is also observed that the alcohol content of the microemulsion phase is greater than would be calculated assuming the alcohol to be distributed between the brine and hydrocarbon phases in the microemulsion in precisely the same proportions found in the excess phases. Thus the quantity ΔV defined as

$$\Delta V = V_A^M \text{ (observed)} - \left[V_B^M \left(\frac{V_A^{EB}}{V_B^E} \right) + V_H^M \left(\frac{V_A^{EH}}{V_H^E} \right) \right] \qquad (1)$$

is generally found to be positive. Here V_A, V_B and V_H are the alcohol, brine and hydrocarbon volume fractions, respectively.

Table 1. Effect of the salinity on the phase behavior of the brine-octane-sulfonate-isopropanol system. WOR = 1.1, sulfonate and isopropanol concentrations are 1.0 and 3.0 wt %, respectively.

Salinity g/l NaCl	Phase Volume (%)			Phase Composition (Vol. %)							
				Lower Phase		Microemulsion Phase				Upper Phase	
	Lower Phase	Middle Phase	Upper Phase	Brine	Iso-propanol	Brine	Iso-propanol	Sulfonate	Octane	Iso-propanol	Octane
26	--	60.6	39.4	None	None	83.7	6.9	1.7	7.7	0.2	99.8
27	39.4	21.2	39.4	92.5	7.5	79.8	6.5	4.9	8.8	0.1	99.9
29	44.2	16.3	39.5	93.1	6.9	56.0	4.8	6.0	33.2	0.1	99.9
31	47.5	14.4	38.1	92.9	7.1	46.9	4.3	7.4	41.4	0.2	99.8
33	49.2	13.7	37.1	92.8	7.2	38.8	3.7	7.8	49.7	0.1	99.9
35	50.5	14.5	35.0	92.8	7.2	29.8	3.3	7.5	59.4	0.1	99.9
37	50.5	15.7	33.8	93.0	7.0	20.3	2.5	7.3	69.9	0.1	99.9
38	50.8	16.2	33.0	93.0	7.0	17.2	2.4	7.2	73.2	0.2	99.8
39	51.0	49.0	--	93.0	7.0	6.8	1.0	2.5	89.7	None	None

V_A^{EB} and V_A^{EH} are the volume fractions of alcohol in excess brine and excess hydrocarbon, respectively. The superscript M refers to the microemulsion phase and E to the excess phase.

Significantly ΔV tends to increase as the salinity is increased. There are several possible explanations. It has been shown that the micellar electrolyte concentration is generally smaller than that of the excess aqueous phase (4) and thus the alcohol solubility would be expected to increase in the water rich regime of the micellar phase assuming, as seems reasonable, that essentially bulk water and hydrocarbon phases coexist in the micellar phase. The data shown by Tosch, Jones and Adamson (4) indicate that the electrolyte concentration of the micellar phase deviated most at about 0.35 moles/l Na_2SO_4 from that of the aqueous phase and the two concentrations rapidly converged thereafter. The NaCl concentration of the micellar phases was not measured, but the differences between this and that of the excess aqueous phase is not expected to be large. Furthermore, small changes in the NaCl concentration do not change the distribution of isopropanol between an aqueous phase and an equilibrium octane phase in the absence of surfactant.

There does appear to be a correlation between the volume fraction surfactant and the volume fraction of excess alcohol (ΔV). The ratio $\Delta V/V$ surfactant is almost exactly 0.1 for all experiments of Table 1. This would seem to imply a degree of interfacial activity exhibited by isopropanol. Such activity should be enhanced by increasing the alcohol chain length, and it will be seen that this is the case.

The effect of water-to-oil ratio: By increasing the overall water-to-oil ratio in a given system, with everything else being kept constant, the alcohol content in each phase can either decrease or increase, depending on the alcohol distribution between hydrocarbon and water. If the partition coefficient, defined as the alcohol concentration in the oil phase divided by its concentration in the aqueous phase, is less than one, an increase in the WOR induces a decrease in the alcohol concentration in both phases. This is the case with isopropanol and the water-octane pair. This is pointed out elsewhere (6).

Table 2 shows that there is a strong decrease in the alcohol concentration in all three phases and, at the same time, an increase in the middle phase volume--from 7 to 21 volume %--when the WOR is increased from 0.27 to 5.23. This result strikingly demonstrates an important effect of the alcohol. Decreasing the alcohol content increases the solubilization parameters and the multiphase zone becomes smaller. This trend has been reported by Salter (7).

Table 2. Water-to-oil ratio effect on the optimum salinity. Sulfonate and alcohol concentrations are 1.0 and 3.0 wt % respectively

Overall Water-Oil Ratio	Optimum Salinity g/l NaCl	Phase Volume (%)			Phase Composition (Vol. %)							
					Lower Phase		Middle Phase				Upper Phase	
		Lower Phase	Middle Phase	Upper Phase	Brine	Iso-propanol	Brine	Iso-propanol	Sulfo-nate	Octane	Iso-propanol	Octane
0.27	41.5	20.0	7.0	73.0	87.2	12.8	40.1	9.6	12.9	37.4	0.3	99.7
1.10	31.5	47.7	14.0	38.3	92.9	7.1	46.9	4.2	7.4	41.5	0.2	99.8
5.23	29.0	74.3	21.0	4.7	95.5	4.5	48.6	3.0	4.7	43.7	0.2	99.8

For the systems studied, the addition of isopropanol increases the optimum salinity which is defined here as that salinity for which the water-to-oil ratio in the middle phase is unity. Thus the optimum salinity for the system recorded in Table 1 is approximately 32 g/l NaCl. This means that the addition of isopropanol increases the solubility of sulfonate in the aqueous phase relative to the oil phase. Thus by increasing the electrolyte concentration appropriately when adding isopropanol, its effect can be compensated thereby maintaining equal volumes of aqueous phase and hydrocarbon in the middle phase.

In Figures 1, 2 and 3, the logarithms of the water-to-oil ratio found in the middle phase are plotted as a function of salinity. The relationship is linear in these three examples [alcohols are isopropanol, 2-butanol and isopentanol (3 methyl-butanol) respectively] for 3 WOR values: 0.27, 1.1 and 5.23. One interesting feature observed in Figures 1, 2 and 3 is that the WOR in the middle phase appears to be close to one when the salinity is equal to the average of the extreme salinities for which middle phases are observed, even though the overall WOR is varied over a wide range. This means that the definition of optimum salinity used by Salager (8) and Wade et al. (9) is almost equivalent to the one used here.

The effect of alcohols: I→III→II transitions can be obtained by adding sufficiently high molecular weight alcohols instead of salt. The results in Table 3 deal with a system containing a fixed wt % of 2-butanol which has been added to speed up the coalescence of macroemulsions, and normal octanol of variable concentration.

The influence of octanol on phase behavior is pronounced. An increase in its concentration from 0.25 to 0.40 g/100 ml is sufficient to produce the complete I→III→II transition.

It should be noted that in alcohol scans

- excess phase compositions do not vary to a significant extent even though the microemulsion composition varies considerably.
- the relationship between the logarithm of WOR in the middle phase and alcohol concentration is linear (Figure 4).
- the higher the alcohol molecular weight the stronger its effect (7,8). However, for molecular weights higher than octanol this effect levels off as far as the water-octane pair is concerned (Figure 5) (to be published).
- the optimum alcohol concentration almost corresponds to the average of extreme values giving a three-phase system.

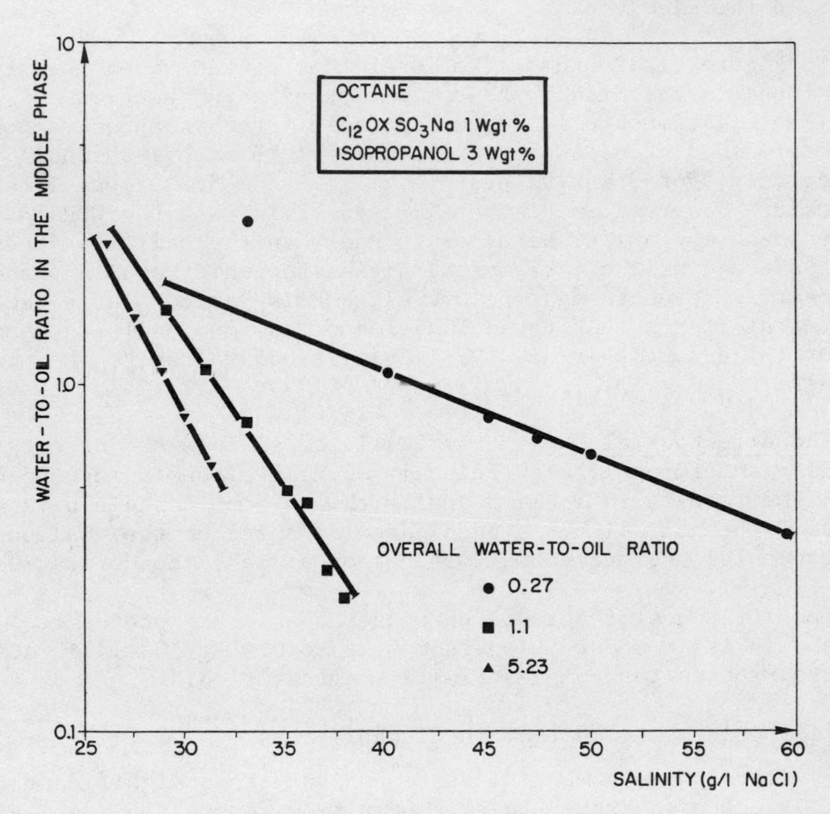

Fig. 1. Water-to-oil ratio in the middle phase as a function of salinity.

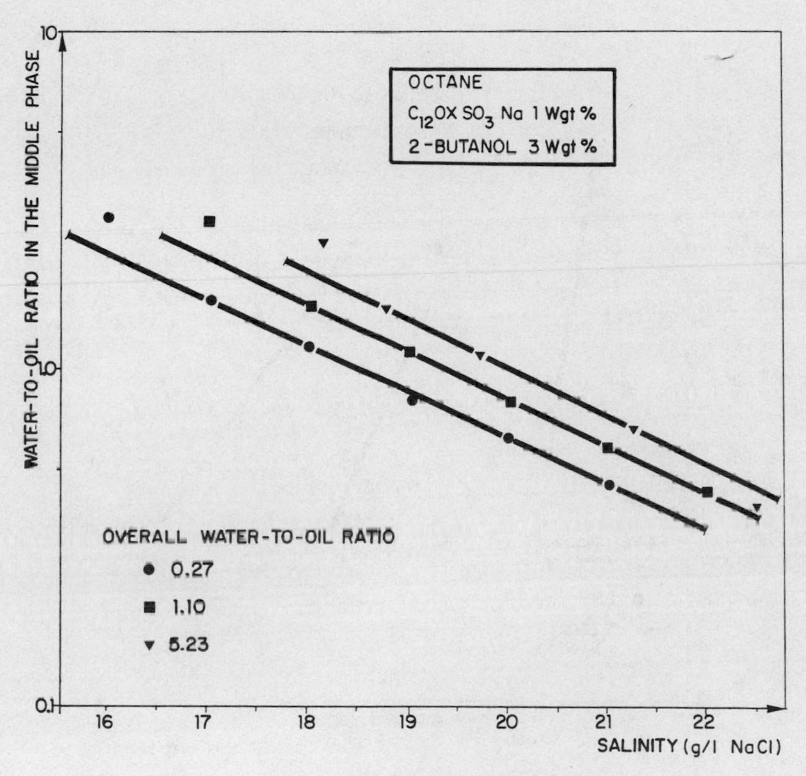

Fig. 2. Water-to-oil ratio in the middle phase as a function of salinity.

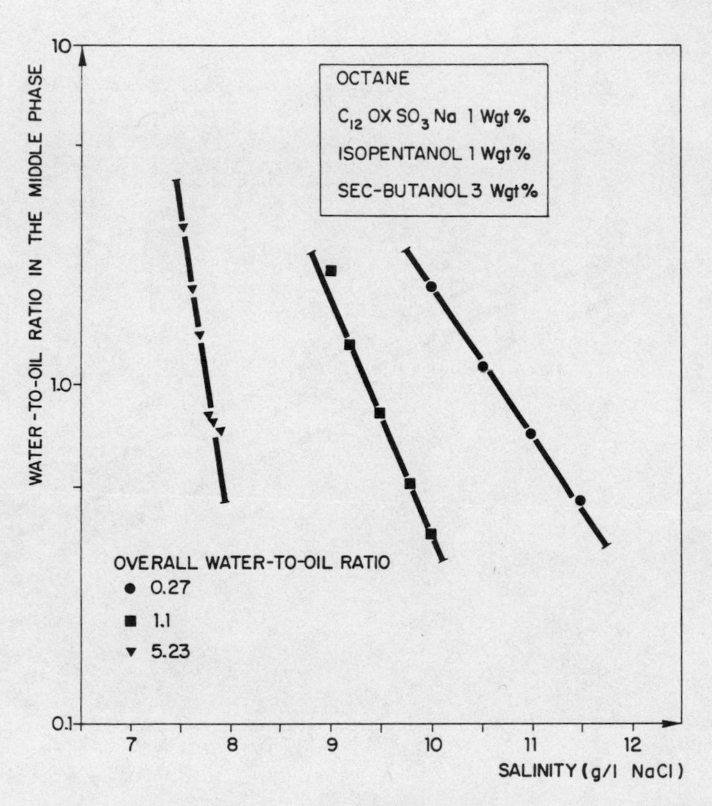

Fig. 3. Water-to-oil ratio in the middle phase as a function of salinity.

Table 3. Effect of the n-octanol content on the phase behavior of the brine-octane-sulfonate-2-butanol system. Brine is 10 g/l NaCl. WOR = 1.1, sulfonate and 2-butanol concentrations are 1.0 and 3.0 wt % respectively.

Octanol Concentration (g/100 ml)	Phase Volume (%)			Phase Composition (Vol. %)									
				Lower Phase		Middle Phase					Upper Phase		
	Lower Phase	Middle Phase	Upper Phase	Brine	2-Butanol	Brine	2-Butanol	Octanol	Sulfonate	Octane	2-Butanol	Octanol	Octane
0.250	--	55.6	44.4	--	--	82.9	5.4	0.1	1.9	9.7	1.6	0.5	97.9
0.275	26.0	31.0	43.0	95.0	5.0	63.6	5.3	0.4	3.3	27.4	1.6	0.6	97.8
0.300	35.0	23.0	42.0	94.9	5.1	51.1	4.9	0.5	4.4	39.1	1.6	0.6	97.8
0.350	37.4	22.2	40.4	94.9	5.1	34.0	4.1	0.7	4.5	56.7	1.7	0.8	97.5
0.375	42.0	24.7	33.3	94.7	5.3	25.6	3.8	0.8	4.0	65.9	1.7	0.9	97.4
0.400	43.0	57.0	--	94.9	5.1	10.2	2.7	0.8	1.8	84.6	---	---	--

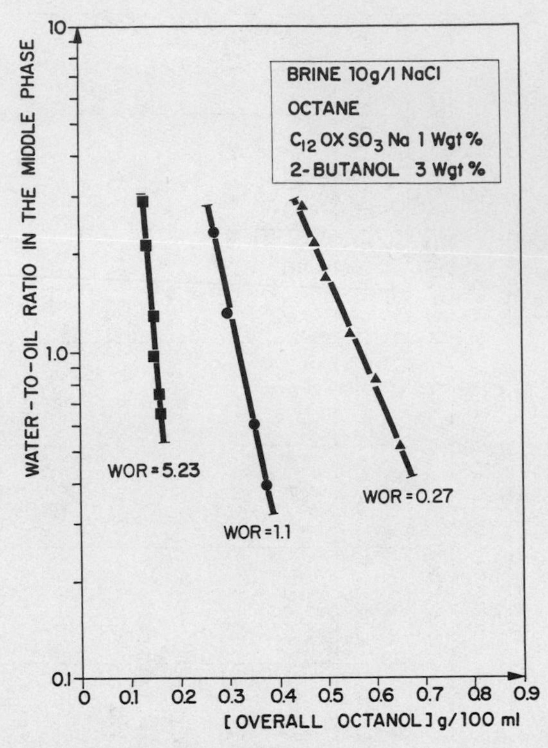

Fig. 4. Water-to-oil ratio in the middle phase as a function of overall octanol concentration.

Perhaps these observations can be generalized to the other optimization parameters, i.e. sulfonate, hydrocarbon molecular weight and temperature.

The decrease in the optimum octanol concentration when the overall WOR is increased, as shown in Table 5, requires further explanation. In these particular compositions, the 2-butanol and octanol partition coefficient between octane and water are less than one and greater than one, respectively. Furthermore, increasing the concentration of either 2-butanol or octanol has the effect of decreasing the optimum salinity. When the overall WOR is increased there is less 2-butanol and more octanol in the middle phase. Thus if the octanol content in all the phases is maintained constant, then it would be necessary to increase the salinity to maintain the optimum formulation. Alternately, instead of increasing the salinity, the same effect can be achieved by increasing the octanol concentration. Since increasing WOR decreases the concentration of 2-butanol in the middle phase microemulsion, less octanol is required to maintain the optimum formulation. The increase in middle phase volume with increasing

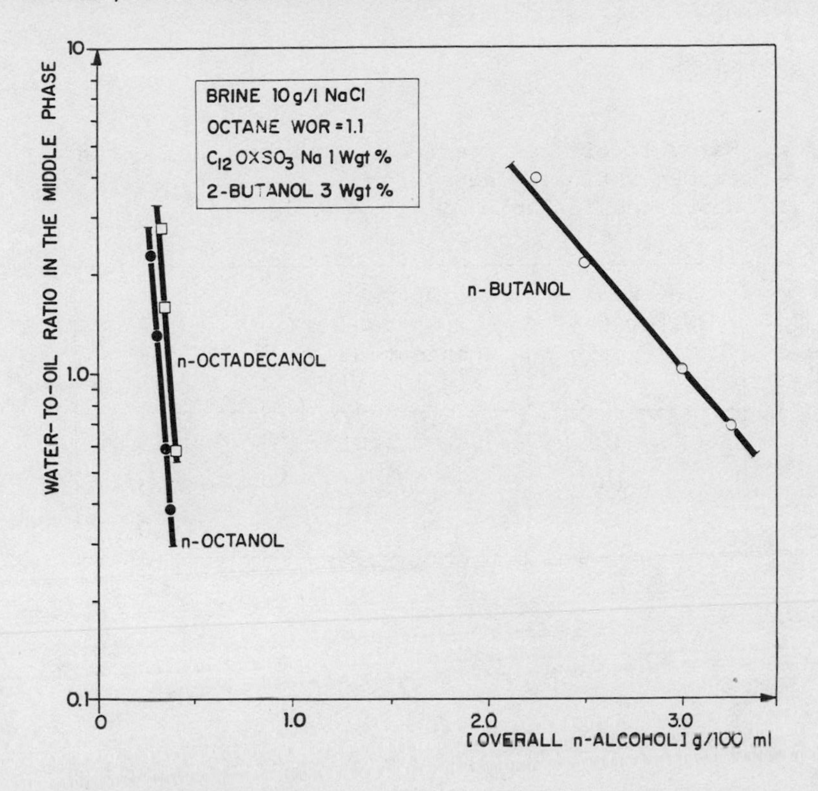

Fig. 5. Water-to-oil ratio in the middle phase as a function of overall n-alcohol concentration.

WOR is due to the decreased amounts of both 2-butanol and octanol used in the optimum formulation (Table 4).

The effect of surfactants: Figure 6 shows that the microemulsion volume is proportional to the surfactant concentration as reported previously (8,10). However, in some cases it appears that the optimum salinity in a given system varies with this concentration when the surfactant used is not an isomerically pure product. It has been pointed out, for instance, by interfacial tension measurements that by increasing the petroleum sulfonate (TRS 10-80 supplied by Witco Chemical Company) concentration, the preferred alkane molecular weight is decreased. This shift occurs apparently because there is a dependence of the mean micellar and monomeric molecular weights on the total surfactant concentration (9,11).

Thus for the sulfonate used here an increase in the sulfonate concentration decreases the optimum salinity (Table 5), a 17.0 g/l NaCl brine gives a type I system at 0.1% sulfonate, but it gives a type II system at 5.0% sulfonate. A salinity scan was made for

Table 4. Water-to-oil ratio effect on the optimum octanol concentration. The system is the same as the one described in Table 3.

Overall Water-to-Oil Ratio	Optimum Octanol Concentration (g/100 ml)	Middle Phase Volume (%)
0.27	0.560	17.5
1.10	0.325	22.5
5.23	0.150	26.0

Table 5. Sulfonate ($C_{12}OXSO_3Na$) concentration effect on salinity range giving a three phase system. The brine-to-octane ratio is 1.1, 2-butanol concentration is 3.0 wt %.

Sulfonate wt %	Salinity for Middle Phase (g/l NaCl)			Middle Phase Vol. at Optimal Salinity (%)
	Mini	Optimal	Maxi	
0.1	17.25	19.75	22.25	1
0.5	16.25	19.00	21.75	7
1.0	15.25	18.00	20.75	15
2.0	14.75	17.50	20.25	31
3.0	14.25	16.40	18.50	48
4.0	14.25	16.00	17.75	67
5.0	14.25	15.50	16.75	77

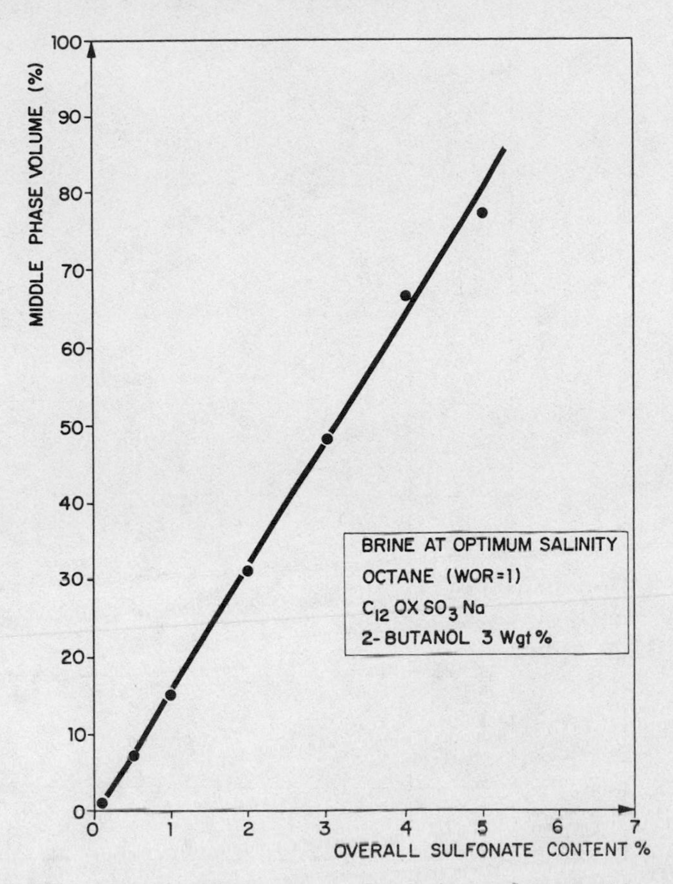

Fig. 6. Middle phase volume as a function of overall sulfonate concentration.

several sulfonate concentrations and all the phases were analyzed to determine the exact composition at the optimum ($WOR_{MP} = 1$). Table 6 shows that an increase in the sulfonate concentration leads to:

- a decrease in alcohol concentration in the excess phases. This supports the assumption that a portion of this component is included in the interfacial structure.
- a decrease in the sulfonate and alcohol concentrations in the middle phase. There is a small change in the sulfonate-to-alcohol ratio.
- an unchanged alcohol partition coefficient between the excess phases.

The shift in optimum salinity with the increase of the sulfonate concentration has not been explained for sulfonate TRS 10-80 and the sodium 4-phenyldodecyl sulfonate. The phase behavior

Table 6. Excess and middle phase compositions at optimum salinity as a function of sulfonate concentration. The brine-to-octane ratio is 1.1, 2-butanol (SBA) content is 3.0 wt %.

Sulfo-nate wt %	Optimum Salinity (g/l NaCl)	Lower Phase Composition		Middle Phase Composition (Vol. %)				Middle Phase Characteristics			Upper Phase Composition	
		Water (Vol. %)	SBA (Vol. %)	Water	SBA	Sulfo-nate	Octane	S+A	R=S/A	WOR_{MP}	Octane (Vol. %)	SBA (Vol. %)
0.1	19.75	94.5	5.5	38.7	5.8	10.1	45.4	15.9	1.74	0.85	98.4	1.6
0.5	19.00	94.7	5.3	42.1	5.6	8.0	44.3	13.6	1.43	0.95	98.5	1.5
1.0	18.00	94.9	5.1	44.0	5.0	7.3	43.7	12.3	1.46	1.01	98.6	1.4
2.0	17.50	95.3	4.7	44.6	4.5	6.9	44.3	11.4	1.53	1.01	98.8	1.2
3.0	16.40	95.5	4.5	44.6	3.9	6.0	45.5	9.9	1.54	0.98	99.0	1.0
4.0	16.00	95.6	4.4	46.7	3.8	6.2	43.3	10.0	1.63	1.08	99.0	1.0
5.0	15.50	96.0	4.0	44.6	3.8	6.5	45.1	10.3	1.71	0.99	99.1	0.9

Table 7. A--Excess and middle phase composition at a given salinity as a function of sulfonate concentration.

Sulfonate	TRS 10-80
Brine	11.5 g/1 NaCl
Octane	WOR = 1/1
SBA (2-butanol)	3 wt %

Sulfonate wt %	Lower Phase		Middle Phase				Middle Phase Characteristics			Upper Phase	
	Water	SBA	Water	SBA	Sulfonate	Octane	S+A	R_{MP}=S/A	WOR	Octane	SBA
0.5	94.7	5.3	62.5	3.9	4.5	29.1	8.4	1.15	2.15	1.3	98.7
1.0	95.1	4.9	63.1	4.0	4.5	28.4	8.5	1.13	2.22	1.3	98.7
3.0	95.4	4.6	62.0	3.4	4.3	30.3	7.7	1.26	2.05	1.1	98.9

B--Excess and middle phase composition at salinity 53 g/1 NaCl as a function of sulfonate concentration.

Sulfonate	Sodium 4 phenyl dodecyl sulfonate
Brine	53 g/1 NaCl
SBA (2-butanol)	3 wt %
Octane	WOR = 1/1

Sulfonate wt %	Lower Phase		Middle Phase				Middle Phase Characteristics			Upper Phase	
	Water	SBA	Water	SBA	Sulfonate	Octane	S+A	R_{MP}=S/A	WOR	Octane	SBA
0.5	95.6	4.4	47.1	5.8	8.3	38.8	14.1	1.43	1.21	98.3	1.7
3.0	96.3	3.7	51.2	4.4	6.3	38.1	10.7	1.43	1.34	98.8	1.2
5.0	96.8	3.2	55.4	3.5	5.6	35.6	9.1	1.60	1.56	99.2	0.8

showed the following results: no shift with petroleum sulfonate and a very small shift with pure sulfonate (Table 7) but in an opposite direction to the one observed with sodium dodecyl ortho-xylene sulfonate. Clearly this confusing situation will require further investigation, but the possibility of a shifting optimum salinity has to be taken into account to predict phase behavior during the oil recovery process.

CONCLUSIONS

(1) During I→III→II transitions associated with a changing salinity or alcohol, there is practically no change in the alcohol distribution between the excess phases.

(2) There is a linear relationship between the logarithm of the water-to-oil ratio in the middle phase and the salinity or alcohol concentration.

(3) In the middle of the range of salt or alcohol concentrations giving a three-phase system, the water-to-oil ratio in the middle phase has a value close to one, a value defined as an optimum, whatever the overall water-to-oil ratio may be.

(4) Alcohol concentration measurements reveal an involvement of alcohol in the interfacial structure of sulfonate-rich micellar phases.

(5) Changing either the sulfonate concentration or the WOR may modify the phase behavior to a great extent. This limits the possibilities of using psuedo-ternary representation.

ACKNOWLEDGMENT

This work was supported by the contract Action Concertée No: 78 7 0739 of the Délégation Générale à la Recherche Scientifique et Technique, France.

REFERENCES

1. P. A. Winsor, Trans. Fara. Soc., 44, 376 (1948).
2. R. L. Reed and R. N. Healy, in "Improved Oil Recovery by Surfactant and Polymer Flooding," D. O. Shah and R. S. Schechter, eds., Academic Press, New York, 383 (1977).
3. J. C. Morgan, R. S. Schechter, and W. H. Wade, in "Improved Oil Recovery by Surfactant and Polymer Flooding," D. O. Shah and R. S. Schechter, eds., Academic Press, New York, 101 (1977).

4. W. C. Tosch, S. C. Jones, and A. W. Adamson, J. Coll. Interface Sci., 31, 297 (1969).
5. E. Vasquez, in "Phase Behavior and Interfacial Tension of Sulfonated Surfactants," Master's Thesis, The University of Texas at Austin (1978).
6. M. Baviere, W. H. Wade, and R. S. Schechter, "The Alcohol Effect on Phase Behavior of Micellar Solutions," to be submitted for publication to the Journal of Colloid and Interface Science.
7. S. J. Salter, SPE 6843, presented at the 52nd Annual Fall Technical Conference and Exhibition, Soc. Pet. Eng., Denver, Colorado, 1977.
8. J. L. Salager, in "Physico-Chemical Properties of Surfactant-Water-Oil Mixtures: Phase Behavior, Microemulsion Formation and Interfacial Tension," Ph.D. Dissertation, The University of Texas at Austin (1977).
9. W. H. Wade, J. C. Morgan, R. S. Schechter, J. K. Jacobson, and J. L. Salager, Soc. Pet. Eng. J., 18, 242 (1978).
10. W. C. Hsieh and D. O. Shah, SPE 6594, presented at the International Symposium on Oil Field and Geothermal Chemistry, Soc. Pet. Eng., La Jolla, California (1977).
11. R. L. Cash, J. L. Cayias, G. R. Fournier, J. K. Jacobson, C. A. Legear, T. Schares, R. S. Schechter, and W. H. Wade, in "Detergents in the Changing Scheme," published by the American Oil Chemist's Society, Champaign, Illinois (1977).

II: MICROSTRUCTURE OF SURFACTANT SYSTEMS

FORMATION AND PROPERTIES OF MICELLES AND MICROEMULSIONS

H. F. Eicke

Physikalisch-chemisches Institut der Universität Basel
Klingelbergstr. 80
CH-4056 Basel, Switzerland

The present view regarding the formation and properties of micelles in aqueous and hydrocarbon media as well as the corresponding microemulsions is discussed. Particular emphasis has been put on new results concerning W/O microemulsions.

INTRODUCTION

The opposing physical properties within typical amphiphilic molecules as, for example, surface active agents (surfactants), which are to be considered exclusively in the following, lead to a competitive situation between adsorption and homoassociation processes. Phenomenologically, the adsorption concept necessarily introduces a second phase (eventually in a dispersed state), i.e. a phase boundary where the surfactant tends to accumulate, thereby reducing the interfacial free energy of the system. Homoassociation can proceed in strictly binary solutions where the molecular arrangement is such that the antagonistic group of the surfactant with respect to the solvent (dispersion medium) is shielded from the latter. This process, of course, is also driven by a decrease of the free energy of the surfactant solvent system. Generally speaking, the formation of a stable emulsion (understood as an adsorption process at the oil/water interphase) has to be achieved by a surfactant concentration smaller than the critical micelle concentration (cmc) (1,2) or by any critical concentration leading to higher organized structures (bilayers).

The situation discussed recently in the literature (2), namely that a particular surfactant prefers to form micelles instead of being adsorbed in the interface, neglects the fact that the interfacial free energy (γ) has not to decrease to zero in order to attain the equilibrium state of the whole system. The entropic contribution due to the dispersion tendency of the surfactant has to be considered. (A negative value of γ, as is occasionally believed to be prerequisite for a stable system, is incompatible with general principles of thermodynamic stability.)

In the following two typical examples of considerable interest are to be discussed, representing the above mentioned two cases of adsorption and homoassociation, i.e. the formation and properties of micelles and microemulsions. In particular the difference between the two mutually antagonistic types (normal and inverted micelles as well as W/O and O/W microemulsions) of these two phenomena will be examined.

Micelles

a) Micelles in Aqueous Surfactant Solutions

Solutions of molecules with pronounced amphiphilic character exhibit unusual concentration dependent properties: dilute solutions behave like normal electrolytes (if ionic surfactants are considered), at higher, rather well-defined concentrations, quasi-sudden changes of several physical properties are observed (see Fig. 1). This phenomenon has been successfully ascribed to the formation of organised aggregates, i.e. micelles. The concentration above which micelles exist in equilibrium with monomers (and eventually small subunits) is the so-called critical micelle concentration (cmc).

Micelles formed by amphiphiles are generally considered to be the result of at least two opposing forces: attractive forces which are favoring the aggregation of monomers and repulsive forces which prevent the (unlimited) growth to very large aggregates and eventually to a continuous phase. As a third condition the cooperativity has to be taken into account. It expresses the fact that a few surfactant monomers are not sufficient to shield the antagonistic moieties with respect to the solvent. In other words a nucleation step is involved in the building up of micelles which is less favorable compared with the growing steps. The micelle formation in aqueous surfactant solutions compares very nicely with this scheme: the so-called hydrophobic effect favors the aggregation and hence an attraction between the paraffin chains of the amphiphiles which simultaneously leads to a relief of the strain on the hydrocarbon-separated water molecules. From this point of view the statement appears reasonable that micelle formation in aqueous media is entropy driven. The water retains the polar (ionic) head groups of the surfactants which are responsible for the repulsive interaction.

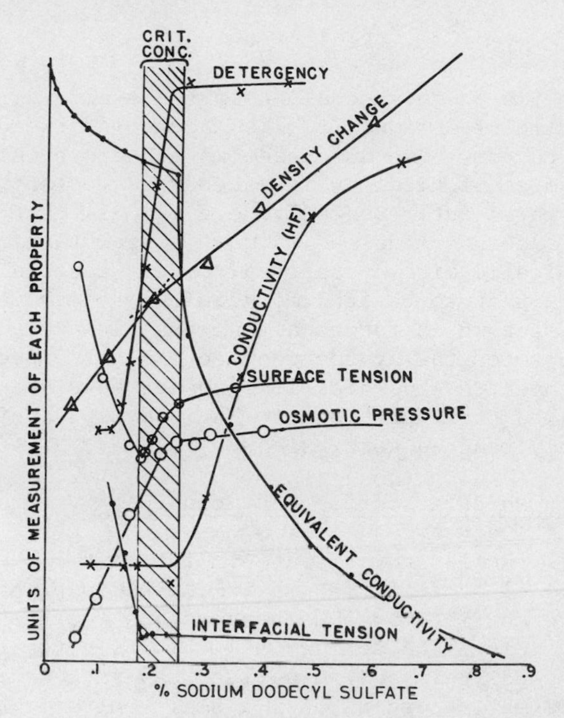

Fig. 1. Changes in some physical properties of an aqueous solution of SDS at the cmc. From W.C. Preston, J. Phys. Colloid Chem. 52, 84 (1948).

According to various experimental information the hydrocarbon core of the "aqueous" micelle has a liquid-like structure (3,4). This has been confirmed, in particular, by spectroscopic probing techniques (5,6). Hence the micelle in aqueous surfactant solutions presents itself to the surfactant monomer as an equivalent with respect to the (macroscopical) oil/water interface. It might be not unreasonable, therefore, to consider this type of micelle formation an "auto-solubilization" to stress the close resemblance between adsorption and homoassociation processes. The hydrocarbon core of a micelle in aqueous surfactant solutions is characterized by its excellent solvent power for crystalline non-polar compounds (7). This latter feature appears remarkable and could serve as a more fundamental distinction between "normal" and inverted micelles than the generally cited apparently more obvious differences. The free energy of micellization is customarily (8) referred to the standard free energy of a monomer in a micelle, i.e. ΔG_n^o represents the free energy of transfer of a monomer from the aqueous solution to a micelle[1] of size n,

---

[1]It has been pointed out by Stigter (27) that besides the integral process of micelle formation also a differential process

$$- RT \ln K_n = n\Delta G_n^o$$

The left hand side of this equation is the standard free energy of forming a micelle of size n. ΔG_n^o is made up of two contributions corresponding to the above discussed main factors which control formation and size of micelles in aqueous surfactant solutions. It has been pointed out repeatedly, e.g. (9-11) that the degree of association of aqueous micelles is much larger than can be accommodated by a micelle with a spherical core. Tanford reasonably attributes this fact to the thermodynamical requirements of the hydrophobic effect which forces an increase in aggregation to prevent contact between the hydrocarbon core of the micelle and the aqueous environment. Thus disk-like shapes have been proposed (9,11). Detailed considerations by Tanford (10) confirm these more intuitively reached conclusions.

b) Micelles in Nonpolar Surfactant Solutions

Apart from the lengthy dispute in the past concerning the existence of micelles in apolar media there was probably agreement as to the different interactions responsible for the stability of such entities. Since the apolar tails of the surfactants which belong to inverted (reversed) micelles stay in contact with the hydrocarbon solvent, there is only a small entropic effect which has to be considered regarding the stability of the micelles. This entropic contribution corresponds to the transition of a monomer from the solution to the micellar pseudo-phase (12). In addition there might be another slight entropic effect accompanying the topological transformation during micelle formation in hydrocarbon solvents (13).

The small effect of the solvent on the aggregational state of inverted micelles is best demonstrated by Fig. 2. Two surfactants, namely AOT (sodium-di-2-ethylhexyl sulfosuccinate) and the analogous POT (sodium-di-2-ethylhexyl phosphate) were correlated. Only the slightly polar ethylacetate or dioxane considerably reduced the aggregational tendency of both surfactants due to the interactions with the polar (hydrated) groups. The slope of the correlation diagram corresponds to the ratio of the smallest aggregate sizes of both surfactants detected. These were considered in the frame of the micellization model as nuclei (14). The enthalpic contributions consist of two parts, i.e. dispersion and electrostatic interactions where the latter include particularly hydrogen bond

---

(describing the condition for reversible micelle growth) has to be considered. The free energy changes for both processes show a gap of about 1 kT/ion, probably due to deficiencies in the Gouy-Chapman model.

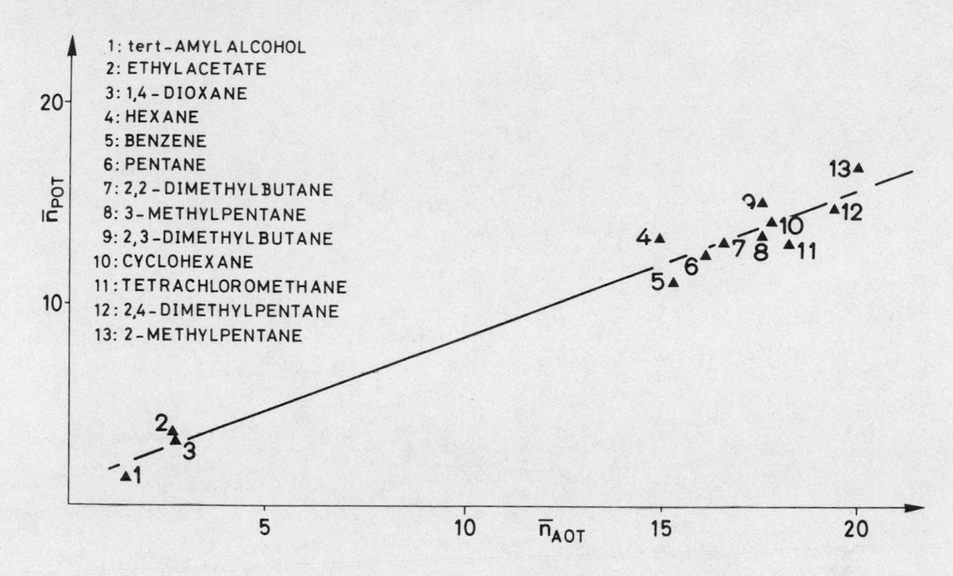

Fig. 2. Correlation diagram between mean aggregation numbers $\bar{n}_{POT}$ and $\bar{n}_{AOT}$ concerning the solvent dependent aggregate sizes of AOT and sodium di-2-ethylhexyl phosphate (POT).

formation. According to the recent experimental results from photon correlation spectroscopy (15), positron annihilation technique (16), IR spectroscopic investigations (17) and vapor pressure osmometric measurements of increasingly hydrophobic (tetra alkylammonium) ions of di-2-ethylhexyl sulfosuccinate (18) it was concluded that hydrogen bond formation (and generally hydration interactions) are decisive regarding the formation and stability of inverted micelles. These conclusions were strongly confirmed by careful IR spectroscopic OH-vibration analyses (17) which resulted in the following model (see Fig. 3). The figure illustrates the case for small amounts of water attached to the ionic groups. This is believed to be the experimental situation generally met with the formation of inverted micelles. The importance of the hydration interaction is seen in Fig. 4 where the average aggregation number $\bar{n}$ is plotted versus the weighed-in concentration c_o of various ionic derivatives of di-2-ethylhexyl sulfosuccinates with increasingly hydrophobic counterions. The effect is quite apparent and shows the expected trend. It appears, therefore, justified to assume that minute amounts of water are essential for the formation, and surprising stability (15), of many ionic micelles in hydrocarbon solvents. This applies in particular to those surfactants with strongly hydrophilic head groups and which are capable to form hydrogen bonds stabilized by polarization.

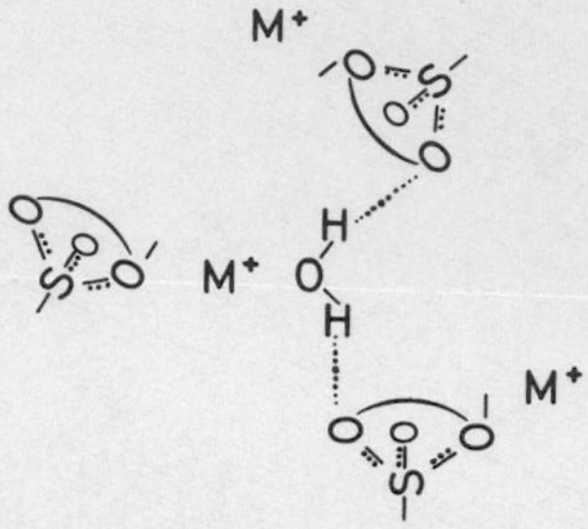

Fig. 3. Formation of trimers of AOT at low degree of hydration according to Zundel's model (17).

With respect to size and shape of inverted micelles the former is, as a rule, considerably smaller than aqueous micellar aggregates. Since no thermodynamically determining factor, like the hydrophobic effect, exists in the case of inverted micelles it appears to be accepted that steric restrictions are important, i.e. the ratio of the cross-sectional areas of the hydrocarbon to the polar moieties essentially determines size and shape of the aggregates (19).

Quite frequently prolate ellipsoids have been suggested according to experimental results (19,20), for example in the case of AOT. For other surfactants, like dinonylnaphthalene sulfonates (21), spherically shaped aggregates seem more probable. Thus, contrary to a general driving force as with association phenomena in aqueous surfactant solutions where the hydrophobic effect forces the micelle to adopt a shape which guarantees the most favorable shielding from water, specific sterical effects seem to dominate the shape of inverted micellar aggregates.

Microemulsions

A phenomenon closely related to the formation of micelles is the building up of thermodynamically stable emulsions, so-called microemulsions. The process giving rise to the final formation of a microemulsion is the so-called solubilization, i.e. the encaging of the respective antagonistic component in the binary oil/water system by the surfactant. It might appear reasonable to consider

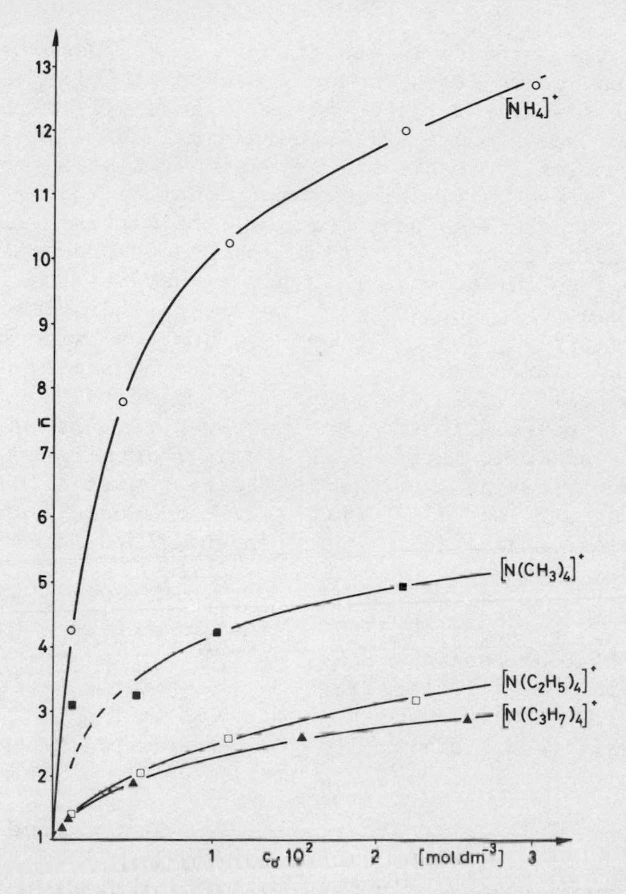

Fig. 4. Number average aggregation number ($\bar{n}$) of AOT with different tetraalkylammonium counterion vs. surfactant concentration in benzene at 25°C.

this solubilization as the leading phenomenon compared to the micelle formation.

Extrapolation to zero solubilizate amount would lead automatically to the micellar aggregate which appears according to this point of view as an auto-solubilization phenomenon. Such a leading role attributed to the microemulsion formation is confirmed by the fact (2) that a thermodynamically stable emulsion can be obtained only if the necessary concentration of surfactant required to build up the microemulsion is smaller (2) than the cmc of the binary surfactant/solvent system. Thus solubilization and micellization are competitive processes. This is nicely demonstrated by Figure 1 where the onset of the solubilization process ("detergency" curve) clearly precedes that of the micellization. Which emulsion type, i.e. W/O-, or O/W- is formed, depends largely on the nature of the surfactant. It is, moreover, possible that a phase inversion occurs depending on the water to oil ratio and the temperature of the system. The latter is generally called the phase-inversion temperature (PIT). Whether such an inversion leads to a thermodynamically stable microemulsion can be inferred from considerations based on classical thermodynamics[2] (22). The properties of both these emulsions, however, are different in many respects according to the respective continuous phases. Comparing both emulsion types it appears that a more natural transition exists between normal micelles and the corresponding microemulsion. These micelles already contain a hydrocarbon core with liquid-like structure. The microemulsion is thus formed by adding up increasing amounts of oil (within the stability region of the microemulsion) which are soluble in the hydrocarbon core. It is interesting to note that also solid material is easily solubilized without the addition of a dissolving agent as is the case with solubilizate taken up by inverted micelles. Hence O/W- microemulsions represent in their simplest form a ternary system while W/O- emulsions are generally to be considered quaternary systems.

With regard to the stability against coalescence there is a noteworthy difference between W/O- and O/W- emulsions. The O/W type is certainly better stabilized by the electrostatic repulsion of the electrical double layer than the W/O microemulsion. The latter is stabilized by sterical repulsion between the micellar "membranes".

In the last few years the interest appears to have strikingly shifted towards W/O- microemulsions. The main reason for this is probably the opportunity to investigate the structure of water in the hydrocarbon environment in the presence of surfactant: in particular its dependence on the water concentration, influence on the micellar membrane by the hydrophobic hydration (23) and the sensitivity of the water structure on additives (salt and hydrocarbon). Moreover, the discovery of a remarkable catalytic activity has considerably fostered the research on this type of emulsion.

Most of the available information stems from thermodynamical investigations. Comparatively little work was directed towards structural details. Only more recently a systematic photon correlation spectroscopic study has been performed (15) on ternary W/O- microemulsions, i.e. H_2O/AOT/isooctane, which revealed generalizable results regarding ternary W/O- microemulsions formed by

---

[2]Naturally, one would expect cosurfactants to be considered in order to produce stable microemulsions with high contents of solubilized material. However, the present discussion is concerned with the proper microemulsion phenomena which can be produced without the addition of cosurfactants.

ionic surfactants. It appears that for the first time experimental evidence has been presented that there actually exists a physically reasonable difference between inverted micelles and W/O micro-emulsions. This could have been inferred already from the well-known phase diagrams of a W/O- microemulsion (Fig. 5).

This is illustrated in more detail by Figures 6 and 7 which show plots of the hydrodynamical radii (r_h) and the incremental surface area covered by one AOT molecule ($\bar{f}_{AOT}$) of the oil/water interface versus the temperature and the amount of added water (w_o), respectively. Both figures mutually support each other, indicating a splitting into two groups of essentially similar curves (Figure 6) within a narrow w_o-range and a clear change in the slope of $\bar{f}_{AOT}$ versus w_o in Figure 7 at about the same w_o-values. The lower group of curves in Figure 6 is attributed to the micellar domain, characterized by the very weak temperature dependence. In contrast with this group is the upper ensemble of curves which strongly depends on temperature, pressure, mutual solubility of the microemulsion components etc., since now the interfacial free energy of the system is the relevant thermodynamical magnitude which determines the stability of the system. This is, accordingly, the microemulsion region. Figure 7 expresses the same fact, as at low w_o-values a steep increase of $\bar{f}_{AOT}(w_o)$ is displayed, whereas above w_o = 15–20 the incremental coverage per AOT molecule of the oil/water interface stays approximately constant. Thus the curve

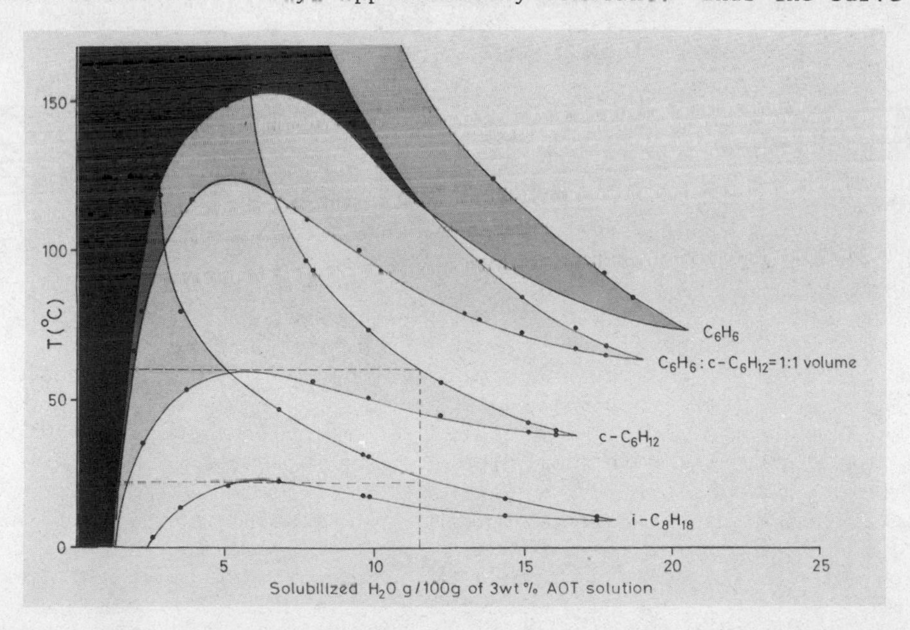

Fig. 5. Phase diagrams of AOT in different solvents and solvent mixtures. Temperature plotted vs. amount of solubilized water. From (24).

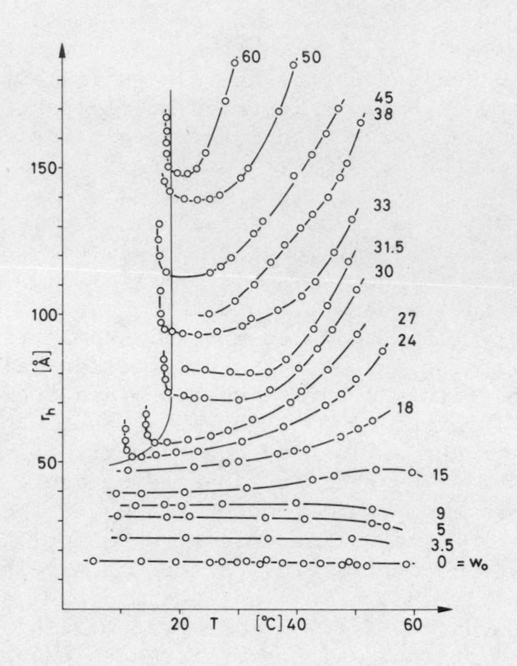

Fig. 6. Stokes radii (r_h) of microemulsion (H_2O/AOT/isooctane) droplets vs. temperature. Parameter: $w_o = [H_2O]/[AOT]$. The vertical line indicates a stability boundary to the left of which spontaneous growth of the aggregates occurs which will ultimately lead to phase separation. From (15).

reflects the transition from a micellar aggregate with a more or less ordered surfactant network to a more defined interfacial monolayer.

An interesting detail was the detection of an optical matching phenomenon which depended on the water content (Figure 8). A recent analysis of this fact revealed a number of molecular details of the microemulsion. The minimum of the scattering intensity versus w_o, i.e. the proper optical matching, could be excellently described by the adsorption model developed earlier on the basis of dipole--image dipole interactions between the water core and the adsorbed surfactant dipoles (25). Its validity is restricted to the microemulsion region which is clearly reflected in Figure 8. The first part of the scattering plot $I_{90°}(w_o)$ is best fitted using the initial part of the experimentally obtained $\bar{f}_{AOT}(w_o)$ plot (Fig. 7), thus identifying the maximum region of the scattered intensity with the micellar domain. The surprising coincidence of the applied equi-partition model with the experimental plot seems to point to a relatively small polydispersity. The physical parameters used to fit the scattering curve indicate that the surface conduction of the microemulsion droplets are negligible, i.e. they appear to the light like dielectric spheres (26).

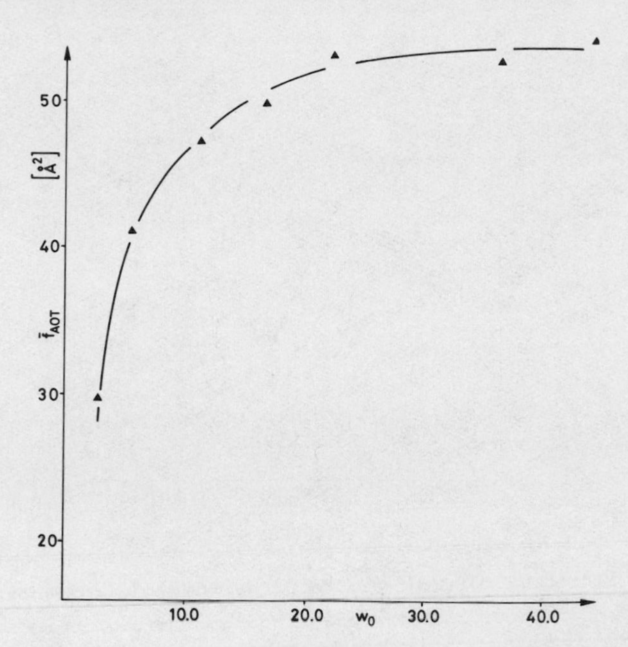

Fig. 7. Average surface fraction ($\bar{f}_{AOT}$) of the H_2O/oil interface covered by one surfactant molecule (AOT) vs. weighed-in amount of water w_o. From H.F. Eicke and J. Rehak, Helv. Chim. Acta 59, 2883 (1976).

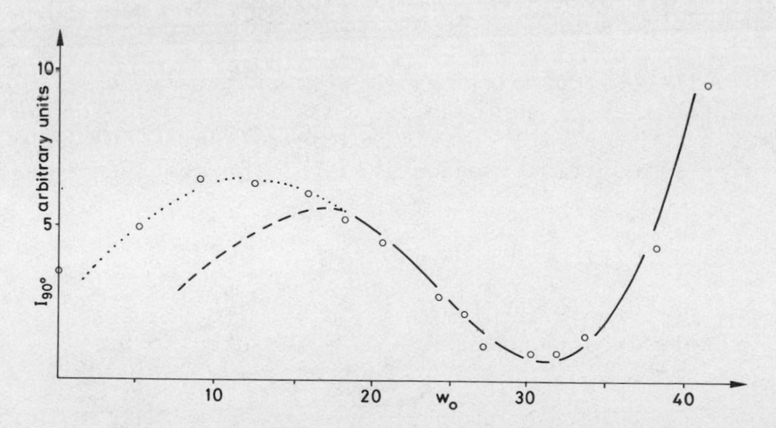

Fig. 8. Scattered intensity $I_{90°}$ vs. $w_o = [H_2O]/[AOT]$ (7) at 25°C. System: H_2O/AOT/isooctane. Solid line: Rayleigh scattering calculated according to monodisperse microemulsion model (26). Dotted line corresponds to initial part of $\bar{f}_{AOT}(w_o)$ curve (Fig. 7).

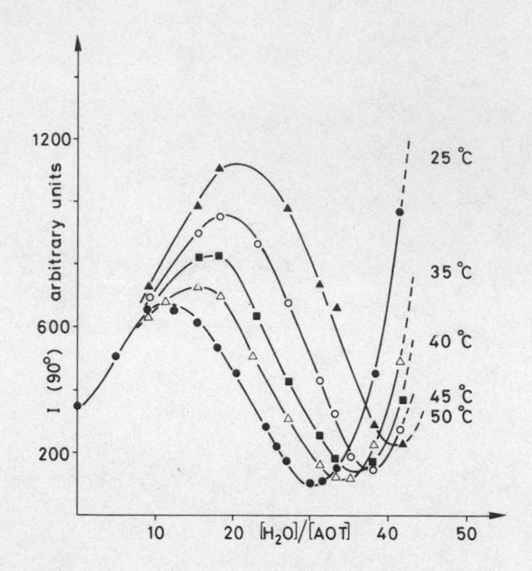

Fig. 9. Extrapolated intensities ($I_{90°}$) vs. w_o = $[H_2O]/[AOT]$ for five temperatures. From (15).

The increasing residual scattering at higher temperatures (see Fig. 9) is straightforward explained by the temperature dependent fluctuations of the interfacial surfactant layer covering the dispersed droplets (26). The fluctuations can be observed since the static contributions of the refractive index increments of the dispersed particles and the solvent (oil) are optically matched. Hence, except for the fluctuations of the surfactant molecules in the monolayer the microemulsions discussed in the present paper show a remarkable monodispersity.

It is believed that our present view of the microemulsion phenomena is already rather detailed on a microscopic (colloidal) level. The next step will certainly acquire deeper insights into the molecular interactions responsible for the observed phenomena.

ACKNOWLEDGMENTS

A grant from the Swiss National Science Foundation (No. 2.014-0.78) is gratefully acknowledged. I am indebted to D. Stigter regarding the comments expressed in footnote 1.

REFERENCES

1. G. S. Hartley, Trans. Farad. Soc., 37, 130 (1941).
2. C. Wagner, Colloid & Polymer Sci., 254, 400 (1976).
3. G. S. Hartley, in "Micellization, Solubilization, and Microemulsion," K. L. Mittal, ed., Vol. 1, p. 23, Plenum Press, New York, (1977).
4. C. Tanford, "The Hydrophobic Effect," John Wiley, New York (1973).
5. M. Shinitzky, A. C. Dianarix, C. Gitler, and G. Weber, Biochemistry, 10, 2106 (1971).
6. A. S. Waggoner, O. H. Griffith, and C. R. Christenson, Proc. Nat. Acad. Sci. USA, 57, 1198 (1967).
7. G. S. Hartley, J. Chem. Soc., 1968, 633 and 1968.
8. C. Tanford, in "Micellization, Solubilization, and Microemulsion," K. L. Mittal, ed., Vol. 1, p. 119, Plenum Press, New York (1977).
9. H. V. Tartar, J. Phys. Chem., 59, 1195 (1955).
10. C. J. Tanford, J. Phys. Chem., 76, 3020 (1972).
11. H. B. Klevens, J. Am. Oil Chemist's Soc., 30, 74 (1953).
12. H. F. Eicke, in "Micellization, Solubilization, and Microemulsion," K. L. Mittal, ed., Vol. 1, p. 429, Plenum Press, New York (1977).
13. H. F. Eicke, R. Hopmann, and H. Christen, Ber. Bunsenges. Phys. Chem., 79, 667 (1975).
14. H. F. Eicke and H. Christen, J. Colloid Interface Sci., 48, 281 (1974).
15. M. Zulauf and H. F. Eicke, J. Phys. Chem., 83, 480 (1979).
16. Y.-C. Jean and H. J. Ache, J. Am. Chem. Soc., 100, 984 (1978).
17. G. Zundel, "Hydration and Intermolecular Interaction," Academic Press, New York (1969).
18. H. F. Eicke and H. Christen, Helv. Chim. Acta, 61, 2258 (1978).
19. J. B. Peri, J. Colloid Interface Sci., 29, 6 (1969).
20. P. Ekwall, L. Mandell, and K. Fontell, J. Colloid Interface Sci., 33, 215 (1970).
21. S. Kaufman and C. R. Singleterry, J. Colloid Sci., 12, 465 (1957).
22. E. Ruckenstein, in "Micellization, Solubilization, and Microemulsion," K. L. Mittal, ed., Vol. 2, p. 755, Plenum Press, New York (1977).
23. G. Hertz, Ber. Bunsenges. Phys. Chem., 68, 907 (1964).
24. H. F. Eicke, J. Colloid Interface Sci., 68, 440 (1979).
25. H. F. Eicke, J. Colloid Interface Sci., 52, 65 (1975).
26. H. F. Eicke and R. Kubik, Ber. Bunsenges. Phys. Chem., 84, 36 (1980).
27. D. Stigter, J. Colloid Interface Sci., 47, 473 (1974).

X-RAY DIFFRACTION AND SCATTERING BY MICROEMULSIONS

S. S. Marsden

Department of Petroleum Engineering
Stanford University
Stanford, California 94305, U.S.A.

Microemulsions are being actively studied as well as field tested as an EOR method in many places. We have learned a great deal about some of their physical properties and their flow in porous media, but we know remarkably little about their physical structure in the size range of colloidal particles. There is some reason to believe that the other properties of microemulsions as well as their performance in EOR are determined by this colloidal structure in the same way that the properties and performance of oil well drilling fluids are determined by the colloidal structure of clay minerals in aqueous dispersions.

One very powerful tool for determining the size, shape, and orientation of colloidal particles in a liquid continuum is through their interaction with X-radiation. The nature of this interaction varies considerably with the colloidal system, and it can often be used to elucidate the structure of the latter when combined with the results of other experimental measurements. X-ray results alone rarely lead to unambiguous conclusions in these kinds of systems.

Microemulsions are comprised of various combinations and concentrations of oils, water (or brine), surfactants, and sometimes co-surfactants. When prepared with colorless materials and equilibrated at a specified temperature, they may be clear, translucent or opaque liquids or gels. Many exhibit birefringence and/or streaming birefringence, and so these either consist of or else contain liquid crystalline phases; others are optically isotropic.

Some microemulsions are thin liquids while others are viscous and still others are gels. Some are simple Newtonian fluids while others are non-Newtonian. The flow properties are determined by chemical composition, and judicious variation of the latter can often be used to get the physical properties desired.

Various procedures have sometimes been followed to orient other liquid crystalline samples before studying them with X-rays. For example, when some smectic liquid crystals are aspirated into a small diameter, thin-walled glass capillary tube, the lamellae become preferentially oriented parallel to the surface of the tube. If this orientation is complete, then the X-ray pattern produced on a film by a thin X-ray beam perpendicular to the axis of the glass tube consists of a series of sharp spots in a line also perpendicular to the axis of the tube. Calculations will show that these are several orders (typically about three) of a single repeating dimension or "spacing" whose numerical value is usually several tens of Angstrom units. This is due to a one-dimensional regularity in such liquid crystals and tells us nothing about any regularity in the other two dimensions.

If the lateral dimensions of the lamellae in this kind of liquid crystals are not particularly large or if Brownian motion disturbs the original orientation relative to the walls of the capillary tube, then instead of spots we will get a series of concentric rings corresponding to the same Bragg spacing. This kind of pattern is analogous to a Debye-Scherrer or "powder" pattern of the sort obtained from a microcrystalline sample, and the interpretation is similarly straightforward. These rings are usually just as sharp as those found on most powder diagrams; that is, they are not the broad, fuzzy haloes obtained with liquids.

The nature of the spots or rings obtained on X-ray film when using a thin X-ray beam of circular cross-section is shown in Figures 1 and 2. The thin-walled glass capillary tubes containing the samples are shown superimposed on these diagrams to illustrate the orientation of the spots relative to the axis of the tube. In a sense, the reader's eye is in the position of the X-ray source; looking toward the page it sees first the sample and then much further on the diffraction results recorded on the film. Those experienced in this field will recognize that these diagrams are not drawn to scale and that the "shadow" of the lead button used to absorb the undiffracted X-ray beam has been omitted.

It is not uncommon to get superposition of the spots and the circles described above when there is partial orientation of the lamellae relative to the walls of the capillary tubes. In theory, it may be possible to learn something about the lateral dimensions of the lamellae from the relative intensities of spots and circles for samples of the same system mounted in tubes of different internal diameters.

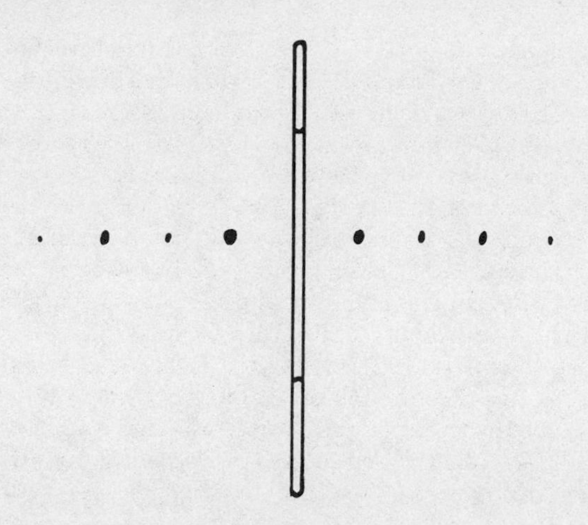

Fig. 1. X-ray diffraction from oriented smectic liquid crystalline phase

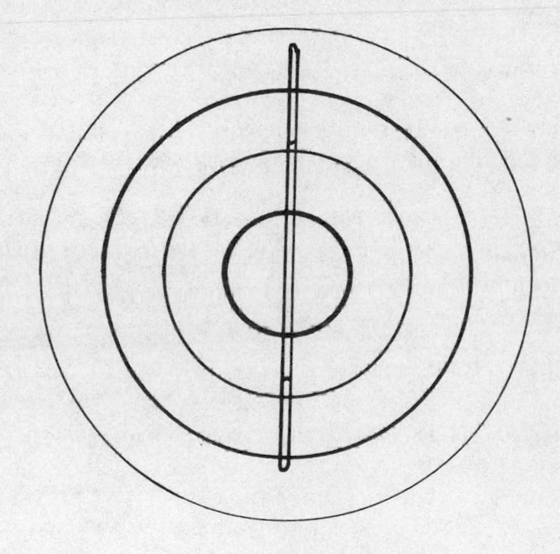

Fig. 2. X-ray diffraction from non-oriented smectic liquid crystalline phase

Besides the lamellar liquid crystals just described, others are known to exist. We shall discuss only one here; namely, the nematic liquid crystals illustrated by the middle soap phase* of a typical soap-water system. An unoriented sample made up of many micro-liquid crystals of this sort will give a series of concentric

---

*This is not to be confused with the "middle phase" of some microemulsion systems.

rings similar in sharpness to those described for the lamellar ones discussed above, but with one significant difference. In the smectic case the calculated Bragg spacings are in the ratios of 1:1/2:1/3:1/4:1/5, etc., because they are different orders of diffraction for the same set of planes in the liquid crystal. In the nematic case the calculated Bragg spacings are in the ratios of $1:1/\sqrt{3}:1/\sqrt{4}:1/\sqrt{7}$, etc., because they are both different orders of the same set of planes, and also diffractions from other sets of planes. The generally accepted interpretation of this data is that it is due to diffraction from a fairly large number of parallel rods whose axes are arranged in a highly regular, two-dimensional, hexagonal pattern with no discernable regularity in the third dimension. By itself, this interpretation has the same kind of ambiguity that arises from similar interpretations of powder diagrams from micro-crystalline samples of solids having three-dimensional regularity.

We are fortunate, however, in that some investigators have been able to get oriented nematic liquid crystals which produce spots rather than simply circles so that the ambiguity in interpretation can be eliminated (1). Not only do these spots have the expected numerical values described above, but they also have the expected orientation relative to each other for this kind of system (Figure 3). These spots represent two-dimensional order rather than the three found in ordinary crystalline solids.

The liquid crystals described above which have been studied with X-rays usually are made from surfactants and water. We expect that those found in microemulsion systems will give similar results when studied by X-rays but this has yet to be demonstrated. Other surfactant-water systems which are not liquid crystalline also give X-ray results which are different from those of the liquid crystalline phases and these will be discussed now. They are of two sorts; as far as we know, neither have been oriented.

The first kind is simply an extension of the concentric rings obtained with non-oriented liquid crystals but with two distinctions. Although the rings can be sharp, they will generally be fuzzy and some even resemble the haloes obtained with ordinary liquids. The interpretation of these results is not as clear-cut as it is for the hydrous liquid crystals. When the rings are sharp, Bragg's law can probably be used again to calculate spacings, but when fuzzy, a simple constant slightly greater than unity has sometimes been incorporated in the equation by some authors. In either event, a numerical value of a spacing or distance is obtained but other evidence is needed to give it real meaning.

The second kind of result is the small angle or Guinier scattering (2) which appears on films as darkening at very small angles relative to the main beam. By using methods described by others,

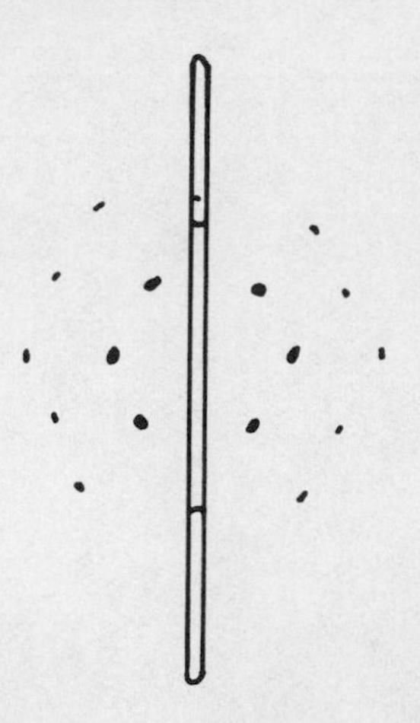

Fig. 3. X-ray diffraction from oriented nematic liquid crystalline phase

this scattering can also give a dimension of individual particles rather than the repeating distance between particles which are obtained by X-ray diffraction. Because we have not had experience with this method, we cannot discuss it more at this point.

A single X-ray diffraction pattern of a liquid crystalline phase or an isotropic solution of a microemulsion will not tell us very much more than the repeating distance within the particles and possibly something about the relative dimensions of the particles. It will be necessary to study a number of related systems under specialized experimental conditions to learn more and these will be described next.

Probably the easiest parameter to vary systematically will be concentration of the several constituents; that is, to study samples having the same components but in different proportions. Then we can get the change in any diffraction pattern or spacing with concentration which can often tell us as much or more than the absolute values of the spacings themselves. This may be illustrated by the earlier work done in several laboratories on simple soap-water systems (3,4). Elementary geometry predicts that for lamellar liquid crystals a double logarithm graph of long X-ray diffraction spacings versus volumetric concentration of the soap should have a slope of minus one if all of the water goes into uniform layers between the sheets of polar heads of the soap molecules. This is indeed found in many but not all cases (Figures 4 and 5). Extension of this modeling to cases where hydrocarbon is solubilized in the detergent-water systems indicates that the hydrocarbon will often lead to additional expansion of the thickness of the liquid crystalline phase by having the hydrocarbon form layers between the non-polar ends of the detergent molecule sheets (4). This has also

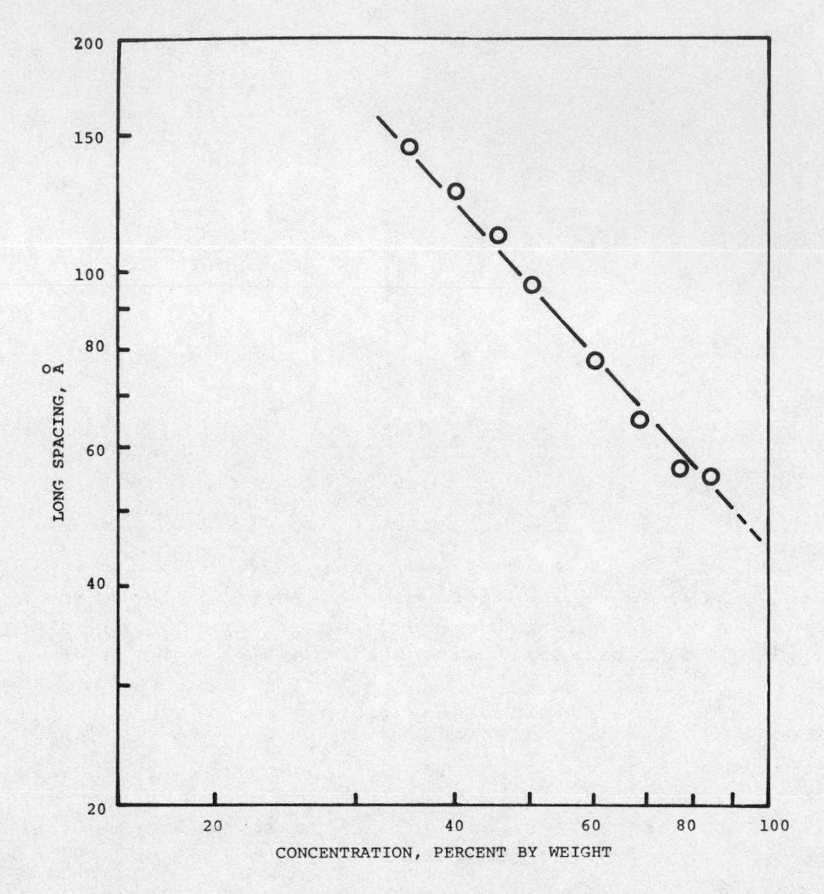

Fig. 4. Long spacing vs. concentration for diglycol monolaurate (Reference 4)

been observed in some but not all cases (Figure 6). When it does happen, we can sometimes speculate as to where the hydrocarbon actually goes in the system when it is solubilized.

For nematic or fibrous liquid crystals whose individual molecular agglomerates consist of rods with centers arranged in a hexagonal array, simple geometry also predicts that a double logarithm graph of the first order of the major X-ray spacing versus the volumetric concentration of detergent should have a slope of minus one-half (1). This together with the numerical ratios of the several diffractions (plus in some cases the oriented spots produced by annealed systems) is rather convincing evidence for this kind of liquid crystal.

Similar geometrical modeling would also predict a slope of minus one-third for a system of isodimensional units such as spheres that expanded uniformly in three dimensions.

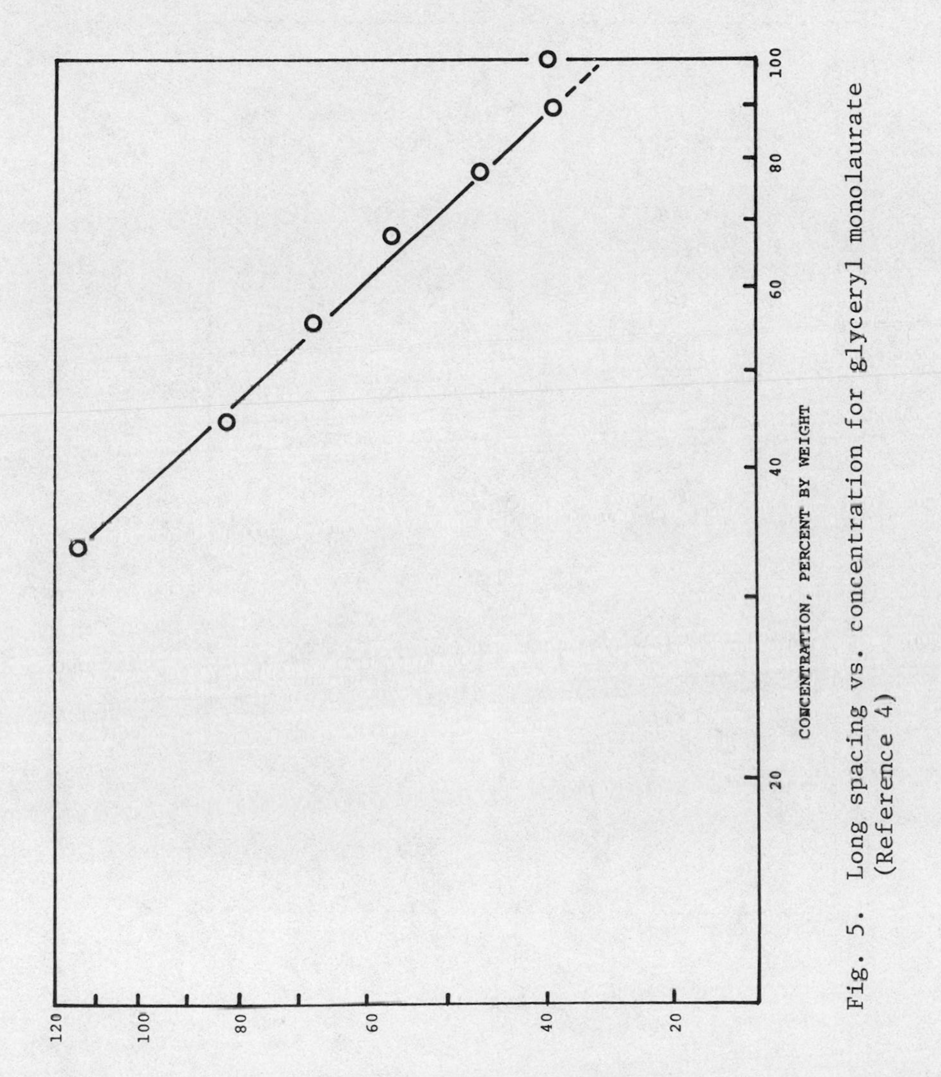

Fig. 5. Long spacing vs. concentration for glyceryl monolaurate (Reference 4)

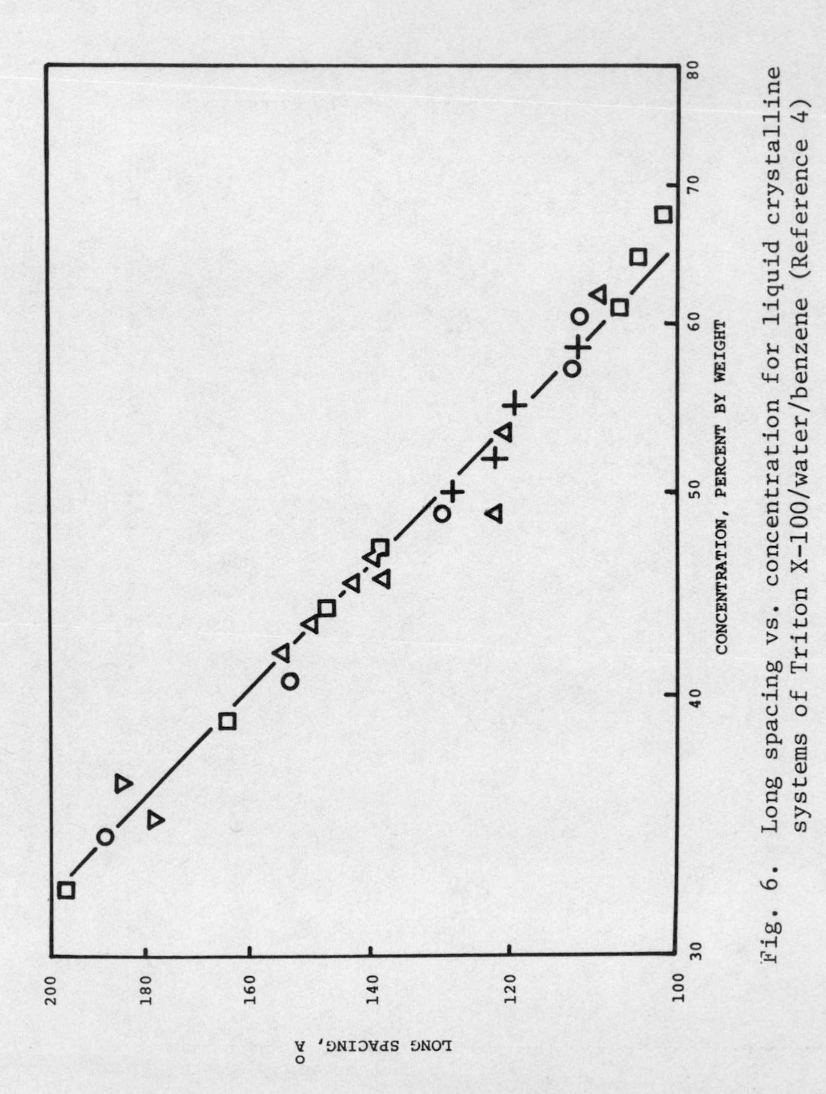

Fig. 6. Long spacing vs. concentration for liquid crystalline systems of Triton X-100/water/benzene (Reference 4)

We have mentioned earlier how some liquid crystalline systems become oriented when they are aspirated into thin-walled glass capillary tubes for study by X-rays and that others do not become oriented. This suggests that we may learn something additional about the overall shape or structure of microemulsions by determining the X-ray patterns of flowing as well as sessile samples. Rheological measurements on the same samples would also be useful in interpreting the X-ray results, as would observations of any streaming birefringence.

Because some liquid crystalline samples become oriented by flow, care must be taken in choosing and using the holder for the samples being studied with X-rays. For example, a flat sample holder oriented perpendicular to the X-ray beam may not work because of liquid crystal lamellae being oriented parallel to the "windows" of the holder, but the same holder may also work well if it is almost parallel to the beam.

Another interesting and important parameter that can and should be studied is the temperature level (5). This is not simply to simulate reservoir temperature but also because it is known to affect the physical properties and phase behavior of microemulsions.

We mentioned earlier that some microemulsions contain water and others brine, so certainly the presence of different electrolytes and their concentration should be a parameter when preparing microemulsions for study. Various surfactants as well as cosurfactants produce microemulsions having different physical properties and so obviously the chemical compositions of these components are still other variables.

The change of colloidal structure as observed by X-ray methods and as related to the physical and chemical properties of microemulsions is the goal of this work. There is a lot to be done and so we would like to encourage those in other laboratories to undertake it as well.

REFERENCES

1. S. S. Marsden and J. W. McBain, J. Chem. Phys., 16, 633 (1948).
2. A. Guinier and F. Fournet, "Small Angle Scattering of X-Rays," John Wiley, 1955.
3. S. S. Marsden and J. W. McBain, Nature, 165, 141 (1950).
4. S. S. Marsden, "Aqueous Systems of Non-Ionic Detergents as Studied by X-Ray Diffraction," Ph.D. Dissertation, Stanford University, 1947.
5. S. S. Marsden and N. R. Sanjana, J. Sci. Instr., 30, 427 (1953).

LIGHT SCATTERING STUDY OF ADSORPTION OF SURFACTANT MOLECULES AT OIL-WATER INTERFACE

A. M. Cazabat, D. Langevin, J. Meunier, and
A. Pouchelon

Laboratoire de Spectroscopie Hertzienne de l'E.N.S.
24, Rue Lhomond 75231 Paris Cedex 05 France

We have studied the following two types of oil-water interfaces by light scattering techniques:

a) Flat Interfaces

Interfacial tension has been deduced from the spectrum of the light scattered by the interface. The results are relative to water-toluene-sodium dodecyl sulfate (SDS)-butanol mixtures either in the two phase, or in the three phase region of the phase diagram. Values down to 10^{-3} dynes/cm have been measured. Measurements down to 10^{-5} - 10^{-6} dynes/cm are expected to be achievable with this technique.

b) Microemulsions

The intensity and the autocorrelation function of light scattered by a microemulsion have been investigated for the water in oil type microemulsions. We studied two mixtures: water-SDS-cyclohexane-pentanol and water-SDS-toluene-butanol. We obtain information about droplet size (radius of the aqueous core, hydrodynamic radius) and about interaction forces between the droplets (from osmotic compressibility and diffusion coefficient data). The role of the nature of oil and the influence of salt on these parameters is also discussed.

INTRODUCTION

The adsorption of surfactant molecules at oil-water interfaces has attracted much interest, in relation to the oil recovery techniques (1). The systems containing oil, water and emulsifier molecules form generally two phases: an aqueous phase containing sometimes solubilized oil in the form of small droplets surrounded by emulsifier molecules and an oil phase which also can contain solubilized water. When the amount of emulsifier is large enough, the system can form only one phase, i.e. all the water (or oil) can be solubilized in the oil (or water). The system is again a dispersion of very small droplets of water (or oil), surrounded by emulsifier molecules, in a continuous medium containing the oil (or water). Such dispersions are currently called microemulsions. The droplet size is usually of the order of 100Å (2).

A very schematic phase diagram showing single and two phase regions is represented by Figure 1(a). Figure 1(b) shows a situation where the system forms three phases. The lower phase is usually pure water, the upper phase pure oil, and the middle phase is thought to be a microemulsion containing most of the emulsifiers.

The interfacial tensions in the two phase region are generally small (1-10^{-2} dynes/cm) and even smaller in the three phase region (10^{-3}-10^{-4} dynes/cm). These properties are expected to be related to the structure of the microemulsions in the bulk phases (3).

Light scattering techniques are useful tools to investigate both interfacial properties such as surface tension and viscoelasticity (4) and bulk properties such as droplet size and interaction forces between these droplets (5,6). It must be pointed out that in each case, light is probing thermal fluctuations in the medium but the fluctuations are of a very different nature: surface roughness in the first case, and droplet concentration fluctuations in the case of bulk scattering.

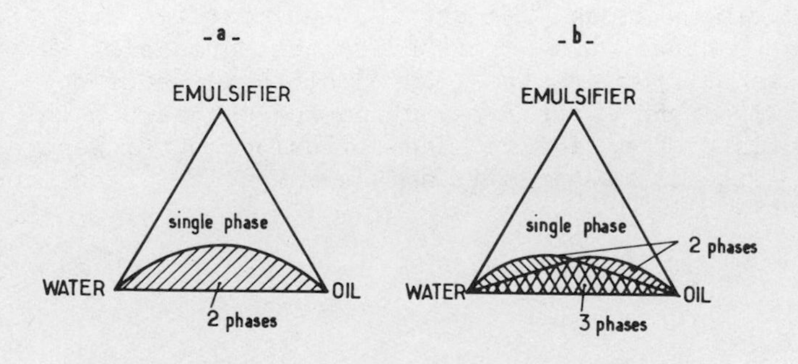

Fig. 1. Schematic phase diagram

In this paper we will present independently the results obtained by the two techniques. We have studied microemulsion bulk properties for samples corresponding to the limit line of the one phase region (Figure 1). On the other hand, the two or three phase samples, for which we studied the interfacial properties, had very small interfacial tensions and are therefore very close to the same demixing line. This preliminary study was intended to determine how much information can be obtained from these two light scattering techniques and how these informations can be coupled.

STUDIES ON OIL-WATER INTERFACES

EXPERIMENTAL

In this paragraph we will briefly recall the main features of the surface light scattering technique. Further details can be found in reference (4).

Thermal motion induces small surface roughness. The amplitude ζ of surface irregularities are of the order of 10Å for usual liquid surfaces, if the surface tension $\gamma \approx 10$ dynes/cm; ζ varies as $\gamma^{-1/2}$ and becomes therefore larger if surface tension decreases. Consequently the intensity of the light scattered by the surface increases. If one chooses a given scattering angle θ (Figure 2), the result will be to select a given wave vector $\vec{q}$ in the spatial Fourier transform of ζ, as in an ordinary diffraction process. This wave vector is related to scattering angle by:

$$|\vec{q}| = \frac{2\pi}{\lambda}\,\theta ,$$

where λ is the incident light wavelength, and provided θ is small and $\vec{q}$ perpendicular to the incidence plane.

As the scattered intensity decreases as θ^{-2} when θ increases, one is limited in practice to very small scattering angles, e.g., $\theta < 5°$. The scattered light is received on a photomultiplier and the photocurrent power spectrum is analyzed in a real time wave analyzer. The spectrum reflects the temporal evolution of the Fourier component ζ_q.

If the surface tension is small,[1] the distorted surface experiences a small restoring force, and owing to the viscosities η of the upper phase and η' of the lower phase, the interfacial motion is strongly overdamped. Hence,

---

[1]The condition for the validity of the above results is:

$$\frac{\gamma(\rho + \rho')}{4(\eta + \eta')^2 q} \lesssim 10^{-2}$$

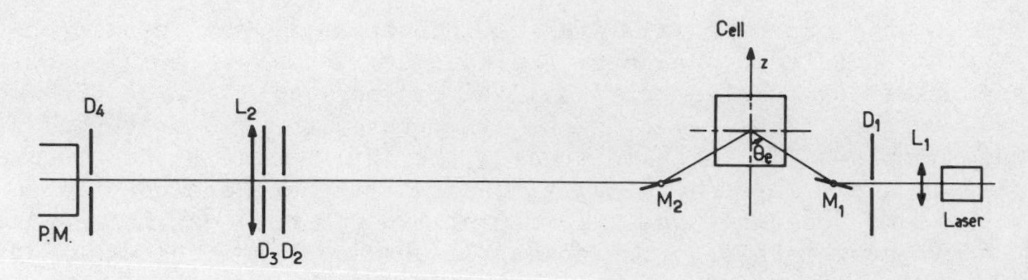

Fig. 2. Experimental setup for surface light scattering

$$\zeta_q(t) = \zeta_q(0)\ e^{-t/t_q}$$

and
$$\tau_q = \frac{2(\eta + \eta')}{[\gamma + \frac{(\rho' - \rho)g}{q^2}]q} \tag{1}$$

where ρ, ρ' are liquid densities and g is acceleration due to gravity. The measured photocurrent power spectrum $P(\omega)$ has a Lorentzian shape:

$$P(\nu) \propto \frac{1}{\nu^2 + \Delta\nu_q^2} \quad \text{where} \quad \Delta\nu_q = \frac{1}{2\pi\ \tau_q} \tag{2}$$

Here, typically $q \sim 1000\ cm^{-1}$, $\eta, \eta' \sim 1$ cp, $\rho, \rho' \sim 1\ g/cm^3$, and eqs. (1) and (2) are valid[1] if $\gamma \lesssim 10^{-2}$ dyne/cm. For larger surface tensions, a slightly more complicated theoretical form of $P(\omega)$ must be used (4).

If the interface exhibits viscoelastic properties, as it often appears when surfactant is adsorbed, the surface tension becomes frequency dependent (7), $\tilde{\gamma} = \gamma + i\omega N$, where N is a surface viscosity associated to the vertical motion of the interface, different from shear (L) and dilatational (M) surface viscosities. These two viscosities are associated with frequency dependent shear S and dilatational K elastic moduli and hence,

$$\tilde{S} = S + i\omega L$$
$$\tilde{K} = K + i\omega M$$

In the case of interfacial deformations at low γ values ($\gamma \lesssim 10^{-2}$), the motion of the interface is purely vertical, and

---

[1] The condition for the validity of the above result is:
$$\frac{\gamma(\rho + \rho')}{4(\eta + \eta')^2 q} \lesssim 10^{-2}$$

is not associated with any shear or dilatation. For this reason, the time τ_q will only depend on γ and N:

$$(2\pi\tau_q)^{-1} = \Delta\nu_q = \frac{1}{1 + \frac{Nq}{2(\eta+\eta')}} \times \frac{\gamma q + (\rho'-\rho)\, g/q}{4\pi\,(\eta+\eta')} \qquad (3)$$

Materials and Methods

The samples were made using toluene as oil and a mixture of sodium dodecyl sulfate (SDS) and butanol as emulsifier. The phase diagram of this system was extensively studied by Lalanne et al. (8). The composition of samples is indicated in Table 1. Their densities, viscosities and refractive indices are reported in Table 2. The three-phase samples were obtained by adding salt into water (the limits of the three-phase domain for the samples of Table 1 were 5.8% - 7.8% of salt in water).

The samples were equilibrated in a thermostated water bath at 20° ± 0.1°C for at least two weeks. The different phases were then taken out separately and each used to fill half of the scattering cell. This cell is also thermostated at 20° ± 0.1°C. Moreover, it has been found necessary to control the temperature of the room itself (within ± 0.5°) in order to avoid thermal gradients in the vicinity of the cell windows, which produce a curvature of the interface. The cell must be filled with great care, in order to avoid soap deposit on the cell glass windows. If the windows are not very clean, the measured spectrum can be distorted in a non-Lorentzian shape, and the results become non-reproducible.

RESULTS

Two samples of the two-phase region were studied, A and B (Table 1). The results for measured half widths have been corrected for the gravity part of the restoring force $(\rho' - \rho)\, g/q^2$ in equations (1) and (3). This leads to apparent half widths of:

$$\Delta\nu_a = \frac{\gamma q}{4\pi\,(\eta+\eta')} \qquad \text{if } N = 0 \qquad (4)$$

$$\Delta\nu_a = \frac{\gamma q}{4\pi\,(\eta+\eta')} \times \frac{1}{1 + Nq/2\,(\eta+\eta')} \qquad \text{viscoelastic interface} \qquad (5)$$

Figure 3 shows the results for $\Delta\nu_a$ as a function of q for sample A. Although a linear fit (full line) is possible and leads to $\gamma = 7.8\ 10^{-2}$ dynes/cm using eq. (4), a fit with an hyperbolic function as indicated by eq. (5) is somewhat better and gives $\gamma = 8.5\ 10^{-2}$ dynes/cm, $N = 4.7\ 10^{-6}$ surface poises.

Table 1

Sample	% toluene	% water	% butanol	% SDS	% salt in water	Interfacial Tension by	
						Light Scattering (γ_{LS})	Spinning Drop (γ_{SD})
A	0.30	0.46	0.16	0.08	--	8×10^{-2}	5×10^{-2}
B	0.55	0.21	0.16	0.08	--	2.5×10^{-2}	1×10^{-2}
						(γ_{WM})	(γ_{MO})
C1	0.47	0.47	0.04	0.02	0.06	3×10^{-3}	9×10^{-3}
C2					0.06	6×10^{-3}	1.25×10^{-2}
C3					0.06	--	1.5×10^{-2}
C4					0.068	3×10^{-3}	7×10^{-3}
C5					0.075	--	2.5×10^{-3}

Percentages are relative to weight proportions.

Table 2

Sample	Phase	η_{cp}	ρ_g/cm^3	n_D
A	upper	0.738	0.852	1.4763
	lower	7.43	0.948	1.3995
B	upper	0.665	0.856	1.4837
	lower	4.139	0.910	1.4429
C1	upper	0.59	0.867	
	middle	3.96	0.979	
	lower			
C2	upper	0.589	0.863	1.4967
	middle	3.479	0.951	1.4301
	lower	1.213	1.039	1.3525
C3	upper	0.588	0.864	1.4973
	middle	3.921	0.974	1.4071
	lower	1.191	1.037	1.3519
C4	upper	0.582	0.859	1.4959
	middle	3.589	0.962	1.4163
	lower	1.202	1.038	1.3518
C5	upper	0.606	0.863	1.4969
	middle	3.636	0.932	1.4492
	lower	1.239	1.067	1.3538

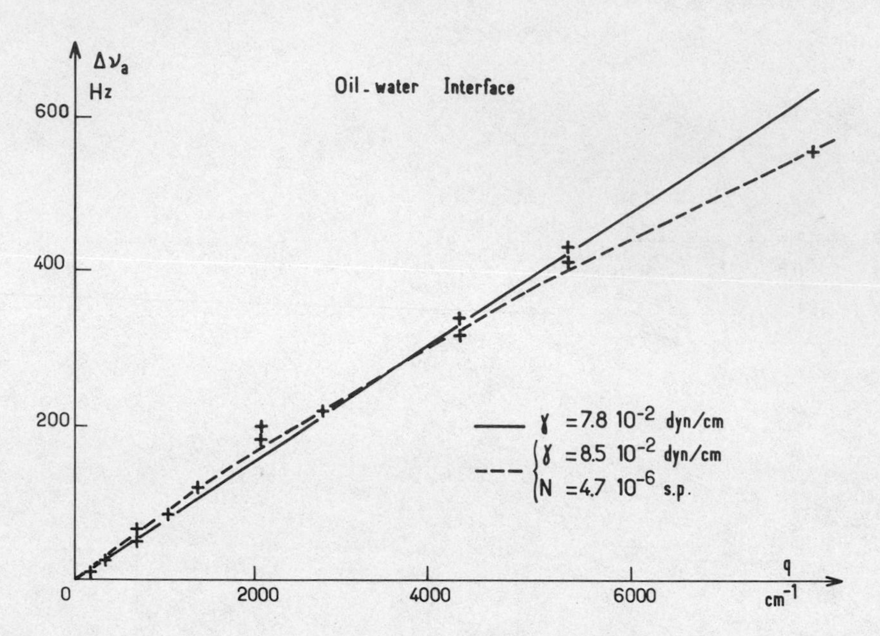

Fig. 3. Apparent spectral half width (corrected for gravity effects) versus scattering wave vector q for sample A of Table 1; + experimental points, — fit with eq. 4; --- fit with eq. 5.

For sample B, the curvature of $\Delta\nu_q(q)$ is still more pronounced, but the precision of the results is poorer, and a precise value of N cannot be determined.

The interfacial tension obtained with light scattering method is compared in Table 1, with values γ_{SD} obtained by the spinning drop technique on similar samples. They are in good agreement.

Several samples have been studied in the three-phase region (C1-C5). The interfacial tensions are smaller in this case and the corresponding spectral width also smaller. A typical spectrum is shown in Figure 4.

The apparent widths $\Delta\nu_a$ for the oil-middle phase interface vary linearly with q as shown in Figure 5. The values of interfacial tension decrease with increasing salt concentration as shown in Table 1. The differences within samples C1-C3 which correspond in principle to same salt concentrations are thought to be due to very small composition differences as indicated in Table 2, and not due to a lack of precision of the technique itself which is, in principle, better than 10%.

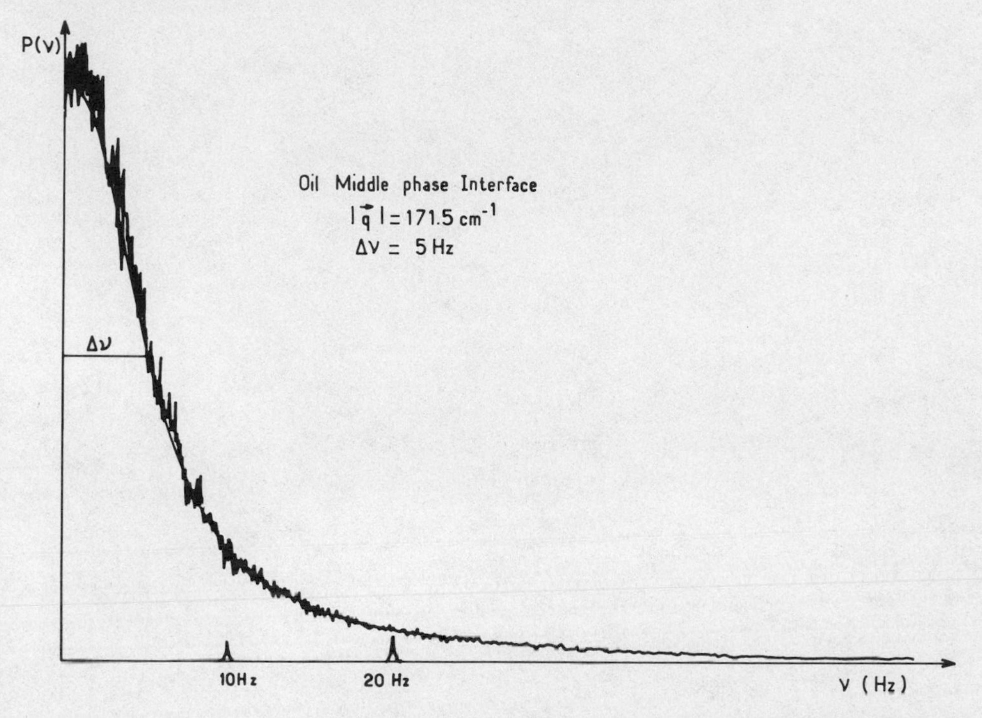

Fig. 4. Typical experimental spectrum of surface scattering.

The water-middle phase interfaces are much more difficult to study. Their behavior varies largely from one sample to the other. The variation of interfacial tension with salt concentration is not clear up to now.

DISCUSSION

The light scattering technique described here happens to be a very precise method for measuring small interfacial tensions. This technique has already allowed the measurements of values of the order of 10^{-4} dynes/cm (4). Measurements down to 10^{-5} - 10^{-6} are expected to be achievable, since the technique allows the determination of very small spectral half widths (9).

Very small values of surface viscosity can also be determined: $N \sim 10^{-6}$ surface poises. However, it is not clear up to now if this viscosity is as large in the systems studied here.

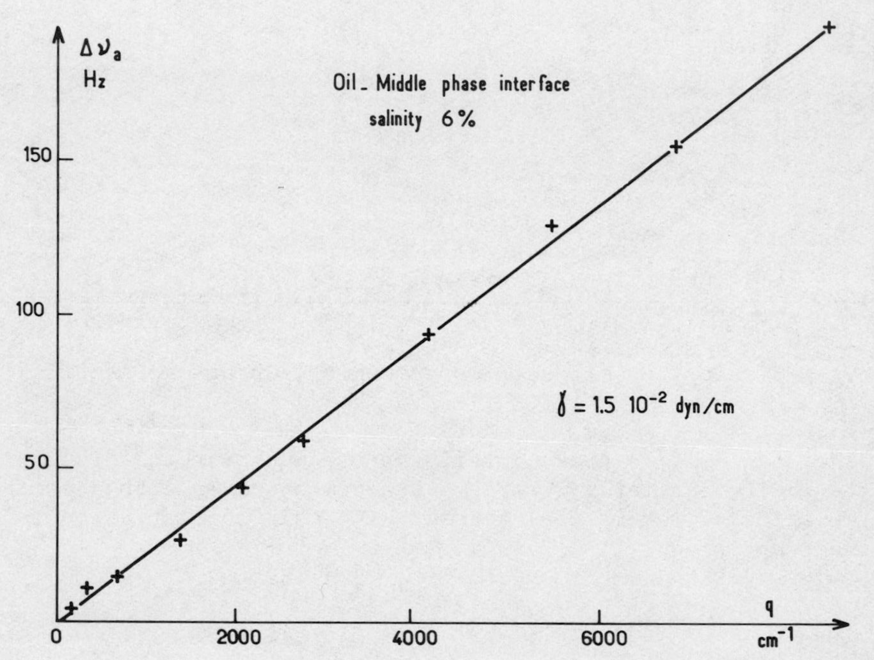

Fig. 5. Apparent spectral half width versus q for sample C2, Table 1 and oil-middle phase interface; + experimental points, - linear fit.

STUDIES ON MICROEMULSIONS

EXPERIMENTAL

We measure both the intensity I and the autocorrelation function $G(\tau)$ of the light scattered by the sample (6). The main contribution to the scattering is due to concentration fluctuations of the droplets. If particle size is assumed to be constant:

$$I \sim \left(\frac{dn}{d\phi}\right)^2 \phi \left(\frac{\partial\pi}{\partial\phi}\right)^{-1} \tag{6}$$

where I is the intensity collected in a fixed scattering solid angle, n the refractive index of the solution, π the osmotic pressure and ϕ the volume fraction occupied by the droplets.

In the limit of small volume fractions:

$$\frac{d\pi}{d\phi} \cong \frac{kT}{V}(1 + 2B\phi) \quad \text{and} \quad V = \frac{4}{3}\pi R^3 \tag{7}$$

where V is the volume occupied by the constituents of the droplets, R the corresponding radius and B the virial coefficient of the osmotic pressure.

The autocorrelation function is[2]

$$G(\tau) = 1 + e^{-\tau/\tau_q} \quad \text{and} \quad \tau_q^{-1} = 2Dq^2 \tag{8}$$

where q is related to the scattering angle θ by $q = \frac{4\pi n}{\lambda} \sin\frac{\theta}{2}$ and D is the diffusion coefficient.

In the limit $\phi \to 0$:

$$D = \frac{kT}{6\pi\eta R_H}(1 + \alpha\phi) \tag{9}$$

[2] The detection is homodyne for $\theta \gtrsim 20°$.

where η is the viscosity of the continuous phase, R_H the hydrodynamic radius, and α a virial coefficient depending on hydrodynamic interactions and interparticular forces (10,11).

If samples containing different volume fractions of droplets of constant size are prepared, this technique allows the determination of sizes and interactions in the system.

Materials and Methods

We studied two types of systems in the water-in-oil region of the phase diagram:

1) water-cyclohexane-SDS-pentanol
2) water-toluene-SDS-butanol

The effect of salt was also investigated. The characterization of the samples can be found in Table 3.

A dilution procedure has been determined for each series of samples in order to prepare different volume fractions of droplets of constant size. Tests of the procedure are:

- linearity of the plot of alcohol amount necessary to obtain a transparent system versus oil volume, the amounts of water and soap in droplets being held constant (12).
- linearity of ϕ/I versus ϕ (6). This is illustrated in Figure 6 for several samples.

A stronger evidence of droplet size constancy is the linearity of Zimm-Plots in small angle neutron scattering (13). But such proof is only available for samples A, B and C of Table 3.

The volume fraction ϕ is defined then as:

$$\phi = \{\text{soap volume} + (\text{water}+\text{alcohol}) \text{ volume in droplets}\}/\text{total volume} \quad (10)$$

Samples are put into 1 cm path length glass cells (Hellma). The light source is an Argon laser "Coherent Radiation" CR2 at a wavelength 5145Å giving an output power of about 0.5w. Most results are relative to a scattering angle of 90°. Refractive indices have been measured using an Abbe refractometer. Intensity measurements are relative. Calibration was made (6) using an aqueous solution of polystyrene spheres (Dow Chemical Company).

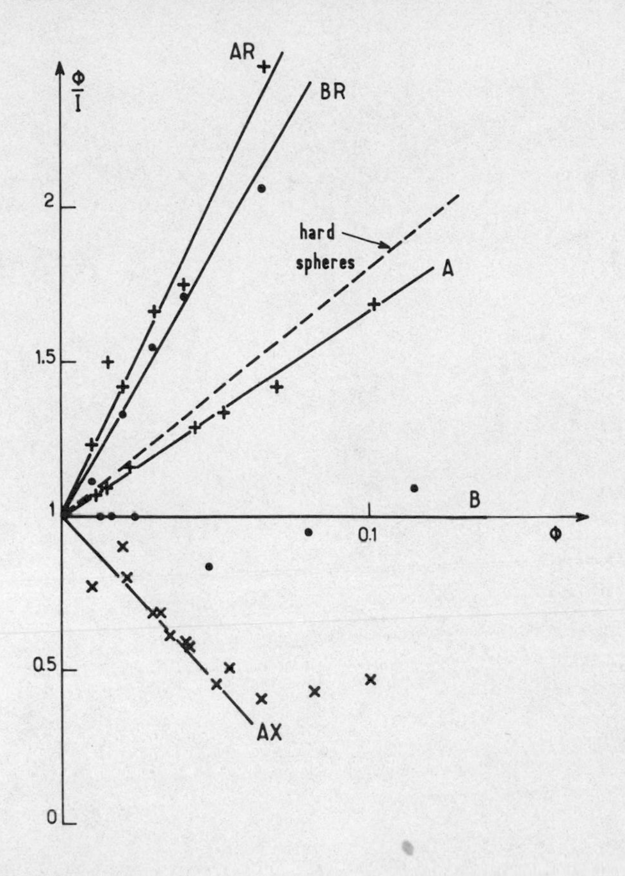

Fig. 6. ϕ/I versus ϕ for $\phi \lesssim 0.1$ and several microemulsions of Table 3.

RESULTS

a) Intensity Measurements: The limit of I/ϕ when $\phi \rightarrow 0$ allows the determination of R and R_w, radius of aqueous core: $R_w = R (\phi_w/\phi)^{1/3}$, where ϕ_w = water volume in droplets/total volume.

The variation of ϕ/I versus ϕ, normalized to 1 for $\phi = 0$, is represented in Figure 6 for several microemulsion series. The slopes of the straight lines are the virial coefficients 2B [from eq. (7)]. All the results are reported in Table 3. If the particles interact as hard spheres, 2B is equal to 8 (5). One can see in Table 3 that this is not the case for microemulsions. Either 2B < 8 indicating the existence of a supplementary attractive potential, or 2B > 8, a case of supplementary repulsive potential.

Table 3

Microemulsion	water/soap volume ratio in droplets	salt in water	R	R_W	R_H	2B (±5%)	α (±10%)
cyclohexane-pentanol							
A	0.69	--	49	36Å	48Å	6.8	3.5
AR	0.69	0.1 M	50	35	49	20	-3
AS	0.69	0.5 M	48	34	48	20	0 to -2
AT	0.69	1 M	48	34	48	20	0 to -2
B	0.345	--	75	65	81	0	-11.5
BR	0.345	0.1 M	84	70	93	17	-4.5
BS	0.345	0.5 M	93	77	94	17	-2
C	0.23	--	92	84	96	≃ -1.8	--
toluene-butanol							
AX	0.86	--	≃ 29	≃ 21	36	-12	-20
AY	0.69	--	≃ 33	≃ 26	≃ 41	-12	≲ -20

Following Vrij (5), these added potentials can be treated as perturbations to hard sphere interactions. This allows to take well into account the shape of I versus ϕ until large volume fractions $\phi \lesssim 0.4$ (see for instance Figure 7). However, such fit cannot give more information about the perturbation potential W than again a value for the virial coefficient 2B. The exact nature of W is still unknown.

b) Autocorrelation Functions: Autocorrelation functions are always exponential within experimental accuracy. Diffusion constants values were deduced using eq. (8) and corresponding hydrodynamic radii and virial term α using eq. (9). The results are reported in Table 3. Again, one sees that α is very different from hard sphere value $\alpha = 1.6$ (10,11).

We found no variation of D with scattering angle ($4° \leq \theta \leq 180°$) for microemulsions A and B. Anomalies seem to appear for microemulsions AY at volume fractions $\phi \sim 0.1$. We attribute this effect to droplet clusters formation leading to large aggregates with sizes comparable to light wavelength λ. Indeed, in such a case, the law $\tau_q^{-1} = 2Dq^2$ is no longer valid (14). Note that in this system the virial coefficient is high and negative, indicating strong attractive interaction between droplets. Such effect has been already

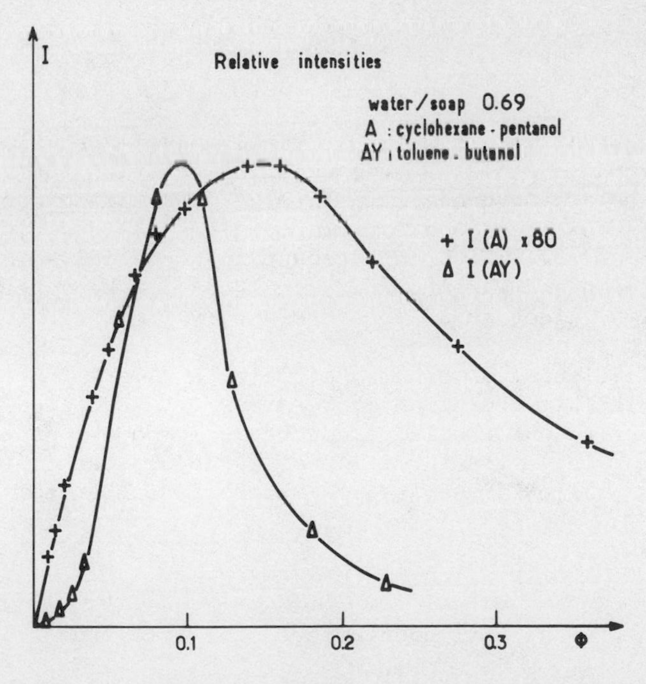

Fig. 7. Relative intensities versus volume fraction ϕ for microemulsions relative to the same water/soap ratio but to different kinds of oil and alcohol.

noted on systems containing copolymers as emulsifiers (15). Further experiments are on the way in order to confirm this observation.

It is possible to compare diffusion coefficient to sedimentation coefficient s. Theory predicts (10,16):

$$D \sim \frac{\partial \pi}{\partial \phi} / f$$

$$S \sim \frac{(1 + \phi)}{f}$$

where f is the friction coefficient. Using $\partial \pi / \partial \phi$ values deduced from intensity data, we have computed the corresponding f values. We find that below $\phi \lesssim 0.1$, f (for microemulsions A and B) has a linear behavior obeying:

$$f_{LS} = 6\pi\eta R_H^{LS} \ (1 + 10\phi)$$

Sedimentation data are available for the same systems (13) and give:

$$f_{SED} = 6\pi\eta R_H^{SED} \ (1 + 3.5\phi)$$

R_H values are very close, but f variation with ϕ is different.

At the present time, we do not understand the origin of this difference.

Adopting a different point of view, we tried to fit the variation of D versus ϕ at low ϕ with existing theories taking into account the role of hydrodynamic interactions and interparticle forces. This has been done for microemulsions A and B, using Felderhof theory (11) with an interaction potential sum of hard sphere repulsion and $W = A(2R/r)^6$, $A = B \frac{kT}{4}$. The agreement with experimental α values is quite satisfactory.

The variation of D versus ϕ at large ϕ is still not understood. All samples exhibit a minimum of D versus ϕ (associated with a maximum of I), but to our knowledge, such minimum does not exist in any other systems (16). The formation of large clusters, if confirmed, would probably allow a qualitative understanding of this point.

DISCUSSION

One can give several conclusions from this study (Table 3, Figures 6 to 9):

1) effect of increasing water/soap ratio: the droplet size increases, the interactions become more attractive.

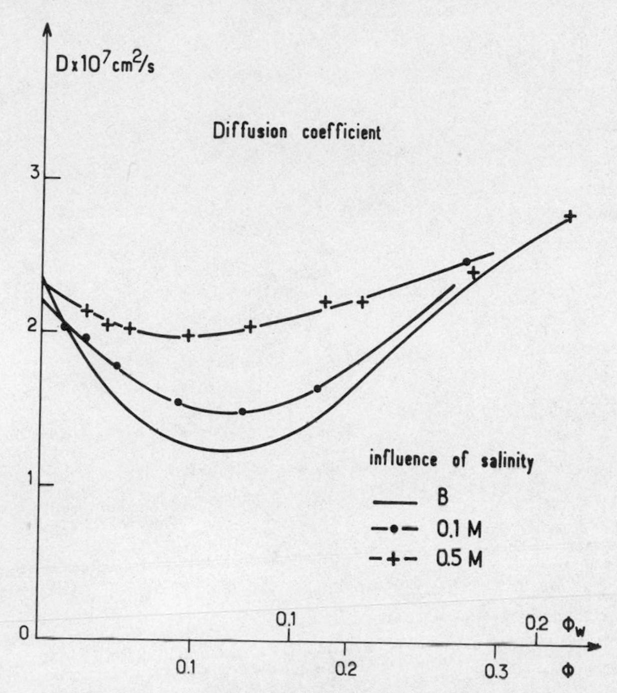

Fig. 8. Influence of salinity on diffusion coefficient

2) effect of increasing amounts of salt: the sizes do not vary significantly, but interactions become more repulsive; we believe that it is due to an increasing mean electric charge per droplet.

3) effect of oil and alcohol: changing cyclohexane into toluene decreases size and increases attraction. This may be due to better solubility of SDS alkyl chain in cyclohexane than in toluene which is a more polar medium.

These ideas can help us to investigate the nature of interaction potential W, electrostatic repulsion W_e and attraction W_a due to the exchange of oil molecules by segments of soap chains when the droplets approach each other (5). The balance between W_e and W_a will be either net repulsion or net attraction.

CONCLUSIONS

Light scattering techniques appear as a very useful tool to investigate the interfacial properties of oil-water mixtures, e.g., interfacial tension, surface viscosity, as well as micellar size and micellar interactions.

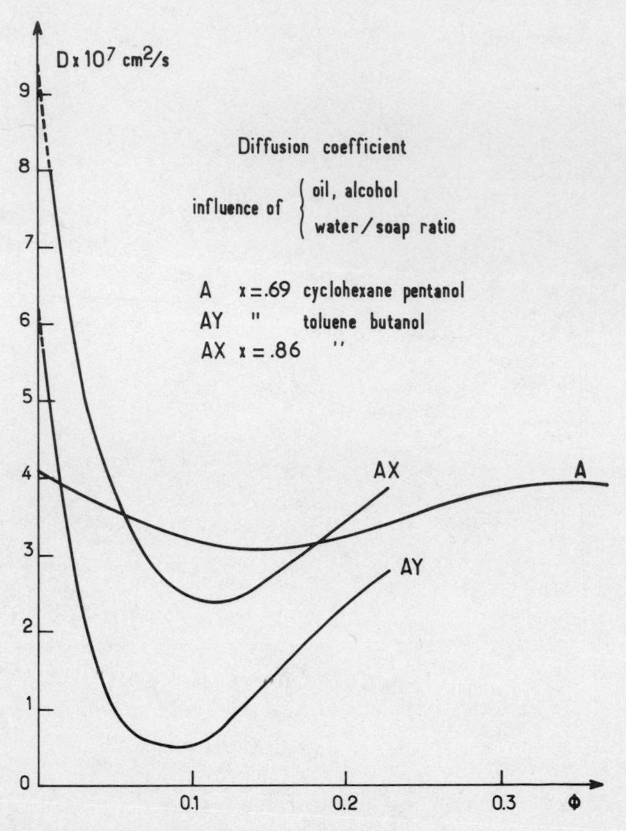

Fig. 9. Influence of water/soap ratio and of oil and alcohol nature on diffusion coefficient.

Further experimental and theoretical investigations are expected to give us a better understanding of the existence of low interfacial tensions and of the thermodynamic stability of the phases. Indeed, interfacial tensions can be related theoretically to micellar interactions in bulk phases (3). The role of surface viscosity is less clear. But it probably influences the phase stability as it happens for ordinary emulsions.

ACKNOWLEDGMENTS

We are greatly indebted to A.M. Bellocq, P. Lalanne, J. Rouch and C. Vaucamps for their help concerning the choice of samples and for many helpful discussions. We also wish to thank C. Taupin, J.P. Le Pesant, R. Ober and M. Lagües for their kind collaboration in the microemulsions study.

REFERENCES

1. R. N. Healy, R. L. Reed and D. G. Stenmark, Soc. Pet. Eng. J., 16, 147 (1976).
2. K. Shinoda and S. Friberg, Adv. Colloid and Interface Sci., 4, 281 (1975).
3. C. A. Miller, R-N. Hwan, W. J. Benton, and T. Fort, J. Colloid and Interface Sci., 61, 554 (1977).
4. D. Langevin and J. Meunier, in "Photon Correlation Spectroscopy and Velocimetry," H. Z. Cummins and E. R. Pike, eds., Plenum Press, New York, 1977.
5. A. A. Calje, W.G.M. Agterof, and A. Vrij, in "Micellization, Solubilization and Microemulsions," Vol. 2, K. L. Mittal, ed., Plenum Press, New York, 1977.
6. A. M. Cazabat, D. Langevin, and A. Pouchelon, J. Colloid and Interface Sci., 73, 1 (1980).
7. C. Griesmar and D. Langevin, Proceedings of the Conference on "Physicochimie des amphiphiles," Bordeaux, 1978.
8. P. Lalanne, J. Biais, B. Clin, A. M. Bellocq, and B. Lemanceau, J. Chim. Phys., 75, 236 (1978).
9. D. Langevin and J. Meunier, Optic Comm., 6, 427 (1972).
10. G. K. Batchelor, J. Fluid Mech., 74, 1 (1976).
11. B. U. Felderhof, J. Phys. (A), 11, 929 (1978).
12. A. A. Graciaa, J. Lachaise, A. Martinez, M. Bourrel, and C. Chamber, CRAS, 282B, 547 (1976).
13. M. Dvolaitzky, M. Guyot, M. Lagües, J. P. Le Pesant, R. Ober, C. Sauterey, and C. Taupin, J. Chem. Phys., 69, 3279 (1978).
14. P. N. Pusey, J. Phys. (A), 8, 1433 (1975).
15. S. Candau, J. Boutillier, and F. Canday, Polymer, 20, 1237 (1979).
16. G. D. Phillies, J. Chem. Phys., 67, 4690 (1977).

AN EXPERIMENTAL INVESTIGATION OF STABILITY OF MICROEMULSIONS AND POLYDISPERSITY EFFECTS USING LIGHT-SCATTERING SPECTROSCOPY AND SMALL ANGLE X-RAY SCATTERING

E. Gulari[1] and B. Chu

Department of Chemical Engineering, University of Michigan, Ann Arbor, Michigan 48109, U.S.A.
Department of Chemistry, State University of New York at Stony Brook, Stony Brook, New York 11794, U.S.A.

Laser light scattering and small angle X-ray scattering were used to study W/O and O/W microemulsions formed by n-tetradecane, sodium dodecyl sulfate, n-pentanol and water. Three of the four W/O systems were stable for long periods of time showing only small changes in size. These systems were fairly monodisperse. The fourth system studied was not stable, changing from a microemulsion to a molecular solution in a few days. The O/W system showed ordering due to strong charge interactions, and also had micelles coexisting with the microemulsion droplets. In all the systems studied, the disperse phase droplets were found to be spherical in shape.

INTRODUCTION

Microemulsions (1,2) [or swollen micellar solutions as some authors prefer to call them (3-5)] have been studied by a variety of techniques including NMR (6), dielectric measurements (7), conductivity (8), X-ray (2) and light scattering (9,10). In this article, we have used the technique of quasielastic light-scattering spectroscopy (QLS) (11,12) to study the formation, stability and polydispersity effects of the dispersed phase in O/W and W/O microemulsions of the n-tetradecane, sodium dodecyl sulfate (SDS), 1-pentanol and water system. Complementary intensity measurements of small angle X-ray scattering (SAXS) and of light scattering are also utilized so that we can estimate the inner core as well as the effective hydrodynamic sizes.

---

[1]Author to whom requests for reprints should be addressed.

QLS has been used routinely to determine the Z-average translational diffusion coefficient and the corresponding hydrodynamic radius of macromolecules in solutions and of colloidal suspensions (11,12); and also to characterize the polydispersity effects by the method of cumulants (13).

Recently, we have developed a new method of data analysis which permits us to obtain an approximate distribution function in terms of translational diffusion coefficients or sizes (14,15). We shall use this new histogram approach which automatically takes into account the effects of polydispersity to study the formation and stability of microemulsions.

MATERIALS AND METHODS

a) Sample preparation: The microemulsions used in this study were prepared from n-tetradecane (Phillips Petroleum 99+ mol % pure), n-amyl alcohol (Fisher certified reagent), gel electrophoresis grade SDS (Bio-Rad Labs) and doubly distilled water. The water in oil samples (2,3 and 4) were prepared by weighing precise amounts of SDS, alcohol and tetradecane and by adding enough water to the resultant mixture in order to obtain a clear solution upon vigorous shaking at 40°C. The remaining portion of the water was added afterwards. Sample 1 was prepared to match the W/O microemulsion studied by Gerbacia et al. (16), with n-hexadecane being replaced by n-tetradecane. Sample number 5 was an O/W microemulsion prepared by mixing water, SDS and tetradecane to form an emulsion which was titrated to clarity with alcohol. Relative compositions of all the samples are listed in Table 1. Light-scattering samples were filtered through 0.22 μm filters to eliminate dust. Table 1 gives the composition of the five samples used in this investigation.

b) Light scattering method: The extrapolated zero-angle excess scattered intensity of a macromolecular solution is related to the solute molecular weight (17). When the scattered intensity is sampled over short time periods, it fluctuates about the mean due to thermal Brownian motions. In applying QLS to macromolecular solutions, the time behavior of concentration fluctuations is expressed quantitatively by the intensity autocorrelation function. The first order (or the field) autocorrelation function of monodisperse particles in a fluid due to only translational motion is given by:

$$g^1(\tau) = [E_s(K,t)E_s^*(K,t+\tau)] = \exp(-\Gamma\tau) \quad (1)$$

where E_s is the scattered electric field, τ is the total delay time and $K = (4\pi n/\lambda_o) \sin \theta/2$ is the scattering vector with λ_o, n, and θ being the wavelength of light in vacuo, the refractive index of

Table 1. Compositions of the Samples (wt %)

Sample No.	n-tetradecane	n-pentanol	SDS	Water	N_w/N_s
1	80.4	14.2	2.45	2.95	19.1
2	51.0	21.0	13.0	15.0	19.1
3	43.0	20.5	10.5	26.0	38.9
4	57.0	23.5	14.0	5.7	6.4
5	17.7	15.3	12.3	54.7	70.9

the scattering medium, and the scattering angle, respectively. $\Gamma = DK^2$ with D being the translational diffusion coefficient of the particle. Experimentally we measure the intensity (or the second order) autocorrelation function. For a Gaussian random process, it is given by:

$$C(\tau) = [I(K,t)I^*(K,t+\tau)] = 1 + |g^1(\tau)|^2 \tag{2}$$

If the particles are polydisperse, equation (1) is replaced by an integral:

$$g^1(\tau) = \int_0^\infty G(\Gamma) \exp(-\Gamma\tau)d\Gamma \tag{3}$$

with

$$\int_0^\infty G(\Gamma)d\Gamma = 1 \tag{4}$$

where $G(\Gamma)$ is the normalized distribution function in the Γ-space. In the method of cumulants (13) equation (3) is expanded about the mean, $\bar{\Gamma}$, of the distribution function $G(\Gamma)$ and the approach is completely general. Unfortunately, it does not provide sufficient information about the shape of the distribution function. We have overcome this difficulty by introducing a histogram method of analyzing the intensity autocorrelation function (14,15). Our method requires no a priori assumption on the form of $G(\Gamma)$. The only approximation is to replace the continuous distribution curve by a histogram of finite steps. In other words, we set

$$G(\Gamma) = \sum_j^M a_j(\Gamma_j) \tag{5}$$

with

$$\sum_j^M a_j(\Gamma_j)\Delta\Gamma_j = 1 \tag{6}$$

where M is the number of discrete steps in the histogram. Then the field autocorrelation function can be approximated by an expression of the form:

$$g^1(\tau) = \int_0^\infty \sum_j^M a_j(\Gamma_j) \exp(-\Gamma_j \tau) d\Gamma \quad (7)$$

where the integration is performed over each discrete step. The measured $g^1(\tau)$ is then fitted to equation (7) by a least squares procedure and the coefficients a_j are determined. If we take the dispersed phase droplets to be spherical in shape, we can use the Stokes-Einstein relation to compute the effective hydrodynamic radius r_h.

$$D = k_B T / 6\pi\eta r_h \quad (8)$$

where k_B is the Boltzmann constant; and T is the absolute temperature. We have taken η to be shear viscosity of the solvent, independent of concentration effects. Using $\Gamma = DK^2$, $G(\Gamma)$ can be converted to G(D). By introducing approximate scattering factors for different spheres we can transform G(D) to a new number distribution function expressed in terms of the effective hydrodynamic radius or the molecular weight (14,15).

The light-scattering spectrometer has been described in detail elsewhere (18). We used a spectra physics model 165 argon ion laser operating at 488.0 nm. The scattered light was detected by an ITT FW-130 photomultiplier tube. A 96 channel single-clipped Malvern correlator was used to measure the intensity autocorrelation function. We also measured the integrated scattered intensity and the incident laser intensity.

c) Small angle X-ray scattering method: At small scattering angles, the excess scattered X-ray intensity of a monodisperse macromolecular solution can be approximated by (19):

$$I(K) = I_o \exp[-(R_g)^2 K^2/3] \quad (9)$$

where R_g is the radius of gyration of the macromolecule and I_o is the extrapolated zero angle intensity. For a polydisperse system, equation (9) becomes:

$$I = \int_0^\infty I_o(R_g) \exp(-(R_g^2/3)K^2) f(R_g) dR_g \quad (10)$$

with

$$\int_0^\infty f(R_g) dR_g = 1 \quad (11)$$

$f(R_g)$ is the normalized distribution function of radii of gyration. While equation (9) is valid for spheres which scatter uniformly, it has to be modified for microemulsion droplets consisting of concentric spheres. If we assume that the W/O droplets consist of an inner core of water surrounded by an interfacial layer, consisting of SDS heads and OH groups of the alcohol, and an outer shell of SDS tails, alcohol tails and solvent. The contribution of each layer has to be summed up so that the excess scattered intensity due to N independent microemulsion droplets in scattering volume, v, has the form (20):

$$I(K) = N \cdot F^2(Kr) \tag{12}$$

$$F(Kr) = (\rho_3-\rho_5)V_3P(Kr_3) + (\rho_2-\rho_3)V_2P(Kr_2) + (\rho_1-\rho_2)V_1P(Kr_1) \tag{13}$$

with ρ_i = the scattering density of the ith layer; $V_i = 4\pi r_i^3/3$, r_i = the outer radius of the ith shell, and $P(Kr_i)$ being the form factor. For a uniform solid sphere,

$$P(Kr) = [3(\sin Kr - Kr\cos Kr)/K^3r^3)] \tag{14}$$

Equation (12) can be used to estimate only the lower limit of the radius of gyration because we cannot vary N without changing the nature of the microemulsion droplets.

The small angle X-ray spectrometer used in this study is shown in Figure 1. A Siemens Kristalloflex-4 generator with a 4CuO/2 tube was used as the X-ray source. The X-rays were monochromatized to Cu-K_α (λ=1.54Å) line by a Ni filter and a Kratky slit system was used to collimate the incident X-ray beam. We used a Tennelec PSD-1100 position sensitive detector which effectively measured the scattered intensity simultaneously and gave an improvement by a factor of about 40 over a slit and proportional counter detection system. The slit geometry was such that the infinite beam length approximation was valid. The sample cells were 2 mm O.D. and 0.01 mm wall thickness glass capillaries. A typical measurement using the position sensitive detector took only half an hour.

RESULTS

a) <u>Mean diffusion coefficients and variances</u>: Table 2 shows the Z average mean diffusion coefficients and their variances for all the samples as determined from the linewidth studies. The means and the variances were obtained by both the cumulants method of data analysis and also by the histogram method. The means are in very good agreement except for the case of sample 5 which is an O/W microemulsion and has a bimodal distribution. The variances as determined by both methods are also in general agreement for all the samples. However, the values determined by the cumulants

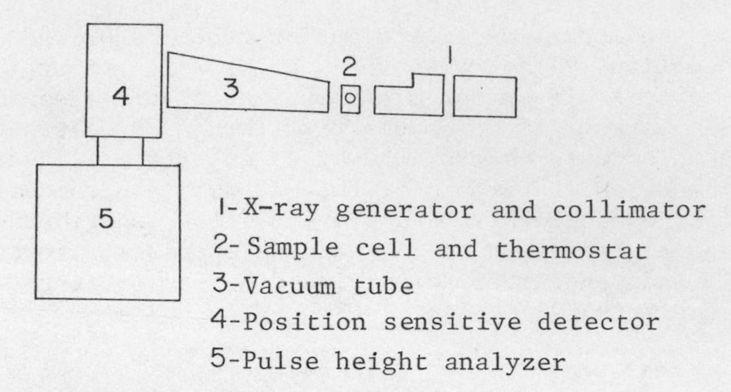

Fig. 1. Schematic drawing of the small angle X-ray spectrometer used in this study.

method show significant changes for the same sample measured at different times; whereas the values obtained by the histogram method are more self-consistent and reliable. This is in part due to the convergence problems associated with the cumulants expansion used in the least squares analysis. The values of the variance will change significantly depending on the number of terms used in the cumulants expansion (14). The variances given in Table 2 were obtained from the highest possible order fit. Convergence also accounts for the error in $\langle D \rangle_z$ by the cumulants method for sample 5 which is highly polydisperse. Consequently, a very large number of terms in the cumulants expansion have to be carried in the fitting procedure in order to obtain reliable diffusion coefficients.

b) Hydrodynamic radii: We have calculated the effective hydrodynamic radii $(r_h)_z$ from $\langle D \rangle_z$ using equation (8). In the computations we have used n-tetradecane viscosity for samples one through four and water viscosity for sample 5. For W/O microemulsions, the surfactant SDS does not dissolve in n-tetradecane but the cosurfactant n-pentanol does. As a result, there is probably some n-pentanol in the solvent phase. If all the n-pentanol was in the solvent phase, the viscosity would change by a maximum of $\simeq$ 15% and decrease the computed effective radius by the same amount. The resultant hydrodynamic radius can be taken as a first approximation to the real hydrodynamic radius of the droplets. There is some evidence indicating a fairly weak concentration dependence of the diffusion coefficient D for hard spheres (21,22). Alternatively, we can compute the real hydrodynamic radius and D_o (the diffusion coefficient at zero concentration) by using the various known theoretical expressions. Unfortunately, for hard sphere interactions (22,23), the hydrodynamic radii obtained are too small to be physically realistic. For example, if we use the following expressions given by Altenberger (23) for the ratio of the effective diffusion coefficient, D_{eff} to D_o,

Table 2. Diffusion Coefficients and the Apparent Hydrodynamic Radii as a Function of Time[a]

Sample	$10^7 \langle D \rangle_Z$ cm^2/sec		$\langle r_h \rangle_Z$ Å		Variance $(\mu^2/\bar{\Gamma}^2)$			
No.	Cum	Hist	Cum	Hist	Cum	Hist	T°C	Time
1	1.598	1.602	64.9	64.8	0.063	0.067	23°	
2	1.860	1.886	55.8	55.0	0.027	0.015	23°	Soon
3	1.374	1.364	75.5	76.1	0.08	0.056	23°	After
4	3.70	---	29.8	--	--	---	26°	Preparation
5[b]	5.65	6.09	40.1	37.2	1.57	1.86	22°	
1	1.608	1.624	64.5	63.9	0.001	0.020	23°	
2	1.894	1.905	54.8	54.5	0.022	0.031	23°	6 Weeks
3	1.406	1.378	73.7	75.3	0.16	0.060	23°	After
5[c]	5.53	6.763	42.9	35.1	1.4	1.95	23°	Preparation
1	1.639	1.632	63.3	63.6	0.04	0.019	23°	
2	1.914	1.919	54.2	54.1	0.029	0.033	23°	10 Weeks
3	1.442	1.386	71.9	74.9	0.031	0.080	23°	After
4	7.31	---	14.2	--	0.08	---	23°	Preparation

a) Calculated from the Stokes-Einstein Relationship and assuming that the viscosity was that of n-tetradecane for 1, 2, 3, and 4. For sample 5, an O/W microemulsion water viscosity was used.

b) This sample is bimodal with $D_1 = 2.07 \times 10^{-7}$ cm^2/sec, and $D_2 = 2.31 \times 10^{-6}$ cm^2/sec for the two peaks. The corresponding r_h's are 109.4Å and 9.8Å.

c) This sample is also bimodal with $D_1 = 2.22 \times 10^{-7}$ cm^2/sec, $D_2 = 2.24 \times 10^{-6}$, $r_h = 104.5$Å and $r_h = 8.8$Å.

$$D_{eff}/D_o = [1-6\phi(1-2.2\phi)]/[1-8\phi + 34\phi^2] \quad (15)$$

where ϕ is the volume fraction of the dispersed phase. We obtain for sample 3 ($\phi \simeq 0.45$) $D_{eff}/D_o = 0.225$ and $(r_h)_o = 0.225\ (r_h)_{eff} = 16.9$Å. This value of the radius is physically impossible due to the fact that the length of one SDS molecule is about 22°A, therefore the hydrodynamic radius of the microemulsion droplets must be larger than 22°A. Thus, the simple hard core potential assumption is not valid for the water in oil microemulsions. It is more lilely that the real potential contains both attractive and repulsive parts. Overall, D_{eff} must increase with the volume fraction in order to be physically acceptable and give a realistic estimate of $(r_h)_o$. For the O/W microemulsion, we did not add any electrolytes to the solution. Consequently, there is strong electrostatic interactions between the microemulsion droplets as well as the hydrated SDS in solution. Due to charge interactions, the diffusion coefficient is known to increase significantly with concentration in SDS solutions (24). Therefore, the hydrodynamic radius computed on the basis of the Stokes-Einstein relation may be too small.

Table 2 also shows repeated measurements made on the same samples 6 weeks after preparation and 10 weeks after preparation in order to make a quantitative check of the stability of the dispersed phase. For the W/O microemulsions 1, 2 and 3, there is a 1-2% systematic increase in the diffusion coefficient and a corresponding decrease in the hydrodynamic radius. This slow drift is beyond the experimental error limits. Interestingly enough, such a trend is opposite to what one would expect. Sample 4 is the only sample that seems to have gone through a drastic change. Unfortunately, sample 4 scatters light very weakly. As a result, the measurements did not have enough precision to warrant a histogram analysis. From the effective r_h values given in Table 2, we see that sample 4 had an initial effective droplet diameter of about 30Å, but after ten weeks, the sample essentially became a molecular solution with an effective r_h of 14Å which is very close to what one can compute from the radius of gyration of the SDS molecule with possibly about 5-6 water molecules attached to each SDS. Thus, we see that not all of the microemulsions are thermodynamically stable and can change into a molecular solution under conditions as those encountered by sample 4. Again, this change is in the opposite direction to the general expectation, but the high surfactant to water molecular ratio makes a stable molecular solution possible.

c) <u>Apparent molecular weights and Rayleigh ratios</u>: As we cannot generally dilute microemulsions without changing the structure of the dispersed phase, we cannot make use of a Zimm plot to obtain the molecular weight and the radius of gyration. The only information we can obtain are the apparent molecular weights and if the asymmetry is large enough, the apparent radius of gyration. For all the microemulsion systems studied, the asymmetry was about 1% or less. Therefore, we were not able to determine the radius of gyration by means of angular distribution of scattered intensity. This finding is in agreement with our hydrodynamic radii computed from the diffusion coefficients.

Table 3 shows the apparent molecular weight for samples 1, 2 and 3. The calculated molecular weights given in the second column were obtained from the hydrodynamic radii listed in Table 2 as follows: The W/O microemulsion droplet was considered to be composed of an inner core of water and an outer shell of surfactant tails and alcohol. The density of the inner core was assumed to be equal to the density of water and an average density of 0.8 gr/cc was used for the outer shell. The molecular weight calculated in this fashion is larger by as much as a factor of two and a half than the apparent molecular weight. However, the disagreement is not surprising because the concentration dependence of the diffusion coefficient on concentration is known to be much weaker than that of the scattered intensity (25) due to cancelling effects. In Table 3 we have also listed the ratio of the core weight to the shell weight. These ratios can be compared with the following values, calculated by assuming that all the alcohol and surfactant are in the shell region; 0.177 for sample 1, 0.450 for sample 2 and 0.832 for sample 3. In sample 1, we should note that there is 3 times more alcohol than SDS and water combined. As most of the alcohol is in the solvent phase rather than the shell region, we should expect a core weight to shell weight ratio greater than 0.177. For samples 2 and 3, the ratios given in Table 3 are smaller than those obtained from a material balance. The difference could be attributed to many factors, such as the use of D instead of D_o in equation (8). A rough estimate of r_{eff}/r_{actual} can be made by comparing the two sets of core weight to shell weight ratios. For samples 2 and 3, agreement is obtained when the hydrodynamic radii are 61Å and 83Å instead of 55Å and 76Å as listed in Table 2, corresponding to a 10% increase. Thus, the use of equation (3) in estimating r_h from D values measured at finite concentrations is reasonably good even at high volume fractions of the dispersed phase.

d) Distribution function of the diffusion coefficient: One of our main aims in this study was to find out how polydisperse the stable microemulsions were and what their distribution functions looked like. A convenient measure of polydispersity is the variance of the distribution function or the second moment divided by the square of the mean ($\mu_2/\bar{\Gamma}^2$). Unfortunately, the variance does not tell us much about the actual distribution function. Questions about the symmetry, the number of peaks in the distribution, etc. cannot be answered. Analysis by the histogram method using equation (7) however yields the distribution function itself directly. Figure 2 shows the normalized distribution functions of

Table 3. Comparison of the apparent molecular weights calculated from the excess scattered intensity and the molecular weights computed from the hydrodynamic radii given in Table 2

Sample No.	Apparent MW From Intensity	Calculated MW Water Core	Calculated MW Outer Shell	Total MW	Core/Shell
1	545000	214091	395435	609586	0.541
2	179700	94487	267558	362045	0.353
3	383279	406845	575709	982554	0.707

the diffusion coefficients obtained in this fashion for samples 1 (-----), 2 (.....) and 3 (———). Sample 1 has the narrowest distribution and sample 3 has the broadest. Thus, the polydispersity seems to increase as the water to surfactant ratio increases. Figure 3 shows the normalized bimodal distribution function of sample 5, an oil/water microemulsion system. The first peak has a $\bar{D}$ of 2.06×10^{-7} cm^2/sec and the second peak has a $\bar{D}$ of 2.31×10^{-6} cm^2/sec. We interpret the first peak being due to the microemulsion droplets and the second peak being due to SDS micelles, or possibly due to individual SDS molecules. In sample 5, we do not have any dissolved electrolytes to act as counter ions. The resultant long range electrostatic interactions could make computation of the hydrodynamic radii by means of the Stokes-Einstein equation unreliable. For example, with $\bar{D} = 2.31 \times 10^{-6}$ cm^2/sec, we get $r_h \simeq 10Å$ which is below the minimum radius, $r_h = 25Å$, of the SDS micelles. Nevertheless, we can probably use the Stokes-Einstein relationship to obtain a reasonable r_h for the microemulsion droplets. In fact, the large number of SDS micelles and single SDS molecules in solution may provide enough shielding to prevent long range electrostatic interactions between the microemulsion droplets. If we then take the Stokes-Einstein equation to be valid, we get $r_h = 109Å$ for the microemulsion droplets.

Figure 4 shows the relative errors between an experimentally measured autocorrelation function and the fitted histogram function. We want to note that the relative error is of the same magnitude as the calculated statistical error.

We also made a check of all the samples to see if there were any deviations from the K^2 dependence in the linewidth. For spheres, the decay time of the autocorrelation function is DK^2. However, for rods and ellipsoids of revolution, there are additional terms. Figure 5 shows a plot of the decay time $\bar{\Gamma}$ (from histogram fits) versus K^2, the square of the scattering vector. The linear behavior predicted for spheres is very precisely obeyed and the intercept is zero. All the samples we studied had similar curves leading us to conclude that in all of them, the dispersed phase droplets were spherical in shape.

e) Small angle X-ray results: Initially, we were hoping to be able to determine the volume ratios of the shell and the inner core region of the droplets as well as the presence of any ordering in the samples from X-ray measurements. Due to the high volume fractions of the dispersed phase, reliable values for the shell and core, radius of gyration could not be obtained. Sample 1 has the smallest volume fraction and the scattering curve for it was closest to that of a dilute suspension.

Sample 5 has a strong nearest neighbor Bragg peak. The desmeared scattering curve of sample 5 is shown in Figure 6. The

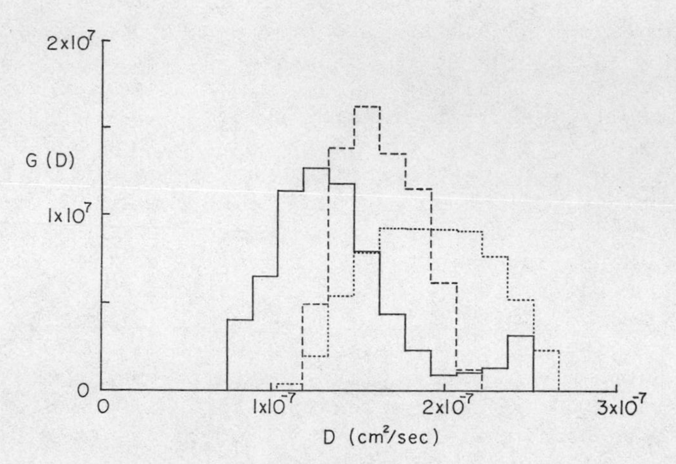

Fig. 2. The normalized histogram distribution functions of the diffusion coefficients for samples 1 ----, 2 ...., and 3 —— .

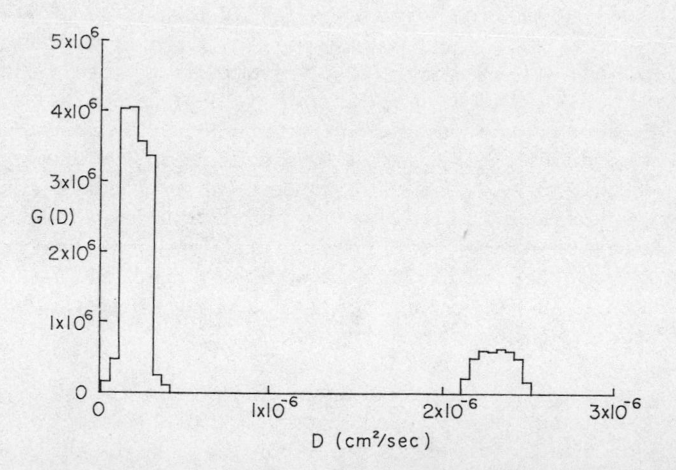

Fig. 3. The distribution function of the diffusion coefficients for sample 5. The peak to the right is the one due to SDS micelles, the other one is due to the microemulsion droplets.

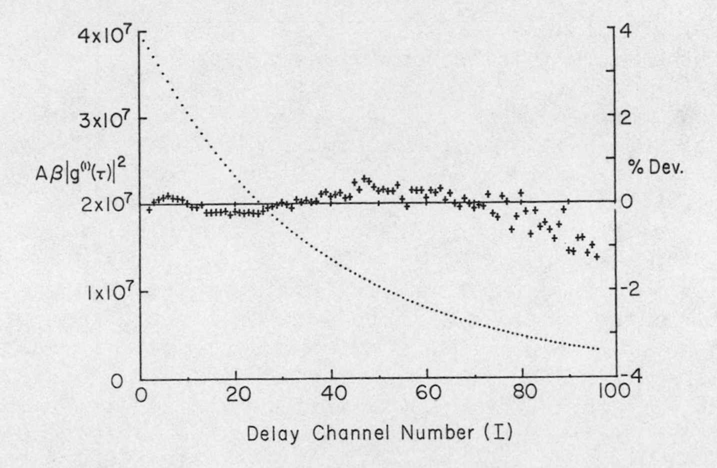

Fig. 4. A typical autocorrelation function and the relative errors. The relative error is:

$$\frac{(|g^1(\tau)|^2_{exp} - |g^1(\tau)|^2_{fitted}) \times 100}{|g^1(\tau)|^2_{exp}}$$

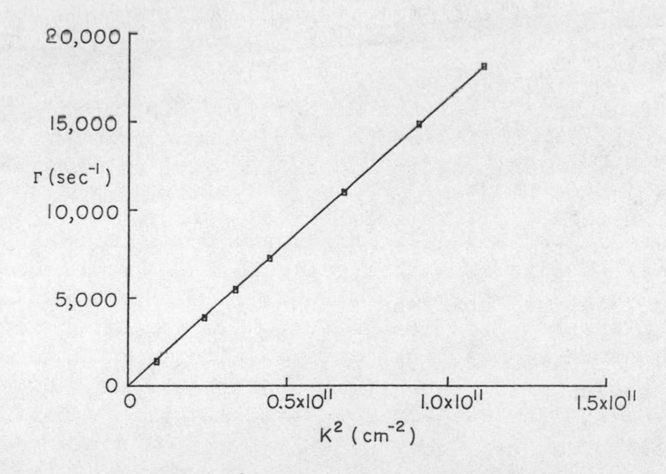

Fig. 5. Linewidth versus K^2 plot for sample 1 covering the angular region 30° to 150°. The slope is the diffusion coefficient $D = 1.602 \times 10^{-7}$ cm^2/sec. The line goes through zero as expected for spherical droplets.

peak corresponds to a spacing of ~ 123Å since the hydrodynamic radius of the microemulsion droplet as determined by quasielastic light scattering is 109Å, the Bragg peak cannot be attributed to interactions between the microemulsion droplets. We speculate its presence as due to interactions between micelles. Schulman and coworkers also report a Bragg spacing less than the hydrodynamic diameter of drops for an O/W microemulsion (2), and suggest that it may be due to the concentrated nature of the solution.

When we examined the X-ray scattering curve as a function of time, the results showed a slight increase in ordering for samples 1, 2, and 3. Sample 5 seemed to remain unchanged but sample 4 showed the most drastic change with time. Figure 7 shows the scattering curve of this sample one day after it was prepared and 10 weeks later. Right after it was prepared, the X-ray results showed that this sample had large microemulsion droplets, $r_g \simeq 200$Å. This is significantly higher than $r_h = 30$Å obtained by light scattering several days later, indicating that the sample underwent significant changes during the first few days. The X-ray intensity data taken ten weeks after shows that the solution is almost a molecular solution, yielding an $r_g \simeq 7$Å (corresponding to an $r_h \simeq 9.2$Å) in reasonable agreement with the corresponding r_h value of 14Å obtained by QLS. This curve also indicates that there is possibly some ordering in the solution, not detectable by light scattering.

DISCUSSION AND CONCLUSIONS

We find that QLS is a very good method of investigating microemulsions because the technique is nondestructive, nondisturbing and highly accurate. The diffusion coefficient of the dispersed phase is measured very precisely (± 0.1-.2%); and due to cancelling effects, the effective hydrodynamic radii can be obtained with reasonable accuracy from the diffusion coefficient. The microemulsion systems studied, with the exception of sample 4, are very stable over long periods of time. There is a very slight change (≤ 2%) over a period of two and a half months but this may be due to external factors or a very slow equilibrium process. We were especially interested in sample 1 because of its similarity to a system studied by Gerbacia and Rosano (16). They report that phase separation started after a few months, with complete separation in six months and no mixing upon subsequent shaking. Our system did not show phase separation and was very stable. Perhaps we can attribute the difference as due to the use of tetradecane instead of hexadecane or due to the fact that a dye was used by Gerbacia and Rosano. We are planning on repeating our measurements with hexadecane. Sample 4 started out as a W/O microemulsion with fairly large droplet size immediately after preparation ($r_h \simeq 200$Å) and changed into a molecular solution after some time. The change

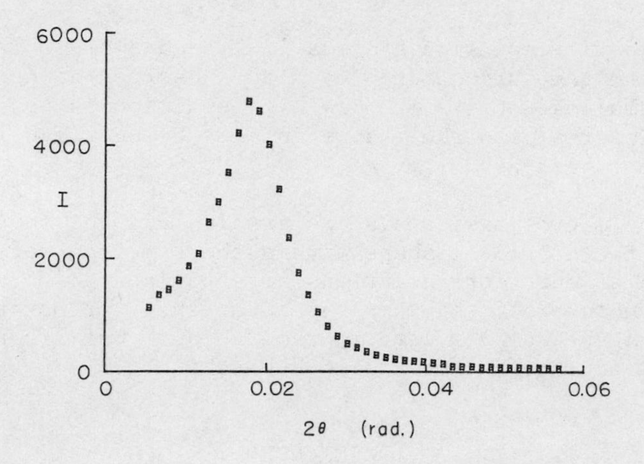

Fig. 6. Desmeared SAXS curve of sample 5 showing the Bragg peak due to strong charge interactions.

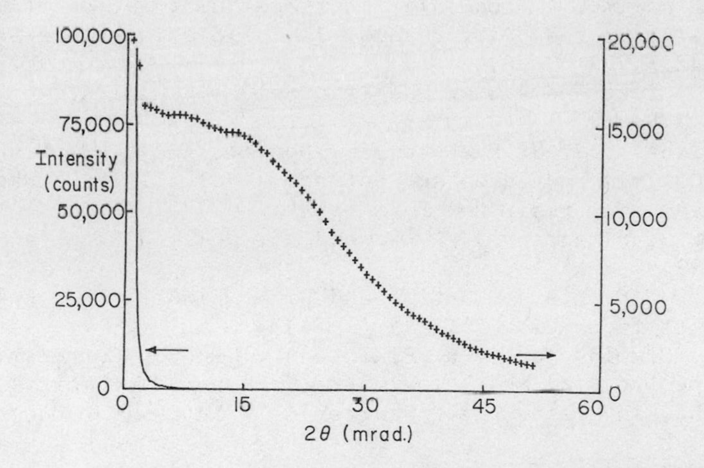

Fig. 7. SAXS curve of sample 4 in the microemulsion state, solid line, and in the molecular solution state ten weeks after preparation, + .

may be reasonable because the number of water molecules per surfactant molecule is only 6 in sample 4.

The W/O microemulsion systems 1, 2, and 3 are fairly monodisperse with the largest droplet sizes observed being only three times the smallest ones. The O/W microemulsion is considerably more polydisperse with the change in size being slightly more than an order of magnitude.

All the microemulsion systems studied did not show any deviations from the spherical shape within the precision of our measurements. All of our stock solutions were accidentally frozen by cooling down to -25°C eight months after their preparation. Upon thawing, all the samples were in two "phases" but after gentle shaking all of them became single "phase".

REFERENCES

1. J. H. Schulman, W. Stoechenius, and L. M. Prince, J. Phys. Chem., 63, 167 (1959).
2. J. H. Schulman and D. P. Riley, J. Colloid Interface Sci., 3, 383 (1948).
3. H. F. Eicke and V. Arnold, J. Colloid Interface Sci., 46, 101 (1974).
4. P. Ekwall, L. Mandell, and K. Fontell, J. Colloid Interface Sci., 33, 215 (1970).
5. A. W. Adamson, J. Colloid Interface Sci., 29, 261 (1969).
6. W. Gerbacia and H. L. Rosano, J. Colloid Interface Sci., 44, 242 (1973).
7. H. F. Eicke and J.C.W. Sheperd, Helvet. Chim. Acta, 57, 1951 (1974).
8. M. Clausse, R. J. Sheppard, C. Boneit, and C. G. Essex, "Colloid and Interface Science," Vol. II, M. Kerker, ed., Academic Press, New York, p. 233, 1976.
9. J. H. Schulman and J. A. Friend, J. Colloid Sci., 4, 497 (1949).
10. D. Attwood, L.R.J. Currie, and P. H. Elworthy, J. Colloid Interface Sci., 46, 249 (1974).
11. H. Z. Cummins and E. R. Pike, eds., "Photon Correlation Spectroscopy and Velocimetry," Plenum, New York, 1977.
12. B. Chu, in "Laser Light Scattering," Academic Press, New York,
13. D. E. Kopell, J. Chem. Phys., 57, 4814 (1972).
14. B. Chu, E. Gulari, and E. Gulari, Physica Scripta, 19, 476 (1979).
15. E. Gulari, E. Gulari, Y. Tsunashima, and B. Chu, J. Chem. Phys., 70, 3965 (1979).

16. W. Gerbacia, H. L. Rosano, and J. H. Whitham, in "Colloid and Interface Science," Vol. II, M. Kerker, ed., Academic Press, New York, p. 245, 1976.
17. M. B. Huglin, ed., "Light Scattering from Polymer Solution," Academic Press, New York, 1972.
18. F. C. Chen, A. Yeh, and B. Chu, J. Chem. Phys., 66, 1290 (1977).
19. O. Kratky, Pure and Applied Chem., 12, 482 (1966).
20. M. Kerker, ed., "The Scattering of Light and Other Electromagnetic Radiation," Academic Press, New York, p. 418, 1969.
21. G.D.J. Phillies, G. B. Benedek, and N. A. Mazer, J. Chem. Phys., 65, 1883 (1976).
22. J. L. Anderson and C. C. Reed, J. Chem. Phys., 64, 3240 (1975).
23. A. R. Altenberger, Macromolecules, 15, 269 (1976).
24. N. A. Mazer, G. D. Benedek, and M. C. Carey, J. Phys. Chem., 80, 1075 (1976).
25. C. Tanford, ed., "Physical Chemistry of Macromolecules," John Wiley, New York, p. 372, 1961.

DIELECTRIC AND CONDUCTIVE PROPERTIES OF WATER-IN-OIL TYPE MICROEMULSIONS

M. Clausse,* C. Boned, J. Peyrelasse and B. Lagourette;
V.E.R. McClean and R. J. Sheppard

Université de Pau et des Pays de l'Adour, Institut Universitaire de Recherche Scientifique, Departement de Physique, Laboratoire de Thermodynamique, Avenue Philippon, F-64000, Pau, France. University of London, Queen Elizabeth College, Department of Physics, Campden Hill Road, Kensington, London W8 7AH, Great Britain

The conductive and dielectric behavior of water-in-oil, (w/o), type transparent isotropic systems using either non-ionic surfactants or a combination of an ionic surfactant with a medium chain-length alcohol used as the cosurfactant was investigated between 400 Hz and 2 GHz or so.

For both types of microemulsions, dielectric relaxations of the Cole-Cole type were in evidence, along with conduction absorptions. The features of the dielectric relaxations were found to depend strongly upon the composition and, even when non-ionic surfactants were used, upon the temperature.

A temperature study of the low-frequency conductivity of water-in-undecane microemulsions using as the surfactant a blend of octylphenylether polyoxyethylenes showed that, as the temperature increases, the conductivity decreases down to a minimum whose temperature coordinate was found to be equal systematically to the solubilization-end temperature corresponding to the system water content. Water-in-dodecane systems exhibited the same behavior.

As concerns microemulsions using ionic surfactants and alcohols, a systematic study performed over $I_{(w/o)}$, the transparent isotropic w/o solubilization area of the pseudo-ternary phase diagram of water-hexadecane systems using potassium oleate and

*To whom correspondence should be addressed.

1-hexanol, put into evidence drastic variations of both the low frequency conductivity and permittivity with system composition and allowed to partition $I_{(w/o)}$ into three adjacent sub-areas, each of which corresponding to a typical dielectric and conductive behavior. The conductivity of these systems proved to be not of the percolative type while that of similar systems incorporating 1-pentanol instead of 1-hexanol was found to fit in with the Percolation and Effective Medium theories.

All these results suggest that the state of solubilization of water is not even throughout the w/o transparent isotropic region and that structural changes occur as the composition varies.

INTRODUCTION

Microemulsions are of great interest to surface and colloid physical chemists, and also because of their numerous potential applications in many industrial fields, (1-6), such as, the manufacturing of health and food specialties, beauty products, paints, varnishes and coatings, the improvement of catalysis processes and the technology of tertiary oil recovery, the microemulsions have been submitted, during the past forty years, to increasingly intensive investigations using more and more varied and sophisticated techniques. Most of the results of these investigations, a great number of which have been published in recently issued books and symposium proceedings (5,7-8), converge to describe microemulsions as resulting from the dispersion of either water in a hydrophobic hydrocarbon or the reverse, in presence of a suitable surfactant or surfactant combination. The disperse phase is considered as consisting of either water or oil (hydrocarbon), minute spherical droplets surrounded by surfactant shells, the commonly reported droplet size range being 50Å - 500Å. In the present state of knowledge, many important points are still to be elucidated as concerns microemulsion formation, stability and structure, in connection with the chemical nature of the oil type phases and of the surface active agents. An interesting review on the subject has been contributed by Shah and co-authors (9), who gave also an account of historical developments in microemulsion science.

Among the numerous techniques available to study the physicochemical properties of microemulsions, conductometry and dielectrometry can provide valuable information as to their structural and phase behavior (10). That is particularly true as concerns w/o type microemulsions, more specifically those using a combination of an ionic surfactant such as an alkaline metal soap with a cosurfactant such as a medium chain length alcohol. These microemulsions have been recognized by Friberg and co-workers (11-14), as being directly related to the ternary solutions formed by mixing in adequate proportions water, an alcohol and an alkaline metal soap such as potassium oleate. Moreover, Sjöblom and Friberg (14) proved through light scattering, density and electron microscopy experiments, that these systems undergo structural changes as the water content is raised in them. Initially, they are in fact hydrated soap molecular dispersions in the hydrocarbon-alcohol phase, their association leading to solutions of swollen inverse micelles taking place at higher water contents. Similar conclusions as to the existence of association processes in microemulsions were arrived at also independently by Clausse and co-workers (15-19) for other systems than those investigated by Sjöblom and Friberg. Aggregation phenomena in ternary compounds analogous to microemulsions have been observed as well by Smith and others (20,21) on the water/2-propanol/n-hexane system, by Rouviere and co-workers (22,23) on the water/AOT/n-heptane or n-decane systems, and by Eicke and collaborators (24-29) who performed numerous and thorough studies of water-apolar solvent systems incorporating either AY or AOT aerosol as the surface-active agent.

In the following, after a brief account of the data available in the literature as concerns microemulsion conductive and dielectric properties, results will be reported and analyzed that show how structural transitions in the transparent isotropic water-in-oil solubilization area can be put into evidence by means of conductometry and dielectrometry. Mention will be made also of the occurrence, in certain w/o microemulsion systems, of percolative conduction phenomena (30) that appear to depend upon the nature of the alcohol used as the cosurfactant, for a given hydrocarbon (31).

HISTORICAL BACKGROUND

Earlier conductivity measurements, such as those performed by Schulman and McRoberts (32) on systems involving water and benzene or paraffin oil, the surfactant being sodium oleate and the co-surfactant a cyclic alcohol, were aimed at determining the nature of the continuous phase and detecting phase inversion phenomena in microemulsions. This approach similar to the one followed in the case of coarse emulsions (33,34) appeared soon to be an over-simplified one as far as microemulsions are concerned. Schulman and co-workers (35,36) remarked that even systems designated as being of the w/o type could be fairly conductive and called them anomalous systems. Later, Dreher and Sydansk (37) and Jones and Dreher (38) reported that isolated conductance experiments could yield ambiguous results as to the nature of microemulsion external phase.

In a series of papers published a few years ago (39-41), Shah and co-workers reported data gained from correlated conductivity, birefringence, interfacial tension and NMR investigations performed on water-hexadecane systems using potassium oleate and hexanol as the surface-active agents. These authors concluded that, upon increasing the water content, these systems undergo a sequence of drastic structural changes, in accordance with the following pattern which describes in fact a phase inversion mechanism: transparent dispersions of water-rich spherical droplets in an oil-phase, turbid birefringent systems of aqueous cylinders dispersed in oil, turbid birefringent lamellar structures of water, oil and surfactants, transparent dispersions of oil-rich spherical droplets in a water-phase and eventually coarse oil-in-water type emulsions. Similar conclusions were reached by Clausse and co-workers (15) who showed that transparent isotropic water-in-hexadecane systems exhibit Cole-Cole type dielectric relaxations whose intensity is strongly dependent upon the water content, the low frequency permittivity exhibiting, as the low frequency conductivity does, a divergent behavior as the water mass fraction, P_w, approaches the critical value, p_c, corresponding to the first transparent-to-turbid transition. A similar behavior was put into evidence later by Senatra and Giubilaro (46,47), on water-dodecane systems using also potassium oleate and hexanol as the surface-active agents. In their preliminary paper on the dielectric behavior of transparent water-in-hexadecane systems, Clausse and co-workers (15) stressed that the theoretical models designed to account for emulsion electrical properties (48) cannot be applied in a straightforward way to the microemulsion case and suggested that the anomalies observed (15,16) could be related to the existence of different solubilization modes of water in hexadecane, in presence of potassium oleate and hexanol. In order to ascertain this suggestion, systematic low frequency conductivity and permittivity determinations were carried out over the entire transparent isotropic water-in-oil type solubilization area of the system pseudo-ternary phase diagram. The results of this study which have been reported at a recent symposium (17) will be analyzed in following sections of the present contribution. It is worth mentioning here that Smith and others (20) put into evidence the existence of different solubilization modes of water in hexane, in presence of short chain length alcohols with no or some hexadecyltrimethylammonium bromide or perchlorate added, by conductance and ultracentrifugation experiments. In a more recent study (21) conductometry was used in conjunction with NMR to investigate the influence of added sodium chloride upon the location and stability of the different solubilization sub-areas of the phase diagram.

Another promising issue of electrical studies is the observation made by different groups of workers (30,31,42,43) that the alcohol chain length has a drastic influence upon the conductive behavior of w/o microemulsions using alkaline metal soap/alcohol combinations as surface-active agents. For instance, Shah and co-workers (42,43) reported that water-in-hexadecane systems involving potassium oleate and either 1-hexanol or 1-pentanol exhibited strikingly dissimilar conductive properties, as concerns the conductivity values which were found to be much larger when 1-pentanol was used and also as concerns the features of the conductivity variations upon increasing water content. On the basis of these observations and of correlated NMR data, these authors proposed to distinguish microemulsions (hexanol as the cosurfactant) from co-solubilized systems (pentanol as the cosurfactant). It will be shown further on that the conductive behavior of the so-called co-solubilized systems can be depicted by means of Percolation and Effective Medium phenomenological theories (49-52) in a way similar to that followed in the case of w/o microemulsions involving toluene, 1-butanol and potassium oleate (30). Percolative conduction has also been reported recently in the case of water-in-cyclohexane microemulsions involving sodium dodecylsulfate and 1-pentanol as the surface-active agents (53). Shah and co-workers also found that the features of the dielectric behavior of microemulsion type systems depend fairly upon the nature of the surface-active agents. Permittivity experiments performed in conjunction with spin-label studies on water-hexadecane systems using sodium stearate as the surfactant and either pentanol, hexanol or heptanol as the cosurfactant led these authors to suggest that the water/oil interface is affected by alcohol chain length as concerns ionization and interfacial polarization (45). In another paper (44) Bansal and Shah reported that the changes observed in the dielectric relaxations exhibited by 5% sulfonate aqueous solutions with 3% isobutanol added reflected fairly well the surfactant formulation modifications when ethoxylated sulfonate was progressively substituted for petroleum sulfonate. The dielectric relaxation increment and critical frequency variations were ascribed to micellar charge increases.

Related works that deserve attention are those devoted to the dielectric properties of ternary micellar solutions build up with water, a hydrocarbon and either AY or AOT aerosols (25,54). For instance, Eicke and Shepherd (25) suggested that the non-linear variations of the dielectric relaxation increment and the sudden increase in conductivity observed upon increasing the water content in water/AY/benzene systems could be interpreted in terms of association process and micelle conformational change.

However interesting and significant the results mentioned above may be, most of them are somewhat lacking of generality since they were derived from experiments limited to some microemulsion compositions. With this in mind, the authors of the present study decided to perform systematic conductivity and permittivity determinations over the entire transparent isotropic water-in-oil type solubilization area of the phase diagram of some microemulsion systems, with a view to gain as thorough as possible information about the structural behavior of the systems.

MATERIALS AND METHODS

a) Sample preparation: Microemulsion samples incorporating non-ionic surfactants were prepared from distilled water and either undecane or dodecane from Fluka A.G. ("Purum" grade). The surfactant used was obtained by blending 10% (w/w) Octarox 1 and 90% (w/w) Octarox 5, which are octylphenylether polyoxyethylenes of different chain lengths supplied by Montanoir (France). Portions of this surfactant blend were mixed at room temperature with either undecane or dodecane so as to obtain "oil-plus-surfactant" phases containing a predetermined surfactant amount and from which microemulsion samples were made by adding the required amount of water and stirring the mixture by means of a magnetic agitator. In order to determine the realm of existence of the transparent isotropic water-in-oil type systems (i.e., $I_{(w/o)}$ region) and its boundaries, namely the haze curve and the solubilization-end curve, the sample phase behavior was studied under conditions of slow heating or cooling, in accordance with the method described by Shinoda and others (55-59).

The materials used to prepare microemulsions incorporating ionic surfactant were distilled water and "Baker Grade" hexadecane, dry pulverulent potassium oleate from Fluka A.G. and either "Baker Grade" 1-hexanol or 1-pentanol. The samples were obtained by stirring together the four constituents whose proportions had been predetermined. All the experiments were performed on systems in which the mass ratio of surfactant to cosurfactant was held equal to 3/5, this value having been chosen as a compromise for the sake of comparison with data available in the literature for similar or close systems. The main solubility areas existing within the pseudo-ternary phase diagram, in particular $I_{(w/o)}$, the transparent isotropic water-in-oil type solubilization domain were delineated through visual observations coupled sometimes with viscosity and dilution tests. For both types of systems, sample mass fractions were determined from weight measurements made by means of a precision balance. Microemulsion stability was checked over several weeks by submitting to regular observations test samples stored at a fixed temperature.

b) Dielectric measurements: In the case of microemulsions using non-ionic surfactants, the relative complex permittivity, $\varepsilon^* = \varepsilon' - j\varepsilon''$ with $j = \sqrt{-1}$, was determined up to 3MHz, with an uncertainty of $\pm$ 0.01 on both ε' and ε'', by means of Schering bridges (General Radio 716C and 716CS1), used in conjunction with a General Radio 1232A detector unit and either a General Radio 1210C or 1211C oscillator, depending on the frequency range investigated. The test cells were Ferisol CS 601 temperature-controlled cylindrico-conical capacitors especially designed for liquid studies. Measurements were also made at 2100MHz by means of a Narda 231N coaxial line powered with a Ferisol OS 301A oscillator connected to a Ferisol SCF 201 power supply unit.

Microemulsions incorporating ionic surfactant were studied over the frequency range [5kHz-100MHz] by means of three admittance bridges, namely a Boonton 75C bridge between 5 and 500 kHz, a Wayne-Kerr B201 bridge between 200kHz and 2MHz and a Boonton 33A bridge between 1 and 100MHz. The measuring cells employed were temperature-controlled capacitors whose solid silver plate electrodes, roughened to decrease electrode polarization, are held parallel to each other by means of a rectangular Perspex spacer containing a cylindrical cavity acting as the sample holder. Microemulsions are fluid enough to allow the cavity to be filled by means of a syringe introduced into a small conduit drilled through the spacer, parallel to the electrode surfaces. Sample extraction is performed through a similar conduit, this twin-hole design being quite convenient for the cleaning and drying of the sample holder between two experiments. Within the frequency range [300-2000MHz], permittivity determinations were carried out by using an automated coaxial line system designed by Sheppard and Grant (60,61). The experimental procedures concerning the use of the bridges and of the coaxial line system have been fully reported in earlier papers (60-64). With these instruments, the permittivity can be determined with a maximum uncertainty of ± 0.2 up to 50MHz, of ± 0.5 at 100MHz, and of ± 0.2 between 300 and 2000MHz.

The low frequency conductivity σ_ℓ was measured for both types of microemulsions by means of Mullard cells (Philips) used in conjunction with either a General Radio 1680 automatically balancing admittance bridge working at 400Hz and 1KHz or a semi-automatic Wayne-Kerr B331 precision conductance bridge working at an angular frequency of 10^4 rad/s. Using the values of the low frequency conductivity, it was possible to compute the values of ε''_R, the dielectric relaxation loss factor, by subtracting the conduction contribution ε''_C from the global loss factor ε''. The uncertainty on σ_ℓ was estimated to be less than 1%. The sample temperature was recorded by means of calibrated thermocouple or thermistor devices, with an uncertainty of ± 0.2°C.

RESULTS

MICROEMULSION PHASE DIAGRAMS

a) <u>Microemulsions using non-ionic surfactants</u>: Figure 1 shows a typical water content versus temperature phase diagram obtained for water-in-undecane systems with 12% surfactant incorporated in the hydrocarbon phase. Such diagrams were delineated by determining, for different values of the water mass fraction p_w, the temperatures corresponding to the haze point and to the solubiliza-

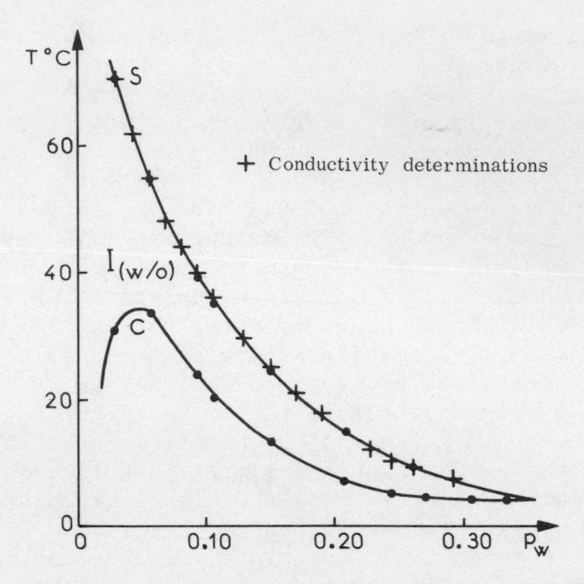

Fig. 1. Water content versus temperature phase diagram of water-undecane systems with 12% (w/w) non-ionic surfactant added to the hydrocarbon phase. $I_{(w/o)}$ is the realm of existence of stable transparent isotropic water-in-undecane microemulsions.

tion-end point. Thus it was possible to draw the haze curve C and the solubilization-end curve S. Similar diagrams were obtained for water-dodecane systems.

The general characteristics of the plots are consistent with those reported by several authors for this type of compound (55-59). All over $I_{(w/o)}$, water solubilizes in the hydrocarbon phase so as to form transparent isotropic w/o type fluid media that remain stable during several weeks when stored at adequate temperatures. These systems will be referred to as water-in-undecane or water-in-dodecane microemulsions. If such a microemulsion, characterized with a fixed value of p_w, is heated above its solubilization-end temperature, turbidity is observed and a separation process takes place that leads to the splitting of the system into two distinct phases. If the microemulsion is cooled down past its cloud point, it becomes hazy and eventually undergoes phase separation as well.

b) Microemulsions using soap-alcohol combinations: Water-hexadecane systems using potassium oleate and 1-hexanol combined in the 3/5 mass ratio exhibit two areas of transparency (Figure 2) that are disjoined and separated by a region within which exist turbid phases of high viscosity, in agreement with previous observations reported by Shah and co-workers (39-41).

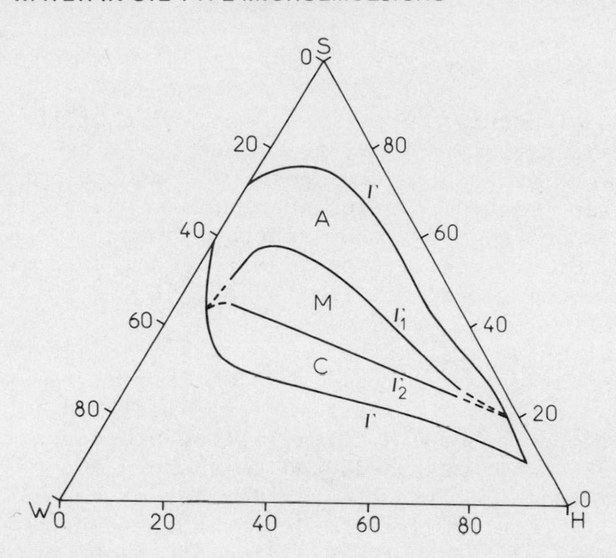

Fig. 2. Pseudo-ternary mass phase diagram of water-hexadecane systems using potassium oleate and 1-hexanol as surface-active agents. Potassium oleate to hexanol mass ratio equal to 3/5. Curve Γ represents the border of the transparent isotropic water-in-oil type solubilization area. Temperature: T = 25°C.

The main transparent area, outlined by curve Γ, represents the realm of existence $I_{(w/o)}$ of water-in-oil type systems. As shown in Figure 2, $I_{(w/o)}$ does not adjoin the no-water side of the pseudo-ternary phase diagram since a minimum amount of water is required to solubilize an ionic surfactant such as a soap in a alkane-alkanol mixture (65). On the contrary, $I_{(w/o)}$ does reach to the no-oil side of the pseudo-ternary phase diagram, which is an indication that the water-in-hexadecane solubility area is connected to the region of inverse micelles existing in the water/soap/alcohol ternary phase diagram. This result conforms to the general statement made by Friberg and co-authors as to the continuity between w/o type microemulsions and inverse micellar solutions of hydrated soap in alkanols (11-14).

As concerns stability, samples with compositions falling within $I_{(w/o)}$ did not exhibit, when stored at room temperature during several weeks, any noticeable changes, as proved from repetitive visual observations and conductivity and permittivity tests. This behavior is remarkably different from that of samples with compositions falling outside $I_{(w/o)}$, though close to curve Γ, in which phase separation takes place after periods ranging from a few hours to several days. Details about these phenomena can be found in references (15) to (17).

DIELECTRIC AND CONDUCTIVE PROPERTIES OF MICROEMULSIONS

a) General features: Experiments showed that, for any composition belonging to $I_{(w/o)}$, systems incorporating either nonionic surfactants or soap-alcohol combinations exhibit dielectric relaxations along with conduction absorptions which are preponderant in the lower frequency range. By subtracting the conduction contribution, ε''_C, determined through low frequency conductivity measurements, from the global loss factor ε'', it was possible to compute at every frequency the value of the dielectric loss factor, ε''_R, and to plot it versus that of ε' so as to obtain Cole-Cole diagrams characteristic for microemulsion dielectric relaxations. Typical Cole-Cole plots thus obtained are displayed in Figures 3 and 4. By using computerized curve-fitting procedures derived from the method proposed by Marquardt and co-authors (66,67), it was possible to check that the frequency variations of the microemulsion relative complex permittivity are depictable by means of the Cole-Cole formula as modified by the addition of an ohmic term:

$$\varepsilon^* = \varepsilon_h + \frac{\varepsilon_\ell - \varepsilon_h}{1 + (j\nu/\nu_c)^{(1-\alpha)}} + \frac{\sigma_\ell}{2\pi j\nu\varepsilon_o} \qquad (1)$$

where σ_ℓ represents the low-frequency conductivity and ε_h and ε_ℓ the relative permittivity respectively at the high frequency end and at the low frequency end of the dielectric relaxation frequency range, ε_o being the absolute permittivity of free space. ν is the frequency of the applied electromagnetic field and ν_c the critical frequency, α being the frequency spread parameter.

Since the constituents used do not exhibit intrinsic dielectric relaxations in the lower and medium radio-frequency range, the occurrence in w/o microemulsions of dielectric relaxations located below 50 MHz or so is to be considered as related to system local heterogeneities created by the dispersion of water in the oil type phase. On the basis of experimental and theoretical results available for coarse emulsions (10), it may be inferred that the Cole-Cole type dielectric relaxations put into evidence arise from migration (or interfacial polarization) phenomena connected to the existence of water/oil interfaces. In that respect, it is gratifying to observe that the relaxation frequency range of microemulsions using ionic surfactant is located at higher frequencies than that of microemulsions using non-ionic surfactant (Figures 3 and 4) and that the critical frequency ν_c decreases when the water content increases (Figure 4). However, other features do not comply generally with the classical models (48,68) that have proved suitable to depict emulsion dielectric behavior (10). In particular, Peyrelasse and co-workers (16,19) obtained rather surprising results as concerns the variations of several dielectric parameters with temperature. For instance, it was found that as the temperature increases from 20 to 40°C, the

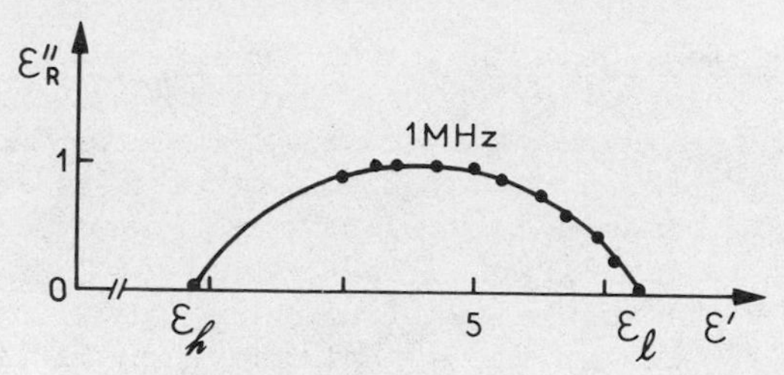

Fig. 3. Cole-Cole plot characteristic for the dielectric relaxation exhibited by microemulsions using non-ionic surfactants. Mass proportion of surfactant in the undecane phase: 12%. Mass fraction of water in the system: $p_w = 0.093$. Temperature: $T = 36°C$.

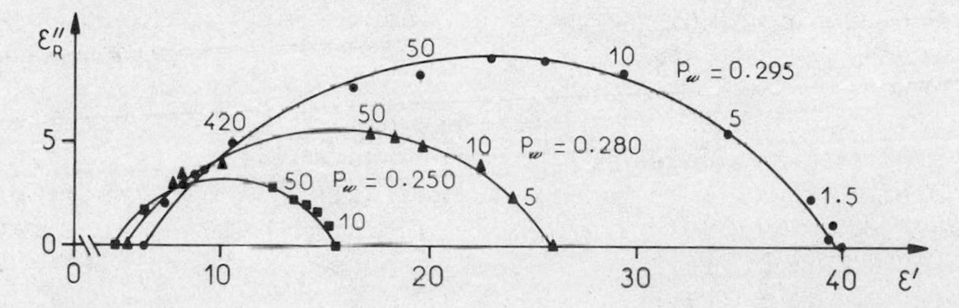

Fig. 4. Cole-Cole plots characteristic for the dielectric relaxations exhibited by water-in-hexadecane microemulsions using potassium oleate and 1-hexanol. Potassium oleate to 1-hexanol mass ratio equal to 3/5. Combined surface active agent to hexadecane mass ratio equal to 2/3. p_W represents the mass fraction of water. Temperature: $T = 25°C$. The frequencies are indicated in MHz.

dielectric relaxation of water-in-hexadecane microemulsions shifts first towards lower frequencies and then back towards higher frequencies whilst the low-frequency permittivity ε_ℓ first increases, reaches a flat maximum centered around 27°C and then decreases (16). Moreover, as it will be reported later, w/o microemulsions, unlike coarse w/o type emulsion systems, are fairly conducting even at low water contents and display upon varying composition very peculiar variations in their low frequency permittivity ε_ℓ and therefore in the dielectric relaxation increment $(\varepsilon_\ell - \varepsilon_h)$, and in their low frequency conductivity σ_ℓ as well. These phenomena suggest that the model of isolated shell-covered aqueous spheres dispersed within an oil-rich phase does not hold all over $I_{(w/o)}$, which is an indication that the state of water "solubilization" is not even throughout $I_{(w/o)}$, owing to association and aggregation processes.

b) Microemulsions using non-ionic surfactants: Systematic conductance measurements (19) showed that the low frequency conductivity of a water-in-undecane or water-in-dodecane microemulsion characterized with a fixed value of the water mass fraction p_w decreases down to a minimum upon varying the temperature T from the cloud point to the solubilization-end point. A further temperature increase past the solubilization-end point induces a conductivity increase. It was found that the temperature at which the conductivity is minimum, T_m, is equal to the solubilization-end temperature, whichever the value of p_w. On the basis of this result, solubilization-end curves have been redetermined quite accurately, as illustrated by Figure 1. Parallel to the σ_ℓ decrease, it was observed that upon increasing T, the low frequency permittivity ε_ℓ decreases as well while the critical frequency ν_c increases, ε_h, the high frequency permittivity, and α, the frequency spread parameter, undergoing no meaningful variations. The influence of the water mass fraction p_w upon the dielectric parameters proved to be rather complex because of the correlation between ε_ℓ and σ_ℓ resulting from the tight p_w-T connection.

c) Microemulsions using soap-alcohol combinations: The pseudo-ternary phase diagram of water-hexadecane systems using potassium oleate and 1-hexanol combined with the mass ratio 3/5 having been determined, permittivity and conductivity measurements were performed all over $I_{(w/o)}$, the transparent isotropic w/o type solubilization area (Figure 2).

Figure 4 shows typical variations of the dielectric relaxation with water content as recorded along a line stretched across $I_{(w/o)}$ and directed towards the 100% water vertex of the pseudo-ternary phase diagram, that is for systems characterized with a fixed ratio of combined surface-active agents to hexadecane and enriched gradually with water. While dielectric relaxation phenomena are hardly detectable at low water contents, systems characterized with higher water contents exhibit striking dielectric relaxations, the dielectric increment $(\varepsilon_\ell - \varepsilon_h)$ increasing drastically as p_w approaches the critical value corresponding to the transparent-to-turbid transition. The increase in $(\varepsilon_\ell - \varepsilon_h)$ results from the drastic increase in the low frequency permittivity ε_ℓ whose variations with p_w are plotted in Figure 5a. While at low water contents, ε_ℓ increases slowly and almost linearly with p_w, it displays a divergent behavior in the vicinity of the $I_{(w/o)}$ border line Γ. Similarly, upon increasing p_w, the low frequency conductivity σ_ℓ increases, reaches a maximum, then decreases down to a minimum and eventually follows a sharply ascending branch, as ε_ℓ does (Figure 5b).

With a view to investigate thoroughly the way in which the dielectric and conductive behavior depends upon system composition, systematic determinations of both ε_ℓ and σ_ℓ were made all over

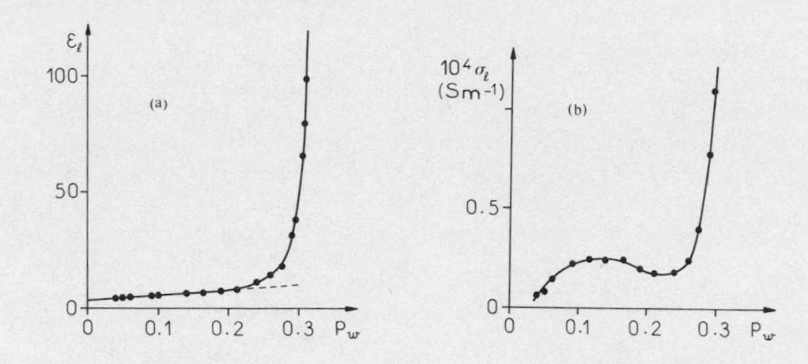

Fig. 5. Variations of the low frequency permittivity ε_ℓ and conductivity σ_ℓ observed upon increasing water content in water-in-hexadecane systems. Specifications identical to those of Figure 4. (a) ε_ℓ versus p_w curve; (b) σ_ℓ versus p_w curve.

$I_{(w/o)}$, according to the experimental procedures indicated thereafter. Runs of experiments were performed along lines parallel to either side of the pseudo-ternary phase diagram, that is for series of systems in which either p_h, the hexadecane mass fraction (Type 1 experiments), or p_w, the water mass fraction (Type 2 experiments), or p_s, the combined surface-active agent mass fraction (Type 3 experiments) were taken as parameters and ascribed fixed values, as sketched on Figures 6a to 6c. By changing step by step the values of either p_h, p_w or p_s, $I_{(w/o)}$ was swept up and down entirely by turns along the three main directions of the pseudo-ternary phase diagram. By combining all the data gained in this way, it was possible to cover $I_{(w/o)}$ with a tight network of permittivity and conductivity values.

Figure 7 shows ε_ℓ versus p_w plots obtained for different values of p_h, the hexadecane mass fraction (Type 1 experiments). For any value of p_h, the curves consist of two branches. The lower branch corresponds to a moderate and almost linear increase in ε_ℓ with p_w, while the upper branch, which is sharply ascending, represents a drastic increase in ε_ℓ upon p_w approaching its critical value located on the $I_{(w/o)}$ border line Γ. The water mass fraction at which both branches are joined has been labelled p_w^i. p_w^i is a decreasing function of p_h. As shown in Figure 8, the low frequency conductivity σ_ℓ undergoes non-monotonous variations as p_w increases. p_h being fixed, σ_ℓ first increases, reaches a maximum to which corresponds the value p_w^c of the water mass fraction, decreases down to a minimum marked with the value p_w^j and then increases drastically as p_w gets closer to its critical value. p_w^c and p_w^j are both decreasing functions of p_h. It was observed that, for a given value of p_h, the values of p_w^j and p_w^i are close to each other, with a discrepancy of less than 0.03 or so. It must be remarked that the features of the ε_ℓ and σ_ℓ curves reported in

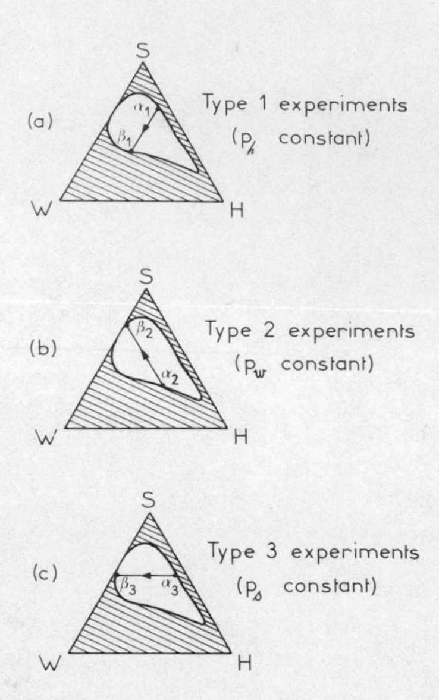

Fig. 6. Schematic pseudo-ternary phase diagrams showing lines along which permittivity and conductivity measurements were performed. (a) Type 1 experiments, p_h constant; (b) Type 2 experiments, p_w constant; (c) Type 3 experiments, p_s constant.

Figures 7 and 8 are similar to those of the corresponding curves in Figure 5. It was checked that the values of p_w^c, p_w^i and p_w^j determined by both methods are consistent.

In Figures 9a and 9b are reported typical curves obtained through Type 2 experiments. For a given value of the water mass fraction p_w, both ε_ℓ and σ_ℓ first decrease to minima, marked respectively with values p_s^i and p_s^j of the combined surface-active agent mass fraction p_s, then increase when the $I_{(w/o)}$ border line Γ is approached. As in the preceding case, it was observed that, for any value of p_w, the values of p_s^i and p_s^j are almost equal. As shown in Figure 10, the ε_ℓ versus p_s curves do not intersect and are regularly ordered, the top curve corresponding to the greatest value of p_w and p_s^i increasing with p_w. Similar features were found for the σ_ℓ versus p_s curves.

No extrema or kinks were found on the ε_ℓ and σ_ℓ versus p_w curves determined through Type 3 experiments, which is quite consistent with the results gained from Type 2 experiments. It can be deduced from Figure 10 that, at a fixed value of p_s, ε_ℓ increases monotonously with the water mass fraction p_w. σ_ℓ exhibits a similar behavior, as illustrated by Figure 11 on which the filled circles represent the variations of σ_ℓ upon water content increasing in water-in-hexadecane systems in which the mass fraction p_s of combined potassium oleate and 1-hexanol was kept equal to 0.40.

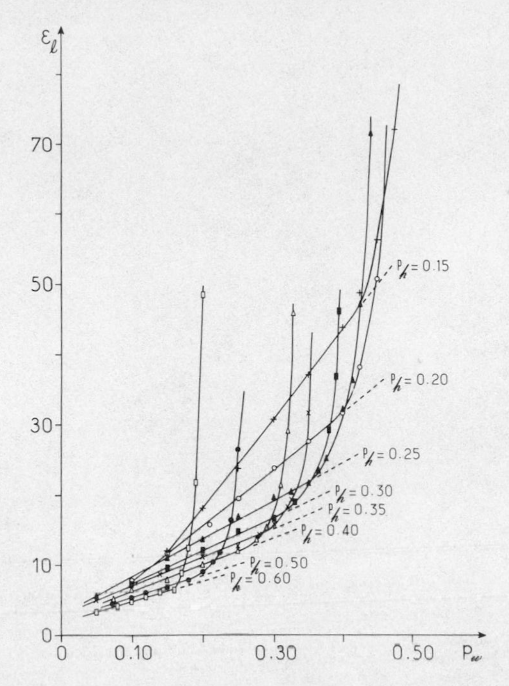

Fig. 7. Variations of the low frequency permittivity ε_ℓ with increasing water mass fraction p_w in water-hexadecane systems using potassium oleate and 1-hexanol combined with the mass ratio 3/5, for different values of p_h, the hexadecane mass fraction. Temperature: T = 25°C.

In the pseudo-ternary phase diagram, the points representative of the system compositions defined by the values of p_w^c, p_w^i, p_w^j, p_s^i and p_s^j are not scattered at random but define two curves, Γ_1 and Γ_2, that divide $I_{(w/o)}$ into three adjacent sub-areas designated A, M and C in Figure 2. Curve Γ_1 is determined from the values of p_w^c that correspond to conductivity maxima as recorded through Type 1 experiments. As shown in Figure 2, for values of p_h greater than 0.15 or so, Γ_1 is a smooth curve stretching across $I_{(w/o)}$ while gradually stepping away from Γ as p_h decreases. For values of p_h smaller than 0.15 or so, Γ_1 exhibits a marked curvature and eventually parallels roughly the no-oil side of the pseudo-ternary phase diagram. As reported in an earlier paper (17), it was found that the values of p_w^i, p_w^j, p_s^i and p_s^j are satisfactorily correlated, which allows to define curve Γ_2 that runs across $I_{(w/o)}$ almost parallel to the lower part of the border line Γ (Figure 2).

Complementary conductance experiments were made with a view to investigate the influence of the cosurfactant upon microemulsion conductive properties. Comparative Type 3 measurements were performed on water-in-hexadecane systems using potassium oleate combined in the 3/5 mass ratio with either 1-pentanol or 1-hexanol.

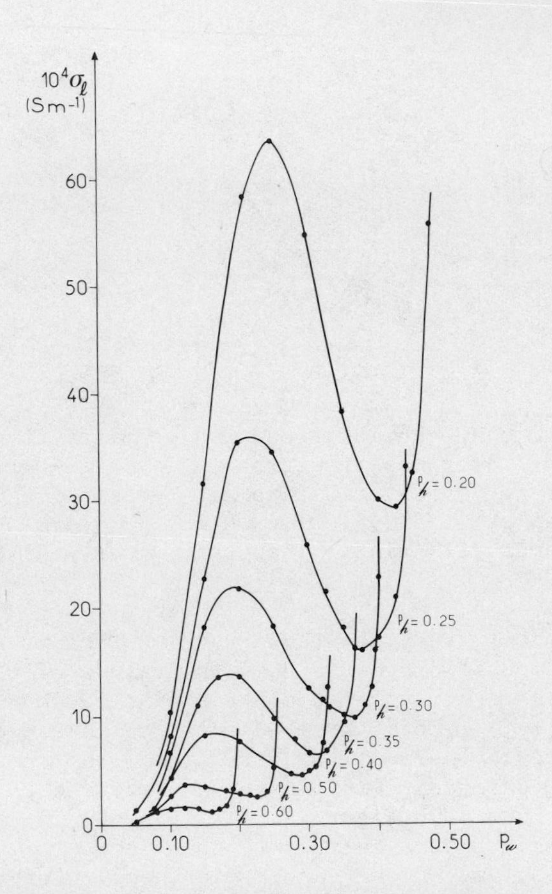

Fig. 8. Variations of the low frequency conductivity σ_ℓ with p_W. Specifications identical to those of Figure 7.

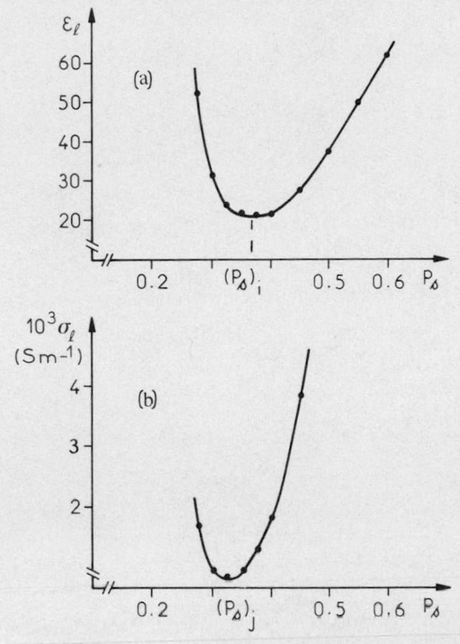

Fig. 9. Variations of the low frequency permittivity ε_ℓ and conductivity σ_ℓ with p_s, p_w being kept equal to 0.35. Other specifications identical to those of Figures 7 and 8. (a) Permittivity curve; (b) Conductivity curve.

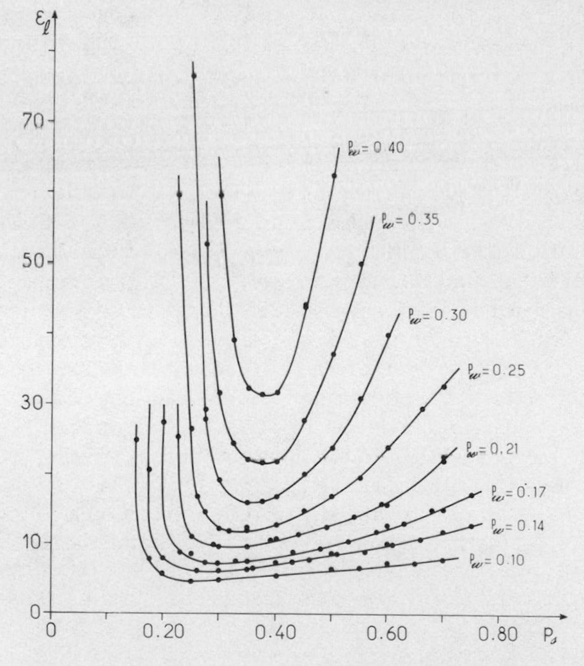

Fig. 10. ε_ℓ versus p_s curves showing the influence of p_w. Specifications identical to those of Figure 9.

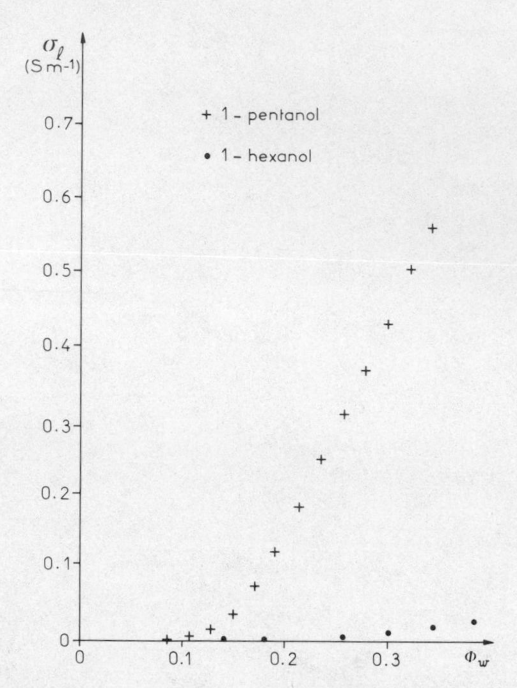

Fig. 11. Comparative conductive behavior of water-in-hexadecane systems using potassium oleate as the surfactant and either 1-pentanol or 1-hexanol. Soap to alcohol mass ratio equal to 3/5. Combined soap and alcohol mass fraction p_s equal to 0.40. Φ_w represents the water volume fraction. Temperature: T = 25°C.

It can be remarked from Figure 11 that the substitution of 1-pentanol for 1-hexanol induces drastic changes in the microemulsions conductive behavior. While the conductivity of 1-hexanol using microemulsions remains relatively small and increases moderately with water content, that of 1-pentanol using microemulsions is much higher and increases drastically upon water content increasing past a critical value. The σ_ℓ versus Φ_w curves recorded on systems incorporating 1-pentanol are quite similar to those obtained by Lagourette and co-workers (30) in the case of water-toluene systems using potassium oleate and 1-butanol and, consequently, can be analyzed in the same way by using Percolation and Effective Medium theories (Figure 12). This result reveals the influence of the cosurfactant nature upon the occurrence in microemulsion type systems of percolative conduction phenomena. In that respect, Clausse and co-workers (31) evidenced the strong influence of alcohol conformation upon the phase behavior and conductive properties of microemulsions using a straight or a branched alcohol belonging to the series of pentanol isomers. In addition, in the case of systems with a unique transparent monophasic domain, it was proved that the arch-like geometry of conductivity plots was characteristic for o/w to w/o microemulsion inversion.

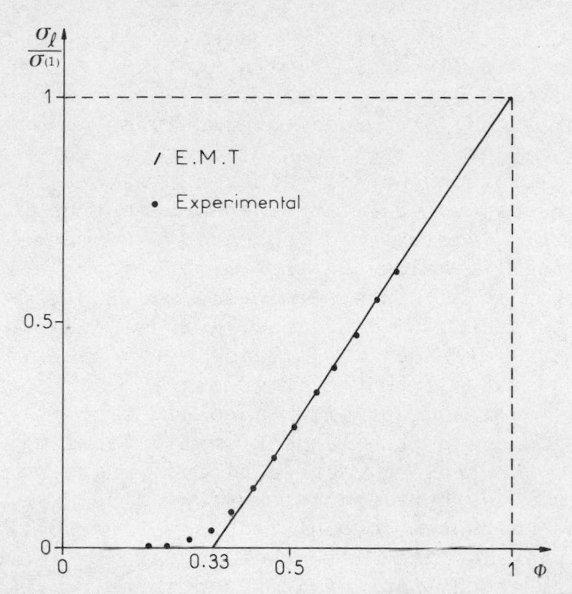

Fig. 12. Reduced conductivity versus disperse volume fraction plot for water-in-hexadecane systems using a potassium oleate and 1-pentanol combination, as obtained from the application of the Percolation and Effective Medium theories. The solid line represents the E.M.T. curve.

DISCUSSION

The division of $I_{(w/o)}$ into three adjacent sub-areas, as revealed from low frequency conductivity and permittivity measurements performed on water-in-hexadecane microemulsions using potassium oleate and 1-hexanol, can be interpreted in terms of conformational and topological changes taking place as the system composition varies. Consequently, each of the sub-areas labelled A, M and C in Figure 2 can be assigned compositions corresponding to distinct modes of water "solubilization" in hexadecane, in presence of potassium oleate and 1-hexanol.

Within A, that is, in the upper part of $I_{(w/o)}$, potassium oleate is in excess compared to the amount of water, and the systems consist of dispersions in the alkane-alkanol phase of hydrated soap molecular aggregates resulting from the preferred association of both water and soap. Thus curve Γ_1 defines compositions corresponding to the onset of the building up of stable water-swollen micelles whose realm of existence is sub-area M. Upon crossing curve Γ_2, the surfactant-to-water ratio becomes too small to secure micellar stability and the inverse micelles tend to coalesce, which results in the formation of micelle clusters, a process that is predictive of system destabilization and transformation into turbid-like structures as Γ is reached to and stepped across.

As reported elsewhere with full details (17), this interpretative scheme is quite consistent with the data gained from conductivity and permittivity measurements. The non-monotonous variations of σ_ℓ with p_w recorded through Type 1 experiments and illustrated by the curves reported in Figure 8 can be explained as follows. The initial conductivity increase with p_w varying up to p_w^c is induced by soap solubilization enhancement upon increasing water content till Γ_1 is reached (sub-area A). Once all the soap is engaged in micelle shells, further water mass fraction increases beyond p_w^c occasion mainly inverse micelle swelling, (sub-area M), and the system conductivity decreases, owing to this dilution type process. The parallel moderate increase in ε_ℓ observed as p_w increases up to p_w^i arises from the progressive enhancement of system global polarization resulting simultaneously from the enrichment with the constituent with the largest static permittivity (i.e., water), and from the growing contribution of migration polarization phenomena due to water/hexadecane interface development (sub-areas A and M). When the system composition is varied so that curve Γ_2 is crossed, further additions of water result in ε_ℓ and σ_ℓ sharp increases because of the formation of non-spherical clusters and conducting paths within the systems. A similar scheme holds for the variations in ε_ℓ and σ_ℓ when the system composition is varied while the mass fraction of water p_w is kept constant (Type 2 experiments). As the proportion of combined surface-active agents is lowered in systems with compositions belonging to sub-areas A and M, ε_ℓ and σ_ℓ decrease, owing to the progressive system enrichment with hexadecane that is the constituent having the smallest permittivity and conductivity. As soon as Γ_2 is reached, ε_ℓ and σ_ℓ increase sharply with p_h because of the spherical to non-spherical topological changes induced in the systems by surface-active agent deficit. The partition of $I_{(w/o)}$ into three adjacent sub-areas, to which are assigned compositions corresponding to molecularly dispersed hydrated soap (sub-area A), water-swollen inverse micelles (sub-area M) and micelle clusters (sub-area C), allow to explain the striking modifications in dielectric behavior observed in water-in-hexadecane systems as the water content is raised in them. At low water contents, that is, for compositions falling in sub-area A, dielectric relaxation phenomena are hardly detectable. Owing to the presence of spherical micelles in systems with compositions belonging to sub-area M, dielectric relaxations arising from migration polarization phenomena are then noticeable, smooth increase of the dielectric increment with p_w being related, as in the case of stable coarse emulsions (10), to the progressive growth of the water/oil interface. By contrast, within sub-area C, the dielectric relaxation intensity is very sensitive to composition and increases drastically when the border line Γ is approached, as illustrated by the Cole-Cole plots reported in Figure 4 and by the ε_ℓ versus p_w curve in Figure 5a. This phenomenon arises from the system migration polarization drastic enhancement induced by micelle clustering and cluster interlinking processes.

Similarly, the dielectric and conductive behavior of either water-in-undecane or water-in-dodecane microemulsions using non-ionic surfactants could be accounted for on the basis of the occurrence within $I_{(w/o)}$, of system conformational changes. In that case, however, the situation is more complicated, owing to the p_w-T correlation. In particular, as suggested already in a previous paper (19) the decrease of σ_ℓ observed upon increasing the temperature T in microemulsions characterized with a fixed value of the water mass fraction p_w could be ascribed to repetitive aggregational processes, the systems becoming the most "emulsion-like" when the solubilization-end curve is reached.

The conclusions arrived at through the present study are quite consistent with those reached by other scientists who used other systems and/or investigation techniques concerning the structural behavior of microemulsion systems of the water-in-oil type. For instance, on the basis of NMR, ultraviolet spectroscopy, light scattering, Fischer titration, ultra-centrifugation and density experiments, Friberg and co-workers (11-14) claimed that w/o type microemulsions using an ionic surfactant such as a soap and a co-surfactant such as a medium chain-length alcohol are structurally identical to inverse micellar solutions obtained from the so-called "structure-forming" components, that is, water, surfactant and co-surfactant (11). Consequently, in the pseudo-ternary phase diagram, the transparent isotropic water-in-oil type solubilization area $I_{(w/o)}$ is the transposition, upon addition of hydrocarbon, of the region of inverse micelles existing in the water-surfactant - co-surfactant ternary phase diagram. In a recent paper (14), Sjöblom and Friberg reported that the intensity of light scattered by ternary systems involving water, potassium oleate and 1-pentanol increases sharply when the water mass fraction exceeds 15% or so. Similar phenomena were observed in microemulsions obtained by adding to the ternary systems different amounts of hydrocarbons such as benzene, decane or phenyldodecane. Density measurements and electron microscopy observations confirmed that, at low contents, the systems consist of molecular dispersions of hydrated soap, association to inverse micelles taking place at higher water contents. The authors reported that the nature of the hydrocarbon has an influence upon the onset of the association process, aromatic hydrocarbons lowering the association concentrations.

A partition of the transparent isotropic water-in-oil type solubilization area has been proposed as well by Smith and others (20,21) who investigated the phase and structural behavior of oil-continuous systems composed of water, hexane, 2-propanol, with or without addition of hexadecyltrimethylammonium bromide or perchlorate. The techniques used were conductometry, ultracentrifugation and, later, NMR. Smith and coworkers (20) put into evidence kinks on conductivity curves as the system composition was varied by increasing 2-propanol content. Plotting in the phase diagram the compositions corresponding to the kinks led to a partition of the transparent w/o type solubilization area into three adjacent sub-areas. From ultracentrifugation experiments, each of these sub-areas was assigned compositions corresponding to either ternary molecular solutions or small aggregates of water and 2-propanol in a hexane-rich medium or w/o microemulsion type systems. The incorporation of surfactant did not affect fundamentally the phenomena and the addition of sodium chloride in the aqueous phase induced a shift in sub-area boundaries, as shown by NMR experiments (21).

Aggregation and association processes have been reported also in the case of ternary systems involving water, a hydrocarbon and either sodium di-2-pentylsulfosuccinate or sodium di-2-ethylhexylsulfosuccinate (AY or AOT aerosol, respectively) (22-29). The behavior of inverse micellar solutions formed with water, AOT and either n-heptane or n-decane was investigated recently by Rouviere and co-workers (22,23) through viscosity, density, diffusivity and Kerr effect experiments. According to these authors, conformational changes can be detected as the water content is raised, which leads to the division of the region of inverse micelles into three distinct sub-areas. Up to a water-to-AOT molar ratio equal to 8 or so, addition of water causes the formation of hydrated surfactant aggregates, as suggested by the sodium self-diffusivity decrease, the viscosity increase and the water proton chemical shift determinations that show water to be only slightly hydrogen-bonded. For water-to-AOT molar ratios ranging from 8 to 35, NMR, viscosity and sodium diffusivity measurements converge to indicate that water is hydrogen-bonded and engaged in inverse micelles whose geometry and mean size varies moderately with the water content. When the water-to-AOT molar ratio exceeds 35, the viscosity increases with the addition of water, which was ascribed by the authors to the deformation of the disperse micelles that were considered as being spheroids since the systems exhibited electrical birefringence. It is worth pointing out that the viscosity increase could be regarded alternatively as resulting from micelle clustering and cluster interlinking, which would be compatible with system electrical birefringence and could explain the sodium diffusivity increase that was not given any interpretation by Rouviere and his co-workers.

In a series of papers (24-29) Eicke and co-workers reported results of thorough studies performed on inverse micellar systems involving hydrocarbons such as benzene or isooctane and surfactants such as AY or AOT aerosols. Dielectric, conductance, ultracentrifugation, NMR, light scattering, fluorescence depolarization and photon correlation spectroscopy techniques were used. The main conclusions arrived at are the following ones, as expressed in (29). For water-to-AOT molar ratios smaller than 10, water-in-isooctane systems consist of dispersions of hydrated soap aggregates behaving like rigid molecules in the temperature range (0–50°C). At temperatures greater than 50°C, the water appears to be not so highly bonded and can be observed as apparently free, owing probably to thermal activation. When the water-to-AOT molar ratio exceeds 10, which seems to be a characteristic limit, the systems incorporate water as a distinct phase, with a surfactant monolayer separating it from the continuous hydrocarbon phase. Discussing the effect of temperature upon system behavior and the reliability of the equipartition model, Zulauf and Eicke (29) proposed two interpretations to account for the discrepancies observed between results gained from ultracentrifugation and from photon correlation measurements. First, a collective analysis of the available data suggests that the microemulsions are polydisperse, the size distribution being of the Gaussian type with a possible 20% width around the radius value predicted from the equipartition model, which implies that the polydispersity increases with the water-to-AOT molar ratio. Alternatively, the authors suggested that micelle radius increase could be ascribed to a coalescence process resulting from constituent partial mutual solubilization at a molecular level and subsequent shell reorganization phenomena. It is worth pointing out that both interpretations do not exclude each other *a priori* and could concur in support of the interpretative scheme designed by the authors of the present paper to account for the dielectric and conductive behavior of w/o type microemulsion systems. In that respect it should be noted that, as reported elsewhere (17), curve Γ_1 defines in the pseudo-ternary phase diagram (Figure 2) for hexadecane mass fractions greater than 0.15 or so, compositions in which the water-to-surfactant ratio remains almost constant and equal to 19 or so, which implies a constant value of 77Å or so for the micelle radius at the aggregate-micelle transition.

By the light of the results reported in the present contribution as to w/o microemulsion type system electrical behavior and of data available in the literature concerning other physico-chemical properties, aggregation and association processes appear to be one of the general clues towards the understanding of w/o type microemulsion formation, stability and structural and phase behavior. In that respect, a very promising issue of conductivity investigations is the discovery, by Lagourette and coworkers (10,30,31), and by Lagues and others (53,69) that certain w/o type microemulsion systems exhibit an electrical conductivity of the percolative type. Percolation is a very general phenomenon encountered either in macroscopic physics, as concerns for instance the displacement of fluids through porous media, or in microscopic physics, as concerns for instance transport properties in media presenting local heterogeneities associated with fluctuations in composition, density or bond configuration. Percolation theory as applied to the case of electrical conduction in locally heterogeneous systems has been given a tractable formulation by different authors (49-52), in particular by Kirkpatrick (49,50) on the basis of numerical simulations on resistor networks. The basic problem is the following one. In a conductor-insulator binary composite, no bulk conduction phenomena can take place as long as the conductor concentration is smaller than a critical value called the percolation threshold. For conductor concentrations exceeding slightly the percolation threshold, the composite exhibits a non-zero conductivity owing to the existence of conducting paths stretching throughout the sample. Then, the conductivity is a sharply increasing function of the conductor concentration. The conductive behavior of the composite can be depicted by the following set of equations (52)

$$\Phi < \Phi^P , \qquad \sigma = 0 \tag{2a}$$

$$\Phi^P < \Phi < 0.4 \qquad \sigma \propto (\Phi - \Phi^P)^{8/5} \tag{2b}$$

$$\Phi > 0.4 \qquad \sigma/\sigma_1 = \frac{3}{2}\left(\Phi - \frac{1}{3}\right) \tag{2c}$$

In these equations, Φ represents the volume fraction of the conductor constituent and Φ^P the critical volume fraction corresponding to the percolation threshold. σ is the binary composite conductivity and σ_1 the conductor constituent conductivity, σ_2 being equal to zero since the other constituent is assumed to be of the insulator type. Equation (2a) states that the percolation phenomenon is a rigorous one, the conductivity being null as long as $\Phi < \Phi^P$. Equation (2b) is a scaling law depicting the conductive behavior in the vicinity of the percolation threshold, the value of the critical exponent γ being 1.6 to within $\pm$ 0.2. Equation (2c) expresses the composite conductivity dependence upon conductor concentration beyond the percolation threshold. Equation (2c) is a simplified form, valid in the case of conductor-insulator mixtures, of a more general equation derived in different ways by Bruggeman (70), Bottcher (71) and Landauer (72) and known as the Effective Medium Theory, (E.M.T.), formula:

$$\frac{\sigma - \sigma_2}{3\sigma} = \frac{\sigma_1 - \sigma_2}{\sigma_1 + 2\sigma}\,\Phi \tag{3}$$

It must be remarked that the simplified E.M.T. equation (2c) can represent a rigorous percolation regime characterized with the critical value $\Phi_c = 0.33$. In fact, numerical simulation data and experimental results show that Φ_c does not represent an actual percolation threshold in the case of three-dimensioned systems. Another remark is that the 0.4 volume fraction value appearing in the inequalities at the left side of equations (2b) and (2c) is only an indicative one and that the validity domains of (2b) and (2c) do overlap. As concerns Φ^P, different values ranging from 0.15 or so to 0.29 or so have been proposed for three-dimensioned systems, depending upon their internal configuration. Details on the subject can be found in references (49) to (52).

Studying the electrical behavior of water-in-toluene microemulsion systems using potassium oleate and butanol, Lagourette and co-workers (10,30) found that the dependency of the low frequency conductivity σ_ℓ upon increasing water content in systems characterized with a fixed mass fraction p of combined surface-active agents was depictable by using the scaling and E.M.T. formulae (2b) and (2c). The conductivity curve fitting procedure used showed that 0.29 could be retained as the value of the percolation threshold Φ^P. It is worth mentioning that, according to Kirkpatrick (50), 0.29 represents the percolation threshold (in terms of volume fraction) of a continuum percolation model in which the percolation sites are assumed to be surrounded by identical spheres permitted to overlap and with centers randomly distributed, which is a geometrical situation particularly suitable to depict microemulsion conformation and internal interactions. Comparing the value 0.29 with the value of the critical water volume fraction yielded 1.18 for the ratio (a/a_c) of the droplet conductive radius, a, to the water core radius, a_c. This figure is in excellent agreement with the 1.17 value reported by Lagües and co-authors (53) who determined it through neutron scattering experiments performed on water-in-cyclohexane microemulsions using sodium dodecylsulfate and 1-pentanol.

As shown in Figure 11, the conductivity σ_ℓ of water-in-hexadecane microemulsions using 1-pentanol exhibits variations with the water volume fraction Φ_w that are characteristic for a percolative behavior. Figure 12 shows the dependency of σ_ℓ upon Φ, the volume fraction of the disperse phase (water plus combined surfactant membrane), as resulting from the application of the Percolation and E.M.T. formulae. It is readily seen from Figure 12 that the experimental data are fitted quite well by equation (2c), for Φ ranging from 0.4 or so up to 0.73 or so, this latter value being almost equal to the limit corresponding to the maximum packing of identical spheres, namely $(\pi\sqrt{2})/6 \simeq 0.74$. An interesting consequence of this result is the following one. Since σ_1, the internal phase conductivity, can be identified with $\sigma(1)$, the value of σ_ℓ obtained by extrapolating at $\Phi = 1$ the conductivity curve, it appears that σ_1 does not depend upon water content although the surfactant proportion is kept constant in the systems. This is an indication that the conductive behavior of such percolating microemulsions does not reflect a volume conduction mechanism but more likely, as suggested by Lagourette and co-workers (30), and Lagues and co-workers (53), an interfacial one related to the number of ions available at the water/oil interface (31). As illustrated by the characteristic toe at the bottom of the diagram reported in Figure 12, the variations of the conductivity σ_ℓ in the vicinity of the percolation threshold are satisfactorily depicted with the scaling formula (2b) in which $\Phi^P = 0.29$. Below Φ^P, σ_ℓ exhibits very low values but is not strictly null as it must be in the case of a rigorous percolation regime. This residual conductivity could result from the contribution of electrophoresis type phenomena affecting the disperse aqueous globules at low water contents (16,30,53).

Topological aspects of percolation phenomena occurring in microscopically inhomogeneous media have been given schematic descriptions, for instance by Kirkpatrick (49,50), and by Cohen and co-authors (51,52). In a conductor insulator binary composite, conducting clusters remain isolated one from another for values of Φ, the conductor volume fraction, smaller than the percolation threshold Φ^P, which inhibits electrical conduction throughout the sample. Upon enriching the system with the conductor component, clusters connect progressively one with another until the formation, at $\Phi = \Phi^P$, of a conducting path stretched across the sample that acquires then a non-zero conductivity. With Φ increasing beyond Φ^P, the cluster association process goes on and the system conductivity increases in proportion to it. As stressed by Lagourette and co-workers (30), this descriptive model seems to be applicable to the case of microemulsions and meets the bicontinuous structure model developed by Scriven (73,74) and the Voronoi tesselation model introduced by Talmon and Prager (75,76).

To add a final touch to this discussion about percolation in microemulsion systems, it is worth pointing out that the occurrence of percolative conduction phenomena in microemulsions appears to depend strongly upon the chemical nature of the constituents (31). The influence of the cosurfactant is strikingly illustrated by the comparative plots in Figure 11 showing the discrepancies existing between the conductive properties of water-in-hexadecane microemulsions incorporating 1-pentanol and of water-in-hexadecane microemulsions incorporating 1-hexanol. This phenomenon has been put into evidence as well by Shah and co-workers (42,43) who proposed to distinguish microemulsions from so-called co-solubilized systems. As far as electrical properties are concerned, this distinction appears to be a sensible one and can be formulated in terms of percolating and non-percolating microemulsions.

ACKNOWLEDGMENTS

Part of this work is the result of a collaboration established between the Physics Department of Queen Elizabeth College, University of London, G.B., and the "Laboratoire de Thermodynamique" of "Université de Pau et des Pays de l'Adour," Pau, France. The authors wish to express their gratitude to both the Institut de France and the British Council for the support they have granted.

Thanks are due also to Miss Trouilh and to Mrs. Anglichau, Bernade and Chaparteguy, all from Pau University, for the great care they have taken for typing of the manuscript and drawing of the figures.

REFERENCES

1. L. M. Prince, in "Emulsions and Emulsion Technology," K. J. Lissant, ed., Part I, M. Dekker, New York, p. 125, 1974.
2. S. Friberg, Informations Chimie, 148, 235 (1975).
3. S. Friberg, Chemtech., 6, 124 (1976).
4. V. K. Bansal and D. O. Shah, in "Microemulsions--Theory and Practice," L. M. Prince, ed., Academic Press, New York, p. 149, 1977.
5. D. O. Shah and R. S. Schechter, eds., "Improved Oil Recovery by Surfactant and Polymer Flooding," Academic Press, New York, 1977.
6. V. K. Bansal and D. O. Shah, in "Micellization, Solubilization, and Microemulsions," K. L. Mittal, ed., Vol. 1, Plenum Press, New York, p. 87, 1977.
7. L. M. Prince, ed., "Microemulsions--Theory and Practice," Academic Press, New York, 1977.
8. K. L. Mittal, ed., "Micellization, Solubilization, and Microemulsions," Plenum Press, New York, 1977.
9. D. O. Shah, V. K. Bansal, K. S. Chan, and W. C. Hsieh, in "Improved Oil Recovery by Surfactant and Polymer Flooding," D. O. Shah and R. S. Schechter, eds., Academic Press, New York, p. 293, 1977.
10. M. Clausse, in "Encyclopedia of Emulsion Technology," P. Becher, ed., Vol. 1, M. Dekker, New York, 1980.
11. S. Friberg, in "Microemulsions--Theory and Practice," L. M. Prince, ed., Academic Press, New York, p. 133, 1977.
12. D. G. Rance and S. Friberg, J. Colloid Interface Sci., 60, 207 (1977).
13. S. Friberg and I. Burasczenska, Progr. Colloid and Polymer Sci., 63, 1 (1978).
14. E. Sjöblom and S. Friberg, J. Colloid Interface Sci., 67, 16 (1978).
15. M. Clausse, R. J. Sheppard, C. Boned, and C. G. Essex, in "Colloid and Interface Science," Vol. 2, M. Kerker, ed., Academic Press, New York, p. 233, 1976.
16. J. Peyrelasse, V.E.R. McClean, C. Boned, R. J. Sheppard, and M. Clausse, J. Phys. D: Appl. Phys., 11, L-117 (1978).
17. C. Boned, M. Clausse, B. Lagourette, J. Peyrelasse, V.E.R. McClean, and R. J. Sheppard, paper presented at the 53rd Colloid and Surface Science Symposium, Rolla, MO, U.S.A., June 11-13, 1979; J. Phys. Chem., 84, 1520 (1980).
18. J. Peyrelasse, C. Boned, P. Xans, and M. Clausse, C. R. Acad. Sc., Paris, Ser. B, 284, 235 (1977).

19. J. Peyrelasse, C. Boned, P. Xans, and M. Clausse, in "Emulsions, Latices and Dispersions," P. Becher and M. N. Yudenfreund, eds., M. Dekker, New York, p. 221, 1978.
20. G. D. Smith, C. E. Donelan, and R. E. Barden, J. Colloid Interface Sci., 60, 488 (1977).
21. B. A. Keiser, D. Varie, R. E. Barden, and S. L. Holt, J. Phys. Chem., 83, 1276 (1979).
22. J. Rouviere, J. M. Couret, M. Lindheimer, J. L. Dejardin, and R. Marrony, J. Chim. Phys. Phys.-Chim. Biol., 76, 289 (1979).
23. J. Rouviere, J. M. Couret, A. Lindheimer, M. Lindheimer, and B. Brun, J. Chim. Phys.-Chim. Biol., 76, 297 (1979).
24. H. F. Eicke and V. Arnold, J. Colloid Interface Sci., 46, 101 (1974).
25. H. F. Eicke and J.C.W. Shepherd, Helv. Chim. Acta, 57, 1951 (1974).
26. H. F. Eicke and J. Rehak, Helv. Chim. Acta, 59, 2883 (1976).
27. H. F. Eicke, J.C.W. Shepherd, and A. Steinemann, J. Colloid Interface Sci., 56, 168 (1976).
28. H. F. Eicke and P. E. Zinsli, J. Colloid Interface Sci., 65, 131 (1978).
29. M. Zulauf and H. F. Eicke, J. Phys. Chem., 83, 480 (1979).
30. B. Lagourette, J. Peyrelasse, C. Boned, and M. Clausse, Nature, 281, 60 (1979).
31. M. Clausse, J. Peyrelasse, B. Lagourette, C. Boned, and J. Heil, paper submitted for publication.
32. J. H. Schulman and T. S. McRoberts, Trans. Faraday Soc., 42B, 165 (1946).
33. S. S. Bhatnagar, J. Chem. Soc., 117, 542 (1920).
34. E. A. Hauser and J. E. Lynn, "Experiments in Colloid Chemistry," McGraw-Hill Book Co., New York, p. 129, 1940.
35. J. H. Schulman and D. P. Riley, J. Colloid Sci., 3, 383 (1948).
36. J. E. Bowcott and J. H. Schulman, Z. Elektrochem., 59, 283 (1955).
37. K. D. Dreher and R. D. Sydansk, J. Pet. Tech., 23, 1437 (1971).
38. S. C. Jones and K. D. Dreher, Soc. Pet. Engr. J., 16, 161 (1976).
39. D. O. Shah and R. M. Hamlin, Science, 171, 483 (1971).
40. D. O. Shah, A. Tamjeedi, J. W. Falco, and R. D. Walker, AIChE J., 18, 1116 (1972).
41. D. O. Shah, Ann. N.Y. Acad. Sci., 204, 125 (1973).
42. D. O. Shah, Proc. 48th National Colloid Symposium, p. 173, Austin, Texas, June 24-26, 1974.
43. D. O. Shah, R. D. Walker, W. C. Hsieh, N. J. Shah, S. Dwivedi, J. Nelander, R. Pepinsky, and D. W. Deamer, Preprint SPE 5815, Improved Oil Recovery Symposium, Tulsa, Oklahoma, March 22-24, 1976.
44. V. K. Bansal and D. O. Shah, J. Colloid Interface Sci., 65, 451 (1978).

45. V. K. Bansal, K. Chinnaswamy, C. Ramachandran, and D. O. Shah, J. Colloid Interface Sci., 72, 524 (1979).
46. D. Senatra and G. Giubilaro, J. Colloid Interface Sci., 67, 448 (1978).
47. D. Senatra and G. Giubilaro, J. Colloid Interface Sci., 67, 457 (1978).
48. M. Clausse and R. Royer, in "Colloid and Interface Science," Vol. 2, M. Kerker, ed., Academic Press, New York, p. 217, 1976.
49. S. Kirkpatrick, Phys. Rev. Lett., 27, 1722 (1971).
50. S. Kirkpatrick, Rev. Mod. Phys., 45, 574 (1973).
51. M. H. Cohen and J. Jortner, Phys. Rev. Lett., 30, 696 (1973).
52. I. Webman, J. Jortner, and M. H. Cohen, Phys. Rev. B, 11, 2885 (1975).
53. M. Lagües, R. Ober, and C. Taupin, J. Phys. Lett. (France), 39, L-487 (1978).
54. T. Hanai and N. Koizumi, Bull. Inst. Res. Kyoto Univ., 45, 342 (1967).
55. K. Shinoda and T. Ogawa, J. Colloid Interface Sci., 24, 56 (1967).
56. K. Shinoda and H. Saito, J. Colloid Interface Sci., 26, 70 (1968).
57. H. Saito and K. Shinoda, J. Colloid Interface Sci., 32, 617 (1970).
58. H. Saito and H. Kunieda, J. Colloid Interface Sci., 42, 381 (1973).
59. K. Shinoda and S. Friberg, Advances Colloid Interface Sci., 4, 281 (1975).
60. R. J. Sheppard, J. Phys. D: Appl. Phys., 5, 1576 (1972).
61. R. J. Sheppard and E. H. Grant, J. Phys. E: Sci. Instrum., 5, 1208 (1972).
62. B. P. Jordan and E. H. Grant, J. Phys. E: Sci. Instrum., 3, 764 (1970).
63. E. H. Grant, G. P. South, S. Takashima, and H. Ichimura, Biochem. J., 122, 691 (1971).
64. C. G. Essex, G. P. South, R. J. Sheppard, and E. H. Grant, J. Phys. E: Sci. Instrum., 8, 385 (1975).
65. P. Ekwall, L. Mandell, and K. Fontell, Mol. Cryst., 8, 157 (1969).
66. D. W. Marquardt, R. G. Bennett, and E. J. Burrell, J. Molec. Spectrosc., 7, 269 (1961).
67. D. W. Marquardt, J. Soc. Ind. Appl. Math., 11, 431 (1963).
68. C. Boned, J. Peyrelasse, M. Clausse, B. Lagourette, J. Alliez, and L. Babin, Colloid and Polymer Sci., 257, 1073 (1979).
69. M. Lagües, C.R. Acad. Sc. Paris, 288, 339 (1979); J. Phys. Lett. France, 40, L-331 (1979).
70. D.A.G. Bruggeman, Ann. Physik (Leipzig), 24, 636 (1935).
71. C.J.F. Bottcher, Rec. Trav. Chim. Pays-Bas, 64, 47 (1945).
72. R. Landauer, J. Appl. Phys., 23, 779 (1952).

73. L. E. Scriven, Nature, 263, 123 (1976).
74. L. E. Scriven, in "Micellization, Solubilization, and Microemulsions," K. L. Mittal, ed., Vol. 2, Plenum Press, New York, p. 887, 1977.
75. Y. Talmon and S. Prager, Nature, 267, 333 (1977).
76. Y. Talmon and S. Prager, J. Chem. Phys., 69, 2984 (1978).

MEASUREMENT OF LOW INTERFACIAL TENSION BETWEEN CRUDE OIL AND FORMATION WATER WITH DISSOLVED SURFACTANTS BY THE SPINNING DROP TECHNIQUE: FACT OR FICTION?

A. Capelle

Chemische fabriek Servo B.V.
Langestraat 167, 7490 AA DELDEN
The Netherlands

The laboratory study of enhanced oil recovery with the aid of surfactants has increased tremendously since the development of the spinning drop apparatus. Almost all recent papers about the measurement of the interfacial tension between crude oil and formation water mention values determined with this apparatus. Modifications as to temperature control and influence of pressure have been suggested. However, all measurements deal with steady state conditions and this fact is always stressed. This aspect of the laboratory study deviates from the real situation where we deal with a dynamic process.

During the laboratory investigation with a spinning drop apparatus several uncontrolled factors are introduced. In the long time necessary for reaching steady state conditions, solubilization of the oil drop (due to the ultra low interfacial tension) will occur. There is also a transport of surfactant from the water phase into the oil phase and its accumulation in the interfacial layer.

A recent paper, confirmed by our investigations, showed that the partition coefficient of a surfactant between water and a hydrocarbon phase is not constant but depends on the surfactant concentration. This leads to the question: are we really measuring the true interfacial tension?

INTRODUCTION

Processes which deal with enhanced oil recovery, especially with the aid of surface active chemicals, have encountered a growing interest during the last few years. From the work of Taber (1), we know that a significant drop in the interfacial tension between the crude oil and the formation water is needed to mobilize the trapped oil from the formation. The range of the required interfacial tensions is between 10^{-3} to 10^{-5} dynes/cm. This drop in interfacial tension can only be achieved by the addition of surface active chemicals to the formation water.

It has been suggested that surfactants to be used in enhanced oil recovery projects should be screened for their ability to decrease the interfacial tension between the crude oil and formation water under investigation. Also the dependence on temperature should be investigated (2).

It has become relatively easy to carry out these measurements since the introduction of the spinning drop apparatus by Wade and colleagues (3). Almost all recent papers on this subject mention values with regard to temperature control and pressure, but all are based on the same principle developed by Vonnegut (4) and refined by Princen (5).

EXPERIMENTAL

For the calculation of the interfacial tension the following quantities are needed.

- density difference between crude oil-formation water: $\Delta\rho$
- time for one period : P
- length of the drop : L
- diameter of the drop : D
- volume of the drop : V

Three methods of calculation are possible:

Case 1: $L > 4D$

Here an approximation described by Vonnegut (4) is used for the calculation $\Delta\rho$, P and D are needed.

$$\gamma = \frac{\Delta\rho\omega^2}{4C} = \frac{\pi^2\Delta\rho D^3}{8P^2}$$

Case 2: $L < 4D$

The method described by Princen et al. (5) is used for the calculation $\Delta\rho$, P, D, L and V are needed.

$$\gamma = \frac{\Delta\rho\omega^2}{4C}$$

with C calculated from $C = \dfrac{4\pi(CR^3)}{3V}$

where CR^3 is taken from a table as function of L/D.

Case 3: L < 4D

The method described by Slattery and Chen (6) for the calculation $\Delta\rho$, P, D and L are needed.

$$\gamma = \frac{\Delta\rho\omega^2}{16}\left\{\frac{D}{r^*_{max}}\right\} = \frac{\pi^2\Delta\rho}{4P^2}\left\{\frac{D}{r^*_{max}}\right\}^3$$

where r^*_{max} is taken from a table as function of D/L.

The readings of P, L and D are taken when a steady state is reached. The times needed to reach this condition varied between 45 to 180 minutes.

Measurements of interfacial tension were carried out with a spinning drop interfacial tension meter according to Wade (Model 300). Glass capillaries obtained from Wildmad Glass Co., with an internal diameter of 2.00 mm were used. The rotating velocity was varied between 4000 and 9600 rpm at a constant temperature of 35°C. Also the influence of the rotating time at constant rotating velocity was investigated. Crude oil and formation water obtained from an oilfield in Germany was used. Crude oil was dehydrated by means of centrifugation prior to usage. The formation water contained 20% salt.

Surfactant, a phosphated ethoxylated alcohol neutralized with a fatty amine or an ethanolamine, was obtained from Chemische Fabriek Servo B.V. and used as received. The critical micelle concentration of this product in the investigated formation water is 20 mg/l. Concentration of the surfactant was varied between 20 mg/l and 40 g/l.

Prior to the measurements, the system under investigation was pre-equilibrated, in some cases measurements were carried out immediately after preparation of the fluids. The results of the measurements are shown in Figures 1 to 3.

DISCUSSION

A dependence of the interfacial tension on time and surfactant concentration at constant angular velocity or on the rotating velocity at a constant surfactant concentration was observed.

The dependence of the measured interfacial tension on the surfactant concentration is most easily explained by the micellization of the surfactant above its critical micelle concentration. However, the difference between the values found at concentrations of

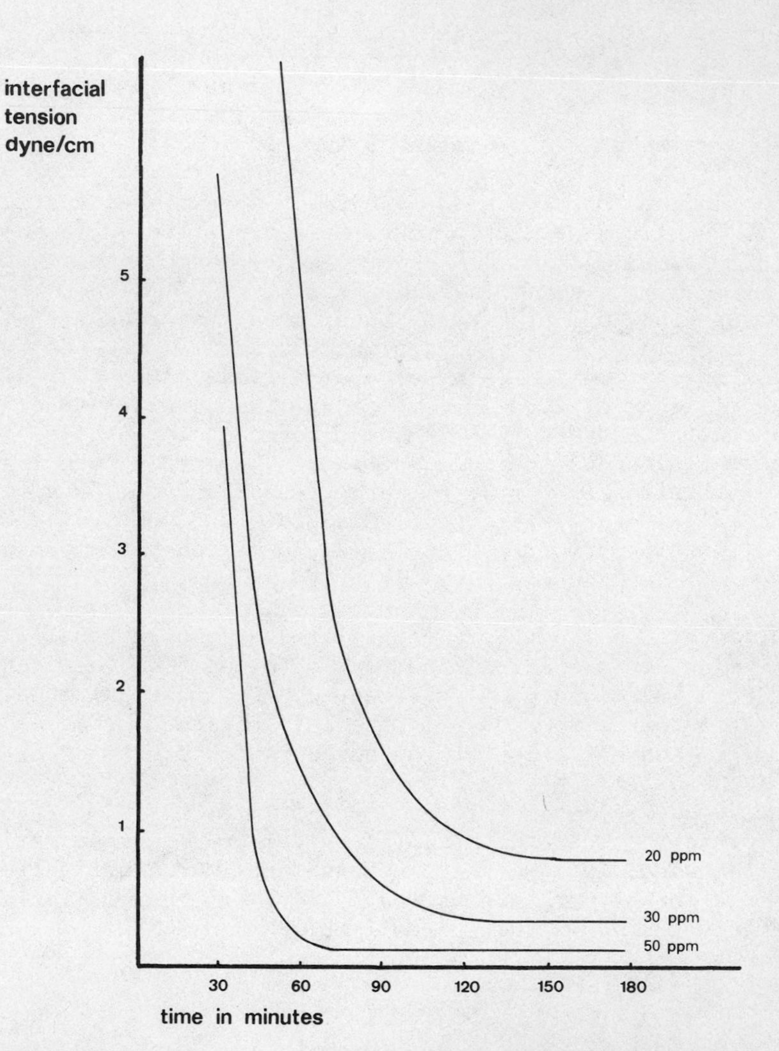

Fig. 1. Interfacial tension as a function of time

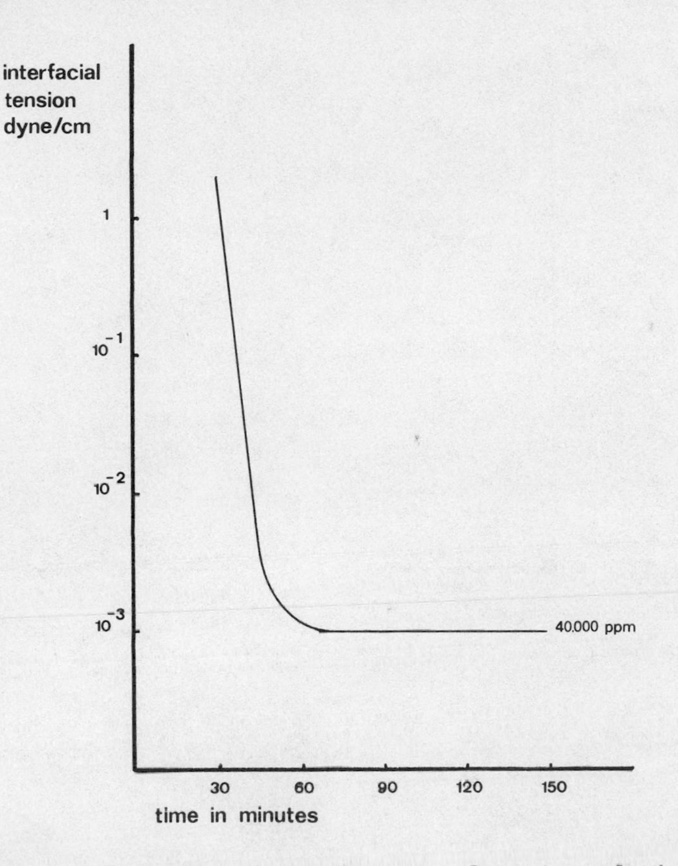

Fig. 2. Interfacial tension as a function of time

50 mg/l and 40 g/l remains unexplained by this effect. At the high surfactant concentration we probably deal with "swollen micelles" or the formation of a middle phase.

The dependence of the measured interfacial tension on time may be due to the diffusion of the surfactant from the formation water into the crude oil. However, if the system was pre-equilibrated this should not take place. An adsorption and accumulation of the surfactant at the interface can occur, resulting in an interfacial layer of different composition than the original phases.

It has been shown (7,8) that the partition coefficient (Co/Cw) both for ionic and nonionic surfactants is not a constant but depends on the surfactant concentration. Greenwald (7) investigated the distribution of a range of ethoxylated alkylphenols in isooctane and water at 26°C and found it to be constant up to cmc. Dupeyrat (8) investigated the distribution of alkylbenzene sulfonates and petroleum sulfonates in paraffin oil or hexane and water with dissolved sodium chloride and found more or less the same results.

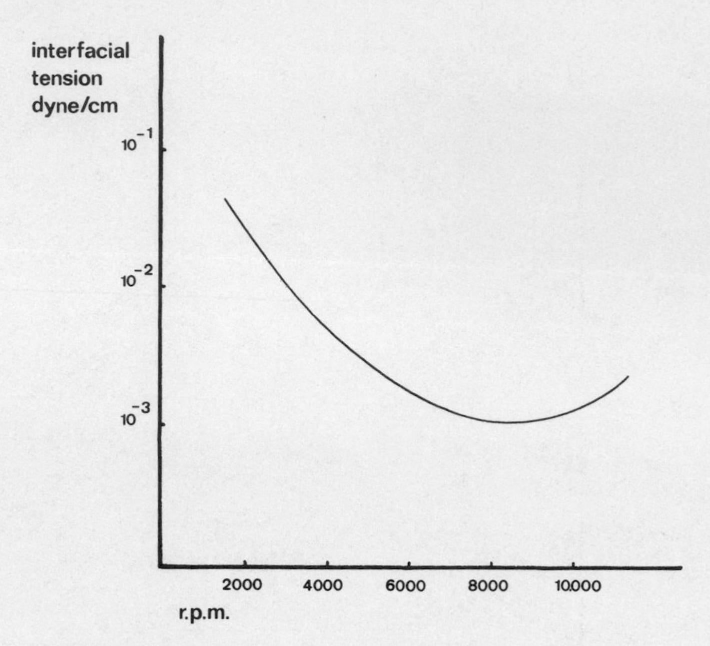

Fig. 3. Interfacial tension as a function of rotation velocity

We have investigated the distribution of a phosphated ethoxylated alcohol, neutralized with either an amine or an ethanolamine in crude oil and formation water (Figure 4) and found also a constant partition coefficient up to a certain value.

The dependence of the interfacial tension on the rotation velocity complicates the measurements enormously. It shows that the system under investigation is unstable and shifts its composition either due to flow or gravitational effects. Anyhow, due to dependence on time, rotation velocity and partitioning, probably an interfacial layer of constantly changing composition builds up. This results in a change of the density difference $\Delta\rho$ and the volume V and consequently in a change of D and L.

If we take also into consideration the work of Shinoda and others (9,10) on the composition of emulsions and microemulsions, we see a drastic change of the composition at a given temperature, the phase inversion temperature PIT. The problems connected with the thermostatting of the measurement apparatus may then easily be imagined.

CONCLUSIONS

Reviewing the above mentioned arguments about the measured value of IFT, or better the quantity resulting from the calculation, we propose that it is a function of:

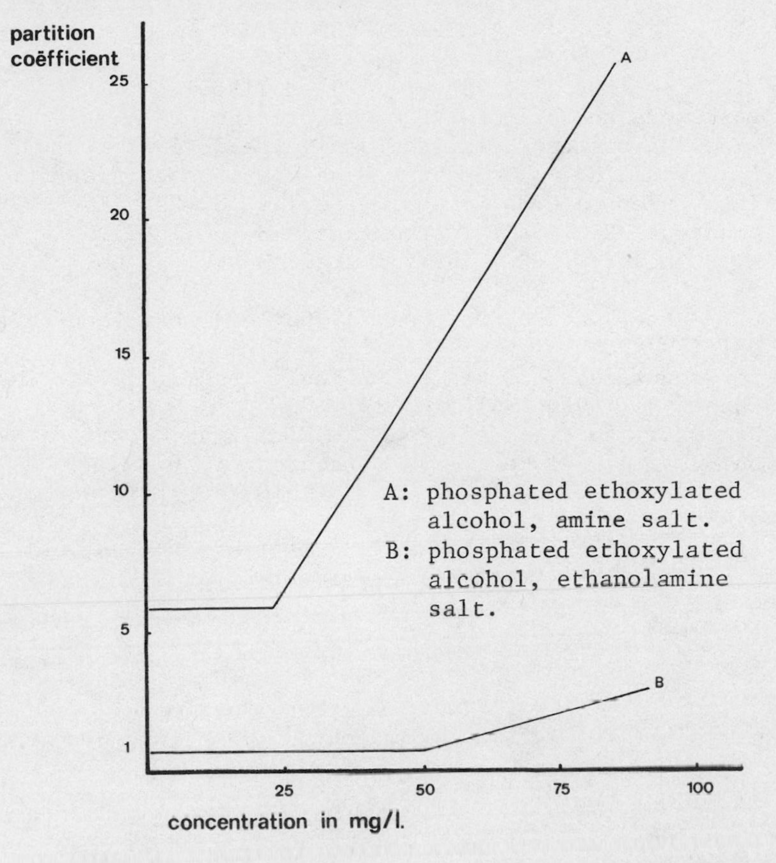

Fig. 4. Partition of surfactant as a function of concentration.

- crude oil
- formation water
- time
- temperature
- rotation velocity
- surfactant

The objective of this measurement technique is to screen surfactants for their ability to decrease interfacial tension and thereby mobilizing trapped oil. If we confine ourselves to this objective, it is not sufficient to measure the interfacial tension between crude oil and formation water only. More relevant information can be obtained by measuring trends (for example temperature and concentration variations) in the effect of other variables. Using this technique to predict behavior in an oil bearing formation, where we deal with totally different processes, leads to the question: what is the real value of measuring ultra low interfacial tension for predicting oil recovery in the reservoir?

REFERENCES

1. J. J. Taber, Soc. Pet. Eng. J., 9, 3 (1969).
2. M. Burkowsky and C. Marx, First Symposium on Interfacial Tension Measurement, Clausthal-Zellerfeld, 1977.
3. I. L. Cayias, R. S. Schechter, and W. H. Wade, Adsorption at Interfaces, A.C.S. Symposium Series, 8, 234 (1975).
4. B. Vonnegut, Rev. Sci. Instruments, 13, 6 (1942).
5. H. M. Princen, I.Y.Z. Zia, and S. G. Mason, J. Colloid Interface Sci., 23, 99 (1967).
6. J. C. Slattery and J. D. Chen, J. Colloid Interface Sci., 64, 371 (1978).
7. H. L. Greenwald, E. B. Kice, M. Kenly, and J. Kelly, Anal. Chem., 33, 465 (1961).
8. L. Dupeyrat, L. Minssieux, and A. El Naggar, European Symposium on Enhanced Oil Recovery, Edinburgh, 1978.
9. K. Shinoda and H. Sagitani, J. Colloid Interface Sci., 64, 68 (1978).
10. S. Friberg, I. Lapczynska, and G. Gillberg, J. Colloid Interface Sci., 56, 19 (1976).

III: EMULSION RHEOLOGY AND OIL DISPLACEMENT MECHANISMS

THE EFFECT OF FILM-FORMING MATERIALS ON THE DYNAMIC INTERFACIAL PROPERTIES IN CRUDE OIL-AQUEOUS SYSTEMS

C. H. Pasquarelli and D. T. Wasan

Department of Chemical Engineering
Illinois Institute of Technology
Chicago, Illinois 60616, U.S.A.

Interfacial behavior of crude oil-aqueous systems depends strongly upon the nature of the crude and of the aqueous displacing fluid. The crude oil contains discrete surface active particles known as asphaltenes in addition to the surface active resins, both of which contain organic acids and bases. The film forming characteristics of these components are exhibited by the interfacial films which exist at the crude-aqueous interface.

This paper discusses the physical characteristics of naturally occurring films formed at the crude oil-aqueous interfaces and the subsequent changes in their structure are characterized by measuring interfacial charge, interfacial tension and interfacial shearing viscosity against brine solutions containing alkaline agents. A California crude oil has been fractionated and the dynamic interfacial behavior of each fraction is examined and compared with the whole crude.

It is concluded that the dynamic properties exhibited by the crude oil-aqueous interfaces are a composite of the interfacial behavior existing between the alkaline aqueous phase and individual crude components. Furthermore, it is observed that the heavy asphaltic components in crude oils are primarily responsible for the high interfacial activity and hence, the resulting dynamic interfacial properties. This suggests the possibility of using these high molecular weight components for improved oil recovery.

INTRODUCTION

Crude oil has been characterized as a colloid (1,2) in which high molecular weight, asphaltic particles are dispersed. Such particulates--known as asphaltenes--are held in solution via peptization by lower molecular weight components--classified as resins. This association forms a micellar type structure (3-5). Classification of the asphaltic constituents in crude oil is based primarily upon each component's solubility in various solvents (e.g., the pentane soluble, propane insoluble resin or maltene fraction and the benzene soluble, pentane insoluble asphaltene fraction) (6,7).

Characterizing the heavy components in crude oil with respect to their molecular structure has been difficult primarily because of the complex mixture of extremely high molecular weight compounds. Work by Yen (8,9) and Pollock and Yen (10) has indicated that asphaltenes occurring in petroleum exist as layered flat sheets of condensed aromatic rings linked by short chain alkanes, kata-condensed naphthenics and attractions between the π-electron clouds of the peri-condensed polynuclear aromatic systems. Yen, Erdman and Saracenno (11) have indicated the presence of stable free radicals associated with these structures. This suggests the presence of gaps or holes within these particles that are capable of complexing heavy metal ions [e.g. Vanadium (IV) or Nickel (II)] found in most crude oils.

Structural analysis of the resinous fraction in crude has indicated the presence of high molecular weight carboxylic acids, esters and porphyrin structures. Cason et al. (12-14) have isolated isoprenoids, cyclic and acyclic carboxylic acids with carbon numbers of C_{10} to C_{20}. Seifert and Teeter (15) have discovered C_{22}-C_{24} steroid carboxylic acids as well as other C_{16} to C_{31} petroleum carboxylic acids in the Midway-Sunset California crude. Jenkins (16) has isolated cyclic monocarboxylic and fatty acids as well as aliphatic esters from petroleum distillates and residues of several crudes. Seifert and Howells (17), through an elaborate extraction and separation scheme, have recovered phenols and carboxylic acids with molecular weights of 300 to 400 from the Midway-Sunset crude. The acidic extracts have been shown to give ultra low ($< 10^{-2}$ dyne/cm) interfacial tensions when contacted with an alkali aqueous phase. Phenolic components isolated from the crude were found to diminish the interfacial activity of the acidic fraction.

Other high molecular weight components in petroleum have been identified by Yen (18) as porphyrin structures that effectively complex with metals such as nickel and vanadium. These metalloporphyrins have been shown to exhibit interfacial activity when contacted with an aqueous phase (19,20). The film forming tendencies that these particular metalliferrous constituents have at an oil-aqueous interface have also been demonstrated (21,22). Early investigations of interfacial behavior and film formation for crude oil-aqueous systems (23) indicated the major role that the heavy asphaltic material played in such activity. Later works by Kimbler, Reed and Silberberg (24) enabled one to study the compressibility and collapse pressure of these interfacial films. Wasan et al. (23,25) developed a technique for measuring the interfacial shear viscosity of these films in an effort to relate film structure to emulsion stability. Strassner (26) and Riesberg and Doscher (27) had also examined the relation between emulsion stability and film formation under conditions of changing pH. Riesberg and Doscher have extended their work to determining the effect that such films have upon adhesion of crude oil to solid surfaces and crude displacement in porous media. Bourgoyne, Caudle and Kimbler (28) have found that interfacial film formation tends to decrease enhanced recovery efficiency in highly heterogeneous porous media only.

Recent work by Lichaa and Herrera (29) and David (30) has shown that the effective permeability to brine decreases as a result of asphaltene precipitation in porous media. Preckshot et al. (31) indicate that the streaming potential of crude oil flowing through porous media is responsible for this precipitation. Radke and Sommerton (32), using an oil phase of high viscosity mineral oil with an acidic additive, were able to model the interfacial behavior of crude oil-aqueous alkaline systems. In view of this, we have been directing our continuing work to determining the individual and combined effects that the asphaltenes and acidic resins in crude oil have upon interfacial tension, interfacial shear viscosity, interfacial elasticity, electrophoretic mobility, emulsion stability and ultimately, tertiary oil recovery for a crude oil-caustic aqueous system.

EXPERIMENTAL

The crude oil system used for our work was the LMZ (Lower Main Zone) S-47 variety from Huntington Beach, California. This crude oil has an acid number of 0.65 mg KOH/gm sample which makes it particularly amenable to caustic flooding techniques. The bulk viscosity measured at room temperature was 108 cp. The API gravity for this crude oil is 23.5°. The standard caustic aqueous phase was 0.15% (wt) sodium orthosilicate plus 0.75% (wt) sodium chloride in double distilled water. At this concentration, the aqueous phase has a pH of 11.7.

A vacuum distillation technique was employed for the separation of the crude into three temperature cuts plus the high boiling asphalt residue. These temperature cuts are:

Cut 1: 70°F - 230°F at 20 mm Hg.
Cut 2: 230°F - 302°F at 0.1 mm Hg.
Cut 3: 302°F - 420°F at 0.1 mm Hg.

Vapor temperatures and operating pressures were minimized to avoid thermal cracking. Using IR and NMR techniques, Farmanian et al. (33) have shown that cracking does not occur under similar conditions with the same crude oil system.

The separation of the saturated oils, polar resins and asphaltene particles from the high boiling residue as well as the whole crude was affected through the use of Institute of Petroleum Method IP143/57 (6). However, normal pentane was substituted for the specified heptane solvent required for asphaltene precipitation. Asphaltene particles and resin fractions were stored under subdued light and a nitrogen atmosphere. Interfacial tension measurements were made using a spinning drop tensiometer that was built in our laboratory. All runs were made at room temperature using a monochromatic light source. Drop size was held constant at 1 $\mu\ell$ through the use of a micro-syringe. All runs were allowed to continue until the interfacial tension reached its equilibrium value. This rarely took longer than one hour. Values reported are the minimum interfacial tensions exhibited for each system.

Electrophoretic mobility measurements were made on a Zeta-Meter Inc. instrument that required using a plexiglass sample cell.

Interfacial shear viscosity measurements were made on the viscous traction shear viscometer developed by Wasan et al. (25) for determining interfacial viscosities of crude oil-aqueous systems.

RESULTS AND DISCUSSION

The LMZ S-47 crude was first examined for asphaltene and resin concentration. Pentane precipitation of asphaltenes resulted in 10.5 (wt) % concentration of these particles in the crude. Resin concentration was determined to be 27.0 (wt) %.

A comparison was made of the dynamic interfacial tensions of cut 2,3 and the LMZ crude against the standard aqueous phase concentration of 0.75% NaCl plus 0.15% sodium orthosilicate (Figure 1). This was done to relate these dynamic responses to the individual and combined resin-asphaltene structures.

Due to the high molecular weights of the asphaltene particles, these species were absent in all temperature cuts with the exception of the high boiling residue. The difference in the dynamic interfacial tensions of cut 2 and 3 can be explained by the corresponding average molecular weights differences for these two cuts.

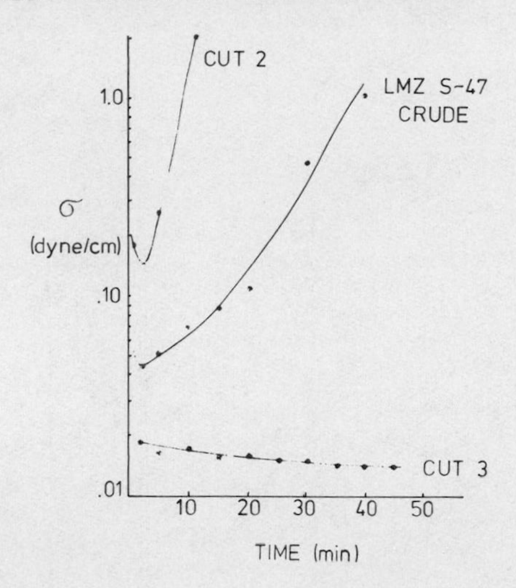

Fig. 1. Dynamic Interfacial Tension for the LMZ Crude, Cuts 2 and 3 vs. 0.15% orthosilicate + 0.75% NaCl solution.

Cut 2, with its lower boiling point and hence lower average molecular weight, gives an interfacial tension response with time that is characterized by diffusion to the interface followed by in situ reaction to form the weak surfactant and mass transfer into the aqueous phase. This final step of surfactant partitioning into the aqueous phase is absent in the Cut 3-caustic case. This is probably due to the inability of the high molecular weight saponified acids to partition into the aqueous caustic phase. The large hydrophobic portion of these species maintains the constant surface concentration suggested by the stable interfacial tension response seen here.

Interfacial behavior exhibited by the LMZ crude is similar to that of Cut 2. However, the adsorption of the asphaltene particles in the crude at the oil/aqueous interface creates an interfacial barrier--or film--that inhibits further saponification of the acidic components that are responsible for lowering interfacial tension.

The effect that the asphaltene particles and resinous components have upon interfacial rigidity and elasticity were examined for the crude oil and temperature cuts when contacted with the aqueous caustic phase. An interface made up only of adsorbed high molecular weight resins might differ significantly from one on which the heavy asphaltene particles were highly concentrated. The results in Table 1 support this hypothesis and show that the Cut 3-aqueous interface is indeed much more rigid than the lighter Cut 2 and whole crude systems.

Crystalline structures were found to develop upon interfacial aging of the Cut 3-caustic system (Figure 2). (This photograph was

Table 1. Interfacial Tension and Shear Viscosity of the Huntington Beach Crude (LMZ S-47)

OIL	AQUEOUS PHASE	INTERFACIAL TENSION DYNES/CM	INTERFACIAL VISCOSITY S.P.
WHOLE CRUDE	0.75% NaCl	28.09	0.25
CUT 2	0.75% NaCl	29.86	4×10^{-3}
CUT 3	0.75% NaCl	26.72	7.10
WHOLE CRUDE	0.75% NaCl + 0.15% ORTHOSILICATE	.045	4.3×10^{-2}
CUT 2	0.75% NaCl + 0.15% ORTHOSILICATE	.729	7.4×10^{-4}
CUT 3	0.75% NaCl + 0.15% ORTHOSILICATE	.014	6.71

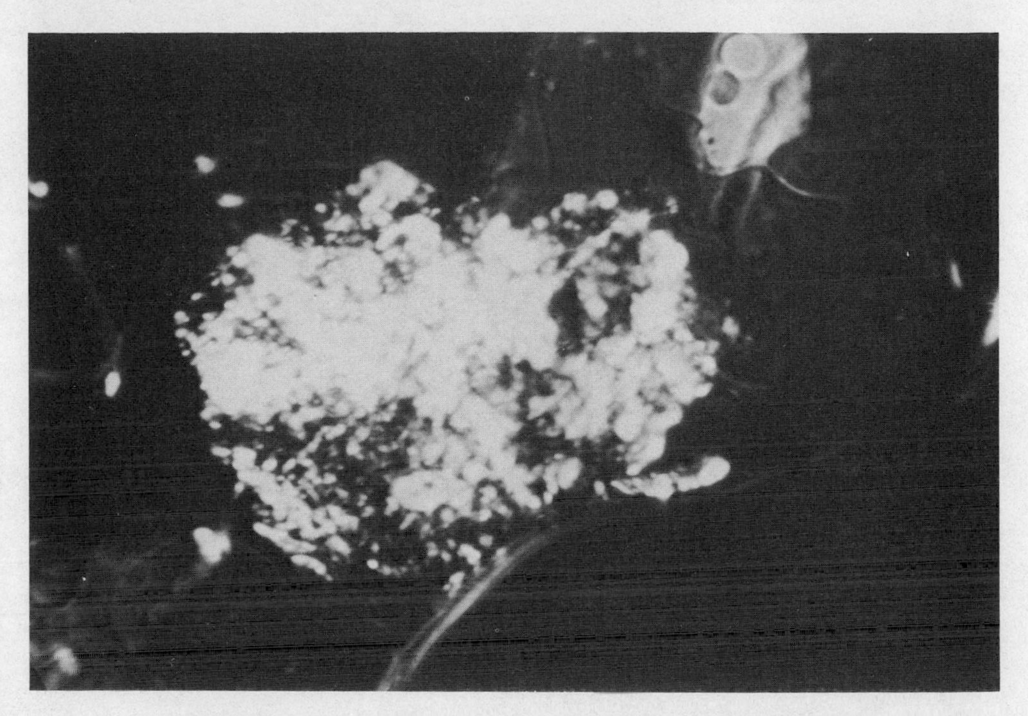

Fig. 2. Photograph of the crystalline structure at the cut 3-aqueous alkaline interface.

taken under a magnification of 400 using cross polarized light.) The sample was aged for a period of three days under an inert atmosphere. A control experiment was conducted with the Cut 3-brine system that was equilibrated for the same three day period. No crystalline formations were observed, even after an aging period of several months.

Dynamic interfacial tension was measured for the Cut 3-caustic system with varying concentrations of asphalt in the Cut 3 oil phase. Figure 3 illustrates the same stable tension response for the asphalt systems that was seen for the pure Cut 3-caustic interface. Figure 4 illustrates the effect that the resinous material in the asphalt has upon interfacial tension and interfacial viscosity. The response of these two parameters to changing asphalt concentration suggests an interface composed of heavy resinous molecules. Previous work by Katz and Beu (34) has shown that oil phases which have a surface tension of less than 21 dynes/cm are capable of depeptizing the asphaltene particles and high molecular weight resins. Measurements of the air-liquid surface tension of Cut 3 with the ring tensiometer indicates a surface tension of slightly less than 20 dynes/cm. Therefore, this behavior indicates adsorption of heavy resinous material at the interface leading to

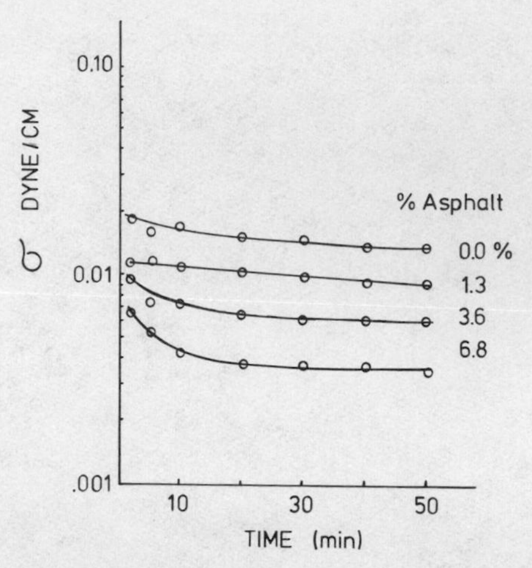

Fig. 3. Dynamic Interfacial Tension of cut 3 with varying amounts of asphalt.

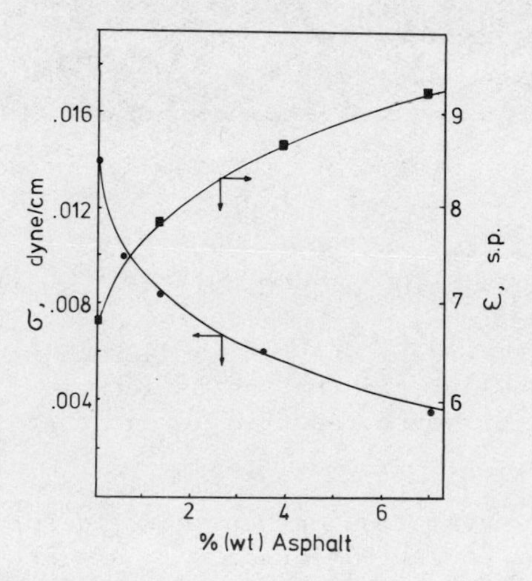

Fig. 4. Variation of Interfacial Tension and shear viscosity for cut 3 with varying amounts of asphalt.

an in situ generation of surfactant with the resulting decrease in tension and increase in interfacial rigidity.

Recently, we have been using electrophoretic mobility measurements in an attempt to support some of our hypotheses. Our experimental plan, however, required the separation of asphaltenes, resins and gas oil as it exists in the crude. To do this we basically used standardized technique #143/57 of the Institute of Petroleum. Electrophoretic mobility measurements were made of the whole crude, Cut 2 and Cut 3 plus asphalt when contacted with the standard caustic phase (Table 2). These measurements indicate a maximum interfacial charge for the Cut 3 system with the asphalt additive. This behavior is the result of asphaltene precipitation occurring in the Cut 3 oil phase that affects the release of the resinous components leading to an increase in resin surface concentration and hence, interfacial charge. The effect that this has on interfacial tension is obvious. The whole crude was also examined for interfacial charge buildup when asphalt concentration was increased. Neither interfacial tension nor electrophoretic mobility were affected by the addition of asphalt to the whole crude. This indicates that the asphaltene particles do not contribute greatly to either surface activity or interfacial charge, although the structure of asphaltene particles may strongly contribute to interfacial elasticity. Electrophoretic mobility measurements indicated that the asphaltene particles are slightly electropositive while the resinous components are electronegative.

Work was done to determine the interfacial activity of the LMZ asphaltenes when dissolved in an oil phase of spectroscopic grade benzene and contacted with the standard caustic aqueous phase. Figure 5 illustrates interfacial activity similar to that of Cut 2 and can be explained in the same way. Solubilizing the solid asphaltene particles in benzene permits the separation of their smaller components, some of which are obviously either surface active or capable of generating such activity when contacted with a caustic aqueous phase. No interfacial activity was seen for the pure benzene oil phase contacted with the caustic aqueous phase.

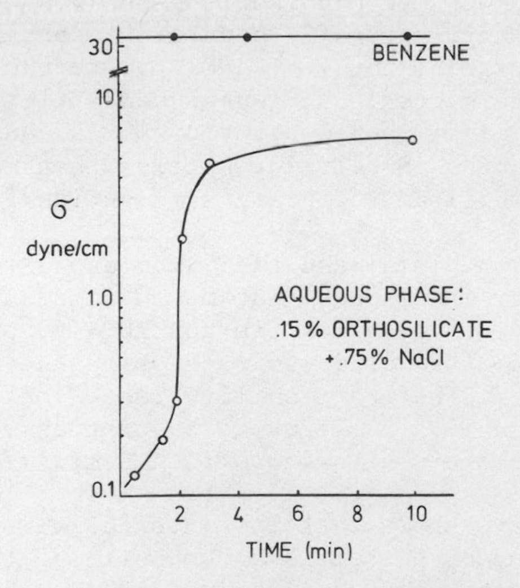

Fig. 5. Dynamic Interfacial Tension of LMZ asphaltenes in benzene vs. 0.15% orthosilicate + 0.75% NaCl.

Table 2. Interfacial Tension and Electrophoretic Mobility of the Huntington Beach Crude (LMZ S-47)

Oil Phase	Aqueous Phase	IFT (dyne/cm)	EM $\left(\frac{\mu}{sec} \cdot \frac{cm}{volt}\right)$
Whole crude	0.15% Orthosilicate 0.75% NaCl	0.045	3.84
Crude + 10% Asphalt	↓	0.049	3.88
Cut 3 + 10% Asphalt	↓	0.004	4.01
Cut 2	↓	0.729	3.73
Asphaltene	↓	--	1.32

In view of this work, we feel that an effort is warranted for choosing a surfactant (e.g. anionic, cationic or nonionic) or co-surfactant that is capable of preventing asphaltene precipitation problems and the subsequent permeability reductions during tertiary oil recovery by chemical flooding methods (29).

SUMMARY

In conclusion, we have observed the individual and combined effects that the heavy, naturally occurring asphaltic components have upon the interfacial activity and structure of an acidic crude oil-aqueous alkaline system. The subsequent film formation that these components exhibit when contacted with a caustic phase suggests the presence of a stabilizing barrier that may impede coalescence of viscous emulsions formed in situ during flooding processes.

Precipitation problems resulting from asphaltene aggregation lead to permeability reductions that may ultimately affect enhanced recovery methods. This suggests the possibility of using acidic resins extracted from the crude to re-peptize the asphaltene particles and prevent further permeability reductions. Core flooding tests in conjunction with a microwave absorption scanning are being conducted in our laboratory to support this claim.

Finally, we have shown that the interfacial phenomena exhibited for a crude oil-aqueous system is a composite of the interfacial activities existing between an alkaline aqueous phase and the individual crude components. This behavior must be considered when employing--or designing--a tertiary technique or aqueous phase.

ACKNOWLEDGMENTS

This study was supported partly by the Department of Energy Grant DE-AC19-79BC10069 and by the National Science Foundation Grant ENG-77-20164.

REFERENCES

1. B. R. Ray, P. A. Witherspoon, and R. E. Grim, J. Phys. Chem., 61, 1276 (1957).
2. I. A. Eldib, H. N. Dunning, and R. J. Bolen, J. Chem. Eng. Data, 5, 550 (1960).
3. S. L. Neppe, Petroleum Refiner, 31, 137 (February, 1952).
4. J. P. Pfeiffer, ed., "The Properties of Asphaltic Bitumen," Elsevier Pub. Co., New York (1950).
5. J. P. Pfeiffer and R.N.J. Saal, J. Phys. Chem., 44, 139 (1940).
6. Institute of Petroleum Standards for Petroleum and its Products, 30th ed., Part 1, Sec. 1, IP 143/57, 576, 1971.
7. R. L. Hubbard and K. E. Stanfield, Analyt. Chem., 20, 460 (1948).
8. T. F. Yen, Energy Sources, 1, No. 4, 447 (1974).
9. T. F. Yen, Nature Physical Science, 233, 36 (1971).
10. S. S. Pollack and T. F. Yen, Analyt. Chem., 42, 623 (1970).
11. T. F. Yen, J. G. Erdman, and A. J. Saraceno, Analyt. Chem., 34, 694 (1962).
12. J. Cason and A.I.A. Khodair, J. Org. Chem., 32, 3430 (1967).
13. J. Cason and K. L. Liauw, J. Org. Chem., 30, 1763 (1965).
14. J. Cason and A.I.A. Khodair, J. Org. Chem., 31, 3618 (1966).
15. W. K. Seifert, E. J. Gallegos, and R. M. Teeter, J. Amer. Chem. Soc., 94, 5880 (1972).
16. G. I. Jenkins, J. Inst. Pet., 51, 313 (1965).
17. W. K. Seifert and W. G. Howells, Analyt. Chem., 41, 554 (1969).
18. T. F. Yen, "The Role of Trace Metals in Petroleum," Ann Arbor Science Publishers, Inc., Ann Arbor, 1975.
19. H. N. Dunning, J. W. Moore, and M. O. Denekas, Industr. Chem., 45, 1759 (1953).
20. H. N. Dunning, Division of Colloid Chemistry, 122nd Meeting, Amer. Chem. Soc., Atlantic City, N.J. (1952).
21. M. O. Denekas, F. T. Carlson, J. W. Moore, and C. G. Dodd, Industr. Eng. Chem., 43, 1165 (1951).
22. C. G. Dodd, J. W. Moore, and M. O. Denekas, Industr. Eng. Chem., 44, 2585 (1952).
23. F. E. Bartell and D. O. Niederhauser, Fundamental Research on Occurrence and Recovery of Petroleum, API, p. 57, 1949.
24. O. K. Kimbler, R. L. Reed, and I. H. Silberberg, Soc. Pet. Eng. J., 153 (1966).
25. D. T. Wasan, J. J. McNamara, S. M. Shah, K. Sampath, and N. Aderangi, J. Rheology, 23, 181 (1979).

26. J. E. Strassner, J. Pet. Tech., 303 (1968).
27. J. Reisberg and T. M. Doscher, Prod. Monthly, 21, No. 1 (Nov. 1956).
28. A. T. Bourgoyne, B. H. Caudle, and O. K. Kimbler, Soc. Pet. Eng. J., 60 (Feb., 1972).
29. P. M. Lichaa and L. Herrera, Soc. Pet. Eng. AIME, Paper No. 5304, 107, 1975.
30. A. David, AIChE Symposium Series, 69, No. 127, 56 (1971).
31. G. W. Preckshot, N. G. DeLisle, C. E. Cottrell, and D. L. Katz, Pet. Tech., 188 (Sept., 1942).
32. C. J. Radke and W. H. Somerton, Oil & Gas J. Proceedings, Vol. 1 - Oil (1977).
33. P. A. Farmanian, N. Davis, J. T. Kwan, R. M. Weinbrandt, and T. F. Yen, ACS Symposium Series 91, p. 103, 1978.
34. D. L. Katz and K. E. Beu, Indust. Eng. Chem., 37, 195 (1945).
35. D. T. Wasan, S. M. Shah, M. Chan, K. Sampath, and R. Shah, Chemistry of Oil Recovery, ACS Symposium Series, No. 91, 115 (Feb., 1979), R. T. Johansen and R. L. Berg, eds.

RECOVERY MECHANISMS OF ALKALINE FLOODING

T. P. Castor, W. H. Somerton and J. F. Kelly

Mechanical Engineering Department
University of California, Berkeley
Berkeley, California 94720, U.S.A.

In the alkaline flood process, the surfactant is generated by the in situ chemical reaction between the alkali of the aqueous phase and the organic acids of the oil phase. The surface-active reaction products can adsorb onto the rock surface to alter the wettability of the reservoir rock and/or can adsorb onto the oil-water interface to lower the interfacial tension. At these lowered tensions (1-10 dyne/cm), surface or shear-driven forces promote the formation of stable oil-in-water emulsions or unstable water-in-oil emulsions; the nature of the emulsion phase depends on the pH, temperature, and electrolyte type and concentration. These different paths of the surface-active reaction products have created different recovery mechanisms of alkaline flooding. The four alkaline recovery mechanisms which have been cited in the recent literature are: (i) Emulsification and Entrainment, (ii) Emulsification and Entrapment, (iii) Wettability Reversal from Oil- to Water-Wet, and (iv) Wettability Reversal from Water- to Oil-Wet. These four mechanisms are similar in that alkaline flooding enhances the recovery of acidic oil by two-stage processes.

Studies on displacement dynamics and interfacial tensions were carried out to establish and improve recovery efficiencies of acidic crudes by alkaline agents. Displacement tests were carried out on restored state oil-field cores and on synthetic Ottawa sand-packs with permeabilities ranging from 100 to 3,500 millidarcies. Concomitant experiments were carried out with a spinning-drop tensiometer and a contact angle goniometer; capillary pressure-determined wettability indices were measured and the type and stability of emulsions were characterized. These experiments indicate that the recovery mechanisms cited in the literature are valid under specific conditions of pH, electrolyte type and concentration. The results also indicated that tertiary and secondary recovery efficiencies could be improved by (v) Partial Wettability Reversal from Water-Wet to Oil-Wet, (vi) Chromatographic Wettability Reversal, and (vii) Emulsification and Coalescence.

INTRODUCTION

During the primary depletion stage, crude oil is produced by the natural energies of the reservoir and the confined fluids. Below the bubble-point, pressure gas percolates out of the oil phase, coalesces and displaces the crude oil. The gas phase, which is much less viscous and thus more mobile than the oil phase, fingers through the displaced oil phase. In the absence of external forces, the primary depletion inefficiently produces only 10 to 30 percent of the original oil in place. In the secondary stage of production, water is usually injected to overcome the viscous resistance of the crude at a predetermined economic limit of the primary depletion drive. The low displacement efficiencies, 30 to 50 percent, of secondary waterfloods are usually attributed to vertical and areal sweep inefficiencies associated with reservoir heterogeneities and nonconformance in flood patterns. Most of the oil in petroleum reservoirs is retained as a result of macroscopic reservoir heterogeneities which divert the driving fluid and the microscopically induced capillary forces which restrict viscous displacement of contacted oil. This oil accounts for approximately 70 percent, or 300 x 10^9 bbl, of the known reserves in the United States.

Enhanced oil recovery processes: In the tertiary or enhanced production of the remaining reserves, an additional energy source is required to microscopically mobilize and macroscopically displace residual and bypassed crude. A thermal or chemical energy source is utilized to alter the mobility of the driving fluid and/or to reduce the restraining capillary forces. Stegemeier (1,2) groups the enhanced oil recovery mechanisms into two predominant types. In the first type the ratio of viscous to capillary forces is altered and in the second type the fluid phase volume is altered. The latter group should be expanded to include processes in which the preferential wettabilities of the porous medium are altered. Thus, the mobilization mechanisms can be generically classified as processes which:

(i) alter the viscous-capillary interaction, and
(ii) alter the residual phase's configuration.

The microscopic mobilization efficiency (or the percentage reduction in the residual oil saturation of a secondary waterflood) of tertiary surfactant floods has been experimentally correlated to be a function of the capillary number.

$$\text{Capillary number, Ca} = \frac{v\mu_w}{\sigma_{ow}\cos\theta} = \frac{\text{viscous forces}}{\text{surface forces}}$$

This dimensionless grouping represents the competitive interaction of the viscous driving forces and the restrictive capillary forces. High recovery efficiencies, > 90%, have been realized for laboratory systems in which the oil-water interfacial tension has been reduced to 10^{-2} - 10^{-4} dynes/cm. Such lowered tensions are difficult to establish and sustain under field conditions. Recently, more attention has been placed on mechanisms which alter the configuration of the residual phase (1,2).

The recovery of naturally acidic oils by alkaline flooding fits into the phase alteration category. The recovery mechanisms of these floods are varied since the surface active salts, which are formed by the in situ acid-base reaction, can adsorb onto the oil-water interface to promote emulsification or can absorb onto the rock surface to alter wettability. The exact recovery mechanism, recently reviewed by Johnson (3) depends on the pH and salinity of the aqueous phase, acidity of the organic phase and wettability of the rock surface (4,5). In this study an additional alkaline recovery mechanism is explored. This mechanism, Emulsification and Coalescence, depends on the valency of the electrolyte as well as the pH and salinity of the aqueous phase. The Emulsification and Coalescence mechanism for the recovery of acidic oils is similar to the Spontaneous Emulsification mechanism suggested by Schechter et al. (6) for the recovery of nonacidic oils with petroleum sulfonate solutions.

Alkaline recovery mechanisms: In a recent article, Johnson (3) reviewed the mechanisms by which alkaline flooding improved the recovery of acidic crudes from partially depleted reservoirs. The mechanisms were:

1. Emulsification and Entrainment (7)
2. Emulsification and Entrapment (8)
3. Wettability Reversal from Oil- to Water-Wet (9)
4. Wettability Reversal from Water- to Oil-Wet (4)

These four mechanisms concur that alkaline flooding enhances the recovery of acidic oil by two-stage processes. In the first and common stage of these alkaline recovery mechanisms, surface active salts are formed by the in situ acid-base reaction between the alkali contained in the floodwater and the organic acid present in the residual oil. The surfactants can adsorb onto the oil-water interface to lower the interfacial tension and thus promote emulsification under the action of surface driven forces (spontaneous) and/or under the action of shear driven forces (external and internal pressure gradients). The surfactants can also react with or adsorb on the rock surface to alter the wettability of the rock and the configuration of the residual ganglia or droplets of crude. Thus, in the first stage of the alkaline recovery processes, residual oil is mobilized as a result of configurational changes (emulsification/wettability alteration). The second stage of the alkaline recovery processes involves macroscopic production of the mobilized oil phase. In this stage, the overall recovery efficiency can be increased by improvement of the displacement efficiency through reduction in the mobility of the floodwater. These two stages, mobilization, and production, are interdependent since the emulsion type, nature, and method of formation are determinants in the incremental oil production and the production efficiency.

EXPERIMENTAL

Dynamic displacement experiments: Dynamic displacement studies were carried out on a well-defined synthetic system to test the hypothesis that the recovery and production efficiencies could be improved by defined and proposed alkaline recovery mechanisms (10). The synthetic system consisted of an Ottawa sandpack with permeabilities from 100-3,500 millidarcies, mineral oil traced with oleic acid and distilled water with reagent grade chemicals; a synthetic system was used to allow the definition and thus to establish the role of each variable in the recovery process.

The multifluid flow displacement apparatus used in the coreflooding experiments is shown schematically in Figure 1. The heart of the apparatus is a hollow lucite cylinder which is packed with sand. This "core" simulates an unconsolidated oil reservoir. The transient pressure drop and electrical resistivity across this core can be monitored continuously via transmitting and recording devices. The core is enclosed in an oven which can be used to simulate reservoir conditions. In this apparatus, a multiple of fluids can be injected into or bypassed around the core. These fluids include gases such as CO_2 and N_2 and as many as five different aqueous and organic phases. The liquid phases are deaerated in their holding chambers and stored under a blanket of N_2 and are collected in the downstream or lefthand portion of the apparatus. The breakthrough concentration profiles of injected chemical species are obtained by batch analysis of the collected fractions.

The porous medium consists of unconsolidated Ottawa sand contained in a cylindrical lucite or lexan polycarbonate core holder. A plastic core holder is used to minimize attrition during dry packing of the sand, to eliminate secondary loss of chemicals at the wall (e.g., reaction to form rust) and to allow flow visualization of the saturation fronts. The Ottawa sand is sieved and thoroughly cleaned in order to obtain reproducible surface characteris-

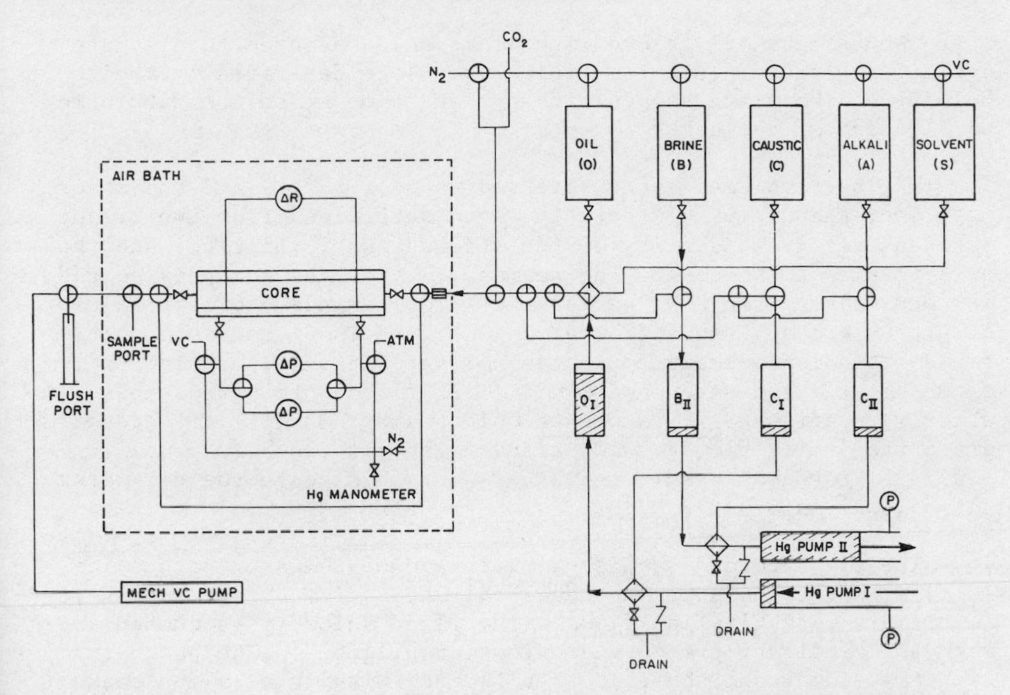

Fig. 1. Process flow diagram of multifluid flow displacement apparatus.

tics. The sieved sand was sequentially washed with (1) 2.0M hydrochloric acid, (2) aqua ammonia, (3) distilled water, (4) sodium tripolyphosphate solution, (5) 1.0M sodium chloride solution, (6) distilled water, (7) 0.1M sodium hydroxide solution, (8) distilled water, (9) warm chromic acid solution, and (10) copious amounts of distilled water. The hydrochloric acid wash leached out a considerable amount of iron; magnetic separation was used to remove the ferromagnetic impurities. This acid wash, which would remove any inorganic oxides, sulfides, and carbonates, was followed by a neutralizing aqua ammonia solution and distilled water. The sand was then treated with saturated sodium tripolyphosphate solution to deactivate any clays present and alternately contacted with distilled water and solutions of sodium chloride and sodium hydroxide to ensure the replacement of polyvalent ions with univalent ions. The final wash with warm chromic acid cleaning solution to oxidize and remove organic impurities was followed by prolonged water washing. The sand was then oven-dried at 150°F. The above treatment renders the Ottawa sand strongly water-wet and allowed reproducible dynamic and equilibrium flow behavior.

Extremely simple fluids are used for the aqueous, alkaline, and organic phases. Deaerated distilled water purified by a Milli-Q nucleopore membrane and CP reagent grade chemicals were used for the preparation of alkaline and saline solutions. A narrow-cut light oil traced with oleic acid was used to simulate the acidic crudes. All the fluids used were deaerated in their holding chambers and capped with a blanket of N_2 before saturation or flooding of the core.

The pore volume (later referred to as PV or τ) and the porosity, ϕ of the dry packed core is first determined from the weight and measured grain density of the Ottawa sand (2.65 g/cc) and the bulk volume of the core. The permeability of the sandpack to gas is then measured in a N_2 permeameter from a minimum of six values of pressure drop versus flowrate. The core is placed in a multi-fluid flow displacement apparatus for saturation with brine or fresh water. The saturated core is oil flooded to irreducible water saturation, S_{wir}, at rates which make the capillary pressure gradients negligible, $v \simeq 10$ ft/day. The oil flood is usually continued for about ten pore volumes; at this stage the oil-water ratio (OWR) is about 200.

The oil-saturated sandpack, with a connate water saturation, is first waterflooded to residual oil saturation before running a tertiary mode alkaline flood. The flood velocity is chosen to make the capillary pressure gradients negligible. The sandpack with a residual oil phase is then flooded with an alkaline phase until no additional oil is produced. During the chemical flood, the core is flooded at velocities which are high enough to result in negligible capillary pressure gradients but low enough to limit the ratio of viscous to capillary forces. This latter requirement prevents incremental production due to the singular action of externally imposed pressure forces since this condition could not be reproduced in the field. The alkaline flood velocity was chosen so as to limit the capillary number to between 10^{-5} and 10^{-4}.

The waterflood stage was replaced by an alkaline flood for a secondary mode alkaline flood. The pressure drop and produced fluids were monitored during the secondary and tertiary floods. The core is removed between floods and weighed in order to determine fluid saturations. The saturation is checked against that obtained from the volume of the produced fluids; in most cases, the volume determination and the mass determination of saturation differed by only 2 to 3 percent in saturation. The volume-determined saturations were used in the reported results. The emulsion characteristics and recovery efficiencies of alkaline floods discussed are listed in Table 1. The end-point saturations, permeabilities and wettabilities (as inferred from contact angle measurement) are listed in Table 2. The dimensions and properties of the unconsolidated Ottawa sandpacks used in these alkaline floods are listed in Table 3. Neutral pH floods are also listed. The neutral pH floods offer a base line for the recovery efficiencies of the alkaline floods.

Table 1. Emulsion Characterizations and Recovery Efficiencies of Alkaline Floods

Run DY-	Flood Type	Alkali/Electrolyte	pH	Acid No.	Velocity (ft/day)	Emulsion Character			% Oil Recovered		
						Type	Stability	Mode	E_{RW}	E_{RC}	E_{RR}
12	S	Distilled water	7.0	0.0	22.0	Nil	Nil	Nil	73.2	--	--
14	S	Distilled water	7.0	5.0	22.5	Nil	Nil	Nil	73.0	--	--
59	S	1.0M NaCl (Brine)	7.0	2.0	16.4	Nil	Nil	Nil	66.8	66.8	0
15	T	0.025M NaOH	12.4	5.0	7.0	O/W	VS	Spon	73.0	73.0	0
18	T	0.025M NaOH	12.4	0.5	55.3	O/W	VS	Shear	55.2	80.0	44.7
20	T	0.025M NaOH/1.0M NaCl	12.4	2.0	8.1	W/O	S	Spon	68.1	68.1	0
25	S	0.025M NaOH/1.0M NaCl	12.4	2.0	23.2	W/O	S	Spon	(75.5)	71.5	(0)
28	S	0.025M NaOH/1.0M NaCl	12.4	2.0	7.9	W/O	S	Spon	(71.8)	71.8	(0)
56	S	0.025M NaOH/1.0M NaCl	12.4	2.0	2.7	W/O	S	Spon	(83.0)	83.0	(0)
102	T	0.05M $Na_2B_4O_7$/1.0M NaCl	8.6	2.0	8.6	W/O	US	Shear	63.0	80.0	43.5
104	T	0.05M $Na_2B_4O_7$/1.0M NaCl	8.6	2.0	8.6	W/O	US	Shear	74.4	79.1	18.2
53	S	0.05M $Na_2B_4O_7$/1.0M NaCl	9.1	2.0	7.4	W/O	US	Shear	(71.4)	71.4	0
106	T	0.05M $Na_2B_4O_7$	8.6	2.0	9.3	O/W	S	Spon	63.0	63.0	0
30	T	0.1M Na_2CO_3/1.0M NaCl	11.4	2.0	8.6	W/O	US	Shear	71.0	78.3	23.1
37	S	0.1M Na_2CO_3/1.0M NaCl	11.4	2.0	8.0	W/O	US	Shear	(62.3)	62.3	(0)
38	T	0.02M $Ca(OH)_2$	12.0	2.0		W/O	US	Spon	72.9	85.4	46.2
44	S	0.02M $Ca(OH)_2$	12.3	2.0	6.0	W/O	US	Spon	(72.0)	83.5	(46.4)

Table 1 (Continued)

Run DY-	Flood Type	Alkali/Electrolyte	pH	Acid No.	Velocity (ft/day)	Emulsion Character			% Oil Recovered		
						Type	Stability	Mode	E_{RW}	E_{RC}	E_{RR}
108	T	0.02M $Ba(OH)_2$	12.2	2.0	9.5	W/O	VUS	Shear	69.8	69.8	0
62	T	0.025M NaOH/0.02M $CaCl_2$	12.4	2.0	8.1	W/O	US	Spon	74.2	79.4	16.4

[a] Parentheses indicate best estimate from same or similar condition.

Table 2. End-Point Saturations, Permeabilities and Wettabilities of Alkaline Floods

Run DY-	Flood Type	Alkali/Electrolyte	S_{wir}	S_{orw}	S_{orc}	k°_{ro}	k°_{rw}	k°_{rwc}	σ_{ow}[a] dyne/cm	θ_{ow}	E_{RR}
12	S	Distilled water	0.10	0.23	--	0.80	0.43	--	30	25°	--
14	S	Distilled water	0.10	0.26	--	0.81	0.46	--	30	25°	--
59	S	1.0M NaCl (Brine)	0.26	0.24	--	0.81	0.43	--	30	20-30°	--
15	T	0.025M NaOH	0.10	0.26	0.26	0.72	0.41	0.0	(0.4)	20-30°	0
18	T	0.025M NaOH	0.33	0.30	0.14	0.68	0.37	0.12[a]	(0.4)	20-30°	44.7
20	T	0.025M NaOH/1.0M NaCl	0.28	0.23	0.23	0.81	0.46	0.39	1.5	120°	0
25	S	0.025M NaOH/1.0M NaCl	0.19	0.23	0.23	0.81	0.46	0.32	1.5	120°	0
28	S	0.025M NaOH/1.0M NaCl	0.27	0.21	0.21	0.90	0.46	0.7[b]	1.5	120°	0
56	S	0.025M NaOH/1.0M NaCl	0.43	(0.24)	0.10	1.00	(0.43)	0.39	1.5	120°	0
102	T	0.05M $Na_2B_4O_7$/1.0M NaCl	0.24	0.28	0.15	0.60	0.4	0.85	1.3	170°	43.5
104	T	0.05M $Na_2B_4O_7$/1.0M NaCl	0.14	0.22	0.18	0.85	0.40	0.42	1.3	170°	18.2
53	S	0.05M $Na_2B_4O_7$/1.0M NaCl	0.30	0.25	0.20	1.00	0.19	0.43	1.3	170°	25.0
106	T	0.05M $Na_2B_4O_7$	0.24	0.28	0.28	0.90	0.46	0.46	0.23	0-30°	0
30	T	0.1M Na_2CO_3/1.0M NaCl	0.27	0.21	0.16	0.90	0.46	0.79	1.39	146°	23.1
37	S	0.1M Na_2CO_3/1.0M NaCl	0.20	0.32	0.30	1.00	0.46	0.61	1.39	146°	6.3
38	T	0.02M $Ca(OH)_2$	0.04	0.26	0.14	0.62	0.55	0.22	NA	120°	44.0
44	S	0.02M $Ca(OH)_2$	0.18	(0.28)[b]	0.15	0.69	0.46	0.19	NA	120°	(46.4)

Table 2 (Continued)

Run DY-	Flood Type	Alkali/Electrolyte	S_{wir}	S_{orw}	S_{orc}	k°_{ro}	k°_{rw}	$k^{\bullet}_{rwc}$	σ_{ow}[a] dyne/cm	θ_{ow}	E_{RR}
105	T	0.02M $Ba(OH)_2$	0.14	0.26	0.26	0.91	0.36	0.86	NA	120°	0
62	T	0.025M NaOH/0.02M $CaCl_2$	0.23	0.20	0.16	1.00	0.67	0.30	NA	NA	20.1

[a] Minimum values which are measured in a spinning-drop tensiometer.

[b] Parentheses indicate best estimate from same or similar condition.

NA--Not available.

Table 3. Dimensions and Properties of Unconsolidated Sandpacks Used in Alkaline Floods

Run DY-	Flood Type	Alkali/Electrolyte	$A_x(cm^2)$	L(cm)	$PV(cm^3)$		S_{wi}	k_g	k_1	k_1/k_g
12	S	Distilled water	2.88	15.88	17.69	0.298	0.99	815	592	0.726
14	S	Distilled water	2.88	15.88	17.69	0.298	0.96	815	592	0.726
59	S	1.0M NaCl (brine)	12.13	21.27	83.27	0.320	1.0	493	285	0.578
15	T	0.025M NaOH	2.88	15.88	17.69	0.298	1.00	815	592	0.726
18	T	0.025M NaOH	3.47	15.10	21.73	0.414	1.50	963	683	0.709
20	T	0.025M NaOH/1.0M NaCl	2.88	20.30	19.35	0.331	NA	NA	3450	NA
25	S	0.025M NaOH/1.0M NaCl	2.88	20.60	18.06	0.304	1.01	1817	1220	0.671
28	S	0.025M NaOH/1.0M NaCl	2.99	21.27	22.04	0.346	0.99	1515	1245	0.822
56	S	0.025M NaOH/1.0M NaCl	12.13	21.27	85.53	0.331	0.99	439	151	0.344
102	T	0.05M $Na_2B_4O_7$/1.0M NaCl	2.88	20.61	19.51	0.329	NA	6100	4290	0.703
104	T	0.05M $Na_2B_4O_7$/1.0M NaCl	2.88	20.61	NA	NA	NA	5800	3920	0.676
53	S	0.05M $Na_2B_4O_7$/1.0M NaCl	2.99	21.27	19.37	0.305	0.99	405	273	0.674
106	T	0.05M $Na_2B_4O_7$	2.88	20.30	20.70	0.354	NA	NA	1020	NA
30	T	0.1M Na_2CO_3/1.0M NaCl	2.88	21.29	22.04	0.346	0.99	1515	1245	0.822
38	T	0.02M $Ca(OH)_2$	2.88	20.30	NA	NA	NA	702	349	0.497
44	S	0.02M $Ca(OH)_2$	2.99	21.27	18.80	0.300	0.99	241	181	0.751
108	T	0.02M $Ba(OH)_2$	2.88	20.30	21.00	0.359	NA	7900	6080	0.770
62	T	0.25M NaOH/0.02M CaCl	2.99	21.27	19.37	0.305	0.97		45	

NA--Not available.

Equilibrium displacement experiments: Equilibrium displacement experiments were carried out to determine pore-size distribution from air-brine curves, capillary pressure versus saturation, and preferential wettabilities of reactive and non-reactive oil/water systems (10). The well-defined synthetic system, described in the previous section and used in the dynamic displacement experiments, was used in the equilibrium displacement studies. The equilibrium displacement characteristics were determined by a constant speed centrifugal technique (11).

The basic unit used in this work consisted of a centrifuge (International, Size 2) with an externally attached voltage regulator for speed control up to 3,100 rpm. The speed was measured by a stroboscopic Xenon phototube driven by an oscillator circuit which was calibrated to the frequency of the house AC cycle before each capillary pressure run. Speeds lower than 600 rpm were measured by isolating double or quadruple images with the stroboscope. A belt driven multispeed attachment with a four place conical head was used with a second centrifuge to develop extremely high speeds required for the determination of residual saturations. The centrifuge tubes used to contain the sandpack during determination of the air displacing water or oil displacing brine capillary pressure curves are very similar to those used by Slobod et al. (11). The capillary pressure curves of water or brine displacing oil were measured in thick glass holders with a graduated nipple which are stoppered and inverted into a slotted brass tube supported by a trunnion ring. This setup is very similar to the one used by Donaldson et al. (12) in the determination of wettability from secondary imbibition and drainage curves.

The fractional wettability of the porous medium, defined as a measure of the fraction of the internal surface of the porous medium in contact with one fluid (13) was determined by the USBM method (14). In the USBM method for determining wettability, the logarithm to the base 10 of the ratio of areas under a secondary drainage (A_3) and an imbibition capillary pressure (A_2) versus water saturation curves is used to define the wettability scale. The scale varies from positive for water-wet conditions, through zero for intermediate-wet conditions, to negative for oil-wet conditions. The scale is believed to be independent of the pore geometry since the influence of the pore geometry, which is similar for the imbibition and drainage curves, cancels out on taking the ratio of areas. This scale, then, is a measure of the number and distribution of pores which are oil-wet and water-wet over the saturation range from irreducible water saturation to residual oil saturation.

The wettability indices, Ω, of the non-reactive and reactive alkaline systems are listed in Table 4. The listed contact angle measurements were made with a contact angle goniometer (NRL Ramé-

Table 4. Wettability Indices and Contact Angles for Reactive Alkaline (Aqueous)-Acidic (Organic) Systems on a Quartz Surface

Run EQ-	Alkali/Electrolyte	Endpoint Sat. S_{wir}	S_{or}	E_R	θ	$\log_{10}\{A_1/A_2\}$	Ω
9b	0.1M NaCl	0.335	0.12	82.0	30°	1.15	0.87
10a	1.0M NaCl	0.20	0.30	62.5	30°	0.76	0.54
10b	1.0M NaCl	0.20	0.30	62.5	30°	0.76	0.54
10c	1.0M NaCl[a]	0.36	0.19	70.8	30°	0.88	0.55
10d	1.0M NaCl[b]	0.36	0.08	87.5	30°	0.88	0.44
17	0.025M NaOH/1.0M NaCl	0.15	0.26	69.8	120°	0.25	-0.023[d]
18	0.025M NaA	0.16	0.27	68.0	120°	0.32	0.045[d]
16	0.05M $Na_2B_4O_7$/1.0M NaCl	0.40	0.21	66.7	170°	0.76	0.48[d]
15,19	0.1M Na_2CO_3/1.0M NaCl	0.15	0.08	90.6	146°	0.21	-0.069[d]
20	0.05M $Ca(OH)_2$	0.30	0.16	77.9	120°	0.01	-0.23[d]
21	0.05M $Ca(OH)_2$	0.20	0.10	87.5	120°	-0.29	-0.56[d]
22	1.0M NaCl	0.11	0.23	74.4	30°	0.65	0.54
23	0.05M $Na_2B_4O_7$/1.0M NaCl	0.25	0.11	85.7	170°	0.79	0.47

A_1--Area under primary drainage curve; A_2--Area under imbibition curve; A_3--Area under secondary drainage curve

Ω Wettability index, $\log_{10}\{A_3/A_2\}$

[a]No acid in displaced oil during imbibition cycle; [b]0.0285 oleic acid in displaced oil during imbibition cycle; [c]0.5M $CaCl_2$ in connate water; [d]In this calculation A_3 is estimated from Run EQ-10, i.e., $A_3 \simeq A_1*(A_3/A_1)$ Run EQ-10.

Hart). The angle between a quartz plate and an acidic oil droplet in an alkaline/electrolyte solution was measured. The oil-water-quartz contact angles are believed to be representative of the oil-water-Ottawa sand contact angles since the Ottawa sand used was 99.7 percent silica. The order of preferential water wettability of the reactive acid-base systems is: (1) $Na_2B_4O_7$/NaCl, (2) NaCl, (3) NaA, (4) NaOH/NaCl, (5) Na_2CO_3/NaCl, (6) $Ca(OH)_2$, and (7) $Ca(OH)_2$ with $CaCl_2$ in the connate water. Multivalent ions such as Ca^{++} or Ba^{++} act as activators for the adsorption of fatty acid on quartz surfaces (15). The surface chemistry of divalent ions supports the observation that calcium hydroxide-Ottawa sand-oleic acid systems are the most oil-wet of the alkaline systems. The presence of calcium chloride increases the wettability to oil by further depressing the electrical double layer on the silica surface and increasing the adsorption of the oleate ions and/or ion molecular complexes. The salt also has the effect of precipitating out the surface-active reaction products. The effects of the salt are evident in the ordering of the increasing water-wettabilities of the remaining alkaline-oleic systems.

Emulsion characterization experiments: There are several documented methods by which the emulsion phase can be characterized. The single and multidrop methods are tedious to perform in that unusually large amounts of samples are required before a statistically determinate result can be obtained. The experimental time of these methods is greatly reduced by placing the emulsified phases in a centrifugal field and measuring the volume reduction in the emulsified phase with time. In this method the reactive solutions are sheared (homogenized) in an electric blender at different speeds in order to control the size of the emulsion drops. The attraction of droplets for each other and thus their stability depends on the mass or volume of the drops and the distance of separation (through Van der Waals' forces of attraction). The third method entails shearing of the two phases through vigorous shaking and allowing the emulsified phase to separate by standing in a gravitational field. In this method, the diameter of the emulsion droplet is determined, for an equal amount of shear, by the type and concentration of the electrolyte and the reacting species. In all of these methods, the influence of capillary forces on the emulsification process and the nature of the emulsified phase are not considered. These comparative methods are then only of qualitative utility when applied to a porous medium.

The constant shear method was used in the present work for the characterization of the emulsion phase because alkaline floods are carried out under analogous conditions of low but constant shear. Using the constant shear procedure, Schulman and Cockbain (16) measured the stability of O/W emulsions by observing the time required for the first visible signs of separation of the two phases. These authors concluded that this method is satisfactory to compare relative stability of emulsions prepared under the same conditions even though stability is based on the efficiency of the emulsification process and the subjective differentiation of the opacity of the emulsified phase. In the experiments, reported in the following, 5.5 cc of the dyed organic phase (Chevron 410 H Thinner laced with 0.0285 moles/l of oleic acid) was gently contacted with 5.5 cc of alkaline solution in a centrifuge tube. The initial contact of the two phases was made with a minimum of physical disturbance in order to establish the occurrence and extent of spontaneous emulsification; the oil was pipetted onto the glass wall about 2 mm above the alkaline surface. After 24 hours, the tubes were vigorously hand-shaken for an equal number of times. The behavior of the reactive phases, before and after shearing, was followed in time. The reactive phases were intermittently photographed. The relative stability of these emulsions was determined by comparing the rates at which the volume of emulsion in the centrifuge tubes decreased. The emulsion type was determined by the addition of emulsion drops to a small pool of water or oil on a glass slide. The fluid (water or oil) which spontaneously attracted the emulsion drop is the continuous phase. The initial and final pH, emulsion type, and relative stability of the emulsions in the systems studied are listed in Table 5.

RESULTS

Mechanistic interpretations: The results of the dynamic and equilibrium displacement experiments are used to evaluate and further define mechanisms by which alkaline floods increase the displacement and recovery of acidic oil in secondary mode and the tertiary mode floods. The data sets used in the mechanistic interpretations of alkaline floods are (a) overall and incremental recovery efficiencies from dynamic and equilibrium displacement experiments, (b) production and effluent concentration profiles from dynamic displacement experiments, (c) capillary pressure as a function of saturation curves and conditions of wettability from equilibrium displacement experiments, (d) interfacial tension reduction and contact angle alteration after contact of aqueous alkali with acidic oil and, (e) emulsion type, stability, size and mode of formation. These data sets are used to interpret the results of the partially scaled dynamic experiments in terms of two-stage phase alteration mechanisms of emulsification followed by entrapment, entrainment, degrees and states of wettability alteration or coalescence.

Emulsification and entrainment: The emulsification and entrainment mechanism was observed to occur during high pH flooding of low-acid number oils in low salinity environments. This mechanism is manifested by partial entrainment and partial entrapment of the emulsion phase. Complete entrainment of the emulsion

Table 5. Characterization of Emulsion Type and Stability

Alkali/Electrolyte	pH		Emulsion Type	Relative Stability
	Initial	Final		
Water	7	7	Nil	
0.025M NaOH	12.4	10.6	O/W	VS
0.025M NaA	12.4	9.85	O/W	VS
0.025M NaOH/0.1M NaCl	12.4	9.81	O/W	VS
0.025M NaOH/0.3M NaCl	12.4	9.15	?	US
0.025M NaOH/0.5M NaCl	12.4	9.8	W/O	VS
0.025M NaOH/1.0M NaCl	12.4	10.1	W/O	VS
0.05M $Na_2B_4O_7$/1.0M NaCl	8.9	8.9	W/O	US
0.025M $Ba(OH)_2$	12.2	11.8	W/O	VUS
0.02M $Ca(OH)_2$	12.2	11.65	W/O	US
0.025M NaOH/0.00005M $CaCl_2$	12.4	10.3	O/W	VS
0.025M NaOH/0.001M $CaCl_2$	12.4	11.1	O/W	VS
0.025M NaOH/0.005M $CaCl_2$	12.4	10	O/W	VS
0.025M NaOH/0.02M $CaCl_2$	12.4	10.45	W/O	US
0.02M $Ca(OH)_2$/0.02M $CaCl_2$	12.2	11.7	W/O	US

VS--Very stable
US--Unstable
VUS--Very unstable

droplets, in contrast to complete entrapment, never occurred in the uniform sandpacks. The emulsification and entrainment mechanism is illustrated in Figure 2 by the pressure, production and concentration history of a tertiary alkaline flood of a low acid # oil (acid # = 0.5). A concentrated emulsion phase broke through at τ = 0.64 PV (around the nominal waterflood value) and stopped after 1.21 PV of caustic was injected. The emulsion phase traveled with the concentration shock front of the depleted caustic slug whose pH jumped 4 units at τ = 0.64 PV. The pressure increase after cessation of emulsion production indicates that redistribution of the entrapped phase continued to occur after the partial entrainment stage. The produced oil fractions were photomicrographed within a few minutes of sampling. The average size of the produced emulsion droplets was around 14 μm (Figure 3). Larger O/W droplets were produced at later times, approximately 25 μm at τ = 0.83 PV and approximately 60 μm at τ = 1.2 PV. The coronas around the droplets in this photomicrograph are speculated to be the interfacial film which limits the coalescence of flocculated O/W emulsions. These droplets are similar to the interference phase contrast micrograph which was used by Wasan et al. (17) to measure a 1.1 micron thick interfacial film around a 16 μm oil droplet for a non-equilibrated 0.02 wt percent NaOH/1.0 wt percent NaCl--acidic crude system.

Entrainment occurred because the stable O/W emulsions were on the average the same size (14 μm) as or smaller than the average pore diameter of the sandpack. The residence time for the sodium

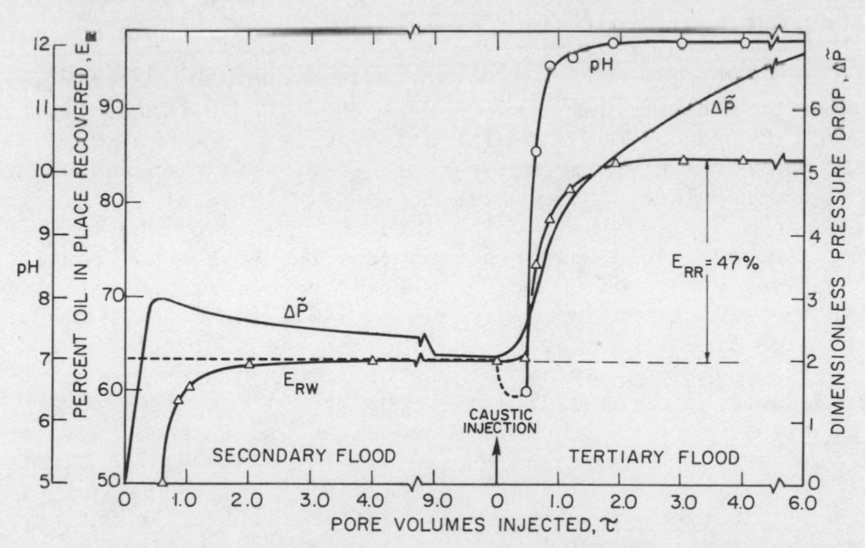

Fig. 2. Production, concentration and pressure histories of tertiary sodium hydroxide flood of Acid Number 0.5 Oil as a function of pore volumes throughput.

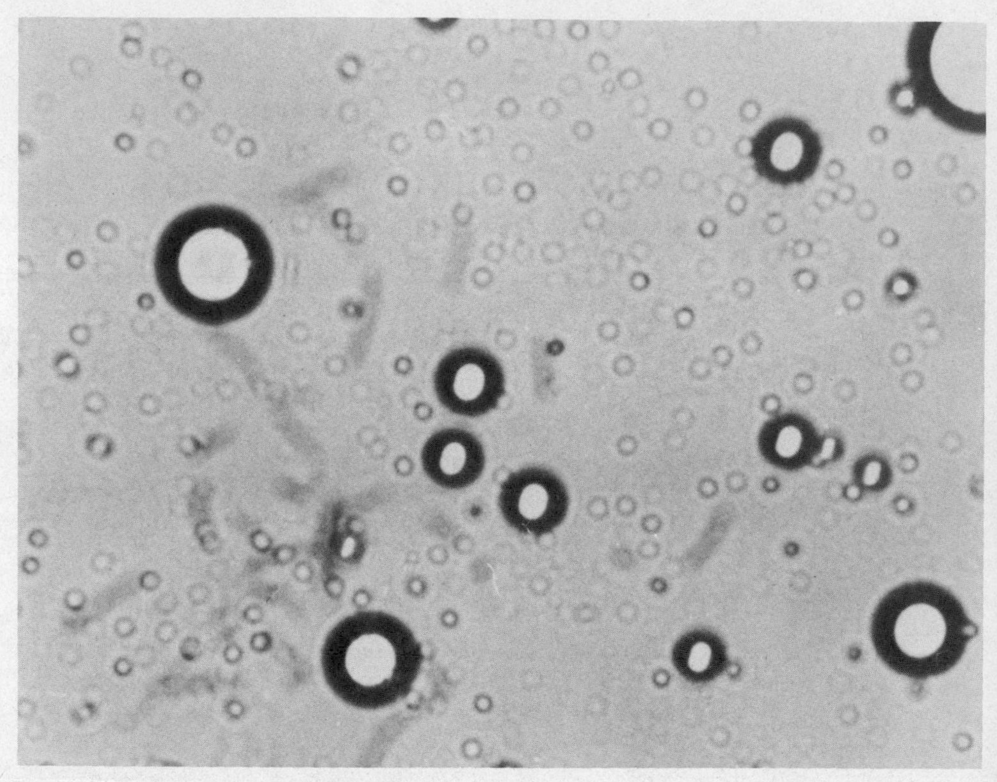

Fig. 3. Stable O/W emulsions produced by shearing an aqueous alkali with a residual acidic oil phase in a porous medium (droplet sizes ~ 14 μm).

hydroxide in the core was much shorter than the residence time required for spontaneous emulsification between the aqueous and organic phases. This suggests that the emulsification was shear induced and resulted in the formation of smaller droplets due to the additional energy input. The shear dominance of the emulsification process could have resulted from the high shear conditions (flooding velocities of 55 ft/day), from the lowered potential for spontaneous emulsification (acid number = 0.5) or from a combination of high flooding velocity and low reaction potential. The high shear conditions may not be the critical parameter since a 0.025M NaOH flood of acid #2.0 oil (Run DY-19) resulted in complete entrapment of the aqueous phase which was injected at a rate of 23 ft/day.

Emulsification and entrapment: The emulsification and entrapment mechanism was observed to occur during high pH flooding of moderate acid number (> 2.0) oils in low salinity environments. This mechanism is manifested by complete plugging of the homogeneous sandpacks which are uniform in porosity and permeability.

The emulsification and entrapment mechanism is illustrated in the pressure and production history of the tertiary alkaline flood, Run DY-15 shown in Figure 4. In this experiment, the flow rate decreased to zero just before breakthrough of the driving fluid (at about τ = 0.6 PV in a nominal waterflood). Substantial differences in the upstream pressure of the flow apparatus (400 psi at τ = 1.2 PV) and the differential pressure across the core (75 psi at τ = 1.2 PV) indicate that most of the emulsification and entrapment occurred in the entrance region of the sandpack. This mechanism was repeatedly observed in high pH, non-saline floods of moderate acid number (> 2.0) oils.

Entrapment occurs if the average drop size of the emulsion phase is greater than the average pore size of the porous medium and persists if the emulsion phase is stable. The stability condition is met since stable O/W emulsions are formed at large planar interfaces under similar conditions of high pH and low salinity. Spontaneous emulsification would favor the emulsification and entrapment mechanism since larger drop sizes are formed in systems which emulsify under limited shear; the droplet size reduces, to minimize the surface free energy, with increase in applied shear or with decrease in interfacial tension. The low saline, high pH systems, which exhibit the emulsification and entrapment mechanism, do spontaneously emulsify with moderate acid number oils. The droplet size of these emulsions ranged from 4-40 micron with an average of 16 μm in an unblended 0.025M NaOH/0.07M oleic acid (acid # 5.0) system (Figure 5). For an identical system which is sheared,

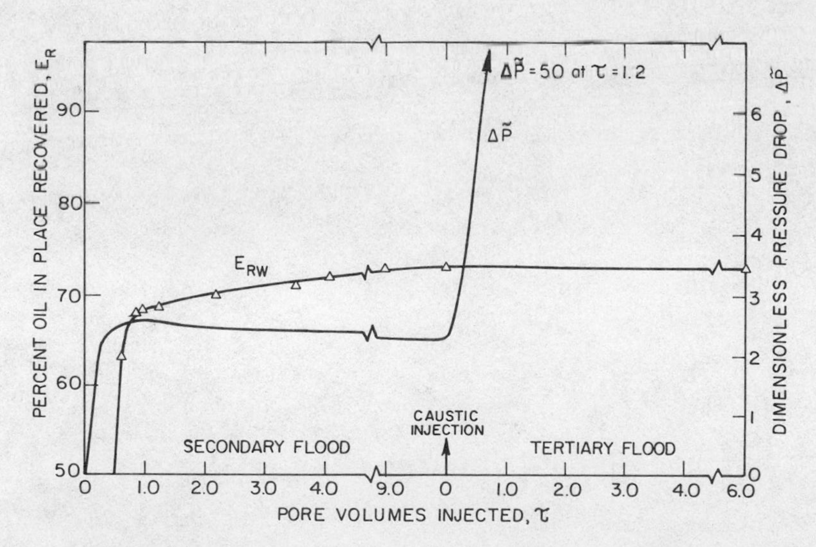

Fig. 4. Production, concentration and pressure histories of tertiary sodium hydroxide flood of Acid Number 5.0 oil as a function of pore volumes throughput.

the droplet size ranged from 3-15 μm with an average of 5 μm. The diameter of the pore constrictions, in the sandpacks which plugged, ranged from 0-30 microns with an average of 10 microns (as determined from air-brine capillary pressure curves). Therefore, the emulsion droplets which are formed under conditions of limited shear, would entrap in the porous medium. These droplets remain entrapped unless the interfacial tension is lowered sufficiently to increase the ratio of the viscous to the surface forces. No displacement of emulsions was observed even though the capillary number for these systems, $Ca \simeq 1 \times 10^{-3}$, was high enough to initiate mobilization of the entrapped emulsion droplets which are water external emulsion droplets.

Emulsification and wettability alteration: Increases in the recovery rate and the microscopic mobilization efficiency of residual oil were observed in high pH/high salinity and moderate pH/high salinity alkaline floods. These increases are due to the formation of W/O emulsions in the presence of a univalent electrolyte. Bulk phase experiments of earlier section indicate that the salt concen-

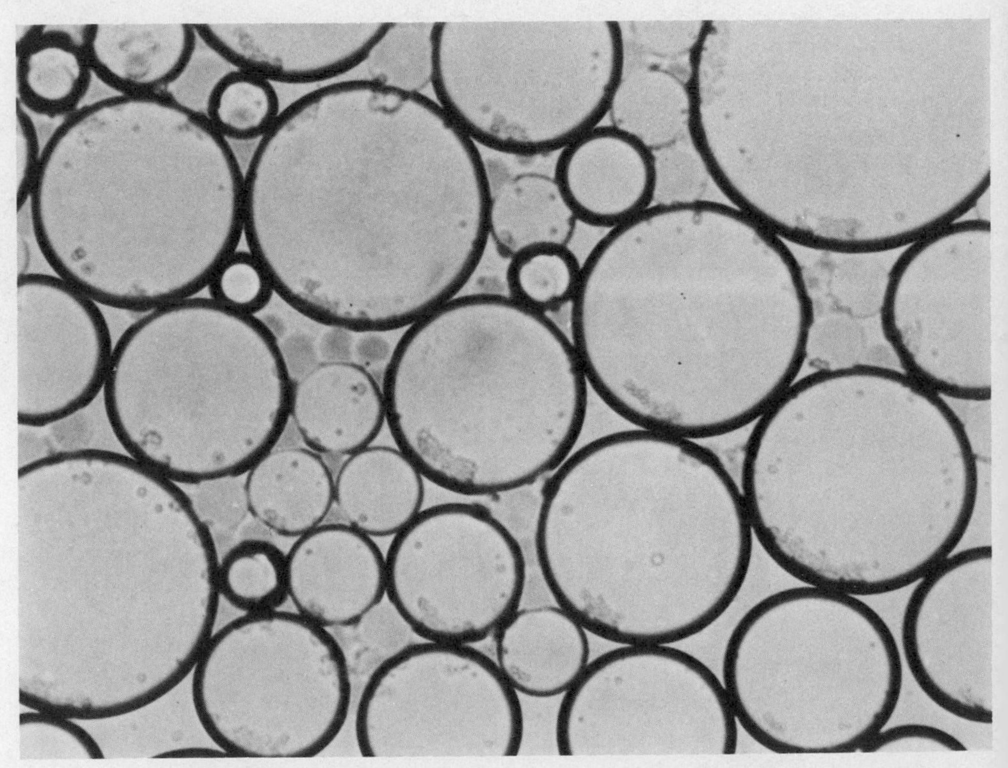

Fig. 5. Stable O/W emulsions which are formed spontaneously at an aqueous alkali-organic acid interface (droplet sizes ~ 4-40 μm).

tration should be greater than 0.3M. The surface active reaction products, which are "salted" out of solution, adsorb onto the rock surface and alter the wettability of the reservoir rock. The several possible mechanisms are partial, complete, and alternate wettability reversal.

High pH/high salt alkaline (such as 0.25M NaOH/1.0M NaCl) systems appear to improve the rate of recovery of acidic oils by an emulsification and partial wettability reversal mechanism. Partial wettability reversal is suggested by the intermediate magnitude of the externally measured contact angle, 120° and the wettability index of the porous media, -0.023 (Table 4).

Figure 6 illustrates the improvement in the production efficiency of secondary alkaline floods (Runs DY-25 and DY-28) relative to secondary waterfloods (Runs DY-12 and DY-14). The efficiency of oil production, as measured by the lowered WOR at breakthrough, is a function of the alkaline flood velocity. Oil production was complete after injection of 1.1 PV of caustic at 24 ft/day in Run DY-25 and 0.68 PV at 8 ft/day in Run DY-28 with equal total recovery. At a lowered rate of 2.6 ft/day shown in Figure 6, the recovery efficiency of the secondary alkaline flood increased from 72 percent at 8 and 24 ft/day to 83 percent at 3 ft/day (Run DY-56). An increase in production efficiency accompanied the increase in recovery efficiency since 72 percent of the oil in phase was recovered after injecting 0.41 pore volumes of NaOH/NaCl solution. The significance of the increased recovery at a flood velocity of 3 ft/day was uncertain because the recovery efficiency of oil from a water-wet medium could very well be a function of the irreducible water saturation, S_{wir}.

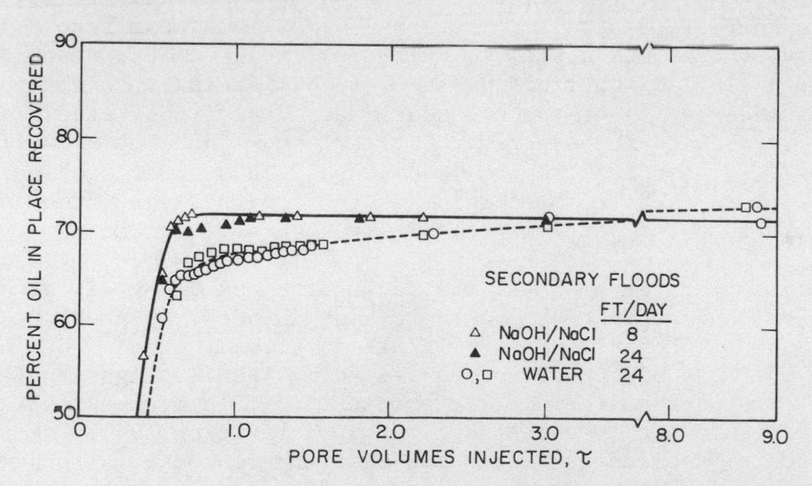

Fig. 6. Production efficiencies of secondary high pH/high salt alkaline floods as a function of pore volume throughput.

The improved production efficiency of the NaOH/NaCl flood resulted from the formation of W/O, water-in-oil, emulsions under the limited shear conditions. The alkaline phase imbibes into the oleic phase as a result of the interfacial chemical reaction. The swollen oil phase, together with its altered configuration, reduces the area available for the passage of the alkaline floodwater. The produced water-oil ratio, WOR, decreases as a result of the lowered water mobility. The consistent presence of a light, emulsion phase at the very end of the oil production stage indirectly confirms the occurrence of the in situ emulsification step. The production efficiency increases with decrease in injection rate because of mass transfer limitations of the interfacial chemical reaction; hydrodynamic effects would act in the opposite direction. The degree of in situ emulsification and wettability alteration, and the correspondent mobility reduction, depends on the residence time of the reactants in the core.

The residual oil saturation is not decreased because the swollen oil phase destabilizes to its original volume after the emulsification step. Such creaming and flocculation of W/O emulsions in a high pH/high univalent salt environment is demonstrated in the emulsion characterization experiments described later. The stable nature of these W/O emulsions, which is classified in Table 5, precludes the formation of local regions of enhanced oil saturation through contacting of the residual oil subsequent to emulsification and partial wettability reversal. The mechanism by which isolated and entrapped ganglia improved the production but not the recovery efficiency is illustrated in Figure 7.

The incremental production of acidic oil by moderate pH (buffered)/high salinity alkaline systems occurred by what is believed to be a complete wettability reversal mechanism. Complete wettability reversal is suggested by the magnitude of the measured contact angle, 170° but not by the wettability index of the porous media. The results of the secondary and tertiary buffered floods appear to confirm the experimental results of Cooke et al. (4); the work of these researchers is discussed in a later section. It will be shown that this mechanism is not a direct extension of the emulsification and partial wettability reversal mechanism.

In the tertiary mode floods, Figures 8 and 9, an oil bank is formed which is preceded by a sharp rise in both the pressure drop and the pH. The sharp pressure peaks in Figures 8 and 9 demonstrate that the oil phase was mobilized by locally high gradients in pressure whereas the diffuse pressure profile of the tertiary high pH/high salinity alkaline flood appears to be caused by the formation of swollen ganglia which restrict aqueous flow and which increase the macroscopic displacement efficiencies.

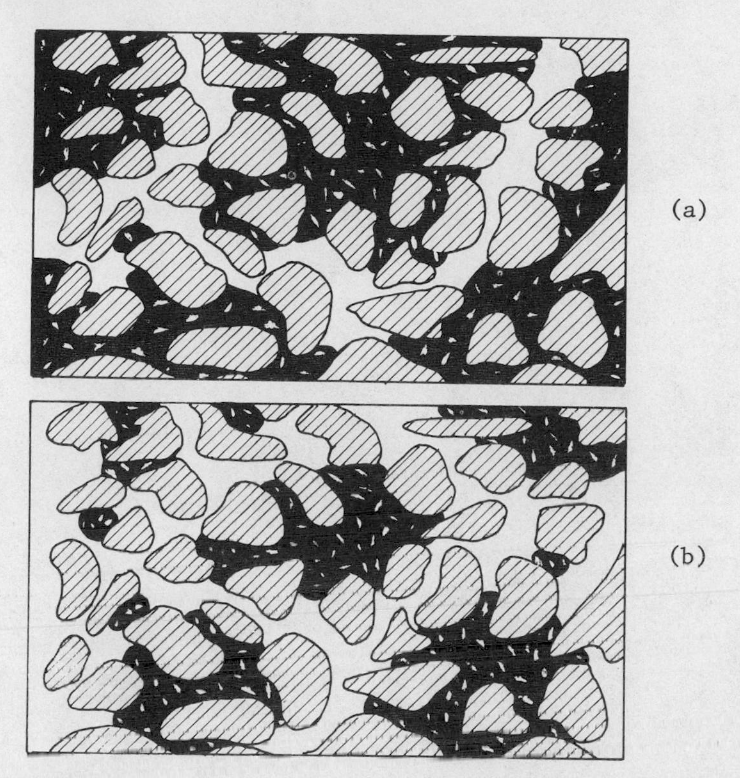

Fig. 7. (a) Swelling which results in an increased production efficiency; (b) destabilization of W/O emulsions to the original oil saturation.

The decreases in the water relative permeabilities of the high pH/high salt alkaline floods are directly contrasted with the increases in the relative permeabilities to water at the end of the moderate pH/high salt flood (compare the end point relative permeabilities column in Table 2). The increased permeability to water is believed to be caused by the formation of rigid interfacial films (which increases the resistance to flow in oil filled pores) and by the oil-wet conditions (under which water flows in the less restrictive flow paths). Such a reduction in permeability, which has been used to indicate the existence of a low tension mechanism (18), is not a valid low tension index since the interfacial tension minimum is only 3.5 dynes/cm and the capillary number is 1×10^{-4} for the buffered alkali/salt-oleic acid system.

High pH/low salt alkaline floods, of the tertiary type, were found to enhance the recovery of acidic oils if the connate water at waterflood residual oil saturation contained a high concentration of a univalent electrolyte such as sodium chloride (19). In these alkaline floods, transients in physico-chemical behavior

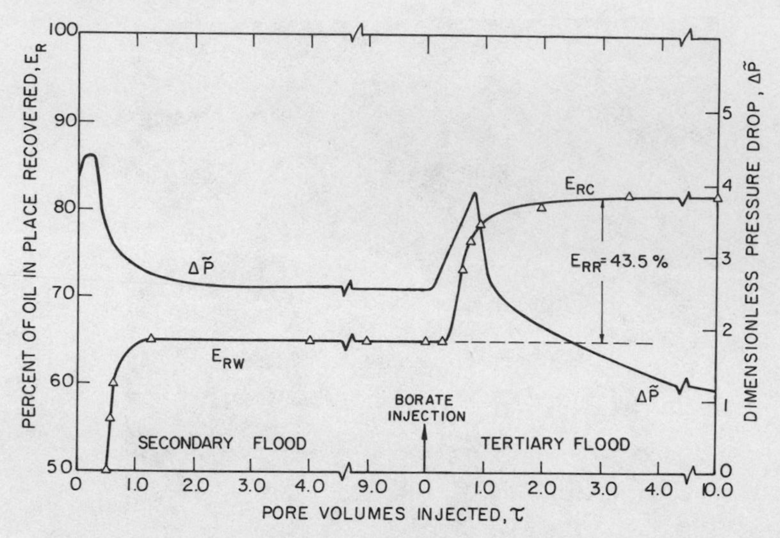

Fig. 8. Production, concentration and pressure histories of secondary waterflood and tertiary sodium borate/sodium chloride flood at pH = 8.6.

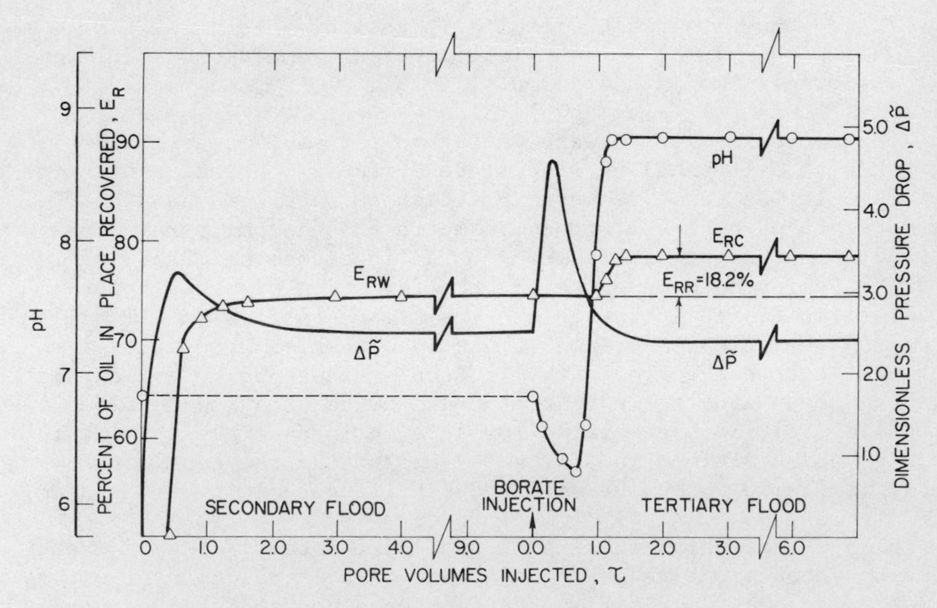

Fig. 9. Production, concentration and pressure histories of secondary waterflood and tertiary sodium borate/sodium chloride flood at pH = 8.6.

would appear responsible for the enhancement in recovery since tertiary production was not stimulated by either of the two constant chemistry processes which are limiting conditions. High pH/low salt alkaline floods of acidic oils in a low salinity environment resulted in emulsification and complete entrapment and high pH/high salt alkaline floods improved the production but not the recovery efficiency of acidic oils in a high salinity environment. The incremental production from high pH/low salt alkaline flooding of acidic oils in a high-salt environment is most probably due to wettability gradients which are established by the step change in electrolyte concentration at the inlet of the core. This recovery mechanism is very similar to the chromatographic wettability reversal mechanism elucidated by Michaels and coworkers (20,21,22).

The enhanced recovery of residual oil, which occurs as a consequence of wettability gradients in a reactive alkaline-acidic system, is illustrated in Figure 10. Incremental production in this tertiary flood begins about 0.6 PV after caustic injection and ends after injection of approximately 1.0 PV of caustic. This production is preceded by a pressure peak with a maximum which is two times the steady-state pressure drop of the secondary waterflood. The pressure drop maximum coincides with the concentration minimum of the hydroxyl ion species which elutes at the rear of the tertiary oil bank.

The chromatographic transport of the high pH/low salt alkaline slug ensures a water-wet trailing edge of the tertiary oil bank since oleate ions promote water-wetness in low salinity environments. The high salinity environment encountered at the leading

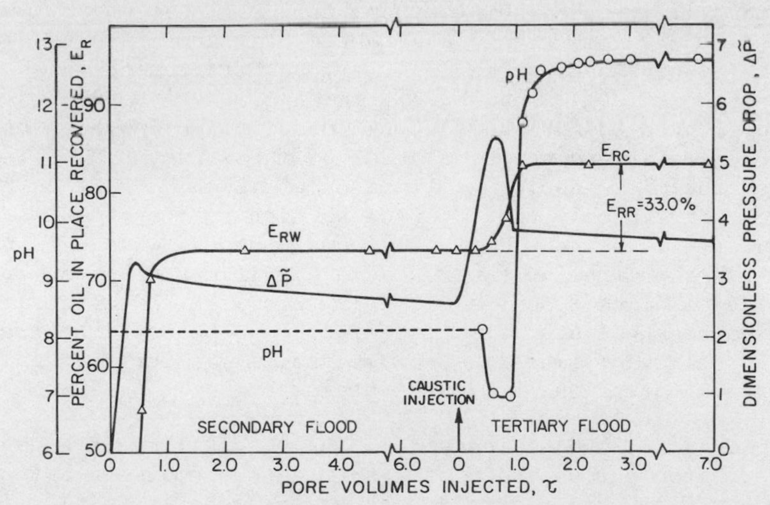

Fig. 10. Production, concentration and pressure histories of secondary waterflood and tertiary sodium hydroxide flood of brine saturated core.

edge of the injected alkaline slug would induce the formation of an oil-wet surface since oleate ions promote oil-wetness in high salinity environments. The residual oil is mobilized since both the capillary forces at the leading and trailing edges act in the direction of the imposed pressure gradient. Enhanced recovery as the direct result of such a favorable wettability gradient is reflected in the elution times of the leading and trailing edges of the tertiary oil bank. The leading edge elutes after 0.6 PV (a nominal breakthrough value for these systems) and the trailing edge elutes after complete displacement of the high salinity connate water at 1.0 pore volumes of the injected high pH/low salt alkali. The decrease in the end-point water relative permeability after the chemical flood (Table 2) supports the contention that the trailing edge of the tertiary oil bank is indeed water-wet.

Ultra-low tension: In the alkaline flooding of acidic acids, some reduction in interfacial tension (from 30 to approximately 10^{-1} dynes/cm) is necessary for the emulsification and subsequent mobilization of waterflooded residual oil by the previously discussed phase alteration mechanisms. The residual oil may also be mobilized and produced by a low-tension displacement process which is similar to surfactant flooding if the interfacial tension can be further reduced to ultra-low values (10^{-4} to 10^{-5} dynes/cm).

Typical capillary numbers of alkaline floods are listed in Table 6; these floods are representative of the different pH/salt systems investigated. The minimum interfacial tension which is measured between the aqueous alkaline drop and the acidic organic phase in a spinning-drop tensiometer is used to calculate the capillary number (23). The capillary number for the secondary waterflood, $Ca \simeq 8 \times 10^{-6}$ is within the ranges of values reported for conventional waterfloods. The capillary numbers for the high pH/high salt systems (e.g., Run DY-25), for the moderate pH/high salt systems (e.g., Run DY-102) and for the moderate pH/low salt flood (Run DY-106) are around the same order of magnitude, i.e., $Ca \simeq 10^{-4}$. Yet, the microscopic mobilization efficiency was significantly increased in only the moderate pH/high salt systems. The noncommonality of recovery efficiencies of these floods eliminates the viability of ultra-low tension as a significant recovery mechanism in the alkaline floods studied in this work. The recovery efficiencies of the listed alkaline floods can be better correlated with the stability of emulsions and wettability alteration than with interfacial tension of these systems.

Emulsification and coalescence: The possibility of enhancing oil recovery from porous media by a spontaneous emulsification mechanism has been examined by Schechter and coworkers (6). These researchers postulated that residual oil, which is entrapped after a conventional waterflood, can be mobilized by spontaneous emulsification and subsequent coalescence of small droplets with other

Table 6. Typical Capillary Numbers Minima Which Can Occur During the Dynamic Displacement of an Acidic Oil by an Alkali

Run No. DY-	Alkali/Electrolyte	Capillary Number
12	Distilled water	7.5×10^{-6}
19	NaOH	1.3×10^{-3}
18	NaOH	4.0×10^{-3}
25	NaOH/NaCl	1.2×10^{-4}
102	$Na_2B_4O_7$/NaCl	7.2×10^{-5}
106	$Na_2B_4O_7$	4.2×10^{-4}

ganglia of oil to form local regions of high oil saturation. Laboratory work was carried out with a 1-3 wt percent petroleum sulfonate solution (89 wt percent active sulfonate, 7-10 wt percent water and less than 1 wt percent inorganic salt). The general conclusions are that the displacement of residual oil is significantly improved if spontaneous emulsification occurs between the bulk phases and that displacement efficiency is favored by the high coalescence rates of the bulk phase emulsions.

Preferential partitioning of the sulfonate into the oil phase was determined as a necessary condition for the spontaneous emulsification of the organic and aqueous phases. Schechter and coworkers (6) suggested that the conditions necessary for spontaneous emulsification are similar to the conclusions of Sterling and Scriven's theoretical analysis (24) on the occurrence of interfacial instabilities across a planar liquid-liquid interface. This analysis suggests spontaneous disruption of a planar-planar interface if the solute simultaneously transfers out of a phase with lower diffusivity and larger kinematic viscosity and decreases the interfacial tension. Ultralow tensions were not, however, a sufficient condition for spontaneous emulsification in the aqueous sulfonate-oil systems; e.g., an organic n-octane aqueous sulfonate system which had an interfacial tension of 0.0011 dynes/cm did not spontaneously emulsify whereas the toluene-sulfonate system with an interfacial tension of 0.4 dynes/cm did spontaneously emulsify. The ultralow tension mechanism was not operative in these systems since the capillary number was approximately 5×10^{-3} in the recovery of 64 percent of the residual oil from a Berea sandstone core. It should be noted that in this particular flood, approximately one-half of the residual oil was recovered as a distinct oil bank before emulsion broke through at 1.02 pore volumes; in fact, early breakthrough in this test suggests a high permeability streak or a fracture in the flow direction.

Spontaneous emulsification in the water-sulfonate-oil system used by Schechter et al. (6) must have occurred by a diffusion and stranding mechanism (24). Diffusion creates local regions of supersaturation with the formation of emulsion droplets (stranding) from phase transformation in these areas. The presence of a third component (sulfonate) which is of common but limited miscibility with the other mutually immiscible phases, is a prerequisite to the diffusion and stranding phenomenon. Wasan et al. (17) have presented photomicrographs of the diffusion and stranding phenomenon occurring between a non-preequilibrated aqueous petroleum sulfonate (3 wt percent active)-crude oil system. The photomicrographs were made from low speed photography of the two phases which were contacted by imbibing the aqueous phase into the oil phase sandwiched between a glass slide and a cover plate. Wasan et al. (17) have also demonstrated that surfactant formulations which exhibited low interfacial tensions and low interfacial viscosities (high coalescence rates) yield good enhanced oil recoveries.

The type and nature of the emulsion which forms at planar liquid-liquid interfaces between an alkaline aqueous solution and an acidic organic phase were investigated theoretically and experimentally in the present work in order to develop an Emulsification and Coalescence recovery process for acidic oils.

Theory of emulsion stability: Emulsions can show instability by creaming, by inversion and by breaking (demulsification). Creaming refers to the separation of an emulsion into a concentrated disperse and a dilute disperse phase. Creaming is considered to be a sedimentation phenomenon which occurs by the interaction of gravity and buoyancy forces. Emulsion instabilities can cause an oil-in-water (O/W) emulsion to change into a water-in-oil (W/O) emulsion and vice versa. This phenomenon is called inversion. The demulsification process occurs by the aggregation of droplets of the dispersed phase (flocculation) and combination of the aggregates to form a single drop (coalescence). Flocculation is often reversible. Coalescence, which is irreversible, decreases the number of droplets and can result in complete demulsification. Flocculation is rate-limiting in dilute emulsions whereas coalescence is rate-determining in concentrated emulsions. Demulsification can be accompanied by creaming and inversion.

The surface of O/W emulsions, which are stabilized by surface active salts of polyelectrolytes, usually carries a negative charge. O/W emulsions are formed if the interfacial film consists of a molecular complex with an oil-soluble component and an ionizable water-soluble component (25). The simple oriented-wedge theory of

Langmuir and Harkins (26) suggests that the polar groups would orient towards the water phase of an O/W emulsion which is stabilized by a monomolecular film of surface active agent such as a sodium oleate soap. The carboxylic heads penetrating the oil phase are mostly ionized so that at the surface of the drop will be the carboxyl ion ($-C\lesssim^{O}_{O-}$) and the droplet will be surrounded by negative charges. The double layer extends only short (10^{-3} - 10^{-2} μm) distances in water-continuous emulsions so that the potential energy is essentially zero at the surface with regard to infinite distance at the surface of the droplets; consequently, the total energy barrier must be overcome before flocculation can occur (25). This rate limitation in flocculation would tend to make O/W emulsions more stable than W/O emulsions.

Geometric considerations, as can be illustrated by the simple oriented-wedge theory, suggest that there should be two or more lipophilic tails to the single hydrophile, i.e., the hydrophile should be at least divalent (27). Indeed, it has long been established that soaps of univalent metals favor the formation of O/W emulsions and those of multivalent ions favor the formation of W/O emulsions. Clowes (28) has suggested that sodium, potassium and lithium soaps will give O/W emulsions. Holmes (29) also listed gelatin, saponin, albumin, lecithin and casein as O/W emulsifiers and gum dammar, rosin, rubber, and cellulose nitrate as W/O emulsifiers. Albers and Overbeek (25) observed that the soaps of polyvalent metals (Al, Ba, Ca, Cd, Cu^{II}, Fe^{III}, Hg^{II}, Mg, Mn, Ni, Pb, Sr, and Zn) promote W/O emulsions which flocculate; some of the flocculated emulsions are stable against coalescence, others are not. The stabilization of coalescence by some oleates of polyvalent metals was found to result from the formation of a thick film of complex hydrolysis products. This observation confirms the findings of Wasan et al. (17) that coalescence rates could be inversely related to interfacial viscosities or thickness of the interfacial film (0.7-0.3 μm).

Under the right conditions of pH, temperature, and concentration, stable O/W emulsions can be converted to W/O emulsions which are also stable. The addition of a calcium salt to a sodium soap-stabilized O/W emulsion can cause the inversion to a W/O emulsion. Clowes (28) observed that this occurred with intense Brownian movement which could be Marangoni in nature. Emulsions can also be inverted by adjustment of the concentration of 1:1 electrolytes. Kremnev and Kuibina (30) report that O/W emulsions of the benzene/0.3M sodium oleate system were broken and inverted by adjusting the aqueous phase to a 0.25-0.3M NaCl concentration. King (31) points out that sodium oleate at a concentration of 1.0 percent normal gives O/W emulsions whereas a W/O emulsion may result at much lower concentrations; in the latter case, the sodium oleate behaves as a normal electrolyte rather than as a colloidal electrolyte as in the high oleate concentration O/W emulsions. It should be noted that the type of emulsion depends on the ratio of the ionic concentrations of uni- and multivalent ions present in solution. Clowes (28) called this an "ion antagonism phenomenon."

Macroemulsion stability between aqueous alkaline and acidic oil systems: The rates of flocculation and/or coalescence can be inferred from plots of the ratio of the emulsion volume to the total volume, V_e/V_T, versus time as shown in Figures 11 and 12. The influence of salt concentration on the emulsion character of univalent surface-active salts is detailed in Figure 11. After shearing, the change in emulsion type with concentration is readily distinguishable by the change in the color of the emulsion phase. O/W emulsions are formed at low electrolyte concentrations, $< 0.3M$ NaCl, and W/O emulsions are formed at high electrolyte concentrations, $> 0.3M$ NaCl. The continuous phase at 0.3M NaCl was indeterminable, by the simple method previously discussed. This emulsion phase collapsed to a value of V_e/V_T of 0.1 between 10,000 minutes (limit of time scale of Figure 11) and 15,000 minutes (last observation point); all the other emulsion phases formed with univalent alkalis at high pH's flocculated to $V_e/V_T = 0.5$ over the observation period. The flocculation rate of the W/O emulsions is faster than the O/W emulsions because of double layer effects (25). The W/O emulsions flocculated but did not coalesce because of the presence of interfacial films discussed by Wasan et al. (17). The relative rates of flocculation of either the O/W or the W/O emulsions appear to depend on the concentration of electrolyte; however, the data are insufficient to make a more definitive statement at this time.

The effects of divalent ion concentration on the emulsion stability at high values of pH are shown in Figure 12. The emulsion type changes from O/W to W/O at high concentrations of $CaCl_2$. The rates of coalescence at the low $CaCl_2$ concentrations, not shown on Figure 12, are very similar to the O/W emulsions with the univalent salt shown in Figure 11; $V_e/V_T = 0.5$ at 15,000 minutes in these systems. The W/O emulsion phase formed by reacting barium hydroxide with oleic acid coalesced much more rapidly than the emulsion phases formed by reacting oleic acid with alkalis containing calcium. This relative ordering of stability is in agreement with the findings of Albers and Overbeek (25). Interestingly, the rate of coalescence of the 0.025M NaOH/0.02M $CaCl_2$-oleic acid system is greater than the 0.02M $Ca(OH)_2$/0.02M $CaCl_2$-oleic acid system. The lower rate in the high calcium environment is probably caused by steric hindrance of the large calcium oleate salts present at the oil-water interface. Sodium borate at pH = 8.6 in a high salt environment forms unstable W/O emulsions with oleic acid. These emulsions are, however, more stable than those formed with divalent alkalis.

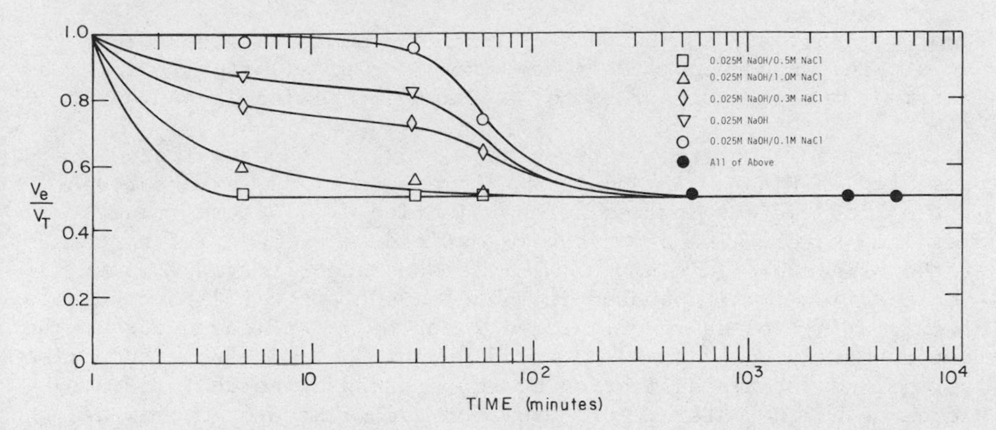

Fig. 11. Flocculation rate of emulsions as a function of time.

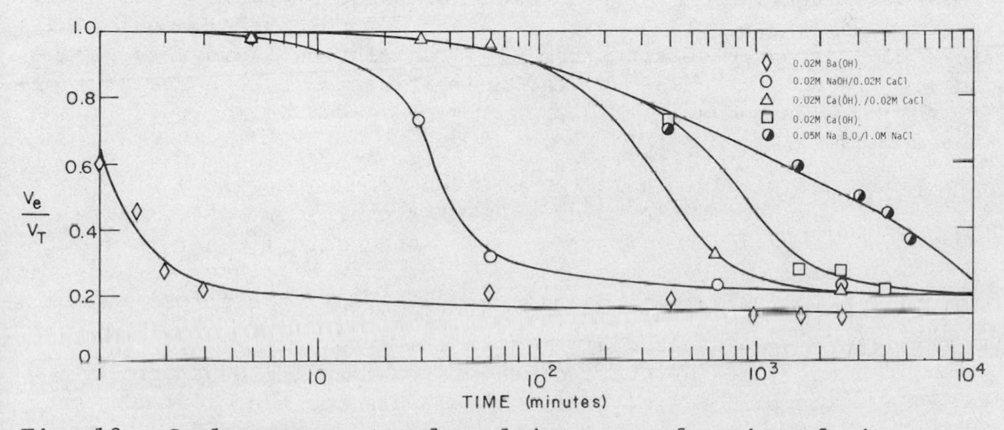

Fig. 12. Coalescence rate of emulsions as a function of time.

<u>Recovery of acidic oils with alkaline agents by an emulsification and coalescence mechanism</u>: Calcium hydroxide [$Ca(OH)_2$] was used to verify the emulsification and coalescence concept since, as suggested by the theoretical and experimental evidence of an earlier section, the carboxylic salts of divalent ions form unstable emulsions of water-in-oil. The emulsification and coalescence concept was quantitatively verified by secondary and tertiary flooding of partially oil-saturated sandpacks. A tertiary chemical flood with $Ca(OH)_2$ (pH = 12) recovered 44 percent of the waterflood residual oil from a 3.5-darcy Ottawa sandpack; the oil had an acid number of 2 and a viscosity of 1.5 cp. A secondary caustic flood with $Ca(OH)_2$ (pH = 12.32) recovered 82.3 percent of the original oil in place from a 0.25-darcy Ottawa sandpack; the oil phase in this secondary flood had the same physical and chemical properties as the oil phase used in the tertiary mode flood. It should be noted that the microscopic mobilization efficiencies of these

floods are not strongly dependent on the geometry of the porous media. This nondependence on geometry is typical of systems which increment the recovery of oil by a phase alteration mechanism.

The results of the tertiary and $Ca(OH)_2$ secondary floods are presented in Figures 13 and 14. In the waterflood, breakthrough of the flood water occurred after injection of 0.6 pore volumes of distilled water. The secondary waterflood recovered 71.7 percent of the original oil in place. In the subsequent tertiary mode alkaline flood, oil appeared in the effluent after 1.2 pore volumes of calcium hydroxide were injected into the waterflooded core. The tertiary oil production was delayed because a finite residence time is required for emulsification of the entrapped residual oil, coalescence of the water-in-oil emulsion and subsequent mobilization of the coalesced droplets into an oil bank.

The tertiary oil bank is preceded by a pressure pulse which is about five times the single, aqueous phase pressure drop. This enormous pressure pulse, which peaks at 1.2 pore volumes of $Ca(OH)_2$ injected, probably results from flow restriction (temporary permeability reduction) caused by the emulsification of the residual oil by the alkaline phase. This probable cause can be inferred from the following:

(i) The pressure pulse which preceded the tertiary oil bank is about three times that of the pressure pulse which preceded the oil bank in the secondary flood. This difference indicates that some mixture of phases (other than two immiscible phases) were mobile in the porous medium.

(ii) The caustic consumption is a maximum at $\tau = 1.2$ PV, i.e., at the start of the tertiary oil production. The position of this maximum suggests that most of the emulsification occurs before any enhanced oil is produced.

(iii) A clear and distinct oil phase was produced throughout the tertiary recovery mode. Traces of an emulsion phase were observed after the tertiary oil bank was produced.

The plot of dimensionless pressure drop versus pore volumes injected for the secondary mode alkaline flood (Figure 15) indicates that breakthrough occurs at the same amount of pore volumes during the waterflood (i.e., $\tau = 0.58$ PV approximately) and during the calcium hydroxide flood. However, in the secondary mode alkaline flood, the pressure drop continues to increase because of the in situ formation of a water-in-oil emulsion phase after breakthrough. As in the tertiary mode alkaline flood, the emulsified phase restricts the flow of the displacing phase and increases the pressure drop. The increased ΔP indicates the formation of an oil bank

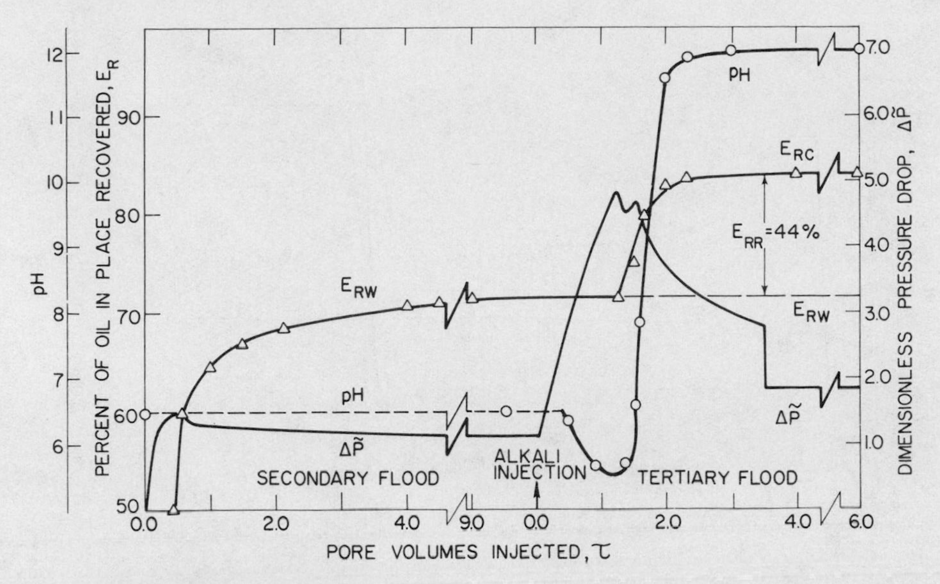

Fig. 13. Production, concentration and pressure histories of secondary sodium chloride flood and tertiary calcium hydroxide flood at pH = 12.0.

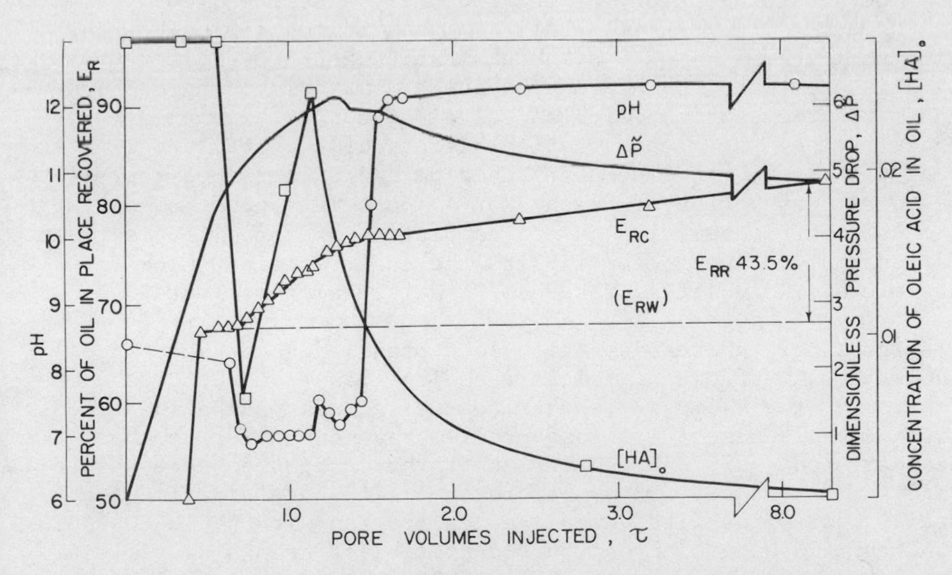

Fig. 14. Production, concentration and pressure histories of secondary calcium hydroxide flood at pH = 12.3.

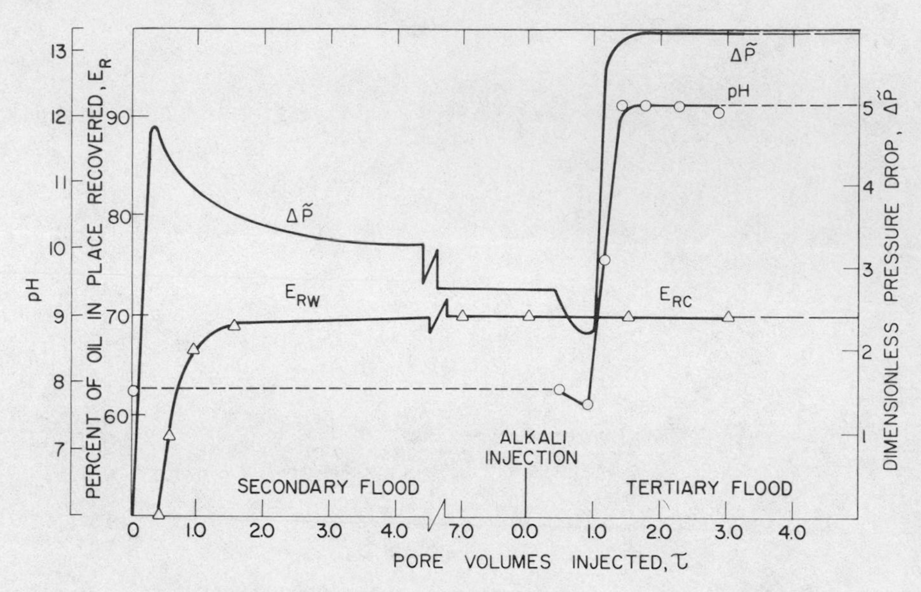

Fig. 15. Production, concentration and pressure histories of tertiary barium hydroxide flood as a function of pore volumes throughput.

which also reaches a maximum saturation at 1.25 PV of calcium hydroxide injected.

The pressure history of the secondary alkaline flood reflects the formation of a secondary oil bank behind the immiscible phase oil bank. This secondary oil bank results in an overall recovery which is above that obtained by secondary waterflooding or by secondary caustic flooding with an univalent ion of high electrolyte concentration. The concentration history of the fractional water production during the secondary calcium hydroxide flood represents the total consumption of the hydroxyl ion. This consumption curve is made up of consumption due to adsorption of the silica surfaces and consumption due to the in situ chemical reaction which forms the more oil-soluble, surface-active salt, calcium oleate.

The single-phase adsorption loss profile can be used to decouple the interfacial chemical reaction loss from the total consumption of the hydroxyl ion during the secondary alkaline flood. A single-phase adsorption loss profile was made prior to the secondary-mode multiphase experiment; the sandpacked core was reconditioned to a neutral pH before the multiphase experiment by partial desorption with water, neutralization with 0.01M HCl solution, and flushing with copious quantities of triple distilled water. The major consumption of alkali by interfacial chemical reaction occurs during the oil production period of the secondary alkaline flood illustrated in Figure 14.

Ninety-two percent of the oil is produced before breakthrough of the hydroxyl ion at τ = 1.45 PV. The first evidence of an acid-base reaction occurs at the trailing edge of the primary oil bank at τ = 0.78 PV. The oleic acid concentration in the oil phase is a minimum at τ = 0.78 PV. The surface active oleates of calcium which are formed by the interfacial chemical reaction stimulate the production of a secondary oil bank. The oleic acid concentration in the oil phase peaks at the oil cut peak of the chemically stimulated oil bank at τ = 1.72 PV during a period of minimum contact. The concentration of the oleic acid displays a diffuse minimum at the trailing edge of the secondary oil bank. Incremental recovery of acidic oil by an Emulsification and Coalescence mechanism in the secondary calcium hydroxide flood can be inferred from the following experimental observations:

(iv) There is a continued increase in dimensionless pressure drop after breakthrough of water in the primary oil bank. This pressure drop peaked at 1.25 PV, immediately after the oil cut production peak at 1.12 PV.

(v) The oleic acid concentration in the oil phase occurs at the start of enhanced oil production. The emulsification, which is partly responsible for the pressure rise, must have occurred before any incremental oil was produced.

(vi) Although a clear and distinct oil phase was produced throughout the production of the oil bank in the secondary waterflood, traces of an emulsion phase were observed after the oil bank was produced in the alkaline flood.

A barium hydroxide flood was carried out to further test the Emulsification and Coalescence concept in the recovery of acidic oils. It was thought that barium hydroxide would be a good test for the Emulsification and Coalescence mechanism, since as discussed earlier, W/O emulsions stabilized by barium oleate were the most unstable of the emulsions studied. The pressure and production history of a tertiary-mode barium hydroxide is graphed in Figure 15. No incremental oil production was obtained although the pressure drop across the core was higher than the pressure peak in the secondary waterflood. The constancy in the pressure drop of this tertiary mode alkaline flood (Figure 15) suggests that some permanent alteration in the fluid distribution occurred within the stimulated reservoir core. Permanent wettability alteration to oil-wet could have occurred since multivalent ions such as barium and calcium are known to adsorb (15,32) on negatively charged quartz surfaces. Oleate ions, with their hydrophilic tails sticking out, would then associate with the chemisorbed calcium ions in the Stern layer (32). It is speculated that the relaxation time for the Emulsification and Coalescence process for barium oleate stabilized emulsions is smaller than the flow time required for the contacting of adjacent ganglia of residual oil which have been mobilized. After coalescence, the surface active agents diffuse away from the oil-water interface through the water phase and adsorb onto the rock surface. The pressure drop in the tertiary-mode alkaline flood increases above that of the secondary-mode waterflood because the wettability alteration causes configurational changes in the ganglia of residual oil.

The production and pressure history of a tertiary-mode 0.025M NaOH/0.02M $CaCl_2$ flood at pH = 12.4, illustrated in Figure 16, are similar to the pressure and production history of the tertiary-mode 0.02M $Ba(OH)_2$ flood depicted in Figure 15. The coalescence rate of the W/O emulsions formed by contacting 0.025M NaOH/0.02M $CaCl_2$ with oleic acid lies between the rate for the emulsions formed by contacting 0.02M $Ba(OH)_2$ and 0.02M $Ca(OH)_2$ (Figure 12). The incremental production of 16.4 percent of the original oil in place also lies between the incremental production of the tertiary-mode barium hydroxide and the tertiary-mode calcium hydroxide flood. This correlation could have resulted because the coalescence rate was low enough to allow some contacting of the residual oil ganglia but high enough to allow redistribution of the oil which is mobilized by the emulsification mechanism. The pressure drop across the core, which remains high and constant, reflects configurational changes within the core which are most likely due to altered states of wettability.

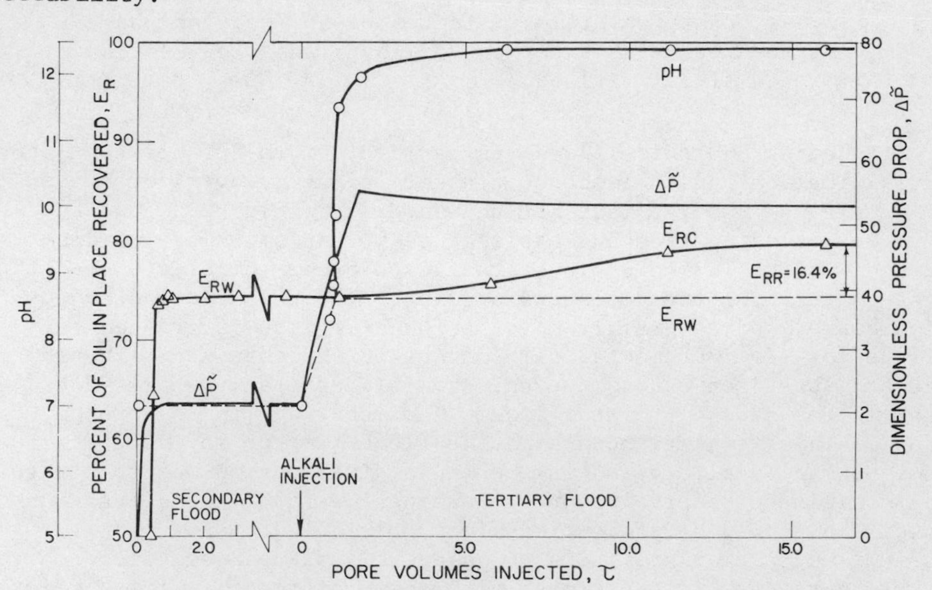

Fig. 16. Production, concentration and pressure histories of sodium hydroxide/calcium chloride flood as a function of pore volumes throughput.

The pattern of incremental recovery with increasing stability of unstable W/O emulsions is maintained by the buffered sodium borate/sodium chloride mixture which is on the right of the W/O emulsion coalescence plot of Figure 12. In the tertiary mode 0.05 $Na_2B_4O_7$/1.0M NaCl flood, Figure 8, an oil bank is preceded by a sharp rise in both pressure drop and pH. Among the buffered alkaline floods (5), the highest tertiary recovery of residual oil was obtained with the sodium borate/salt floods at a pH of 8.6. This result confirms the experimental work of Cooke et al. (4). However, Cooke et al. (4) alleged that the high recovery efficiencies of the residual oil occurred predominantly by an emulsification and wettability reversal (from water-wet to oil-wet) mechanism. Equilibrium displacement experiments of 0.05 $Na_2B_4O_7$/1.0M NaCl-oleic acid traced mineral oil systems in similar sandpacks indicate that no permanent alteration in wettability occurs after displacement between these reactive systems (5), c.f. values of wettability indices in Table 4. Contact angle measurements do indicate alteration of contact angle from 30° for the nonreactive oil-water system to 180° for the reactive oil-water system (19). Thus, transient changes in wettability with some emulsification could have been responsible for the mobilization of the residual oil. Cooke et al. (4) also correlated the maximum in recovery efficiency with the highest interfacial viscosity which purportedly increased the stability of emulsion droplets and residual oil lamellae and subsequently caused the locally high pressure gradients. Somasundaran et al. (33) obtained a good correlation between the pH (7.8) of maximum concentration of an ion-molecular reaction product A_2HNa and the pH (8.5-9.5) of maximum surface activity of oleate solutions and the maximum floatability of hematite ore; it was suggested that the small differences in these maxima were caused by the difference in surface pH and bulk solution pH. The high concentration, along with the steric effects, of the ion-molecular complex at the oil-water interface would increase the thickness of interfacial films and the interfacial viscosity. This interfacial film can cause interfacial rigidity and promote local increases in pressure drop; the ion-molecular complexes in the film can also adsorb onto the rock surface and alter the configuration (through a wettability change) of residual oil ganglia.

Thus, it appears that the high recovery of the buffered borate/salt system was obtained by an Emulsification and Coalescence mechanism in which the mobilization step was promoted by an emulsification and transient wettability alteration process and in which the production step was initiated by a coalescence process and aided by interfacial film formation. Mobilization of residual oil in the calcium hydroxide floods appear initiated by an emulsification mechanism which swells the residual oil and increases its apparent saturation. Contacting of adjacent ganglia of residual oil is made as the result of swelling of the residual oil with concomitant wettability alteration. The mechanisms by which an emulsification and coalescence process increase the production and recovery efficiencies of oil from a porous medium is illustrated in Figure 17. The actual oil saturation and permeability increase locally after coalescence of swollen ganglia into a volume or saturation which was less than the waterflood's residual volume or saturation (as illustrated in Figure 12). Emulsification and partial wettability alteration of residual ganglia in the univalent alkaline (high pH/high salt) floods decrease the water permeability as the result of flow restriction. The reduced permeability results in an increased production efficiency. The recovery efficiency of the high pH/high salt floods are not increased because the swollen residual oil ganglia flocculate into a volume or saturation which is equal to the waterflood's volume or saturation (as illustrated in Figures 7 and 11). Thus, the data and theory presented suggest that production and recovery efficiencies of acidic oils from porous media can be improved by alkaline flooding via an Emulsification and Coalescence mechanism.

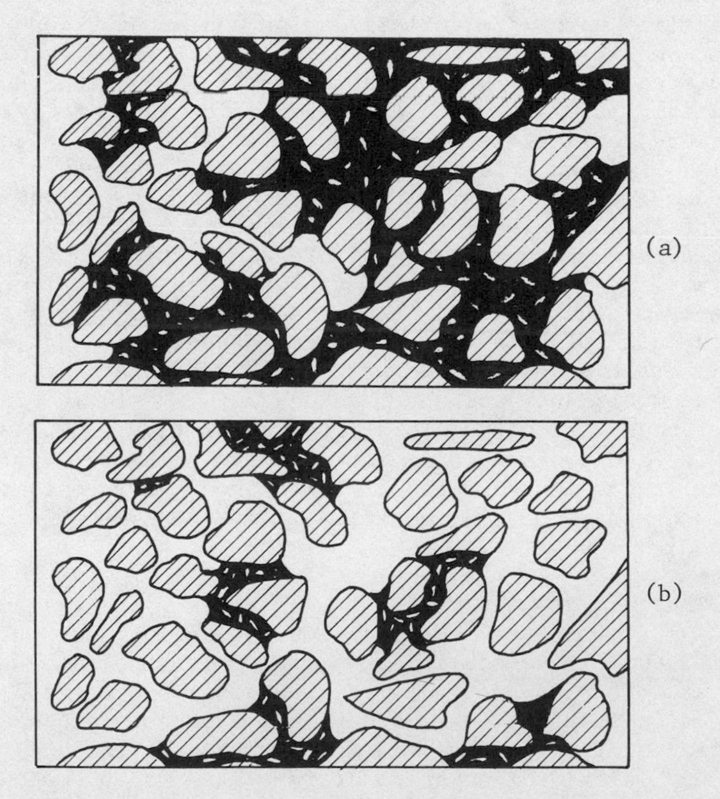

Fig. 17. (a) Emulsification or swelling which results in increased production efficiency; (b) coalescence which results in increased recovery efficiency.

CONCLUSIONS AND SIGNIFICANCE

(i) EMULSIFICATION AND ENTRAINMENT, which increases the recovery, occurs during increased-rate, high pH flooding of low acid-number oils in low salinity environments.

(ii) EMULSIFICATION AND ENTRAPMENT, which improves the sweep efficiency, occurs during the high pH flooding of moderate acid-number oils in low salinity environments.

(iv) COMPLETE WETTABILITY REVERSAL (Water- to Oil-Wet) increases the recovery of acidic crudes in moderate pH flooding in high salinity environments.

(v) PARTIAL WETTABILITY REVERSAL (Water- to Oil-Wet) increases the rate of production in high pH flooding of acidic crudes in high salinity environments.

(vi) CHROMATOGRAPHIC WETTABILITY REVERSAL (Water- to Oil-to-Water-Wet) increases the recovery of acidic crudes in high pH flooding of systems with salinity gradients.

(vii) EMULSIFICATION AND COALESCENCE increases the rate of production and recovery in systems which spontaneously form unstable W/O emulsions.

It was determined from dynamic displacement experiments that alkaline flooding of acidic oils with hydroxides of certain divalent cations increased the production and recovery efficiencies above that obtained by alkaline floods with hydroxides of univalent ions with or without high electrolyte concentration. The increased efficiencies resulted from an Emulsification and Coalescence mechanism.

The first stage, Emulsification, is accomplished by the in situ formation of surfactants which can promote emulsification through shear driven forces (external and/or internal pressure gradients) and/or by surface driven forces (spontaneous emulsification). Emulsification, with concomitant wettability alteration, reduces the permeability of the aqueous phase as a result of flow restriction and increases the production efficiency of the improved waterflood. The second stage, Coalescence, involves the recombination of the emulsified droplets to form local regions of high oil saturation with a correspondent increase in the permeability of the oil phase. The local regions of high oil saturation further combine to form a mobile oil bank. The formation and production of such an oil bank, under externally imposed pressure gradients, improves the recovery efficiency of oil from waterflooded cores above that obtained by conventional processes. Emulsification and Coalescence of capillary retained oil are interdependent processes. The emulsions will recoalesce if they are unstable in nature; the nature of this instability depends on the emulsification process. Emulsification and Coalescence also partially utilizes the Emulsification and Entrapment and the Emulsification and Entrainment mechanisms to improve the overall recovery efficiency.

The recovery of acidic oils via an Emulsification and Coalescence mechanism is very similar to the recovery of nonacidic oils by petroleum sulfonate solutions (6). Although high coalescence rates are prerequisites to the success of both systems, the recovery efficiencies of alkaline-acidic systems appear to be inversely related to the rate of coalescence. It is speculated that the relaxation time for coalescence must be greater than the flow time required for the contacting of adjacent ganglia of residual oil which have been mobilized by emulsification and wettability alteration processes. Yet coalescence was a necessary condition for enhanced recoveries since systems (such as hydroxides with univalent cations) in which the emulsion phase flocculated into the original oil volume did not improve the recovery efficiency. Some reduction in interfacial tension (10^{-1}-10^{0} dynes/cm) is required for the emulsification step although ultralow tension (10^{-2}-10^{-4} dynes/cm) was not necessary to the alkaline recovery processes. This observation concurs with an identical result of Schechter and coworkers (6). The two steps to this recovery mechanism, Emulsification and Coalescence, appear to be necessary and sufficient conditions for the enhanced production and recovery of nonacidic (6) and acidic oils.

NOMENCLATURE

A	Area, cm^2
Ca	Capillary number, ratio of viscous to interfacial forces, $\mu_w v/\sigma_{ow} \cos\theta$
E_{RW}	Recovery efficiency of waterflood, $(S_{oi}-S_{orw})/S_{oi}$
E_{RC}	Recovery efficiency of chemical flood, $(S_{oi}-S_{orc})/S_{oi}$
E_{RR}	Residual recovery or microscopic mobilization efficiency, $(S_{orw}-S_{orc})/S_{orw}$
k	Permeability, darcy
L	Length of flow field, cm
M	Molarity, moles/liter
NaA	Sodium oleate
O/W	Water external oil emulsions
pH	Negative logarithm to the base 10 of hydrogen ion concentration
PV	Pore volume, cc
S	With subscript, saturation
S	Without subscript, stable emulsion phase
US	Unstable emulsion phase
V	Volume, cc
v	Velocity, cm/sec
W/O	Oil external water emulsions

Greek Letters

θ Contact angle measured through water phase
μ Viscosity, centipoise
τ Pore volume, cc
σ Interfacial tension, dynes/cm
ϕ Porosity, pore volume/bulk volume

Subscripts

g Gas
l Liquid
o Oil
oi Initial oil
orc Residual oil to chemical flood
orw Residual oil to water flood
wir Irreducible water
ro Relative to oil during water flood
rw Relative to water
rwc Relative to water during chemical flood
w Water

Superscripts

0 Endpoint

REFERENCES

1. G. L. Stegemeier, "Mechanisms of Entrapment and Mobilization of Oil in Porous Media," preprint prepared for the 81st Am. Inst. Chem. Eng. Meeting, Missouri (April, 1974).
2. G. L. Stegemeier, SPE 4754 presented at Improved Oil Recovery Symposium, Oklahoma (April, 1974).
3. C. E. Johnson, Jr., J. Pet. Tech., 85 (January, 1976).
4. E. E. Cooke, Jr., R. E. Williams, and Kolodzie, J. Pet. Tech., 1365 (December, 1974).
5. T. P. Castor and W. H. Somerton, SPE 7143 presented at the SPE-AIME Annual California Regional Meeting, San Francisco, California (April, 1978).
6. R. S. Schechter, G. D. Jones, M. Hayes, J. L. Cayias, and W. H. Wade, SPE 5562 presented at the 50th Annual Fall Meeting of the Society of Petroleum Engineers of AIME, Dallas, Texas (September 28-October 1, 1975).
7. P. Subkow, U.S. Patent No. 2,288,857 (July, 1942).
8. H. Y. Jennings, Jr., C. E. Johnson, Jr., and C. D. McAuliffe, J. Pet. Tech., 1344 (December, 1974).
9. P. Raimondi, B. J. Gallaguer, G. S. Bennett, R. Ehrlich, and J. H. Messmer, SPE 5831 presented at the Improved Oil Recovery Symposium of the SPE-AIME, Tulsa, Oklahoma (March, 1976).

10. T. P. Castor, "Enhanced Recovery of Acidic Oils by Alkaline Agents," Ph.D. Thesis, University of California, Berkeley, California (1979).
11. R. L. Slobod, A. Chambers, and W. L. Prehn, Jr., Trans. Am. Inst. Min. Eng., 192, 127 (1951).
12. E. C. Donaldson, R. D. Thomas, and P. B. Lorenz, Soc. Pet. Eng. J., 13 (March, 1969).
13. R.J.S. Brown and I. Fatt, Trans. Am. Inst. Min. Eng., 207, 262 (1956).
14. N. Y. Jennings, Jr., Prod. Monthly, 20 (May, 1957).
15. A. M. Gaudin and D. W. Fuerstenau, Trans. AIME, Mining Eng., 66 (January, 1955).
16. J. H. Schulman and E. G. Cockbain, Trans. Far. Soc., 36, 651, 661 (1940).
17. D. T. Wasan, S. M. Shah, M. Chan, K. Sampath, and R. Shah, presented at the Symposium on Chemistry of Oil Recovery, Anaheim, California (March 12-27, 1978); Am. Chem. Soc. Symposium Series Preprints 23(12), 705 (March, 1978).
18. R. Ehrlich and R. J. Wygal, Jr., Soc. Pet. Eng. J., 263 (August, 1977).
19. J. F. Kelly, "Tertiary Displacement of Oleic-Acid Oils with Alkaline Agents," M.S. Thesis, University of California, Berkeley, California (1979).
20. A. S. Michaels and M. C. Porter, Am. Inst. Chem. Eng., Vol. II, No. 4, 617 (July, 1965).
21. A. S. Michaels, A. Stancell, and M. C. Porter, Soc. Pet. Eng. J., 231 (September, 1964).
22. A. S. Michaels and R. S. Timmins, Am. Inst. Min. Eng., 219, 150 (1960).
23. C. J. Radke and W. H. Somerton, "Enhanced Recovery with Mobility and Reactive Tension Agents," Fourth DOE Symposium on Enhanced Oil and Gas Recovery and Improved Drilling Methods, Tulsa, Oklahoma, B-2 (August, 1978).
24. C. V. Sterling and L. E. Scriven, Am. Inst. Chem. Eng. J., 514 (December, 1959).
25. W. Albers and J.Th.G. Overbeek, J. Coll. Sci., 14, 501, 510 (1959).
26. W. D. Harkins, in "The Physical Chemistry of Surface Gums," Reinhold Publishing Corp., New York, p. 83 (1952).
27. P. Becher, in "Emulsions: Theory and Practice," Second Edition, Reinhold Publishing Corp., New York, p. 97 (1966).
28. G.H.A. Clowes, J. Phys. Chem., 20, 407 (1916).
29. H. H. Holmes and R. H. Bogue (eds.), Colloidal Behavior, McGraw-Hill Book Co., New York, 1, 227 (1924).
30. L. Y. Kremnev and N. I. Kuibina, Kolloid Zhur., 9, 292 (1947); Chem. Abs., 43:525h.
31. A. King, Trans. Far. Soc., 37, 168 (1941).
32. M. R. Kruyt, Colloid Science I, Chapters 4 and 5, Elsevier Publishing Co., New York (1952).

33. P. Somasundaran, K. P. Ananthapadmanabhan, and R. D. Kulkarni, "Flotation Mechanism Based on Ion Molecular Complexes," presented at the Twelfth International Mineral Processing Congress, Sao Paulo, Brazil (1977).

THE ROLE OF ALKALINE CHEMICALS IN OIL DISPLACEMENT MECHANISMS

T. C. Campbell

The PQ Corporation
Research & Development Center
Lafayette Hill, Pennsylvania 19444, U.S.A.

Alkaline inorganic chemicals such as sodium silicates, sodium hydroxides, sodium carbonate, and sodium phosphates have been added to injection fluids used in enhanced oil recovery systems. These chemicals can, in varying degrees, affect various rock and fluid parameters such as interfacial tension, interfacial viscosity, emulsion stability, rock wettability, hardness-ion content, ion-exchange capacity or equilibria, surfactant adsorption, phase equilibria, etc., in order to improve recovery efficiency for residual oil remaining after waterflooding.

This paper discusses the solution properties of the various alkaline chemicals which are important in chemical flooding systems. Data on the comparative efficiencies of highly alkaline sodium silicate solutions and sodium hydroxide solutions in recovering a heavy California crude oil from Berea sandstone cores by alkaline waterflooding are presented. A study on the use of various alkaline chemicals as a preflush for the removal of hardness ions from reservoir brine in Berea sandstone cores was made, and part of these findings are presented. Finally, some preliminary results on the increased recovery obtained with combinations of petroleum sulfonates and less-alkaline sodium silicates are given.

INTRODUCTION

The addition of alkaline inorganic chemicals to injection fluids to promote increased recovery of crude oil had been suggested as early as 1917 by Squires (1). Other workers, such as Atkinson (2), Nutting (3), Beckstrom and Van Tuyl (4), and Subkow (5) described the benefits of various alkaline salts such as sodium silicates, sodium hydroxide, sodium carbonate, sodium phosphates, etc., in the recovery process. A large body of laboratory studies and several field trials in recent years have employed alkaline chemicals, either as the primary enhanced recovery agents or in conjunction with various types of surfactants to augment the recovery efficiency of the surfactants. An excellent review of alkaline waterflooding has been published by Johnson (6), which covers the laboratory studies and field trials performed prior to 1976. In the area of surfactant flooding, many processes have been developed which utilize alkaline salts as a preflush solution to condition the reservoir prior to injection of the surfactant. Also, the alkaline salts can be added to the surfactant solution. Examples of these processes have been patented by Holm (7) and Chang (8). Many other patents and articles on these types of processes have been published, which are too numerous to list.

CHEMICAL PROPERTIES NEEDED FOR IMPROVED OIL DISPLACEMENT

Several processes can be employed to modify the chemical environment of the reservoir fluids in order to improve the relative mobility of the crude oil. Among these are removal of hardness ions from the brine by precipitation or sequestration, changes in surface forces at the fluid interfaces and between the fluids and rock surfaces, emulsification, changes in adsorptivity of the rock surfaces, and reaction with acidic substances in the crude oil to form surfactants at the fluid interface.

The ability of the various alkaline chemicals to affect the above processes is summarized in Table 1 which compares sodium orthosilicate, sodium hydroxide, sodium carbonate and sodium tripolyphosphate.

ALKALINE FLOODING

The chemicals which have been used, or are being used, for alkaline flooding are sodium orthosilicate and sodium hydroxide. Sodium carbonate and sodium tripolyphosphate are not alkaline enough to react with the acids in crude oils. Johnson (6) and Mungan (9) have published reviews of alkaline flooding in which they summarize the proposed mechanisms of oil displacement (Table 2) and some of the important parameters for selection of alkaline flooding candidates (Table 3). These topics are covered in these two publications and will not be discussed further in this paper.

In a recent publication (10) a comparison was made between sodium orthosilicate and sodium hydroxide in recovering residual crude oil after waterflooding in Berea sandstone cores. The crude oil was a sample from a coastal California reservoir, for which an

Table 1. Comparison of Chemical Properties of Alkaline Salts

Sodium Orthosilicate	Sodium Hydroxide	Sodium Carbonate	Sodium Tripolyphosphate
Na_4SiO_4	NaOH	Na_2CO_3	$Na_5P_3O_{10}$
• Reacts with acids in crude oil	• Reacts with acids in crude oil	• Does not react with acids in crude oil	• Does not react with acids in crude oil
• Precipitates Ca^{++} and Mg^{++}	• Precipitates Ca^{++} and Mg^{++}	• Precipitates Ca^{++} but not Mg^{++}	• Sequesters Ca^{++} and Mg^{++}
• Good emulsifier	• Fair emulsifier		• Good emulsifier
• Reduces surfactant adsorption on rock surfaces	• Changes oil wettability		• Hydrolyzes over time to orthophosphate
• Interfacial properties less affected by hardness ions than other alkaline salts			
• Hydrolyzes in solution to equilibrium mixture of OH^-, SiO_4^{-4} and complex silicate oligomers			
• Changes oil wettability			

Table 2. Proposed Mechanisms for Alkaline Waterflooding

1) Emulsification and Entrainment
2) Wettability Reversal (Oil-wet to Water-wet or vice versa)
3) Emulsification and Entrapment
4) Mobility Control

Table 3. Important Parameters for Selection of Alkaline Flood Candidates

1) Acidity of Crude Oil
2) Viscosity and Density of Crude Oil
3) Reservoir Temperature
4) Permeability and Porosity
5) Rock Reactivity
6) Salinity and Hardness of Reservoir Brine

alkaline field trial is in progress. Sodium orthosilicate will be used in this trial. The results of the laboratory oil displacement studies are presented in Tables 4 and 5. Table 4 presents the results for continuous flooding with sodium orthosilicate and sodium hydroxide at concentrations of 0.15%, 0.25% and 0.50% by weight in injection fluids also containing 0.75% NaCl. Table 5 gives the results for the runs made with 0.5 PV of alkali injected, followed by brine injection.

The data from these tests show that sodium orthosilicate is more effective than sodium hydroxide in recovering residual oil under the conditions studied, both for continuous flooding and when 0.5 PV of alkali was injected. The mechanisms through which sodium orthosilicate produced better recovery than sodium hydroxide in this system have not been completely elucidated. Reduction in interfacial tension is similar for both chemicals, so other factors must play a more important role. Somasundaran (26) has shown that sodium silicates are more effective than other alkaline chemicals in reducing surfactant adsorption on rock surfaces. Wasan (27,28) has indicated that there are differences in coalescence behavior and emulsion stability which favor sodium orthosilicate over sodium hydroxide. Further work is being done in this area in an attempt to define the limits of physically measurable parameters which can be used for screening potential alkaline flooding candidates.

Table 4. Continuous Alkaline Floods of Huntington Beach Field Crude in Berea Sandstone Cores

RUN	1	2	3	4	5	6
Alkali	Na_4SiO_4	NaOH	Na_4SiO_4	NaOH	Na_4SiO_4	NaOH
Conc., WT %	0.15	0.15	0.25	0.25	0.5	0.5
Total PV Inj.	3.2	3.4	3.5	3.4	3.1	3.3
PV, ml	236.0	277.8	253.2	255.9	232.4	243.1
So_i	0.627	0.590	0.608	0.664	0.633	0.646
So_{wf}	0.415	0.385	0.403	0.434	0.417	0.415
% WF Recovery	33.8	34.8	33.8	34.7	34.0	35.7
So_{af}	0.322	0.331	0.288	0.340	0.318	0.337
% Residual Oil Recovered by AF	22.4	14.0	28.4	21.6	23.7	18.8
% PV Recovered by AF	9.3	5.4	11.5	9.4	9.9	7.8

Table 5. Alkaline Floods - Huntington Beach Field Crude in Berea Sandstone Cores Using 0.5 PV of Alkali Injection

RUN	1	2	3	4	5	6
Alkali	Na_4SiO_4	NaOH	Na_4SiO_4	NaOH	Na_4SiO_4	NaOH
Conc., WT %	0.15	0.15	0.25	0.25	0.5	0.5
PV Alkaline Inj.	0.5	0.5	0.5	0.5	0.5	0.5
Total PV Inj.	3.3	3.1	3.5	3.3	3.1	3.4
PV, ml	237.8	239.3	251.2	247.5	247.0	245.3
So_i	0.716	0.711	0.632	0.651	0.669	0.656
So_{wf}	0.474	0.465	0.410	0.435	0.428	0.433
% WF Recovered	33.7	34.6	35.1	33.2	36.0	34.0
So_{af}	0.415	0.421	0.319	0.359	0.355	0.401
% Residual Oil Recovered	12.4	9.4	22.2	17.4	17.1	7.5
% PV Recovered by AF	5.9	4.4	9.1	7.6	7.3	3.2

ALKALINE PREFLUSH AGENTS IN MICELLAR-POLYMER FLOODING

The use of preflush systems for conditioning oil reservoirs to provide an optimum environment for surfactant-based enhanced recovery systems has been extensively documented (11-22). These preflush systems are based upon three different approaches to reservoir conditioning, which are as follows:

- Design the preflush and surfactant systems with balanced concentrations of monovalent cations and divalent cations which would minimize the ion-exchange from the clays in the reservoir.

- Preflush the reservoir with slugs of high monovalent cation salinity so that the divalent cations are exchanged from the reservoir clays and eluted through the reservoir ahead of the surfactant slug.

- Inject alkaline chemicals which can disrupt or minimize the ion-exchange process by various mechanisms, such as adsorption, precipitation or ionic effects.

A study to compare alkaline and soft saline preflush systems for removal of hardness ions from reservoir brines was presented recently (23). In this study, Berea sandstone cores were saturated with a simulated reservoir brine containing 2.84% NaCl, 1110 ppm $CaCl_2$ and 490 ppm $MgCl_2$ by weight. The hardness level of the effluent from the cores was measured by EDTA titration. The initial brine used had a hardness ion content which was equivalent to 1500 ppm of $CaCO_3$.

Figure 1 presents the hardness ion profile of the effluent from Berea cores injected with one of the following fluids:

- Continuous injection of 3% NaCl solution to greater than three pore volumes.

- Injection of 0.25 pore volume of 0.5% Na_4SiO_4 by weight in 3% NaCl solution, followed by 3% NaCl solution to completion.

- Injection of 0.25 pore volume of 0.5% NaOH in 3% NaCl solution, followed by 3% NaCl solution to completion.

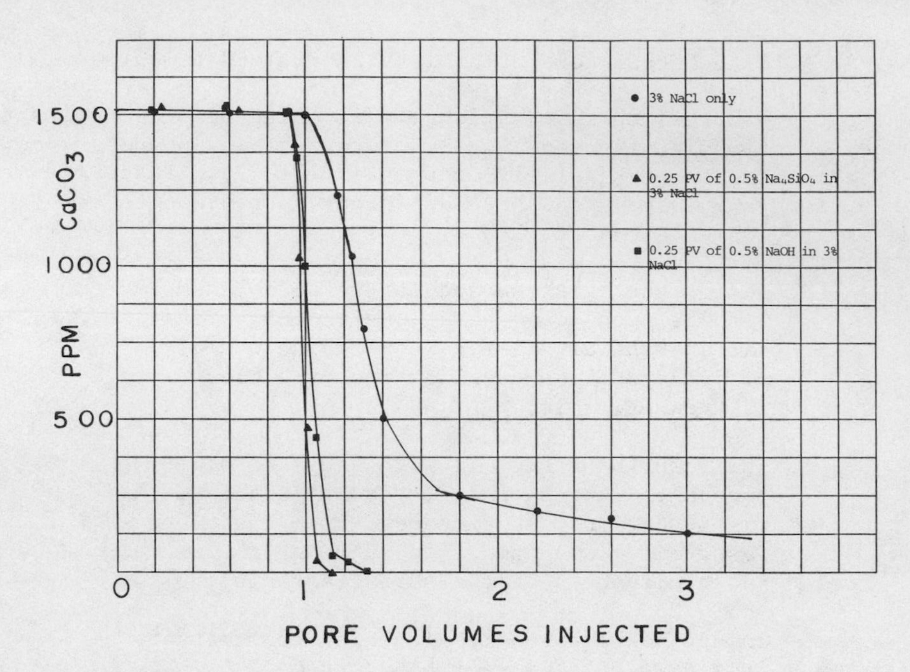

Fig. 1. Elution of hardness ions by 0.25 pore volume of Na_4SiO_4 or NaOH, followed by 3% NaCl

The hardness ion level of the core injected with 3% NaCl solution showed a rapid drop-off after 1.0 pore volume of fluid had been injected. The level did not reach zero hardness due to ion-exchange of hardness ions from the clays in the sandstone cores. The injection of 0.25 pore volume of 0.5% Na_4SiO_4 or NaOH in 3% NaCl produced a sharp reduction in hardness ion level with no additional elution beyond 1.2 pore volumes of injection. The alkaline salts effectively suppressed ion-exchange from the clays, probably by precipitation of the hardness ions. There was no significant reduction in permeability when the alkaline salts were injected.

Figure 2 shows the effect of injecting sodium carbonate as a preflush chemical. In this case, the elution of hardness ions followed a similar pattern when only 3% NaCl solution was injected. The elution of hardness ions following injection of 0.1 pore volume or 0.5 pore volume of 0.5% Na_2CO_3 in 3% NaCl solution, followed by 3% NaCl solution to completion, showed significant increases in hardness ion levels after one pore volume of fluid had been injected. These high levels of hardness ion eluted from the core suggest that the removal of Ca^{++} ion by precipitation does not occur under these conditions. High levels of hardness ion are exchanged from the clays, as shown by the increase in concentration above that of the resident brine.

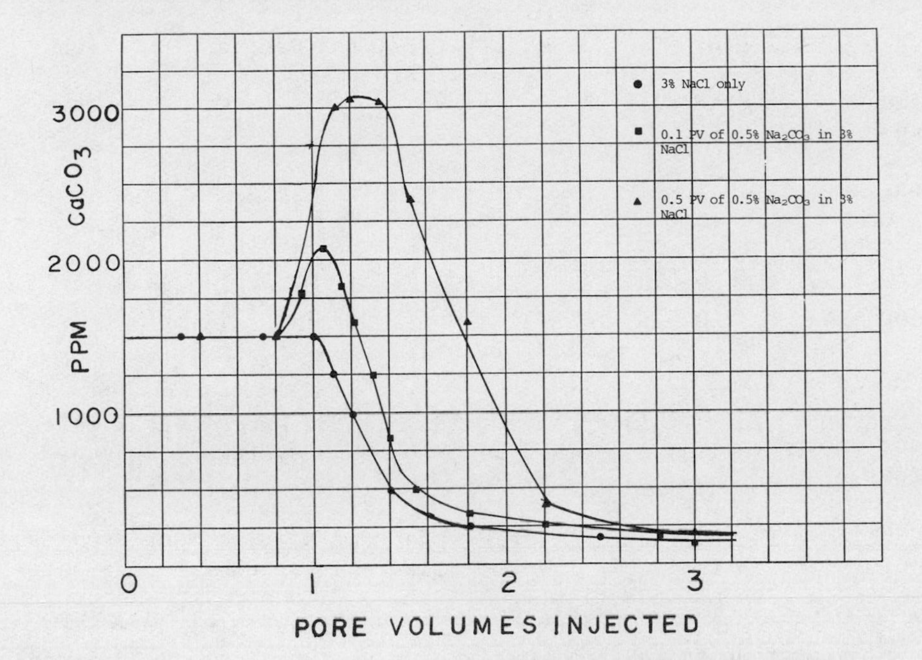

Fig. 2. Elution of hardness ions by sodium carbonate injected for 0.1 PV or 0.5 PV, followed by 3% NaCl

Although sodium orthosilicate and sodium hydroxide exhibit similar behavior in reducing the hardness ion levels in the brine eluted from cores which do not contain oil, there is a significant increase in oil recovery efficiency when sodium orthosilicate solution is used as the preflush for a micellar-polymer recovery system. These results have been documented by Holm and Robertson (19) and some of the comparative data are shown in Table 6.

More recently, a field trial in the Bell Creek Field in Montana of a micellar-polymer recovery system was reported by Goldburg and Stevens (24). For this project, the preflush was 0.16 pore volume of 0.45% sodium orthosilicate solution injected over a period of 130 days.

SILICATE-SURFACTANT SYSTEM FOR OIL RECOVERY

Considerable effort has been expended to develop a surfactant-based oil recovery system that would be less complex, and potentially less costly than the various micellar-polymer systems proposed by many major oil companies, examples of which are given in an excellent review by Gogarty (25). One system which shows considerable promise , based on initial laboratory oil displacement studies, is a combination of dilute petroleum sulfonate and sodium

Table 6. Preflush Effect on Micellar/Polymer Recovery Efficiency

Preflush Type	Concentration WT %	% PV Slug	pH	Oil Recovered % OIP
None	--	--	--	28.2
NaCl	5.0	10	8.3	79.8
Na_4SiO_4	0.5	10	13.0	82.8
NaOH	0.5	10	13.1	65.4

silicate injected in fairly large volumes. The experimental conditions for the oil displacement tests on this system are given in Table 7.

The results from the silicate-surfactant displacement studies are summarized in Table 8. The slug size was varied from 0.5 pore volume to continuous injection, following a 0.25 pore volume preflush with soft water. The surfactant concentration was 0.10% to 0.25% by weight on an "as is" basis. The percent oil recovery was calculated from the residual oil saturation after waterflooding. The recovery figures are given for tests in which surfactant alone was injected compared to the same surfactant concentration with 1.0% by weight of 3.22 ratio sodium silicate solution (37.6% solids) added to the injection fluid. The total injection volumes are given for each experiment. In the cases when the slug size was 0.5 or 1.0 pore volume, the surfactant slug was followed by brine until the total volume indicated was injected. The recovery of residual oil from the watered-out Berea cores by the silicate-surfactant combinations was significantly higher than when the surfactant was used alone, especially for the lower surfactant concentration levels.

The mechanisms by which the addition of sodium silicate improves oil recovery have not been completely defined. However, Somasundaran (26) has shown that the addition of sodium silicate can significantly reduce the adsorption of surfactants on mineral surfaces. Also, recent tests in our laboratory, completed by P.H. Krumrine subsequent to this symposium, which are summarized in Table 9, show a 75% decrease in surfactant adsorption in Berea cores when no oil is present. Another factor, which may be of importance, is the lower interfacial tension values obtained between the crude oil and the surfactant solution. The data in Table 10 show that the addition of 1.0% by weight of 3.22 ratio sodium silicate solution (37.6% solids) produced a significant reduction in interfacial tension values compared to the values measured with surfactant alone.

Table 7. Silicate-Surfactant System for Oil Recovery

SILICATE - 3.22 Ratio SiO_2/Na_2O, N[†] Sodium Silicate, PQ Corporation

SURFACTANT - Stepan PETROSTEP[†] 420 Petroleum Sulfonate

OIL - Kansas Crude, 34.5° API Gravity

CORE - Berea Sandstone, 150 - 250 md Permeability, 5 cm OD by 60 cm

TEMPERATURE - 22° - 23°C

FRONTAL VELOCITY - 30 cm/day

PREFLUSH - 0.25 PV of 0.1% NaCl

SALT CONCENTRATION - 1.0% NaCl

BRINE COMPOSITION - 3% NaCl
1000 ppm Ca^{++}
500 ppm Mg^{++}

Table 8. Effect of 3.2 Ratio Silicate on Surfactant Flooding

		% Recovery of OIP (PV Inj)*	
Slug Size	Surfactant Concentration	No Silicate	1.0% 3.22 Ratio Sodium Silicate Added**
Continuous	0.10 WT %	31.4 (4.0 PV)	43.2 (4.0 PV)
Continuous	0.25 WT %	45.6 (4.0 PV)	52.3 (4.0 PV)
1.0 PV	0.10 WT %	--	31.9 (2.5 PV)
1.0 PV	0.25 WT %	33.1 (2.7 PV)	--
0.5 PV	0.10 WT %	8.8 (2.0 PV)	22.0 (2.5 PV)

*Total PV injected to obtain recovery indicated, including preflush

**PQ Corporation's "N"[†] sodium silicate (37.6% solids)

[†]Trade name

Table 9. Surfactant Adsorption in Berea Cores

Surfactant System	Adsorption mg/g rock	% Surfactant Recovered
0.25% Petrostep 420 and 1% NaCl	0.64 mg/g	15.5%
0.25% Petrostep 420 and 1% NaCl and 1.0% 3.22 ratio sodium silicate solution*	0.15 mg/g	80.0%

*PQ Corporation's "N"† sodium silicate (37.6% solids)

INJECTION SEQUENCE:

1) 0.25 PV of 1% NaCl solution
2) 3.0 PV of surfactant solution
3) 0.25 PV of 1% NaCl solution
4) 1.0 PV of simulated reservoir brine

Table 10. Interfacial Tension Values for Silicate-Surfactant Systems

PETROSTEP† 420	3.22 Ratio Sodium Silicate Solution*	Interfacial Tension dyne/cm at 22°C
0.1 WT %	--	2.12
0.1 WT %	1.0 WT %	0.157
0.25 WT %	--	1.76
0.25 WT %	1.0 WT %	0.088

- Solutions contained in 1% NaCl
- IFT measured by spinning drop technique, not equilibrated, 15 minute spinning time
- Kansas Crude Oil

* PQ Corporation's "N"† sodium silicate solution (37.6% solids)

†Trade name

CONCLUSIONS

• In chemical flooding processes for enhanced oil recovery, alkaline chemicals can be useful for hardness ion suppression or removal, reaction with acidic crude oils to generate surface-active species, reduction in surfactant adsorption on reservoir rock surfaces, changes in interfacial phase properties, mobility control and increased sweep efficiency, oil wettability reversal and increased emulsification.

• In alkaline waterflooding, sodium orthosilicate is the most effective reagent for recovering oil remaining after waterflooding.

• For micellar-polymer systems, a highly alkaline preflush is more effective than a saline preflush in conditioning a reservoir for maximum recovery of residual oil. Sodium orthosilicate is more effective than sodium hydroxide for this use.

• The addition of sodium silicate to dilute surfactant solutions promotes increased recovery of residual oil compared to the use of surfactant alone.

ACKNOWLEDGMENTS

The author wishes to thank the PQ Corporation for permission to publish this work. The author also wishes to thank P.H. Krumrine for providing the data on silicate-surfactant combinations, and acknowledge the work of S. Maliar, R. Bauer and S. Gingold in obtaining the experimental data.

REFERENCES

1. F. Squires, U.S. Patent No. 1,238,355 (August 28, 1917).
2. H. Atkinson, U.S. Patent No. 1,651,311 (November 29, 1927).
3. P. G. Nutting, Oil & Gas J., 25, 32, 106 (1927).
4. R. C. Beckstrom and F. M. Van Tuyl, Bull., AAPG, 223 (1927).
5. P. Subkow, U.S. Patent No. 2,288,857 (July 7, 1942).
6. C. E. Johnson, Jr., J. Pet. Tech., 85 (1976).
7. L. W. Holm, U.S. Patent No. 4,011,908 (March 15, 1977).
8. H. L. Chang, U.S. Patent No. 3,977,470 (August 31, 1976).
9. N. Mungan, "A Review and Evaluation of Alkaline Flooding," Applications Report AR-4, Petroleum Recovery Institute, Calgary, Alberta, November 1979.
10. T. C. Campbell and P. H. Krumrine, SPE Paper No. 8328, presented at the 54th Annual Technical Meeting, Las Vegas, Nevada, September 23-26, 1979.
11. C. S. Chiou and H. L. Chang, Paper 376 presented at AIChE 84th National Meeting, Atlanta, Georgia, February 26-March 1, 1978.

12. H. J. Hill, F. G. Helfferich, L. W. Lake, J. Reisberg, and G. Pope, J. Pet. Tech., 1336 (1977).
13. G. A. Pope, L. W. Lake, and F. G. Helfferich, SPE Preprint 6771, presented at the 52nd Annual Fall Technical Conference, SPE, Denver, Colorado, October 9-12, 1977.
14. H. J. Hill and L. W. Lake, SPE Preprint 6770, presented at the 52nd Annual Fall Technical Conference, SPE, Denver, Colorado, October 9-12, 1977.
15. L. W. Lake and F. G. Helfferich, SPE Preprint 6769, presented at the 52nd Annual Fall Technical Conference, SPE, Denver, Colorado, October 9-12, 1977.
16. H. L. Hill, J. Reisberg, F. G. Helfferich, L. W. Lake, and G. A. Pope, U.S. Patent No. 4,074,755 (February 21, 1978).
17. T. G. Griffith, SPE Preprint 7587, presented at the 53rd Annual Fall Technical Conference, SPE, Houston, Texas, October 1-3, 1978.
18. F. W. Smith, J. Pet. Tech., 959 (1978).
19. L. W. Holm and S. D. Robertson, SPE Preprint 7583, presented at the 53rd Annual Fall Technical Conference, SPE, Houston, Texas, October 1-3, 1978.
20. L. W. Holm, U.S. Patent No. 4,011,908 (March 15, 1977).
21. L. W. Holm and D. H. Ferr, U.S. Patent No. 4,036,300 (July 19, 1977).
22. L. W. Holm, SPE 7066, SPE, 5th Symposium on Improved Oil Recovery, Tulsa, Oklahoma, April 1978.
23. T. C. Campbell, SPE Paper No. 7873, presented at the 1979 SPE Symposium on Oil Field and Geothermal Chemistry, Houston, Texas, January 22-24, 1979.
24. A. Goldburg and P. Stevens, Volume 1, pp. A-4/1-4/22, Proceedings of the 5th Annual DOE Symposium on Enhanced Oil and Gas Recovery, Tulsa, Oklahoma, August 22-24, 1979.
25. W. B. Gogarty, in "Improved Oil Recovery by Surfactant and Polymer Flooding," D. O. Shah and R. S. Schechter, eds., Academic Press, New York, New York, 27, 1977.
26. P. Somasundaran and H. S. Hanna, SPE 7059, presented at the Society of Petroleum Engineers 5th Symposium on Improved Oil Recovery, Tulsa, Oklahoma, April 1978.
27. D. T. Wasan, J. Perl, F. Milos, P. Brauer, M. Chang, and J. J. McNamara, presented at Symposium on Chemistry of Oil Recovery, ACS Division of Petroleum Chemistry Meeting, Anaheim, California, March 12-17, 1978.
28. D. T. Wasan, S. M. Shah, M. Chang, K. Sampath, and R. Shah, Volume 1, pp. C-2/1-C-2/19, Proceedings of the 5th Annual DOE Symposium on Enhanced Oil and Gas Recovery, Tulsa, Oklahoma, August 22-24, 1979.

INTERFACIAL RHEOLOGICAL PROPERTIES OF CRUDE OIL/WATER SYSTEMS

E.L. Neustadter, K.P. Whittingham and D.E. Graham

New Technology Division, BP Research Centre
Chertsey Road, Sunbury-on-Thames
Middlesex, England

It has been shown that a study of crude oil/water interfacial rheology can be used to investigate crude oil emulsion stability and rationalize the effect of chemical demulsifiers. Both kinetic factors relevant to droplet coagulation and the mechanical resistance to coalescence which gives rise to permanent emulsion stability can be studied and defined by these techniques. The effects of crude oil type, temperature and aqueous phase changes can be followed and used to pinpoint emulsion problems that may arise in a practical situation.

Measurements of interfacial dilatational elasticities and viscosities at crude oil/water interfaces have been carried out. The dilatational viscosity at fixed frequency does not change materially with time over a period of $\frac{1}{2}$ - 3 hours. This strongly suggests that the nature and extent of the film relaxation process is not a function of time. The fall in dilatational viscosity with frequency indicates that the relaxation process involves mainly bulk to interface diffusion interchange. The increase of dilatational elasticity with time reflects the irreversible adsorption of slowly adsorbing high molecular weight species.

INTRODUCTION

The crude oil/water interface possesses very marked rheological properties. This can be demonstrated by the formation of wrinkled skins around an oil drop as its volume is reduced in water. These rheological characteristics have a marked effect on the resolution of water-in-crude oil emulsions which occur in the production and processing of crude oil. The incidence of water in crude oil emulsions in the production of crude oil is on the increase as water injection to boost oil recovery is practiced more widely. The efficient resolution of these emulsions is becoming more important as more oil is produced in offshore locations. In the refinery desalting operation, water is added to the crude oil stream. The coalescence of this added water with the existing highly saline water is necessary to adequately reduce the salt content. We have shown (1,2) how the oil/water interfacial rheology affects crude oil emulsion stability and how added demulsifiers modify the rheology of the interface. A study of the relevance of these interfacial rheological characteristics to the oil/water immiscible displacement process is also being carried out (3).

The rheological properties of the crude oil/water interface are thus of considerable importance to the oil industry. In this paper we describe the techniques we have employed to study the interface and the conclusions we have drawn from this.

The rheological characteristics of the crude oil/water interface arise from adsorption of crude oil surfactants at that interface. These surfactants are amphoteric and their adsorption at the liquid/liquid interface depends markedly on the properties of the aqueous phase. Attempts to produce model crude oil surfactants (4) showed that though the surface wetting properties of crude oils could be produced, the emulsion forming characteristics could not. We have therefore chosen to examine real crude oil/water interfaces in spite of the difficulties in interpretation that this sometimes entails.

INTERFACIAL SHEAR VISCOSITY

Interfacial shear viscosity has been measured using a biconical bob torsion pendulum as previously described (4). We have examined whether the apparatus allows a reliable torque transmission property of the interface to be measured. It has been shown that slip does not occur at the extremities of the interfacial annulus (i.e., at the bicone bob or at the glass container wall) (6). The measurement was also not unduly sensitive to precise location of the knife edge of the bob at the interface. Whilst it has been pointed out (7) that such a device fails in the region of low interfacial viscosity, it is very effective for measuring high interfacial viscosities. It is our belief that only high interfacial viscosities are of practical significance in crude oil/water systems and that the biconical bob pendulum viscometer is suitable for this study.

Our earlier work (1) highlights problems in determining whether or not interfacial shear viscosity is a measure of the ease of emulsification or emulsion resolution. Thus, it was shown that Iranian Heavy crude oil forms extremely stable water-in-crude oil emulsions, whereas those formed by Kuwait and Forties oils are not so intractable. The interfacial shear viscosity for all these crude oil/water interfaces is, however, very similar (1).

The interfacial viscosity of some crude oils can show considerable batch to batch variation. Such crude oils are also sensitive to loss of light end components by degassing of the crude. It is considered that the extent of this gas removal is primarily responsible for the viscosity changes, reflecting changes in the crude oil solvency character for the crude oil surfactants (1).

The viscosity measurement probes the whole "interphase" volume and responds to small changes in adsorption in that region. In contrast the film compression experiments described below show less variation and reflect the fact that all but the most surface active species are squeezed out during the compression experiment.

Highly viscous interfacial films will retard the rate of oil film drainage during coagulation of the water droplets. This can lead to greatly reduced rates of emulsion breakdown. Viscoelastic films act as strictly mechanical barriers to coalescence and hence emulsion stability is high. The time taken to build up these viscous or viscoelastic films is important in the formation of emulsions, particularly in the high water content water-in-oil mousse emulsion. Thus, in cases where little interfacial compressibility resistance to coalescence occurs, stable mousses will be formed if high η_i is observed and if time is given for these viscous films to build up.

Low interfacial viscosities do not, however, imply that an emulsion will be unstable. The coalescence of water droplets is governed by the compressibility of the interfacial films formed. The extent of film viscoelasticity cannot be determined for interfaces where $\eta_i < 1$ surface poise with the present rheometer. When η_i is high, however, an analysis of the initial part of the creep compliance curve (1) (dln J(t)/dln t) can be used to determine the extent of film viscoelasticity. Figure 1 shows the increasing viscoelastic nature of the interface across the mid pH range. This maximum in film viscoelasticity agrees well with the observed maximum in emulsion stability at this pH, and with the extent of rigid film formation (Figure 2) (1).

It is interesting to note that in many of those cases where interfacial viscosity is low, high η_i can frequently be generated by a compression of the interfacial film. Figure 3 shows one such effect for Iranian Heavy/distilled water interfaces. Increases of η_i over several orders of magnitude occur for relatively small area compressions. Even interfaces that exhibit low η_i and compressible relaxing interfacial films may show increases on compression. This again must have an effect on emulsion stability, particularly those

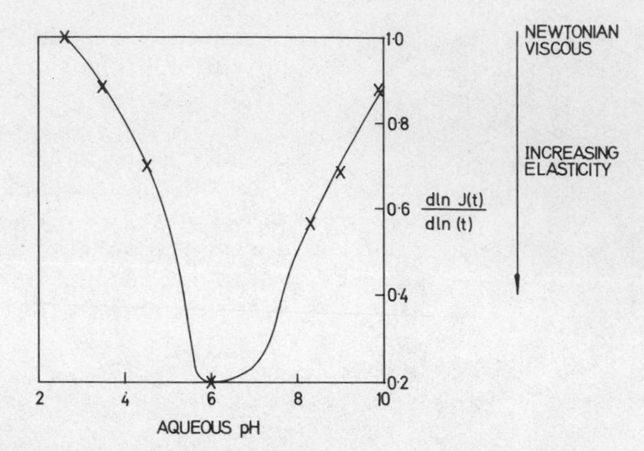

Fig. 1. Extent of film viscoelasticity as a function of aqueous pH.

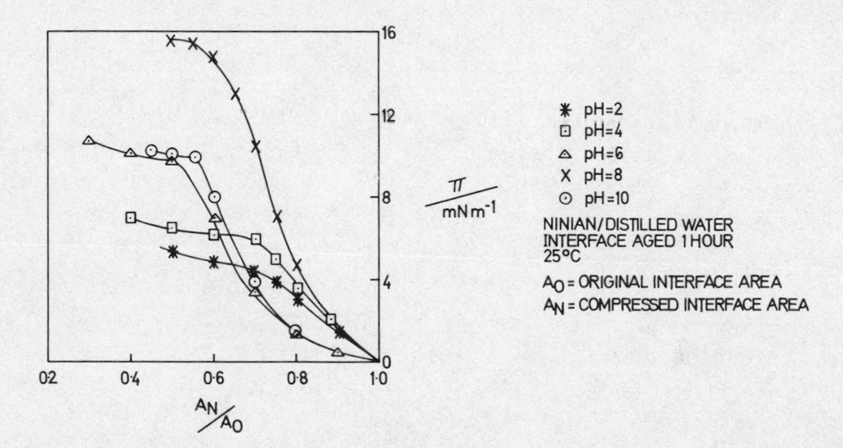

Fig. 2. The influence on film compressibility of variation of aqueous phase pH for Ninian crude oil-distilled water interfacial film.

involving small droplet size distributions (and hence large decreases of surface area during coalescence/separation). The rate of thinning of the oil film between water drops will be greatly reduced.

However, though interfacial shear viscosity (η_i) can influence the kinetics of droplet coagulation, it is not possible to rely solely on this measurement to predict emulsion behavior. Specific effects can occur which affect monolayer compressibility and film elasticity and these are not registered by shear viscosity measurements.

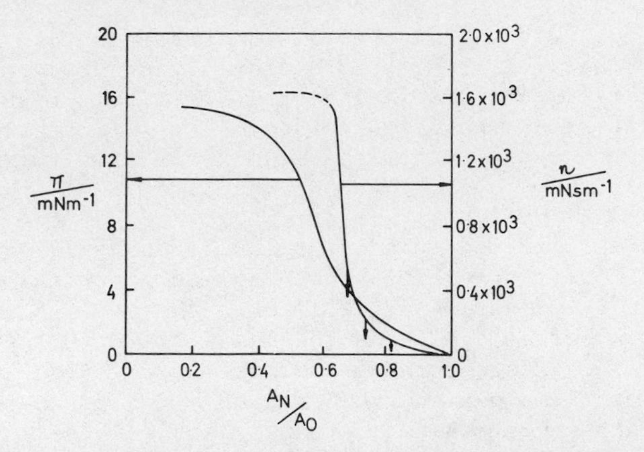

Fig. 3. Effects of compression on interfacial film pressure and viscosity.

PSEUDOSTATIC INTERFACIAL FILM COMPRESSIBILITY

The rigid oil/water interfacial films formed by adsorption of crude oil surfactants have previously been investigated. Their importance in stabilizing emulsions has been suggested (8,9) but few precise measurements have been carried out.

Hydrodynamic interaction between interfacial film and underlying liquid means that drainage of oil from the layer between two approaching water drops creates a local expansion of the interface. The resulting surface tension gradient then creates a force which opposes the thinning (Marargoni effect). In addition the coalescence of water droplets in a water/oil emulsion is accompanied by an initial increase in surface area as the drops form a connecting neck in a dumb bell configuration and a subsequent reduction in surface area and hence a compression of the crude oil/water interface as the drops coalesce. The relevance of film compression characteristics to emulsion stability is hence evident.

We employed a Langmuir type oil/water interfacial film balance (10). The relative film pressure during compression and expansion of the oil/water interface was determined. This was measured using a hydrophobic Wilhelmy plate, which remained oil-wet even in the presence of added demulsifiers. During film compression, solvent molecules may be squeezed out of the film, surfactant molecules not tightly held in the interface may redissolve into the bulk phases and molecular rearrangements of the adsorbed species may occur. The behavior of interfacial films will depend on the rate of compression. Up to speeds of compression of 1.80 cm/min (initially 16% area reduction/minute) no marked differences in film behavior were observed. Hence, all compression expansion cycles were performed at this speed. The compressibility of crude/oil water interfacial films show varying characteristics. Figure 4 shows four types of film behavior observed. The film pressure (in mN m^{-1}) is plotted against the fractional reduction in interfacial area (A_N/A_O).

Incompressible non-relaxing films (curve I, Figure 4) give rise to the most stable emulsions, whereas compressible relaxing films (curve IV, Figure 4) give rise to the least stable ones. Increasing interface age leads to greater film incompressibility. In general, the greatest change occurs in the first three hours, although some films can change detectably for up to one day before final equilibrium is reached. These changes reflect the time dependency of surfactant adsorption to/and final rearrangements and association at interfaces. Changes in crude oil solvency due to loss of light ends will also affect film compressibility.

Specific ion effects are important because formation water that can be associated with oil field emulsions may vary greatly with respect to salt content. The most pronounced effects observed, particularly with North Sea crudes, is the condensing effect of Ca^{++} on the interfacial film rendering these more incompressible, resulting in more stable emulsions (1).

In many instances emulsions are heated to assist resolution. The rate of droplet coagulation being increased by a reduction in the viscosity of the bulk oil phase. The effect of heat on the interfacial film characteristics has not been adequately studied. We have performed some tests at higher temperatures. For example, at 65°C many incompressible non-relaxing films start to relax, indicated by the fall in film pressure for films held at constant compressed area, which would still represent a kinetic barrier to coalescence. In many cases the rate of buildup of the resistance to film compression also increases with temperature. Hence, many emulsions need demulsifier treatment to overcome these rapidly developing kinetic barriers even at high temperatures.

It has also been noted (1) that where interfacial film behavior exhibits a sensitivity to loss of light end components, subsequent degassing (further stabilization of the crude by light end removal under N_2 atmosphere) leads to significant changes in the interfacial film behavior at high temperature. Often the films generated by this process remain incompressible and non-relaxing even at high temperatures and then emulsion resolution is not achieved by heating in spite of the low bulk oil viscosity.

In certain cases, e.g., with Magnus crude oil and formation water, the interfacial film at low temperature gives rise to a

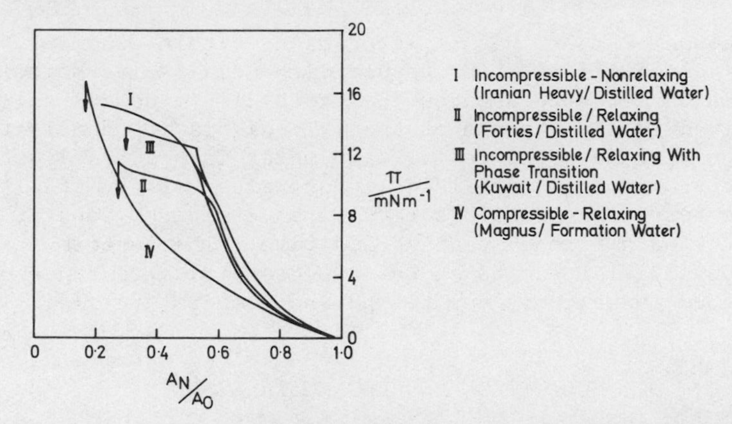

Fig. 4. Crude-oil/water interfacial compressibility characteristics.

weak compressible film. By degassing the crude, a rigid incompressible non-relaxing film is formed and this remains temperature-insensitive. Hence, the film properties are to a considerable extent governed by the changes in the solvent properties of the degassed crude for the crude oil surfactants. The influence of chemical demulsifiers on crude oil/water interfaces and their role in destabilizing emulsions can be clarified by studying their effect on interfacial films in compressibility experiments described above.

Using these techniques we have been able to distinguish between two types of demulsifier action. Some chemical demulsifiers are able to displace aged interfacial films. We designated those as film displacing demulsifiers. With others when added to the crude oil prior to contacting with the aqueous phase, the additive significantly reduced the rate of buildup of the interfacial film, this we designated as film inhibiting behavior.

It is easy to confirm this difference in behavior. Two stable Iranian Heavy/2.5% w/w sodium chloride solution emulsions (90% oil/ 10% salt solution v/v) were prepared by high-shear homogenizing. These were left for several days, with further homogenizing, to allow rigid interfacial films to build up at the oil/water droplet interface. Demulsifier A* (shown to be film displacing and inhibiting) or B* (only film inhibiting) were added, along with a further 5% v/v distilled water, with further homogenizing. Demulsifier A, within one hour, resulted in a separation of 70% of the salt and 90% of the distilled water. Demulsifier B gave a separation of 5% of the original salt and 50% of distilled water

---

*A = an ethoxylated phenol in an aromatic solvent.
B = a carboxylic acid containing a small amount of alkoxylated ester also in an aromatic solvent.

within one hour. There was no separation, within one hour, in the absence of demulsifiers. The result with demulsifier B indicated that little coalescence between the added distilled and salt solution occurred in the absence of the film-displacing demulsifier. Thus, the distilled water (e.g., wash water in practice) acts as a scavenger to increase collision frequencies between itself and the salt water droplets. This results in efficient separation in the presence of a film-displacing additive. The performance of a chemical demulsifier can be shown to be sensitive to crude oil type, aqueous phase, salt nature, pH and temperature (1).

MULTISTAGE DROPLET COALESCENCE

The formation of smaller secondary droplets by coalescence of the primary drop at an oil/water interface can be of great practical significance. These smaller drops are most difficult to remove from w/o or o/w emulsions. For oil and water drop settling rates are decreased as the droplet diameter is reduced.

The coalescence times for oil and water drops at the crude oil/water interface using the three crudes Tia Juana Medium, Put River and Zakum are shown in Table 1. Multistage coalescence was observed for oil and water drops with Zakum at pH 11, in all the other cases multistage coalescence was not observed. The Zakum oil/water interface at pH 11 also exhibited the lowest (hardly measurable) interfacial viscosity of all the systems studied.

Mason (11) explained the occurrence of multi and single stage coalescence on the basis of Raleigh perturbations. According to Raleigh, for an infinitely long cylindrical column of an incompressible liquid of radius R, surface disturbance of λ greater than $2\pi R$ will grow in amplitude and cause the column to break up. For a column of finite length, as in the draining of a drop, if the length exceeds the circumference, the column will start to break up. Once this has occurred a race develops between the necking down process and drainage of liquid in the drop; if the former wins, multistage coalescence will occur. The presence of surfactant in sufficient amount will not only stop the growth of the instabilities damping interfacial waves, but will also stabilize the column itself so that the drop can drain completely.

The results obtained to date show that multistage coalescence with crude oils will occur only when the interfacial stability is decreased to a marked extent. Extremely fluid interfacial films with low η_i only occur at high pH values in the absence of additives. The addition of demulsifiers results in a fluid interface which could give rise to multistage coalescence and hence pose additional problems with respect to emulsion resolution. The dilatational modulus technique (see below) will be applicable to the

Table 1. Drop Lifetimes of Water and Oil Drops at Various Crude Oil/Water Interfaces for Aqueous Phase pHs of 2, 5 and 11

Oil	pH	Drop Lifetimes seconds	
		O/W	W/O
Tia Juana	2	180	> 600
	5	40-50	54
	11	180	> 600
Put River	2	120	> 600
	5	37	48
	11	> 600	> 600
Zakum	2	131	594
	5	145	> 480
	11	1st stage 240	1st stage 120-600
		Total > 600	Total > 600

study of this problem. The effect of demulsifier on interfacial instabilities at different frequencies can be determined.

INTERFACIAL DILATATIONAL PROPERTIES

The components of the surface stress tensor depend upon the extent and the rate of surface deformation, in a relationship involving the resistance of the surface to both changes in area and shape. Either of these two types of resistance can be expressed in a modulus which combines an elastic with a viscous term. This leaves us with four formal rheological coefficients which suffice for a description of the surface stress. Two of these, viz., the surface dilatational elasticity, ε_d, and viscosity, η_d, measure the surface resistance to changes in area, the other two, viz., the surface shear elasticity, ε_s, and viscosity, η_s, describe the resistance of the surface to changes in the shape of a surface element.

Since it is commonly (5) held that the occurrence and magnitude of surface tension gradients (Marangoni effects), whether due to spatial variations in temperature or concentration or compression/expansion of the interface, are important to many colloid problems, some consideration has been given to methods of determining the dilatational rheological parameters.

A longitudinal wave propagation approach has been adopted in view of the problems of applying capillary wave techniques to liquid/liquid interfaces. These arise from two quarters. Firstly the variation of damping coefficient with elastic modulus can be shown (12) to be negligible at sufficiently high and low elastic moduli.

Secondly for capillary (transverse) wave propagation, when the liquid viscosities and densities become equal, there is no net tangential motion of the interface and here the elastic modulus plays no role in the motion of transverse waves. Since many organic liquid/water pairs have characteristics not far from this condition, it places a severe restriction on applicability to liquid/liquid systems. Since the longitudinal wave is accompanied by a horizontal rather than a vertical interfacial motion, such waves are often associated with liquid flows of a highly dissipative character causing rapid damping. It is necessary therefore to determine conditions where this effect is minimized. This effectively requires operating under conditions where the wavelength of the propagated wave is long compared with the interfacial length under observation. A theoretical examination of the propagation of longitudinal waves leads to a representation of the wave characteristics [that is wavelength (λ) and damping coefficient (β)] in terms of the dilatational properties of the liquid/liquid interface (13).

Thus it can be shown that:

$$\beta\ (\mathrm{cm}^{-1}) = \frac{[(\omega^3\rho_o\eta_o)^{\frac{1}{2}} + (\omega^3\rho_w\eta_w)^{\frac{1}{2}}]^{\frac{1}{2}}}{|\varepsilon|\ \frac{1}{2}} \sin\left(\frac{\pi}{8} + \frac{\theta}{2}\right) \qquad (1)$$

and

$$\chi\ (\mathrm{cm}^{-1}) = \frac{2\pi}{\lambda} = \frac{[(\omega^3\eta_o\rho_o)^{\frac{1}{2}} + (\omega^3\eta_w\rho_w)^{\frac{1}{2}}]^{\frac{1}{2}}}{|\varepsilon|\ \frac{1}{2}} \cos\left(\frac{\pi}{8} + \frac{\theta}{2}\right) \qquad (2)$$

where β (cm^{-1}) = the damping coefficient
λ (cm) = wavelength
ω(H$_z$) = angular frequency
η_o, η_w = liquid bulk viscosities
ρ_o, ρ_w = liquid densities
θ = phase angle (between γ and surface area variation)
$|\varepsilon|$ = dilatational modulus

The dilatational elasticity (ε_d/mN m^{-1}) and the dilatational viscosity (η_d/mN s m^{-1}) can be related to $|\varepsilon|$ and θ by

$$\varepsilon_d = |\varepsilon| \cos \theta \tag{3}$$

$$\eta_d = \frac{|\varepsilon| \sin \theta}{\omega} \tag{4}$$

A given amplitude and frequency wave is propagated by sinusoidal oscillation of one barrier of the film balance. Details of the dilatational modulus apparatus can be found elsewhere (9). If we arrange for every small element of area to be subjected to the same change, then measurement of the variation of interfacial tension can be made anywhere along the interfacial length to determine the dilatational properties. To do this, use is made of the cooperative effect of multiple reflections with the outgoing wave. This can be achieved provided the interfacial length used is small compared with the wavelength (13).

The procedure adopted is to estimate values of wave frequency (ω) and interfacial lengths (L) such that the wavelength (λ) of the longitudinal wave is long compared to L. Since λ depends upon the dilatational properties this requires an iterative procedure. That is, estimate a suitable L and check for uniformity of $|\varepsilon|$ and θ along the interfacial length. If this is uniform then the dilatational properties may be calculated. If not, alter either ω or L (or both) and repeat the process until uniformity is obtained. From the measured $|\varepsilon|$ and θ we may determine by equations (3) and (4) the dilatational properties of the interface.

The amplitude and frequency of barrier oscillation and hence wave propagation are variable. Frequencies between 0.005 - 1 Hz can be obtained; and amplitudes up to ± 0.5 cm can be used. The other barrier can be smoothly adjusted to alter the extent of the interfacial area, as in a conventional Langmuir type film balance. The change of interfacial tension produced by the area variation is monitored continuously. Use is made of Wilhelmy plate suspended from one arm of a microforce balance (Beckman microforce balance). The output of this feeds one arm of an X-Y recorder. A position transducer monitors the movement of the oscillating barrier and this feeds the Y axis of the recorder. Lissajou figures are therefore produced on the X-Y recorder. Here θ is defined as being negative when the change in surface pressure follows the change in area and positive when the surface pressure precedes the change in area. When conditions have been set to achieve uniform measurement of $|\varepsilon|$ down the interfacial length, if θ is also uniform then the whole film is relaxing by a single uniform cooperative process. When θ varies along the interfacial length, a check must be made by further reducing the interface length to see whether the problem is one of viscous damping, i.e., the trough length is too long with respect to λ.

RESULTS OF PRELIMINARY DILATATIONAL MODULUS MEASUREMENTS

Iranian Heavy crude oil/distilled water interfaces were aged for set intervals on the dilatational modulus apparatus. The interfacial length was initially set as 13.5 cm and amplitudes of area oscillation of ± 0.5 cm were used. The Wilhelmy plate (platinum, hydrophobic) was always aligned parallel to the oscillating barrier. The X-Y recorder was used to produce Lissajou figures of interfacial tension-interfacial area variations. The Wilhelmy plate was located at points $\frac{L}{5}$, $\frac{2L}{5}$, $\frac{3L}{5}$ and $\frac{4L}{5}$ along the interface length (L). At each location, Lissajou figures were produced over a range of wave frequencies. From these the modulus $|\varepsilon|$ and phase angle θ were determined. Table 2 summarizes the dilatational elasticity and viscosity results calculated from these. It can be seen that for the oldest interface (3 h) (most cohesive film with highest dilatational elasticity) using the longer trough length that the conditions sought (interface length << wavelength of longitudinal wave) was not met because variation of η_d down the length of the trough was observed. For this interface uniform behavior could only be obtained by reducing the interfacial length (and its amplitude). For the shortest aged interface (½ hour), although data is not given, there was a slight increase of ε_d measured along the interface length. This reflects the fact that the ε_d increase with film age was comparable with the time scale taken to make the measurements along the trough length. Figure 5 shows the change of ε_d (the dilatational elasticity) with film age, and Figure 6 shows η_d (the dilatational viscosity) against interface age (full curve). For the dilatational viscosity the dotted line shows the result at 3 hours, obtained where viscous damping was evident. In an attempt to minimize the damping effect, the η_d values were obtained by extrapolating back to the oscillating barrier. Clearly such a procedure leads to large errors.

Figures 7 and 8 show the dilatational elasticity and viscosity over the limited frequency range so far studied.

Table 3 gives the dilatational properties of 10 ppm of Bovine serum albumin (Sigma Chemical Company) in 0.1 M NaCl solution/toluene oil phase made for comparative purposes. In order to

Table 2. Dilatational Properties of Iranian Heavy Crude Oil/Distilled Water Interfaces

Oscillation Frequency ω/Hz	Interface Age/Hr.	Interface Length L/cm	ε_d/mN m^{-1}				η_d/mN sm^{-1}			
			$\frac{L}{5}$	$\frac{2L}{5}$	$\frac{3L}{5}$	$\frac{4L}{5}$	$\frac{L}{5}$	$\frac{2L}{5}$	$\frac{3L}{5}$	$\frac{4L}{5}$
0.005	3	13.5	<—— 11 ± 1 ——>				160	150	140	130
0.01	3	13.5	<—— 13 ± 1 ——>				90	70	55	35
0.005	3	6.8	<—— 11 ± 1 ——>				<——110 ± 15——>			
0.01	3	6.8	<—— 13 ± 1 ——>				<——50 ± 7——>			
0.02	3	6.8	<—— 15 ± 1 ——>				<——25 ± 5——>			
0.005	1½	13.5	<——7.5 ± 1 ——>				<——100 ± 15——>			
0.01	1½	13.5	<—— 9 ± 1 ——>				<——50 ± 7——>			
0.005	½	13.5	<—— 7 ± 1 ——>				<——90 ± 15——>			
0.01	½	13.5	<—— 8 ± 1 ——>				<——40 ± 7——>			

Table 3. The Dilatational Properties of BSA at the Toluene/0.1N NaCl Solution Interface

ω/Hz / Plate Location	0.01		0.05		0.2	
	ε_d/mNm^{-1}	$\eta_d/mNsm^{-1}$	ε_d/mNm^{-1}	$\eta_d/mNsm^{-1}$	ε_d/mNm^{-1}	$\eta_d/mNsm^{-1}$
$\frac{L}{5}$	51	57½	53	< 3	48	11
$\frac{2L}{5}$	47	52½	51	< 3	50	13
$\frac{3L}{5}$	46	51	52	< 3	49	14
$\frac{4L}{5}$	48	52½	49	< 3	48	14

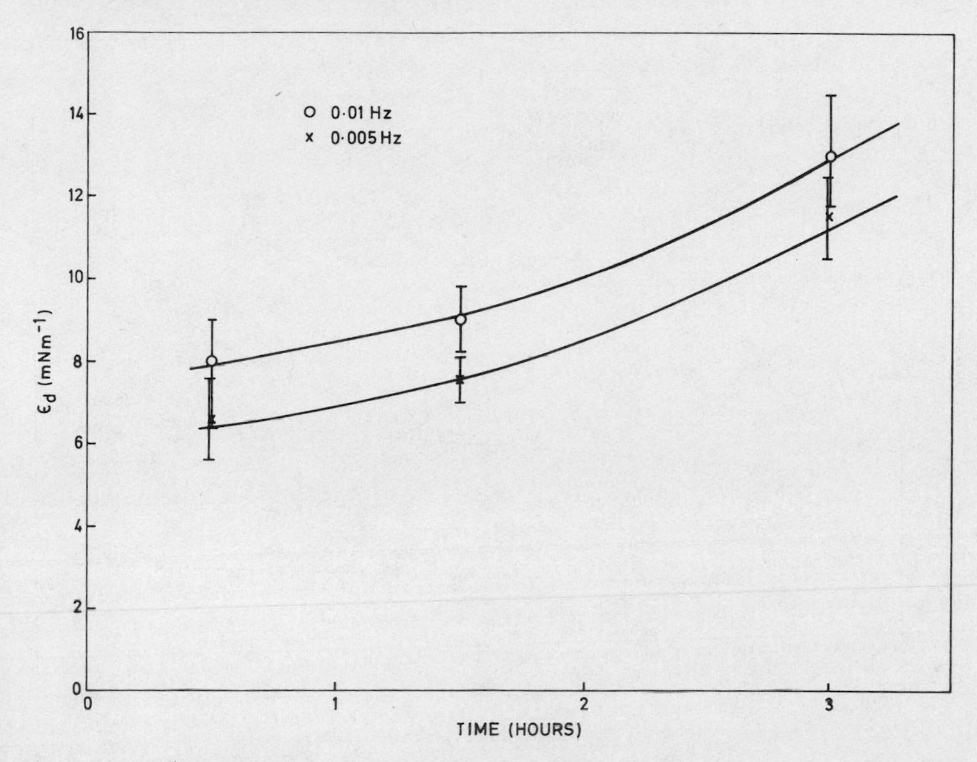

Fig. 5. Dilatational elasticity vs. age of interface for Iranian Heavy/distilled water interface.

prevent contact angle problems that were experienced with this system a fresh hydrophobic glass Wilhelmy plate was used for measurements at each location.

The values of the dilatational properties will clearly depend upon the molecular relaxation processes that occur during the timescale of film compression/expansion. The relaxation processes occurring will have their own characteristic rates and it is possible to recognize, if not identify, the different relaxation rates involved.

Molecular relaxation processes envisaged to occur within an interfacial film are not clearly understood but they are likely to include the following:

1. Increased packing (surface diffusion) of relatively low molecular weight species.

2. Molecular conformational changes of high molecular weight, irreversibly adsorbed, species.

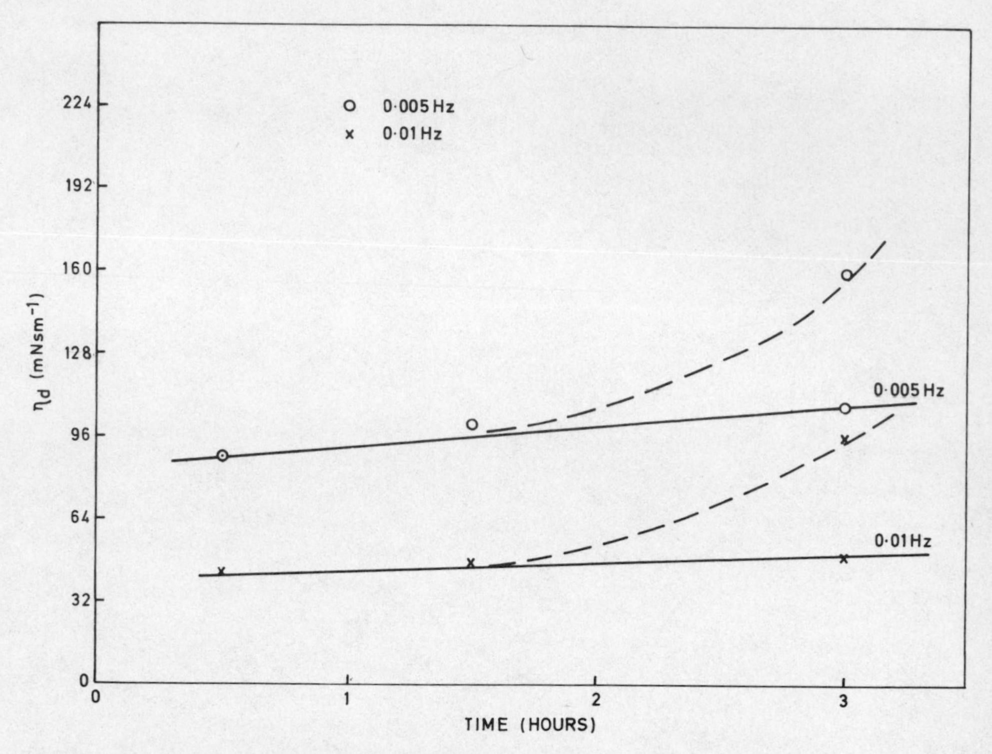

Fig. 6. Dilatational viscosity vs. age of interface for Iranian Heavy/distilled water.

3. Desorption (and readsorption) of low molecular weight reversibly adsorbed molecules during compression/expansion. This will also include solvent losses from molecular films.

The rates of these relaxation processes (characteristic relaxation times) are not well known for liquid/liquid systems. This is especially true with monolayers where the interfacial concentrations and molecular adhesion/cohesion forces within the monolayer are high. Experimental evidence for relaxation processes (rates) can be derived from a study of the dilatational properties of adsorbed film as a function of frequency of the propagated longitudinal wave. Where bulk-to-interface interchange occurs during the compression/expansion of the film, the interfacial tension gradient is essentially "short-circuited". Lower 'apparent' values of interfacial compressional moduli are observed. There is a phase difference between the imposed strain (change in area) and the resultant stress (change in interfacial tension).

Alternatively, with an irreversibly adsorbed monolayer during quick expansions or compressions the value of the interfacial tension may be significantly different from the same surface at rest

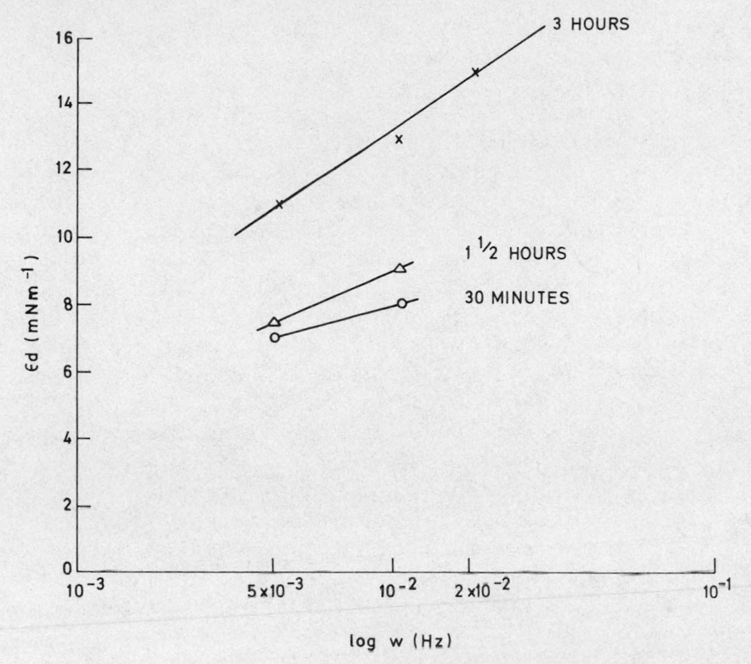

Fig. 7. The dilatational elasticity crude oil/water interface as a function of wave frequency (and interface age).

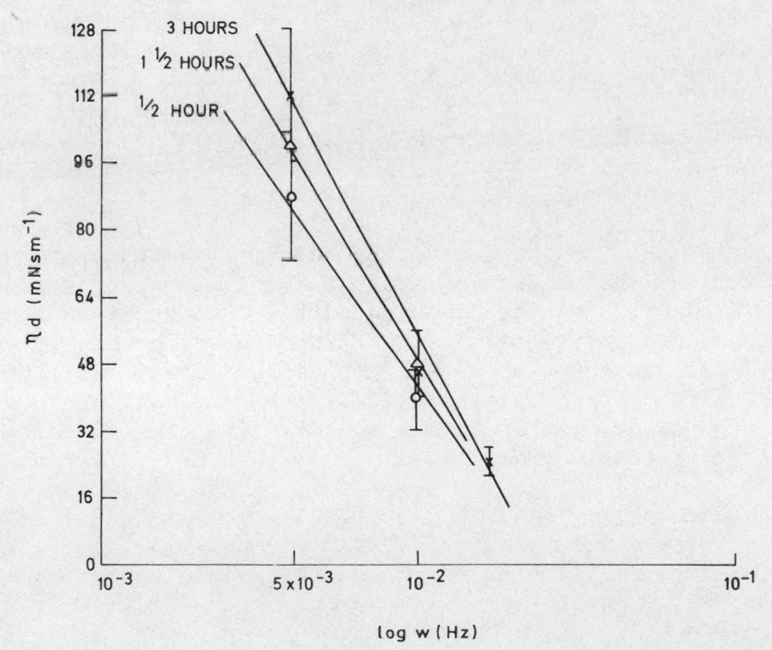

Fig. 8. The dilatational viscosity of an Iranian Heavy/distilled water interface as a function of frequency (and age).

with the same interfacial coverage. This arises as a consequence of molecular reorientational (conformational) changes of irreversibly adsorbed species.

In the case of bulk-to-interface interchange, the effect should be less marked as the frequency of oscillation increases. In the second case (conformational changes), the effect will increase with increase of frequency. The relaxation times of the two processes should be very different. Slow reorientations are generally only observed for long chain, practically insoluble (irreversibly adsorbed) components. These are not subject to diffusional interchange. Conversely diffusion will be more important the greater the solubility and the shorter the chain length.

For the Iranian heavy crude oil/distilled water interface it seems that the dilatational elasticity, ε_d, is not particularly high (compared with subsequent BSA values) or when compared with very cohesive insoluble interfacial films at the liquid/air interface. However, there is a measurable phase angle (θ) and hence dilatational viscosity and this confirms that relaxation processes will play an important part in the behavior of the crude oil/water interface.

The dilatational viscosity seems high; however again there is little with which to compare it. The dilatational elasticity of the interfacial film increases with film age, particularly at the longer interfacial contact times (i.e., 2-3 hours). This behavior is as would be expected from previous knowledge on the pseudostatic elasticity of these interfacial films (1).

There seems to be little change (within experimental accuracy) in the dilatational viscosity of the interfacial film over the period of time investigated (Figure 6). This strongly suggests that the nature and extent of the film relaxation process is hardly affected as a function of time. The fall of η_d with frequency is indicative that the relaxation process involves predominantly bulk-to-interface diffusion interchange; with low molecular weight surfactant species adsorbing/desorbing during the timescale of the compression/expansion cycle. The changing dilatational elasticity with time (Figure 7) probably reflects the increase in the proportion of the high molecular weight species with time (i.e., increasing surface concentration).

Adsorbed monolayers of BSA at the liquid/liquid interface give rise to a phase difference, at certain frequencies, between the sinusoidal area oscillation and the interfacial tension variation. This contrasts sharply with the behavior at the liquid/air interface, where a previous investigation found no phase angle (14) for both spread and adsorbed films. Thus, relaxation processes can be shown to be important at the liquid/liquid interface for adsorbed

BSA films. The film relaxation processes are however complex.

It is not possible to directly compare the moduli values resulting from our investigation with those found at the liquid/air interface (14). This is because of viscous damping effects observed at the liquid/air interface; the $|\varepsilon|$ values being sensitive to the location along the interface length. As has been shown for a crude oil/water interface, large errors result (especially in η_d) when back extrapolating the behavior to the oscillating barrier. Indeed as has been pointed out elsewhere (15) close to the oscillating barrier, the surface film is moved a large distance compared to the size of the adsorbed molecules and in so doing introduces other problems (i.e., shearing between the film/subphase and the film/trough wall). The dilatational moduli (liquid/air) obtained by this back extrapolation technique for BSA at the liquid/air interface are in the order of 150 mN m^{-1} for the nearest comparable bulk phase concentration. It has been shown by Lucassen et al. (13) that for the liquid/air interface, measurements made at an interfacial length $x = 0.423$ L (the interface length) should give the correct dilatational properties. At this point the local and uniform area displacements should be equal. Where this was done for BSA at the liquid/air interface (14) values of ε_d of around 45 mN m^{-1} (compare with 46-53 mN m^{-1} for BSA at the liquid/liquid interface) were obtained at the comparable bulk phase concentration.

From the dilatational behavior observed for BSA, this globular protein seemingly gives rise to highly cohesive interfacial films at the liquid/liquid interface. In marked contrast to the behavior at the liquid/air interface adsorbed BSA interfacial films at the liquid/liquid interface show relaxation processes within the time-scale at the compression/expansion cycles. At the lowest frequencies these seem to arise from bulk-to-interface diffusional interchange whilst at the higher frequencies there may be some evidence of conformational changes occurring (i.e., amino acid train-to-loop configurational changes).

REFERENCES

1. T. J. Jones, E. L. Neustadter, and K. P. Whittingham, J. Canadian Pet. Tech., 17, 100 (1978).
2. T. J. Jones, E. L. Neustadter, and K. P. Whittingham, Proceedings of the International Conference on Colloid and Surface Science, E. Wolfram, ed., Budapest, Hungary, September 15-20, 1979.
3. C. E. Brown, E. L. Neustadter, and T. J. Jones, 3rd International Conference on Surface and Colloid Science, Stockholm, Sweden, August 1979.

4. R.J. R. Cairns, D. M. Grist, and E. L. Neustadter, SCI Symposium Theory and Practice of Emulsion Technology, Brunel University, September 16-18, 1974.
5. P. Sherman, J. Colloid Interface Sci., 45, 427 (1973).
6. K. P. Whittingham, D. M. Grist, and E. L. Neustadter, Paper SPE 7819, Soc. Pet. Engr. of AIME, Dallas, Texas, 1978.
7. F. C. Goodrich, "Progress in Surface and Membrane Science," Vol. 7, Academic Press, New York
8. J. Reisberg and T. M. Doscher, Prod. Monthly, 21, 43 (1956).
9. C. E. Brown, E. L. Neustadter, and K. P. Whittingham, Symposium on Enhanced Oil Recovery by Displacement with Saline Solutions, Britannic House, London, May 20, 1977.
10. J. H. Brookes and B. A. Pethica, Trans. Faraday Soc., 60, 208 (1964).
11. G. E. Charles and S. G. Mason, J. Colloid Interface Sci., 15, 105 (1960).
12. J. Lucassen and R. S. Hansen, J. Colloid Interface Sci., 22, 32 (1966).
13. J. Lucassen, Trans. Faraday Soc., 64, 2230 (1968).
14. D. E. Graham, "Structure of Protein Films and Their Role in Stabilising Foams and Emulsions," Thesis, 1976.
15. E. H. Lucassen-Reynders and J. Lucassen, in "Advances in Colloid and Interface Science," Vol. 2, Elsevier, Amsterdam, March 1970.

POROUS MEDIA RHEOLOGY OF EMULSIONS IN TERTIARY OIL RECOVERY

S. Vijayan,[1] C. Ramachandran,[2] H. Doshi and D. O. Shah

Department of Chemical Engineering
University of Florida
Gainesville, Florida 32611, U.S.A.

Emulsions containing 5 wt % TRS 10-410, 3 wt % isobutanol, sodium chloride (X %), water and equal volume of dodecane oil were prepared by sonication and by hand-shaking. The coalescence behavior of emulsions was studied for hand-shaken as well as sonicated systems. In general, sonicated emulsions required a longer time for phase separation as compared to hand-shaken systems. It was observed that for both the cases, the coalescence rate at room temperature (25°C) was maximum at the optimal salinity (1.5% NaCl) while interfacial tension was minimum at this salinity.

For the flow of emulsions through a sand pack at room temperature (25°C) the emulsions at most of the salinities studied exhibited a non-Newtonian behavior. However, at and near the optimal salinity the flow behavior was Newtonian. Moreover, the apparent viscosity of the emulsions in the porous media was minimum at the optimal salinity, hence it is proposed that ultra low interfacial tension causes minimum resistance to fluid flow and thus aids in the process of maximum oil recovery obtained at this optimal salinity.

Porous media studies of sonicated emulsions in a EM-Gel packed bed at 35°C formed with different aqueous to oil phase ratios for different salt concentrations indicate that pressure drop across the bed increases with an increase in the amount of the dispersed

---

[1]Presently with the Whiteshell Nuclear Research Establishment, Atomic Energy of Canada Ltd., Pinawa, Manitoba, Canada R0E 1L0.

[2]Presently at the School of Pharmacy, University of Wisconsin, Madison, Wisconsin 53706, U.S.A.

phase. Thus, for water-external macroemulsions the pressure drop increases with an increase in the amount of oil, while for oil-external macroemulsions pressure drop increases with an increase in the amount of water.

Sonicated emulsion results at 35°C as a function of phase ratio and salt concentration further show that the behavior of these emulsions in the porous media is different from those at 25°C. Even three phase systems give relatively more stable emulsions and possess non-Newtonian behavior. Hysteresis effect shown by some emulsions in porous media experiments is highlighted. The nature of emulsions produced by sonication and by the shearing of liquids in the porous media is contrasted. Extensive physical property data for these emulsions are also given. The applicability of Darcy's law for the present situation is discussed.

The results of our earlier investigation using spin-labelling technique to understand the structural aspects involved in the various emulsions, support the theory that water-external macroemulsions exist below optimal salinities and oil-external types exist beyond optimal salinity. In addition it was found that microemulsions coexisted with macroemulsions and were of the same type in the sonicated emulsions. These findings are further complemented by electrical conductance and bulk viscosity data.

INTRODUCTION

In tertiary oil recovery process, similar to primary and secondary recovery processes, one encounters problems associated with the formation of emulsions. Evidently such emulsions may either impede or accelerate the recovery depending on their physicochemical characteristics under given conditions. For example, if the formation of emulsion is considered adverse in operating the recovery scheme then selective methods of demulsification should be invoked.

A review of the literature shows very little mentioning of flow of emulsions through porous media. Some interesting observations favoring the presence of macroemulsion in the oil recovery process have been reported by McAuliffe (1). The ease with which emulsions flow at high pressure gradients was proposed by McAuliffe to be an advantage in the field applications because in a radial flow system the highest pressure decrease per unit distance occurs near the well-bore.

The unfavorable aspect of having an emulsion arises outside the reservoir conditions. Although it is well known that macroemulsions are thermodynamically unstable, the presence of multiple surfactants in the flood can effectively prolong the emulsion stability to time periods of the order of weeks or months. The formation of such relatively stable emulsions has been identified by Boneau and Clampitt (2).

Lissant (3) in a recent review indicated that the occurrence of emulsion is a natural and all pervasive phenomenon in petroleum production and emulsions are to be avoided if possible. He further reported that a considerable amount of literature on the development of techniques for resolving emulsions can be found and that this technology should be taken into consideration when emulsification is deliberately employed to achieve additional oil recovery.

It appears from these investigations that we have two opposing effects in having macroemulsions in the oil recovery process--one a possibly favorable situation in the reservoir and the other an adverse effect off-the reservoir. The obvious questions to be posed are: (1) Can a compromise for having the emulsion considering the two effects in terms of its efficiency, be reached? (2) For a given surfactant formulation and conditions, how would one go about assessing emulsion formation and separation? (3) What is the influence of oil-external, water-external and middle phase microemulsions on the formation of macroemulsions and emulsion phase separation? (4) With an impure surfactant system such as TRS 10-410, can we characterize the macroemulsion containing oil, cosolvent and brine? (5) If a relatively stable macroemulsion is produced can one fundamentally investigate its flow behavior in a porous system with respect to pressure drop, permeability, shear degradation, droplet population, etc.?

With all these factors in mind, we have attempted to carry out the emulsions aspect of the investigations at the University of Florida Improved Oil Recovery Research Program (4,5). The emulsion systems contain TRS 10-410, isobutanol, sodium chloride, dodecane and water. Extensive physical property data and microstructural studies of the aqueous surfactant formulations have been already reported by Vijayan et al. (6). Also, the structural aspects of the emulsions containing the same species with aqueous to oil ratio of 1:1 as well as various physical property data as a function of salt concentration have been reported by Vijayan et al. (7). A detailed study of the middle phases formed by the same surfactant formulation with dodecane oil with respect to microstructural changes and microemulsion (swollen micelle) phase inversion has been reported by Ramachandran et al. (8).

With this background knowledge of the behavior of the aqueous surfactant formulations and 1:1, aqueous : oil emulsions and, middle phases as a function of sodium chloride content, the objectives of the present investigation are:

(1) to experimentally measure the physical properties of relatively stable emulsions of a specific surfactant formulation consisting of TRS 10-410 + isobutanol (5:3% W/W), sodium chloride (X%) + water and dodecane oil (Y%).

(2) to investigate the flow behavior of these emulsions in an unconsolidated sand pack and in an unconsolidated porous bed containing EM-Gel packing.

(3) to obtain the phase separation characteristics of various sonicated and hand-shaken emulsions by batch settling tests, and

(4) to correlate physical property data and structural state of emulsions with phase separation data, pressure drop results, shear rate effects and permeability factors.

The concentration of brine and aqueous to oil ratio are the variable parameters considered throughout the investigation.

EXPERIMENTAL

Chemicals: The surfactant formulation consisted of fixed amounts of the surfactant TRS 10-410, a petroleum sulfonate (5 wt %) and isobutanol (3 wt %) in brine solutions of different sodium chloride concentrations up to 8.0 wt %. The oil used was dodecane. Double distilled water with conductivity less than 2 μS/cm was used throughout the experiments. Dodecane oil was of technical grade (95 mole %) supplied by Phillips Petroleum Company (Lot N-919).

The petroleum sulfonate TRS 10-410 (~ 60% active) was obtained from Witco Chemical Company. Isobutanol (IBA) and sodium chloride were of high purity grade (> 99%) from Chemical Sample Company and Fisher Scientific, respectively.

Porous media: The following two porous media were employed in the present study. (1) A sand packed bed (400 mesh), (0.25 inch diameter and 2 inches long) with column made of stainless steel, and (2) a second porous media system with column made of stainless steel (0.25 inch diameter and 6 inches long and EM-Gel SI 2500Å packing material with particle size 0.063-0.125 mm). Mean pore diameter measured with a mercury porosimeter gave a value of 2800Å. The material was obtained from Merck, Cat. No. 9364, batch no. YE 94, 3949803.

Packing mode: dry packing
Porous media state: Unconsolidated porous bed

Porous media (1) was exclusively used for emulsions containing equal volumetric amounts of oil and aqueous phases and salt levels from 0 to 8.0 wt %.

Preparation of solutions: Aqueous solutions were prepared by dissolving 5% (W/W) of TRS 10-410 and 3% (W/W) isobutanol together with constant stirring. Desired concentration of sodium chloride in water was prepared separately and then known amounts of the surfactant mixture and the brine solution were mixed volumetrically. It should be noted that while the concentrations of surfactant and alcohol are expressed in percent weight based on total weight of aqueous phase, the concentration of sodium chloride is based on the total weight minus the weight of surfactant and alcohol.

Experimental set-up and procedure: Macroemulsions were produced by constant sonication for a period of about forty-five minutes in a thermostated bath at desired temperatures. Kinematic viscosity and specific conductance data of emulsions were obtained using standard Cannon-Fenske viscometer and conductivity meter, respectively. Bulk density and Screen factors of emulsions and equilibrated phases were determined by standard specific gravity bottles and screen viscometer, respectively. The interfacial tension values of oil/aqueous systems were measured by spinning drop technique. The details of measurement procedures are described elsewhere (4,5).

Porous media experiments: Flow experiments in sand-packed beds were carried out using 1:1, aqueous : oil emulsions at 25°C for salt concentrations of 0 to 8 wt %. Porous media experiments in EM-Gel-packed columns were done at 35°C for emulsions produced by several aqueous to oil ratios (4:1, 2:1, 1:1, 1:2, 1:4) and different salt concentrations varying from 1 to 3.5 wt %.

Procedure: Appropriate aqueous-oil systems were taken in glass bottles with a capacity of 150 ml and were placed in a thermostated bath. (The experimental set-up is shown in Figure 1.) Sonication was effected by an ultrasonic vibrator. Simultaneously, constant stirring was maintained by a magnetic stirrer system. The emulsion thus formed, was pumped from the bottle by a Constametric 1 HPLC Pump (Milton Roy Co., Serial No. 7804-03). The flow rate was accurately controlled by an in-built potentiometer setting. The pressure drop (against atmosphere) was measured using three pressure transducers (for higher accuracy in measurement) depending on the range of their calibration. Calibrations of 0-50 psi range, 0-100 psi range and 0-500 psi range were used. For a given emulsion, pressure drop was measured as a function of flow rate and the emulsion was recycled continuously.

During the experiment, five different columns were used so that the pressure drop of water, which was used as a standard for judging the condition of the regenerated bed, remained within reasonable limits of the pressure drop obtained for fresh beds. The procedure to fill the columns with the porous material was followed consistently.

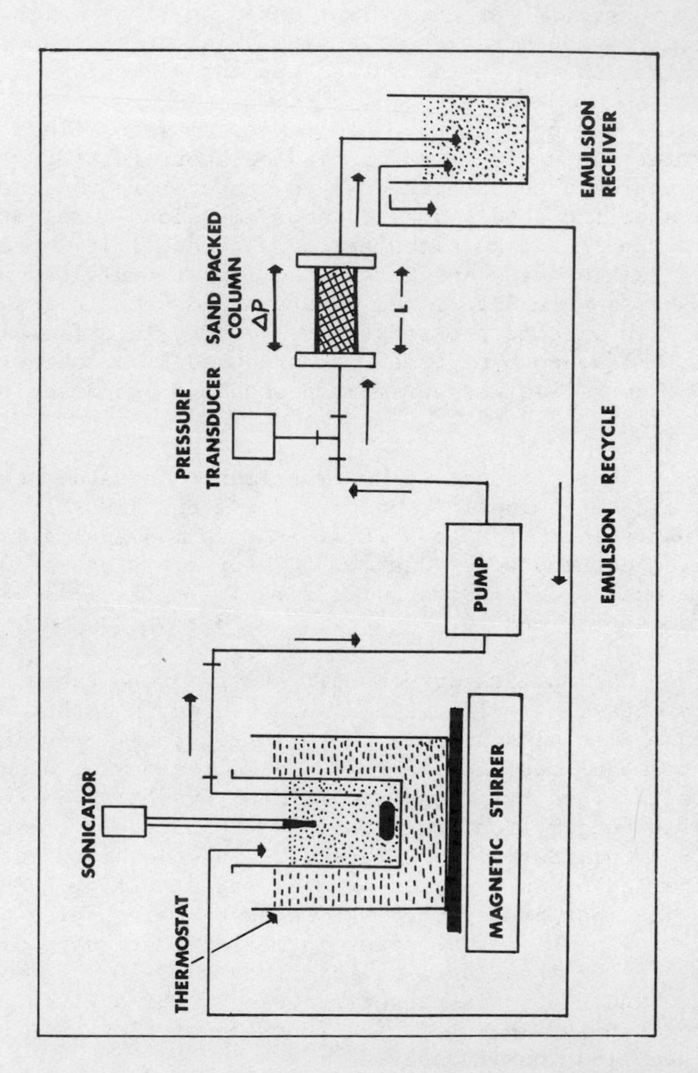

Fig. 1. Schematic flow diagram of laboratory emulsion flow through porous media studies.

It is obvious that the bed characteristics alter slightly once an emulsion is pumped through the bed. Regeneration was attempted by flushing the bed thoroughly with acetone and distilled water. In order to account for the variations in the bed behavior, if any, the following procedure was used:

Pressure drop of water (ΔP_{WM}) was measured and plotted as a function of flow rate for the regenerated bed. Then the pressure drop of the emulsion (ΔP_{EM}) was measured; and the values at the particular flow rate were normalized by dividing them by the ΔP_{WM} values, obtained from the experimental plot and at the corresponding flow rate (i.e. the ratio $\Delta P_{EM}/\Delta P_{WM}$ was thus obtained.) A generalized pressure drop (ΔP_{WG}) curve for water obtained by taking an average of about 30 calibrations was used to multiply the above mentioned ratio. This gave the normalized pressure drop of the emulsion (ΔP_E), that is, $\Delta P_E = [\frac{\Delta P_{EM}}{\Delta P_{WM}}](\Delta P_{WG})$.

Figure 2 is used as the generalized curve of the pressure drop of water (ΔP_{WG}) in the 6 inch long porous bed. Figure 3 illustrates typical pressure drop versus flow rate curves for water flowing through a 2 inch long sand-packed column after passage of various emulsions.

Our contention is that by using the above method, any permeability changes taking place due to the flow of the emulsions, can be accounted and therefore a meaningful comparison of the pressure drop results of different systems can be achieved.

Batch Settling Experiments for Hand-Shaken and Sonicated Emulsions for Oil to Aqueous Phase Volume Ratios of 1:1 and 2:1

Appropriate volumes of the aqueous formulation and dodecane oil were taken in 50 ml mixing cylinders. For hand-shaken experiments, the cylinders were hand-shaken for one minute and then the time required for complete separation was followed by noting the volume change of separated phase with time. Similar experiments were also carried out using sonicated emulsions at constant sonic power input of 30 watts for a period of 5 minutes.

The objective in these experiments was to collect separation-rate data for the various emulsions as a function of sodium chloride concentration. As the time taken for complete separation of sonicated emulsions containing low (less than 1 wt %) and high (greater than 2%) salt concentrations was in the order of days, we restricted our measurements to time taken for 50% phase separation.

Reproducibility check was made by using different batches of the same formulation. It was estimated that the data are reproducible within ± 5%. It should be emphasized that our primary aim

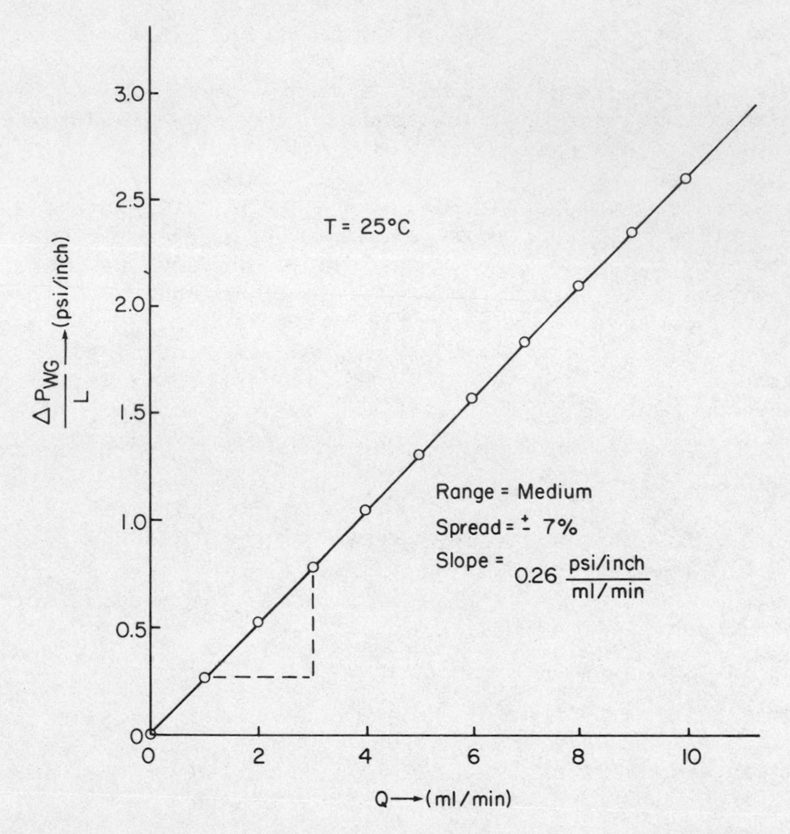

Fig. 2. Generalized curve for the pressure drop of water with flow rate.

was not to obtain accurate phase separation data, but to understand the relative ease with which the emulsions separate for several salt concentrations. Phase separation experiments were all carried out at room temperature (23 ± 1°C).

RESULTS AND DISCUSSION

1. Physical Properties

Density, viscosity and interfacial tension data of the equilibrated phases corresponding to an aqueous to oil ratio of 1:1 are presented in Table 1. Table 2 is the summary of the number of equilibrium phases present at 35°C for different aqueous to oil ratios and sodium chloride concentrations. It should be noted that the 1:1 system having 2% NaCl at 35°C represents a three phase system. However, at 25°C the same system gave only two phases. Upon prolonged storage (of the order of weeks) at 25°C, this system

Table 1. Physical property data at 30°C and property group controlling phase separation for conditions of negligible double layer (repulsive) and van der Waal (attractive) forces

NaCl Concn. (wt %)	Aqueous : Oil = 1 : 1						
	ρ_{aq} (g/ml)	ρ_{oil} (g/ml)	μ_{aq} (cp)	μ_{oil} (cp)	σ_i (dyne/cm)	$\frac{\sigma_i^2}{\mu_{aq} x \Delta\rho}$	$\frac{\sigma_i^2}{\mu_{oil} x \Delta\rho}$
0	0.9833	0.7410	1.26	1.2	0.422	57	60
0.5	0.9851	0.7410	1.21	1.24	0.105	3.7	3.6
1.0	0.9729	0.7422	2.33	1.26	0.0178	0.058	0.109
1.2	0.9612	0.7434	6.86	1.25	0.0056	0.002	0.011
1.5*	1.0018[1]	0.7417[1]	0.89	1.24	0.0008[2]	0.00028	0.0002
2.0*	1.0490[1]	0.7518[1]	0.96	--	0.0097[2]	0.033	--
3.5	1.0174	0.7664	0.84	2.77	0.083	3.26	0.99
6.0	1.0362	0.7581	0.83	1.56	0.243	25.7	13.6
8.0	1.0490	0.7664	0.93	1.72	0.39	57.8	31.3

*These systems have three equilibrium phases

1) For systems showing three phases ($\Delta\rho$) refers to maximum density difference in system.
2) Interfacial tension values refer to the lowest value in the three phase system. Here it is the value at the middle phase/oil interface.

Table 2. Equilibrium number of phases at 35°C for different aqueous to oil ratios and sodium chloride concentrations

NaCl Concn. (wt %)	Aqueous : Oil				
	4 : 1	2 : 1	1 : 1	1 : 2	1 : 4
1.0	2*	2	2	2	2
1.3	3*	3	3	2	2
1.4	3	3	3	3	2
1.5	3	3	3	3	3
1.6	3	3	3	3	3
1.8	3	3	3	3	3
2.0	3	3	3	3	3

*Numbers 2 and 3 indicate that the system has two and three equilibrium phases, respectively.

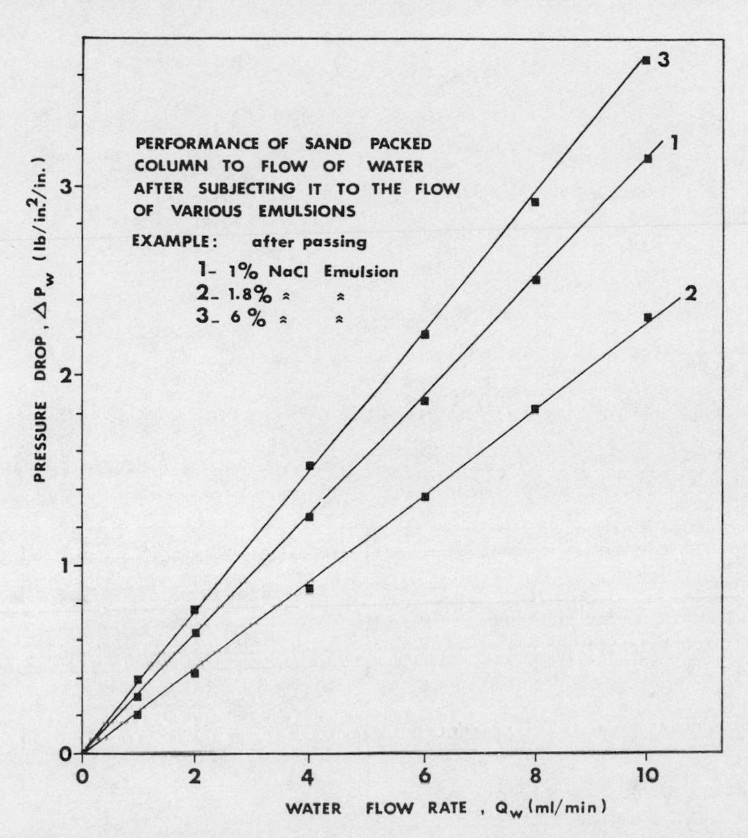

Fig. 3. Pressure drop vs. flow rate curves for water flowing through sand-packed column (1/4" diameter & 2" inch long) under various column conditions.

gradually produced three phases. Electrical conductivity and kinematic viscosity data of the 1:1 emulsions as a function of sodium chloride concentration are given in Figure 4. Extensive data of bulk density, absolute viscosity and electrical conductivity of the emulsions for various aqueous to oil ratios and sodium chloride concentrations are presented in Tables 3, 4 and 5, respectively.

a) Temperature effects: Figure 4 shows that electrical conductivity values of 1:1 emulsions are about 2000 μS/cm lower than those reported in Table 5 at low salinities (< 1.6 wt %) and similarly lower by about 400 μS/cm at 2% NaCl. Allowing for experimental errors, the differences in these two sets of data arise because of the 10°C difference in the temperature of measurements. Temperature effects can be further illustrated by considering the optimal salinity concentration, ease of formation of emulsion and equilibrium number of phases. Recent phase equilibria studies (8) have indicated that even 3.5% NaCl system with aqueous to oil ratio 1:1, separates out into three phases at room temperature. (Note

Table 3. Bulk density in g/ml of the sonicated emulsions at 35°C

NaCl Concn. (wt %)	Aqueous : Oil				
	4 : 1	2 : 1	1 : 1	1 : 2	1 : 4
1.0	0.9549	0.9150	0.8659	0.8292	-
1.5	0.9512	0.9147	0.8652	0.8253	-
1.6	0.9525	0.9183	0.8699	0.8262	-
2.0	-	-	-	0.8190	0.7824
3.5	0.9620	0.9229	0.8762	0.8299	0.7926

Table 4. Absolute viscosity (cp) of sonicated emulsions at 35°C

NaCl Concn. (wt %)	Aqueous : Oil				
	4 : 1	2 : 1	1 : 1	1 : 2	1 : 4
1.0	2	3	20.8	70.5	–
1.5	8.2	3.5	8.7	12	–
1.6	7.3	5.2	7.7	13.6	–
2.0	–	–	12.8	14.6	46.5
3.5	138	36	9.2	3.8	2.3

Table 5. *Specific electrical conductivity (μS/cm) of the sonicated macroemulsions at 35°C

NaCl Concn. (wt %)	Aqueous : Oil				
	4 : 1	2 : 1	1 : 1	1 : 2	1 : 4
1.0	13700	11300	8300	4700	< 1 μS/cm
1.3	18000	12600	8400	4900	"
1.4	18200	14000	8800	4700	"
1.5	18700	15000	9000	5500	"
1.6	19700	16000	10000	6100	"
1.8	590	740	530	485	"
2.0	400	430	620	700	"
3.5	500	15	< 1	< 1	< 1

*For values greater than 5000 μS/cm error is about 10%. For values less than 1000 μS/cm error is considerably greater. However, the order of magnitudes are highly reproducible in both cases.

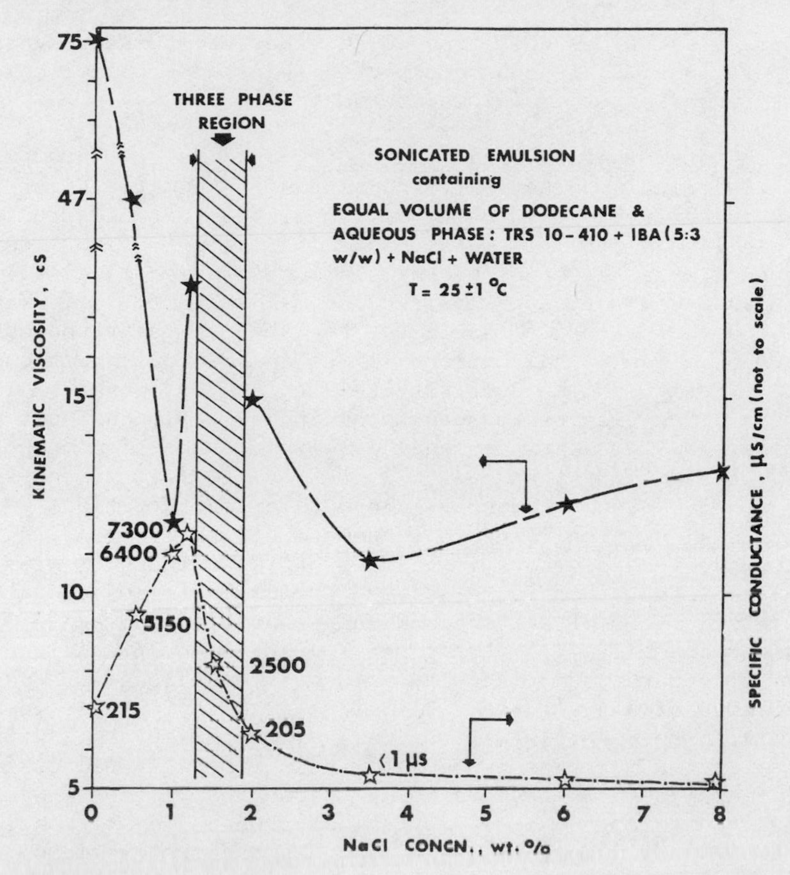

Fig. 4. Variation of kinematic viscosity and specific conductance of sonicated emulsions (aqueous : oil = 1:1) with NaCl concentration.

that till now it was believed that the 1:1 system separates into two phases at room temperature (25°C) for NaCl concentration greater than 1.9% at room temperature upon prolonged standing for periods of the order of three to four weeks.) However, in recent experiments we have found that the time necessary for the formation of three phases in these systems can be accelerated from 'weeks' to 'hours' by increasing the temperature from 25 to 45°C. Once a three-phase was formed, lowering the temperature back to 25°C did not alter the number of phases. When making sonicated emulsion at 1.5% NaCl with aqueous:oil, 1:1 we found that relatively stable emulsions could be produced at 35°C by prolonged sonication (about one hour). However, sonication of the system at 25°C did not give any emulsion at all. Evidently this experiment reveals the temperature sensitivity of the emulsions. In the later discussions of pressure drop results of emulsions at 25°C and 35°C, we will further elaborate the temperature effects. In the context of reservoir temperature conditions where one would expect temperature variations up to 60-70°C, it can be expected that the problems associated with the formation of emulsions are unavoidable.

b) <u>Effect of phase ratio and addition of salt</u>: Interpretation of electrical conductivity and viscosity results (Figure 4) reported in Reference (7) together with microstructural studies using spin-labels for 1:1 emulsions at 25°C seem to favor the coexistence of water-external microemulsions (swollen micelles) and macroemulsions (O/W type) below the optimal salinity (~ 1.5% NaCl). However, above the optimal salinity, the results favor the coexistence of oil-external microemulsions and W/O type macroemulsions. Viscosity and conductivity results at 35°C given in Tables 4 and 5 as a function of aqueous to oil phase ratio and salt concentration show two distinct behavior with a transition occurring around 2 wt % NaCl. At and below 2 wt %, viscosity increases as aqueous to oil ratio decreases. Above this concentration, viscosity decreases drastically up to aqueous : oil ratio of 1:1 and thereafter decreases gradually. Emulsions containing 1% NaCl show an opposite trend, that is, viscosity gradually increases up to aqueous : oil ratio of 2:1 and suddenly increases thereafter. If we consider the viscosity variation as a function of salt concentration it appears that viscosity values of emulsions with aqueous to oil ratios of 1:4, 1:2 and 1:1 decreased with increase in salt concentration up to 1.6 wt %, gradually increases up to 2 wt % and decreases thereafter. As opposed to this, viscosity values of 2:1 and 4:1 emulsions increase gradually up to about 1.5 wt % NaCl and after passing through a local optimum at around 1.5 wt % NaCl, increase abruptly for further increase in salt concentrations. Electrical conductivity results of emulsions as a function of sodium chloride concentration (Table 5) at 35°C follow essentially the pattern shown in Figure 4 for 1:1 emulsions at 25°C. The only difference is that the salinity at which a maximum in the conductivity occurs is shifted from 1.2% (Figure 4) to 1.6% NaCl in Table 5 for all phase ratios. In terms of the variation of conductivity with decrease in aqueous to oil ratio, it can be readily seen that up to 1.6% NaCl and above 2%, conductivity values abruptly decrease when aqueous to oil ratio is decreased. A transition appears to exist in the 1.8 - 2.0% NaCl region. Regarding the state of emulsion with respect to the effect of addition of salt it is evident that up to 1.6% NaCl the emulsions are of water-in-oil type for all phase ratios studied. Above 2% NaCl, a decrease in conductivity results show the presence of oil-in-water type emulsions for all phase ratios. The state of emulsion in the transition region (2.0% > [NaCl] > 1.6%) is not clearly understood presently.

Although the presence of higher volume percent of aqueous phase is expected to produce oil-in-water emulsions, such a conclusion is not obvious in the present situation because of the interfering effect of sodium chloride. The viscosity and electrical conductance behavior of 4:1 (aqueous:oil) and 2:1 emulsions present some interesting problems: The magnitude and the relative increase in the viscosity values with the addition of sodium chloride tend to favor the existence of oil-in-water emulsions. However, the conductivity values do not seem to support this deduction. With this uncertainty, in the absence of other relevant data, it can be speculated that the abrupt increase in the viscosity values could at best be the result of a drastic reduction in the emulsion droplet size.

2. Batch Settling Tests

a) Hand-shaken emulsions of 1:1 aqueous:oil system at 25°C: Typical batch settling curves for the emulsions at 25°C as a function of concentration, are shown in Figure 5. Four major differences in the emulsion behavior with respect to salt concentration were observed. There were two types of emulsions where the first phase boundary leading to phase separation was formed after a certain time. We have designated this time as the "initial time taken for the formation of an observable phase boundary" (ITFOPB). The four types are:

I. Gradual phase separation with time with ITFOPB of the order of 5 to 15 minutes.

II. Gradual phase separation with ITFOPB of the order of 1 to 3 hours.

III. Gradual phase separation leading to "creaming" of emulsion.

IV. Rapid phase separation.

Sodium chloride concentrations in the emulsions falling under the above types are:

Type I. [NaCl] = 1.2% [ITFOPB = 10 min.],
2% [ITFOPB = 5 min.], 3.5% [ITFOPB = 15 min.],
6% [ITFOPB = 5 min.],

Type II. [NaCl] = 1% [ITFOPB = 3 hrs.],
8% [ITFOPB = 1 hr.].

Type III. [NaCl] = 0 and 0.5%

Type IV. [NaCl] = 1.5%

With Type I emulsions, the volume *versus* time curves (Figure 5) are S-shaped. In the case of Type III emulsions it should be emphasized that although the time taken for 50% phase separation is lower than that of the emulsion containing 1% NaCl, in fact, "creaming" of these emulsions takes place approximately after 50% phase separation. As observed in the systems without isobutanol (see ref. 4), "creaming" gives rise to a "denser white texture"

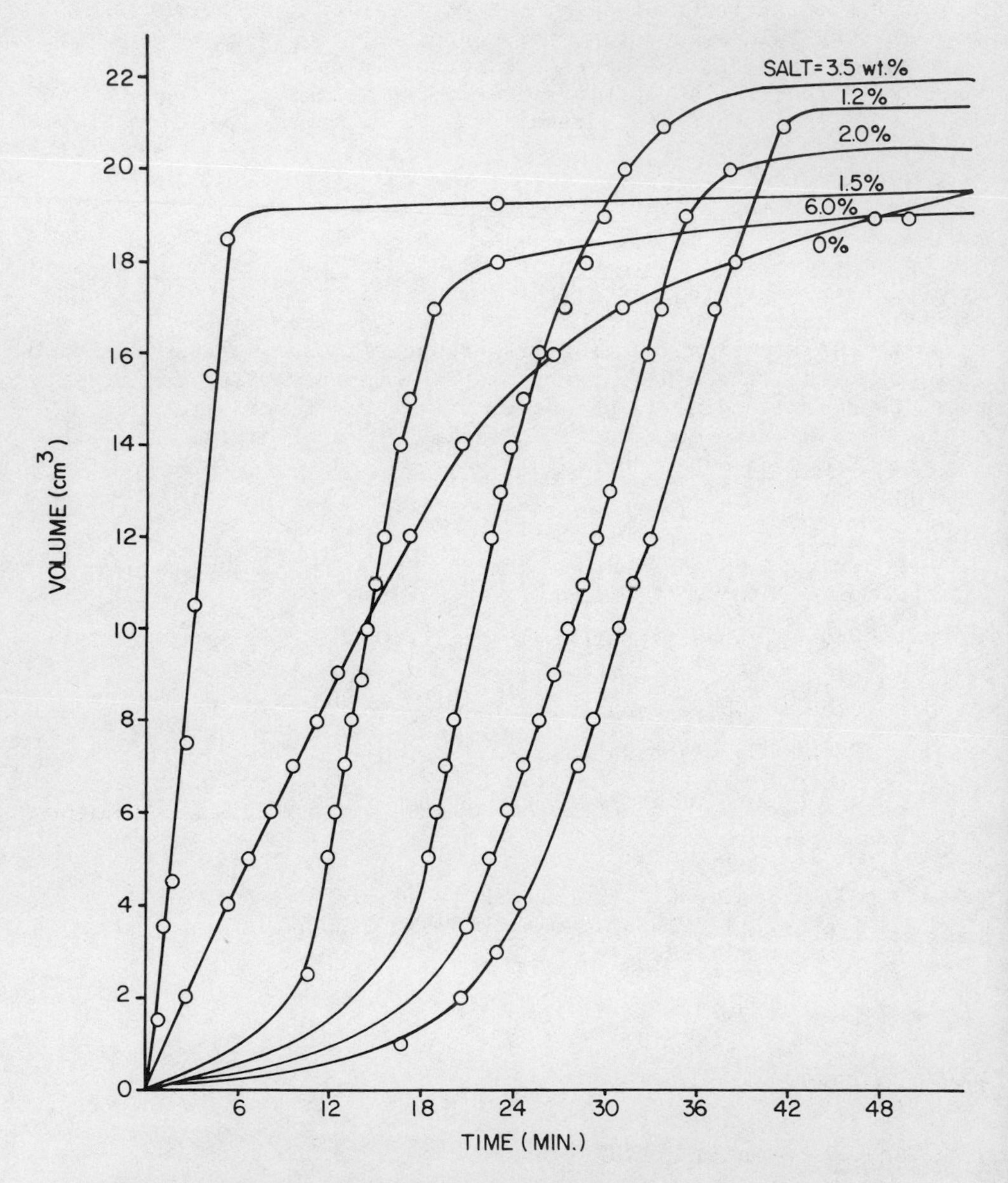

Fig. 5. Batch settling curves for TRS 10-410 (5%) + IBA (3%) + NaCl (varying wt %) + equal volume of dodecane.

near the phase boundary and also gives a greater stability to the emulsion in the batch settling test. Taking this effect into consideration the total time taken for complete phase separation of these emulsions would be very large.

Characteristic Phase Separation Behavior of Emulsions at Optimal Salinity

Type IV system yields three phases after complete phase separation. That is, when the emulsion is being separated, movement of two boundaries can be observed. The results are shown in Figure 6. It is evident from the figure that both the boundaries move at almost identical rates.

Recalling the observation (Table 1) concerning the occurrence of low and almost the same interfacial tension values between the middle phase/oil and the middle phase/aqueous phase systems, we were interested in the study of the phase separation behavior of these two phase systems. Therefore, three different combinations of "two phase systems" out of one three phase system were made by selectively deleting the third phase (Figure 6). Macroemulsions were produced as described in the experimental section.

It is interesting to note that the phase separation behavior of all of the three combinations is nearly the same. The behavior of the oil/middle phase emulsion is similar to that of the three phase system. Likewise, the two emulsion systems constituting oil/aqueous and middle phase/aqueous phase appear to behave identically within the experimental error.

Macroscopically, the important observation is the occurrence of rapid separation of the emulsion at optimal salinity in comparison with that of other emulsions. This effect is best illustrated in Figures 6 and 7. When one considers the time taken for 50% phase separation, the time at which an abrupt change in slope of the volume versus time curve occurs and the reciprocal separation rate with NaCl concentrations as criteria, three pronounced anomalies are obvious:

1) a maximum at 1% NaCl

2) a pronounced minimum at 1.5% NaCl and

3) a local minimum at 6% NaCl

Alternatively, if the total time taken for complete separation of these emulsions is considered, then we are left with the anomalies (2) and (3). Anomaly (1) drops out because of creaming effects.

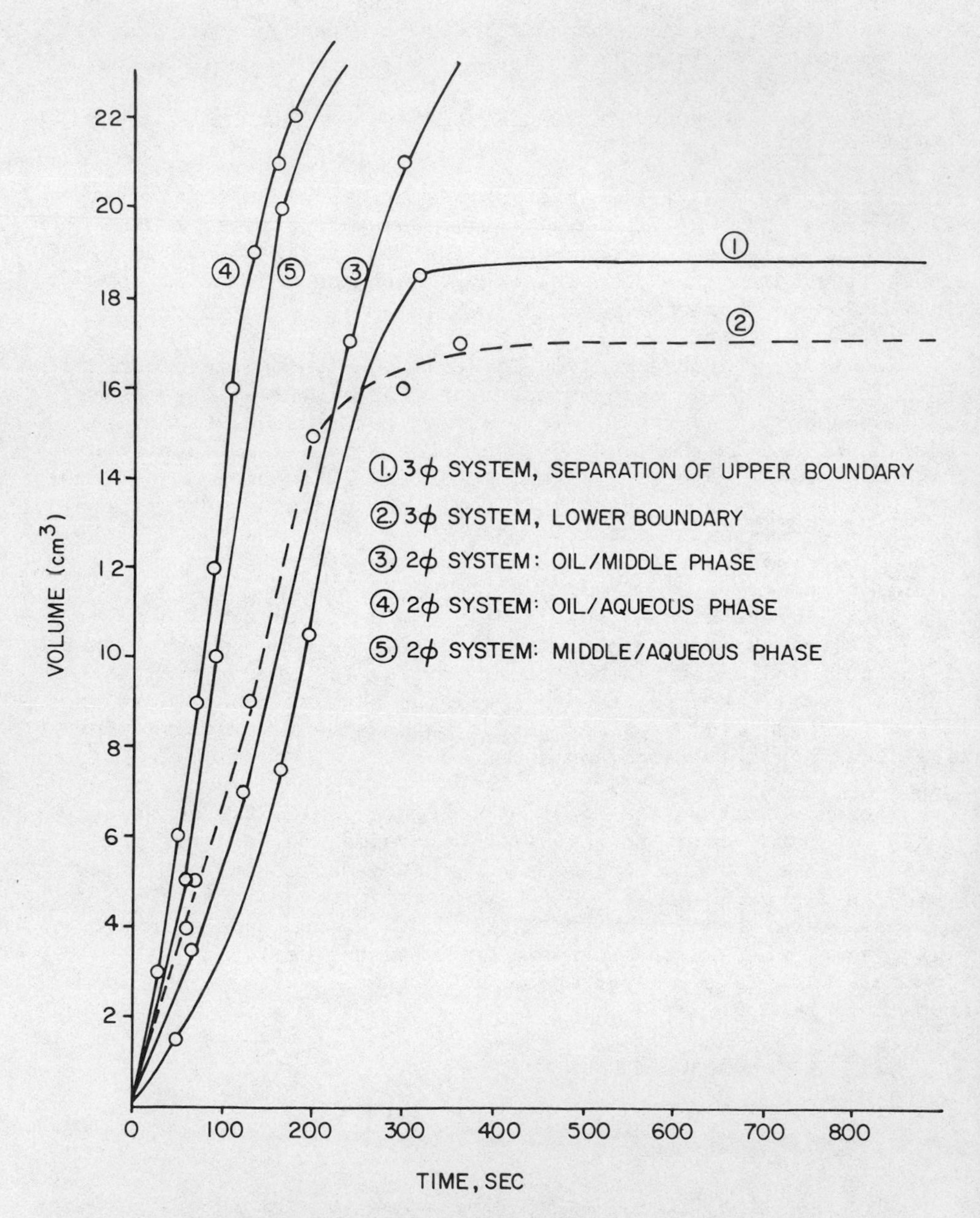

Fig. 6. Batch settling curves for 1.5% NaCl + TRS 10-410 (5%) + IBA (3%) three phase system with dodecane.

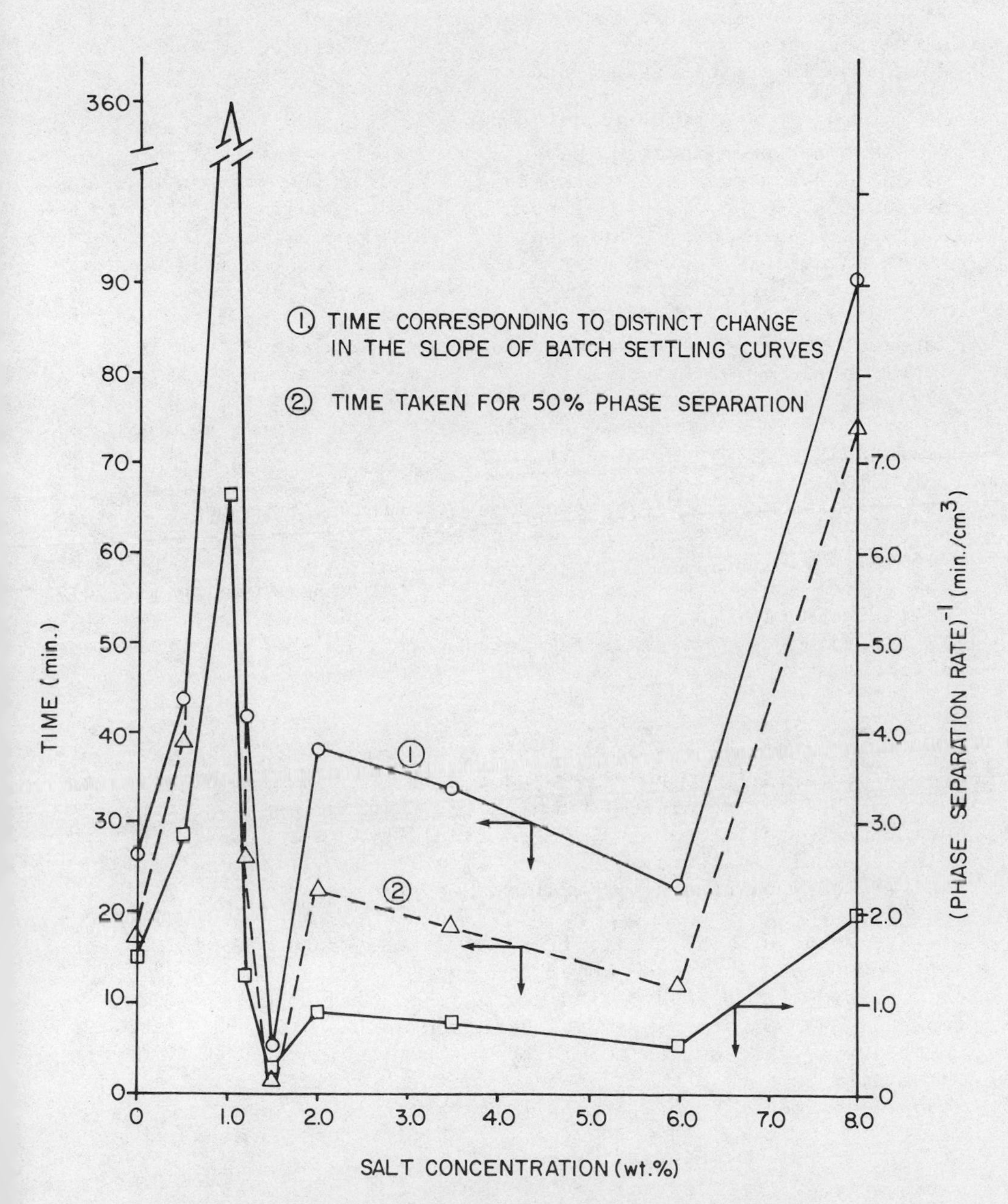

Fig. 7. Comparison of phase separation data

Drop stability: It is well known that the phase separation of the emulsions is controlled by the factors which influence the coalescence of drops. Therefore, we will now attempt to apply the developed Theory of Coalescence to the present situation.

Although the stability of emulsions containing micron-size droplets has been studied using a variety of emulsions in the past no unified and quantitative approach based on the physico-chemical properties of the system has been achieved. Obviously, this lack of understanding can be attributed to the complexity of the systems which invariably contain several species. A substantial contribution to the factors affecting emulsion stability has been made by Boyd et al. (9) using Spans or Tweens with Nujol and water. Some aspects of energy barriers associated with emulsion stability have been considered by several authors. The details can be found in the reference, Davies and Rideal (10). We will consider the electrical barrier-effects encountered with the oil-water and water-oil emulsion systems in a later section.

For the prediction of emulsion stability conditions, it is necessary to have some criteria in terms of the measurable quantities. Evidently, the role of equilibrium physical properties of the system would give some insight into the understanding of the operative mechanisms causing stability. Although a large amount of literature exists depicting this aspect, no satisfactory generalization can be made presently.

For illustrative purposes, we will now consider the role of interfacial tension on the stability of emulsions. A majority of researchers [Bickerman (11)] agreed that since emulsions possess very high interfacial area when the interfacial tension is reduced one expects an increased emulsion stability. Adamson (12) stated, "interfacial tension criterion is unassailable. It is also relatively useless except for qualitative arguments."

Yasukatsu Tamai et al. (13) showed that both the oil separation curve and the interfacial tension (σ_i) curve have a minimum at the same value of HLB. In the case of blended emulsifiers, they did not observe any marked minimum and, at the same time the rate curves also showed no minimum. They indicated that emulsion stability is indeed ruled by many factors but interfacial tension should be one of them and appears to be a useful measure of getting a stable emulsion. They suggested that low tension means energetically a stable interface and consequently may represent a stable structure of the adsorbed emulsifier at the interface.

Davis et al. (14) studied the stability of fluorocarbon emulsions. They reported that although coalescence times are expected to decrease with increased interfacial tension their results for the perfluorinated oils did not support any such trend.

Having realized the complexity of micrometer-droplet systems and the lack of reported literature dealing with the modeling of such systems we thought it would be interesting to see as to how the fluid properties would influence liquid/liquid dispersions containing larger drops of the order of 1 mm diameter or greater. It is interesting to note that a considerable amount of information is available with regards to the fluid dynamic behavior of droplets. In view of the many similarities that exist between micrometer and millimeter size droplets we believe that the "film drainage" concept developed for millimeter droplets can be equally applied to the case of micron droplet dispersions.

In the past, investigations involving the phenomenon of coalescence have been largely concerned with the study of single droplets resting on plane liquid/liquid interfaces. The main objective was to ascertain the influence of various physical parameters on this process. In the earlier works, experiments were directed mainly towards measuring the rest-time of drops. Although rest-time is a measure characterizing the drop residence and hence the coalescence behavior of the drop at a liquid/liquid interface, the direct factor which controls the coalescence step has been considered to be of hydrodynamic origin, namely "film thinning". Much attention was focused on this problem with the result that investigations were extended so that instead of measuring the rest-time, the variations of film thickness during thinning were determined. Extensive details of this process for a number of systems have been reported by many researchers, for example, Jeffreys and Davies (15), Woods and Burrill (16), Vijayan and Ponter (17,18).

It should be emphasized here that all of the earlier-developed models and experiments have been limited to systems containing low surfactant concentrations (below the CMC) and for droplet sizes of about 1 mm diameter or greater. We would like to restate some of the major assumptions in these models and will use them as the basis for extending the models to micron-size dense dispersions (or emulsions) containing high concentrations of one or more surfactants and electrolytes.

The major assumptions are:

(1) The effects due to electrical double-layer repulsion forces, electroviscous effects and the physical presence of adsorbed film are neglected.

(2) Effects due to attractive forces of molecular origin are assumed to be negligible.

(3) All external body forces are absent except for buoyancy (or gravity) force.

(4) Among surface forces, only surface (or interfacial) tension contribution is included. Other probable effects arising from surface rheological parameters are neglected.

Following the suggestions of various authors (19), we will now examine the influence of viscosity ratio of phases, density difference and interfacial tension values.

Drop size and density difference between phases: It is known that the size of drops and the difference in density between the oil and water phases determine the force which a drop exerts on an interface. The greater the force pushing a drop against an interface, the shorter the lifetime of the drop. Generally, large differences in density give rise to deformation of the drop. Thus the drop tends to flatten and the area of the draining film is increased. But at the same time the hydrostatic force causing drainage does not increase proportionately. In fact it was confirmed by many earlier workers that the rest-time increased with increase in the density difference between the phases.

Viscosity ratio of the phases: An increase in the viscosity of the continuous phase relative to the drop phase increases the rest-time as would be expected because the resistance to drainage of the film is increased.

Interfacial tension: A high interfacial tension results in the drop resisting deformation so that the area of the drainage film will tend to decrease as the interfacial tension increases. Thus coalescence time tends to decrease with increase in the interfacial tension. However, increase in interfacial tension also tends to inhibit the flow of film itself. Thus, here again the physical property produces two opposing effects. In general, however, coalescence time decreases with increase in interfacial tension.

Application of the theory of coalescence to the observed phase separation anomalies: For the "even mode of drainage" of the liquid film separating the drop from another drop or from an interface, several models have predicted a physical property group as the controlling group in addition to functions characterizing the drop dimensions and the rupture film thickness.

As a first approximation let us now assume (1) the average drop diameter in the emulsions to be a constant and (2) the magnitude of rupture film thickness to be a constant throughout the NaCl concentration range. Based on these assumptions let us see whether this physical property group ($\sigma_i^2/\mu_c \Delta\rho$) would adequately explain the anomalous phase separation behavior at 1.5% and 6% NaCl concentrations. This group was evaluated using the physical property data given in Table 1. The results are presented in Figure 8. It is

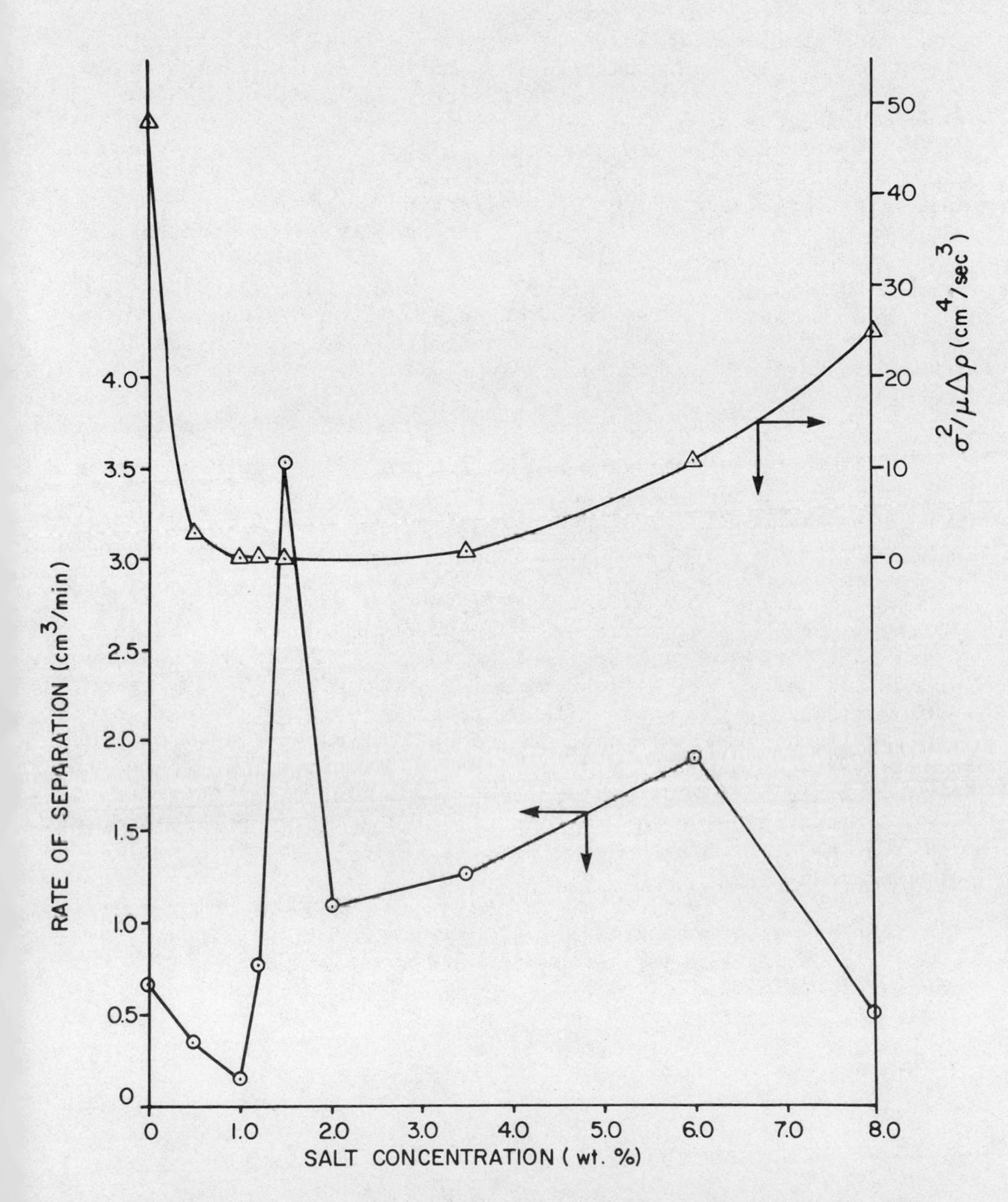

Fig. 8. Rate of separation in batch settling experiment for TRS 10-410 (5%) + IBA (3%) + NaCl (varying from 0% to 8.0%).

evident from the figure that the variation of this group with increase in NaCl concentration does not reveal a parallel behavior with the phase separation curve. Instead, we get a minimum for $(\sigma_i^2/\mu_c \Delta\rho)$ at 1.5% NaCl compared to a maximum in the phase separation rate. In the case of 6% NaCl system, although the phase separation curve shows a local minimum no such trend is found in the physical property group. It is obvious that the low values of interfacial tension in these systems play a dominant role in determining the magnitude of the physical property group as compared to the effect of either bulk viscosity or density difference variations.

When it was realized that the equilibrium physical property group just considered was not sufficient to explain the observed phase separation anomaly, one is left with the following questions:

(1) Are the assumptions in the coalescence modeling realistic?

(2) For concentrated mixed surfactant system such as the one under investigation, would it be possible to apply the models developed primarily for "even film" drainage mode?

In an earlier investigation Vijayan and Woods (20) studied the stability of oil (benzene) drops at oil oil/water interface containing mixed surfactants (sodium dodecyl sulfate and dodecanol) around and below the critical micelle concentration. They measured both coalescence times and film drainage rates using interferometer-cinematography. For the range of concentration studied, they found uneven film drainage modes which finally lead the drops to coalescence. Also, anomalous coalescence time behavior was found at some concentration of the surfactant. (Note the comparison between this observation with the anomaly at 1.5% NaCl concentration in the current study.)

For uneven draining films, Liem and Woods (21) extended the even film drainage model and proposed a new expression for the coalescence time, t, given by

$$t = \frac{\phi \eta_G^2}{16} \left[\frac{\mu_c \Delta\rho \; g}{\sigma_i^2} \right] b^5 \left[\frac{1}{h^2} - \frac{1}{h_0^2} \right] \qquad (1)$$

where

ϕ = a surface mobility parameter defined as

$$= \frac{4 - 2n_M}{1 + n_M}$$

n_M = number of moving surfaces which can take fractional values suggesting a local mobility in the interface.

n_G = dimensionless geometrical parameter that relates the radius of curvature of the film to the radius of curvature of the undeformed drop of radius b.

It is interesting to note that this model basically contains the same physical property group as the other models but now has, in addition, a "surface mobility parameter". (If the drop and the planar interfaces are relatively motionless, then $n_M \simeq 0$.) The use of the model (eqn. 1) to study mixed-surfactant system by Vijayan and Woods (20) partially explained some of the coalescence time anomalies.

Furthermore, the evaluation of the quantity ϕ (or n_M) is tedious and needs extensive film thinning rate data.

In spite of the partial success in the application of the model one is left with some ambiguities as to the magnitude and to the physics and chemistry of the parameter, n_M. A more fundamental problem would be to understand the factors which would directly influence the mobility of surfaces. Evidently factors such as attractive and repulsive forces of molecular origin between surfaces, and the mechanical properties of the liquid film separating the surfaces have to be understood. Quantitative experimental measurements of these parameters are not readily available.

b) Sonicated emulsions: effects of phase ratio and salt concentration at 25°C: Similar to hand-shaken emulsion studies discussed in section (2a), three sonicated systems containing 2:1, 1:1 and 1:2 aqueous to oil ratios were investigated as a function of salt concentration. From extensive data of volume separated versus time, results expressed as time taken for 50% separation as a function of sodium chloride concentrations were obtained. Typical results are shown in Figures 9 and 10 for 2:1 and 1:1 systems, respectively.

From the results obtained using the three phase ratios, it can be seen that the general tendency of these emulsions has been to give fastest separation at 1.5% NaCl, and also give comparable separation at 1.2% NaCl. For all the phase ratios studied, separation of 0% and 0.5% NaCl emulsions was not measured because these emulsions were very stable (of the order of days) when sonicated. Emulsions with 1:2 aqueous to oil ratio provide the most stable emulsions at several salt concentrations. In this case, separation of 6% and 8% NaCl in addition to 0% and 0.5% was not measured because of the large times associated with the separation process (of the order of days).

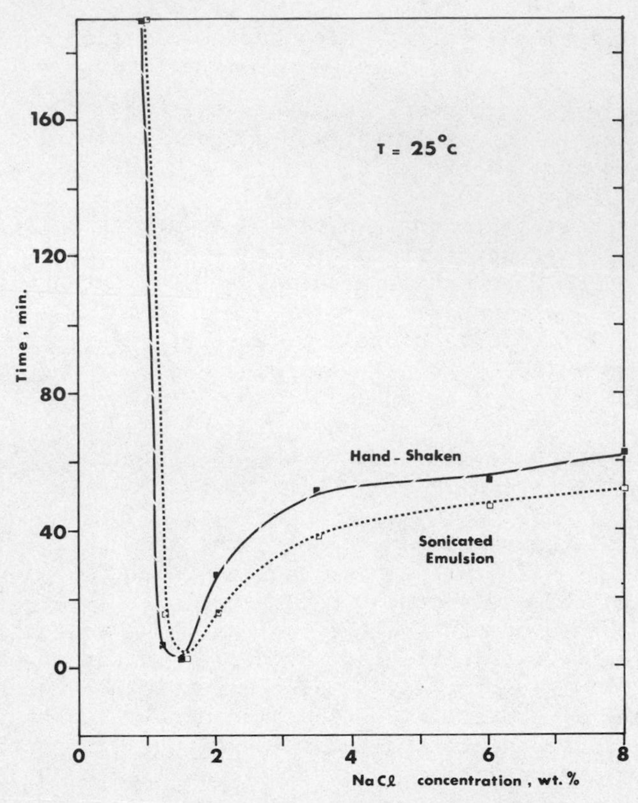

Fig. 9. Time taken for 50% phase separation for 2:1 sonicated and hand-shaken emulsions versus sodium chloride concentration.

Note on 1:1 emulsions: When constructing the settling curves for these systems an interesting observation was recorded. For all salt concentrations, with the exception of 1% and 8% NaCl, all the emulsions showed a gradual phase separation behavior with time. However, the emulsions containing 1% and 8% NaCl revealed a large initial lag before any visible separation commenced. This initial time taken for the formation of an observable phase boundary (ITFOPB) is of the order of hours. Furthermore, a comparison of the macroscopic structures of these emulsions by scanning electron microscopy (7) surprisingly reveal similar polyhedral structures. The physico-chemical aspect of the presence of polyhedral structures in relation to the initial stagnation period (at least for visual observations) is an important consideration in the area of emulsions. Obviously the next step is to see the microscopic changes taking place during the macroscopic stagnation period.

Separation behavior of emulsions: hand-shaken versus sonicated systems: For the different aqueous to oil ratio systems considered presently, the general trend in the settling curves (volume of phase separated versus time) remains essentially the

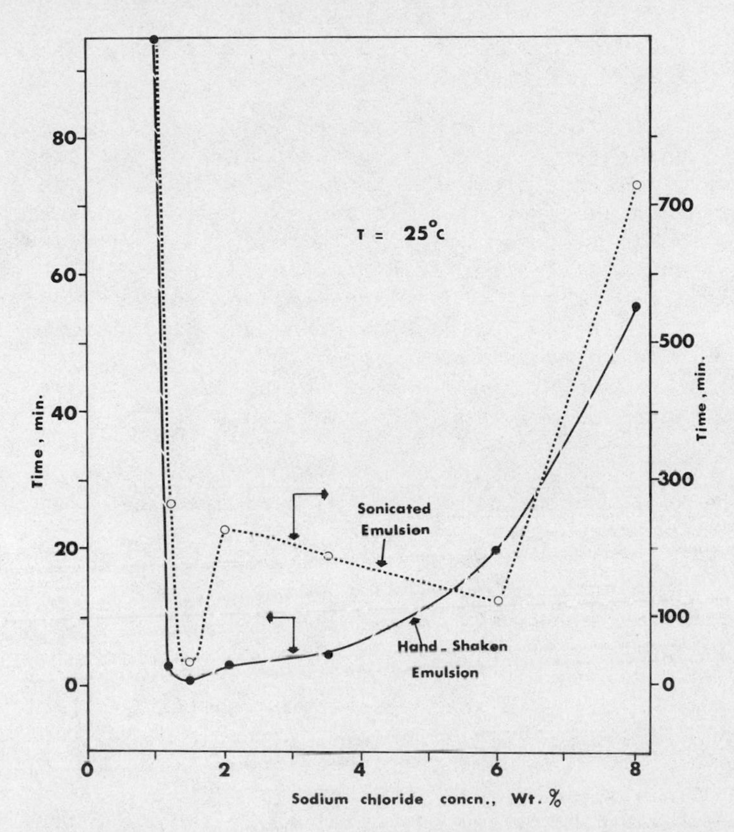

Fig. 10. Time taken for 50% phase separation for 1:1 sonicated and hand-shaken emulsions versus sodium chloride concentration.

same for the hand-shaken and the sonicated emulsions for a given aqueous to oil ratio but differs slightly from one ratio to the other. However, a comparison of the time taken for 50% separation does show appreciable differences in the separation behavior between the two modes of emulsion production. This difference is pronounced in the case of 1:1 emulsion system. The 2:1 emulsion system does not reveal any such differences between the hand-shaken and sonication modes.

Globally speaking, one can reason out the difference in the behavior of hand-shaken and sonicated emulsions in terms of the production of stable drop sizes. One would expect the sonicated emulsions to provide a stable equilibrium drop population. In the case of 2:1 emulsions, it is surprising as to why a difference in the separation behavior, in spite of the two modes of emulsion-making used, is not observed. Probably the power input by the hand-shaking mode is sufficient to produce equilibrium drop population.

3. Flow Through Porous Media Studies in Unconsolidated Sand-Packed Beds at 25°C

a) Water: Pressure drop results in the sand-packed bed using water (see apparatus, Figure 1) for several flow rates and different column conditions are presented in Figure 3. It can be seen from the figure that the pressure drop values increase linearly with flow rate passing through origin. That is, the flow curves demonstrate evidently Newtonian behavior as expected. Small variations in the expanded pressure drop diagram arise from small changes occurring as the result of passing other fluids and emulsions. Obviously it is an anticipated effect. Nevertheless, it is small compared to the overall increase in the pressure drop values associated with aqueous solutions and macroemulsions.

b) Aqueous solutions of TRS 10-410: isobutanol (5:3% w/w) containing sodium chloride (0 to 2 wt %): Pressure drops produced by the flow of aqueous solutions for several NaCl concentrations and flow rates are given in Figures 11 and 12. Aqueous solutions with NaCl concentration 0%, 0.5%, 1% and 1.2% follow Newtonian type flow behavior over the flow rates investigated. At low shear rates (less than 2 ml/min. flow rate) aqueous solutions containing 1.5% and 1.8% NaCl demonstrate a non-linear pressure drop-flow rate relation. However, at higher shear rates the ΔP-Q relation appears to be linear. It has been demonstrated from molecular order parameter studies (5,6) and physical property data (see for example, bulk viscosity, Figure 12) that maximum structural effect and liquid crystalline behavior are revealed by these solutions. On this basis, it is not surprising to observe a non-Newtonian type of behavior at low shear rates. At high shear rates it is possible that this structural effect is destroyed, in a relative sense, and the pressure drop increases proportionally with flow rate. Aqueous solution containing 2% NaCl seems to lie in between ΔP and Q still persists but the intensity is considerably reduced. At flow rates higher than 2 ml/min., we have essentially a Newtonian type fluid behavior.

The study of flow of aqueous solutions through unconsolidated sand bed thus demonstrates the influence of structural aspects of surfactant formulation on pressure drop values at low shear rate flows.

c) Sonicated macroemulsions consisting of aqueous phase (5:3% w/w TRS 10-410: IBA) + NaCl (X%) and dodecane oil (oil: aqueous = 1:1 w/w): Pressure drop data with flow rate for sonicated macroemulsions as a function of sodium chloride concentration is given in Figure 13. Here, as discussed in the earlier section, we have both two phases and three phases under equilibrium conditions. That is, the macroemulsions upon standing give rise to equilibrated clear phases which could be two or three phases

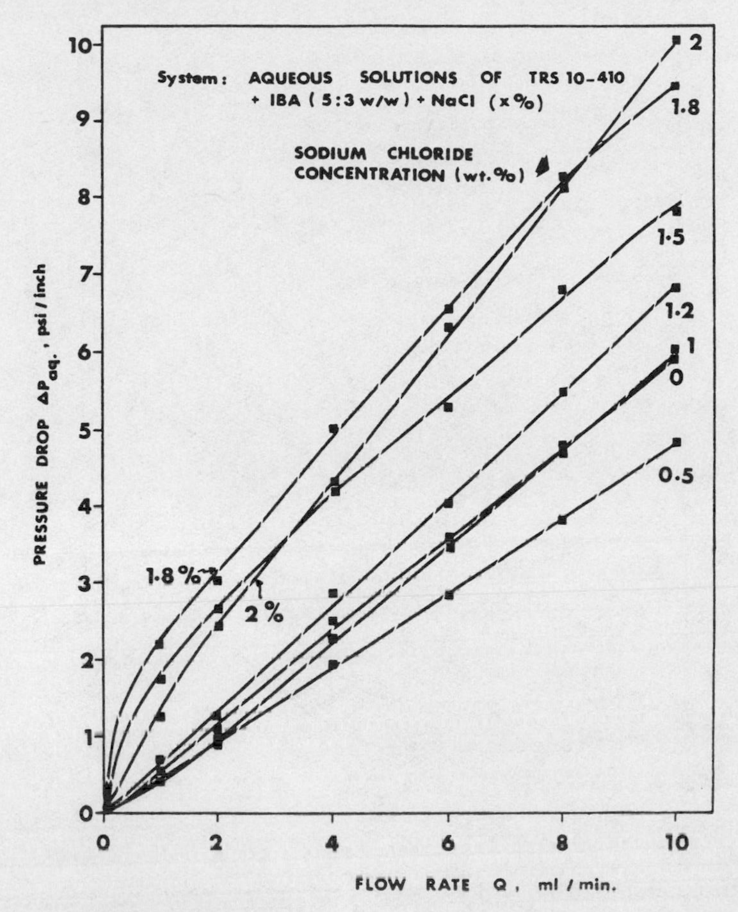

Fig. 11. Pressure drop versus flow rate curves for aqueous surfactant formulations flowing through a sand pack as a function of sodium chloride concentration.

depending on the salt concentration. In particular, we have two salt concentrations, 1.5 and 1.8 wt %, where three phases are produced whilst other concentrations (0, 0.5, 1, 1.2, 2, 6, 8 wt %) give two phases. Unlike aqueous phase pressure drop behavior, macroemulsions containing 0%, 0.5%, 1%, 1.2%, 2% and greater salt levels show non-linear ΔP-Q relation at low flow rates. This effect is more pronounced for 0 and 0.5 wt % NaCl emulsions. Again at high flow rates, similar to aqueous solution behavior, proportional variation in ΔP values with emulsion flow rate is observed. The very high pressure drop values of 0% and 0.5% emulsions can be readily attributed to their viscosity values (see Figure 4). The kinematic viscosity of these emulsions are 75 cS and 47 cS, respectively. With the exception of 1.5% and 1.8% NaCl emulsion, the kinematic viscosity of all other emulsions fall in the range 12 to 17 cS. Being very unstable, viscosities of 1.5% and 1.8% NaCl

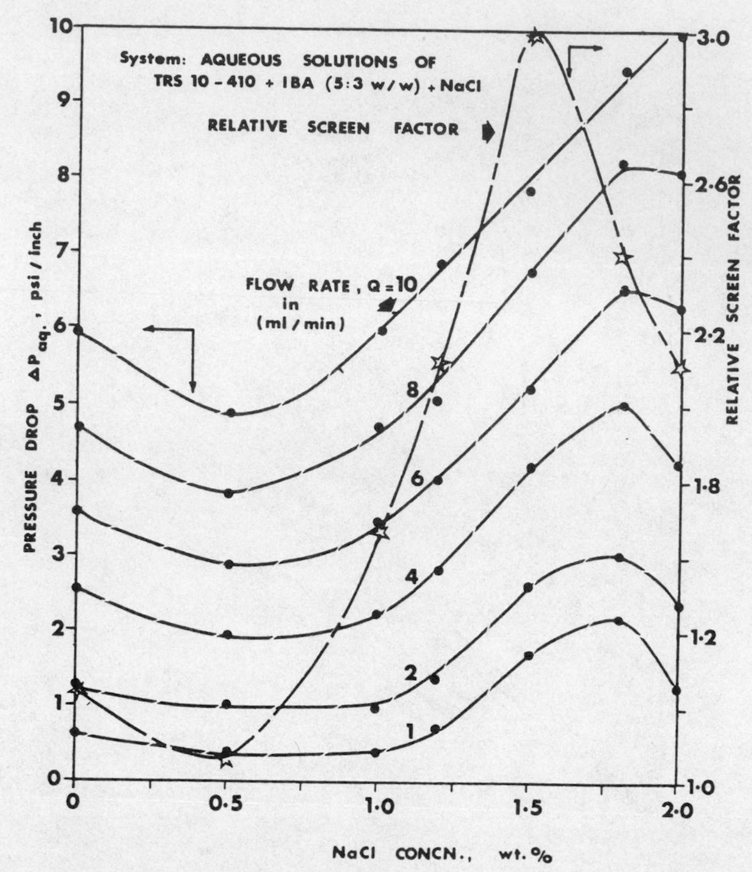

Fig. 12. Pressure drop and screen factor vs. sodium chloride concentration for aqueous solution of TRS 10-410 + IBA (5:3 w/w) + NaCl (X%).

emulsions could not be measured. (See Experimental section.) It is interesting to note that while equilibrium two phase systems with dodecane oil forming macroemulsions show higher pressure drops for a given flow rate and non-Newtonian type (probably pseudoplastic) of behavior at low shear rates (less than 2 ml/min.), three phase systems with dodecane oil forming macroemulsions exhibit very low pressure drops and a perfectly Newtonian type of behavior. This reasoning can be deducted directly by a comparison of water pressure drop in the porous media (see Figure 3). One obvious reason is that the effective viscosity of these emulsions is very low, of the order of a few centipoise.

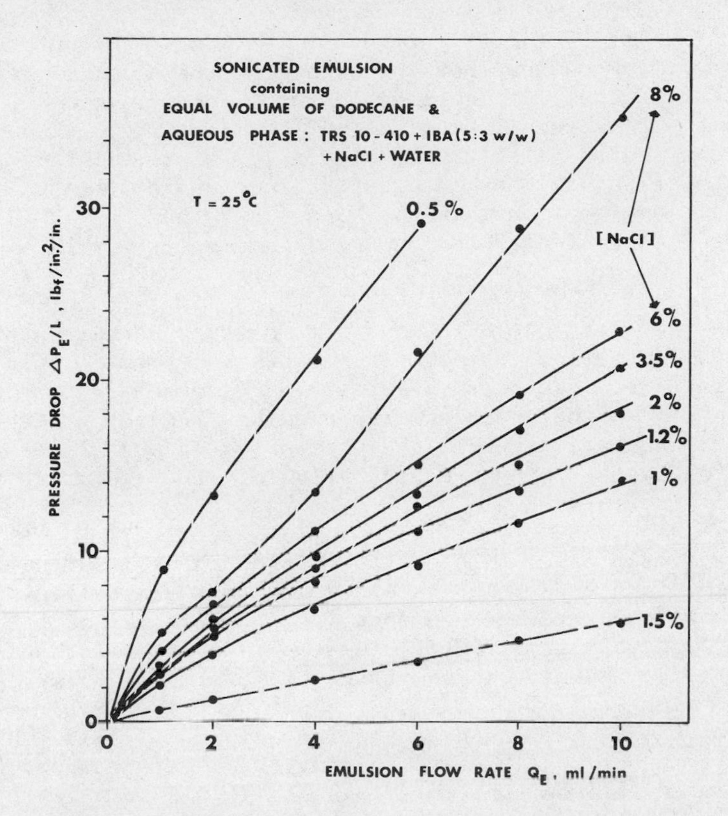

Fig. 13. Pressure drop values of sonicated emulsions in a sand packed column as a function of flow rate and NaCl concentration.

Effect of Addition of Sodium Chloride to Aqueous Surfactant Formulation Containing TRS 10-410 + IBA (5:3% w/w) on Pressure Drop in a Sand-Packed Bed and on Screen Factor

Figure 12 depicts that the $(\Delta P)_{aq}$ (aqueous phase pressure drop) remains steady up to 1% NaCl, gradually increases reaching a maximum at about 1.8% NaCl and then decreases. At higher flow rates (between 2 to 8 ml/min.) $(\Delta P)_{aq}$ decreases from 0% NaCl reaching a minimum at 0.5% NaCl and then increases up to 1.8% NaCl and again decreases. At 10 ml/min. flow, $(\Delta P)_{aq}$ - minimum is more pronounced at 0.5% NaCl but at the same time the maximum in $(\Delta P)_{aq}$ occurring at 1.8% NaCl with low flow rates disappears. For all NaCl concentrations, increasing the flow rate increases the pressure drop. When the relative screen factor values of these solutions are superimposed, two optima are found to occur--a minimum at 0.5% NaCl and a maximum at 1.5% NaCl. It is interesting to see that both screen factor results and pressure drop results reveal a minimum at 0.5% NaCl. However, the maximum screen factor occurring at 1.5% NaCl does not correspond to the maximum in the pressure drop results. It is probable that a correspondence for maxima between the screen factor and $(\Delta P)_{aq}$ might have occurred at much lower flow rates (less than 1 ml/min.). Microstructure studies also indicate structural changes at these concentrations (5,6). Among the aqueous solutions, it has been found that 0.5% NaCl solutions form the most clear solution and remain so upon prolonged standing.

The important outcome of this study is that under the conditions of experimentation, the occurrence of a maximum in bulk viscosity and in screen factor at a certain NaCl concentration does not correspond with the pressure drop results. Except for this disagreement, the variations in the screen factor and viscosity are faithfully followed by pressure drop values in the flow rate range of 4 to 8 ml/min.

d) Comparison of emulsion pressure drop in a sand-packed bed with batch settling separation data and emulsion viscosities: Figure 14 shows the agreement between $(\Delta P_E/\Delta P_W)$ [ratio of pressure drop of emulsion to that of water] and the time taken for 50% separation of the emulsion. The pronounced minima in these two factors occurring at 1.5% NaCl is worth noting. The significance here is that at this temperature, the formation of stable emulsions at 1.5% NaCl was not possible and this may be the reason why the pressure drop as well as the emulsion stability is very low. Viscosity data of the emulsions again show a similar trend. An increase in the emulsion viscosity at 1.2% NaCl from 1% NaCl system (12 cS to 18 cS) is reflected directly on the pressure drop values. However, this effect is not revealed by the phase separation results. It should be emphasized that all these measurements were taken at 25 ± 1°C. Later measurements in our laboratory indicated that by increasing the temperature to 35°C and by prolonging the sonication for periods of hours, it was possible to form relatively stable emulsions at 1.5% NaCl. This effect together with the oil to aqueous phase ratio effect on the physical properties of emulsions and pressure drop in a 6 inch column are discussed in the following section.

4. Flow of Sonicated Emulsions in a 0.25 Inch Diameter, 6 Inch Long EM-Gel Packed Unconsolidated Bed at 35°C

a) Pressure drop as a function of phase ratio and salt concentration: The pressure drop results of the systems, described in Table 2, across the 6 inch bed as a function of salt concentration and aqueous to oil ratios, are presented in Figures 15-20 and in Table 6. In particular, the pressure drop results for the three phase systems are somewhat different from that of the results for the flow of macroemulsions through a sand-packed column at 25°C

Table 6. Pressure drop of sonicated emulsions ($\Delta P_E/L$ in psi/inch) flowing at Q = 4 (ml/min) across the bed at 35°C

NaCl Concn. (wt %)	Aqueous : Oil				
	4 : 1	2 : 1	1 : 1	1 : 2	1 : 4
1.0	3.4	4.3	11.4	19.3	31.6
1.3	6.6	7.0	7.3	9.9	11.3
1.4	6.0	5.7	6.3	10.7	14.8
1.5	4.3	7.0	8.6	11.4	11.2
1.6	7.1	6.1	9.0	12.7	14.5
1.8	4.0	4.4	10.1	13.3	12.8
2.0	14.8	6.5	3.9	14.4	23.3
3.5	48	18.4	8.5	4	2.7
Mean for 1.3 to 1.8%	5.6	6	8.3	11.6	12.9

Mean for 1.3% to 1.8% salt concentrations has a maximum spread = ± 25%.

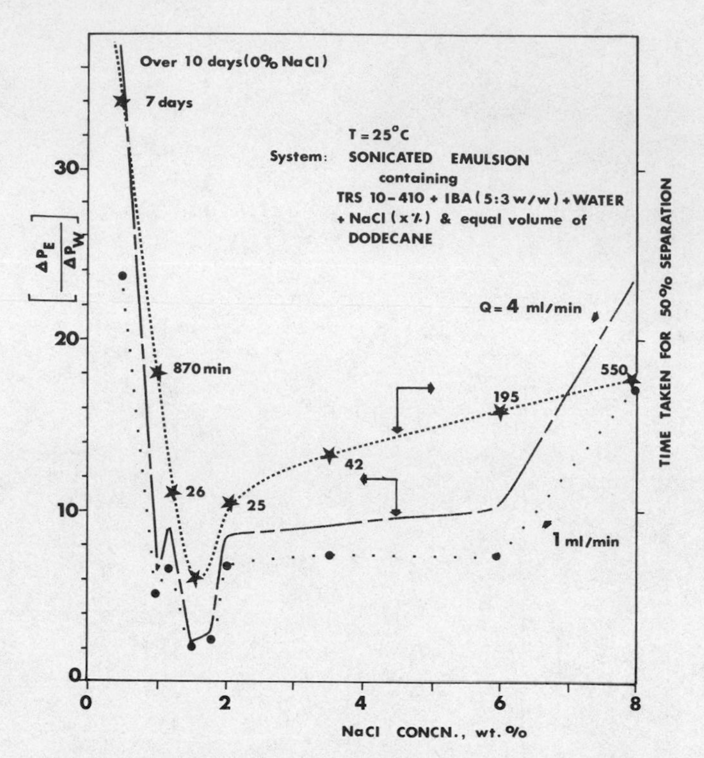

Fig. 14. Ratio of pressure drop of emulsion to water in a sand pack and phase separation data plotted against NaCl concentration.

(see section 3) where a Newtonian type of characteristic linear plot was obtained between ($\Delta P_{E/L}$) and Q for 1:1 emulsions containing 1.5% and 1.8% sodium chloride. However, the results shown in Figures 16-18 do not entirely reveal Newtonian type of behavior. It should be noted that our recent observations (8) indicate that even 3.5% salt concentration system with aqueous to oil ratio of 1:1 separate out into three phases upon standing at room temperature (~ 25°C) for a few weeks. Evidently this system did not reveal any Newtonian type behavior in the sand-pack experiments at 25°C where this system presumably gave only two phases at the experimental conditions. Therefore, it is incorrect to conclude that three phase systems exhibit, in general, Newtonian type characteristics.

Interestingly, results of the system containing 1.5% salt and 1:1 aqueous to oil ratio (Figure 16) do not show a Newtonian type behavior (especially at low shear rates). The reason being that we could form a relatively stable emulsion this time as opposed to the previous attempt (~ 35°C as opposed to 25°C in the sand-pack experiments). The higher temperature operation thus seems to have

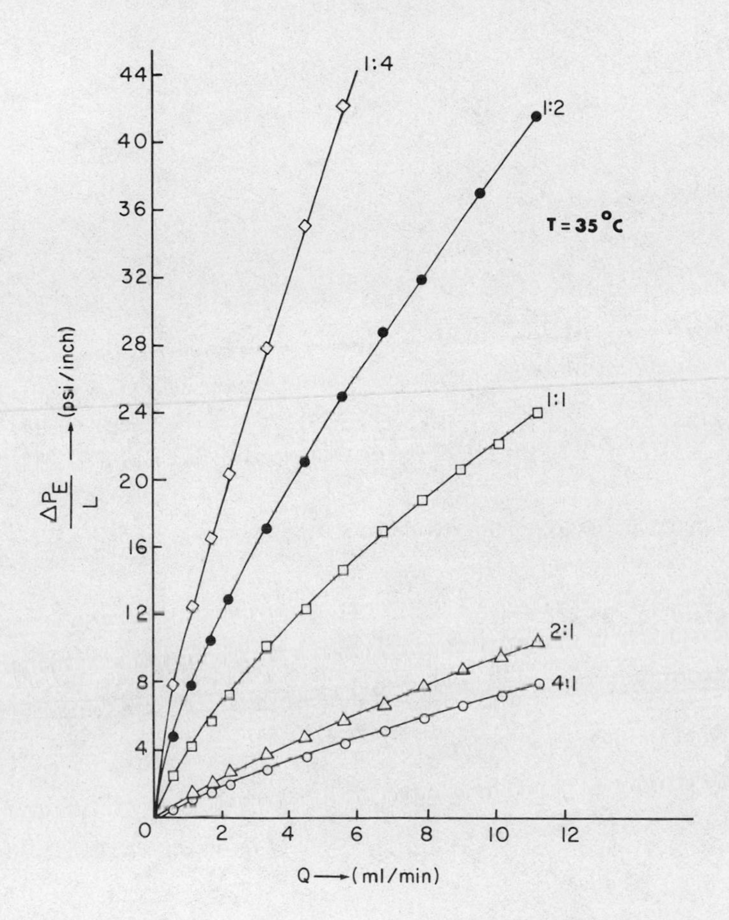

Fig. 15. Pressure drop variation with flow rate for 1% salt emulsion system as a function of aqueous to oil ratio.

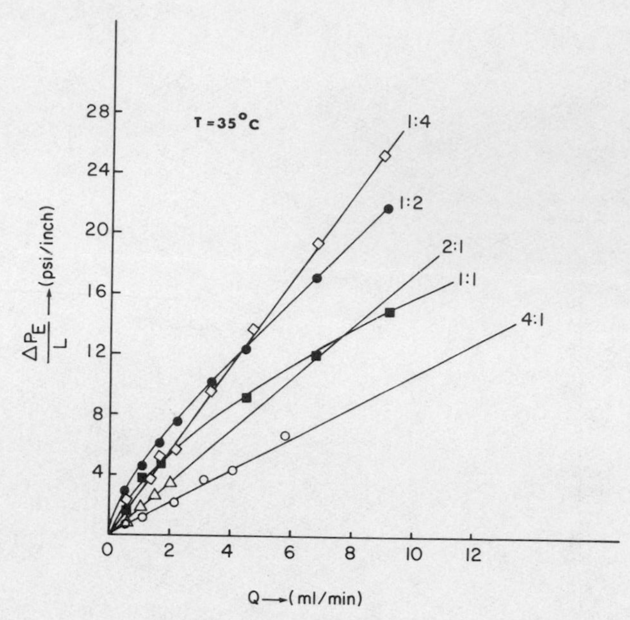

Fig. 16. Pressure drop variation with flow rate for 1.5% salt emulsion system at different aqueous to oil ratios.

made this profound difference in the state of the emulsion. Therefore, it may be concluded that higher temperatures and continuous sonication for about one to two hours can lead to the formation of relatively stable emulsions.

In general, most of the 1:4 aqueous to oil ratio systems did not produce stable emulsions even after prolonged sonication (~ three hours). However, by passing these systems, well-mixed, through the porous bed at very high flow rates, relatively stable emulsions were obtained. This observation leads us to conclude that sonication alone is not capable of producing emulsions in 'certain' systems while mere shearing at high flow rates is capable of emulsifying such systems.

The 2% NaCl system with aqueous to oil phase ratio of 2:1 showed hysteresis phenomena as shown in Figure 17. With emulsion systems having other phase ratios very long time period (of the order of five to six hours) was required for achieving steady state in the pressure drop values at a particular flow rate. Once steady state conditions were reached at higher flow rates, the time taken for the attainment of steady state at lower flow rates was considerably reduced (about 15 minutes). This system again reveals non-Newtonian type characteristics. It appears that 2% NaCl systems show a transition state with respect to pressure drop values. Comparison of ($\Delta P_{E/L}$) values for different salt concentration at

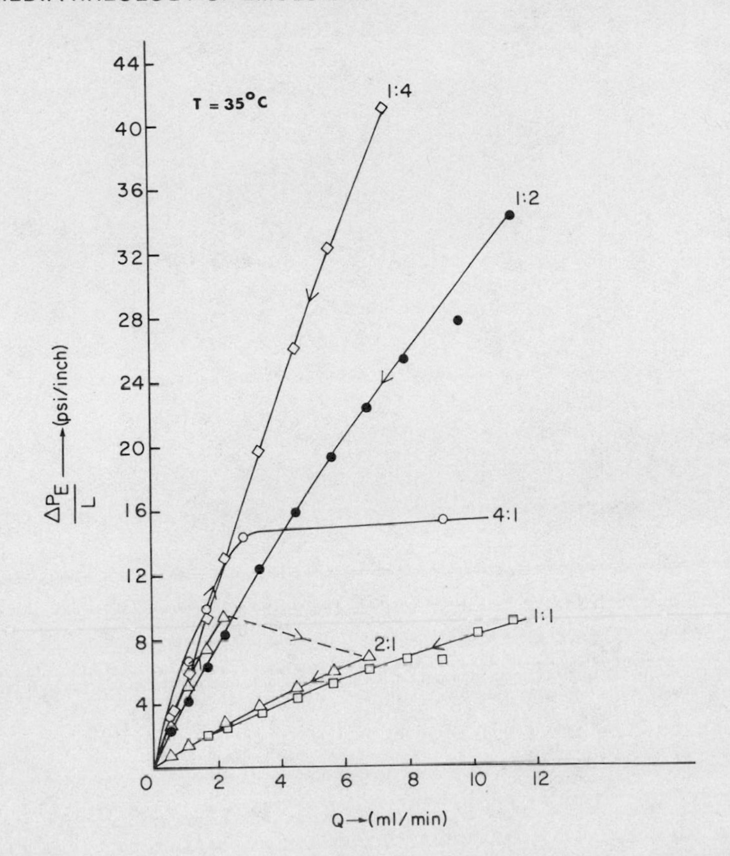

Fig. 17. Pressure drop variation with flow rate for 2% salt emulsion system at different aqueous to oil ratios.

particular aqueous to oil ratios (Figure 19) shows that (a) for 4:1 and 2:1 systems pressure drop increases considerably beyond 2% NaCl, (b) for 1:2 and 1:4 systems pressure drop decreases beyond 2% NaCl and (c) for 1:1 systems, although the changes in the pressure drop with salt concentration are not significant, a discontinuity with an optimum at 2% NaCl can be seen. In general, the discontinuities present in the vicinity of 2% NaCl region correspond with the discontinuous changes as revealed by the conductivity data in Table 5.

Comparison of the pressure drop values as a function of aqueous to oil ratios at particular salt concentrations (Figure 20) shows that: (a) ($\Delta P_{E/L}$) increases almost exponentially with phase ratio for 1% NaCl systems, (b) ($\Delta P_{E/L}$) decreases exponentially with phase ratio in the case of 3.5% NaCl systems, (c) for 2% NaCl systems ($\Delta P_{E/L}$) decreases with phase ratio, going through a minimum at aqueous to oil ratio = 1:1 and increases abruptly thereafter, (d) for emulsion systems containing salt concentration in the range of 1.3 to 1.8 wt %, ($\Delta P_{E/L}$) increases gradually but linearly with

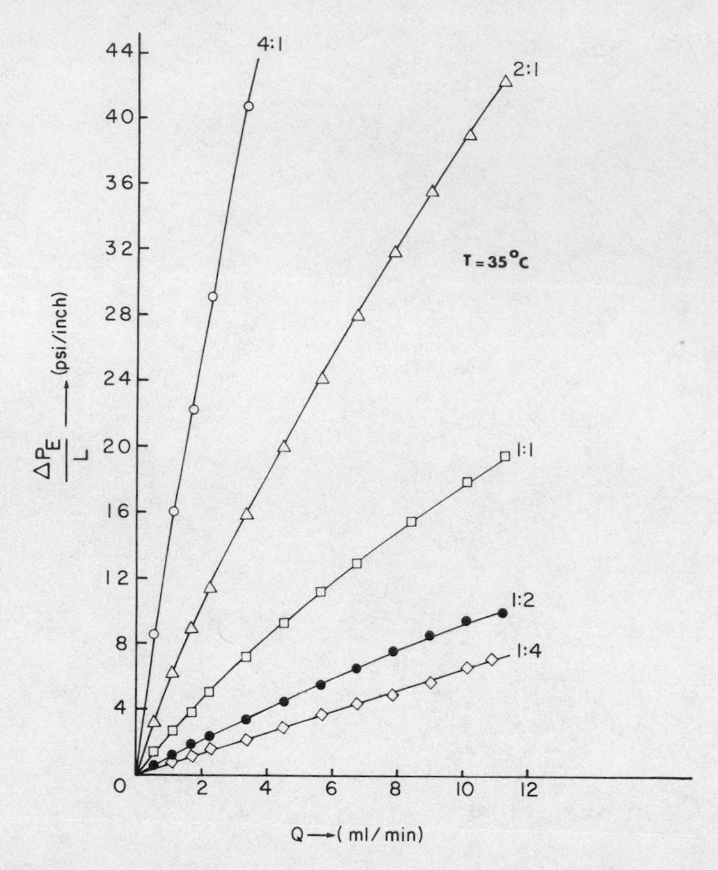

Fig. 18. Pressure drop variation with flow rate for 3.5% salt emulsion system at different aqueous to oil ratios.

phase ratio. For prediction purposes, various pressure drop results in the region are grouped together at a given flow rate. The curve shown in Figure 20 is estimated to be accurate within ± 30%. It should be noted that although Figures 19 and 20 illustrate the behavior of emulsions at Q = 4 ml/min., the interpretations presented above remain essentially unaltered for other flow rates as well.

The electrical conductivity results discussed in section 1 have shown that the macroemulsions seem to exsit as W/o type below 1.8% NaCl and as O/W type above 1.8% NaCl for all aqueous to oil phase ratios. This result together with the pressure drop data thus suggest that pressure drop associated with the flow of macroemulsions increase with the increase in the amount of the dispersed phase irrespective of whether it is oil or aqueous phase. It is suggested that the anomalous behavior at 2% NaCl could only be attributed to 'optimal salinity' effects. Rheological data for this composition and for those compositions in the transition region

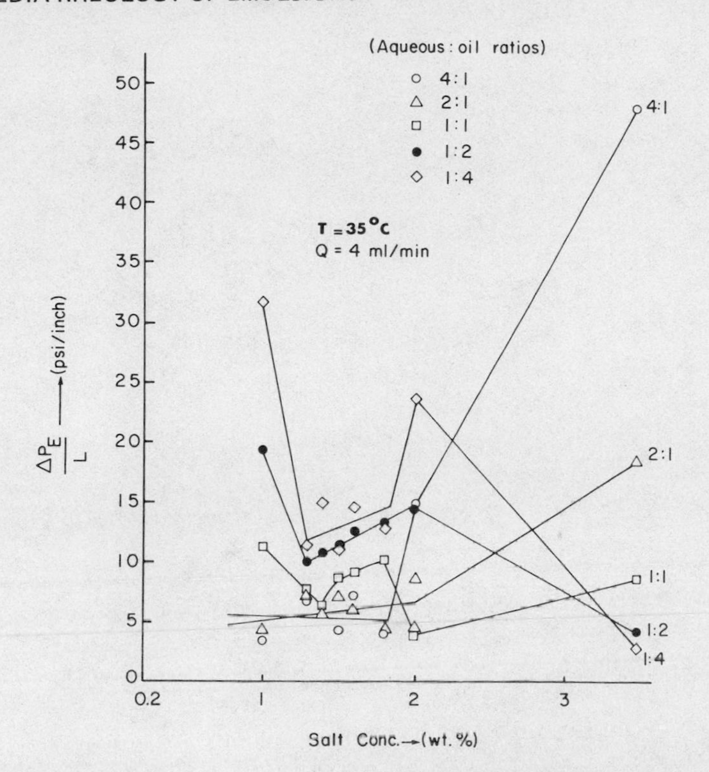

Fig. 19. Pressure drop variation with salt concentration of sonicated emulsions for different aqueous to oil ratios at 4 ml/min. flow rate.

indicate that the discontinuity must be due to structural changes present within the fluid which is in good agreement with our microstructural results published elsewhere (8). Similar transition states associated with phase inversion and structural properties as reflected by rheological properties of some microemulsion systems, have been studied by Dreher et al. (22). Their flow experiments using microemulsions (swollen micelles) showed that oil-external microemulsions exhibit Newtonian flow behavior whilst systems in those composition range that are in the oil-external to water-external transition region, exhibit non-Newtonian behavior. Our results for the transition region appear to corroborate these findings. In aqueous systems (Figure 11) the birefringent characteristics (liquid crystalline) of 1.2% - 2% NaCl systems evidently show non-Newtonian flow behavior of pseudoplastic type characteristics.

b) Applicability of Darcy's law: For steady laminar viscous flow in horizontal beds, Darcy's law is:

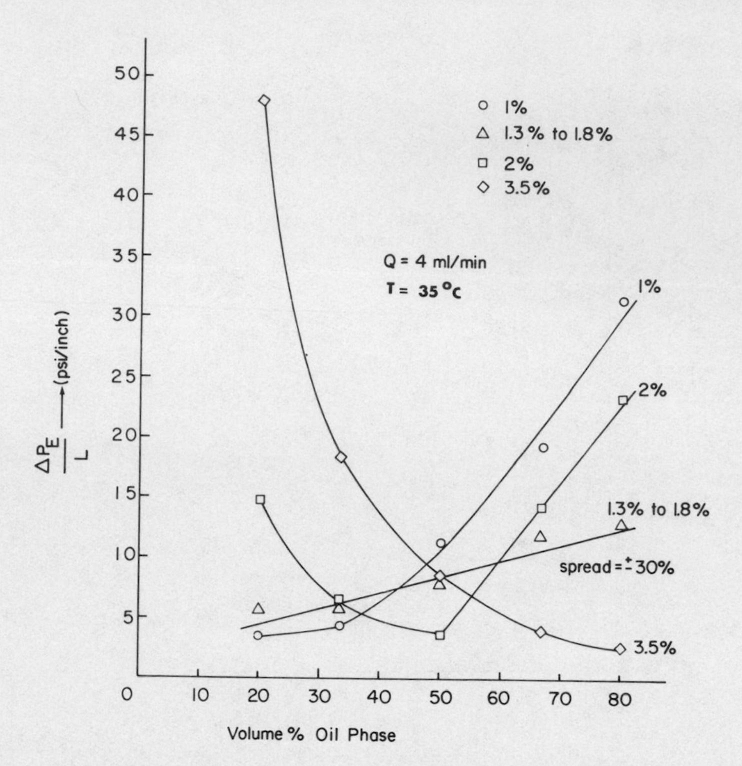

Fig. 20. Variation of pressure drop with percent volume of oil phase for different salt concentrations at 4 ml/min. flow rate.

$$Q = -\frac{KA\ \Delta P}{\mu L} \tag{2}$$

where the permeability K is determined by the structure of the porous material and has the unit: 'Darcy' (d) [1 darcy = 1 (cm^3/s). 1 (cp)/1 (cm^2). 1 (atm/cm)]. The transition from laminar to turbulent flow occurs, say in sands and sandstones, gradually in the range of Reynolds number from 1 to 10. The Reynolds number is defined as

$$Re = \frac{Q\cdot\rho\cdot\delta}{\mu\cdot A\cdot\phi} \tag{3}$$

where δ and ϕ are the sand-grain diameter and porosity of the bed respectively. Since porous systems possess a distribution of pore sizes, the transition from laminar to turbulent flow is not abrupt at a critical Reynolds number as in the case for flow through pipes.

To verify the applicability of eqn. (2), it is possible to calculate the maximum Reynolds number from eqn. (3) that could have been encountered in the present investigation. Substituting values for the parameters in eqn. (3) except for the porosity [δ = 0.063 - 0.125 mm, A = 0.317 cm^2, μ = 0.95 cp (lowest value), ρ = 1.0 g/cm^3 (maximum value), Q = 11.3 ml/min. (maximum value)], we have Re_{max} = 0.78/ϕ. If a percentage porosity as low as 10% is substituted, it would only give a maximum Reynolds number of 7.8 which is well within the range of 1 to 10 for the transition from laminar to turbulent flow in sands and sandstones. Although the porous media used in the present study are composed of different materials, the order of magnitude of the maximum Reynolds number is quite within reasonable limits, should be sufficient for the applicability of Darcy's law. It should be noted that the viscosities of emulsions are, in most cases, much higher than that of water. Consequently, Reynolds numbers would go even below 1. This condition would rule out the possibility of a change from laminar to turbulent flow in the present studies.

c) Permeability and viscosity considerations: In the absence of reliable viscometric data (relating viscosity as a function of shear rate) for the emulsion systems under study, an attempt is made presently to calculate the ratio of permeability to viscosity parameter of the emulsions in the porous bed using Darcy's law. The calculated values are presented in Table 7 for three aqueous to oil ratios, four different salt concentrations and five flow rates. With increase of emulsion flow rate, the (k/μ) ratio increases for salt concentrations less than 2% for 1:1 and 1:4 emulsions. At and above 2% NaCl, the ratio remains essentially constant as the flow rate is increased. For 4:1 emulsion system, with NaCl concentrations of 2% and higher the ratio increases with the flow rate. These observations appear to reveal the existence of a 'transition region' in the phase ratio. More likely the transition point is expected to lie in between 2:1 and 1:1 emulsions.

The determination of absolute permeability values as a function of shear rate, salt concentration and phase ratio, poses some problems because of the complex interaction of the parameters. Although permeability is the property of the porous medium, it has been recognized by investigators (see for example, 23 and 24) that the nature of the fluid in terms of its wetting characteristics with the packing material, would equally contribute to permeability effects. Therefore, a strict evaluation of this quantity from Darcy's law requires extensive viscosity-shear rate data from independent viscometric measurements. Our data from low and constant shear rate experiments (Table 4) although indicate variations with respect to phase ratio and salt concentration (paralleling the observed pressure drop values) are not sufficient to be able to give realistic values of permeability. Order of magnitude calculations show that K can vary anywhere from 5 to 16 darcy. It should be noted that the flow of water through the same porous system gives an average permeability of 7 darcy. If the fluid is assumed to be Newtonian, then it is possible to use the water permeability to evaluate the 'apparent viscosity' of the various emulsion systems from the calculated (k/μ) values in Table 7. Evidently, this approach is not entirely correct for the present situation. However, this basis can be useful to understand the relative behavior of the various systems. From such a calculation it can be seen that: (a) for 4:1 emulsions having 2% and 3.5% NaCl, μ_a decreases as the flow rate is increased while μ_a of 1% and 1.5% NaCl emulsions remain constant at about 3-3.5 cp. (b) Apparent viscosity values of 1%-2% emulsions with 2:1 aqueous to oil ratio are essentially independent of flow rate. However, the 3.5% NaCl emulsion shows a decrease in μ_a with an increase in flow rate. (c) Both 1:1 and 1:4 emulsions behave identically. Here, for 1% and 1.5% NaCl, μ_a decreases with flow rate and for 2% and 3.5% NaCl the apparent viscosity values are independent of flow rate.

The observations presented above evidently show that by increasing the oil phase volume, a phenomenological change occurs around 2:1 aqueous to oil ratio in the apparent viscosity values as a function of sodium chloride concentration. For lack of viscometric data, further analysis of the rheological behavior of the emulsions is not carried out.

In summary, the authors wish to point out that all of the results reported in this investigation of emulsion behavior under various conditions show evidently the complexity of the problem. Simplified correlations are difficult to propose at this stage. Nevertheless, this investigation highlights a number of pertinent problems caused by the presence of emulsion in tertiary oil recovery systems. This is only the beginning and obviously more detailed studies are inevitable relating microstructure of micellar systems with macrostructure of emulsions and their physical and rheological properties before analyzing their fluid dynamic behavior in porous media.

CONCLUSIONS

1. Discontinuities in electrical conductivity values demonstrate the existence of a phase inversion process in macroemulsions. For 1:1 (aqueous:oil) emulsions at 25°C, phase inversion of O/W to W/O occurs around optimal salinity (1.5% NaCl).

2. For a given phase ratio, the phase inversion process parallels the observed (ref. 7) phase inversion in microemulsions (swollen micelles) by electron spin resonance studies.

3. Emulsions containing optimal salinity concentration of sodium chloride are very unstable at 25°C. By increasing the temperature to 35°C, relatively stable emulsions can be formed by sonication.

4. Increasing the system temperature favors the production of

Table 7. Values of the ratio of permeability to viscosity (k/μ) from Darcy's equation using experimental emulsion pressure drop and flow rate data at 35°C

Flow rate, Q (ml/min)	(k/μ) in $(cm^3/sec)/(cm^2 \cdot atm/cm)$											
	aq:oil = 4:1				aq:oil = 1:1				aq:oil = 1:4			
	Salt concn. (wt %)				Salt concn. (wt %)				Salt concn. (wt %)			
	1	1.5	2	3.5	1	1.5	2	3.5	1	1.5	2	3.5
0.5	2.9	1.9	0.3	0.11	0.5	0.5	1.96	0.98	0.12	0.33	0.33	1.96
2.0	2.5	1.9	0.3	0.13	0.6	0.8	1.96	0.82	0.19	0.56	0.36	3.9
4.0	2.3	1.8	0.5	0.16	0.7	0.9	2.02	0.92	0.24	0.7	0.34	2.9
6.0	2.5	1.8	0.8	–	0.84	1.05	2.4	1.0	0.26	0.7	0.35	3.1
10.0	2.8	1.8	1.3	–	0.93	1.24	2.3	1.1	–	0.7	–	3.3

3-phases (otherwise 2-phase systems at low temperatures). Once three phases are produced reverting to room temperatures makes the phase change irreversible.

5. Discontinuities in emulsion viscosity values reconfirm the phase inversion process as revealed by electrical conductivity results.

6. At 25°C, hand-shaken as well as sonicated 1:1 emulsions at optimal salinity exhibited maximum phase separation rate.

7. Phase separation results for hand-shaken and sonicated emulsions with aqueous to oil ratios of 1:1 and 2:1 as a function of sodium chloride concentration follow very closely to each other although the time taken for separation of sonicated emulsions are much higher than that of hand-shaken emulsions.

8. Application of the simplified theory of coalescence (developed for millimeter-droplets and for relatively high interfacial tension values) to micron-droplet system (present case) having ultra low interfacial tension values, has not been successful.

9. For low shear rate experiments in sand packs for the flow of aqueous surfactant formulations containing 1.5% NaCl and more, a non-linear pressure drop-flow rate relation exists. Other formulations having less than 1.5% NaCl, ΔP-Q plots show linearity. High shear rate experiments reveal linear variations in ΔP and Q.

10. Non-linear behavior of aqueous formulations containing 1.5-2.0% NaCl at low shear rates, is attributed to liquid crystalline nature of these phases. Screen factor measurements further corroborate these findings.

11. Flow of sonicated emulsions in sand packs at 25°C having equilibrium 3-phases, produce low pressure drops. Also, they possess Newtonian type flow behavior. However, 1:1 emulsions with 0-1.2% NaCl give high viscosities and consequently high pressure drops. Reasons for this behavior are attributed in terms of emulsion droplet size and structure.

12. Results of the flow of aqueous surfactant formulations in sand packs together with the physical property data demonstrate that the occurrence of a maximum in bulk viscosity and in screen factor at 1.8% NaCl, does not correspond to the pressure drop results. Otherwise, pressure drop values, in general, follow the bulk viscosity and screen factor variations.

13. Emulsion pressure drop results as a function of NaCl concentration in sand pack experiments follow closely the separation rate results from batch settling tests at 25°C with a pronounced minimum at 1.5% NaCl.

14. Emulsion flow in a EM-Gel packed bed at 35°C as a function of phase ratio and salt concentration demonstrates the presence of a transition region in the vicinity of 2% NaCl. This, in turn,

corresponds to the discontinuous change as revealed by the conductivity data.

15. Pressure drop results of emulsions containing 1.3 to 1.8% NaCl in a EM-Gel packed bed reveal a linear variation with phase ratio.

16. In general, flow experiments in EM-Gel packed beds show that pressure drop of emulsions increase with the increase in the amount of dispersed phase, irrespective of whether it is oil or aqueous phase. The anomalous behavior at 2% NaCl is attributed to optimal salinity effects.

17. Reynolds number calculations rule out the possibility of a change from laminar to turbulent flow in the present studies. The shear rate magnitudes and flow regime favor the applicability of Darcy's law.

18. The average permeability of EM-Gel packed beds using water is 7 darcy and the emulsion systems give permeabilities in the range of 5 to 16 darcy.

19. In the absence of reliable viscometric data, the apparent viscosity values (evaluated using Darcy's law with the permeability of water) show a discontinuity around aqueous to oil ratio of 2:1. For several salt concentrations and phase ratios, the apparent viscosity variation with emulsion flow rate is essentially constant.

ACKNOWLEDGMENTS

It is with pleasure that we acknowledge that this project was financially supported by a grant from the Department of Energy, Washington, D.C., #EY-77-S-05-5341 and by a consortium of twenty major oil and chemical companies. Facility given to S. Vijayan by the Department of Chemical Engineering, McMaster University, during the preparation of this paper is gratefully acknowledged.

NOMENCLATURE

A	cross-sectional area of porous bed, cm^2
B	a constant with dimension of reciprocal length squared which is characteristic of the pore geometry. B is also proportional to the specific surface
g	acceleration due to gravity
h	liquid film thickness at any time t (eqn. 1)
h_o	initial liquid film thickness (eqn. 1)
K	permeability ($= \phi_{/B}$), a constant characteristic of the medium, darcy [1 darcy = $(cm^3/s) \cdot cp/cm^2 \cdot (atm/cm)$]
L	length of porous bed
P	pressure, psi
ΔP_E	pressure drop across porous bed by the flow of emulsion, psi

ΔP_W	pressure drop of water across porous bed, psi
Q	volumetric flow rate of fluid, ml/min
ρ	bulk density, g/cm^3
$\Delta\rho$	density difference between aqueous and oil phases, g/cm^3
μ_a	apparent viscosity, cp
$\mu_{aq.}$	aqueous phase viscosity, cp
μ_{oil}	oil phase viscosity, cp
μ_c	continuous phase viscosity, cp
ϕ	porosity of bed
σ_i	interfacial tension, dyne/cm

REFERENCES

1. C. D. McAuliffe, J. Pet. Tech., 25, 721,727 (1973).
2. D. F. Boneau and R. L. Clampitt, J. Pet. Tech., 29, 501 (1977).
3. K. J. Lissant, in "Improved Oil Recovery by Surfactant and Polymer Flooding," D. O. Shah and R. S. Schechter, eds., Academic Press, New York, 1977, p. 23.
4. S. Vijayan, H. Doshi, and D. O. Shah, in Improved Oil Recovery Research Program, University of Florida, Gainesville, Semi-annual Report, Dec. 1977, p. B.17.
5. S. Vijayan, C. Ramachandran, H. Doshi, and D. O. Shah, in Improved Oil Recovery Research Program, University of Florida, Gainesville, Semi-annual Report, June 1978, p. B.11.
6. S. Vijayan, C. Ramachandran, and D. O. Shah, J. Amer. Oil Chemists' Soc. (in press).
7. S. Vijayan, C. Ramachandran, and D. O. Shah, J. Amer. Oil Chemists' Soc. (in press).
8. C. Ramachandran, S. Vijayan, and D. O. Shah, J. Phys. Chem., 84, 1561 (1980).
9. J. Boyd, C. Parkinson, and P. Sherman, J. Colloid Interface Sci., 41, 359 (1972).
10. J. T. Davies and E. K. Rideal, "Interfacial Phenomena," 2nd ed., Academic Press, London, 1963, p. 366.
11. J. Bickerman, "Foams," Springer Verlag, New York, Ch. 10, 1973.
12. A. W. Adamson, "Physical Chemistry of Surfaces," 2nd ed., Inter. Sci., New York, 1967, p. 509.
13. Y. Tamai, S. Ebina, and K. Hirai, in "Colloid and Interface Science," Vol. II, M. Kerker, ed., Academic Press, New York, 1976, p. 257.
14. S. S. Davies, T. S. Purewal, R. Buscall, A. Smith, and K. Choudhury, in "Colloid and Interface Science," Vol. II, M. Kerker, ed., Academic Press, New York, 1976, p. 265.
15. G. V. Jeffreys and G. A. Davies, in "Recent Advances in Liquid-Liquid Extraction," C. Hanson, ed., Pergamon Press, Oxford, 1971, p. 495.

16. D. R. Woods and K. A. Burrill, J. Electroanal. Chem., 37, 191 (1972).
17. S. Vijayan and A. B. Ponter, Chem.-Ing.-Tech., 47, 748 (1975).
18. S. Vijayan and A. B. Ponter, Tenside (Germany), 13, 193 (1976).
19. S. Vijayan, "Coalescence in a Laboratory Continuous Mixer/ Settler Unit," D.Sc. Thesis, No. 190, Swiss Federal Inst. Tech., Lausanne, 1974.
20. S. Vijayan and D. R. Woods, paper presented at the 51st Colloid and Surface Science Symposium, June 19-22, 1977, Grand Island, New York.
21. A.J.G. Liem and D. R. Woods, Can. J. Chem. Eng., 52, 222 (1974).
22. K. D. Dreher, W. B. Gogarty, and R.D. Sydansk, J. Colloid Interface Sci., 57, 379 (1976).
23. J. J. Taber, Soc. Pet. Eng. J., 335 (1966).
24. J. J. Taber, J. C. Kirby, and F. U. Schroeder, A.I.Ch.E. Symp. Ser., 69, 53 (1970).

RHEOLOGICAL PROPERTIES OF INTERFACIAL LAYERS FORMED BY IRON AND ALUMINUM SOAPS, USED IN REVERSIBLE EMULSIONS IN PETROLEUM INDUSTRY

S. Shalyt, L. Mukhin, L. Ogneva, G. Milchakova and V. Rybalchenko

Moscow Institute of Petrochemical and Gas Industry named after I. M. Gubkin, Leninski prospect, 65, Moscow, USSR

In connection with the application of water-in-oil concentrated emulsions in the petroleum industry, emulsified hydrocarbon films, rheological properties of interfacial layers formed by aluminum and iron soaps of synthetic fatty acids (SFA), C_{17}-C_{20}, and stabilities of model systems were studied. It was found that sufficiently thick (700Å) structured interfacial layers having mechanical properties similar to those of solid-like bodies were the basic stabilizing factors for water-in-oil concentrated emulsions.

INTRODUCTION

The greater are the drilling depths in complicated geological formations, the more urgent is the necessity of creating and improving the drilling muds. In this respect, hydrocarbon-based muds are very promising, in particular, water-in-oil concentrated emulsions which are complex multicomponent systems which are gaining wider and wider application. In spite of the fact that many scientists work on the formation of stable water-in-oil concentrated emulsions, insufficient attention is paid to the inter-relation between the physico-chemical properties of surfactants and their emulsifying and stabilizing actions.

According to Rhebinder (1), stable concentrated emulsions may be obtained only when there is a structural-mechanical barrier at the interface. The structural-mechanical barrier may result from adsorption of surfactants forming a structure of sufficiently high mechanical strength at the interface. In previous studies (2-4), it was shown that SFA aluminum and iron soaps (C_{17}-C_{20}) can be used to obtain stable water-in-oil concentrated emulsions as models of hydrocarbon-based drilling muds.

In our work we have attempted to determine the relationship between interfacial layer properties and water-in-oil emulsion stability. For this objective, physico-chemical properties of surfactants (stabilizing agents), properties of emulsified hydrocarbon films and rheological properties of interfacial layers must be studied. The results of the study may lead to recommendations for preparing high-stability systems for commercial well-drilling.

EXPERIMENTAL

SFA aluminum and iron soaps (C_{17}-C_{20}) synthesized by a double exchange reaction were studied. Soaps were freed of impurities by washing with hot water and extracting with methanol. A water/n-decane system was taken as a model of water-in-oil emulsion. Sodium dodecylsulfonate was purified by means of recrystallization from an alcoholic solution.

Formation and properties of emulsified hydrocarbon films were determined by visual observation and measurements of contacts between a small drop of water submerged into oil and the larger water phase. The cell within which the film was formed was a small glass cylindrical vessel having a flat parallel bottom (1.5 cm diameter and 3 cm high). By means of a microsyringe water drops were injected into the upper oil phase, time was given for an equilibrium adsorption layer to be formed, then the drop was spread upon the interface. To obtain drops of a similar radius, a glass capillary (0.04 cm diameter) was used.

The films obtained were observed from below using a "Zeis-Epitip" microscope. Monochromatic light was supplied by a mercury lamp with a filter passing a 5770-5790Å band. The method was utilized to study kinetics of emulsified film thinning-out and to determine the half-lives ($\tau_{1/2}$) of water drops at flat interfaces (5).

Water-in-oil emulsions were prepared in a mixer conventionally, by stirring at a speed of 1000 rpm for 30 min. at 20°C. Emulsions contained 50% water and 50% oil. Stabilities of model emulsions were estimated by the time (τ) within which a 50% dispersed phase was settled. To accelerate the measurements the samples were placed within a field of centrifugal forces of a centrifuge making 3000-6000 rpm.

Rheological properties of interfacial layers were determined by means of a device based on twisting a disk suspended on an elastic wire and placed at interface (6).

Two methods were used in our work:

Method I - Measurement of kinetics of shearing strain development, $\varepsilon = \varepsilon(\tau)$, where P is constant. The method allows to study deformation with time at constant loading of the system, and reversible release of deformation after instantaneous loading (P = 0).

Method II - Study of kinetics of establishing a steady flow. The method allows to study shearing stress variations, P_s, with time upon gradually increasing deformations and constant speeds of shearing ($\dot{\varepsilon}$ is constant).

Application of the two methods made it possible to plot the entire rheological curve, speed of deformation vs. equilibrium stress $\dot{\varepsilon} = f(P_{ss})$, as well as to determine rheological properties of interfacial layers.

RESULTS AND DISCUSSION

It has been shown (7-9) that properties of thin liquid films are of great importance for stabilization of dispersed systems. We have used thin hydrocarbon films as models of water-in-oil emulsions and their properties may be characteristic of reversible emulsion stabilities. We investigated hydrocarbon films stabilized by aluminum and iron soaps with concentrations from 0.001 to 0.1%.

Upon the contact of a water drop with the flat water/soap solution in n-decane interface, a liquid hydrocarbon film was formed which is surrounded by Newton rings that may be seen in reflected light. Upon thinning, the film itself changes its color from white to grey and then to black.

It was found that up to 0.004% concentration of the iron soap in n-decane, which corresponds to the formation of a saturated adsorption layer at the interface, black spots are formed. However, these black spots do not produce a black film and interfacial film is broken quickly. Soap concentrations of up to 0.1% produce a film whose thinning-out stops when a thick stable grey film is created. Similar results have been observed for films stabilized by aluminum soap. During the formation of the film, monochromatic light showed alternative dark and bright bands corresponding to the interference maxima and minima. By measuring the parameters of the latter, the film thickness could be estimated.

In the case of thick films, their thermodynamic properties may be considered to be practically the same as those of the bulk phase. Using the topographic technique for determining the angle of contact between a thick film and the larger phase (10), the angle of contact was taken to be equal to zero.

Estimation of hydrocarbon film thickness was carried out in relation to soap concentrations and the time from interface formation. Soap concentrations from 0.01 to 0.1% increased the film thickness from 600 to 1000Å, the time from interface formation influencing the film stability. Similar results were obtained when film thickness was determined by the ellipsometry. Thus, the study of emulsified hydrocarbon films stabilized with aluminum and iron soaps showed that the process of film formation did not lead to stable black films.

Film thinning is completed when a grey thick film is formed. Its stability increases with time, apparently on account of a structure formation process, the latter may be estimated by the rheological parameters of interfacial layers. Rheological investigations were carried out for an interfacial layer formed by a 0.5% SFA iron soap solution in n-decane at water interface. Kinetics of shearing strain development was studied within the shearing stress range from 0.099 to 0.396 dyne/cm.

Curves for shearing strain development are shown in Figure 1. As seen from Figure 1, immediately upon loading, fast instantaneous deformation is developed in the layer, the former increasing with the load growth according to Hooke's law. Fast instantaneous deformations are followed by elastic deformations which are increasing with time. As the shearing stress applied exceeds the yield point, in addition to elastic deformations, residual deformations are observed, the magnitude of the latter increasing with constant speed. Residual deformations in the layer may be seen upon unloading.

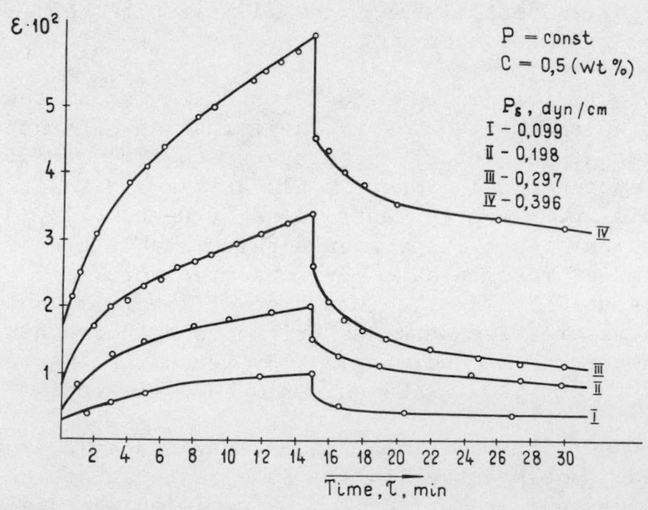

Fig. 1. Development of deformation with time in the interfacial layer formed by SFA iron soap solution in n-decane/water interface. I--P_S = 0.099 dyne/cm; II--P_S = 0.198 dyne/cm; III--P_S = 0.297 dyne/cm; IV--P_S = 0.396 dyne/cm.

The data obtained were used to estimate the moduli of fast and slow elastic deformations, the equilibrium modulus of elasticity and viscosity of elastic after-effect.

To obtain rheological characteristics in the range of high shearing stress, the method $\dot{\varepsilon}$ = constant was used. The speed of deformation was taken in the range from 0.064 to 1.28 sec^{-1}.

Figure 2 shows that the curves obtained have a clearly seen maximum corresponding to the critical shearing stress, P_{rs}, achieving its equilibrium value, P_{ss}, one minute after the beginning of deformation. The maximum on the curves, $P_s = f(\tau)$, is associated with the elastic properties of the interfacial layer. The greater the speeds of deformation, the larger the differences between the values of P_{rs} and P_{ss}, which is due to the relationship between the time of relaxation and speed of deformation.

Rheological curves for speed of deformation and interfacial layer viscosity vs. equilibrium shearing stress, $\dot{\varepsilon} = f(P_{ss})$ and $\eta = f(P_{ss})$, are given in Figure 3. Rheological curves have several sections of different nature. In the elastic area, at $P_s < P_{K_1}$, deformations are fully reversible, the interfacial layer structure is characterized by infinitely great viscosity values and absence of broken links. In the $P_s > P_{K_1}$ range of shearing stresses there is a short shoulder of slow creep characterized by constant plastic viscosity, η_o^*, equal to 10^4 dyne sec/cm. Within this area the number of restored linkages exceeds that of broken ones. Further on, at shearing stresses higher than P_{K_2} (Bingam yield point) the number of broken linkages begins to exceed, the layer structure is broken and viscosity drops by 4-5 orders. The presence of two yield points on the rheological curves proves a solid-like nature of the interfacial layer formed, whose structure represents a 10-15% gel. Rheological characteristics are listed in Table 1.

In the previous work (3), the disintegrating action of sodium dodecylsulfonate on the critical shearing stress of the interfacial layer formed by iron soap solution in decane was shown, when the interfacial layer in question was used in drilling muds as a hydrophobic agent for a solid surface.

In this work the authors studied the effect of sodium dodecylsulfonate on fast elastic deformation, ε_o, of the interfacial layer formed by a 0.05% iron soap solution in n-decane at water interface.

As seen from Figure 4, small concentrations (5.10^{-5} mol/l) of water-soluble sodium dodecylsulfonate reduce fast elastic deformation (ε_o). This fact may be associated with a greater degree of interfacial layer filling. Gradually increasing concentrations result in a sharp growth of ε_o, which is due to the disintegration of the layer structure (Curve I). Variations in sodium dodecyl-

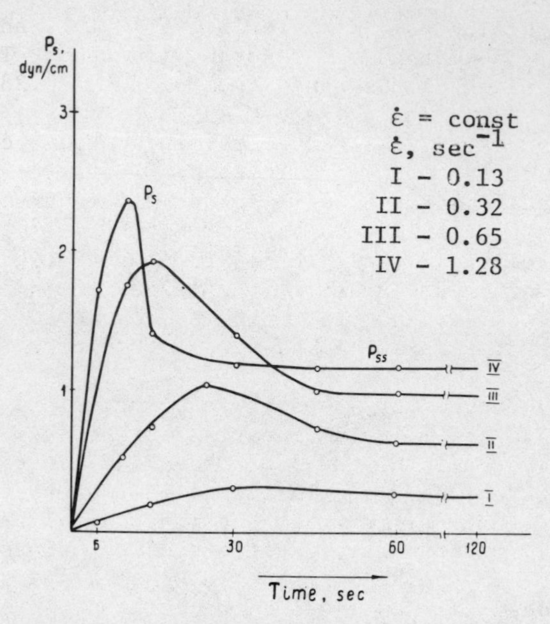

Fig. 2. Variations of shearing stress in the interfacial layer formed by SFA iron soap solution in n-decane/water interface. Conc. = 0.5%. I--$\dot{\varepsilon}$ = 0.13 sec^{-1}; II--$\dot{\varepsilon}$ = 0.32 sec^{-1}; III--$\dot{\varepsilon}$ = 0.65 sec^{-1}; IV--$\dot{\varepsilon}$ = 1.28 sec^{-1}.

Table 1. Rheological characteristics of a 0.5% SFA iron soap solution in n-decane/water interface

E_{1s}	E_{2s}	E_{ss}	η_{2s}	Q_2	P_{K_1}	P_{K_2}	η_o^*	η^*
dyne/sec / cm			dyne/sec / cm	sec	dyne/sec / cm		dyne/sec / cm	
34,3	23,1	13,8	3170	140	0,09	0,80	9880	0,31

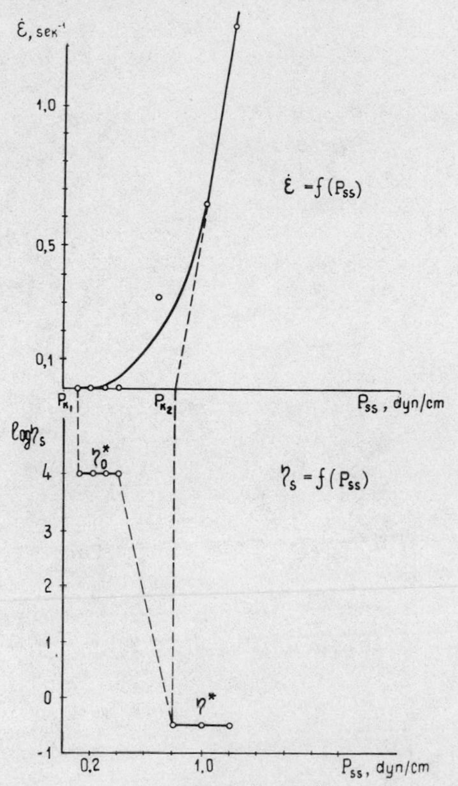

Fig. 3. Curves showing dependence of speed of deformation ($\dot{\varepsilon}$) and viscosity (η) on equilibrium shearing stress (P_{ss}) for the interfacial layer formed by SFA iron solution in n-decane/water interface. Conc. = 0.5%.

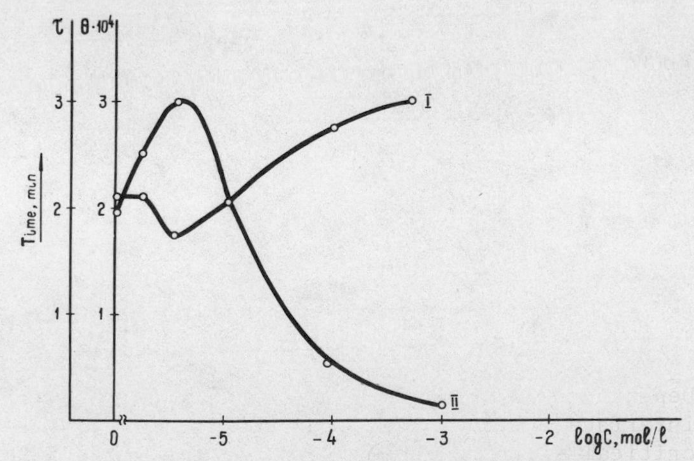

Fig. 4. Effect of sodium dodecylsulfonate concentration on fast elastic deformation θ (I), and stability of concentrated emulsions τ (II).

sulfonate concentrations produce similar effect on emulsion stability (Curve II).

To study the inter-relationship between rheological characteristics of interfacial layers and stability of water-in-oil concentrated emulsions, model emulsions containing 50% water and 50% soap in n-decane were prepared and their properties studied. In our previous work (3,4) high stabilities of emulsions stabilized by aluminum and iron soaps were observed, rheological properties of stabilizing interfacial layers being compared with the stability of drops.

In Figure 5 are given relations of water drop stability at the flat interface (Curve I), adsorption of SFA aluminum soap on emulsion drops (Curve II) and the critical shearing strength of the interfacial layer (Curve III) depending on the equilibrium soap concentration in solution. Greater equilibrium soap concentrations in solutions lead to higher stabilities due to adsorption and formation of a stable structure at the interface.

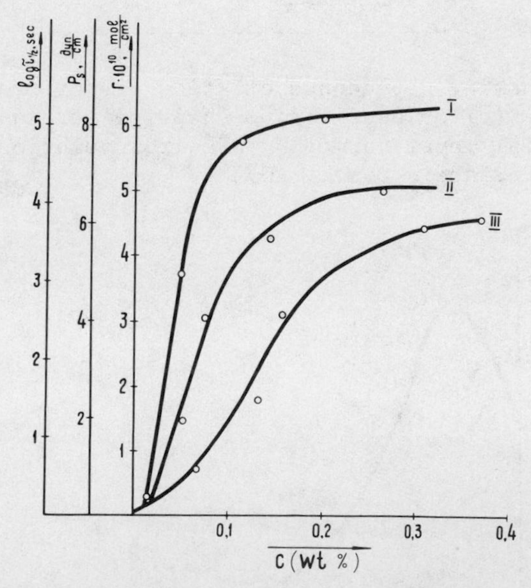

Fig. 5. Dependence of half-lives ($\tau_{1/2}$) of water drops at flat interface (I), adsorption on emulsion drops (II) and critical shearing stress (P_{ss}) of the interfacial layer (III) on equilibrium concentration of SFA aluminum soap in n-decane.

In our work we also compared stabilities of model emulsions with those of thin hydrocarbon films. Soap concentrations corresponding to black spot formation did not lead to stable emulsions. To obtain them, concentrations much higher than those at which stable films are formed are necessary. This is due to the fact that during emulsification a sharp increase of the interfacial adsorption surface occurs, which results in considerably lower concentration of surfactant in solution.

CONCLUSIONS

The correlation between stability of water-in-oil concentrated emulsions and rheological characteristics of interfacial layers allows us to make a conclusion that the basic stabilizing factor at interface is the presence of a colloid-adsorption layer having a gel-like structure of sufficient strength.

ACKNOWLEDGMENTS

It is our pleasure to thank Professor A. D. Sheludko, Dr. D. Platikanov and Professor G. Sonntag for reading and discussing the manuscript.

REFERENCES

1. V. N. Ismailova and P. A. Rhebinder, "Structure Formation Processes in Protein Systems."
2. L. K. Mukhin, S. J. Shalyt, L. G. Ogneva, and I. I. Kasakova, Trans. of the IV-th International Conference on Surfactants, Berlin (1974).
3. L. K. Mukhin, S. J. Shalyt, L. G. Ogneva, E. P. Fedorova, and I. I. Kasakova, Trans. of the International Conference on Colloids and Surfactants, Budapest (1975).
4. L. K. Mukhin, S. J. Shalyt, L. G. Ogneva, I. I. Kasakova, and G. F. Milchakova, Trans. of the VII-th International Conference on Surfactants, Moscow (1976); S. J. Shalyt, L. K. Mukhin, L. G. Ogneva, and G. F. Milchakova, Trans. of the IV-th Symposium on Surfactants, Yugoslavia (1977).
5. P. A. Rhebinder and E. Wenstrem, Coll. J., 53, 145 (1930).
6. P. A. Rhebinder and A. A. Trapesnikov, J. Phys. Chem., 12, 573 (1938) (Rus.).
7. A. D. Sheludko, Usp. Koll. Khim. (Rus.), Adv. Colloid Interface Sci., 1, 391-464 (1967).
8. H. Sonntag and K. Strenge, "Coagulation and Stability of Dispersion Systems," L., "Khimia" (Rus.) (1973); H. Sonntag and H. Klare, Tensick, 4, 104-108 (1967).

9. P. M. Kruglyakov and Y. G. Rovin, "Physico-Chemistry of Black Hydrocarbon Films," Isd. "Nauka," M. (1978) (Rus.).
10. A. Sheludko, B. Radoev, and T. Kolarov, Trans. Faraday Soc., 64, 2213 (1968).

IV: WETTABILITY AND OIL DISPLACEMENT MECHANISMS

EFFECT OF VISCOUS AND BUOYANCY FORCES ON NONWETTING PHASE TRAPPING IN POROUS MEDIA

N. R. Morrow and B. Songkran

Petroleum Recovery Research Center
New Mexico Institute of Mining and Technology
Socorro, New Mexico 87801, U.S.A.

The trapping of residual nonwetting phase in packings of equal spheres has been investigated for a wide range of capillary numbers, $v\mu/\sigma$ (ratio of viscous to capillary forces) and Bond numbers, $\Delta\rho g R^2/\sigma$ (ratio of gravity to capillary forces). For vertical displacement, residual saturations varied from normal values of about 14.3% when the Bond number was less than 0.00667 and the capillary number was less than 3×10^{-6}, down to near zero when either the Bond number exceeded 0.35 or the capillary number exceeded about 7×10^{-4}. Precise correlation of residual saturation with a linear combination of Bond and capillary numbers showed that the effects of gravity and viscous forces on trapping are superposed. The equivalence of gravity and viscosity forces is used to determine relative permeability of the displacing fluid at the flood front. Relative permeabilities to the wetting phase behind the flood front at less than normal residual saturations are reported. For displacement at various dip angles, the effect of gravity is greater than that predicted by using results for vertical displacement with the Bond number resolved according to angle of dip.

The conditions required for prevention of trapping are compared with those required for mobilization of trapped blobs of residual nonwetting phase. It is estimated that mobilization of trapped blobs is about five times more difficult to achieve than prevention of trapping. For low interfacial tension systems, buoyancy forces may sometimes have significant influence on the microscopic mechanism of displacement in reservoir rock.

INTRODUCTION

The entrapment and mobilization of residual saturations is pertinent to many aspects of enhanced recovery (1-4). Entrapment mechanisms determine what proportion of oil is recovered from the swept zone of a waterflood, and, consequently, how much oil remains for possible recovery by tertiary methods. The volume of oil remaining in swept zones and the fraction of this oil that is recovered are critical economic factors in application of enhanced recovery processes.

Entrapment and mobilization mechanisms at low flow rates and low interfacial tension can also control the recovery obtained by tertiary methods. In micellar flooding, for example, high ratios of viscous to capillary forces arise at field flow rates when interfacial tensions are very low. Development of a continuous oil bank having significant mobility requires that discontinuous oil be mobilized to form a continuous bank which gathers more residual oil as it advances. Interfacial tensions may exist or develop between the micellar bank and the oil, or between the micellar fluid and the aqueous polymer bank used to push the micellar fluid. Entrapment of oil by the micellar bank, or of micellar fluid by the polymer bank would eventually cause the process to fail.

Taber (2) showed, for water-wet sandstones, that a critical value of viscous to capillary forces was required for production of residual oil. For economically significant recovery of oil, this critical value, which is an intrinsic property of a given rock, must be exceeded by a factor of about ten. Because of practical limitations on pressure drop and distances of separation between injection and producing wells, it was estimated that extremely low interfacial tensions of the order of 1/100 of a dyne/cm must be achieved for mobilization of significant quantities of residual oil. At normal oil-water interfacial tensions, capillary forces which retain residual oil far exceed buoyancy forces. However, comparatively little attention has been given to the distinct possibility (4) that when interfacial tensions are lowered, as in surfactant flooding, so that viscous forces can compete with capillary forces, buoyancy forces may also have significant effect on both trapping and mobilization.

BOND NUMBER AND CAPILLARY NUMBER

The factors which determine the microscopic mechanism of trapping are: geometry of the pore network; fluid-fluid properties such as interfacial tension, density difference, viscosity ratio, and phase behavior; fluid-rock interfacial properties which determine wetting behavior; and applied pressure gradient and gravity.

Results presented in this paper concern the effect of applied pressure gradient and gravity on residual nonwetting phase saturations for systems in which there is complete wetting, the fluid-fluid properties are held constant, and the porous media (random packings of equal spheres) are geometrically similar.

The capillary forces which cause trapping or resist mobilization can be overcome by viscous pressure gradient or buoyancy forces. The ratios of gravity to capillary forces and viscous to capillary forces are expressed as dimensionless groups, known respectively as the Bond number (N_B), and capillary number (N_{Ca}).

$$N_B = \frac{\Delta\rho g R^2}{\sigma} \tag{1}$$

$$N_{Ca} = \frac{v\mu}{\sigma} \tag{2}$$

where: $\Delta\rho$ = fluid density difference (gm/cm^3)
g = acceleration due to gravity (cm/sec^2)
R = particle radius (cm)
v = displacing fluid velocity (cm/sec)
μ = displacing fluid viscosity (poise - gm/cm-sec)
σ = interfacial tension (dyne/cm)

The particle radius R is chosen as a convenient characteristic length which is of the same order of magnitude as the pore dimensions. The main reason for choosing unconsolidated packings of close-sized particles as the porous media is that particle radius can be varied at will to give systems of different permeabilities but essentially similar geometries which scale according to particle radius. The absolute permeability, in C.G.S. units, can be expressed in terms of particle radius by using the Kozeny-Carman equation (5):

$$k = \frac{\phi^3}{K_z(1-\phi)^2 A_s^2} \tag{3}$$

where: A_s = specific surface area per unit solid volume
= $3/R$
K_z = Kozeny constant, which by experiment has been shown to be approximately equal to 5 for well-sorted sands or sphere packings
ϕ = porosity, which is usually equal to about 0.38 for random packings of equal spheres

Substitution of these values gives an approximate relationship between permeability and particle radius for random packings of equal spheres:

$$k = 0.00317R^2 \tag{4}$$

It follows that the relationship between the Bond number, expressed as, $\Delta\rho g R^2/\sigma$, and that expressed as $\Delta\rho g k/\sigma$ can be written:

$$N_{B(k)} = 0.00317\ [N_{B(R^2)}] \qquad (5)$$

where k and R^2 indicate the basis of the Bond number. Unless stated otherwise, the values of Bond number, N_B, or the reciprocal of the Bond number, are based on the particle radius, R.

EXPERIMENTAL

In most of the experiments, an aliphatic oil (Soltrol-130) was used as the wetting phase and air was the nonwetting phase. This choice of fluids ensured that the contact angle was zero through the oil phase for all experiments. The Soltrol had a viscosity of 1.42 cp, density of 0.7484 gm/cm^3, and surface tension of 23.19 dynes/cm. All measurements were made at room temperature (23±0.5°C).

Packings of close-sized glass beads were used as the porous media. The spheres were first separated into the sizes listed in Table 1. Bead packs were held in a modified burette (see Figure 1) which had a disc of fine stainless steel screen of about 450 mesh size at its base. The screen was supported in a hollow rubber plug or in a stainless steel cylinder with an o-ring which provided a seal with the inside wall of the burette. After adding oil to the burette, it was attached to a vibrator and beads were added slowly in order to avoid entrapment of air during packing. This procedure gave a reasonably dense and consistent packing which was initially 100% saturated. After adding the beads, a plastic washer was forced against the upper surface of the pack to prevent its movement during subsequent displacement tests. The oil above the bead pack was then drained to the level of the upper surface of the saturated pack and porosities were determined from subsequent weight and volumetric measurements. All weighings were made on a top loading balance having an accuracy of ± 0.01 gms. Errors in saturation measurement were estimated to be less than one-tenth percent of the pore volume. Permeability measurements were made with equipment set up as shown in Figure 1(a). A plot of change in height, Δh, versus time, t, on semi-logarithmic graph paper yielded a straight line relationship because pressure drop is directly proportional to Δh, and flow rate is proportional to rate of change in Δh. The permeability of the rock was determined from the overall permeability by correcting for the resistance to flow in the apparatus caused mainly by the stainless steel retaining screen (6). After measuring absolute permeability, the oil was drained from the column into a burette connected as shown in Figure 1(b). The height, H, in Figure 1(b), was calculated to provide for drainage to the irreducible saturation of the bead pack but was kept

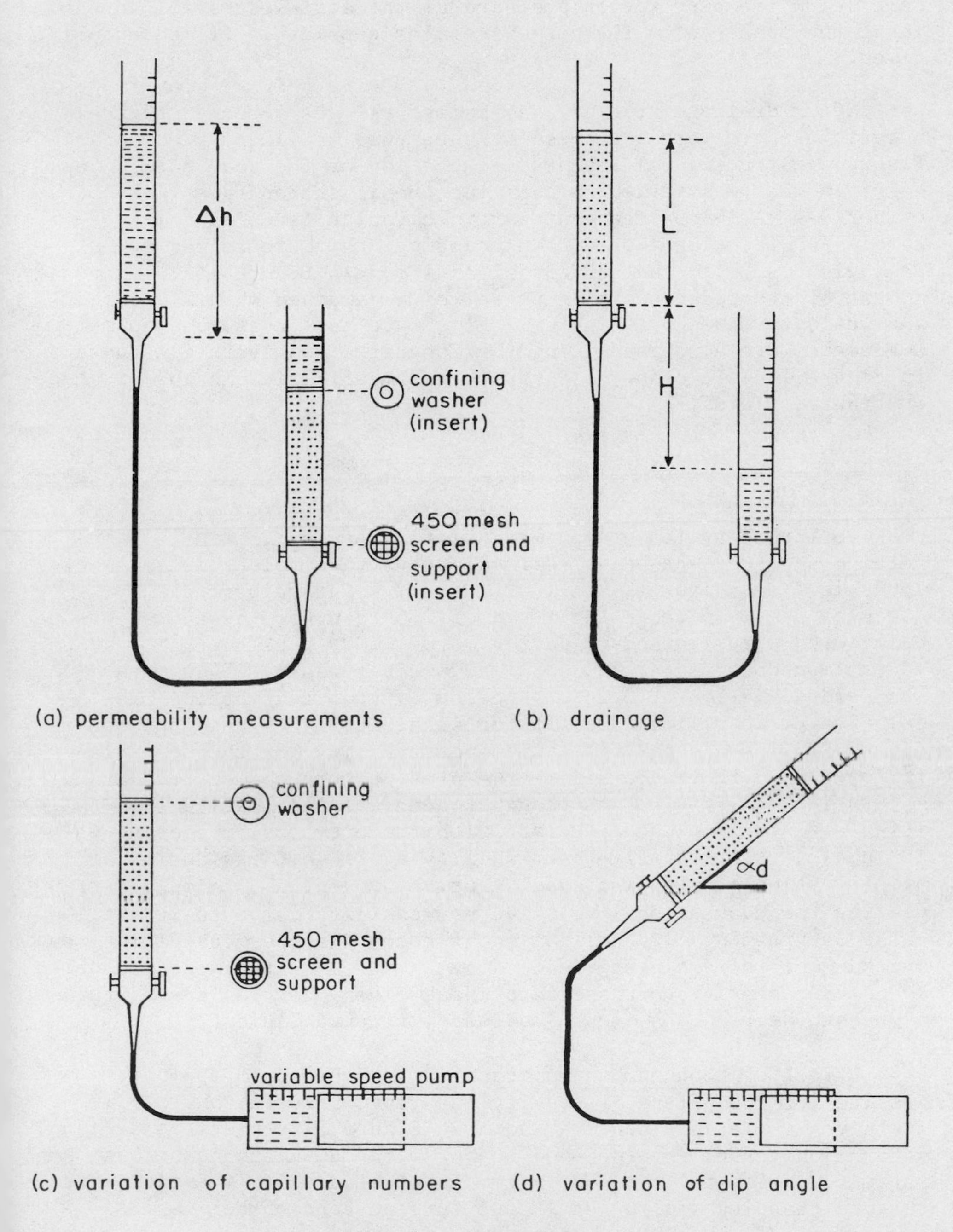

Fig. 1. Apparatus

well below the penetration pressure of the stainless steel screen. Thus, the space below the screen remains completely filled at all times.

After drainage, the oil was pumped via the screen into the burette using a variable speed syringe pump as illustrated in Figure 1(c), until oil appeared at the top of the bead pack. Capillary number was varied by changing the injection rate. The Bond number was varied by changing particle radius (see Table 1). In studies of the effect of dip on trapping, the burette was set at an angle, α_d, with the horizontal as shown in Figure 1(d). The amount of trapped nonwetting phase which remained within the packing was determined by weighing. Flow rate and pressure drop measurements were also made using the apparatus configuration shown in Figure 1(a) in order to determine permeabilities in the presence of trapped fluids.

RESULTS

Permeability measurements: Relative permeabilities to the wetting phase at and below residual saturation values are presented in Table 2 and Figure 2. All of the end point values for five sizes of sphere were in the range of 0.6 to 0.65, the average value being 0.63 with a standard deviation of 0.02. The average residual saturation obtained at low capillary number was 14.25% with a standard deviation of 0.25%. Previously reported wetting phase relative permeability data (7) for sphere packings are included in Figure 2. The end point relative permeability for these results is lower than that obtained in the present work, being 0.45 at a saturation of 16.5%. Even though the end point relative permeabilities are not in close agreement with the previously reported value, it is clear that relative permeability at reduced residuals obtained by prevention of trapping should not be predicted by simply extrapolating the conventional relative permeability curve to 1.0 at 100% phase saturation. It should also be noted that permeabilities, obtained when residuals are reduced by change in entrapment mechanism, do not necessarily correspond to those given when residual saturations are decreased by mobilization of trapped fluid.

Effect of bond number and capillary number on trapping: Results for the effect of gravity and viscous forces on trapping are given in Table 1 and are shown in Figure 3 as plots of residual saturation versus inverse Bond number with capillary number as parameter. For the type of system under study, it is estimated that no trapping will occur if the inverse Bond number is less than about 3. When the inverse Bond number is greater than about 200, buoyancy forces have no effect on the amount of trapped residual nonwetting phase saturation, and residual saturation depends only on the capillary number. Above the critical capillary number

Table 1. Nonwetting phase saturations trapped in random packings of equal spheres at various Bond and capillary numbers. ($\Delta\rho$ = 0.748, σ = 23.19 dyne/cm).

Mesh Size	Mean Dia. (mm)	Bond Number	Inverse Bond Number	$\%S_{nw}$ at Capillary Number				
				2.82×10^{-6}	2.82×10^{-5}	7.07×10^{-5}	1.41×10^{-4}	2.82×10^{-4}
12-14	1.55	0.1895	5.277	2.97	2.56	1.90	1.62	1.21
14-16	1.29	0.1313	7.619	4.83	4.28	3.41	2.76	--
16-18	1.09	0.0937	10.671	6.31	5.22	4.35	3.37	--
18-20	0.925	0.0675	14.818	7.79	6.68	5.45	3.74	--
25-28	0.655	0.0338	29.551	10.30	9.19	6.65	4.78	2.11
32-35	0.462	0.0168	59.399	12.36	10.60	8.19	5.14	2.64
42-48	0.327	0.00843	118.568	13.95	11.72	8.51	6.14	3.07
48-60	0.275	0.00596	167.647	14.03	11.94	9.27	6.54	3.68
65-80	0.191	0.00288	347.532	14.06	11.92	9.21	6.92	4.07
80-100	0.165	0.00215	465.686	13.99	11.90	9.39	6.94	4.05
100-115	0.137	0.00148	675.492	14.06	11.88	9.28	6.96	4.06

Table 2. Relative permeabilities at reduced residual saturations

Mesh Size	$\%S_{rr}$	Relative Permeability	Mesh Size	$\%S_{rr}$	Relative Permeability
42-48	Run No. 1		65-80	Run No. 1	
	3.07	0.882		2.36	0.906
	6.14	0.841		4.57	0.851
	8.51	0.768		6.78	0.799
	11.72	0.684		9.14	0.703
	13.95	0.656		12.09	0.642
				14.60	0.607
	Run No. 2			Run No. 1	
	1.70	0.963			
	2.97	0.928		4.05	0.916
	4.24	0.874		6.94	0.859
	5.94	0.790		6.39	0.790
	8.77	0.726		11.99	0.719
	12.16	0.655		14.31	0.600
	14.14	0.627			
				Run No. 2	
48-60	Run No. 1				
				3.72	0.918
	9.15	0.704		6.84	0.822
	11.75	0.671		11.90	0.676
	13.93	0.649		14.43	0.641
	Run No. 2		100-115	Run No. 1	
	1.98	0.923		6.96	0.853
	3.40	0.904		9.28	0.815
	6.66	0.806		11.88	0.726
	8.92	0.712		14.06	0.726
	12.18	0.648			
	14.59	0.623			

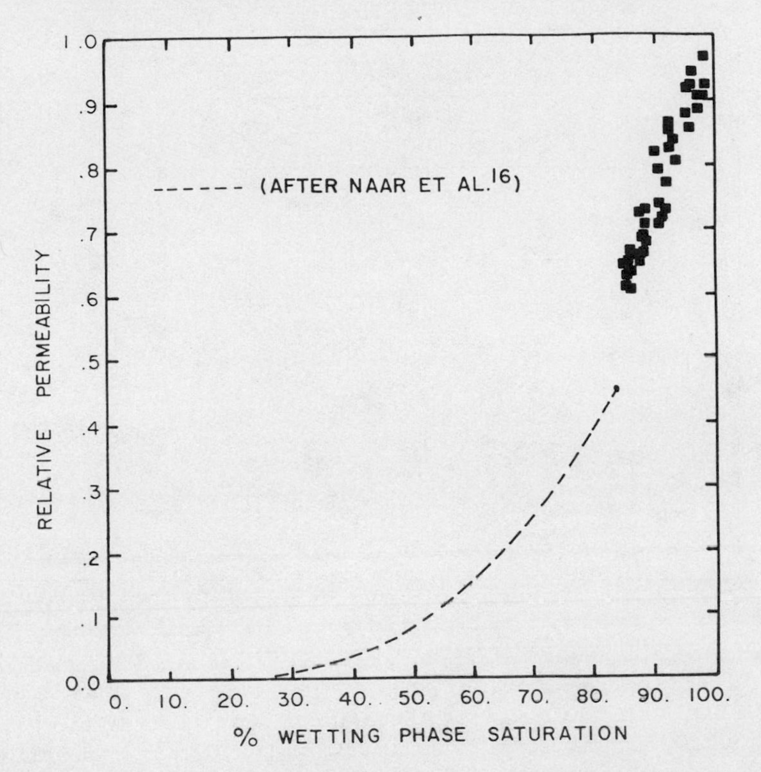

Fig. 2. Relative permeability to the wetting phase at less than normal nonwetting phase residual saturations. (Detailed results given in Table 2.)

and Bond number for trapping, the residual saturation depends on the combined effect of viscous and gravity forces. The plot in Figure 4 shows percent residual saturation versus capillary number, with inverse Bond number greater than 200; at capillary numbers of about 10^{-6} and lower, the trapping mechanism is dominated by capillary forces and the residual saturation is constant.

Correlation of combined effect of gravity and viscous forces: Attention was next turned to correlating the combined effect of gravity and viscous forces. The results presented in Figure 3 were used to obtain the plot shown in Figure 5 of capillary number versus Bond number at given residual saturations. The straight line relationships show that a given saturation can be achieved for a complete range of linear combinations of Bond and capillary numbers. Furthermore, each line in Figure 5 is parallel to any other line given by different choice of saturation. Thus, residual saturation can be correlated by simple linear combination of Bond number and capillary number as shown in Figure 6. If the capillary and Bond numbers of the system are given, the amount of nonwetting phase that is trapped can be predicted for any combination of these two numbers.

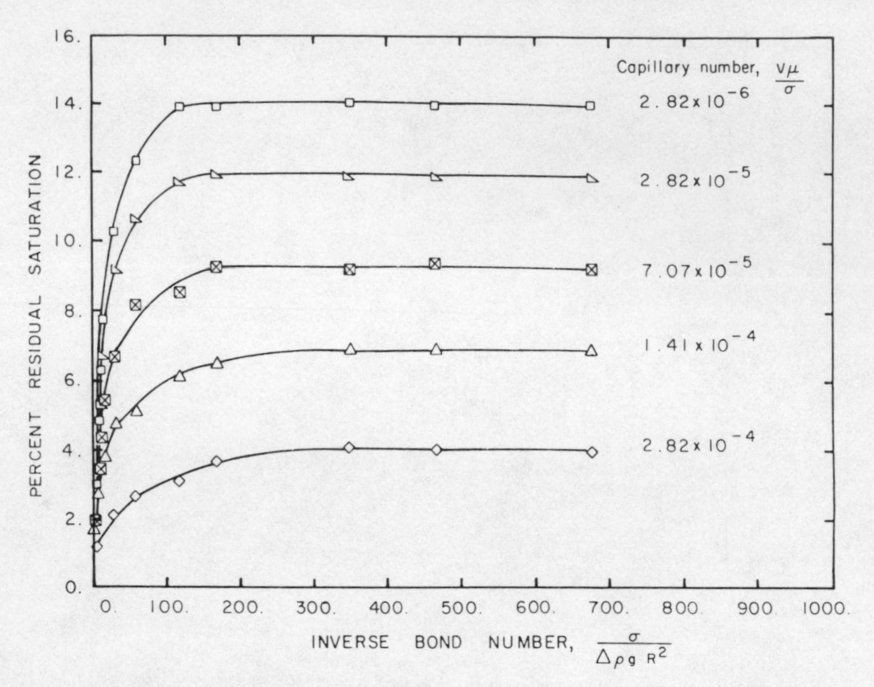

Fig. 3. Trapped residual saturation versus inverse Bond number for a range of capillary numbers.

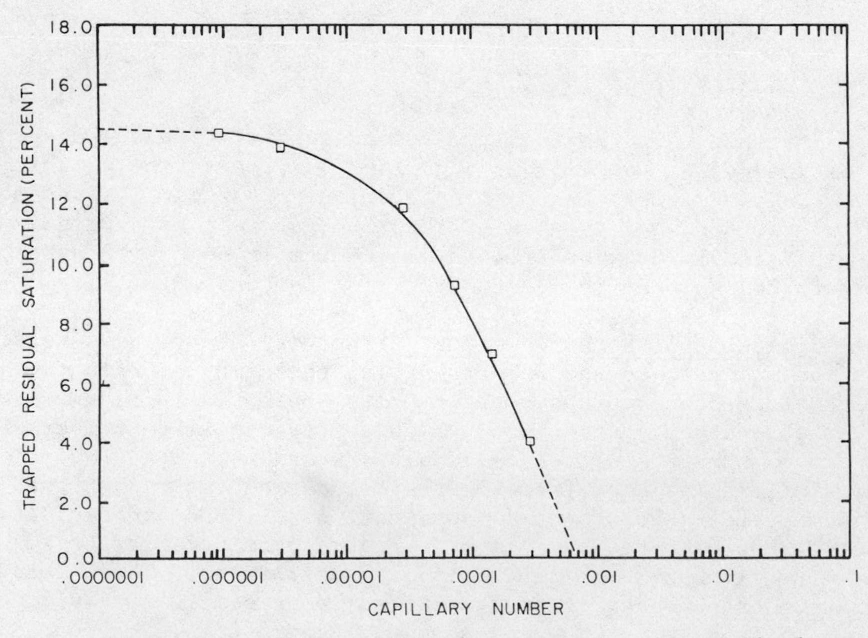

Fig. 4. Effect of capillary number on trapping of residual saturation at Bond numbers greater than 0.00667.

Table 3. Effect of dip angle on residual nonwetting phase saturation at low capillary number (2.82×10^{-6}).

Dip Angle (Degrees)	% Residual Saturation						
	N_B 0.1895	N_B 0.1313	N_B 0.0937	N_B 0.0675	N_B 0.0338	N_B 0.00288	N_B 0.00215
90	2.97	4.75	6.31	7.68	10.20	14.45, 14.60	14.29
60	3.14	5.10	6.79	8.19	10.83	15.04	15.03
45	3.81	5.77	7.51	8.85	11.25	15.05, 14.45	14.88, 14.29
30	4.90	6.93	8.67	9.36	11.67	14.90	14.88
20	5.59	-	9.25	9.87	12.52	15.04	15.03
15	6.40	7.79	9.54	10.45	12.80	15.19	15.18
10	7.49	8.92	9.82	10.89	13.01	-	-
5	9.13	10.25	11.56	11.62	13.64	14.60	14.58, 14.43
0	-	-	-	-	-	14.60, 14.75	14.43

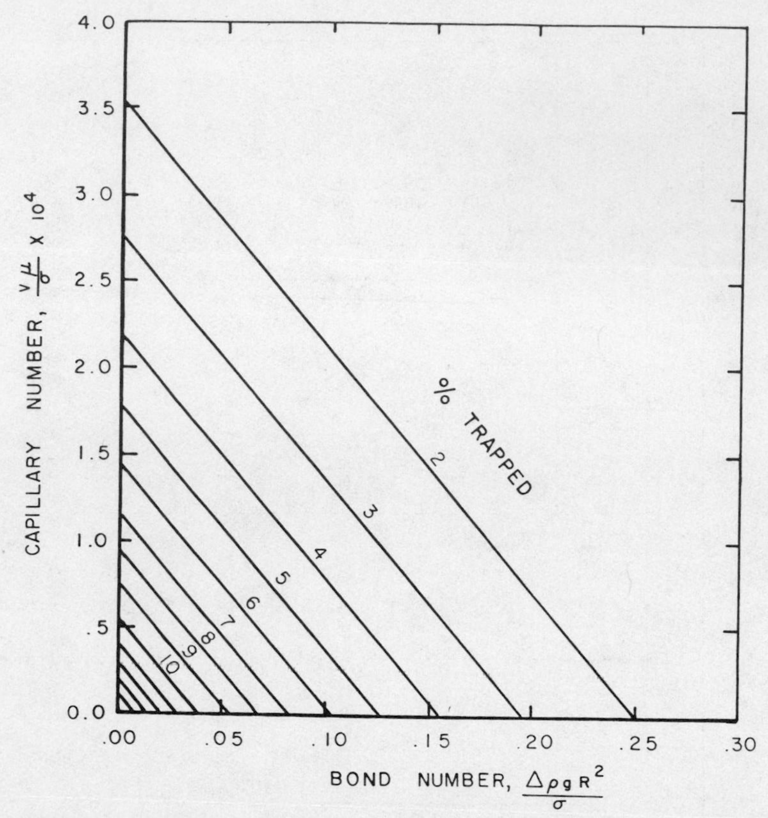

Fig. 5. Plots of capillary number versus Bond number with percent trapped residual saturation as parameter.

Residual saturations at various Bond numbers were also measured using water as the wetting phase and air as the nonwetting phase. The relationships between residual saturation and Bond number were in good agreement with those obtained by using air and oil. A comparison is presented in Figure 7. Recently, similar relationships have also been obtained using aqueous and oleic liquid pairs.

Effect of dip angle on trapping: All of the experiments described so far were for vertical displacement with the more dense (wetting) fluid advancing upwards. When results are sensitive to Bond number, it is reasonable to expect that the amount of trapping will decrease if the angle of dip, α_d, measured from the horizontal is increased. Results for the effect of dip angle are presented in Table 3. They are plotted in Figure 8 as residual saturation versus Bond number with dip angle as parameter.

When a modified Bond number, $N_B \sin\alpha_d$, was used to allow for the effect of dip angle, the reduction in residual saturation

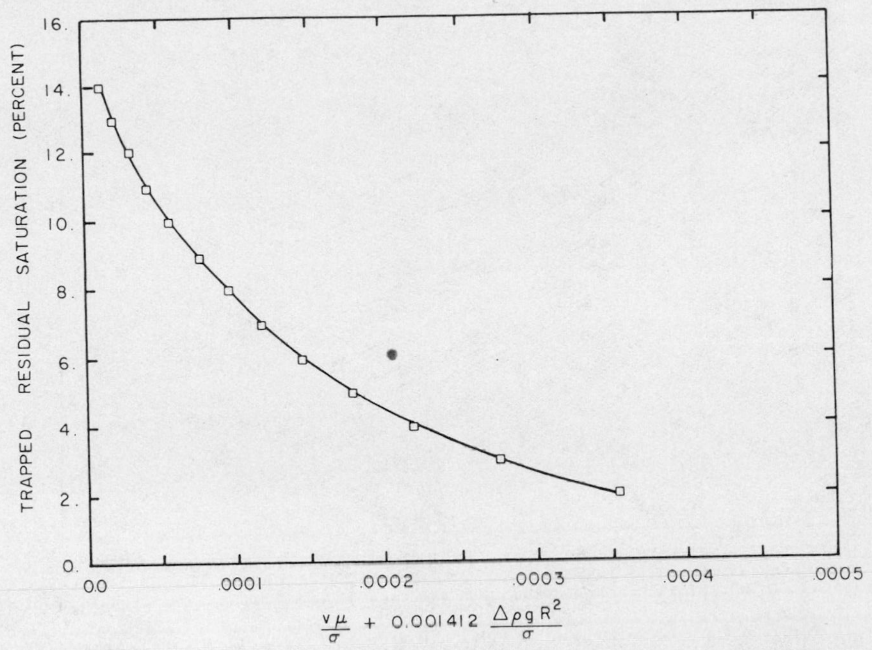

Fig. 6. General correlation of trapped residual saturation with a combined linear function of Bond number and capillary number.

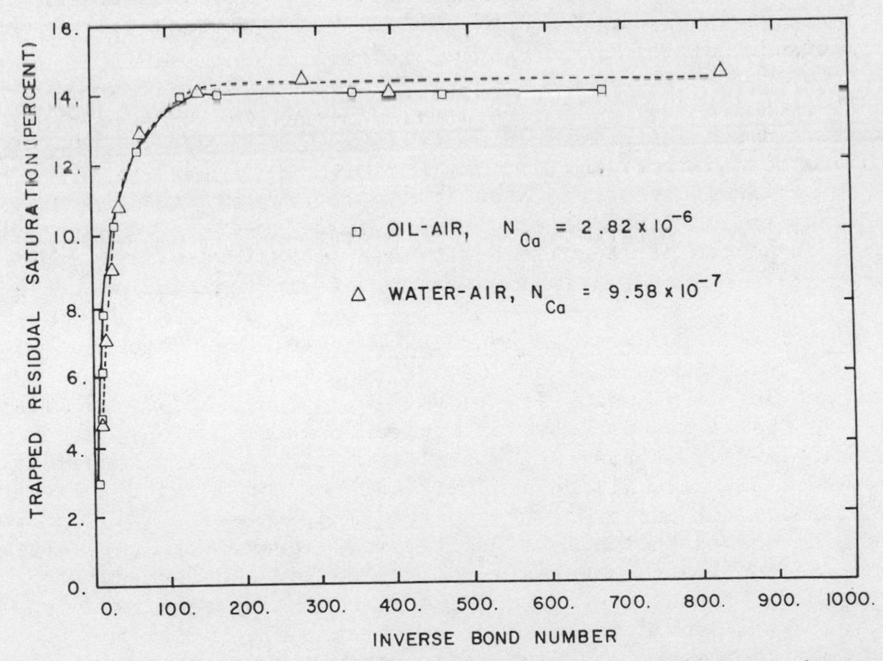

Fig. 7. Comparison of trapped residual saturation versus inverse Bond number for oil-air and water-air as fluid pairs.

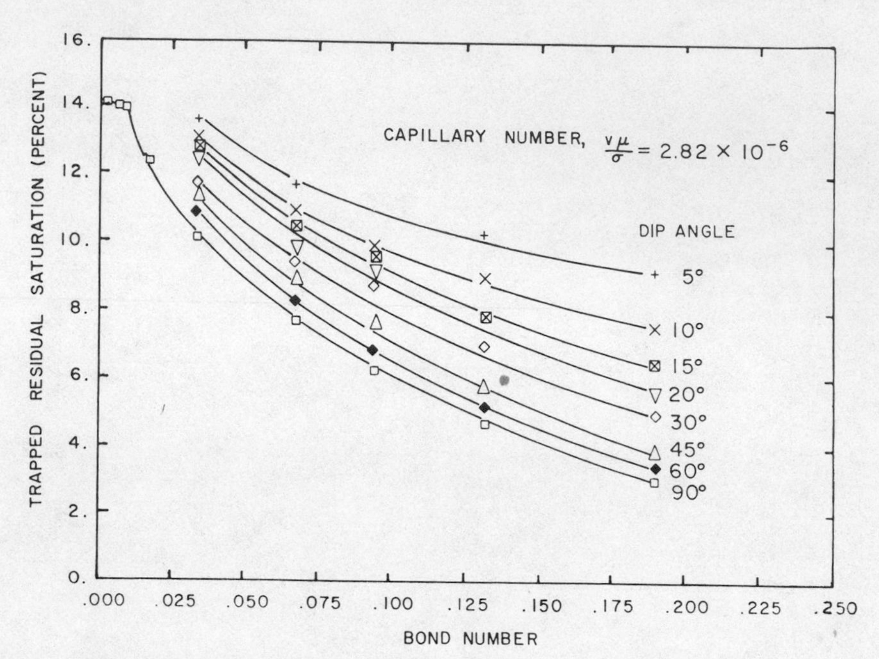

Fig. 8. Effect of Bond number on trapping of residual saturation at different dip angles ($v\mu/\sigma = 2.82 \times 10^{-6}$).

always tended to be higher than predicted from measurements obtained for vertical displacement. Results are plotted in Figure 9 as residual saturation versus ($1/N_B \sin\alpha_d$).

The results for trapping at low capillary and Bond numbers presented in the last two columns of Table 3 were obtained as a check against the possibility that the measured effect of dip angle might be an artifact of the experiment. It is seen that when capillary number and Bond number are well below their critical values so that displacement is dominated by capillary forces, residual saturations are essentially independent of dip angle.

Mobilization of trapped nonwetting phase: Laboratory studies show that trapped residual oil can be recovered if the pressure difference due to viscous flow is sufficient to overcome capillary forces so that trapped blobs, or at least parts of blobs, are mobilized (2-4). If the ratio of viscous to capillary forces is raised sufficiently, almost complete recovery of residual oil can be achieved. For packs of very coarse sand, Leverett (8) reported data which showed indications of slightly improved recovery efficiency for capillary numbers (in this case, of the form $\mu v/\sigma\phi$) as low as 10^{-5}. In the present work, some of the systems in which trapping was measured were later subjected to high flow rate in an attempt to measure the capillary number required for mobilization of trapped nonwetting phase so that results could be compared with capillary numbers for trapping.

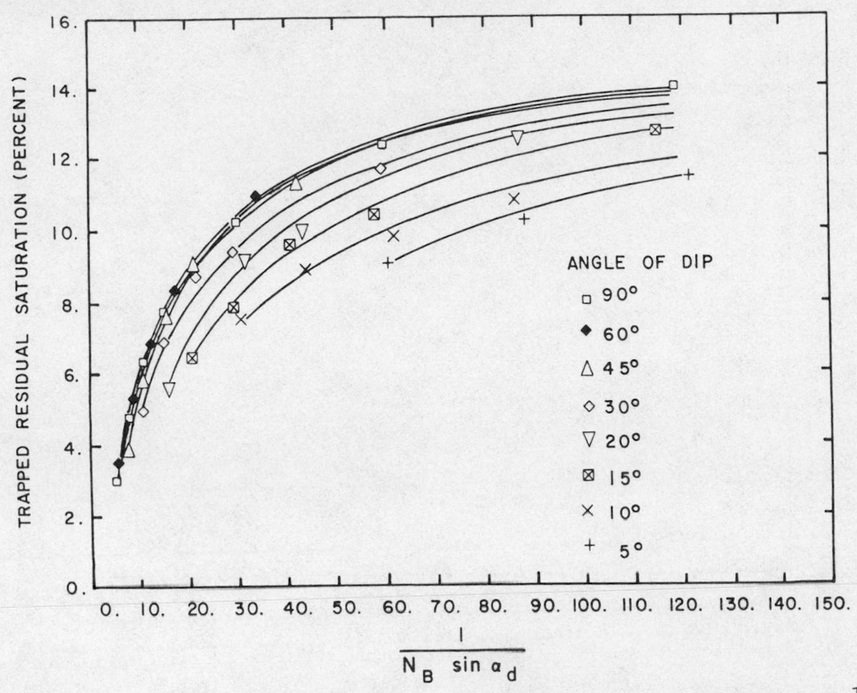

Fig. 9. Trapped residual saturation versus $[N_B \sin \alpha_d]^{-1}$

The amount of injected fluid (number of pore volumes) was found to have significant effect on the residual saturation (see Figure 10). It was apparent that in these experiments, results were strongly influenced by gas passing into solution because of the pressure gradient in the column. During the experiment, the residual gas appeared to develop a shock front. Behind the front, the gas saturation was very low. Ahead of the front, the gas saturation appeared to be uniform and at least as high as the normal residual values. Consideration of pressure gradients and gas solubilities shows that solution effects can account for the observed results. The severe solution effects encountered in an attempt to mobilize trapped gas by increasing the capillary number are in distinct contrast to trapping behavior where solution effects proved to be insignificant. For example, in the experiments on trapping saturations appeared uniform throughout the column and relative permeabilities at reduced residual saturations were independent of time and flow rate.

The observed solution effects present a distinct limitation of using gas-oil systems to model mobilization in completely wetted oil-water systems. However, the mobilization results obtained do at least provide a lower limit to the capillary number required for mobilizing residual fluid. The residual saturation remaining after injection of one pore volume was used to estimate a lower bound for

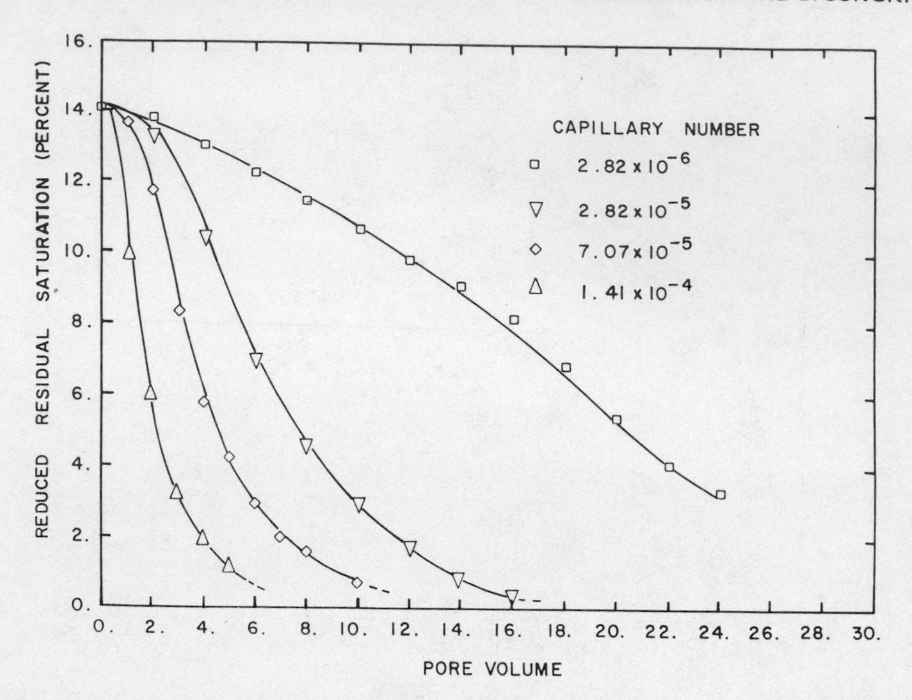

Fig. 10. Effect of amount of injected fluid on reduced residual saturation at various capillary numbers.

the capillary number relationship for mobilization of residuals in sphere packs.

Oil recovery from a swept region can be expressed in terms of the ratio of oil removed to that originally in place (3). This ratio is the microscopic displacement efficiency, E_m, defined for normal nonwetting phase residual saturations, S_{nwr}, as

$$E_m = \frac{1-S_{nwr} - S_{wi}}{1-S_{wi}} \tag{6a}$$

where S_{wi} is the initial wetting phase saturation. When a reduced residual saturation, S_{rr}, is achieved, E_m is defined as

$$E_m = \frac{1-S_{rr} - S_{wi}}{1-S_{wi}} \tag{6b}$$

Results for reduction of residuals after one pore volume injection are plotted in Figure 11 as E_m versus capillary number. Previously reported typical capillary number relationships for trapping of residuals in rocks of narrow and wide pore size distributions are included in Figure 11.

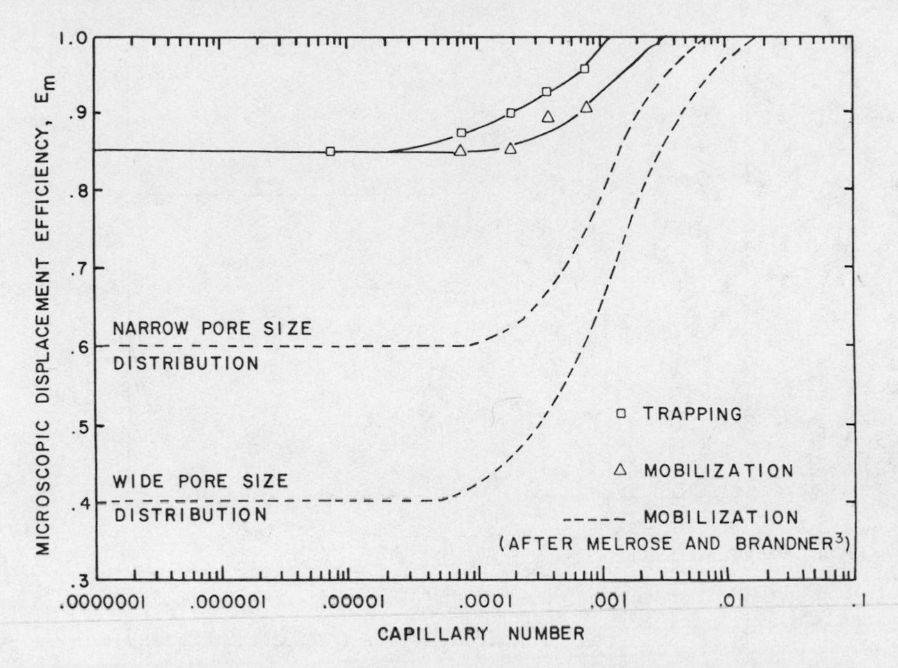

Fig. 11. Plots of microscopic displacement efficiency versus capillary number [$v\mu/\phi\sigma$].

DISCUSSION

Mechanism of trapping: The results obtained for the effect of Bond number on residual saturation provide insight into the mechanism by which trapping occurs and how it can be avoided. Obviously, if the ratio of gravity to capillary forces is very high, for example, a trough containing bowling balls in which air is displaced by water, the interface will be essentially plane and no trapping will occur. However, for the range of Bond numbers of 0.006 to 0.35 over which residual saturation is reduced, the small changes of interface curvature that would occur are apparent from the profiles (9) shown in Figure 12 for Bond numbers of 0, 1, 10, 1000. Local changes in interface shape within individual pores will be small and are not likely to account for the large changes in residual saturation that were measured. It is believed that reduction in residual saturation is due to change in imbibition mechanism caused by small hydrostatic pressure differences. When capillary forces are dominant ($N_C < 10^{-6}$, $N_B < 0.0067$), potential blobs become isolated once an imbibition event occurs whereby the blob separates from the main body of continuous fluid as illustrated in Figure 13. Each blob will have a number of pore openings across which an imbibition capillary pressure is maintained. With increase in Bond number, the tendency for imbibition to occur into the lower region of the blob before the upper region is increased

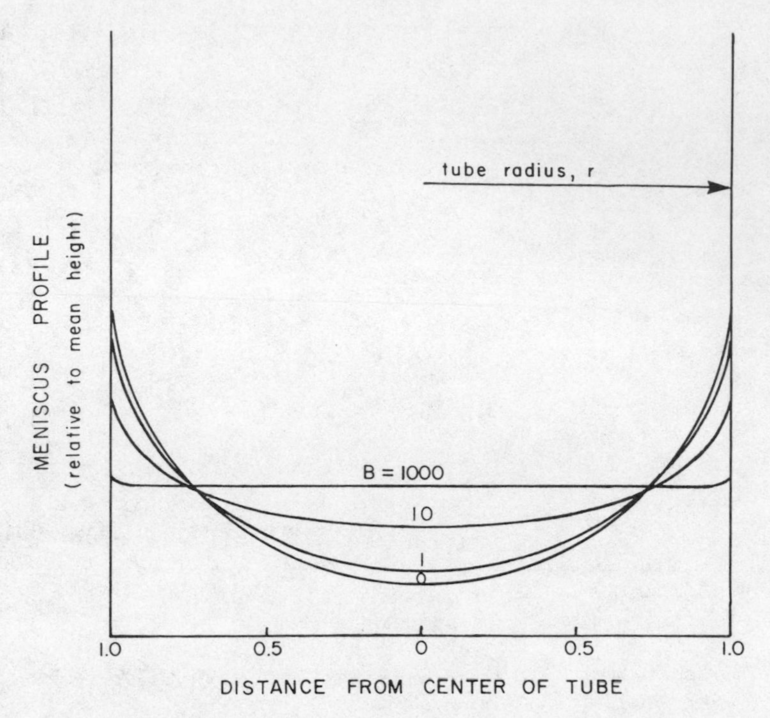

Fig. 12. Interface profile at various Bond numbers (based on tube radius, r) [after Concus (9)].

because the hydrostatic pressure between the regions is increased. If the supplementary hydrostatic pressure is sufficient to cause imbibition into the lower region first, the potential blob is recovered. A similar mechanism can be expected to apply to reduction of residual saturation by viscous forces except that the required supplement in pressure at the trailing edge of the potential blob is provided by the viscous pressure gradient. The correlation obtained, between residual saturation and a linear combination of Bond number and capillary number, demonstrates that the effects of viscous and gravity forces on supplementation of imbibition pressure are additive for vertical displacement, and points to the equivalence of the effect of gravity and viscous forces on the trapping mechanism. It may be noted that the proposed mechanism is distinct from pore doublet theory (1) which predicts that both the location and, consequently, the magnitude of trapped residual oil changes with viscous pressure gradient. For the proposed mechanism, the space occupied by reduced residual oil saturations will, in general, be a sub-set of the space occupied by residual oil under conditions where capillary forces are dominant.

When displacement was carried out with the packing set at various dip angles, α_d, the reduction in trapping was distinctly greater than that predicted by resolving gravity according to the

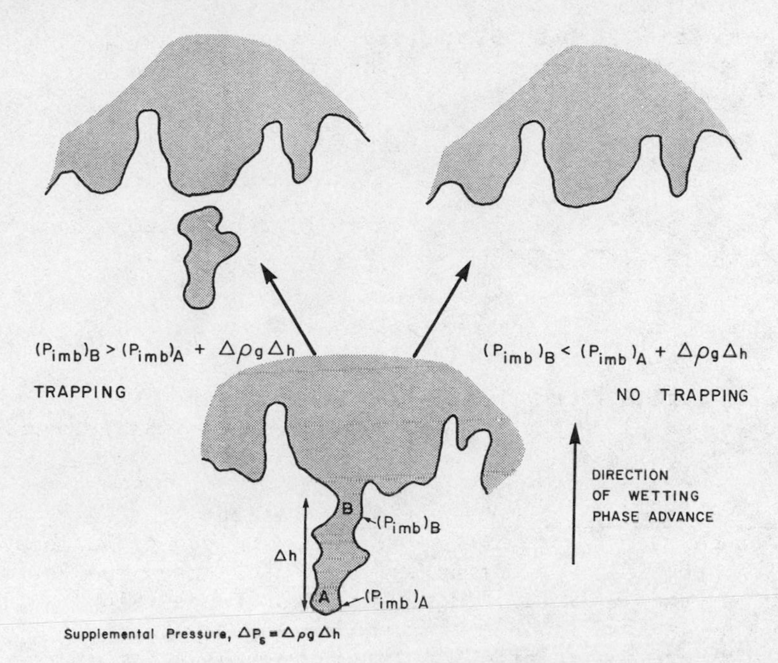

Fig. 13. Illustration of change in trapping mechanism caused by hydrostatic contribution to imbibition capillary pressure.

angle of dip. This is believed to reflect the influence of the three dimensional structure of a potential blob. On a microscopic scale, gravity can supplement imbibition pressures because of local vertical height of the potential blob and this can be sufficient to cause reduction in trapping.

Magnitude of supplemental imbibition pressures: Raimondi and Torcaso (10) showed that entrapment of nonwetting phase tends to occur as the invading wetting phase approaches high saturations. From capillary pressure measurements for sphere packings (11), the imbibition pressure at 70% wetting phase saturation is given by:

$$P_{imb} = \frac{3.9\sigma}{R} \tag{7}$$

Preliminary estimates based on an examination of blobs formed in sphere packings indicate that when residual saturation is reduced to one-half of its normal value of about 14%, the larger blobs will tend to be decreased in size and potential blobs will have maximum lengths of about 7R. The supplementary hydrostatic pressure, ΔP_S, which develops within a potential blob of this length is given by:

$$\Delta P_s = 7\Delta\rho g R \tag{8}$$

Thus, the ratio of the supplementary pressure to imbibition pressure can be expressed in terms of the Bond number as

$$\frac{\Delta P_s}{P_{imb}} = \frac{1.8\Delta\rho g R^2}{\sigma} \tag{9}$$

From the experimental results given in Figure 5, the Bond number for 50% reduction in trapping is 0.08, which gives:

$$\frac{\Delta P_s}{P_{imb}} = 0.14 \tag{10}$$

Thus, if the imbibition pressure at the trailing edge of a potential blob is supplemented by 1/7 of the average imbibition pressure, entrapment is halved.

Comparison of entrapment and mobilization: The foregoing result can be used to provide a comparison of the conditions required for prevention of entrapment with those required for mobilization of trapped liquid. Movement of a trapped blob involves drainage at the leading edge of the blob and imbibition at the rear (see Figure 14). For a completely wetted random sphere pack, the pressure drop required for mobilization, ΔP_m, is given by the difference between drainage and imbibition displacement pressures (3,12,13). At 70% water saturation, these pressures are $6.7\sigma/R$ and $3.9\sigma/R$, giving a value for ΔP_m of $2.8\sigma/R$. The value of ΔP_s is:

$$\Delta P_s = 0.546 \frac{\sigma}{R} \tag{11}$$

Hence, the ratio of ΔP_s to ΔP_m is 0.2. The conclusion that it is about 5 times more difficult to mobilize entrapped fluid than to prevent entrapment is in fair agreement with the experimental results presented in Figure 11. With respect to interpretation of experimental results, it would seem reasonable to expect that the correct relationship for mobilization of residuals in sphere packs will lie somewhere between the sphere pack results, which, because of solution effects represent a lower limit, and the curve obtained for rocks of narrow pore size distribution.

Relative permeability at the flood front: Capillary number relationships are often expressed in terms of pressure gradient, $\Delta P/L$, where ΔP is the pressure drop over a core of length L. Reed and Healy (14) point out that capillary number relationships should be based on the pressure gradient and effective permeability to the displacing phase at the flood front. Because the hydrostatic pressure gradient is constant, the equivalence of hydrostatic and viscous pressure gradients obtained in the present work indicates that the viscous pressure gradient at the flood front is close to constant. This leads to a method of estimating the relative

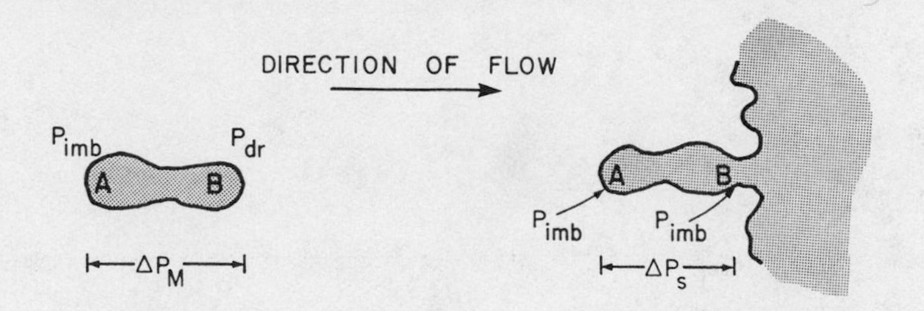

Fig. 14. Conditions required for (a) mobilization of a trapped blob, (b) prevention of entrapment of a potential blob.

permeability of the wetting phase at the flood front, k_{wf}. It is convenient to relate k_{wf} to the fraction, F_d, of the normal residual nonwetting phase saturation, S_{nwr}, that is displaced. F_d is defined as:

$$F_d = \frac{S_{nwr} - S_{rr}}{S_{nwr}} \tag{12}$$

It has been shown that for a given reduction in residual saturation, achieved at a given capillary number, there is a corresponding Bond number which will give the same reduction. At low capillary numbers, recovery is determined by the interplay of gravity and capillary forces, and the displacement process is rate independent. For vertical displacement, the pressure gradient through the flood front $(\Delta P/L)_f$ is then given by

$$\left(\frac{\Delta P}{L}\right)_f = \rho g \tag{13}$$

Application of Darcy's Law to the flood front gives

$$v = \frac{k_{wf}k}{\mu} \left(\frac{\Delta P}{L}\right)_f \tag{14}$$

This relationship can be written as:

$$\left(\frac{v\mu}{\sigma}\right) = k_{wf} \left(\frac{k\rho g}{\sigma}\right) \tag{15a}$$

or from Equations (4) and (5),

$$N_{Ca} = 0.00317 k_{wf} \, (N_B)_{R^2} \tag{15b}$$

Values of N_{Ca} and $N_{B(R^2)}$ for given reductions in residual saturations and calculated values of k_{wf} are presented in Table 4. A plot of k_{wf} vs. F_d is shown in Figure 15 together with relative

Table 4. Capillary numbers, Bond numbers and relative permeability at flood front for given increases in recovery of nonwetting phase.

S_{rr} (% pore volume)	% Reduction of S_{nwr}	N_{Ca}	N_B	k_{wf}
14.20	0	2.82×10^{-6}	0.00667	0.13
12.78	10	1.60×10^{-5}	0.0150	0.34
11.36	20	3.60×10^{-5}	0.0240	0.47
9.94	30	6.00×10^{-5}	0.0378	0.50
8.52	40	9.40×10^{-5}	0.0556	0.53
7.10	50	1.36×10^{-4}	0.0794	0.54
5.68	60	1.95×10^{-4}	0.110	0.56
4.26	70	2.65×10^{-4}	0.150	0.56
2.84	80	3.60×10^{-4}	0.194	0.59

permeabilities determined at reduced residual saturation (see Table 2 and Figure 2), which pertain to flow conditions behind the flood front. The reduced relative permeability at the flood front arises because of interference to flow caused by commingled fingers of displacing and displaced fluids in this region. The ratio of the relative permeability behind the flood front to that at the flood front is inversely proportional to the ratios of the corresponding pressure gradients. A plot of the ratio given by the pressure gradient at the flood front to that behind the front versus the fraction of normal residual nonwetting phase displaced is included in Figure 15. The pressure gradient ratio is seen to decrease sharply with initial increase in recovery, but then remains close to constant for a wide range of further increased recovery.

Comparative effects of viscous and gravity forces: At ordinary waterflood conditions, capillary number is about 10^{-7}. This corresponds to a flow rate of about 1 ft/day for water of one centipoise viscosity and an oil-water interfacial tension of 30 dyne/cm. The capillary number for 50% reduction of residual saturations in sphere packings is close to 10^{-4} (Figure 4), and would require that interfacial tension be reduced to 0.3 dyne/cm for the above flow rate and viscosity. For vertical displacement at very low capillary numbers, the Bond number at which saturation is reduced 50% by buoyancy forces is about 0.08. If the fluid density difference is 0.2, the sphere radius of the packings would be 35

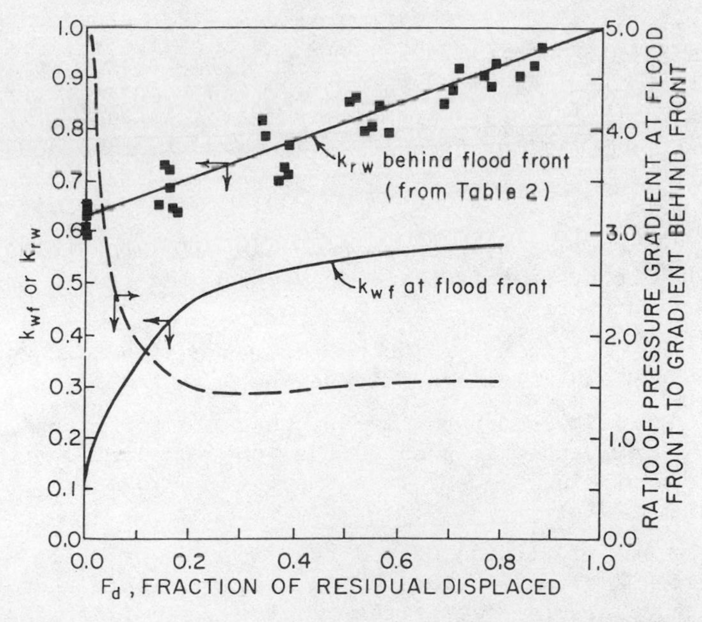

Fig. 15. Relative permeabilities of wetting phase and ratio of pressure gradient at and behind the flood front vs. fraction of residual nonwetting phase displaced.

microns with corresponding pore radii of about 10 microns. Although the permeability (about 4 Darcys) and porosity (about 38%) are high relative to most reservoir rocks, the pore radius of 10 microns is of the same order as that for good producing zones. With respect to the present discussion, pore size is the more relevant parameter. Results for the effect of dip angle (Figure 8) indicate that even for horizontal displacement, buoyancy forces can have a favorable effect on recovery. It follows that in any low tension displacement, the influence of buoyancy forces on the displacement mechanism may well be significant.

CONCLUSIONS

1. Rate independent variation in residual nonwetting phase saturation in bead packs can be achieved by varying Bond number (ratio of gravity to capillary forces) over the range 0.00667, below which there is no effect on trapping, to 0.35, above which there is no trapping.
2. Decrease in residual saturation is ascribed to changes in microscopic displacement caused by differences in hydrostatic pressure; local changes in surface curvature over the above range of Bond number are small and are not likely to have significant effect on the displacement mechanism.
3. Trapping is sensitive to capillary number (ratio of viscous to capillary forces) over the range from 10^{-6}, below which there is no effect on trapping, to 10^{-3}, at which there is no trapping.
4. The amount of trapped nonwetting phase can be correlated with a linear combination of capillary number and Bond number which accounts for the combined effect of gravity and viscous forces on trapping.
5. The space occupied by reduced residual oil saturations will generally be a sub-set of the space occupied by the normal residual saturation.
6. Prevention of trapping is about 5 times easier than mobilization of residual fluid once it has become trapped.
7. The effect of buoyancy forces on prevention of trapping with change in dip angle is greater than that predicted from results for vertical displacement by resolving Bond number according to angle of dip.
8. Relative permeabilities behind the flood front at less than normal residual saturations are generally higher than those predicted by extrapolating the wetting phase imbibition relative permeability curve.

9. Relative permeability to the wetting phase at the flood front is significantly lower than the relative permeability behind the flood front.

ACKNOWLEDGMENTS

The authors are pleased to acknowledge support for this work provided by the United States Department of Energy, Contract No. DE-AC03-ET-3251 and the New Mexico Energy and Minerals Department, Contract No. 77-3301.

REFERENCES

1. T. F. Moore and R. L. Slobod, Prod. Monthly, 20, 20 (1956).
2. J. J. Taber, Soc. Pet. Eng. J., 9, 3 (March 1969).
3. J. C. Melrose and C. F. Brandner, J. Can. Pet. Tech., 13, 42 (1974).
4. W. R. Foster, J. Pet. Tech., 25, 205 (1973).
5. P. C. Carman, Trans. Inst. Chem. Eng., 15, 150 (1937).
6. B. Songkran, M.S. Thesis, New Mexico Tech. (May 1979).
7. J. Naar, R. J. Wygal, and J. H. Henderson, Soc. Pet. Eng. J., 2, 13 (March 1962).
8. M. C. Leverett, Trans. AIME, 132, 149 (1938).
9. P. Concus, J. Fluid Mech., 34, 481 (1968).
10. P. Raimondi and M. A. Torcaso, Soc. Pet. Eng. J., 4, 49 (March 1964).
11. W. B. Haines, J. Agric. Sci., 20, 97 (1930).
12. K. M. Ng, H. T. Davis, and L. E. Scriven, Chem. Eng. Sci., 33, 1009 (1977).
13. N. R. Morrow, J. Can. Pet. Tech., 19, 35 (1979).
14. R. L. Reed and R. N. Healy, in "Improved Oil Recovery by Surfactant and Polymer Flooding," D. O. Shah and R. S. Schechter, eds., Academic Press, New York, 1977.

DEPENDENCE OF RESIDUAL OIL MOBILIZATION ON WETTING AND ROUGHNESS

J. P. Batycky

Petroleum Recovery Institute
3512 33 Street, N.W.
Calgary, Alberta, Canada T2L 2A6

Following a waterflood, the capillary or trapping forces, that limit oil displacement from a porous medium, can be overcome by reducing interfacial tension (σ), by increasing the displacing fluid viscosity (μ), and by accelerating the flooding velocity (V). However, the influence of surface wettability and roughness on the mobilization of residual oil is not adequately understood and it is to this question that the present work is addressed. Experiments were performed to quantitatively relate receding and advancing contact angles to the surface roughness. Roughness was quantified by using a stylus trace of the surface and an integrated centre-line-average (CLA) roughness was employed as a correlating parameter.

It was observed that both receding and advancing surface contact angles generally declined with increasing surface roughness and increasing contact angle. This behavior was characteristic of composite surfaces formed as the result of incomplete drainage of the wetting phase as it receded.

The prior exposure of test surfaces to the wetting phase that produced the composite surface effect would be consistent with the saturation history that is normally imposed on porous media during two-phase displacement studies. Consequently, composite surfaces may indeed occur in porous media.

Mobilization of single ganglia within a regular geometric pack was studied experimentally. The mobilization process was observed to involve expansion and movement of the leading surface through a pore throat and a corresponding movement of the trailing surface as displacement into a downstream pore occurred. The necessary pressure gradients required to induce mobilization would be proportional to the difference between reciprocal radii of curvature that exist at the leading and trailing surfaces of a ganglion, i.e.

$$\Delta P = 2\sigma \left(\frac{1}{r_L} - \frac{1}{r_T}\right)$$

where r_L is the radius of curvature of the leading surface, r_T is the radius of curvature of the trailing surface, and σ is the interfacial tension. For given regular geometries, and by employing above equation with the experimentally determined contact angle behavior, it was possible to predict mobilization pressure drop and its dependence on contact angle and surface roughness. Pressure drop was not predicted to increase with contact angle, a result contrary to studies reported in the literature and from experiments performed here with the sphere-pack model. It would seem reasonable that, by improving the water wetness of a reservoir and thereby decreasing ganglion adhesion through the use of chemical additives, increased oil production would result.

INTRODUCTION

After the completion of waterflooding of an oil-bearing stratum, the oil which remains unrecovered is left behind for two reasons, both of which would appear to be equally significant. Firstly, due to inhomogeneities, portions of the reservoir would not have been contacted by the waterflood. This lack of penetration can as a result be accounted for by what are known as poor sweep efficiencies (1). Secondly, within the swept zones and, if sweep efficiencies can be improved in previously unswept zones, significant saturations of oil also remain trapped due to inefficiencies in the displacement mechanism (2). While it can be considered speculative as to whether this latter oil becomes trapped as a result of heterogeneities on a more microscopic scale, and/or unfavorable levels of capillary forces causing snap-off, it is important to examine requirements that are necessary for mobilizing and thus recovering this immobilized oil.

To address this problem, two types of experiments are normally envisaged--displacement studies utilizing reservoir-type porous media and fundamental studies aimed at quantifying the relevant parameters that incorporate idealized geometries. Both types of studies have obvious disadvantages. Results from displacement work using porous media can sometimes require high levels of interpretation and this can invalidate some of the findings. Examples of displacement studies in porous media include those performed by Moore and Slobod (3), McCaffery and Bennion (4), Lefebvre du Prey (5), Abrams and Taber (7). On the other hand, idealized studies such as those of Batycky and Singhal (8), Morrow (9,10), Oh and

Slattery (11), Payatakes et al. (12), or Ng et al. (13), can be criticized for not adequately reflecting reservoir rock properties and the realities associated with fluid movement within porous media. However, only when the exact nature of the mobilization process is understood can the possibility of new technologies for oil recovery be assessed.

A mathematical model which can predict mobilization of entrapped ganglia under the influence of both gravity and flow has been recently developed (8). The work, which was based on suitable force balances and included the effects of contact angle, was supported by measurements performed using a fluid-fluid-solid system and a simple cubic geometry. This model differed from that of Ng et al., in that their analysis of "blob" movement was semi-quantitative for what was assumed to be a uniformly wetted system (13). Ganglion mobilization experiments which included various intrinsic or smooth surface contact angles were conducted by Morrow, but were not quantitatively assessed (9).

Using the previously developed model, it is possible to predict mobilization criteria from knowledge of the surface properties--interfacial tension and contact angles--for a specified geometry. Based on theoretical aspects of that study it was confirmed that for strongly non-wetting residuals, one suitable criterion for correlating mobilization of residual oil was the capillary number in the form proposed by Melrose and Brandner (2), i.e.,

$$N_{ca} = \frac{V\mu}{\sigma\phi} , \qquad (1)$$

a dimensionless quantity which does not depend on rock permeability (or equivalently on pore throat size). The parametric dependence of oil recovery on pore size distribution was found to be directly related to the contrast between pore throat and pore body dimensions which are determined by the local packing geometry and the degree of cementation. It was also shown that for a given geometry, there exists a theoretical maximum in capillary number which ensures complete mobilization of the oil.

Kinetic surface properties such as interfacial viscosity have been suggested as perhaps playing a significant role in mobilization (14,15). Based on the model it has been possible to assess the very minor role which this property can assume in aiding mobilization of residuals by immiscible displacement.

One aspect of the model which remained poorly understood and which is explored in the present work is that related to interpreting the effects of wettability as reflected by contact angle measurements and the related influences which surface texture or surface roughness can exert on contact angles.

The dependence of mobilization upon two types of surface texture were considered in the earlier study. For a smooth surface which displayed no hysteresis in contact angle, i.e.,

$$\theta = \theta_R = \theta_A = \theta_o \tag{2}$$

the maximum force necessary to obtain mobilization within any given geometry was shown to decrease with increasing contact angle. On the other hand, the effects of type III surface roughness as defined by Morrow (10) were found to produce modest increases in this same mobilization force with increasing contact angle. For type III roughness as pictured in Figure 1, unique values of advancing and receding contact angle are determined for any value of the intrinsic or smooth surface contact angle. Before proposing tertiary recovery schemes which may include wettability alteration, an attempt to understand these divergent results is essential. While type III roughness represented the maximum roughness which was attainable inside TFE capillary tubes, the roughness was not quantified. There is also no certainty that the results obtained from capillary measurements can adequately reflect displacement phenomena in even idealized pore geometries. An attempt has therefore been made in the present study to quantify actual surface textures.

MOBILIZATION THEORY

Macroscopic droplets of oil (the less wetting phase) are contained within one or more interconnected pore bodies that are formed by solid phases of the porous medium. These immobilized portions of the oil phase, which can be referred to as ganglia or blobs, are surrounded with continuously connected reservoir brine (the more wetting phase). At the three-phase contact lines which are formed on the solid surfaces where the two immiscible liquids meet, the apparent contact angles reflect the relative affinities of the three phases for each other, surface textures, surface compositions, and solid surface saturation histories.

It is assumed that a porous medium can be adequately represented by a packing of nearly spherical particles, the radii of which can be suitably modified to reflect solid surface buildup as a consequence of cementation or surface removal due to dissolution, etc. While the geometric configuration of particles is random reflecting a distribution of sizes, it suits our purpose to consider a single particle size arranged in one configuration. The packing arrangement to be emphasized is that described previously by Morrow (9)--a cubical array of spheres as shown in Figure 2. An entrapped ganglion is then contained in a well-defined geometry between the spheres and an overlying plate. For the multilobed ganglion pictured, a quantitative analysis has been

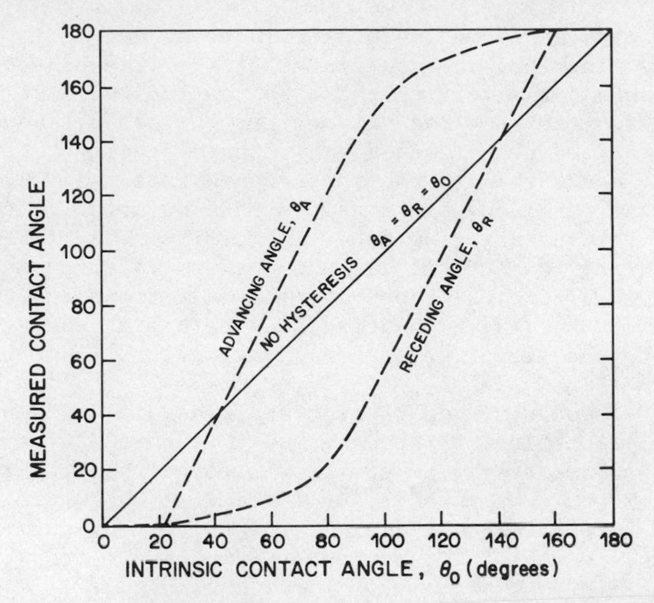

Fig. 1. Type III contact angle hysteresis (10).

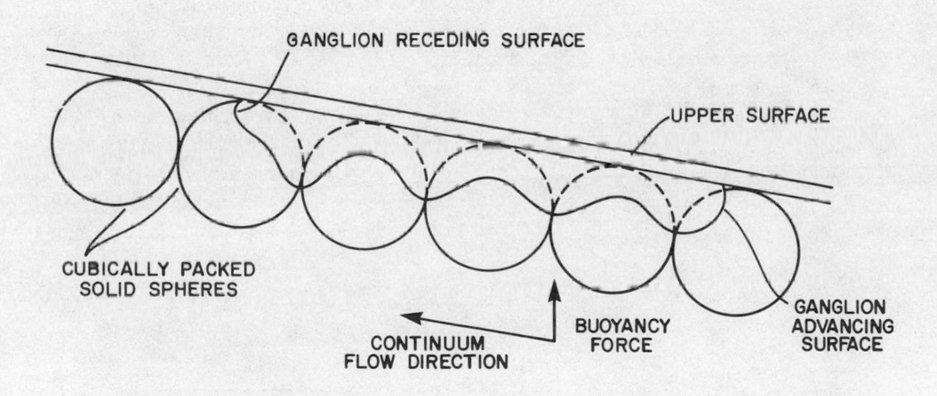

Fig. 2. A ganglion entrapped in simple geometry.

performed and it is in agreement with the proposed theory within experimental error at low contact angles. Such an analysis would not be straightforward when using a random geometry such as that employed by Ng et al. (13). In Figure 2, the ganglion occupies the pore bodies pictured and is interconnected through adjacent pore throats. Depending on the direction of flow or gravity forces with respect to the geometry, the ganglion can be made to move through any of the pore throats. However, for simplicity it will be considered that the forces operate so as to permit movement to the left.

The entrapped ganglion is considered to be non-wetting or rather less wetting than the continuous phase. The contact angle is measured through the wetting phase so that at the pore throat on the left which the ganglion will enter, the contact angle that results is the receding contact angle with respect to the wetting phase, θ_R, and hence this is the receding surface. At the other extremity of the ganglion, there exists and advancing contact angle, θ_A, at the advancing surface, again with respect to the continuous wetting phase. All surfaces deform in response to the imposed pressure gradients. However, the only two which result in ganglion movement are the two surfaces which are at the extremities with respect to the direction of mobilization.

Mobilization occurs when the imposed pressure gradient is sufficient to cause flow of the ganglion through the array to the left. For the geometry given here it was found that regardless of the number of lobes on the ganglion, movement of both the receding and advancing surfaces occurred simultaneously. By assuming that all curvatures are spherical the necessary pressure drop can be given by (2),

$$\Delta P = 2\sigma\left(\frac{1}{r_R} - \frac{1}{r_A}\right) \tag{3}$$

This equation specifies that the mobilization pressure gradient is independent of the number of ganglion lobes and is only sensitive to the two end surface radii. The range of these curvatures can, in turn, be determined by pore size and surface tension. By changing the ganglion volume it is possible to obtain a maximum in mobilization pressure which corresponds to the minimum receding radius at the pore throat, $(r_R)_{min}$, and a maximum advancing radius within the pore body, $(r_A)_{max}$. For any ganglion size, as indicated by the number of lobes,

$$\Delta P_{max} = 2\sigma\left(\frac{1}{(r_R)_{min}} - \frac{1}{(r_A)_{max}}\right) \tag{4}$$

By assuming spherical geometry it is possible, knowing the solid surface contact angles to determine suitable radii of curvature, i.e.

$$r = R \cdot \frac{\sin\varepsilon\sqrt{\cos^2\theta_1\sin^2\varepsilon + 2\cos\theta_1\cos\theta_2 + 1 - \cos\theta_1\sin^2\varepsilon - \cos\theta_2}}{\sin^2\varepsilon - \cos^2\theta_2} \tag{5}$$

The angle ε from the horizontal, centered at the point where two spheres contact each other, permits determination of the center of the radius of curvature of any surface being considered. Limits on ε correspond to limits on the radii of curvature.

Within the pore throat at $\varepsilon = \pi/2$,

$$(r_R)_{min} = R \cdot \frac{\sqrt{\cos^2\theta_1 + 2\cos\theta_1 \cos\theta_2 + 1} - \cos\theta_1 - \cos\theta_2}{\sin^2\theta_2} \quad (6a)$$

and in the pore body,

$$(r_A)_{max} = R \cdot \frac{\sqrt{(\cos\theta_1 + \cos\theta_2)^2 + 2\sin^2\theta_2} - \cos\theta_1 - \cos\theta_2}{\sin^2\theta_2} \quad (6b)$$

For the geometry pictured in Figure 2, θ_1 is the contact angle occurring on a sphere surface and θ_2 is the contact angle at the overlying plate. When $\theta_2 = \pi/2$, the geometry becomes a simple cubic packing of spheres. The values of θ_1 and θ_2 applied in equation (6) must reflect contact angle hysteresis and the direction of three-phase line movement.

The pressure drop that acts to deform and possibly mobilize an entrapped ganglion does so across planes that are located within each surface and which are aligned perpendicularly to the direction of movement which the locus of each center of curvature follows. These planes, called characteristic planes, are located with respect to each origin of the radius of curvature and the angle θ_c, the characteristic angle, as in Figure 3. The receding surface angle is given by θ_{cR} and that for the advancing surface is θ_{cA}. When there is no contact angle hysteresis, the two angles are equal. For a hemisphere in a uniform velocity field $\theta_c = 30°$. The applied pressure drop must operate across these panes which are separated by the characteristic length. At the maximum mobilization condition, this would be:

$$L_c = (r_R)_{min} \sin\theta_{cR} + (r_A)_{max} \sin\theta_{cA} + (2n-1)R \quad (7)$$

where n is the number of ganglion lobes. The pressure drop requirement which is specified by interfacial tension, surface composition, geometry, and sphere radius in equation (3) can be obtained from viscous and buoyancy forces but is reduced by the energy dissipation necessary to maintain internal circulation, i.e.:

$$\Delta P = \Delta P_F + F_V + \Delta P_B - \Delta P_i \quad (8)$$

For the maximum pressure drop, as specified by the geometry of the ganglion ends, i.e., $\Delta P_{max} = \Delta P$, the hydrostatic pressure drop (as measured normal to the flow direction) is evaluated by:

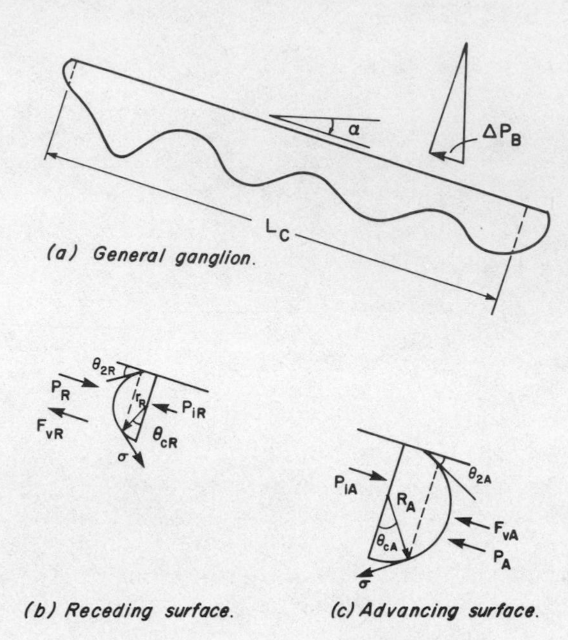

Fig. 3. Force balanced schematic

$$\Delta P_F = C_1 \mu q \sum_{L_c} \frac{\Delta x_i}{(A_{av} R_h^2)_i} \tag{9a}$$

The term representing the distortion in velocity profile due to the presence of the ganglion between the extremities and the characteristic planes is added in equation (8) and is given by:

$$F_V = 3\mu q \left[\frac{(1-\sin\theta_{cR}) C_2}{(r_R)_{min} \cos^2\theta_{cR} (A_{av})_R} \left\{ \frac{2\mu + (r_R)_{min} \cdot f}{3\mu + (r_R)_{min} \cdot f} \right\} + \frac{(1-\sin\theta_{cA}) C_3}{(r_A)_{max} \cos^2\theta_{cA} (A_{av})_A} \left\{ \frac{2\mu + (r_A)_{max} \cdot f}{3\mu + (r_A)_{max} \cdot f} \right\} \right] . \tag{9b}$$

The hydrostatic pressure drop due to the presence of gravity forces is:

$$\Delta P_V = \rho g [L_c \sin\alpha + \{(r_A)_{max} \cos\theta_{2A} - (r_R)_{min} \cos\theta_{2R}\} \cos\alpha] \tag{9c}$$

where α represents the inclination of a layer of spheres from the horizontal. For the experimental situation to be included here, the ganglion fluid was considered inviscid (air), hence no description of ΔP_i was pursued and it was taken to be zero.

SURFACE ROUGHNESS AND CONTACT ANGLE

The relative affinity of two immiscible liquids for a solid surface at the three-phase contact line is given for a clean, smooth, flat plate by Young's Equation (16) as,

$$\sigma^{ba} \cos\theta = \sigma^{sa} - \sigma^{sb} . \tag{10}$$

Because of the inherent restrictions imposed, this expression is rarely representative of actual behavior. Instead, a range of contact angle behavior is possible and hysteresis in the measured contact angle occurs depending on whether the three phase line has advanced or receded with respect to the phase through which θ is measured.

Hysteresis in static contact angle arises because of the presence of surface roughness and surface impurities or heterogeneities. A useful model consisting of sinusoidally-shaped concentric grooves on a flat surface has been used to account for observed hysteresis behavior (17). Chemical heterogeneity can also be accommodated by considering concentric rings of different materials. Both heterogeneity and surface roughness are normally considered to be important causes of hysteresis when considering the surface textures and minerology of porous media (18,19).

In considering a roughness model which is composed of concentric rings (17) or other shapes (20), it is assumed that the local contact angle is in fact the intrinsic contact angle that would be measurable on a smooth surface. Deviation in the observed value is then a result of variation in the angle that is formed by the undulating surface on a microscopic scale and the plane of the surface when viewed macroscopically. The model proposed by Johnson and Dettre (17), who used spherical bubbles, accounted for the movement of the three-phase contact line during increases or decreases in drop volume. They postulated that the preferred configurations were those that were stable or metastable according to the evidence of minima or plateaus in the Helmholtz free energy. In the Johnson and Dettre model, the radius of curvature of the droplet was taken to be much greater than the distance between the sinusoidal peaks representing roughness. Wenzel's roughness ratio, defined by

$$r^* = \frac{A \text{ actual}}{A \text{ observed}} , \tag{11}$$

was in turn related to the ratio of peak height to peak frequency for the hypothetical surface. It was determined that for a simple surface, deviation of both receding and advancing contact angles from the intrinsic angle increased initially with surface roughness as in Figure 4.

Subsequent increases in roughness were shown to favor the formation of composite surfaces and are indicated in Figure 4 to start forming at a roughness near that which corresponds to the maximum in advancing surface contact angle. Composite surfaces are created when the more wetting phase does not totally drain from the surface region occupied by the less wetting phase and thus occupies recesses present on the roughened surface. The formation of these types of surfaces is favored by the occurrence of lower free energy minima than would be possible with a simple surface. Also, depending on the saturation history of the surface, the non-draining wetting phase can itself trap elements of non-wetting fluid. The behavior pictured in Figure 4 illustrates composite surface formation as reported experimentally by Dettre and Johnson (21) and by Bartell and Shepard (22).

As mentioned earlier, work illustrating the dependence of contact angle hysteresis on roughness has also been performed by Morrow (10). Measurements were performed using capillary rise experiments in TFE tubes which had been roughened internally by rolling the tubes while they were filled with various mesh sizes of crushed dolomite. The range of roughness extended from smooth (Type I), through slightly roughened (Type II), to a maximum roughness (Type III) which resulted in the hysteresis pictured in Figure 1. No differences in hysteresis were perceived by increasing the grit size from 60-80 mesh up to 20-32 mesh. There was also no evidence of composite surface formation when compared with the advancing surface behavior pictured in Figure 4. This would be reasonable if the experimental procedure involved starting with a zero initial

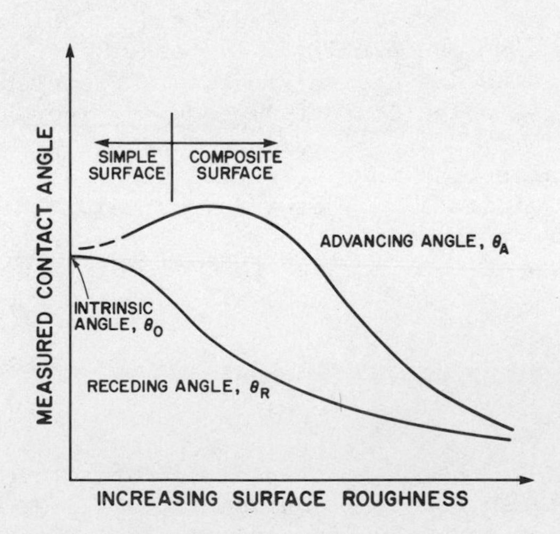

Fig. 4. Dependence of contact angle on surface roughness when composite surfaces occur.

wetting-phase saturation. Then there would be little possibility of trapping non-draining residuals which are required for composite surface formation. Consequently it would seem that, in theory, a range of behavior is possible depending on another aspect of saturation history, i.e., the duration permitted for wetting phase contact and wetting phase removal. Work to be reported here represents an attempt to relate the contact angle hysteresis to a semi-quantitative measure of surface roughness and then to explain the mobilization experiments. In order to better mirror the reservoir situation, experiments are to be conducted slowly and as near equilibrium conditions as is practical.

EXPERIMENTAL

Three types of experiments were performed at room temperature (21° ± 1°C). They included mobilization of entrapped ganglia contained within the cubic array as in Figure 1, measurement of contact angle hysteresis behavior using both flat plates and spheres, and a determination of surface roughness for spheres and plates. In all studies, air formed the entrapped or bubble phase and one of two liquids, a mineral oil (Oil) or ethylene glycol (ETG), with properties as listed in Table 1, were used as continuous phases. The three solid surfaces employed allowed coverage of a range of contact angles from near 0° to greater than 90°. These low energy surfaces included polytetrafluoroethylene (TFE), acrylic, and nylon with majority of work being performed with the first two.

Table 1. Physical Properties of the Fluids Employed[+]

Fluid	Density (g/cm^3)	Viscosity (mPa•s)	Surface Tension (mN/m)
Oil*	0.8765	171.9	30.6
ETG**	1.108	--	46.4

+ Room Temperature: 21±1°C
* Kaydol (White Mineral Oil) - Witco Chemical
** Ethylene Glycol (certified) - Fisher Scientific Company

Surface tension measurements reported in Table 1 were performed using a du Nouy ring tensiometer. They include appropriate corrections for surface deformation (23). Cleaning of the surfaces was accomplished with alternate washings of isopropanol and methanol.

Contact angle measurements: Contact angle measurements were performed using a sessile drop apparatus. Buoyant bubbles were formed on the underside of the prepared plates or spheres as shown in Figure 5. The air was introduced from a syringe through a hole drilled in the surface being studied. The measurements which were recorded, also indicated in Figure 5, included the height, H, the major diameter, D_m, and the base diameter, D_b. These data were obtained using a Wild M5 stereomicroscope that could be fitted with both a filar micrometer eyepiece and either a polaroid or 35 mm SLR camera. The best reproducibility and sensitivity was obtained using the micrometer eyepiece since the resolution was superior to that obtainable on either the negatives or positives produced using the cameras. The results presented for contact angles of less than 90° were obtained using the eyepiece measurements.

Contact angle and interfacial tension measurements were determined with the aid of a computer program. The program, developed by E. Lefebvre du Prey iteratively solves the Bashforth-Adams equation matching the three dimensions indicated in Figure 5. Sensitivity studies were performed using both low interfacial tension (0.058 mN/m) and high interfacial tension (43.5 mN/m) systems to determine the necessary measurement precision and to obtain estimates of the errors associated with both interfacial tension and contact angle. These results are summarized in Table 2. For the high tension measurements it can be seen that the highest level of precision is required in the measurement of D_m and also that the

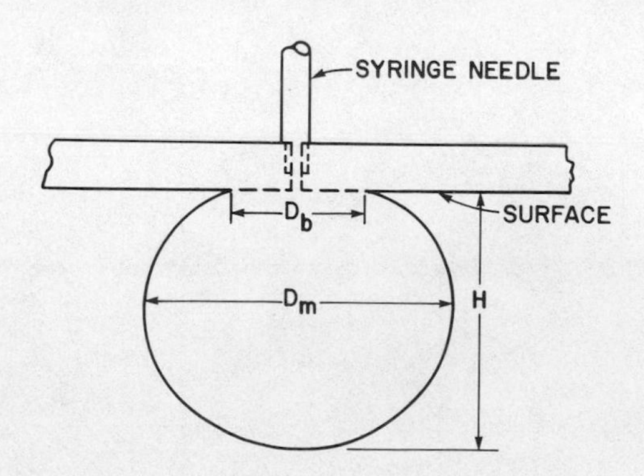

Fig. 5. Sessile drop measurement

Table 2. Sensitivity of the Sessile Drop Measurement

1. High Interfacial Tension Measurements

Variables	Reference Condition	Perturbation in H	Perturbation in D_m	Perturbation in D_b
H (cm)	0.97	± 0.01	--	--
D_m (cm)	1.29	--	± 0.01	--
D_b (cm)	1.04	--	--	± 0.01
σ (mN/m)	43.5	± 9.0 (20.6%)	±10.7 (24.5%)	± 5.3 (12.1%)
θ (Deg.)	50	± 0.7	± 1.2	± 1.0

2. Low Interfacial Tension Measurements

Variables	Reference Condition	Perturbation in H	Perturbation in D_m	Perturbation in D_b
H (cm)	0.37	± 0.01	--	--
D_m (cm)	0.95	--	± 0.01	--
D_b (cm)	0.81	--	--	± 0.01
σ (mN/m)	0.058	± 0.004 (7.7%)	± 0.001 (2%)	± 0.001 (1.4%)
θ (Deg.)	19	± 1.8	± 3.8	± 3.7

contact angle results are relatively insensitive to measurement errors when compared to the sensitivity of the surface tensions. In the contact angle results that are to be presented later the interfacial tensions which were also measured independently were used to infer the accuracy of the contact angle results. With the low interfacial tension system it can be seen that the dependence on the measurement precision of σ is much lower and that of θ is higher than was the case for the high tension system. Consequently, the errors in σ and θ appear to be in the same range. It should also be emphasized that the use of this method of analysis for determining contact angle is more reproducible and hence superior to our attempts at measuring contact angles visually.

In measuring hysteresis, two techniques for approaching the maximum and minimum contact angles were attempted. Once a bubble was formed on a plate it could be expanded or contracted until the base moved and measurements could be taken. Alternatively the bubble volume could be slowly increased or decreased and measurements taken until movement occurred. Both methods produced similar results but, when differences occurred, the greater level of hysteresis was associated with the latter procedure, owing probably to the duration necessary to complete the movement of the wetting phase along the solid surface.

A mild dependence of contact angle on bubble volume was discernible for volumes of less than 0.07 cm^3 perhaps due to the relative size of surface texture inhomogeneities. However, by constantly forming and measuring the dimension of bubbles from 0.11 to 0.18 cm^3, no such effect could be detected.

Roughened surfaces: Surfaces with a range of textures were prepared by roughening them with various grades of sanding papers using grit sizes extending from 600 up to 320 mesh. Since it was not possible to purchase sufficiently smooth TFE sheet, polishing was necessary to obtain surfaces with low levels of roughness. A circular motion was imparted to all roughening (and polishing) operations so that unidirectional lines that could influence interpretation of the results would be minimized. Polishing was performed either by rubbing the required surface against a glass slide or by extruding the TFE against a suitable template. Attempts at melting the TFE in an oven between glass surfaces were fruitless.

There are many techniques available to aid in the quantification of surface roughness. The method chosen for use here was that of a tracing stylus instrument for which the operation is straightforward. Attendant disadvantages are related to the possibility of leaving a permanent impression on the sample and the lack of resolution associated with the stylus size.

The instrument used was a Rank Organization Talysurf 4 which embodies a diamond stylus of 2.5 μm diameter that exerts a normal force of 100 mg (25). Various traverse distances could be employed to produce a recording of the surface. Quantification in the form of center-line-average (CLA), roughness which is a standard measure of surface texture (ASA B46:1955), could be obtained.

For the surface roughness measurements that were performed using flat plates, the CLA value of roughness is reported. This measurement does not reflect the horizontal spacing of the surface irregularities but it does divide the scan into segments and for each it electronically obtains averages for each linearized sample as illustrated by the shaded portions of Figure 6(a). The CLA value reported is then the average of all samples which for a selected sample length of 0.25 mm would involve 7 samples. The CLA values obtained for the spheres were determined by superimposing a spherical pattern on the recorded traces and manually determining the CLA roughness. The reported roughness levels represent an average of scanning results obtained for mutually perpendicular directions. In checking to see if there was a change in surface texture because of the tracing operation it was found that at least the hysteresis behavior did not change so that further contact angle measurements could, if necessary, be performed.

Mobilization experiments: Bubbles of various specified volumes were introduced to the liquid-filled cubic array of spheres using a gas-tight syringe inserted through a hole in the side of the cell. An injected bubble was permitted to rise so that it occupied the pore volume immediately below the upper cell surface. This surface was always perspex regardless of the sphere material employed. In order that the continuum fluids could be given time to drain from the surfaces, a bubble was allowed to sit statically for 15 to 20 minutes. The mobilization forces were then imposed either by tipping the cell or by causing the continuum fluid to flow so that the bubble would deform in response to the forces. The forces were applied in a stepwise manner every 30 seconds. It was found that the results produced with this time interval were very similar to those using longer times but that the kinetics of fluid movement were noticeable when the size of a time step was lowered to 15 seconds. This would agree with the experience obtained during contact angle measurements. The step size used during buoyancy mobilization for example was 1° for high angles (> 45°) and 1/2° for low inclinations. The reader is referred to the previous study for further discussion of this experimental procedure (8).

RESULTS AND OBSERVATIONS OF CONTACT ANGLE HYSTERESIS

Figures 6(b) to 6(e) illustrate profiles typical of those recorded using the Talysurf instrument. While the size of the hemispherically-shaped stylus reduces the resolution that can be obtained, the instrument provides an excellent response to various levels of roughness thus enabling a comparison of various textures. The smoothest surface used, A1 in Figure 6(c), had a glossy surface that was easily able to reflect images. The smooth TFE surface, T1 in Figure 6(b), could also be described as shiny but it was less so than that of A1. The well-roughened surfaces represented by Figures 6(d) and 6(e) did not, on the other hand, provide any discernible reflection when viewed in a lighted laboratory environment.

CLA roughness values are reported in Table 3 for the various flat surfaces employed for contact angle measurement. The reported CLA values represent averages that were determined electronically by the instrument from a number of scans. Included in Table 4 are CLA's obtained manually for the three types of spheres used in the mobilization experiments. The averaging process was performed by superimposing smooth profiles obtained from steel bearings and then calculating the deviation as pictured in Figure 6(a). For illustration, Figure 7 shows, respectively, photographs of the surfaces of a smooth TFE, a roughened TFE and a smooth nylon sphere.

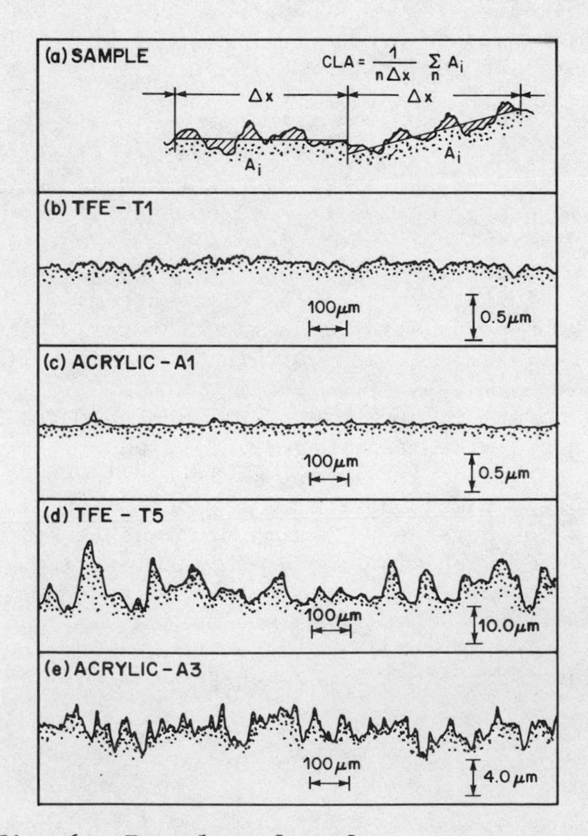

Fig. 6. Examples of surface texture traces

Table 3. Surface Roughness and Contact Angle Hysteresis: Plates

Plate	CLA Roughness μm	Receding Contact Angle θ_R (Minimum, Maximum)	Advancing Contact Angle θ_A (Minimum, Maximum)
A. Air-Acrylic-Oil			
A1	0.017	6.5 (6.2, 7.0)	7.3 (6.5, 9.9)
A2	0.094	7.3 (6.1, 8.4)	6.6 (3.3, 8.7)
A3	1.45	7.7 (7.5, 7.8)	7.2 (6.8, 7.7)
B. Air-Acrylic-ETG			
A1	0.017	23.8 (18.9, 26.9)	42.7 (28.8, 53.9)
A2	0.094	16.0 (6.4, 25.3)	43.3 (30.0, 53.7)
A3	1.45	7.9 (6.8, 9.7)	22.5 (17.0, 26.4)
C. Air-TFE-Oil			
T1	0.058	40.3 (35.3, 44.6)	57.3 (45.8, 60.7)
T3	0.361	27.4 (24.9, 29.9)	50.5 (48.4, 52.7)
T4	1.57	14.5 (7.7, 21.0)	29.8 (14.5, 35.6)
T5	2.21	6.5 (2.4, 17.1)	5.1 (1.8, 10.7)
D. Air-TFE-ETG			
T1	0.058	71.5 (62.0, 75.4)	92.8 (89.8, 94.8)
T2	0.091	69.6 (67.1, 72.2)	87.5 (73.0, 99.3)
T3	0.361	57.0 (53.3, 63.3)	93.7 (93.1, 94.3)
T4	1.57	39.0 (32.1, 48.2)	88.0, 69.8, 44.4 (42.5, 46.0)
T5	2.21	{45.1 (41.6, 58.0) 26.7 (24.2, 28.9)	106.3 (103.7, 109.3)

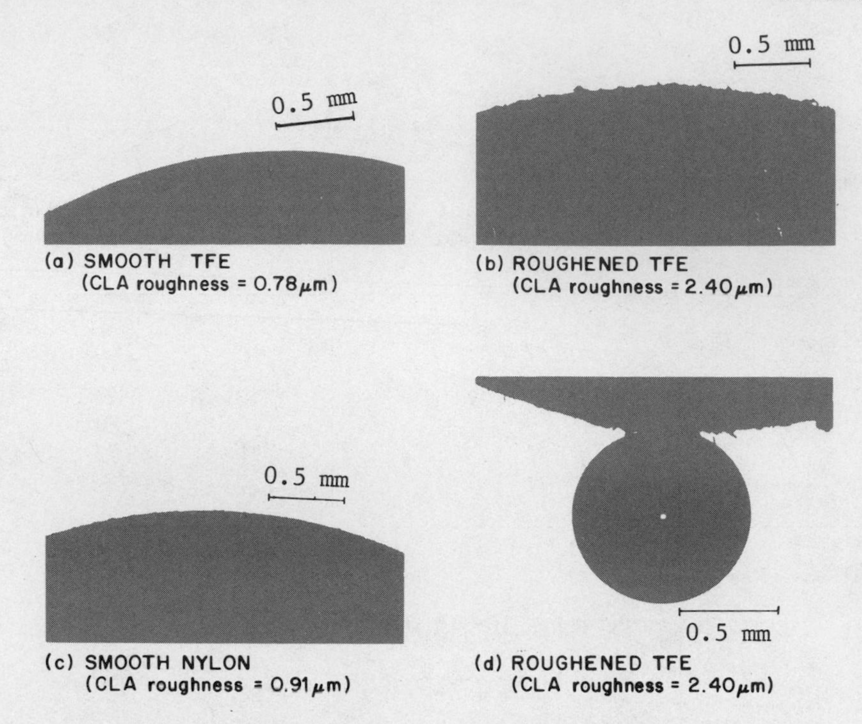

Fig. 7. Photographs of sphere surfaces

Receding and advancing contact angles were measured for the four systems utilized. Four combinations of low surface energy solids and two liquids are listed in Table 4. Figure 8 shows the receding contact angle behavior as measured for each of the combinations as a function of the CLA surface roughness. The relationships pictured are characterized by a decrease in receding contact angle with increasing roughness and are qualitatively reflected in Figure 4.

It was possible to extrapolate the curves to low levels of roughness in order to obtain estimates of the intrinsic contact angle for each of the combinations. The contact angles so determined are presented in Table 4 and range from 7° to 90°. The 90° determination of intrinsic contact angle for air-ETG-TFE is in agreement with the same value obtained by Fox and Zisman (26) who incidentally also experienced difficulties in preparing smooth TFE surfaces.

In Figure 8, for the neutrally-wetted system there is evidence of a widening of the range of possible contact angles and the development of a bimodal functionality for the roughest surfaces investigated. This behavior could have depended on the surface saturation history or on the presence of local surface nonuniformities and was not pursued.

Table 4. Surface Roughness and Contact Angle Hysteresis: Spheres

Sphere*	CLA** Roughness μm	Receding Contact Angle θ_R (Measured)	θ_R (Figure 8)	Advancing Contact Angle θ_A (Measured)	θ_A (Figure 10)
A. Air-Oil-Solid					
N	0.91	9.2	--	17.5	--
RT	2.40	18.5	13.3	26.0	13
T	0.78	19.5	30.7	50	43
A	0.015	--	6.5	--	7.3
B. Air-ETG-Solid					
N	0.91	16	--	18	--
RT	2.40	20	26.5	45	27
T	0.78	57	46.5	~90	60
A	0.015	--	24	--	44

*Abbreviations: N - nylon
T - TFE
RT - roughened TFE
A - acrylic (overlying plate)

**Manually computed.

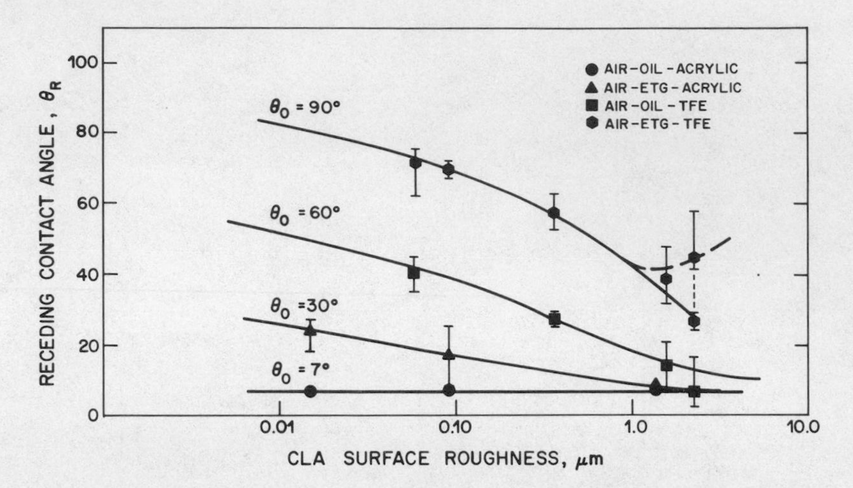

Fig. 8. Effect of surface roughness on receding surface contact angle.

A cross plot of the data from Figure 8 is shown in Figure 9 illustrating the dependence of receding contact angle on intrinsic contact angle for four levels of roughness. For reference there is shown a line with a slope of 45° which shows the behavior expected for a "perfectly" smooth surface displaying no hysteresis. The dashed line represents the contact angle behavior recorded by Morrow for type III roughness. The measurements recorded here indicate that the offset from the 45° line measurable for any surface roughness increases with increasing roughness and that as contact angle increases, the dependence of the receding angle on intrinsic angle becomes parallel with the non-hysteresis line. The line representing type III behavior is qualitatively similar to the CLA value of 2 μm but it also tends, rather than lying parallel to the 45° line, to show a slope characteristic of simple surface behavior. This will be illustrated more fully in Figure 11.

The dependence of advancing surface contact angle on CLA roughness is pictured in Figure 10 for the same systems as those used for receding angle determinations in Figure 8. For the lower surface roughnesses, the measured contact angles are higher than the intrinsic contact angles and it is likely that these surfaces behave as simple surfaces. As roughness is increased, the contact angles decrease--a behavior that is consistent with the formation of composite surfaces as shown in Figure 4. For the extremely rough surfaces, the advancing contact angle declines due to the surface filling with the wetting phase and the angles approach the receding surface angles which in turn appear to go to 0°. An exception to this description exists for the intermediately wetted (θ_0 = 90°) system. Besides displaying the above behavior a wide range of contact angles occur and appear to reflect a behavior

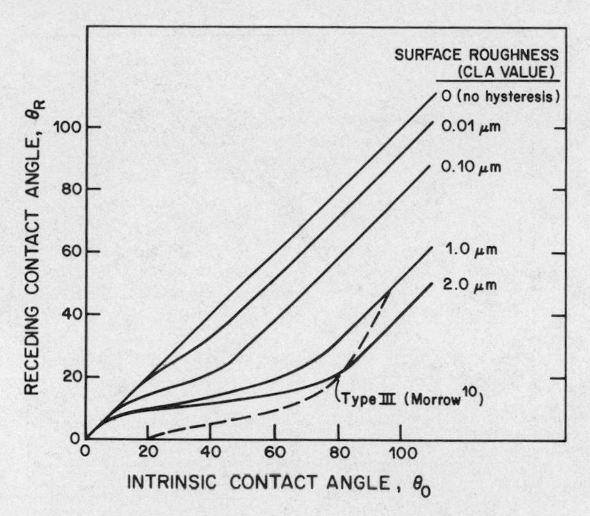

Fig. 9. Dependence of receding contact angle on intrinsic contact angle for various levels of surface roughness.

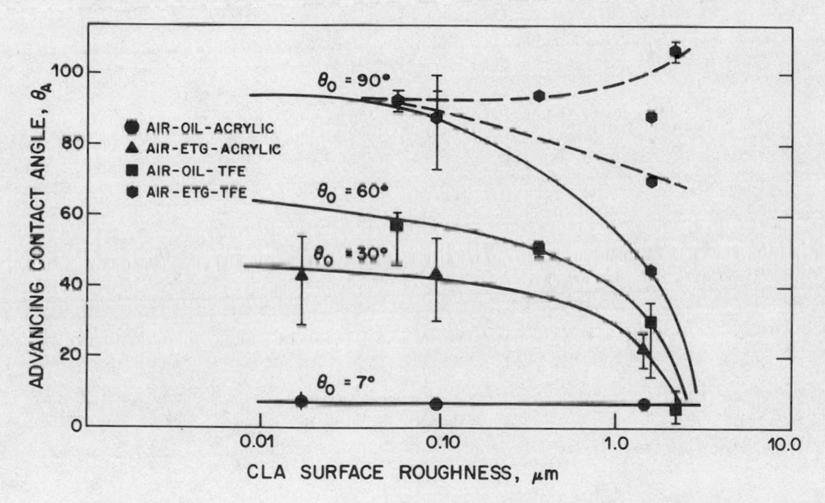

Fig. 10. Effect of surface roughness on advancing surface contact angle.

similar to, but more magnified than, that determined at the receding surface.

Figure 11 represents a cross plot of Figure 10 and shows the dependence of advancing contact angle on surface roughness at four roughness levels. At low intrinsic contact angles and at low levels of roughness the behavior illustrated is parallel to that determined by Morrow on a simple surface. As surface roughness levels increase and at higher contact angles the dependence of

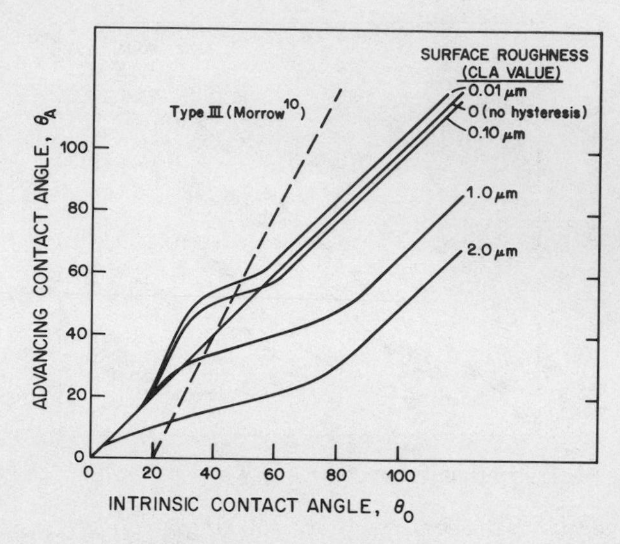

Fig. 11. Dependence of advancing surface contact angle on intrinsic contact angle for various levels of surface roughness.

advancing contact angle on the intrinsic angle begin to parallel the 45° datum and also tends to approach the respective receding angles as shown in Figure 9.

In Figure 7(d) is pictured a bubble in the oil continuum supported by a roughened TFE sphere. It can be suggested for this system, in which the intrinsic contact angle is 60°, that near-zero contact angles result from an inability to drain wetting phase from the sphere surface and composite surface formation is assured. The fluid retained by the surface in this manner illustrates one type of microporosity described by Kieke and Hartmann (19).

The results shown in Figures 8 to 11 appear to be internally consistent and reflect a generalized interpretation of contact angle hysteresis dependence on contact angle and surface roughness. In fact it would appear reasonable to measure contact angle hysteresis and obtain a value of CLA roughness in order to predict effects for different wettabilities. It is recognized that CLA values do not reflect the roughness frequency but only give deviation averages. Preliminary indications do show, for example, that at lower frequencies, both receding and advancing contact angles behave in a manner consistent with lower levels of roughness and as long as it is remembered that CLA values are relative, this is a useful comparison basis.

GANGLIA MOBILIZATION PREDICTIONS

The next step in the analysis is to predict mobilization requirements by coupling the theory presented earlier with the contact angle results of the previous section. The simple cubic geometry will be utilized although the results are broadly applicable to any other geometrical description. Two situations are examined.

In the first, the situation would correspond in porous media to single or multilobed oil globules that had become isolated due to bypassing of the drive water in a manner similar to the doublet analogy presented by Moore and Slobod (3). Mobilization would be characterized by a receding surface at the most downstream pore throat and an advancing surface within the pore body at the other extremum. Here contact angle behavior would be expected to be reflected by the situations presented in Figures 8 to 11 for various surface roughnesses. Mobilization maxima, i.e. $(R\Delta P/\sigma)_{max}$, as computed by combining equations (4), (5), and (6) are given in Figure 12(a) for CLA surface roughness levels depicted in Figures 9 and 11. For comparison, when no prior contact with the wetting phase has occurred, the behavior indicated as Type III for simple surfaces shows increases in mobilization requirements with contact angle.

At low contact angles, hysteresis, due to surface roughness, can cause higher pressure drops than those for ideally smooth spheres. The higher the level of roughness, the larger this deviation becomes, however this tendency becomes muted by the occurrence of composite surfaces which form at successively lower contact angles for increasingly rougher surfaces. The net conclusion to be drawn from examining Figure 12(a) is that there would be peaks in the necessary mobilization force maxima for each of the surface roughness levels investigated. Either increases or decreases in the contact angle, away from the values yielding the peaks, would result in lesser mobilization force requirements.

Another scenario would result when considering the mobilization of small ganglia of oil which had become trapped after flowing temporarily. Stoppage might have occurred due to snap-off or the encountering of a reduced pore throat diameter. In either case the movement of the ganglion would be characterized by the oil phase displacing a previously water-contacted surface and wherever the three-phase contact angle occurred it would have to have been established by the receding contact angle condition. For this reason, ganglion geometry could be defined with both upstream and downstream surfaces being established when the three phase contact line was receding. The corresponding mobilization force behavior would be, depending on roughness and contact angle,

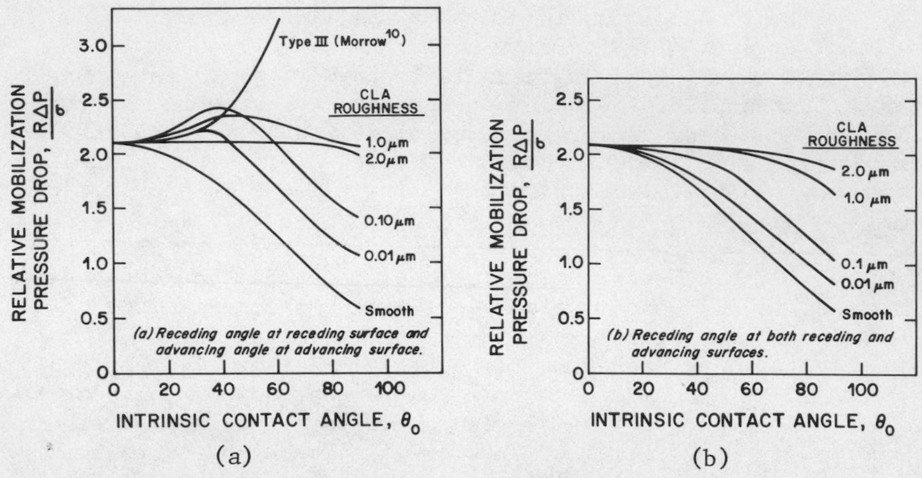

Fig. 12. Dependence of relative maximum mobilization pressure drop on contact angle using equation (4) with roughened surfaces.

predicted as in Figure 12(b). In this case, mobilization would appear to become easier with increasing contact angle. Also, when compared with the results in Figure 12(a), it can be seen that once mobilized, a ganglion might remain in motion even as the pore throat (or R) decreases in radius. It remains now to explore the mobilization behavior obtained experimentally in order to evaluate the applicability of the foregoing predictions. It is important to note that the predicted results which show decreases in the mobilization force with contact angle do not reflect core displacement results obtained by Lefebvre du Prey (5).

GANGLIA MOBILIZATION RESULTS

Contact angle measurements, as listed in Table 4, were determined for three types of spheres--TFE, nylon, and roughened TFE, an upper acrylic plate and two fluid pairs of air--oil and air--ETG. The sphere contact angles to be used in mobilization calculations correspond to those which were determined using sessile drop measurements and the actual spheres. The contact angles reported for the acrylic plate were taken from Figures 8 and 10 for the surface roughness of 0.015 μm. For comparison, data obtained from Figures 8 and 10 for plates with surfaces roughened to the same extent as those of the two types of TFE spheres is also included. The tendency for both sets of data to reflect a lowering of contact angle with increases in roughness indicate composite surface formation. Thus, the mobilization behavior would be expected to reflect a decline as intrinsic contact angle is increased to 90°.

Before presenting the ganglia mobilization results it is necessary to decide whether the advancing surface is best represented by either an advancing or a receding angle in the experiment. A ganglion is initially formed by displacing the continuous phase and all contacts correspond more closely to a receding rather than an advancing three-phase line saturation. For advancing conditions to apply, the ganglion phase should have occupied the pore with a much larger radius of curvature [as determined from equation 6(b)] in order to display a diameter consistent with measured values of θ_A. For the largest contact angle system, air-ETG-TFE, in which the value of $\theta_R = 57°$ and $\theta_A \simeq 90°$, the corresponding maximum advancing surface diameter would be either 0.443 cm or 0.656 cm. The maximum diameter that was measured at this maximum mobilization condition was about 0.47 cm. Consequently, it is assumed that receding contact angles apply and they will be used to evaluate conditions at the sphere surfaces. On the other hand, the upper acrylic plate behaves similarly to a simple capillary in that the encroaching oil at the advancing surface is following a previously unwetted surface and there are no geometrical restrictions at the surface. Hence, for the upper plane, receding angles will be applied to the receding surface and advancing angles will be used at the advancing surface.

Maximum angles of inclination as measured for each ganglion volume are presented in Table 5 for the six systems. These results are ordered with respect to the receding surface sphere contact angle for each of the two liquids. For any air-liquid-solid system, the maximum inclination, α, decreases as the number of ganglion lobes is increased. Also, for each of the two fluids, the inclinations increase with the sphere contact angles, θ_{1R} and θ_{1A}.

The reported inclinations and volumes usually represent averaged data collected for ganglia located at four different locations. The system 3 data are identical to that reported earlier for both buoyancy and flow mobilization. The measurement accuracy for any pore would be in the neighborhood of ± 1/2 degree for three- and four-lobed ganglia and up to ± 1 to 2 degrees for single-lobed ganglia. Table 6 contains a summary of the calculations that were performed employing the contact angles from Table 5 in the evaluation of equations (5), (6a), and (6b). Values of the characteristic angles located for the receding and advancing surfaces are given as θ_{cR} and θ_{cA}, respectively. By assuming the formation of spherical surfaces, it is thus assumed that the applied force field causes no deviation of the end surfaces from spherical. Hence, the positioning of the characteristic planes as indicated by values of θ_{cR} and θ_{cA} that depart from 30° is caused by the removal from the hemispherical ends of portions of these spheres that are intersected by either the solid spheres or by the overlying flat plate. Calculations to obtain these values were performed using standard

Table 5. Ganglia Mobilization - Buoyancy Data

System*	Number of Lobes	Volume (ml)	Inclination α, (deg.)	θ_{1R}	θ_{1A}	θ_{2R}	θ_{2A}
1. A-N-O	1	.030	49	9.2	9.2	6.5	7.3
	2	.083	15.5				
	3	.148	10				
	4	.200	7				
2. A-RT-O	1	.033	54	18.5	18.5	6.5	7.3
	2	.080	17.5				
	3	.140	10.5				
	4	.190	8				
3. A-T-O	1	.029	60.2	19.5	19.5	6.5	7.3
	2	.083	18.5				
	3	.135	11				
	4	.181	8				
4. A-N-E	2	.080	24	16	16	24	44
	3	.135	15				
	4	.185	11				
5. A-RT-E	2	.080	30.3	20	20	24	44
	3	.145	16.3				
	4	.195	11				
6. A-T-E	2	.095	29	57	57	24	44
	3	.155	17.5				
	4	.214	13.5				

*Abbreviations: A - Air
N - nylon
T - TFE
RT - roughened TFE
O - oil
E - ETG

equations of spherical geometry (27) but, since only the receding surface for System 6 departs mildly from 30° and the effects on L_c are minor, the calculations were not included here.

There are two mobilization pressure drops given in Table 6. ΔP_{max} represents the prediction of mobilization pressure drop calculated from equation (4). These values of ΔP_{max} clearly decrease with increases in sphere contact angle θ as listed in Table 5. Buoyancy mobilization pressure drops, ΔP_B, over the characteristic lengths, L_c, are presented for the experiments with various ganglia sizes and air-fluid-solid systems. Within experimental error, ΔP_B does not change for various sizes of ganglia. The standard errors (± the estimate of standard deviation) lie between 0.9% and 2.4% of the average values with the exception of System 5 for which the standard error was 8.6%. The larger deviation occurred with the roughened TFE and may have reflected non-uniformities of roughness. It is important to note for each air-fluid system, that rather than decreasing with increases in contact angle like predicted values of ΔP_{max}, the experimental ΔP_B values tended to increase.

Before continuing to discuss the buoyancy results, the data for flow induced mobilization, which will be compared with the buoyancy data, are presented in Table 7 and represent experiments performed with Systems 1, 2, and 3. Reported in columns (3), (4), and (5) are the measurements of volume, mobilization flow rate, and the mobilization cell pressure drop which was measured across 12 spheres. Next, it is necessary to evaluate terms in equation (8). This can be obtained from columns (4) and (5), and from the dependence of pressure drop on flow rate for a ganglion-free cell, i.e.:

$$\Delta P = q/0.0005302 \tag{12}$$

Reapplying equation 9(a) between the characteristic planes as illustrated earlier, for which C_1 was determined to be 0.1160, the mobilization pressure drop acting due to hydrostatic pressure, ΔP_F, was linearly related to flow rate, q, by:

$$\Delta P_F = P^+ - C^+ q \tag{13}$$

Since the cell was maintained horizontally during the flow tests, a buoyancy pressure drop existed due to the height differences between advancing and receding surface centers only and ΔP_B could be evaluated for use in equation (8). Next, looking at the F_V contribution to ΔP, as given by equation 9(b), it can be seen that, if θ_{cR} and θ_{cA} are relatively invariant for systems 1, 2, and 3, then for the three tests:

$$F_V = C'q \tag{14}$$

Thus, by substitution, equation (8) becomes:

Table 6. Ganglia Mobilization - Buoyancy Summary

System*	r_R (cm)	r_A (cm)	θ_{cR}	θ_{cA}	ΔP_{max} (Pa)	No. of Lobes	L c (cm)	ΔP_B (Pa)
1. A-N-O	.0801	.1602	30.5	29.5	38.20	1 2 3 4	.437 1.072 1.707 2.342	32.84 31.21 32.21 31.31 <31.89>
2. A-RT-O	.0817	.1633	31.8	28.4	37.45	1 2 3 4	.438 1.073 1.708 2.343	34.58 34.39 33.61 34.93 <34.38>
3. A-T-O	.0819	.1638	32.0	28.2	37.34	1 2 3 4	.438 1.073 1.708 2.343	36.17 35.90 34.88 34.95 <35.48>
4. A-N-E	.0837	.1751	32.8	27.7	57.89	2 3 4	1.079 1.714 2.349	52.62 53.41 53.99 <53.34>
5. A-RT-E	.0846	.1770	33.5	28.2	57.20	2 3 4	1.083 1.718 2.353	64.07 57.62 54.13 <58.61>
6. A-T-E	.1068	.2216	45.6	32.7	45.00	2 3 4	1.149 1.784 2.419	66.39 64.70 67.90 <66.33>

*Abbreviations:
A - air
N - nylon
T - TFE
RT - roughened TFE
O - oil
E - ETG

Table 7. Ganglia Mobilization - Viscous Flow Data

System*	No. of Lobes	Volume (ml)	Mobilization Flow Rate, q (cc/s)	Mobilization (12-sphere) ΔP_{Total} (Pa)	Mobilization (Characteristic Plane) ΔP_F, (Pa)
1. A-N-O	1	.0300	.0992	189	13.6
	2	.0775	.0553	106	18.3
	3	.1375	.0397	76	20.8
	4	.1875	.0317	60	22.8
2. A-RT-O	1	.0325	.1032	198	14.1
	2	.0775	.0582	112	19.3
	3	.1400	.0403	78	21.2
	4	.1950	.0322	64	23.1
3. A-T-O	1	.0290	.1131	218	15.5
	2	.0830	.0622	121	20.6
	3	.1300	.0434	85	22.8
	4	.1850	.0332	65	23.9

*Abbreviations: A - air
N - nylon
T - TFE
RT - roughened TFE
O - oil

**Pressure drop measured over a distance of 12 sphere diameters for no ganglion present in cell $\Delta P = q/.0005302$.

$$\Delta P = \Delta P^+ - C^+q + C'q + \Delta P_B \tag{15}$$

and, at zero flow rate:

$$\Delta P = \Delta P^+ + \Delta P_B \tag{16}$$

Since it can be argued that the geometry is not affected by flow rate, equation (16) is valid when flow occurs, and thus $C' = C^+$. Assuming that equation (13) is valid for the three air-oil-solid systems, a tabulation of the resulting flow-induced maximum mobilization force data is shown in Table 8. The differences between the buoyancy-induced mobilization results and the flow-induced data are only between 0.6% and 4.3%. Both experiments therefore result in equivalent conclusions about mobilization pressure drop in a static system. It can be seen, as in the buoyancy experiment, that the maximum pressure drop necessary to initiate mobilization increases with increases in contact angle.

Table 8. Ganglia Mobilization - Viscous Flow Summary

System*	ΔP^+ (equation 12) (Pa)	C^+ (equation 12) (Pa•s/cc)	ΔP_B (Pa)	ΔP (equation 8 or equation 12) (Pa)
1. A-N-O	24.87	114	6.82	31.69
2. A-RT-O	25.97	114	6.96	32.93
3. A-T-O	27.65	114	6.98	34.63

*Abbreviation: A - air
N - nylon
T - TFE
RT - roughened TFE
O - oil

A summary of all the experimentally derived results for zero flow rate is presented in Table 9, and a plot of reduced mobilization pressure drop, $R\Delta P/\sigma$, as a function of receding surface contact angle is given for the two air-fluid pairs in Figure 13. It can be seen that the mobilization requirement is increased as the receding surface contact angle is increased (either for an air-fluid pair as the sphere material is varied, or for a given air-solid pair for which the fluid is changed from oil to ETG). This experimental behavior is in qualitative agreement with oil recovery experiments using consolidated media and a series of fluid

Table 9. Maximum Mobilization Pressure Drops - Comparison

System*	Sphere CLA μm	θ_R	Required ΔP_{max} (equation 4)	Buoyancy ΔP (equation 9c)	Flow ΔP (equation 15)
1. A-N-O	.91	9.2	38.20	31.89	31.69
2. A-RT-O	2.40	18.5	37.45	34.38	32.93
3. A-T-O	.78	19.5	37.34	35.48	34.63
4. A-E-E	.91	16	57.86	53.34	--
5. A-RT-E	2.40	20	57.20	58.61	--
6. A-T-E	.78	57	45.00	66.33	--

*Abbreviations: A - air
N - nylon
T - TFE
RT - roughened TFE
O - oil
E - ETG

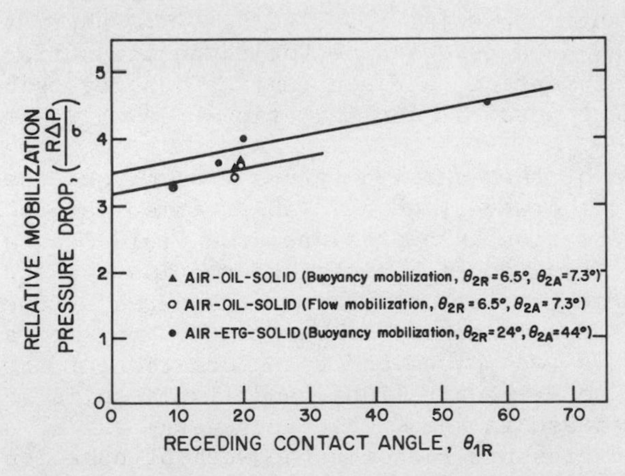

Fig. 13. Experimental dependence of maximum mobilization pressure drops on contact angle from the sphere pack model.

pairs that covered a range of intrinsic contact angles that increased from 0° to 180° (4,5). In addition, for each air-fluid-TFE system, it can also be seen that a reduction of mobilization pressure drop accompanied an increase in roughness of from 0.78 to 2.40 μm for intrinsic contact angles of 60° and 90°. This dependence on roughness violates the expected behavior for Type III surfaces and also does not appear to be supportive of equation (4), (ΔP_{max} in Table 9). Instead, there are consistent increases in mobilization pressure drop with increases in contact angle.

An observation that enables interpretation of the increasing pressure drop behavior was obtained by viewing the experiment conducted with system 6, air-ETG-TFE, for which the highest receding contact angle conditions prevailed. Here, mobilization occurred in two slightly discrete steps and was easily followed. Firstly, deformation to the conditions given by equation (4) occurred; then, the receding surface began to enter the pore throat and, as the ganglion began to lengthen, the advancing surface jumped from contacting the two right-hand spheres directly into the pore body and the entire ganglion was mobilized. In other words, movement of the advancing surface was not a smooth process whereby the curvature underwent gradual reduction as the receding surface deformed into the pore throat. The only way in which the ganglion can move "smoothly" in response to the pressure drop existing across the advancing surface, which would result in a reduced radius of curvature, would be if contact angles down to 0° could occur and equation (3) would totally describe the necessary pressure drop behavior. However, the receding contact angle represents a limiting situation and the associated radius of curvature is the smallest that can maintain contact with the solid spheres, as given for spherical systems by equation (6). Due to the sphere packing geometry and the range of possible contact angles, only larger radii of curvature can exist or, alternately, spherical geometry can no longer be used to describe the advancing surface. At some point in the deformation, a rheon (29) or non-reversible jump must occur to the next possible configuration (i.e. 0° contact angle).

Prediction of the adhesion forces lies beyond the scope of the present work; however, certain observations are possible. Initially, deformation in a pressure drop field occurs, satisfying equation (4). Then, non-spherical deformation can occur whenever a ganglion is touching either the solid spheres or the overlying surface since there is adhesion between the ganglion and the solid. Separation of the two phases and transformation of the air-solid interface into separate air-liquid and liquid-solid interfaces requires an increase in the Gibbs free energy which, at constant temperature and pressure, becomes the work of adhesion per unit area, i.e.,

$$\Delta G = \omega_A = \sigma(1 - \cos \theta_1) \qquad (17)$$

As the surfaces near both right-hand sphere contacts (Figure 2) deform non-spherically, the contact angles increase and, presumably, the three-phase contact line advances. The value attained prior to spontaneous separation is then influenced by the possible levels of hysteresis due to surface roughness and intrinsic contact angle (see Figures 9 and 11). Consequently, no information on contact area at the point of separation is available and equation (17) cannot be readily evaluated.

Another way of looking at the mobilization condition can be approached by recognizing that mobilization is related to the pressure drop that exists across the advancing surface at the right-hand sphere contact regions, ΔP_A, which can be evaluated as follows:

first,
$$\Delta P_A = P_{iA} - P_A \tag{18}$$

then, if there is negligible internal circulation,

$$P_{iA} = P_{iR} \tag{19}$$

and, if at the receding surface,

$$\Delta P_R = P_{iR} - P_R \tag{20}$$

then,
$$\Delta P_A = \Delta P_R - \Delta P_B \,. \tag{21}$$

It is next possible to compare the increase in experimental pressure drop across the advancing surface, ΔP_A, with that which would exist for spherical geometry, $(\Delta P_A)_c$, using the maximum radius of curvature given by equation (6b), i.e.,

$$(\Delta P_A)_c = (P_{iA} - P_A)_c \tag{22a}$$

$$= 2\sigma \frac{1}{(r_A)_{max}} \tag{22b}$$

The extra pressure drop due to adhesion can be determined from:

$$\Delta P_A^* = (P_A - P_{iA})_{experiment} - (P_A - P_{iA})_c \tag{23}$$

or, substituting equations (20) and (21):

$$\Delta P_A^* = \Delta P_B - \Delta P_R + (\Delta P_A)_c \tag{24a}$$

$$= \Delta P_B - 2\sigma\left(\frac{1}{(r_R)_{min}} - \frac{1}{(r_A)_{max}}\right) \tag{24b}$$

A resulting plot of ΔP_A^* versus θ_{1R} is similar to that in Figure 13. This expression, contained in equation (23) suggests that, at low contact angles ($\theta = 0°$), ΔP_A^* would be zero since ΔP_B would be given by equation (4). Unfortunately, at low contact angles, it was seen that measured values of ΔP_B were lower than theoretically anticipated but, due to their systematic nature, it is likely that this occurred for experimental reasons such as the occurrence of a slight bow in the upper plate leading to slightly larger radii of curvature. On the other hand, the increase in ΔP_A^* that was obtained with increasing values of contact angle, θ_i, lends credence to the proposed mechanism. Problems are still associated with the high sensitivity of ΔP_R at the minimum radius of curvature. Also, realization that different mobilization conditions occur must be reflected in the various limits of contact angle, contact angle hysteresis, and areas of adhesion that precede separation.

APPLICATION TO MOBILIZATION OF RESIDUAL OIL

A consistent interpretation which relates contact angle hysteresis to surface roughness has been experimentally obtained with the aid of sessile drop measurements on spheres and plates and from correlations with surface roughness measurements. It was found that, as surface roughness was developed, the tendency for composite surface behavior was increased. Due to the presence of clays and other minerals, and the processes of cementation and dissolution, it is generally agreed that reservoir rocks can possess high degrees of surface roughness (10,18,19). Also, most laboratory tests first involve saturating the cores with wetting phase (brine) before introducing oil so that simple surface formation, which depends on no previous wetting-phase exposure, should not occur. Therefore, rather than reflecting simple surface hysteresis, radii of curvature and contact angle behavior would likely reflect the formation of composite surfaces. This might provide a decrease in the anticipated pressure drop necessary to mobilize trapped oil with increasing contact angle, using equation (3). However, the adhesion of the ganglion material (oil) to the surfaces of the pores must be overcome in order to detach the rear of the ganglion surface from the solid that is preventing its further movement. Experimental results from the sphere model show that reductions in contact angle decrease the mobilization force requirement. That is, the more strongly non-wetting the oil phase tends to be, the smaller the area contacted by the ganglion, and the lower the force level necessary to recover this oil. Also, increases in surface roughness, which tended to reduce contact angles (see Figure 13 and Table 9), provide for lower mobilization forces than those of smoother surfaces.

While the occurrence of simple surfaces has been questioned, it is recognized that they could occur. When this is so, higher advancing contact angles would be expected when compared with those of composite surfaces, and the mobilization forces would be higher.

Increases in pressures around the advancing surface, rather than supplying energy for a 'jump', might alternately provide sufficient energy to cause snap-off of the ganglion when the pore throat diameter separating any two lobes is sufficiently small. Or perhaps, when the receding surface expands, the ganglion can break into two in a manner similar to that described by Roof (30), rather than producing a movement of the advancing surface. Then, since necessary mobilization pressure drops become higher in shorter ganglia, for similar end curvatures and geometries, there would appear to be another reason for the increasing difficulty of mobilization of the non-wetting phase with increasing contact angle. This adhesion phenomenon might also play an important role in establishing residual oil levels in weakly water-wet media.

In either a tertiary recovery scheme or during a waterflood, reduction of the contact angle and hence reduction of adhesion forces, as a result of the use of chemicals in the flood water, could lead to improved recoveries since lower mobilization threshholds would need to be overcome and reduced oil wetting could lower snap-off tendencies.

CONCLUSIONS

1. Receding and advancing radii of curvature can be correlated successfully as functions of roughness with the aid of a stylus-type tracing instrument which enables determination of center-line-average (CLA) roughness.
2. Receding surface contact angle behavior is characterized by reductions in contact angle and by increasing departures from a non-hysteresis behavior as contact angle and roughness are increased.
3. Advancing surface contact angle behavior, at low contact angles and low levels of roughness (surfaces which one would consider polished), show simple surface behavior in which increases in contact angle mirror receding contact angle behavior. However, at angles over about 45° and on mildly roughened to highly roughened surfaces, this tendency was found to collapse and composite surface behavior occurred in which the advancing angle began to approach the receding angle.

4. Using hysteresis behavior outlined and the simplified description of mobilization requirements, equation (4), a tendency to reduce the necessary mobilization force is predicted at any level of roughness as the intrinsic contact angle is increased. This prediction does not, however, reflect the reality of ganglia mobilization experiments.

5. Experimentally measured mobilization behavior in the spherepack model shows an increase in the necessary force levels as the receding surface contact angle is increased. This is in qualitative agreement with porous media displacement studies.

6. For weakly wetted systems, the mobilization process appears to occur as follows. The receding surface first begins to move through the throat increasing the pressure drop across the advancing surface; then, a 'jump' of the advancing surface (which requires a significant threshold force) rapidly follows. This behavior could not have been identified from capillary-type experiments.

7. The effects of increasing surface roughness were experimentally determined to give reductions in non-wetting phase contact angles and reductions in the mobilization forces requirement. When composite surfaces form, roughened surfaces reduce the mobilization requirements when compared with those for smooth surfaces. This behavior could occur in porous media that has first been exposed to the wetting brine phase.

8. The mobilization theory predictions of pressure drop embodied in equation (4) are applicable only to strongly non-wetting residuals or blobs which are not in contact with the rear pore surfaces.

ACKNOWLEDGMENTS

The author would like to sincerely acknowledge the dedication and effort which both Mrs. K.G. Parker and Miss G. Milosz extended in obtaining the measurements contained in this report and who as a result have made this report possible. The author would also like to thank Dr. F.G. McCaffery who readily provided valuable discussions throughout the course of this work. In addition, the contributions of Mr. N.R. Smith, Mr. C.H. Jackson and Mr. V. Masata who assisted at various times are sincerely appreciated. Ms. F. Martens and Ms. B. Moore, who have typed the report as it has taken shape, are sincerely thanked. Also appreciated are Mr. A. Moehrle and Mr. J. Holdsworth who were kind enough to permit us to borrow the Talysurf 4 instrument from the Department of Mechanical Engineering, University of Calgary.

LIST OF SYMBOLS

A	area
C_1	constant, equation 9(a)
C_2	constant, equation 9(b)
C_3	constant, equation 9(b)
C	constant, equation (13)
C^+	constant, equation (12)
D_b	base diameter (sessile drop), cm
D_m	major diameter (sessile drop), cm
F_V	viscous deformation term, equation (8), Pa
f	coefficient of friction
ΔG	Gibbs Free Energy Difference
g	acceleration due to gravity, (9.8083 m/s^2)
H	height (sessile drop), cm
L_c	characteristic length, cm
N_{ca}	capillary number
n	number of lobes on ganglion
P	pressure, Pa
ΔP_B	pressure drop from buoyancy forces, Pa
ΔP_F	pressure drop from viscous forces, Pa
ΔP_i	pressure drop due to internal circulation, Pa
ΔP_{max}	maximum mobilization pressure drop, equation (4), Pa
ΔP_R	deformation pressure drop, equation (16), Pa
ΔP_{TOT}	total pressure drop, equation (17), Pa
$\Delta P^{\ddagger}$	pressure drop (measured) with viscous flow, Pa
ΔP^*	pressure drop necessary for jump, Pa
q	flow rate, cc/s
R	sphere radius, cm
r	radius of curvature, cm
r^*	Wenzel roughness ratio
ω_A	work of adhesion
x	distance, cm
α	cell inclination angle
Δ	difference operator
ε	inclination angle
θ	angle
θ_c	characteristic angle
θ_1	sphere contact angle
θ_2	upper surface contact angle
μ	continuum viscosity, mPa·s
ρ	continuum density, gm/cm^3
σ	surface or interfacial tension, mN/m

Subscripts

A	advancing
av	average
B	buoyancy
c	characteristic

F flow or viscous
h hydraulic
i internal
max maximum
min minimum
R receding
v velocity distortion
o intrinsic or zero-contact angle

Superscripts

a air or liquid
b liquid
s solid
* adhesion

Note: Since ΔP is in pascals and σ is in mN/m, and r is in cm, conversion is assumed to occur, i.e.,

$$\Delta P = 2 \cdot \frac{\sigma}{10} \left(\frac{1}{r_R} - \frac{1}{r_A} \right) .$$

REFERENCES

1. F. F. Craig, Jr., in "The Reservoir Engineering Aspects of Waterflooding," published by the Society of Petroleum Engineers of AIME, Dallas, Texas (1971).
2. J. C. Melrose and C. F. Brandner, J. Can. Pet. Tech., 13, No. 4, 54 (1974).
3. T. F. Moore and R. L. Slobod, Prod. Monthly, 20, No. 10, 20 (1956).
4. F. G. McCaffery and D. W. Bennion, J. Can. Pet. Tech., 13, No. 4, 42 (1974).
5. E. J. Lefebvre du Prey, Soc. Pet. Eng. J., 13, 39 (Feb. 1973).
6. A. Abrams, Soc. Pet. Eng. J., 15, 437 (October, 1975).
7. J. J. Taber, Soc. Pet. Eng. J., 9, 3 (February, 1969).
8. J. P. Batycky and A. K. Singhal, Research Report RR-35 of Petroleum Recovery Institute, Calgary (Oct. 1977).
9. N. R. Morrow, paper No. 78-29-24 presented at 29th Annual Technical Meeting of the Petroleum Society of CIM, Calgary, June 13-16 (1978).
10. N. R. Morrow, J. Can. Pet. Tech., 14, No. 3, 42 (1975).
11. S. G. Oh and J. C. Slattery, paper no. D-2, Proc. of ERDA Symposium on Enhanced Oil and Gas Recovery, Tulsa, Oklahoma, Sept. 9-10 (1976).
12. A. C. Payatakes, C. Tien, and R. M. Turian, Amer. Inst. Chem. Eng. J., 19, 53 (1973).
13. K. M. Ng, H. T. Davis, and L. E. Scriven, Chem. Eng. Sci., 33, 1009 (1978).

14. R. W. Flumerfelt, paper no. D-5, Proc. of ERDA Symposium on Enhanced Oil and Gas Recovery, Tulsa, Oklahoma, Sept. 9-10 (1976).
15. J. C. Slattery and A. A. Kovitz, paper no. B-4, Proc. of DOE Symposium on Enhanced Oil and Gas Recovery, Tulsa, Oklahoma, Aug. 29-31 (1978).
16. P. C. Hiemenz, in "Principles of Colloid and Surface Chemistry," Marcel Dekker Inc., N.Y. (1977).
17. R. E. Johnson, Jr. and R. K. Dettre, Advances in Chemistry Series, No. 43, American Chemical Society, p. 112, Washington, D.C. (1964).
18. B. F. Swanson, J. Pet. Tech., 31, 10 (1979).
19. E. M. Kieke and D. J. Hartmann, J. Pet. Tech., 26, 1080 (1974).
20. C. Huh and S. G. Mason, J. Colloid Interface Sci., 60, 11 (1977).
21. R. H. Dettre and R. E. Johnson, Jr., Advances in Chemistry Series, No. 43, American Chemical Society, p. 136, Washington, D.C. (1964).
22. F. E. Bartell and J. W. Shepard, J. Phys. Chem., 57, 211 (1953).
23. H. H. Zuidema and G. W. Waters, Ind. Eng. Chem. Anal. Ed., 13, 312 (1941).
24. E. Lefebvre du Prey, Dev. Inst. Fran. Pétrole, 24, 701 (1969).
25. Rank Organization, Rank Taylor Hobson Division, Leicester, England: Operating Instructions for 'Talysurf 4' - Surface Measuring Instrument.
26. H. W. Fox and W. A. Zisman, J. Colloid Sci., 5, 514 (1950).
27. C. D. Hogman, ed., in "Mathematical Tables from Handbook of Chemistry and Physics," 11th edition, p. 355, Chemical Rubber Publishing Co., Cleveland, Ohio (1969).
28. N. R. Morrow and F. G. McCaffery, in "Wetting, Spreading and Adhesion," J. F. Padday, ed., Academic Press, London, 1978.
29. N. R. Morrow, Ind. Eng. Chem., 62, 32 (1970).
30. J. G. Roof, Soc. Pet. Eng. J., 10, 85 (March 1970).

EXPERIMENTAL INVESTIGATION OF TWO-LIQUID RELATIVE PERMEABILITY AND DYE ADSORPTION CAPACITY VERSUS SATURATION RELATIONSHIPS IN UNTREATED AND DRI-FILM-TREATED SANDSTONE SAMPLES

P. K. Shankar and F.A.L. Dullien

Department of Chemical Engineering
University of Waterloo
Waterloo, Ontario, Canada N2L 3G1

Drainage and imbibition relative permeabilities of brine and Soltrol 160 were determined in a water-wet Berea sandstone sample and in one treated with 2.5% Dri-Film Solution in hexane. Drainage relative permeabilities were also measured in two other Berea sandstone samples, one treated with 1% and the other with 0.02% Dri-Film solution. The Penn State method was used throughout. The dye adsorption capacity of the sample as a function of increasing brine saturation was also determined in every test. A water-wet sample, initially saturated with brine, was oil-flooded, and then water-flooded; two samples, one treated with 5% and the other with 0.02% Dri-Film Solution, were initially saturated with Soltrol and then water flooded.

INTRODUCTION

The primary objective of the investigation, the results of which are reported below, has been to measure the amount of a water-soluble dye (methylene blue) adsorbed on the pore surface of untreated and Dri-Film treated Berea sandstone samples as a function of the (increasing) brine saturation of the sample (1).

The total amount of dye adsorbed at a given saturation may be regarded as the amount of dye adsorbed per unit area of solid surface in contact with the dye solution times the area of the solid surface in the sample which has been reached by the dye solution at that saturation.

The amount of a polar dye which is adsorbed from an aqueous solution of a given dye concentration by unit area of a solid depends on the magnitude of the affinity of the solid surface for the dye, compared with its affinity for water. As a solid surface of high polarity has a high affinity for both the polar dye and water, whereas a surface of low or zero polarity has a low affinity for both, no dramatic difference in dye adsorption capacity can be expected between the untreated polar sandstone and the less polar (or, nonpolar) surface covered with Dri-Film (2).

The surface area of the solid reached by the dye solution at any particular stage of a relative permeability experiment is determined by a number of factors, such as the polarity of the solid surface, the saturation history, the saturation, and the length of time of exposure.

For example, let us consider a sample with a perfectly water-wet surface (i.e., the contact angle measured through the water is equal to zero) which has been saturated with brine and, subsequently, driven down with oil to irreducible water saturation. It is usually assumed that at this stage there is still left a film of water covering the entire solid surface. Let us suppose that an imbibition-type relative permeability test is started at this point. The question may be asked whether in the vicinity of the irreducible water saturation all, or only a portion of the water contained in the sample will flow.

The answer to this question may be found by using a dyed brine solution in the relative permeability experiment. The amount of dye adsorbed is independent of the saturation of the sample if all the connate water were to flow, but it will be found to be an increasing function of the brine saturation if the brine were to flow only in those capillaries from which the oil has been displaced by the imbibing brine.

Let us consider now a preferentially oil-wet sample, but not necessarily one with a contact angle measured through the oil equal to zero degree, which has been saturated with oil. Let us suppose that a drainage relative permeability test is performed on this sample. The experimental evidence reported in the literature seems to indicate that varying portions of the pore surface are covered with oil after being penetrated by brine. In case all or most of the solid surface remains covered with oil, practically no dye adsorption would be expected throughout the entire (drainage-type) relative permeability test. On the other hand, if a significant portion of the pore surface is exposed to the penetrating brine, then a corresponding amount of dye is going to be adsorbed, and the adsorption capacity of the sample will be found to increase with increasing brine saturation.

In this paper the behavior of samples that were treated with Dri-Film solutions of different concentrations was investigated.

BRIEF REVIEW OF PREVIOUS WORK

Measurement of dye adsorption with the purpose of determining wettability of sandstone cores was first proposed by Holbrook and Bernard (3). These authors defined "relative water wettability" as the fraction of the total surface area of a core that is contacted by injected water. There follows from this definition that if this fractional area were found to be a function of the saturation of the core then "relative water wettability" is also saturation dependent.

Holbrook and Bernard (3) showed that Berea sandstone cores and 250-mesh sand adsorbed about 0.4-0.7 mg dye/gm core and 0.45 mg dye/gm core, respectively, of methylene blue from 0.01 percent dye solutions, whereas the amount of adsorption was nil when the cores, or the sand, were treated with 5 percent Dri-Film SC-87 solution in hexane, oven dried and saturated with Soltrol. One of the untreated sandstone samples which was first saturated with water, driven to irreducible water with oil and then flooded to residual oil saturation with the dye solution adsorbed the same amount of dye as the untreated sample not containing oil, thereby indicating that in the process of water-flooding a water-wet core, the entire pore surface is contacted by the flood water.

In other tests, Holbrook and Bernard (3) treated three adjacent Berea cores with hexane solutions containing 0.01, 0.1 and 5.0 percent Dri-Film SC-87. The dye adsorption capacities of the three samples were determined (0.53, 0.57, and 0.46 mg dye/gm core) and compared with a fourth, adjacent untreated sample whose adsorption capacity was 0.71 mg dye/gm core. The relatively modest decrease in adsorption capacity observed in the treated cores may be due to a number of causes, including a possibly diminished adsorption capacity of unit area of pore surface contacted by the dye solution and the inability of the dye solution to penetrate the clays clogged by Dri-Film.

Holbrook and Bernard (3) determined the dye adsorption capacities also in a second set of four Berea cores which were processed as follows. The first core was initially saturated with water and then driven down to irreducible water with Soltrol, the second core was initially saturated with Soltrol, driven with water to irreducible oil and then with Soltrol to irreducible water. The third and the fourth cores were treated with 0.1 percent and 5 percent Dri-Film, respectively, dried and then subjected to the same sequence of operations as the second core. On all four samples water imbibition tests were run and subsequently the dye adsorption capacities were determined. The third and the fourth cores imbibed only a few percent pore volume of water, the second core about 30 percent, the first core, however, imbibed about 55 percent pore volume of water. The dye adsorption capacities decreased from the first to the fourth core as follows: 0.76, 0.59, 0.18, and 0.05 mg dye/gm core. The most important lesson to be learned from these experiments is that even though the second core was preferentially water-wet according to the usual definition of wettability, it adsorbed only 78 percent of the amount of dye it would have, had it been initially saturated with water, rather than with Soltrol. It is evident that under these conditions some of the solid surface remained covered by Soltrol throughout the subsequent operations and, therefore, the dye could not adsorb on this (intrinsically water-wet) surface. As it would take a very long time (4) before the water could penetrate between the Soltrol and the solid surface and establish capillary equilibrium, the method of Holbrook and Bernard (3) is of great value in determining the "status quo" of wettability.

EXPERIMENTAL

Equipment and Materials

The Penn State or three-core-section method of measuring relative permeabilities was used in this work (e.g., Morse et al. (5); Geffen et al. (6); Caudle et al. (7); Richardson et al. (8); Schneider and Owens (9,10); McCaffery (11)). The schematic diagram of the relative permeability equipment is shown in Figure 1. In the present study the two fluids injected simultaneously at a fixed ratio of flow rates were Soltrol 160-oil (1.5 cp viscosity) and 2% NaCl brine. The cores were mounted horizontally in a modified Hassler-sleeve core holder as shown in Figures 2a and 2b. All experiments were run at room temperature. All of the materials used in the flow system were inert to the test fluids (1). The Ruska constant rate proportionating pumps, used by most of the earlier investigators to displace the two liquids were replaced, for reasons of economy, with hydraulic actuators driven mechanically by D.C. electric motors via variable transmission controls (1). In order to prevent rust problems the hydraulic actuator for brine and dyed brine was primed with Soltrol-oil, thus avoiding contact of the actuator with brine. The Soltrol-oil displaced by the actuator was led into the brine or the dyed-brine reservoir. Both reservoirs were kept under the same pressure so as to avoid altering the steady-state saturation of the sample when switching over from one reservoir to the other. Flow rates were measured by monitoring the input RPM with the help of photoelectric counter system (electronic pulse counter). The pressure drop across the test core section was measured with a differential pressure transducer with a range of 0 to 1000 mm Hg and calibrated with a precision dial gauge, and a digital transducer indicator. The pressure drop was continuously recorded on a strip chart recorder. The pressure at the interface of the test core section with the end section was measured by means of a transducing cell. The pressure

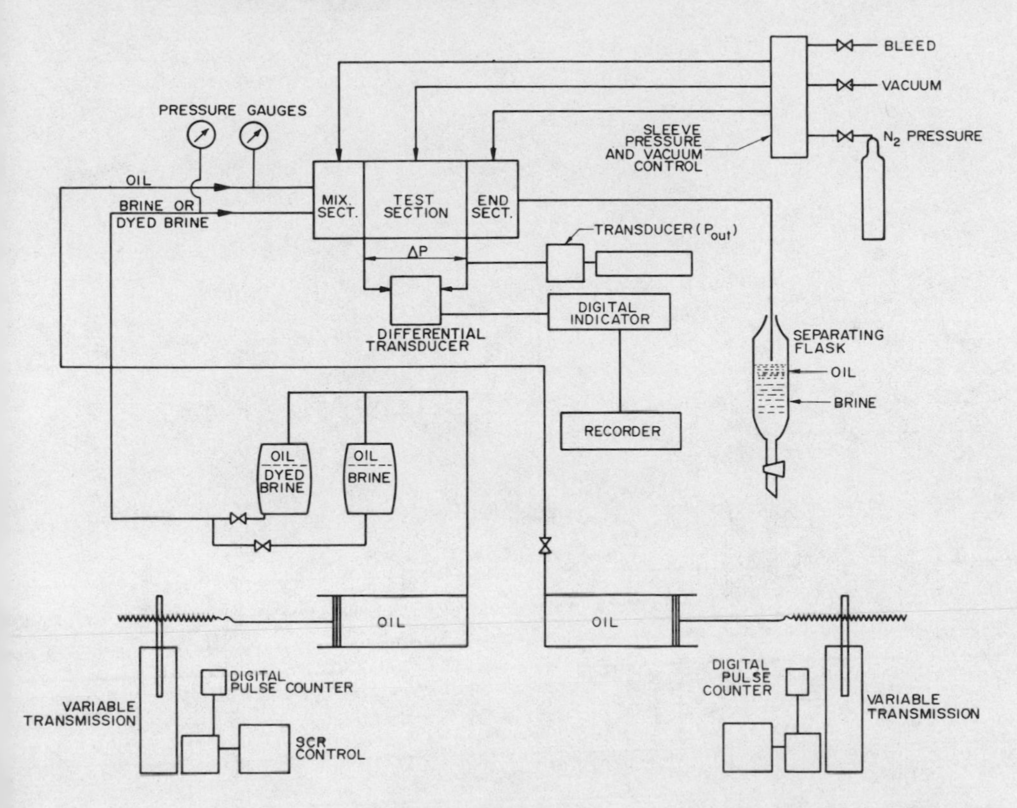

Fig. 1. Layout of two-phase relative permeability equipment

taps communicated with a 0.01 inch wide annular recess around the core periphery, serving as a piezometer ring. After attainment of steady-state conditions, the saturations were determined by weighing the test core section wrapped tightly in tin foil.

0.1 percent solution of methylene blue in brine was used in the dye tests. The effluent dye concentration was determined using a "spectronic" calorimeter calibrated with standard dye solutions of known concentration at the wavelength corresponding to the absorbance maximum (see Figure 3).

Experimental Procedure

The Berea outcrop sandstone cores were supplied as three-section cylindrical samples of 1.5" diameter; the test section was 3.00" long and the end of each section was 1.00" long. The faces were squared off to ensure good contact and the cores were heated at 575°F for forty-eight hours to decompose any organic matter. After cooling, the cores were weighed. The core permeabilities ranged from 440 to 647 md and the porosities were in the range from 0.209 to 0.230.

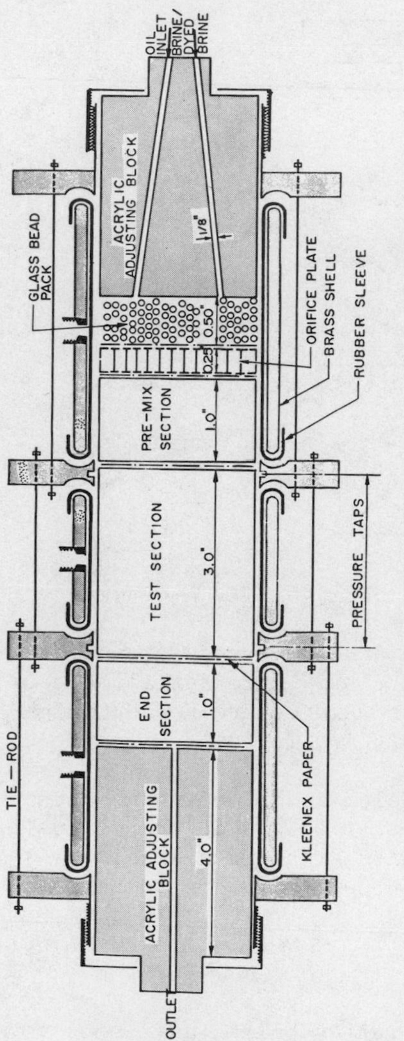

Fig. 2a. Cross-sectional drawing of modified Hassler-sleeve core holder used for Berea 11, Berea 12, and Berea 13 samples.

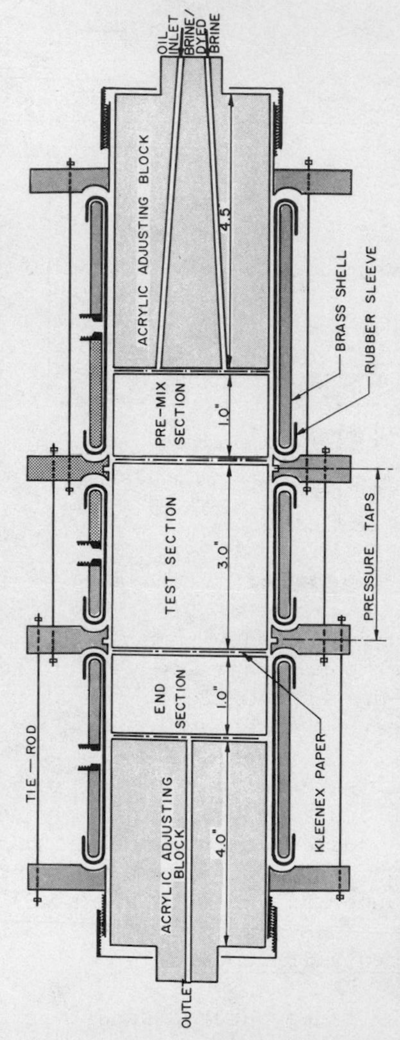

Fig. 2b. Cross-sectional drawing of modified Hassler-sleeve core holder used for Berea 18 sample.

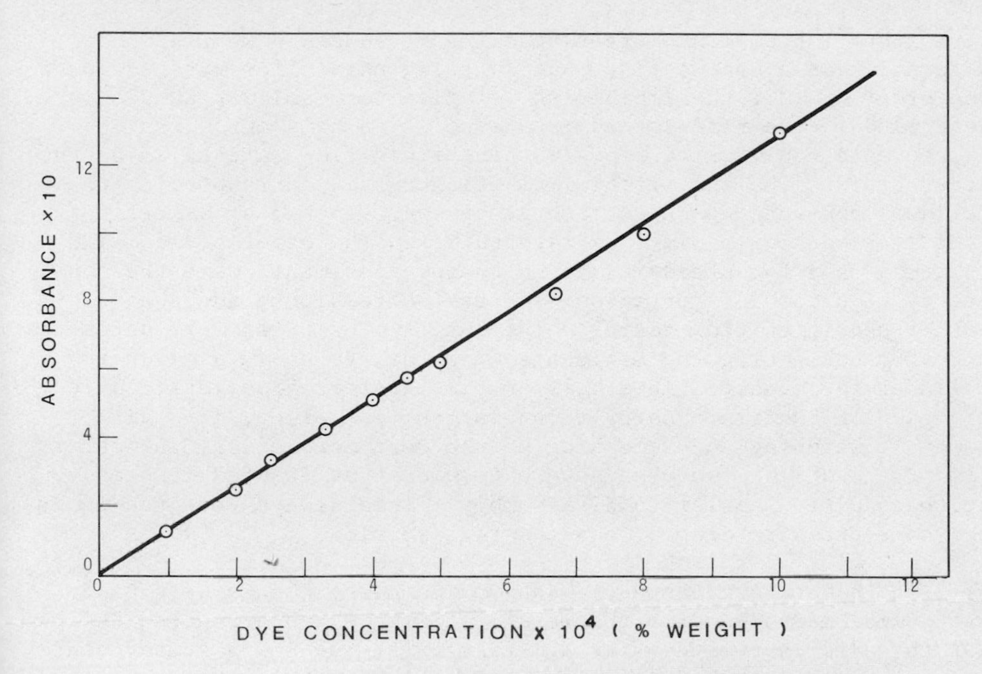

Fig. 3. Calibration curve for spectronic - 20

The sample to be treated with Dri-Film SC-87 solution in hexane was loaded into the Hassler-sleeve core holder and about 60 pore volumes of Dri-Film solution was passed through it in both directions and, subsequently, it was flushed with about 25 pore volumes of N_2 gas. On removal from the core holder the sample was dried at 225°F for about four hours and after cooling it was weighed.

The untreated cores were saturated with brine, whereas the Dri-Film treated cores were saturated with Soltrol in a core saturator. The saturated cores were weighed, and with the help of the known density of the liquid the porosities were calculated.

Detailed description of the loading procedure of the three-piece core into the Hassler holder has been given by Shankar (1). Kleenex paper tissue was used regularly between the two faces of the test-piece and the two end pieces, and the three sections were held pressed tight against each other.

Before commencing two-phase flow operations, single-phase permeability determinations were made at several flow rates, using brine in the water-wet sample, and Soltrol-oil in the samples that had been treated with Dri-Film. In every case, first the drainage relative permeability curves were determined. The first steady-state point was obtained typically at a wetting fluid/non-wetting fluid flow rate ratio of about 10. The filter velocities used were about 1.5 cm/min. After reaching irreducible wetting phase saturation at a non-wetting phase/wetting phase flow rate ratio on the order of 100, the imbibition relative permeability curves were determined. The first steady-state point on this curve was obtained at a non-wetting fluid/wetting fluid flow rate ratio on the order of 10. The end of the imbibition curve, corresponding to residual non-wetting phase saturation, was reached at a wetting fluid/non-wetting fluid flow rate ratio on the order of 10. The criteria used for steady state in these experiments were the constancy of pressure drop across the test core piece, and equal outlet and inlet flow ratios. The outlet flow rates were determined by collecting the effluents from the system in a graduated cylinder for measured intervals of time. After gravimetric determination of the fluid saturations in the test piece, involving removal, weighing and reloading of the test core piece, the same flow rates of the two phases were resumed that existed at steady state conditions prior to disassembly. This always resulted in the same pressure drop across the test section.

In those tests where dye adsorption measurements were made, i.e., imbibition run for the water-wet sample and drainage runs for the samples treated with Dri-Film, after reaching steady state, injection was switched over from brine to dyed brine, and oil and dyed brine were simultaneously injected until the effluent dye concentration nearly equaled the inlet concentration and remained essentially constant. The amount of dye adsorbed was determined, by difference, from the measured dye concentration of the effluent.

After this stage, injection was switched back from dyed brine to brine and the brine/oil flow rate ratio was increased to determine the next point on the relative permeability curve. In this process desorption of the dye took place. The amount of dye desorbed was determined colorimetrically in the effluent brine. After reaching steady state, the new saturation and the dye-adsorption at this point were again determined. The process was continued point-by-point until reaching the saturation where the oil flow rate became immeasurably slow. For a water-wet sample it took about 20 hours to determine each steady-state point. For oil-wet samples this time was shorter.

RESULTS AND DISCUSSION

Relative Permeabilities

All relative permeabilities have been based on the absolute permeability measured before the drainage test. The results obtained with the water-wet sample are shown in Figures 4 and 5. The first oil flow in the drainage experiment was measured at 31% pore volume Soltrol saturation, the irreducible water saturation

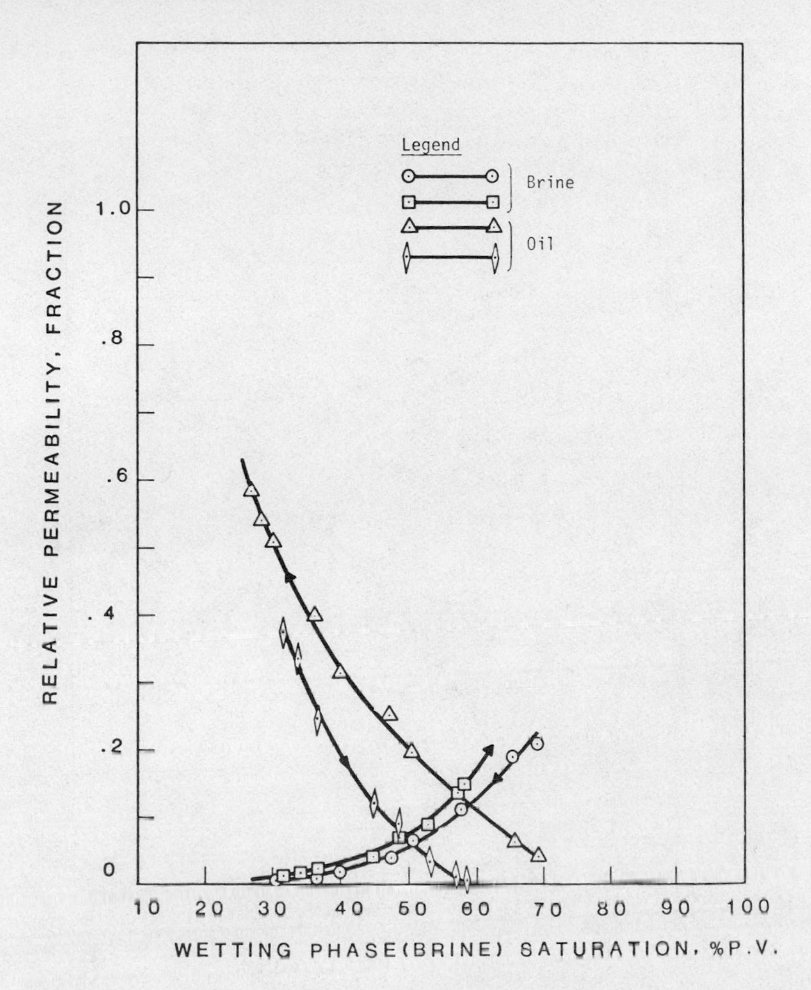

Fig. 4. Drainage and imbibition relative permeabilities of Berea 11 (water-wet sample).

was 28.5% pore volume, and the residual oil saturation 42% pore volume. The oil curve extends up to a relative permeability value of 0.54. The imbibition curve of the oil lies underneath the corresponding drainage curve, and the percent difference between points on the two curves corresponding to the same saturation keeps increasing steadily with increasing brine saturation. The imbibition curve for the water lies only slightly higher than the corresponding drainage curve and it ends at 0.145 relative permeability. Qualitatively speaking, the observed relative permeability characteristics of the water-wet sample are the same as those reported previously in the published literature (4).

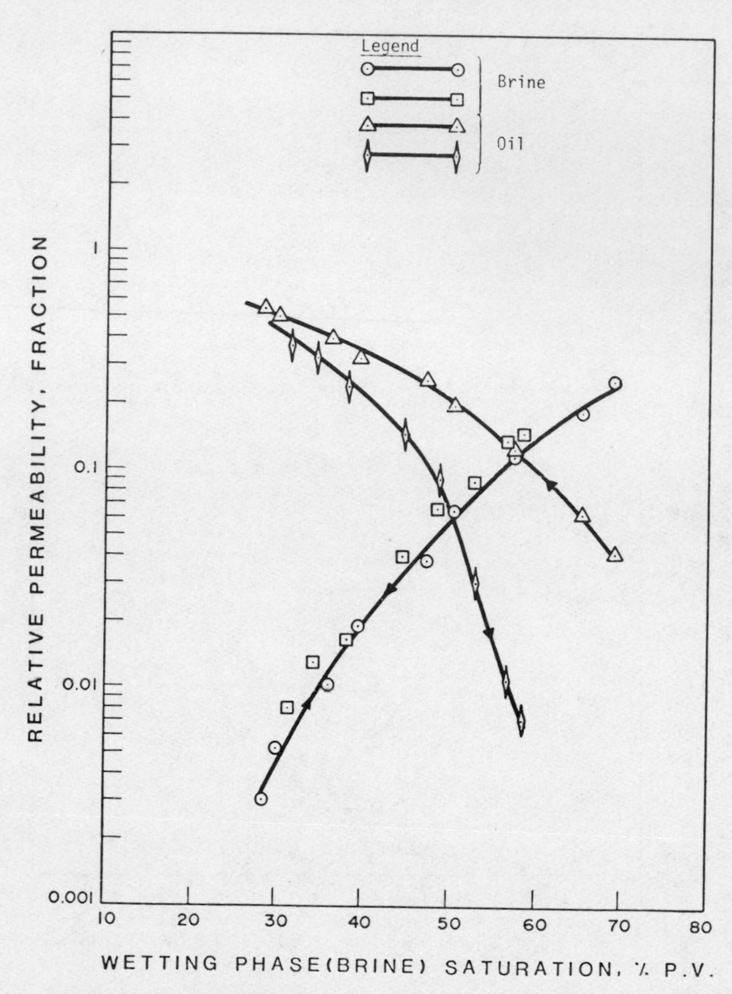

Fig. 5. Drainage and imbibition relative permeabilities of Berea 11 (water-wet sample).

One point meriting discussion is the magnitude of the minimum saturation at which the non-wetting phase will flow, sometimes referred to as "critical saturation". According to Geffen et al. (6) the critical saturation to gas or oil is approximately 1 percent. The lowest reported non-wetting phase saturations at which measurable relative permeabilities were found in drainage tests, however, have been between about 5% and 35% pore volume.

Theoretical studies of Chatzis and Dullien (12,13) on random networks of capillaries have indicated that it is extremely unlikely to have a continuous phase present in a network at saturations less than 10-15% pore volume. Hence the very low "critical concentrations" reported in the literature may be best explained in terms of a mechanism in which the non-wetting phase is flowing as a dispersed phase (14). Under such conditions, however, there is no pressure gradient in the (discontinuous) non-wetting phase and, hence, Darcy's law cannot be applied to this phase.

The results of the relative permeability tests performed on the sample which had been treated with 2.5% Dri-Film solution are shown in Figures 6 and 7. In this case the first flow of the non-wetting phase (brine) in the drainage test was measured at a saturation of 11% pore volume, the wetting phase (oil) relative permeability was 0.004 at 54% pore volume wetting phase saturation, and the residual brine (connate water) saturation was 13.5% pore volume. The water curve extended to a value of 0.37 relative permeability units, compared with the highest point on the oil curve (in the imbibition test) of 0.185 relative permeability units. Some of these features agree qualitatively with the typical water-oil relative permeability characteristics of strongly oil-wet rock (4), but the dramatic shift of all the relative permeability curves by about 0.2 saturation units in the direction of higher wetting phase saturations as compared with the water-wet sample, is in need of a satisfactory explanation.

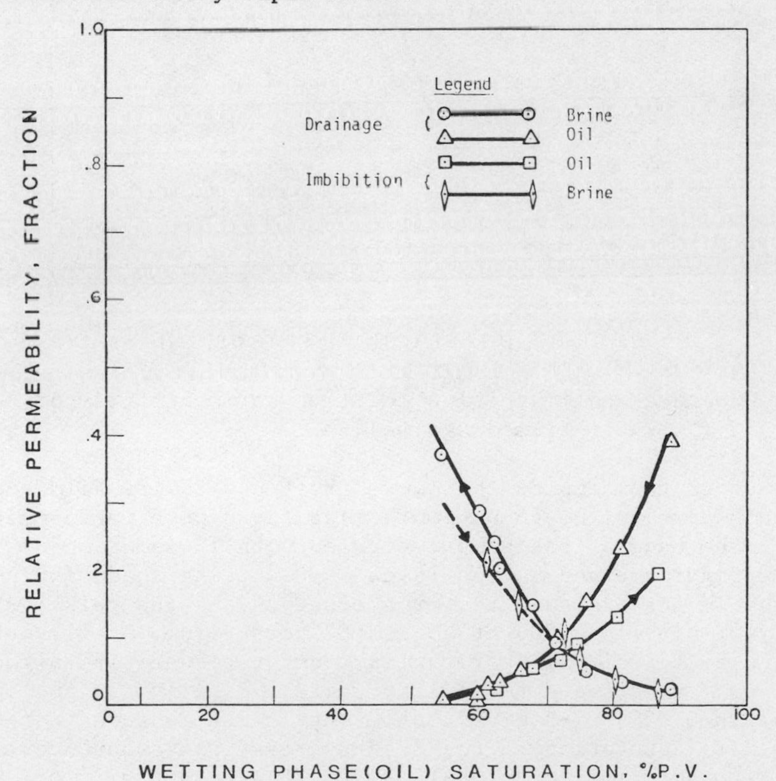

Fig. 6. Drainage and imbibition relative permeabilities of Berea 12 (2.5% Dri-filmed sample).

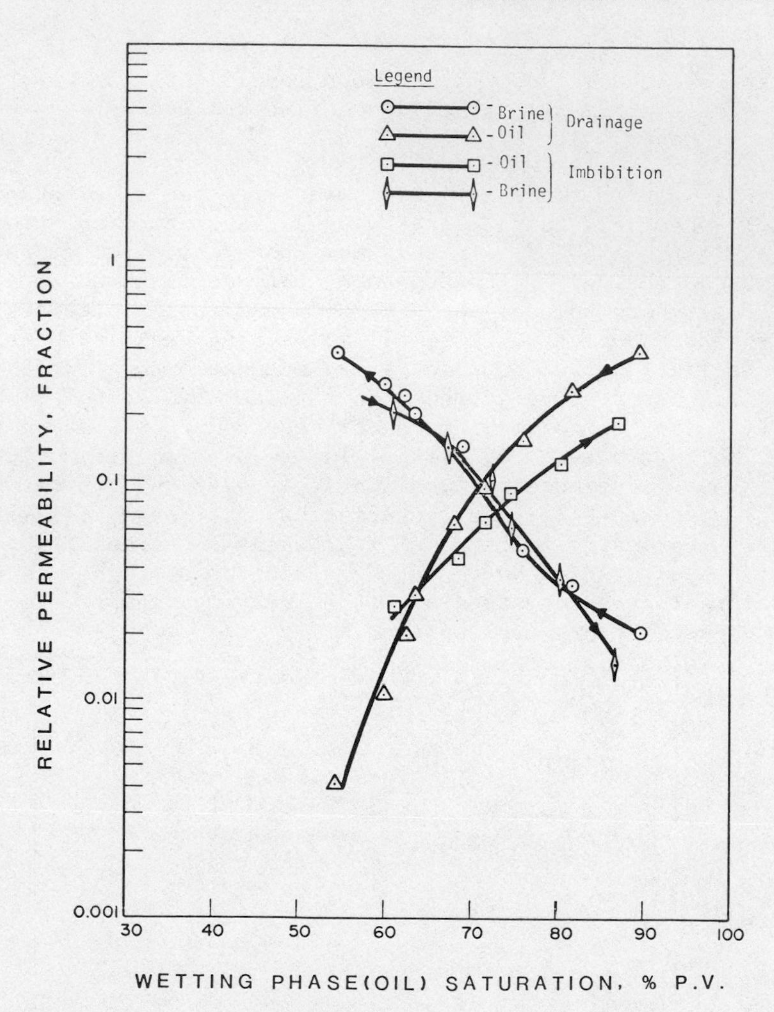

Fig. 7. Drainage and imbibition relative permeabilities of Berea 12 (2.5% Dri-filmed sample).

The other two samples that were treated with Dri-Film showed similar behavior, although no imbibition tests were performed on them. In the case of the sample treated with 1% solution of Dri-Film (see Figures 8 and 9) the lowest non-wetting phase (brine) saturation in the drainage test was about 5% and the relative permeability of oil was 0.006 at about 60% pore volume of oil saturation. The brine relative permeability curve reached its highest point at 0.37 relative permeability units. Finally, in the sample treated with 0.02% Dri-Film solution the lowest measured non-wetting phase (water) saturation in the drainage test was about 6.5% and the relative permeability of oil was 0.025 at about 61% pore volume oil saturation (see Figures 10-12). In this case, however, the

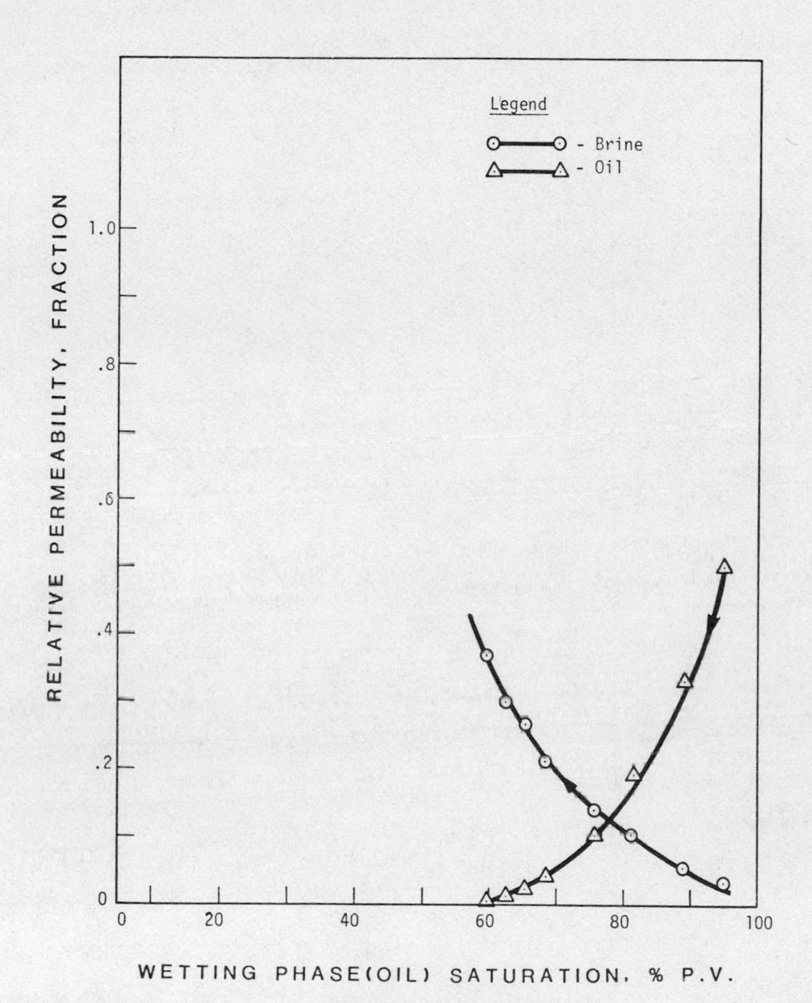

Fig. 8. Drainage relative permeabilities of Berea 13 (1.0% Dri-filmed sample).

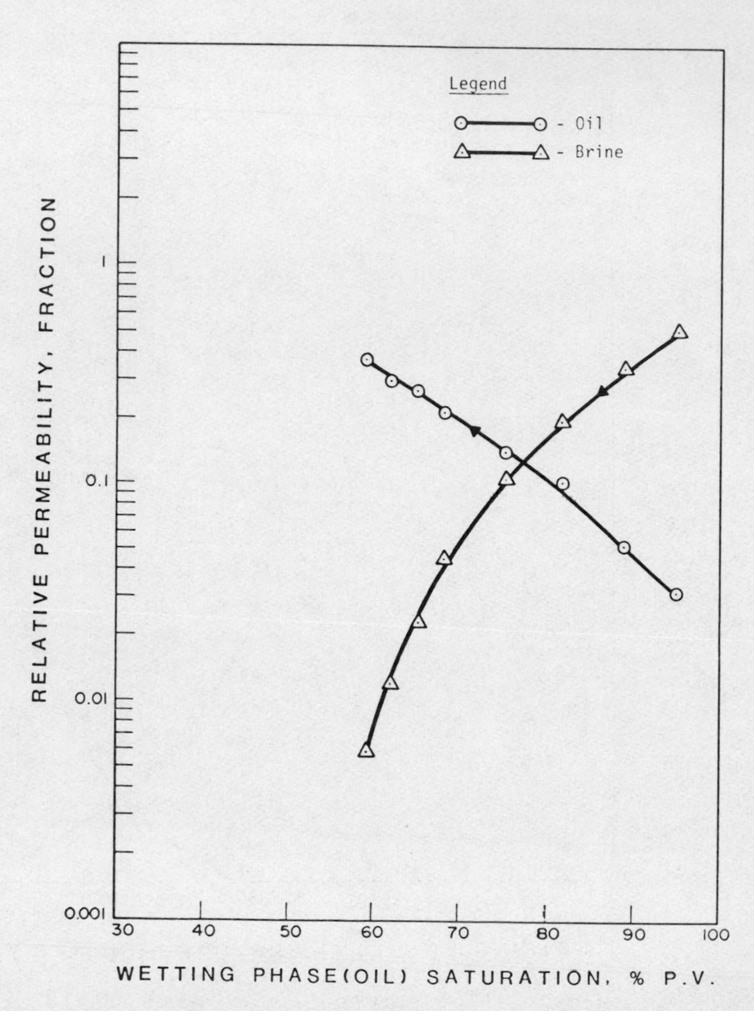

Fig. 9. Drainage relative permeabilities of Berea 13 (1.0% Dri-filmed sample).

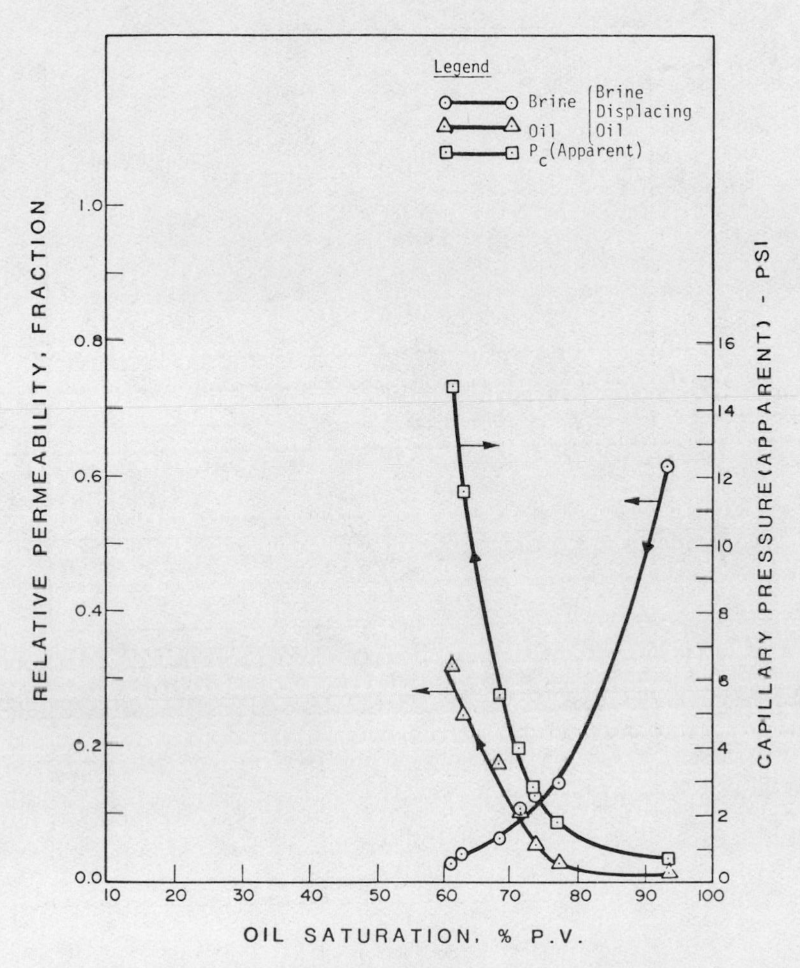

Fig. 10. Drainage relative permeabilities and apparent capillary pressure of Berea 18 (.02% Dri-filmed sample).

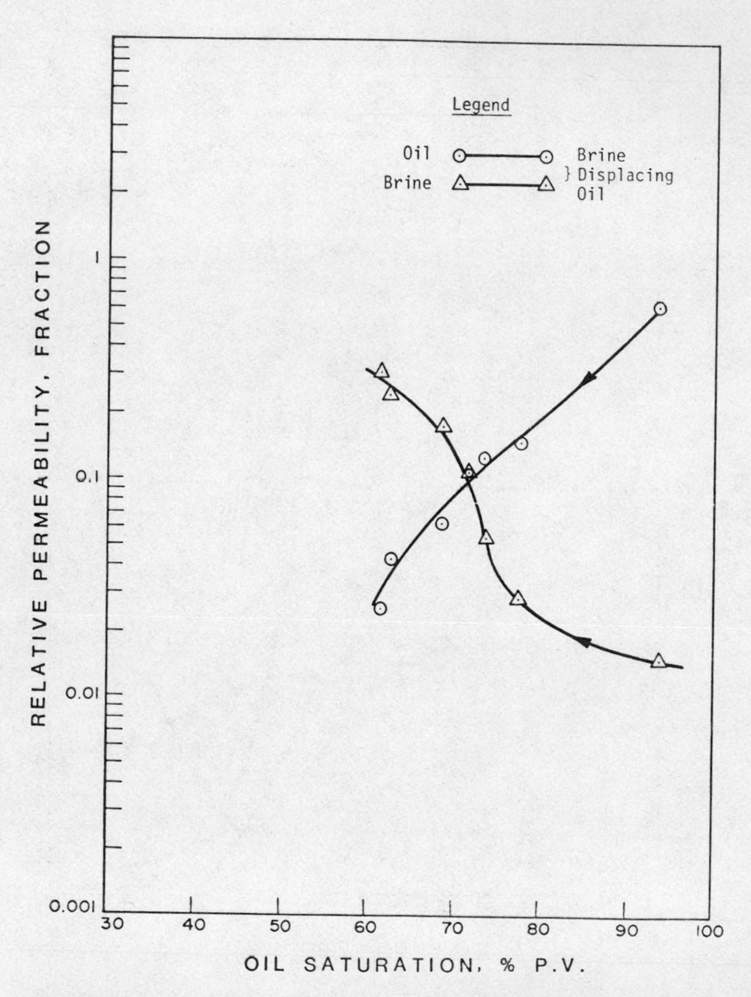

Fig. 11. Drainage relative permeabilities of Berea 18 (.02% Dri-filmed sample).

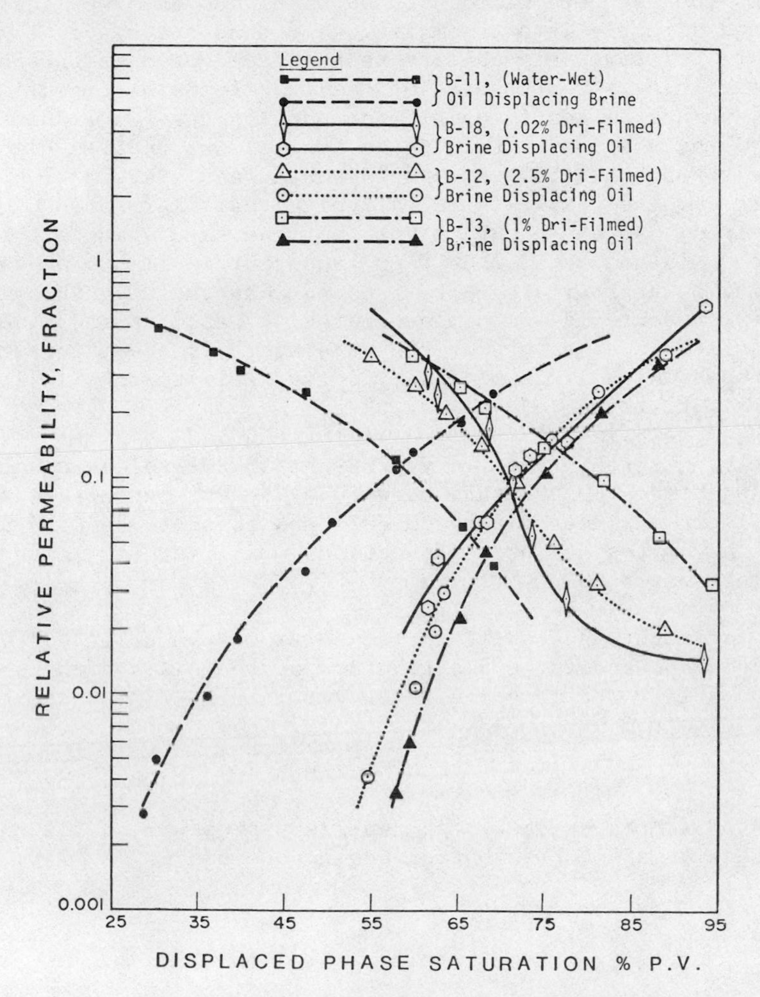

Fig. 12. Comparison of drainage relative permeability vs. saturation relationships obtained for water-wet and Dri-filmed Berea sandstone samples.

brine curve extended as high as 0.5 relative permeability units.

It is noted that in the last test there was an important departure from the usual way of introducing the two phases into the core (see Figure 2b). The usual high permeability mixing section upstream from the core was dispensed with, and the two liquids were introduced directly into the mixing sandstone end piece. When the Penn State relative permeability test is conducted in the usual manner the injection pressures of the two fluids are identical, however, under the conditions of introducing the two liquids shown in Figure 2b, the non-wetting phase (brine) was under a higher pressure throughout the test and, by using a differential pressure transducer at the sample inlet, a typical capillary pressure curve was obtained. As the accurate value of the saturation at the inlet of the core is not known, the term "apparent capillary pressure curve" may be appropriate. It is noted that the breakthrough capillary pressure (displacement pressure) was measured at 0.8 psi. This quantity is a measure of the "diameter" of the thinnest capillary through which it is necessary for a fluid to pass if it is to reach the outlet face of the core.

The breakthrough capillary pressure of Soltrol in a water-wet Berea sandstone sample saturated with brine was measured quasi-statically in a porous diaphragm cell and found to be 0.55 psi, whereas the corresponding value obtained with mercury in an evacuated sample was 8 psi.

The breakthrough capillary pressures may be utilized to make some deductions about the distribution of the oil and the water phases on a microscopic scale. Letting

$$(P_c)_b = \frac{4\sigma \cos \theta}{D_e} \tag{1}$$

where $(P_c)_b$ is the breakthrough capillary pressure, σ the interfacial tension and θ the contact angle, one can write

$$\left(\frac{\sigma \cos \theta}{D_e}\right)_{w/o} \Big/ \left(\frac{\sigma \cos \theta}{D_e}\right)_{o/w} = \frac{0.8}{0.55} \tag{2}$$

whence, as $\cos \theta_{o/w} = 1$ and $\sigma_{w/o} = \sigma_{o/w}$,

$$\cos \theta_{w/o} = 1.45 \frac{(D_e)_{w/o}}{(D_e)_{o/w}} \tag{3}$$

As $\cos \theta_{w/o} \leq 1$, there follows from Eq. (3)

$$\frac{(D_e)_{o/w}}{(D_e)_{w/o}} \geq 1.45 \tag{4}$$

indicating that the effective neck size was severely reduced in the Dri-Film treated sample.

Assuming that this reduction is due to the presence of oil on the surface it can be understood why, for the same wetting phase saturation, the relative permeability of oil was found very much smaller in the oil-wet samples than the relative permeability of water in the water-wet one, where there was much less wetting fluid (water) left on the walls of those pores that have been penetrated by the non-wetting phase. The wetting fluid left on the pore walls will contribute much less to flow than the same amount of fluid when filling the entire pore cross-section.

It is interesting in this context to examine the results of the water flood and oil flood tests conducted with oil-wet and water-wet Berea samples respectively, and shown in Figure 13. These results are in qualitative agreement with results published in the literature (15). The floods were always started with the sample initially saturated with the wetting phase. Only 20-25% pore volume of the oil (wetting fluid) was recovered after injecting the same number of pore volumes that resulted in 60% pore volume recovery of brine (wetting fluid). Assuming that when water is displacing oil from an oil-wet capillary there is an oil film left on the solid surface, the average thickness of which is much greater than that of the water film left on the capillary surface when oil is displacing water from a water-wet capillary, one obtains for the ratio of effective average capillary diameters $(D_e)_{o/w}/(D_e)_{w/o}$ the range of values of $\sqrt{60/25}$ to $\sqrt{60/20}$ (1.55-1.73), which is in good agreement with the values calculated above from the breakthrough capillary pressure of penetration. The physical reason for the startling lack of symmetry with respect to water and oil in the behavior of the water-wet core and the cores treated with Dri-Film is not yet proven, but it seems likely that it is related to the fact that the contact angle measured through the oil in the samples treated with Dri-Film may have been greater than zero.

Dye Adsorption

The cumulative dye adsorption capacities of the samples treated with 2.5% and 1.0% Dri-Film solutions and the water-wet sample are shown in Figure 14 as a function of the saturation. It can be seen that the samples treated with Dri-Film and subsequently saturated with Soltrol adsorbed much less dye than the water-wet sample.

The dye-adsorption capacity of a core treated with 2.5% Dri-Film solution, saturated with Soltrol and then driven down to irreducible oil with the dye solution was found to be 0.024 mg dye/gm core, which is only a small fraction of the dye adsorption capacities shown in Figure 14. A possible explanation of this dis-

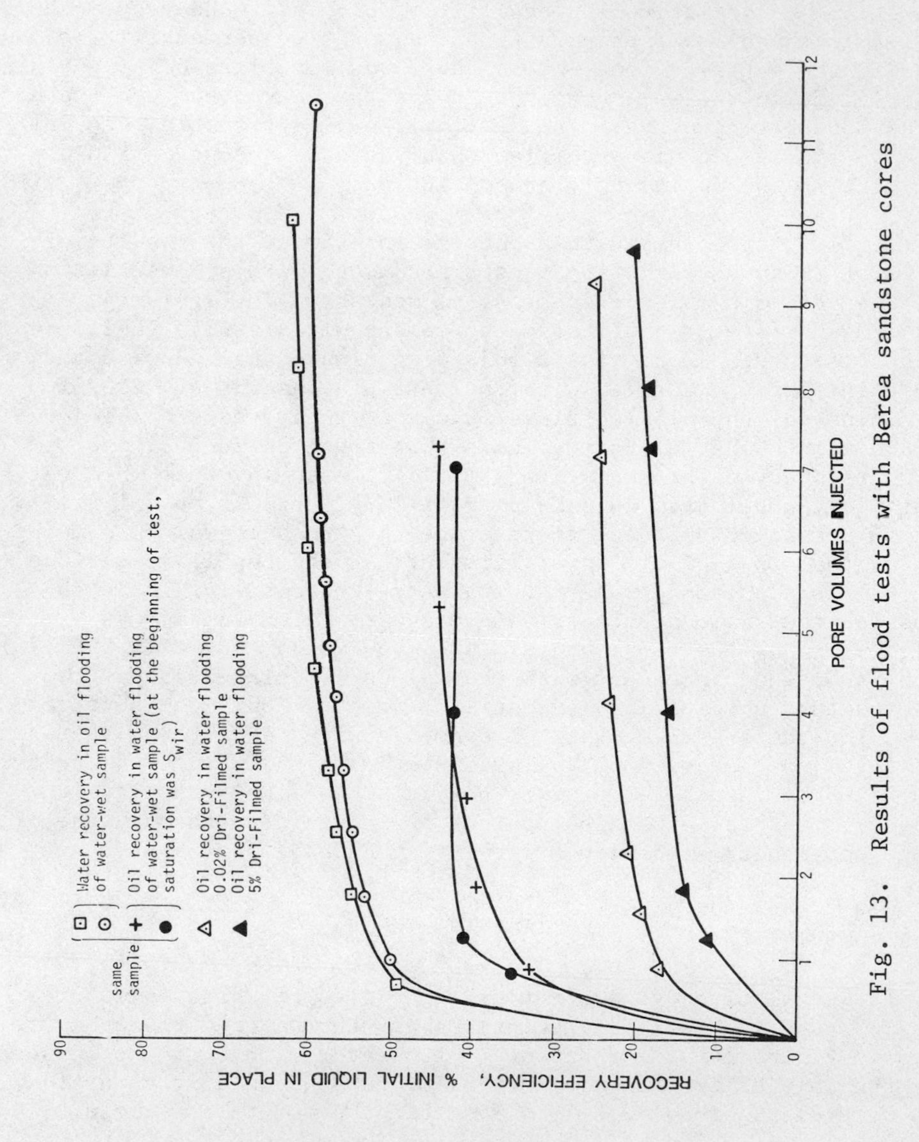

Fig. 13. Results of flood tests with Berea sandstone cores

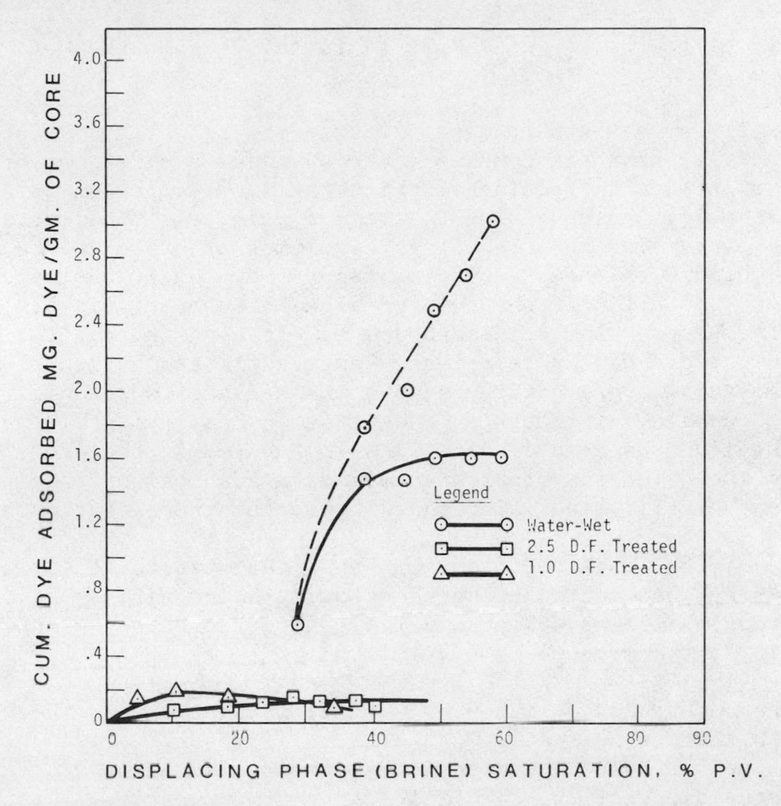

Fig. 14. Cumulative dye adsorption capacity vs. brine saturation.

crepancy is that in the course of the relative permeability tests, which lasted several days, more of the pore surface might have become exposed to the dyed brine than in the direct dye adsorption test, which lasted only a few hours. This explanation would not be admissible if the contact angle measured through the oil had been zero, because in that case the oil would cover the whole surface. Another possible explanation is that the excess adsorption was due to the glass bead pack mixing section present in these experiments.

The dye adsorption capacity of water-wet Berea sandstone cores was determined by saturating two samples with brine and then flooding them with about 700 pore volumes of 0.1% solution of dye in brine. The values obtained were 1.59 mg dye/gm core and 1.47 mg dye/gm core. In two other tests, water-wet samples were driven down to irreducible brine saturation with Soltrol and then flooded with 0.1% solution of dye in brine. The adsorption capacities obtained in these tests were 1.59 mg dye/gm core and 1.60 mg dye/gm core, in excellent agreement with the first two results reported above. Hence there can be little doubt that the dye adsorption capacity of the water-wet sample is in the neighborhood of 1.6 mg dye/gm core.

As it can be seen from the plot in Figure 14, the first point, obtained very near the connate water saturation already represents a very significant fraction of the total dye adsorption capacity of the sample. In obtaining the second point and the rest of the points, every time hundreds of pore volumes of brine had to be passed through the sample before steady flow conditions were reached. The effluent was colored blue throughout this process, owing to the presence of desorbed dye. It appears that the amount of desorbed dye was systematically underestimated by roughly the same amount at every point, because the effluent was too dilute for an accurate calorimetric concentration measurement. The straight line shown in dashed lines in Figure 14, therefore, represents the erroneous apparent cumulative adsorption capacities, whereas the solid line is presumed to be the true relationship.

It can be concluded from this test that practically the entire pore surface was contacted by the flowing brine already at a brine saturation as low as 40% pore volume where the brine relative permeability was a mere 2%.

The core treated with 0.02% Dri-Film solution adsorbed more dye than the cores treated with 2.5% and 1.0% Dri-Film solution, respectively. As can be seen in Figure 15, in the case of this sample the slope of the cumulative adsorption versus saturation curve kept increasing steadily with increasing brine saturation. This trend may reflect the fact that the surface of pores increases approximately as the square, whereas their volume increases approximately as the cube of the pore size. As shown in the figure, the shape of the dye adsorption curve is similar to that of the apparent capillary pressure curve. In the tests conducted with this sample steady state was reached at every point much sooner than in the water-wet sample and desorption of dye was much less, too.

CONCLUSIONS

1) The relative permeabilities in the three Berea sandstone samples treated with different concentrations of Dri-Film solution lay fairly close together, and were displaced by about 20% pore volume in the direction of higher displaced phase saturation, as compared with the relative permeabilities in the water-wet Berea sandstone.

2) The breakthrough capillary pressure (displacement pressure) of brine was 1.5 times higher in a sample treated with 0.02% Dri-Film solution than the value obtained with Soltrol in a water-wet Berea core.

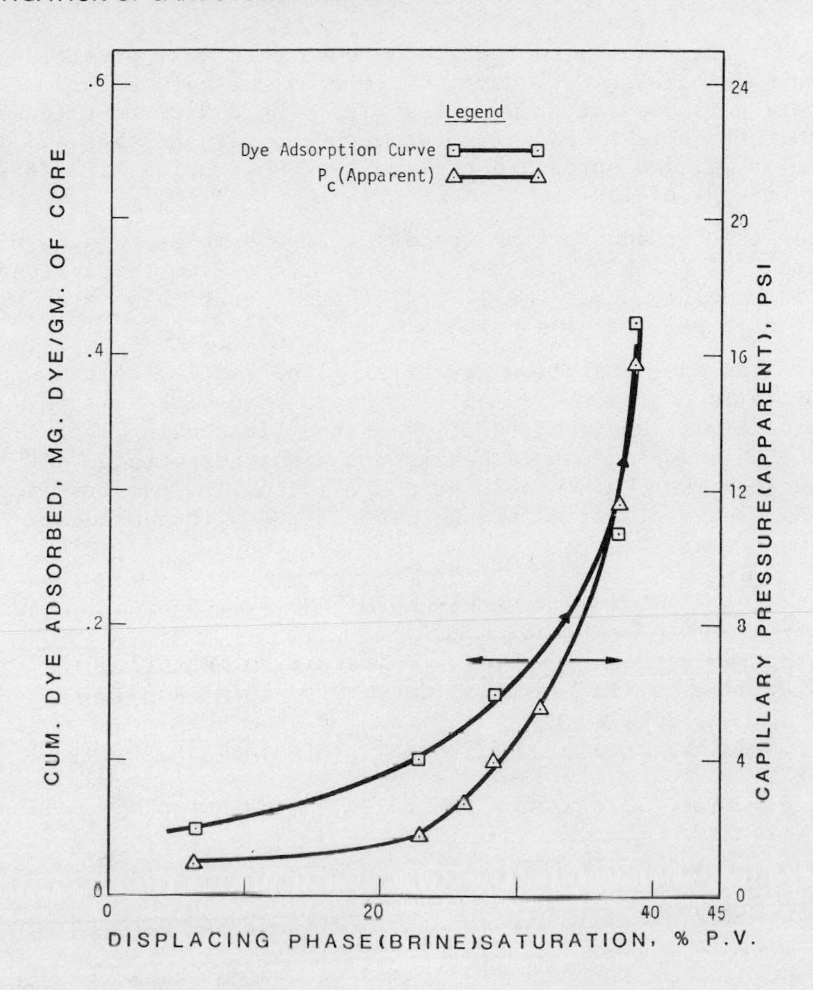

Fig. 15. Cumulative dye adsorption capacity and apparent capillary pressure for Berea 18 (.02% Dri-filmed).

3) The water recovery at the same number of injected pore volumes, from a water-wet core, initially saturated with brine, was about three times higher than the oil recovery from samples treated with Dri-Film solutions and initially saturated with Soltrol 160.

4) Conclusions 1) - 3) are consistent with the assumption that the average thickness of the oil film left on the surface of oil-wet capillaries, after the penetration of the brine into them, is much greater than that of the water film left on the water-wet capillary surface when oil was displacing water.

5) The shape of the dye adsorption capacity vs. brine saturation curve in the water-wet sample initially saturated with brine,

shows that practically the entire pore surface was contacted by flowing brine already at about 40% pore volume saturation. This observation is consistent with the widely held view that in water wet media there is a brine film between the oil and the capillary surface in all the pores and that most of this brine film is flowing in two-phase flow.

6) The dye adsorption capacity in the samples treated with 2.5% and 1.0% Dri-Film solution, respectively, and initially saturated with Soltrol, was small, and behaved erratically probably because of experimental error.

7) The dye-adsorption capacity of the sample treated with 0.2% Dri-Film solution and initially saturated with Soltrol, was approximately a quadratic function of the (increasing) brine saturation. This dependence is consistent with the picture that in a drainage process the non-wetting phase (brine) penetrates increasingly smaller capillaries as the saturation is increased.

REFERENCES

1. P. K. Shankar, Ph.D. Thesis, University of Waterloo (1979).
2. A. W. Adamson, in "Physical Chemistry of Surfaces," Interscience Publishers, New York, 1967.
3. O. C. Holbrook and G. G. Bernard, Petroleum Trans. AIME, 213, 261 (1958).
4. F. F. Craig, Jr., Monograph Series, 3, SPE of AIME, Dallas (1971).
5. R. A. Morse, P. L. Terwilliger, and S. T. Yuster, Oil & Gas J., 109 (August 1947).
6. T. M. Geffen, W. W. Owens, D. R. Parrish, and R. A. Morse, Petroleum Trans. AIME, 192, 99 (1951).
7. B. H. Caudle, R. L. Slobod, and E. R. Brownscombe, Petroleum Trans. AIME, 192, 145 (1951).
8. J. G. Richardson, J. K. Kerver, J. A. Hafford, and J. S. Osoba, Petroleum Trans. AIME, 195, 187 (1952).
9. F. N. Schneider and W. W. Owens, Soc. Pet. Eng. J., 10, 75 (1970).
10. F. N. Schneider and W. W. Owens, Soc. Pet. Eng. J., 16, 23 (1976).
11. F. G. McCaffery, Ph.D. Thesis, University of Calgary (1973).
12. I. Chatzis and F.A.L. Dullien, J. Can. Pet. Tech., 16, 97 (1977).
13. I. Chatzis and F.A.L. Dullien, International Symposium "Fluid Mechanics and Scale Effects on the Phenomena in Porous Media," IAHR. Thessaloniki, Greece, August 29-September 1, 1978.

14. G. L. Langes, J. O. Robertson, Jr., and G. V. Chilinger, "Secondary Recovery and Carbonate Reservoirs," American Elsevier Publishing Co., Inc., New York, N.Y., 1972.
15. E. C. Donaldson, R. D. Thomas, and P. B. Lorenz, Soc. Pet. Eng. J., 9, 13 (1969).

WETTABILITY OF, AND OIL FILM STABILITY ON GLASS CAPILLARIES AS BASIC PROCESSES IN TERTIARY OIL RECOVERY

J. Pintér and E. Wolfram

Department of Colloid Science
Lorand Eötvös University
P.O. Box 328, Budapest 8, Hungary

A study of immiscible liquid/liquid displacement on model systems, paraffin oil/aqueous surfactant solutions in glass capillary tubes led to the conclusion that flow and wetting properties cannot be treated separately. Even though there is a correlation between static and dynamic contact angles, the validity of this relation is restricted to relatively low surfactant concentrations.

Phenomena, like change of the shape of the liquid/liquid interface and of the apparent contact angles, finger forming and the process of displacement that occur during, and as a result of, the flow, can be interpreted by means of a simplified energetic consideration. This possibility, however, is limited by the stability of the oil film on the solid surface in the sense that a too high stability represents a kinetic hindrance for the above phenomena to occur.

Results of measurements on surfactants of different nature but of the same surface activity revealed the fact that even their effect on lowering the liquid/liquid interfacial tension and, on the other hand, on modifying the wettability of the solid, are interdependent and strongly connected with each other. An ultra low interfacial tension of the liquid/liquid interface is only a necessary but not a sufficient condition for high efficiency of oil displacement.

INTRODUCTION

The microscopic processes of crude oil displacement by immmiscible surfactant solutions and, therefore, the efficiency of tertiary oil recovery are determined by static and dynamic wetting as well as by the stability of the liquid film adjacent to the solid surface. All these properties are connected also with the liquid/liquid interfacial tension and the flow conditions. The two-liquid flow influences dynamic wetting via the shear force exerted on the liquid/liquid interface (1-5), and it affects also the film stability by those forces that act directly at the film of liquid _B_/bulk liquid _A_ boundary (6-7).

It is known that, unlike static contact angles, dynamic contact angles strongly depend on the geometry of the system (8-9). Hoffman showed (3) that there is a correlation between dynamic and static contact angles if inertia forces, gravity and the adsorption in the vicinity of the moving three-phase contact line (TPL) are so small that their effect can be neglected. According to his results obtained by studying a number of liquids, the relation between the static contact angle, Θ_s, and the dynamic contact angle, Θ_d, can be written as

$$\frac{\cos \Theta_s - \cos \Theta_d}{\cos \Theta_s + 1} = \tanh (4.96\ Ca^{0.702}) \qquad (1)$$

where $Ca = \eta v/\gamma$ is the capillary number, that is, the ratio of viscous forces to interfacial forces. Slattery et al. (4) have confirmed the validity of Eq. (1) and concluded that the measured dynamic contact angles depend only on the properties of the flowing liquid and the flow rate.

From the assumption that the capillary flow is laminar, it follows that the flow rate is very low near the capillary wall and becomes zero at the wall. As a consequence the TPL should be at rest and the equilibrium contact angle should develop according to the ratio of the corresponding wetting energies (see curves denoted by 1 of Figure 1). The greater the distance of separation from the wall, the higher the deformation of the fluid interface, the extent of which is given just by the capillary number.

Let the capillary wall be wetted better by phase _A_ than by phase _B_. As a result of the deformation of the fluid interface, phase _A_ will move faster than the TPL and will reach those areas of the wall that are covered by phase _B_. If there is sufficient time left for the film that had previously formed from phase _B_ at the wall to break down, then displacement can occur. It is to be noted to this point that microscopic changes within the small distance of the order of 1 μm (δ in Figure 1) cannot be detected and,

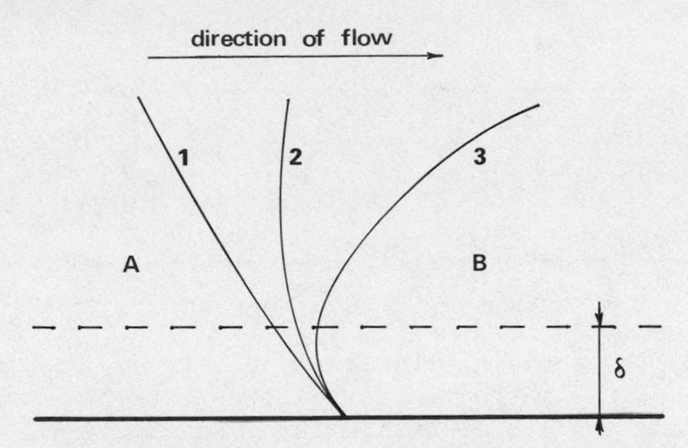

Fig. 1. The steps of deformation of a convex liquid-liquid interface in immiscible liquid displacement.

therefore, the intrinsic microscopic contact angle is not available to direct measurements.

If we disregard that the displacement is hindered by the stability of the adhering film and we consider this process from the thermodynamic rather than from the kinetic point of view, then it is obvious that both the area of the liquid/liquid and the solid/liquid interface can undergo changes (10). The energy change connected with the variation of the liquid/liquid interfacial area, assuming the interface to be a segment of a sphere, is (see Figure 2)

$$E\,(\Theta)_h = 2R^2\,\Pi\gamma_{L_AL_B}\left(\frac{1}{1+\sin\Theta_2} - \frac{1}{1+\sin\Theta_1}\right) \tag{2}$$

Similarly, the energy change for the solid/liquid interface is given by

$$\Delta E(h)_\Theta = 2R\Pi\Delta h(\gamma_{SL_B} - \gamma_{SL_A}) = 2R\Pi\Delta h\gamma_{L_AL_B}\cos\theta \tag{3}$$

The total energy change is the sum of Eqs. (2) and (3). If $\Delta E(h)_\Theta < \Delta E(\Theta)_h$, then change of contact angle and hence of the liquid/liquid interfacial area is preferred, whilst if $\Delta E(\Theta)_h <$ $\Delta E(h)_\Theta$ is true, oil displacement is likely to occur.

For two-liquid flow in capillaries, the behavior of the interface has to be taken into account, too. According to Huh and Scriven (1), the rate profile is, even for laminar flow, not a parabolic one. In the vicinity of the interface, the flow field has, according to Dussan (2) the shape as shown in Figure 3, allowing for phase A to penetrate into phase B. This will occur at point 1 of Figure 3 where the flow has a stagnation point. This is the fountain effect which occurs if the ratio of viscosity of phase A to that of phase B is either too high or too low. The

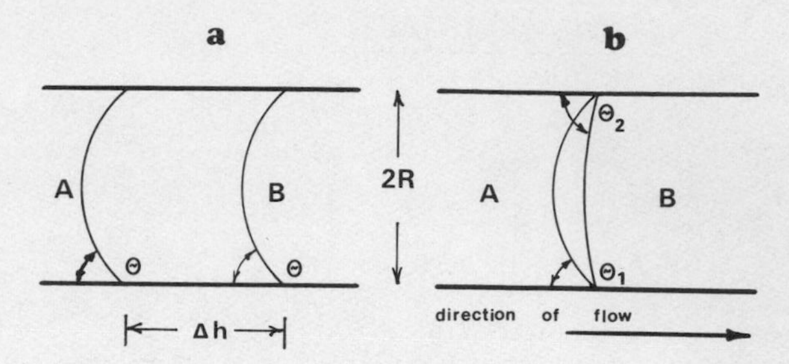

Fig. 2. Immiscible liquid displacement in a capillary tube: (a) change of the wetted area; (b) change of both the contact angle and the liquid-liquid interfacial area.

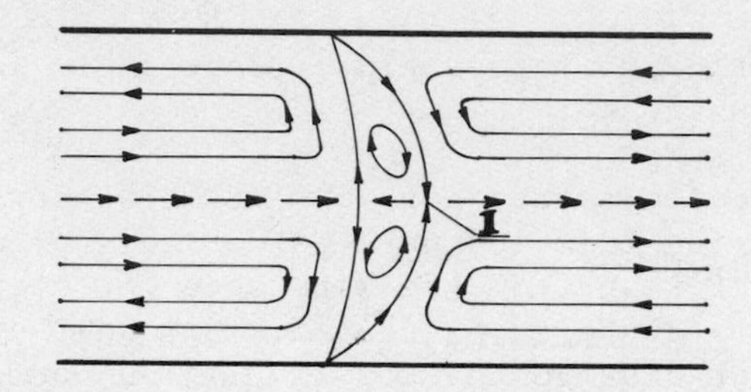

Fig. 3. Flow field near the liquid-liquid interface.

fountain effect results in the dispersion into small droplets of the liquid plug.

For hindered displacement, particularly if the interfacial tension is low and the diameter of the capillary tube is not constant, the moving liquid front often shows the phenomenon of finger-forming (11).

EXPERIMENTAL

The materials used in this study were as follows: Paraffin oil: $\rho = 8.6 \cdot 10^2$ $kg \cdot m^{-3}$; $\eta = 0.164$ $kg \cdot m^{-1} \cdot s^{-1}$; $n = 1.4749$; $\gamma_{LV} = 34.5$ $mN \cdot m^{-1}$, sodium dodecylsulfate (SDS) (product of Merck Ltd., Darmstadt, laboratory grade) purified by crystallization from benzene-ethanol mixtures ($M = 288.3$; $c_M = 8$ $mol \cdot m^{-3}$), dodecyltrimethylammonium bromide (DTAB) (product of Schuchardt, Müchen, laboratory grade) ($M = 308.4$; $c_M = 16$ $mol \cdot m^{-3}$), hexadecyltrimethylammonium bromide (HTAB) (product of Schuchardt, München, laboratory grade) ($M = 364.5$; $c_M = 1$ $mol \cdot m^{-3}$), nonylphenyldekaglycolether (Lerolat-100) (product of Bayer Inc., laboratory grade) ($M = 660.9$; $c_M = 0.07$ $mol \cdot m^{-3}$). The symbols are: ρ density, η viscosity, γ

surface tension, M relative molecular mass, c_M critical micelle concentration.

Capillaries were freshly made from Pyrex-capillaries using a Hewlett-Packard GC glass capillary column maker apparatus. They were pretreated by letting oil stream through them for 30 minutes.

Dynamic measurements were carried out by taking photographs with a Reichert ME-F type universal camera microscope and evaluating them to get the contact angles. The experimental setup is shown in Figure 4. The oil and aqueous phases were mutually saturated with each other prior to the measurements.

Static contact angles were determined separately by measuring the contact angle of 100 mm^3 (±2 mm^3) drops of water and the solutions, respectively, that were placed on to a glass plate immersed in oil using a HVS microsyringe.

The film stability was measured either directly in the capillaries, by observing the time needed for the breakdown of (oil or water) film, or in separate measurements carried out on glass plates of the same material. The latter method which is more accurate, consists of determining the time that elapses from placing water (or solution) drops of 100 mm^3 volume on the surface until a sudden deformation of the drop shape takes place. This time is indicative of the rupture of the oil film due to gravity. The same device was used for the contact angle measurements.

RESULTS AND DISCUSSION

Water/oil systems: In this case, the flow is considered to be laminar, as $Re_{eff} = 0.4$ is much less than $Re_{crit} = 1160$ ($Re = v \cdot R \cdot \rho/\eta$ is the Reynolds number). The linear flow rate as given by the Hagen-Poiseuille equation is

$$v = \frac{1}{8\eta} \frac{\Delta P}{L} R^2$$

From this, it follows that the relation v vs. 1/L and v vs. $\Delta P/L$ should be a straight line, as shown in Figures 5 and 6.

The advancing contact angle at the moving water front depends on how long the capillary wall was wetted by the oil phase prior to the measurement. The longer this time, the more probable is the attainment of adsorption and wetting equilibrium. This is shown by the experimental fact that the advancing contact angle increases with increasing time of previous contact. No time dependence of the contact angle was observed if this "pre-wetting" time exceeded 25-30 minutes.

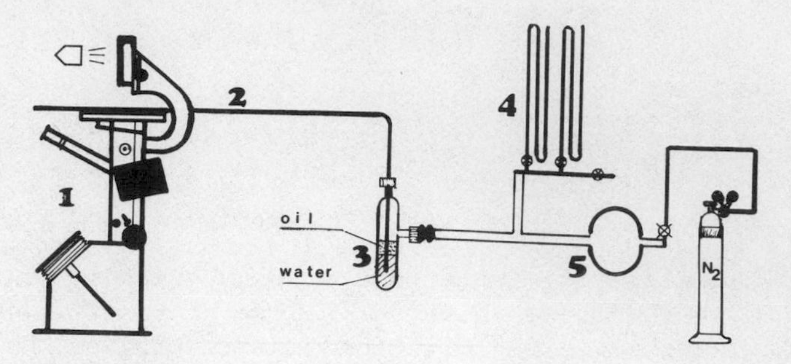

Fig. 4. Experimental setup: (1) microscope with camera; (2) glass capillary tube; (3) pressure vessel; (4) pressure gauge; (5) buffer vessel.

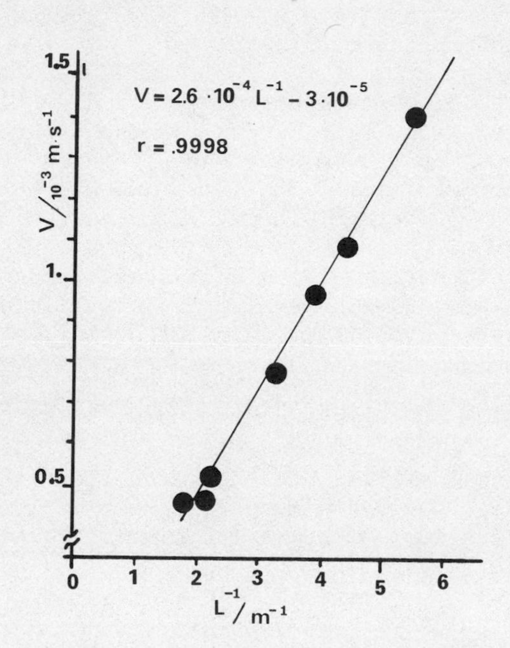

Fig. 5. Linear velocity as a function of the inverse of the capillary length (R = 150 μm, liquid: oil).

At the receding side of the moving water front, one can almost always observe small oil droplets dispersed in the water plug as a result of the fountain effect. This effect, which is a consequence of the flow pattern demonstrated in Figure 3, is shown in Figure 7.

Aqueous SDS solutions/oil systems: The interfacial tension of the solutions against oil and the corresponding capillary numbers are summarized in Table 1. Static contact angles and the advancing dynamic contact angles measured in capillaries as a function

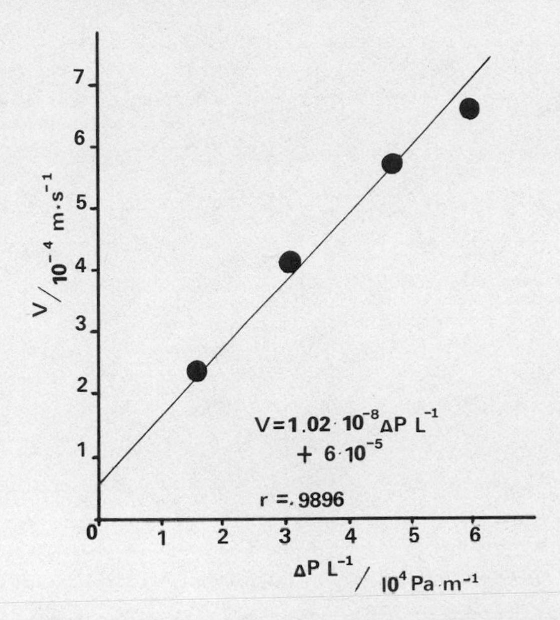

Fig. 6. Linear velocity as a function of pressure gradient ($r = 150$ μm, liquid: oil).

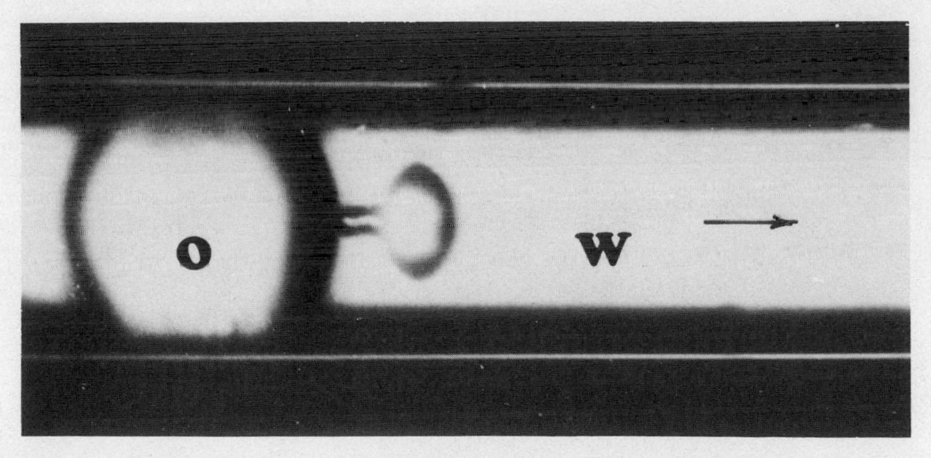

Fig. 7. Fountain effect in the receding water-oil interface ($v = 1{,}5$ mm$\cdot$s^{-1}), $R = 150$ μm, direction of flow is shown by the arrow).

of the SDS concentration are shown in Figure 8. The values for the advancing contact angle measured in capillaries and those calculated from Equation (1) agree with each other for SDS concentrations less than 0.5 c_M within 14-24 percent. This corresponds to an error of 3-30 percent of data given in Reference 4. At higher

Table 1. SDS solution/paraffin oil interfacial tension (γ_{WO}) and capillary number (Ca) as a function of SDS concentration relative to the critical concentration of micellization

SDS Conc. (c/c_M)	γ_{WO} $mN \cdot m^{-1}$	$10^3 \cdot Ca$ --
0	56.0	1.2
0.1	28.5	2.3
0.2	24.3	2.7
0.5	17.0	3.9
1.0	11.4	5.8
2.0	11.6	5.7
5.0	11.0	6.0
10.0	10.4	6.3

surfactant concentrations this disagreement increases up to 67% and then, at a concentration of 10 c_M, drops to 32%.

The observed difference is likely to be due to the stability of the oil film. Figure 9 shows the advancing and receding contact angles measured in capillaries as well as the time of the breakdown of the oil film obtained from separate measurements on glass plates as described above. (The data are mean values from 70-100 individual measurements.) As can be seen from Figure 9, the receding contact angle is practically independent of the surfactant concentration, and it has a value of 27 ± 1 degree.

Aqueous DTAB solutions/oil systems: This surfactant was chosen, as the interfacial tensions of its solutions against oil are practically identical with those of SDS at the same concentration (see Table 2), and this circumstance enables us to make conclusions as to the effect of wettability, independently of other factors.

The static contact angles measured on glass plates, the contact angles that were calculated from these values using Equation (1) and the advancing contact angles measured in capillary tubes are shown in Figure 10. The values for the "capillary" angle and the calculated ones are practically identical at concentrations of 0.2 and 1.5 c_M. For c/c_M being less than 0.2, the difference is 24%, it raises up to 125% between 0.2 and 1.5 c/c_M, amounts to 60% for $1.5 < c/c_M < 10$, but drops to 16% at $c = 10\ c_M$ as was found for SDS, too.

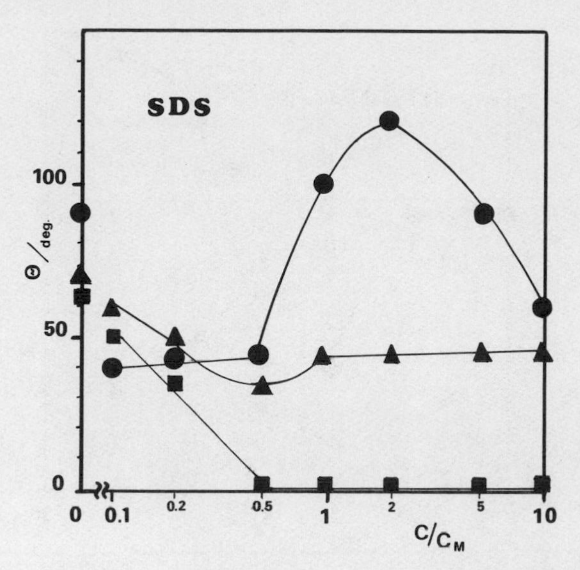

Fig. 8. The contact angle as a function of the SDS concentration; (●) advancing contact angle measured in capillary tube (R = 150 μm); (■) static contact angle; (▲) contact angle calculated from Equation (1).

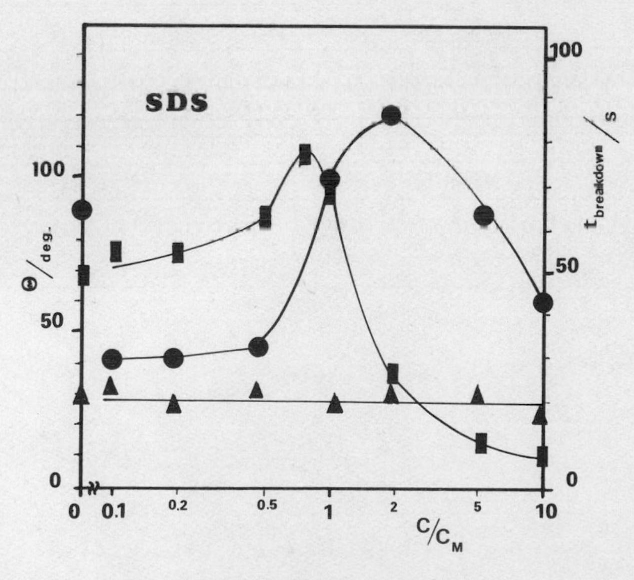

Fig. 9. Advancing (●) and receding (▲) contact angle as well as the time of the breakdown of the oil film (■) as a function of the SDS concentration.

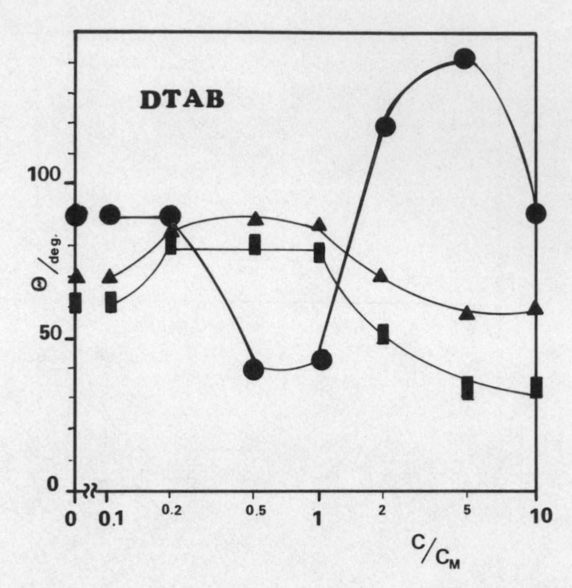

Fig. 10. The contact angle as a function of the DTAB concentration; (●) advancing contact angle measured in capillary tube (R = 150 μm); (▲) static contact angle; (■) contact angle calculated from Eq. (1).

Table 2. DTAB solution/paraffin oil interfacial tension (γ_{WO}) and capillary number (Ca) as a function of DTAB concentration relative to the critical concentration of micellization

DTAB Conc. (c/c_M)	γ_{WO} $mN \cdot m^{-1}$	$10^3 \cdot Ca$ --
0	56.0	1.2
0.1	29.0	2.3
0.2	28.0	2.4
0.5	18.0	3.6
1.0	11.0	6.0
2.0	10.4	6.3
5.0	10.0	6.6
10.0	9.6	6.8

We believe that the differences of the measured and calculated values arise from differences in the film stability. As is apparent from Figure 10, the two curves intersect each other at c/c_M = 1.5 which is very close to the concentration at which the film stability is at maximum. Figure 11 shows also the receding contact angles and one sees that these values are, as was found for SDS solutions, surprisingly invariant (38 ± 2 deg.).

Systems with HTAB and Lerolat-100: Measurements with these surfactants were carried out in order to get information on whether the finger-forming can be attributed primarily to the wettability, or to the interfacial tension. At a concentration of c = 10 c_M, HTAB renders the glass surface more hydrophobic than does DTAB, but the interfacial tension against oil of both solutions is practically the same (9.6 mNm^{-1} for DTAB, and 9.0 mNm^{-1} for HTAB). Lerolat-100 in a concentration of 100 times the c_M wets glass very well, but the interfacial tension against paraffin oil is 2 mNm^{-1} only.

Comparing the photos a and c of Figure 12, it is obvious that a low interfacial tension is the necessary condition for "fingers" to form. On the other hand, the latter can occur to a considerable extent only if the force needed for the adherent oil film to break down is small enough and so is the time available for the displacement. For instance, increasing of the flow rate results in a decrease of the contacting time, which leads to completion of finger-forming, see photo 12b.

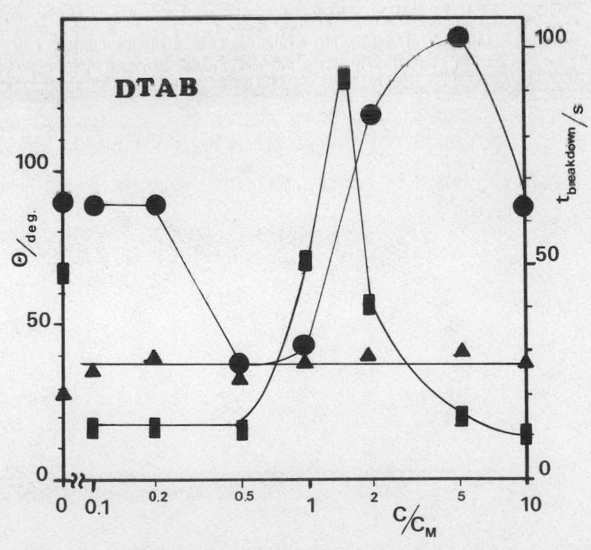

Fig. 11. Advancing (●) and receding (▲) contact angle as well as the time of the breakdown of the oil film (■) as a function of the DTAB concentration.

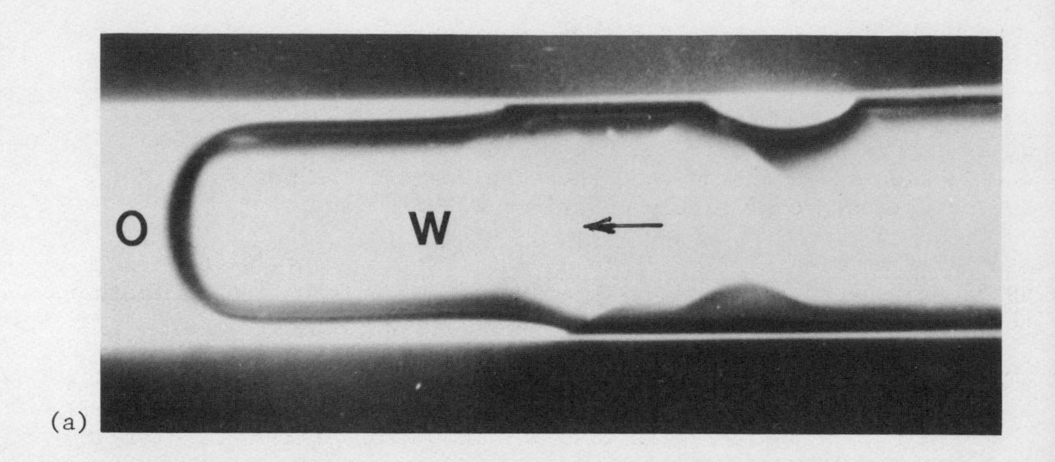

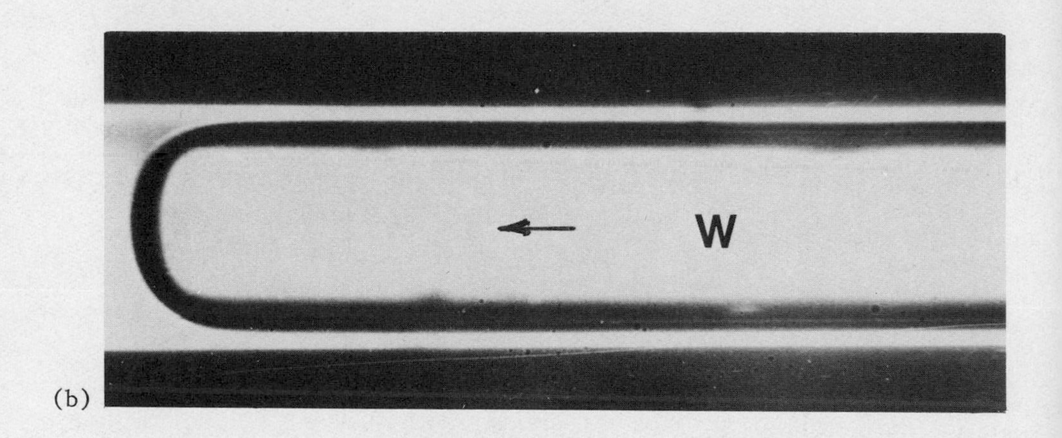

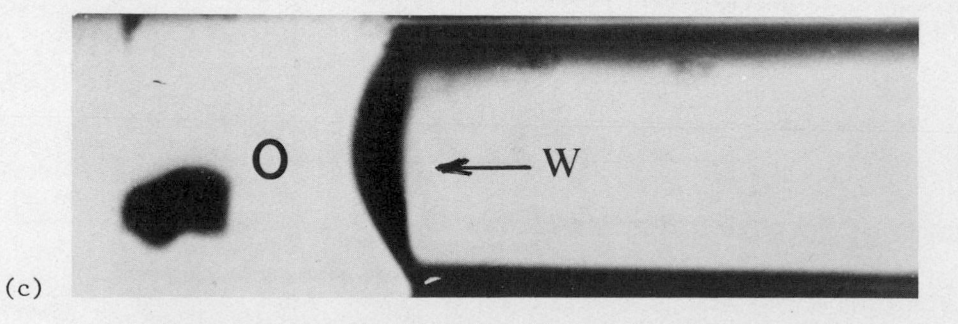

Fig. 12. Finger forming at the advancing water-oil interface: (a) 10 c_M HTAB, v=0.5 $mm \cdot s^{-1}$; (b) 10 c_M HTAB, v=0.8 $mm \cdot s^{-1}$; (c) 100 c_M Lerolat-100 v=0.4 $mm \cdot s^{-1}$ (R=150 μm).

For a HTAB solution of $c = 10\ c_M$, $\cos\Theta$ is negative, and so the value for $\Delta E(h)_\Theta$ calculated from Equation (3) is negative as well. On the other hand, $\sin\Theta$ is positive and, therefore, $\Delta E(\Theta)_h$ calculated by using Equation (2) is positive, too. This means that $\Delta E(h)_\Theta$ being less than $\Delta E(\Theta)_h$, the change in the liquid/liquid interfacial area is preferred to the displacement of oil by solution from the solid surface.

In the case of Lerolat-100, both $\Delta E(h)_\Theta$ and $\Delta E(\Theta)_h$ have positive values but, since $\gamma_{L_AL_B}$ is low, these values are small. In other words, for this surfactant, the probability of the two changes is equally high. This explains why the finger forming is not so predominant as in the case of the rest of surfactants investigated in this study.

As to the fountain effect, it was found for HTAB and even for DTAB that a rather big difference of the viscosities of the two liquids is a necessary but not sufficient condition for this effect to occur. The necessary condition is given by a strongly hindered displacement of the oil in which case a practically complete disintegration of the front can take place as shown in Figure 13.

Finally, the dependence on the surfactant type of the advancing and receding contact angles in capillary tubes of various diameter, but at constant Reynolds number, has been studied as well (Figure 14). For distilled water, solution of SDS and HTAB, both of concentration of $10\ c_M$, contact angle values were obtained as expected from Equation (1), i.e., the higher the capillary number, the greater the contact angle, except for Lerolat-100 of $c = 100\ c_M$ for which the opposite is true. This cannot be ascribed

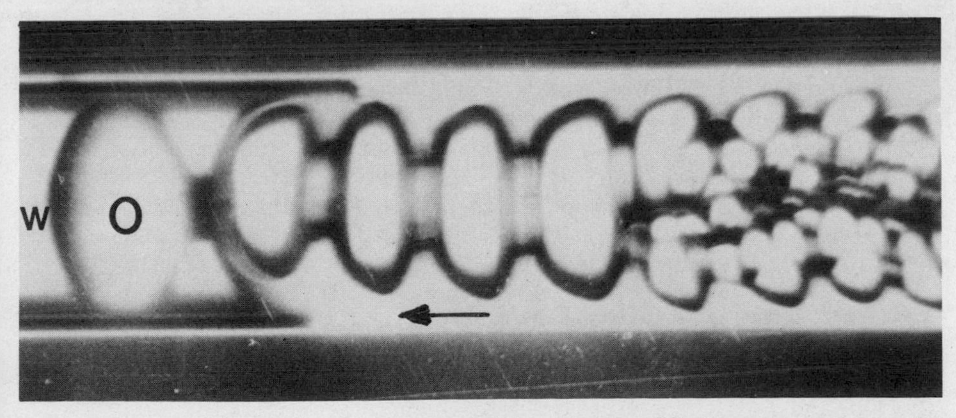

Fig. 13. Dispersion of the receding water-oil front as a result of the fountain effect. (10 c_M HTAB, $v=0.5\ mm \cdot s^{-1}$, $R = 150\ \mu m$.)

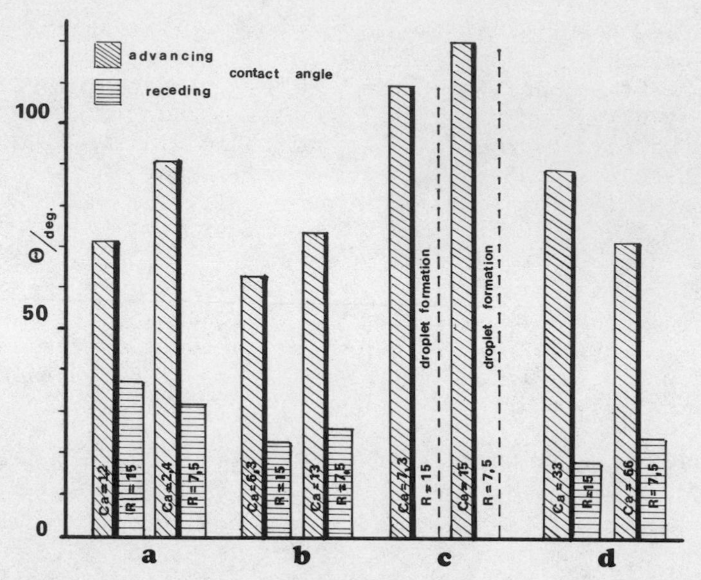

Fig. 14. Dependence of the advancing and the receding contact angle on the radius of the capillary tube at the same Reynolds number ($Re = 3.2 \cdot 10^{-4}$): (a) distilled water; (b) 10 c_M SDS solution; (c) 10 c_M HTAB solution; (d) 100 c_M Lerolat-100 solution. ($Ca = Ca \cdot 10^3$; $R = R \cdot 10^5 m$).

to the fact that the capillary number is in this case at maximum, because the range of capillary number studied in Reference 3, viz., $4 \cdot 10^{-5}$ to 36, includes this value.

REFERENCES

1. C. Huh and L. E. Scriven, J. Colloid Interface Sci., 35, 85 (1971).
2. E. B. Dussan V, AIChE J., 23, 131 (1977).
3. R. L. Hoffman, J. Colloid Interface Sci., 50, 228 (1975).
4. T. S. Jiang, S. G. Oh, and J. C. Slattery, J. Colloid Interface Sci., 69, 74 (1979).
5. R. J. Hansen and T. Y. Toong, J. Colloid Interface Sci., 37, 196 (1971).
6. R. L. Chuoke, P. van Meurs, and C. van der Poel, Petrol. Trans. AIME, 216, 188 (1959).
7. V. G. Levich, Physico-chemical hydrodynamics (Hung.), Akadémiai Kiadó, Budapest (1958).
8. E. B. Dussan V, J. Fluid Mech., 81, 665 (1976).
9. C. Huh and S. G. Mason, J. Fluid Mech., 81, 401 (1977).
10. E. Wolfram and J. Pintér, Acta Chim. Acad. Sci. Hung., 100, 434 (1979).

11. S. G. Oh and J. C. Slattery, ERDA Symp., Oklahoma, 1976. Proc. Vol. 1.D/2, pp. 1-27.
12. K. Henger, M.Sc. Thesis, L. Eötvös Univ., 1978.
13. E. Kiss, M.Sc. Thesis, L. Eötvös Univ., 1978.

THE INFLUENCE OF INTERFACIAL PROPERTIES ON IMMISCIBLE DISPLACEMENT BEHAVIOR

C. E. Brown, T. J. Jones and E. L. Neustadter

New Technology Division
BP Research Centre
Sunbury-on-Thames, Middlesex, UK

In order to obtain better understanding of displacement mechanisms of importance to oil recovery, a technique has been developed to study the immiscible displacement of a liquid/fluid system in capillaries of model geometry. Called the Interfacial Displacement Tensiometer, this technique enables measurement of the pressure changes accompanying displacement of a single interface. The effects of discrete events, such as changes in pore geometry, can therefore be determined. From the pressure measurements, together with determination of the wettability, dynamic interfacial tensions can be calculated. The results presented are for a range of pure oil/aqueous phase systems; in particular effects due to the presence of water soluble and oil soluble surfactants are examined. The results demonstrate that the surfactant systems possess high (dynamic) interfacial tensions during displacement, and the factors controlling their values are discussed. In addition, it is demonstrated that highly visco-elastic interfacial films can contribute an additional resistance to displacement.

INTRODUCTION

In the development of Enhanced Oil Recovery (EOR) systems based on water flooding, the generation of a very low interfacial tension is often considered to be of paramount importance (1,2). However, it is also recognized that a system with a low interfacial tension is not by itself certain of success as an EOR process (3,4). Hence the measurement of an equilibrium (or steady state) interfacial tension is a valuable screening test, but other parameters are also relevant to the dynamic process of oil recovery. This paper focuses attention on these dynamic processes. Its scope is limited to consideration of the local microscopic immiscible displacement mechanisms in capillary size pores; thus interfacial properties dominate the overall system behavior.

The development and optimization of chemical EOR systems can best be achieved on the basis of a good understanding of the mechanisms by which oil is displaced from a porous matrix, and of the parameters which control those mechanisms. Better insight into the displacement processes is required than is afforded by "black box" experiments such as displacement tests in sand packs. It is therefore necessary to carry out laboratory experiments concerned with specified system parameters, and to link the results of these experiments with displacement tests. The most commonly applied measurement used in this way is of interfacial tension, which is linked to displacement tests via intuitive conceptual models of the displacement process. Only relatively recently have systematic studies been undertaken (5) to test these models.

Other properties have also been discussed in terms of their impact on oil recovery efficiency. Thus Reisberg and Doscher (3) demonstrated that formation of rigid interfacial film material may occur during displacement of a crude oil/water interface through a capillary of irregular bore, and suggested that such material may oppose displacement. In the same paper, the strong dependencies upon pH of crude oil/water interfacial tension and of the wettability of silica were shown, and a possible correlation between wettability and rigid film formation was suggested. Most importantly, the necessity of considering interfacial rheological effects in the context of the relevant wetting conditions within a porous medium was recognized.

Possible effects of rigid interfacial films upon the displacement process have been discussed by Bourgoyne, Caudle and Kimbler (6), who concluded that oil recovery would be reduced by the presence of rigid films if the porous medium were highly heterogeneous, but not if it were highly homogeneous. This result was rationalized on the grounds that the films inhibited imbibition. Imbibition is of key importance in a heterogeneous water-wet porous matrix at slow rates of displacement of oil by water.

More recently, Slattery (7) has presented a theoretical description of displacement through irregular shaped pores taking account of the interfacial tension and of the interfacial rheology. He concluded that the interfacial viscosities (dilatation and shear) do not influence entrapment of residual oil, but will be of importance in the displacement of residual oil by surfactant solutions (i.e. when the interfacial tension is low).

The potential importance of hydrodynamic effects in the interfacial region and the paucity of information concerning them has been highlighted by Dussan (8), who emphasized the importance of this topic to surfactant behavior. Blake, Everett and Haynes (9) had previously pointed out that immiscible displacement in even a single cylindrical capillary is associated with radial flow near the interface, so that a dynamic interfacial tension will apply. Blake et al. (9) reported measurements of displacement rate as a function of dynamic contact angle and pressure gradient for pure oil/water systems in a single cylindrical glass capillary. Capillaries of the same geometry were used by Hansen and Toong (10), who reported displacement phenomena other than the piston-like mode. Templeton (11) described the breakup of a crude oil/water interface during displacement in a single capillary, and confirmed the applicability of Poiseuille's law in micron size capillaries. The present authors (12,13) have also studied displacement in single capillaries with crude oil/water systems.

None of the above studies were able to give an instantaneous measurement of displacement pressure, and hence were unable to examine the effects of changes in capillary geometry. Studies using more complex geometries, such as thin section bead packs (14-16), have given information on average pressure gradients and displacement patterns. The present study has developed a technique to measure the pressure changes accompanying discrete events as a single interface is displaced through a model capillary of varying dimensions. Preliminary results have been presented previously (17) for a number of crude oil/water systems; this paper describes the development of the technique to determine dynamic interfacial tensions during displacement, and presents results for a range of pure oil/aqueous phase systems both in the absence and the presence of surfactants.

EXPERIMENTAL

Interfacial displacement tensiometry (I.D.T.): The I.D.T. apparatus used is shown schematically in Figure 1. It consists of a capillary system mounted on the stage of a microscope (Vickers Instruments, model M41). The present capillary system is formed from three glass tubes fused together to form a single cylindrical tube. Thus this capillary comprises of one 50 mm length of 0.25 mm bore radius section fixed between two 80 mm lengths of 0.5 mm bore radius sections. The choice of this particular geometry is explained later in this paper. The capillary system is housed in a rectangular cell with glass lid and base to enable transmitted light illumination and observation of the capillary via the microscope optics. The cell is filled with glycerol to compensate for refractive index effects of the curvature of the glass walls of the capillary (18). The cell is thermostatted to ± 0.05°C by means of an external circulating control and heat exchanger within the cell.

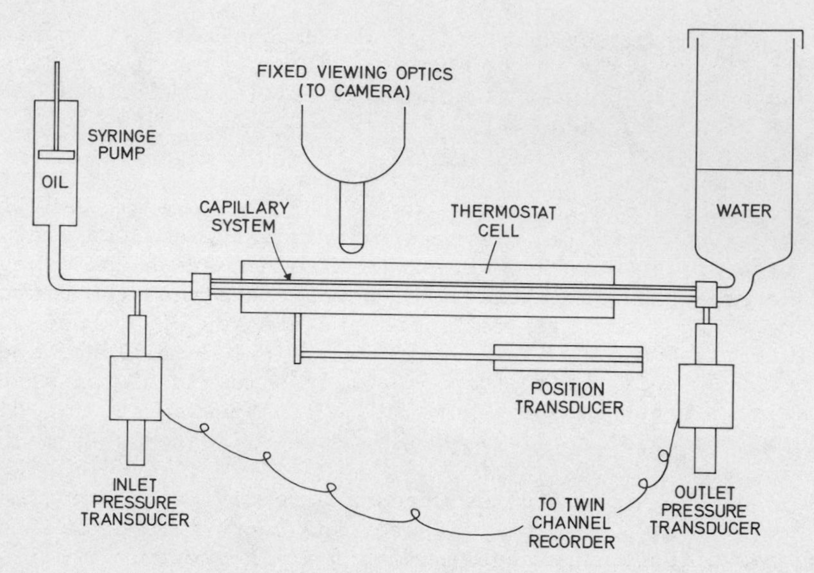

Fig. 1. Interfacial displacement tensiometer apparatus

Lateral movement of the microscope stage, and hence the capillary system, is recorded via a position transducer. By observation of an interface within the capillary, and by maintaining it in the view of the fixed viewing optics, the displacement velocity can be directly determined. The interface can be photographed through the microscope optics, and the dynamic (or static) wetting angle measured (12,18).

Fluid displacement through the capillary system is achieved by use of a syringe pump (Braun, Unita I) driving a 1 ml (Hamilton series 1000) syringe. This arrangement gives flow rates of 10^{-5} to 0.036 ml s^{-1}, constant to ± 0.1%. The use of a larger syringe enables faster flow rates to be achieved. These flow rates correspond to linear displacement velocities in the narrow capillary from 0.003 mm s^{-1} upwards; hence the Capillary Numbers ($N_{ca} \equiv \eta v/\gamma$) were in the range 10^{-5} to 10^{-7} for most runs. This compares with the value $N_{ca} \approx 10^{-6}$ associated with residual water flood saturation in an oil reservoir. The syringe is connected to the capillary system by "Cheminert" PTFE tubes and fittings.

An essential feature of the I.D.T. apparatus is the measurement of pressure at each end of the capillary system. The pressure at the outlet end is held constant, as it is connected to a reservoir of effectively constant hydrostatic head. The pressure measurements are made by all stainless steel variable inductance transducers (EMI model SE 1150/WG) energized by oscillator/demodulator units. The pressure transducer outputs are fed to a twin channel flat bed potentiometric recorder. The pressure measurements have a sensitivity better than 1 Nm^{-2} (0.1 mm H_2O).

Hence the I.D.T. apparatus enables measurement of the volumetric flow rate (Q), the displacement velocity (v), the dynamic contact angle (θ) and the pressure changes (Δ) for the process of immiscible displacement in the capillary system.

Interfacial tension measurement: Interfacial tensions were measured using the pendant drop method (19), and by use of a Wilhelmy plate in conjunction with an interfacial film balance (20).

Materials: The water was triply distilled from all glass equipment. The oils used were "Analar" or of equivalent standard. The toluene was treated by activated alumina to remove trace surface active impurities. The surfactants used were sodium dodecyl benzene sulfonate (SDDBS) supplied by Pfaltz and Bauer, sodium lauryl sulfate (SLS) ("specially pure") supplied by BDH, and sorbitan trioleate (Span 85) supplied by Honeywill-Atlas. The surfactants were used as supplied. Two proteins (Bovine Serum Albumin, BSA, and α-lactalbumin) were used as supplied by Sigma Chemicals.

THEORETICAL

For a single capillary of cylindrical symmetry with the geometry as indicated in Figure 2, the volumetric flow rate (Q) is related to the pressure difference ($P_1 - P_2$) by the expression:

$$P_1 - P_2 = \frac{8Q\eta_j}{\pi}\left(\frac{x[(\eta_i/\eta_j) - 1] + \ell_1}{r_1^4} + \frac{\ell_2}{r_2^4}\right) + \frac{2\gamma(A_1)\cos\theta_1}{r_1} \quad (1)$$

where the symbols are defined by Figure 2. The interfacial tension ($\gamma(A_1)$) is a dynamic quantity, not necessarily equal to the equilibrium value; it is represented as being a function of the interfacial area (A_1). Equation (1) is based on the following assumptions:

(i) The flow is laminar, and can be described by the Poiseuille equation. This has been shown to be a good assumption, even in micron size capillaries (11).

(ii) The only pressure drop associated with the presence of the interface is that expressed by the Laplace equation. However, the interfacial tension is not necessarily the equilibrium one.

(iii) The radius of curvature of the interface can be directly related to the bore of the capillary and the (dynamic) contact angle. This is so if the interface is of constant curvature, i.e., it is not distorted by the displacement, and capillary forces dominate grativational

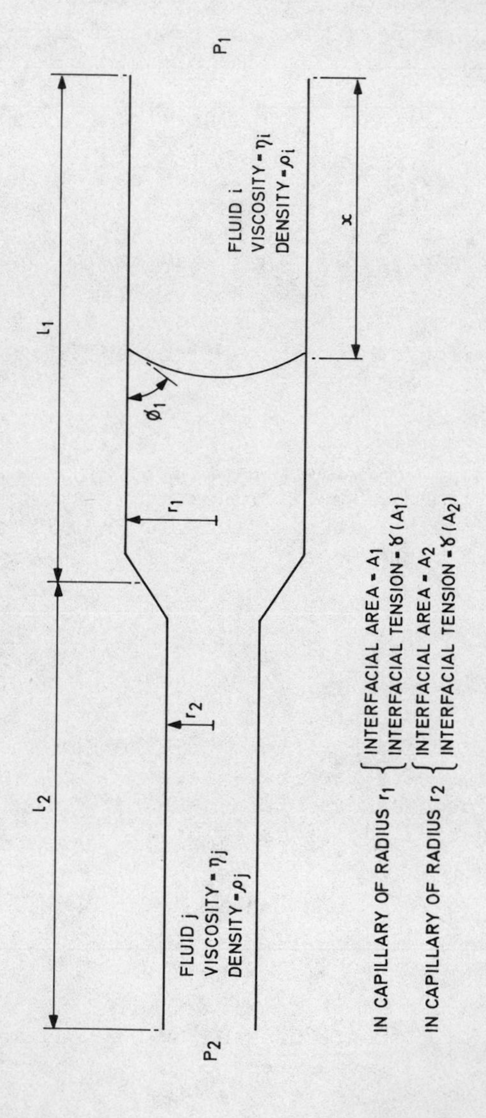

Fig. 2. Capillary system geometry

effects. The latter can be represented as:

$$\gamma \operatorname{Cos} \theta \gg r^2(\rho_i - \rho_j)g \tag{2}$$

The condition expressed by equation (2) is met if the fluids are of equal density ($\rho_i = \rho_j$), or if r is sufficiently small. This condition dictates the maximum bore capillary which may be used for a given pair of immiscible fluids.

(iv) Pressure variation due to acceleration of fluid at the change of bore (Venturi effect) is neglected. This can be shown to be legitimate for the slow flow rates used in the present study.

(v) Effects of interfacial rheology are not explicitly included, except insofar as a dynamic interfacial tension is a function of the dilatational viscosity and elasticity.

(vi) The possibility that the interface may give rise to an intrinsic resistance to displacement (21) of the three phase line is not allowed for.

(vii) Energy dissipation due to non-linear flow in the region of the interface is neglected. Again, this condition is appropriate to the slow flow rates employed in this study. If at these slow flow rates the capillary forces are far greater than the viscous forces as well as the inertial effects, equation (1) can be simplified to the Laplace equation:

$$P_1 - P_2 = \frac{2\gamma(A_1) \operatorname{Cos} \theta_1}{r_1} \tag{3}$$

Hence direct measurement of P_1, P_2 and θ enables calculation of $\gamma(A_1)$.

An expression equivalent to equation (1) can be written for the interface located in the capillary of bore radius r_2. Elimination of Q by combining these expressions, both taken in the limit $\chi \to \ell_1$, gives the pressure drop (ΔP) associated with the movement of interface from one bore to the other:

$$\Delta P = 2\left(\frac{\gamma(A_2) \operatorname{Cos} \theta_2}{r_2} - \frac{\gamma(A_1) \operatorname{Cos} \theta_1}{r_1}\right) \tag{4}$$

The I.D.T. apparatus enables determination of each of the variables in equation (4) as described in the following section. Hence the validity of the assumptions listed above can be tested, and values of the effective interfacial tension obtained during displacement of an interface both into and out of a constriction.

RESULTS OF INTERFACIAL DISPLACEMENT TENSIOMETRY EXPERIMENTS

Pure oil/water systems: In order to evaluate the I.D.T. paparatus and confirm the applicability of equation (4) a number of model systems were examined. In these pure oil displaced distilled water to give a reproducible and stable dynamic contact angle ($\theta = 0°$, measured through the aqueous phase). A typical I.D.T. trace is shown in Figure 3, which can be seen to be simple in shape, featuring pressure jumps as the interface passes through the changes in capillary bore. For ease of reference, displacement of the interface from the wider to the narrower bore is designated "narrowing flow," and displacement in the opposite direction "widening flow." For the model systems, both widening flow and narrowing flow gave rise to the same pressure jump within experimental accuracy.

In the absence of surfactant species the interfacial tension will be constant (i.e., $\gamma(A_1) = \gamma(A_2) = \bar{\gamma}$, say). Therefore, $\bar{\gamma}$ can be calculated directly from the pressure jump using equation (4). These values are compared in Table 1 with directly measured interfacial tensions. It can be seen that the values agree generally to within a spread of $\pm$ 0.5m Nm^{-1} for each of the separate techniques. No differences could be detected between the ΔP values corresponding to narrowing and widening flows.

Direct measurement of the individual pressures (P_1 and P_2) to enable use of equation (3) necessitates accurate absolute calibration of the pressure transducers. A difficulty was that density differences of the oil and water phases affected the hydrostatic pressures within each transducer. Consequently the procedure adopted was to use the pressure transducer outputs for the pure oil/water system as the calibration points for subsequent runs with surface active agents present, as outlined in the following section.

In order to confirm that the bulk viscosity has no appreciable effect, a paraffin oil of viscosity 10 cp was used. The shape of the I.D.T. pressure trace was of the standard form and the interfacial tension calculated from the pressure jump was in agreement with the pendant drop value. The I.D.T. traces were independent of displacement velocity for the displacement rates used (0.03 to 0.3 mm s^{-1}).

Pure oil/surfactant systems: In each case oil displaced the aqueous phase. Displacement in this direction ensured a constant zero contact angle (as measured through the aqueous phase) and corresponded to the advancing interface of an oil blob, as is of

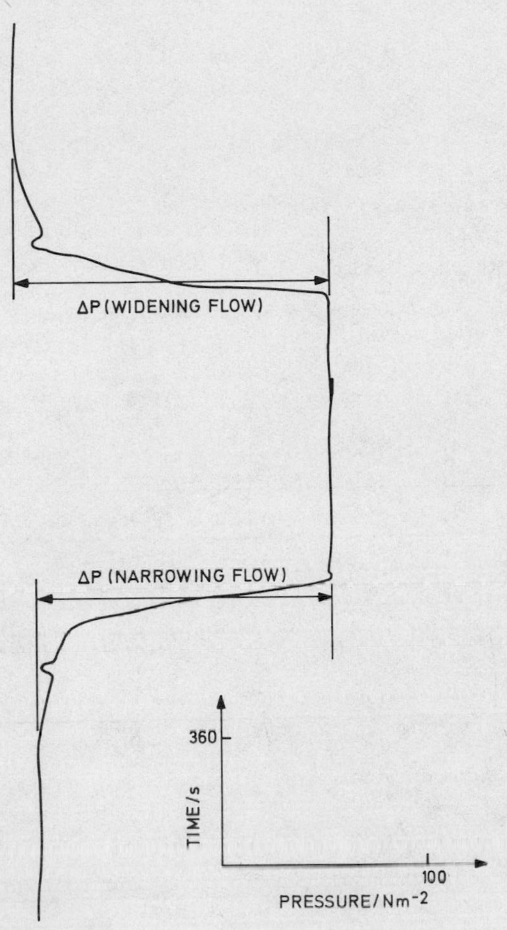

Fig. 3. I.D.T. pressure trace; toluene displacing water.

interest in mobilization of residual oil in enhanced oil recovery. The surfactants used were as follows:

(i) Sodium dodecyl benzene sulfonate (SDDBS)--a selection of I.D.T. traces are shown in Figure 4 for toluene displacing SDDBS solutions over a range of concentrations. The effect of changes of displacement velocity was minimal over the range studied (0.02 mm s^{-1} to 2 mm s^{-1}) as shown by the traces in Figure 5. Values of interfacial tension calculated from the steady state values (or the maximum value attained in the sloping regions) of the I.D.T. traces are shown in Figure 6, together with equilibrium values measured by the pendant drop method. The sloping of the I.D.T. traces indicates that the interfacial tension was not a constant in these runs; thus equation (4) cannot be applied with the simplifying assumption $\gamma(A_1) = \gamma(A_2)$, as was the case for the pure oil/water systems,

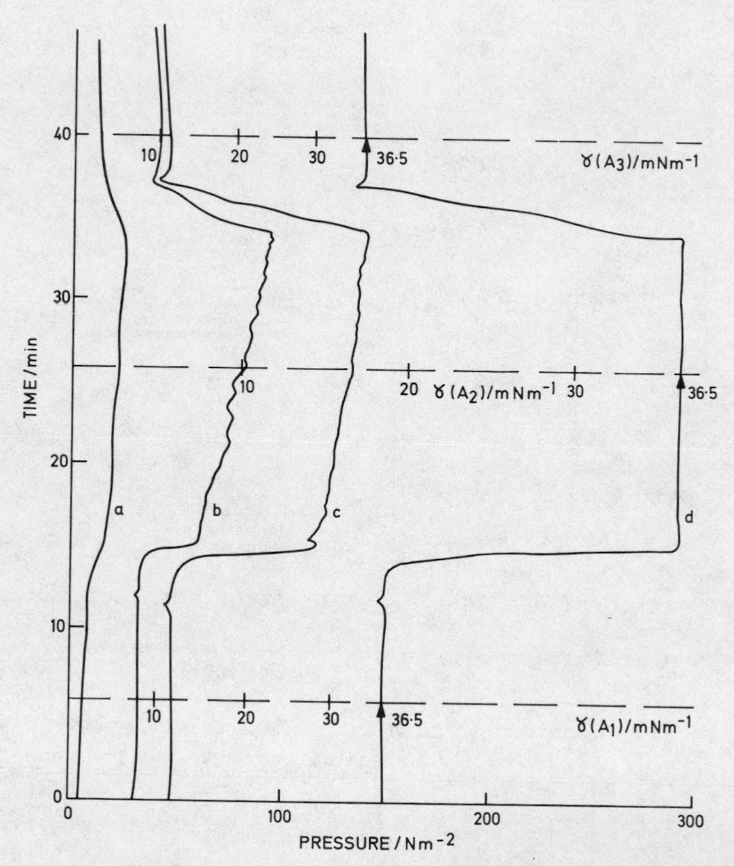

Fig. 4. I.D.T. pressure trace; toluene displacing sodium dodecyl benzene sulfonate solutions.

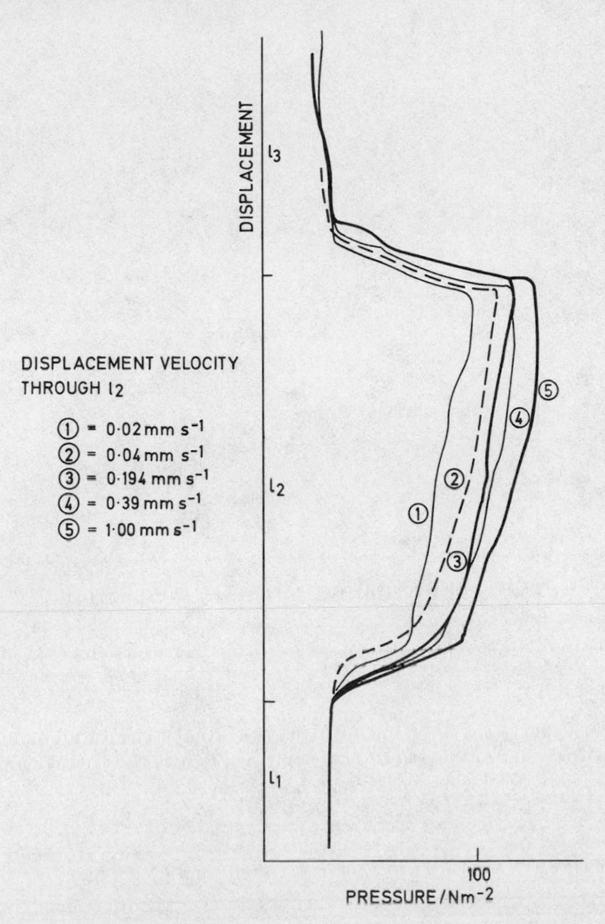

Fig. 5. Toluene displacing 500 ppm sodium dodecyl benzene sulfonate solution; effect of velocity.

described in the previous section. Direct application of equation (3) demands accurate absolute calibration of the pressure transducers. The procedure adopted was to adjust the reservoir to give a constant hydrostatic head (P_2) for all runs. The variation in reading of the inlet pressure transducer between the reference and the test systems could then be used to calculate the interfacial tension at all points along the test system I.D.T. trace, using the toluene/water I.D.T. trace as a base line.

Attempts to measure dynamic interfacial tensions for the SDDBS/toluene system using the film balance were unsuccessful, due to the appreciable solubility of the surfactant in the oil phase. Thus the values of interfacial tension measured were dependent on the factors governing partitioning of the surfactant (principally phase volume

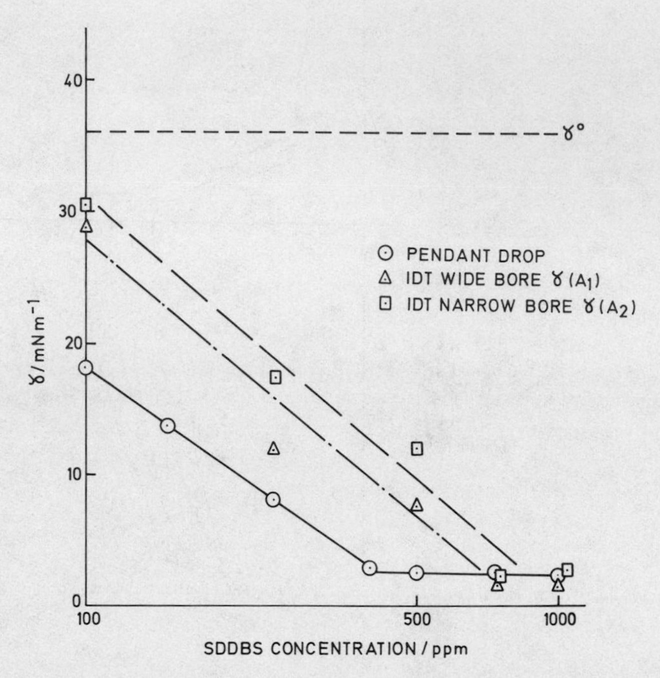

Fig. 6. Interfacial tensions of toluene/sodium dodecyl benzene sulfonate solutions by pendant drop and I.D.T.

Table 1. I.D.T. Results - Pure Oil/Water Systems

Oil	γ/mN m^{-1} (Pendant Drop)	γ/mN m^{-1} (I.D.T.)
Toluene	36	36
_n_-heptane	50	51
_n_-nonane	52	53
_n_-decane	53	53
Carbon tetrachloride	45	45

ratio and aging) rather than upon the expansion/compression cycles imposed upon the interface.

(ii) Sodium Lauryl Sulfate (S.L.S.)--I.D.T. traces for toluene displacing S.L.S. solutions over a range of concentrations are shown in Figure 7. Values of interfacial tension obtained from the I.D.T. traces are compared in Figure 8 with the equilibrium pendant drop values.

(iii) Sorbitan trioleate (Span 85)--I.D.T. traces for toluene displacing Span 85 solutions are shown in Figure 9, and values of interfacial tensions obtained from these traces are compared with equilibrium pendant drop values in Figure 10.

Pure oil/protein solution systems: The Interfacial Displacement Tensiometer was used to examine the effect of additives known to promote an appreciable interfacial rheology. The systems selected were Toluene/α-lactalbumin (100 ppm), toluene/BSA (100 ppm), and nonane/BSA (100 ppm); these systems were expected to produce highly visco-elastic interfacial films. The BSA solution when displaced by either toluene or nonane (see Figure 11) gave rise to pressure jumps on the I.D.T. traces at points equivalent to entry and exit of the narrower capillary. These peaks indicate an increased resistance to displacement of the interface associated with its deformation and changes of extent. The maximum pressures are considerably in excess of those given by the pure oil/water reference system, also shown in Figure 11.

The toluene/α-lactalbumin system did not show the same characteristic peaks on the I.D.T. trace as did the BSA systems. However, the pressure values were constantly in excess of the pure toluene/water case.

DISCUSSION

Immiscible displacement in single capillaries: The Interfacial Displacement Tensiometry results for the pure oil/water systems described previously demonstrate that the simple theory outlined earlier is capable of describing the experimental results satisfactorily. This approach extended to the surfactant solution systems results in measured values of interfacial tension, as described before. These tensions differ from their equilibrium values, so that a better understanding of their behavior is dependent on as full as possible analysis of the factors involved:

Wettability

In considering the dynamic behavior of an oil/water interface during a displacement process, the wettability of the solid

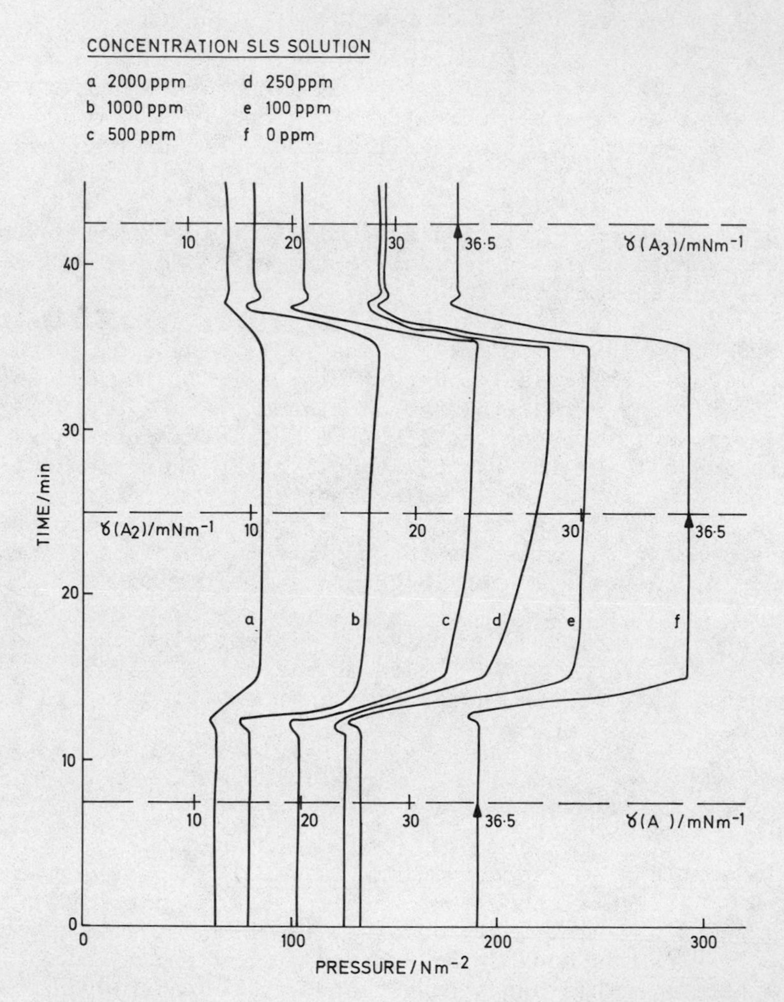

Fig. 7. I.D.T. pressure traces; toluene displacing sodium lauryl sulfate solutions.

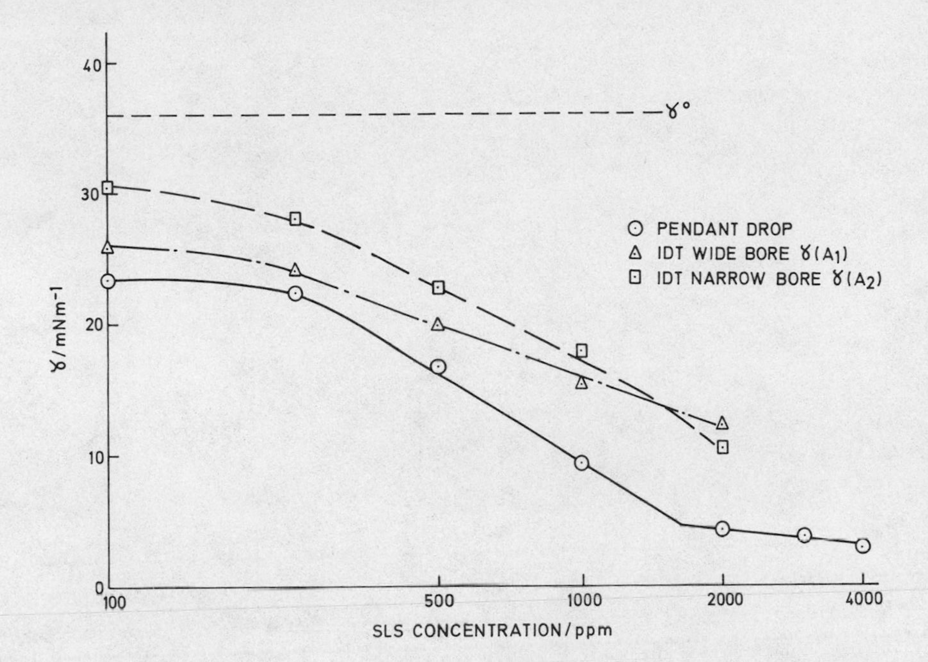

Fig. 8. Interfacial tensions of toluene/sodium lauryl sulfate solutions by pendant drop and I.D.T.

forming the pore matrix is of prime importance. It is the wetting condition prevailing at the oil/water/solid junction which forms a boundary condition for both the interfacial area and extent. Hence, it controls the interface's rheology and the capillary pressure acting across it.

For systems of simple geometry, with smooth solid surfaces, the wettability can often conveniently be characterized by a contact angle. The angle relevant to the present consideration is that which defines the radius of curvature of the interface. This may not correspond to the contact angle as considered at the sub-microscopic level, i.e., in the interphase region above the solid surface. The macroscopic contact angle has been shown (9,18) to be a function of displacement velocity, although the molecular basis for the behavior is not well understood. In the present experiments the displacement of the aqueous phase by the oil phase in clean glass capillaries ensured a constant zero contact angle, as measured through the aqueous phase.

Changes in interfacial extent: In order to follow the geometric interfacial area changes associated with the changes of capillary tube radius, it is necessary to define the wettability of the system. Two classes of behavior can be defined, as follows:

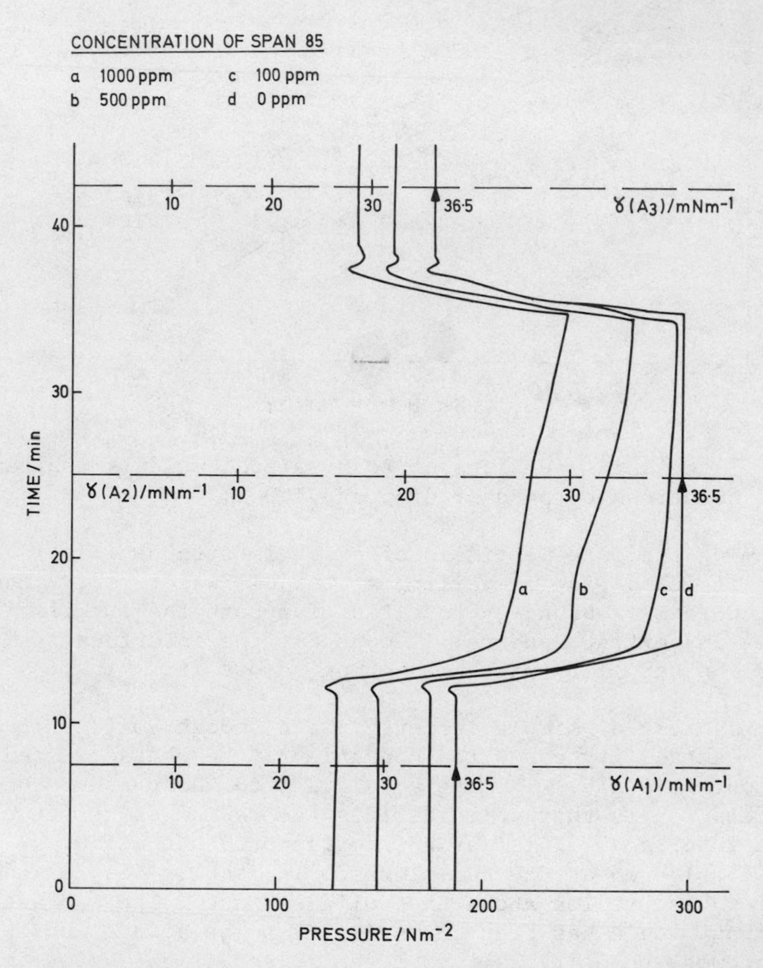

Fig. 9. I.D.T. pressure traces; toluene solution of Span 85 displacing water.

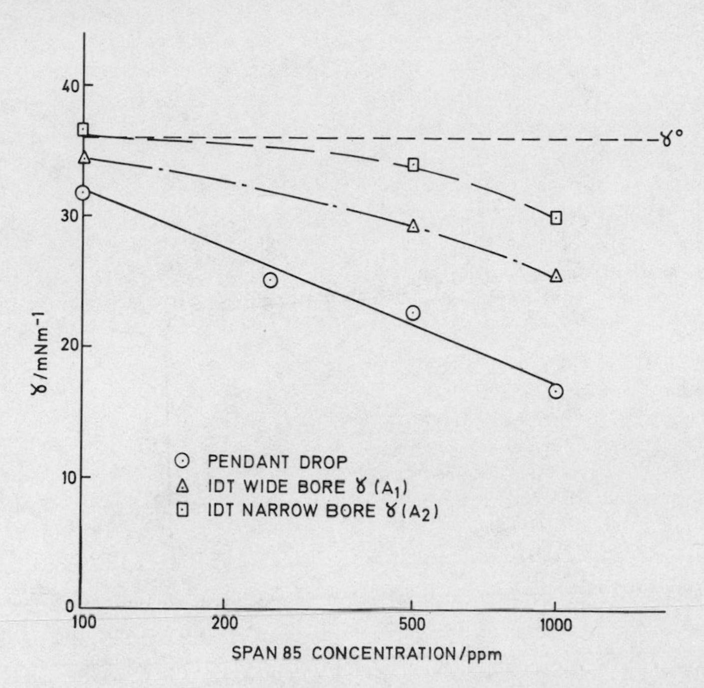

Fig. 10. Interfacial tensions of toluene Span 85 solutions/water by pendant drop and I.D.T.

(i) Intermediate Wetting:

This class considers cases in which the oil/water interfacial extent is bounded by the three phase contact line, even if the contact angle is zero. Thus widening flow results in interface dilation, with the interfacial tension either remaining constant (for systems with no surfactant components) or increasing,

$$\gamma(A_1) \geq \gamma(A_2)$$

Narrowing flow results in compression of the interface so that,

$$\gamma(A_2) = \gamma(A_1) - \pi_i(A_1,A_2)$$

where the film pressure (π_i) is a function of the area reduction and (probably) of time. It has a zero value for pure systems with no surfactant components.

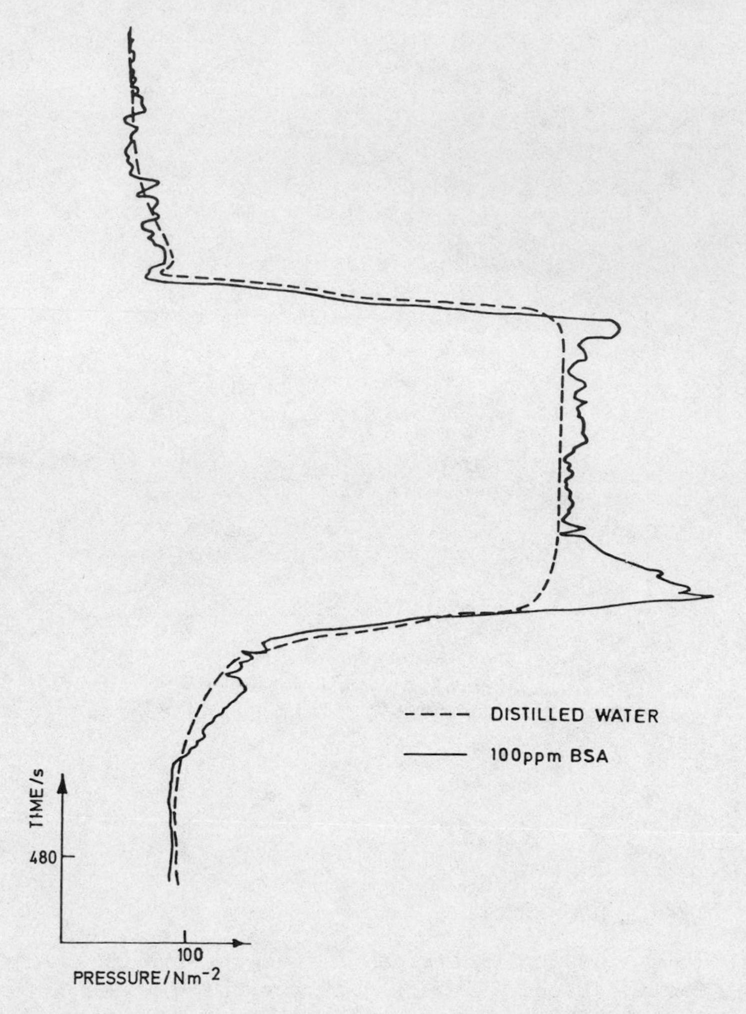

Fig. 11. I.D.T. pressure traces; nonane displacing 100 ppm Bovine serum albumin solution.

(ii) Completely Water Wet:

In this case the interface is no longer bounded by the apparent three phase line of contact, since a thin aqueous film is interposed between the oil and the silica surface. Hence the effective oil/water interface is of much larger extent than the hemispherical cap. This leads to the important and somewhat counter-intuitive result that widening flow results in reduction of the geometric interfacial extent which forms the boundary of a given volume of oil. Conversely, narrowing flow results in interfacial expansion.

Hydrodynamic and interfacial flow: In addition to the wettability, the interfacial state is determined by hydrodynamics of the system. As noted previously (12) the flow profiles in the region of an interface during its displacement are not well known. It is these flows which affect the supply and/or depletion of surfactant to the interface, as well as the flow of the interface itself. Two extreme cases can be envisaged: (a) a fully flowing interface, described by Dussan (8), which results from shear forces imposed on the interface by the interaction of the the two laminar flow profiles in the two liquid phases, (b) a non-flowing interface (12). A non-flowing interface would exhibit interfacial properties dependent on the time lapse since its formation and subsequent deformations. A flowing interface would exhibit a dynamic value of interfacial tension, since fresh interface would be formed continuously.

At fast displacement velocities the aging of a flowing interface would be minimal, and the interfacial tension would approach a pure dynamic value. This would not be strongly influenced by deformation of the interface. However, at rest such an interface would age, so that the aged properties would determine its initial mobilization.

A key factor in the consideration of the interfacial movement is the mechanism by which the three phase line moves. The easiest mechanism to envisage is when the displacing phase moves over a thin film of the displaced phase, i.e., the three phase line is only apparent, as in the definition of completely water wet introduced above. For a real three phase contact line, molecular pictures have been attempted (22). Still uncertain is whether the displacement of the three phase contact line results in a resistance to the flow. Measurements reported by Jacobs (21) have suggested that the presence of surfactants is associated with meniscal resistance to flow, although the mechanism is obscure.

The extent to which interfacial material is conserved during displacement is therefore unknown. Possible behavior ranges from a fluid and flowing interface with material being continuously depleted and replaced, to a rigid interfacial structure. For the latter case, anchoring of interfacial material at the triple phase line may aid "plating out" of interfacial material onto the solid surface and loss of the rigid film material. For fluid interfacial films Marangoni effects may also result in loss of interfacial material.

Liquid/liquid interfaces where surface active agents are present will display interfacial rheology different from that of the bulk phases. An analysis of the rheological properties of liquid/liquid interfaces, as opposed to the static properties such as interfacial tension, is essential in describing the behavior of fluid interfaces where interfacial motion is involved. The dynamic properties of interfacial films can be divided into two classes. Firstly there is the ability of an interfacial film to resist being compressed or expanded (i.e., to resist the area of the film being changed). Secondly, there is the ability of the film to resist being sheared (i.e., to resist the shape of the film being deformed). The first class gives rise to dilatational viscosities and elasticities whereas the second gives shear viscosities and elasticities.

Clearly the factors briefly discussed above are interlinked, and a full analysis presents a formidable task even for the most simple of model geometries. The experimental results will therefore be discussed essentially in qualitative terms at this time.

Displacement of surfactant systems: The three surfactants selected do not give the ultra-low interfacial tensions of greatest interest to EOR, reflecting the limitations of the present I.D.T. equipment; lower interfacial tensions would require concomitantly smaller capillary dimensions. The three surfactants do cover a range of behavior, in that SLS is water soluble, SDDBS is water soluble with an appreciable oil solubility, and Span 85 is oil soluble.

For each of the three surfactant systems, the interfacial tension values calculated from the I.D.T. traces are higher than their equilibrium counterparts as measured by the pendant drop method. This holds true even when the interface is subject only to displacement, rather than expansion or compression; this is shown in Figures 6, 8 and 10, where steady state dynamical interfacial tension values are shown. These values are obtained from the constant linear parts of the I.D.T. traces, or from the last part of the central (small capillary) section of the trace in those cases where it is not certain that the trace had yet reached its plateau (steady state) value.

A notable feature for each of the systems studied is that the interfacial tension increases as the interface progresses along the smaller section of capillary. The rate of increase is largely independent of displacement velocity (Figure 5); hence the conclusion that this increase is a function of the displacement within the capillary, rather than being a relaxation towards a steady state following the deformation of the interface upon the change of capillary bore. None of the I.D.T. traces show features directly identifiable as being a function of the actual changes in bore, i.e., the γ values immediately before and after each pressure jump are in agreement as far as can be determined. It can therefore be concluded that the dynamics of interface displacement dominate any effects of interfacial expansion or compression imposed by changes in geometry.

The dynamic interfacial tension can be related to the fractional change in interfacial area by the dilatational modulus (ε), which, in the absence of relaxation processes, is simply given by (23):

$$d\gamma = \varepsilon\left(\frac{dA}{A}\right) \quad (5)$$

where $d\gamma$ denotes the change in tension from the equilibrium value. Relaxation processes which could be present include changes in configuration of adsorbed molecules, flow of surfactant in the interface along the cylindrical film, and control of supply of surfactants to the interface by diffusion and hydrodynamics.

For the present geometry of a hemispherical interface and zero contact angle, equation (5) can be rewritten as:

$$d\gamma = \frac{\varepsilon V}{r}(dt) = \frac{\varepsilon d\ell}{r} \quad (6)$$

assuming that generation of fresh interface occurs only at the hemispherical cap. This relationship indicates that the deviation from equilibrium interfacial tension is an inverse function of the capillary radius, and is not directly dependent on the displacement velocity (V) but rather upon a perturbation length, $d\ell$. The results for the present systems support these general conclusions that the dynamic interfacial tension is dependent on the capillary bore, but is not strongly dependent on the displacement velocity. The latter point is demonstrated by Figure 5, in which scanning a fifty fold change in displacement rate produces a systematic but gradual change in interfacial tension. The results in Figure 5 are for 500 ppm SDDBS, at which concentration variations can be most sensitively detected. Hence the sloped sections of the I.D.T. traces do not solely arise because of the increased displacement rate in the smaller capillary.

Definition of the so-called perturbation length ($d\ell$) immediately prompts re-evaluation of the assumptions made in constructing equation (6); for complete water wet systems the interface is not limited to the hemispherical cap, but extends as a cylindrical film. The effect of surfactant migration along this film should be distinct for the water soluble surfactant case (surfactant being depleted from the hemispherical part of the interface along the cylindrical film) and for the oil soluble case (surfactant being supplied to the hemispherical region of the interface from the cylindrical film). The possible range of behavior is schematically shown in Figure 12. However, the I.D.T. traces for the systems presented do not reveal any differences which can be clearly ascribed to such an effect at present.

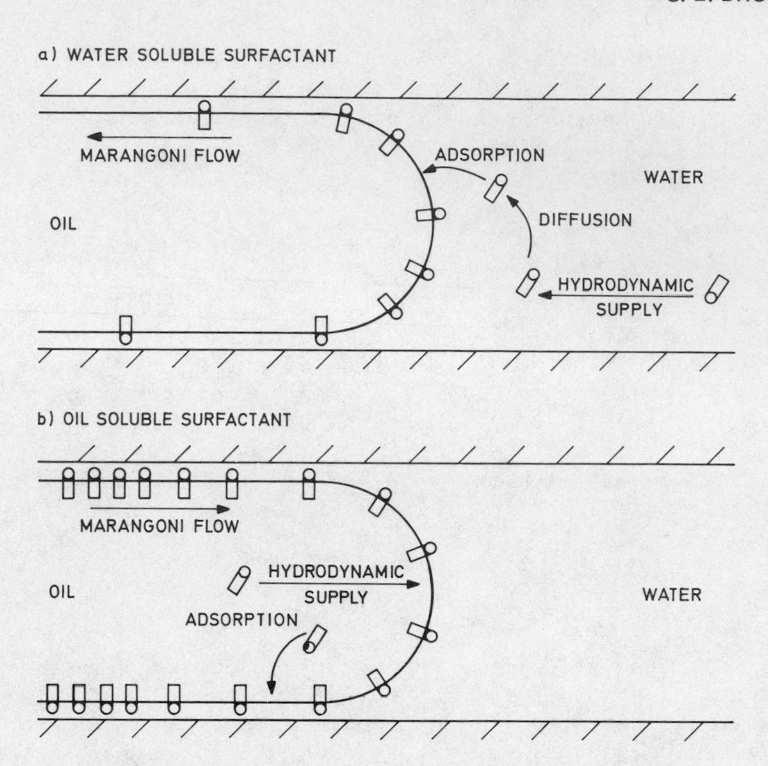

Fig. 12. Surfactant supply and depletion; oil displacing water. (a) water soluble surfactant; (b) oil soluble surfactant.

The I.D.T. traces are sensitive to surfactant concentration. It therefore appears that the value of interfacial tension most directly depends on the supply of surfactant from the bulk to the interface, with a dynamic balance being obtained between the hydrodynamic supply and the surfactant depletion as fresh interface is formed. The similarity of the I.D.T. traces for the three surfactant systems suggest that the effects of non-equivalent flow in the displaced and displacing phase are of secondary importance. However, the sloped I.D.T. traces in the smaller bore capillary do indicate that hydrodynamic supply/depletion of surfactant is not the only factor. A hydrodynamic steady state would be expected to be attained quickly; this conclusion is supported by the constancy of γ values on each side of the changes of bore. Therefore at least one other controlling mechanism is operating, such as depletion of the surface subphase by surfactant diffusion, although no major changes are observed when the surfactant bulk concentration exceeds the c.m.c. In the present geometry the interfacial area to volume ratios are too small to result in a significant reduction of the bulk surfactant concentration.

Displacement for enhanced oil recovery: The results for the model systems presented in this paper demonstrate that the effects of dynamic interfacial tension and of interfacial rheological resistance to deformation induced by displacement can occur during immiscible displacement at flow rates relevant to oil recovery. For displacement in real systems of complex geometry, interfacial movement would proceed episodically, rather than in the smooth manner in the present experiments; thus individual interfacial rearrangements would be significantly faster than the average displacement rate. For this type of movement it is to be expected that dynamic interfacial tension effects are of greatest importance during the flowing or displacement stage. Interfacial rheology will be significant, on the other hand, during initial mobilization, since the film structure required to give rheological resistance to flow will be most developed after aging. Results obtained using the I.D.T. technique (17) with crude oils have shown that the interfacial tension value applicable to displacement corresponds to that of a freshly formed interface, and that crude oil/water systems of markedly different interfacial rheological characteristics display no detectable differences in behavior.

The effect of a dynamic interfacial tension will be to increase the probability of a moving oil blob being trapped. The overall kinetics of blob entrapment and mobilization which depend on the dynamics of interfacial tension variation will determine whether or not the blobs will aggregate; this will be a key factor in formation and stabilization of an oil bank. In addition, any interfacial rheological resistance will reduce the probability of mobilization, and of drop coalescence during oil bank formation. The quantitative assessment of interfacial dynamic properties is therefore of major importance in the development and optimization of chemical EOR systems.

CONCLUSIONS

1. A novel method enables determination of dynamic interfacial tension values for a system undergoing immiscible displacement.

2. For toluene displacing water in the presence of surfactants dynamic interfacial tensions have been measured. The dynamic tension values are dependent on the concentration of surfactant, are inversely dependent on the capillary bore, but show only a weak dependence on displacement rate.

3. The dynamic interfacial tensions for surfactant systems are not appreciably varied by changes in the geometric extent of the interface per se; that is, the interfacial dynamics due to displacement dominate the changes induced by geometry variation.

4. Highly visco-elastic interfacial films can reveal additional resistance to the deformations associated with displacement.

5. The dynamic interfacial tensions for the surfactant systems studied show the same behavior patterns both for oil soluble and water soluble species. This suggests that the effects due to non-equivalent flow patterns in the displaced and displacing phases are of secondary importance.

6. Slow attainment of steady state values of dynamic interfacial tension indicates that diffusion as well as hydrodynamic factors determine the value attained.

7. A complete screening of an EOR surfactant must include determination of the kinetics of interfacial tension changes in addition to their equilibrium values. Considerable work remains to be done to characterize dynamic processes such as oil droplet mobilization, entrapment and oil bank formation.

ACKNOWLEDGMENT

Permission to publish this work has been granted by The British Petroleum Company Limited.

LIST OF SYMBOLS

A	Interfacial area (m^2)
P	Pressure (Nm^{-2})
Q	Volumetric flow rate ($m^3\ s^{-1}$)
V	Displacement velocity ($m\ s^{-1}$)
N_{ca}	Capillary number
g	Gravitational constant ($m\ s^{-2}$)
ℓ, χ, r	Dimensions (m), see Figure 2
γ	Interfacial tension ($mN\ m^{-1}$)
η	Bulk viscosity ($g\ m^{-1}\ s^{-1} \equiv cP$)
ε	Dilatational modulus ($mN\ m^{-1}$)
ρ	Density ($g\ m^{-3}$)
θ	Dynamic contact angle (degrees)
π_i	Interfacial film pressure ($mN\ m^{-1}$)

REFERENCES

1. J. J. Taber, Soc. Pet. Eng. J., 9, 3 (1969).
2. A. Abrams, Soc. Pet. Eng. J., 15, 437 (1975).
3. J. Reisberg and T. M. Doscher, Prod. Monthly, 21, 43 (1956).
4. D. T. Wasan, S. M. Shah, M. Chan, K. Sampath, and R. Shah, preprints to "Chemistry of Oil Recovery Symposium," Am. Chem. Soc., 23, 705 (1978).

5. K. M. Ng, H. T. Davis, and L. E. Scriven, Chem. Eng. Sci., 33, 1009 (1978).
6. A. T. Bourgoyne, B. M. Caudle, and O. K. Kimbler, Soc. Pet. Eng. J., 12, 60 (1972).
7. J. C. Slattery, AIChE Journal, 20, 1145 (1974).
8. E. B. Dussan, AIChE Journal, 23, 131 (1977).
9. T. D. Blake, D. H. Everett, and J. M. Haynes, S.C.I. Monograph No. 25, 164 (1967).
10. R. J. Hansen and T. Y. Toong, J. Colloid Interface Sci., 36, 410 (1951).
11. C. C. Templeton, Petrol. Trans. AIME, 201, 162 (1954).
12. C. E. Brown, E. L. Neustadter, and K. P. Whittingham, "Enhanced Oil Recovery by Displacement with Saline Solutions," pub. British Petroleum, p. 91, 1977.
13. C. E. Brown, Chem. Industry, Nos. 22, 875 (1978).
14. A. Chatenever, Petrol. Trans. AIME, 195, 149 (1952).
15. O. K. Kimbler and B. H. Caudle, Oil & Gas J., 55, 85 (December 16, 1957).
16. J. A. Davies and S. C. Jones, J. Pet. Tech., 1415 (1968).
17. C. E. Brown, E. L. Neustadter, and K. P. Whittingham, paper presented at the "European Symposium on Enhanced Oil Recovery," Edinburgh, 1978.
18. R. L. Hoffman, J. Colloid Interface Sci., 50, 228 (1975).
19. C. E. Stauffer, J. Phys. Chem., 69, 1933 (1965).
20. J. H. Brooks and B. A. Pethica, Trans. Faraday Soc., 61, 571 (1965).
21. H. R. Jacobs, Biorheology, 1, 229 (1963).
22. T. D. Blake and J. M. Haynes, J. Colloid Interface Sci., 30, 421 (1969).
23. J. Lucassen and M. Van den Temple, Chem. Eng. Sci., 27, 1283 (1972).

EFFECT OF ALCOHOLS ON THE EQUILIBRIUM PROPERTIES AND DYNAMICS OF MICELLAR SOLUTIONS

R. Zana

C.N.R.S., Centre de Recherches sur les Macromolécules
6 rue Boussingault
67083 - Strasbourg Cedex, France

The addition of alcohol (ethanol to hexanol) to micellar solutions of alkyltrimethylammonium bromides results in a decrease of cmc and of micelle molecular weight, and in an increase in the degree of ionization of micelles, at detergent concentration close to the cmc. Moreover the relaxation times for the exchange of detergent ions between micelles and surrounding solution and for the micelle formation-dissolution become much shorter (labilization of the micelles) upon addition of the alcohol. The degree of ionization of alkyltrimethylammonium bromide micelles in water-alcohol mixtures goes through a minimum when plotted as a function of the alkyl chain length. The various results have been interpreted in terms of the effect of solubilization of alcohol in the micelle palissade layer on the micelle surface charge density.

INTRODUCTION

Microemulsions have recently aroused considerable interest owing to their potential uses, most notably in tertiary oil recovery (1,2). These systems are generally made of water (or brine), surfactant, cosurfactant and oil. They appear as monophasic, stable and transparent or slightly translucent systems. The nature of microemulsions and origin of their stability is still a matter of discussion (3-7). Likewise, the role of the cosurfactant which, in most instances, is a short chain alcohol (propanol to hexanol) is still unclear (7).

We have tried to contribute to the understanding of the role of alcohol in microemulsions by systematically studying the effect of alcohol on the equilibrium properties (critical micellization concentration, the degree of ionization of micelles, and micelle molecular weight) of micellar solutions as well as on their dynamics (8) (kinetics of the micelle formation-dissolution, and of the exchange of amphiphilic ions between micelles and surrounding solution). The purpose of this paper is to report the results obtained in the course of this study, and the preliminary conclusions reached thus far.

MATERIALS AND METHODS

The measurements were performed on alkyltrimethylammonium bromides, $RN^{+}(CH_3)_3\ Br^{-}$ with R = octyl, decyl, dodecyl, tetradecyl and hexadecyl. These detergents were selected because they are easily synthesized and because commercially available bromide ion specific electrodes can be used to probe their micelle forming properties.

The cmc and micelle ionization degrees near the cmc (9) were determined by emf measurements using a specific bromide ion electrode (Orion 9435), in conjunction with a double junction reference electrode (Orion 9002) and a millivoltmeter (Orion 701A). Some determinations were also performed by means of conductivity. It was noted that emf measurements yielded cmc values slightly lower than those obtained from the equivalent conductivity vs. $(\text{concentration})^{1/2}$ plots. This, however, is of no importance in this work where we are only interested in relative changes.

In aqueous solutions the micelle molecular weights were determined by light scattering using a Fica 50 instrument and a Brice Phenix refractometer for the index of refraction increments. In the presence of added salt the micelle molecular weights were determined by membrane osmometry (High speed osmometer Mechrolab type 502). As in studies by other investigators (10,11) a micellar solution with a concentration well above the cmc was used in the solvent compartment, in order to slow down diffusion. The cellulose acetate membranes (B19, from Sartorius, Gottingen, W. Germany) used in the present investigation did not permit measurements of micellar weights below 10,000.

The dynamic studies involved measurements of the relaxation times of the systems by means of the ultrasonic absorption, T-jump and shock-tube apparatuses used in previous investigations (8,12, 13).

RESULTS

Figure 1 shows the changes of cmc and ionization degree α near the cmc, for tetradecyltrimethylammonium bromide (TTAB) in mixtures of water and 1-alcohols of increasing chain length (ethanol to 1-hexanol). The primary effect of alcohol additions is always a decrease of cmc, even for ethanol. For short chain alcohols, however, the cmc goes through a minimum and increases at higher alcohol content. Similar results have been reported by other workers, for other detergents, such as sodium dodecylsulfate (14-17) and hexadecyltrimethylammonium bromide (17). The changes of α show a simpler behavior than that of the cmc: in all instances α increases with the alcohol content. The results however demand three remarks.

First, the change of cmc is opposite to what is expected on the basis of the observed increase of α upon addition of alcohol. Indeed, for a given alkyl chain length the cmc of detergents decreases as the ionic character decreases. Here, α increases, i.e., the ionic character increases and the cmc decreases.

Second, α increases to values close to 1 at sufficiently large alcohol concentration c_{ROH} (3.5$\underline{M}$ propanol or 0.88$\underline{M}$ butanol). Such a result cannot be understood if the micelle retained the same aggregation number as in the absence of alcohol because the electrostatic repulsions would become much too strong for the micelle to be stable. This result suggests that micelles undergo a drastic reduction of molecular weight. This conclusion is confirmed by the results presented below.

Finally, it can be seen that the α-curves run almost parallel. Each curve can be obtained from the preceding one by a translation along the concentration axis corresponding to a concentration ratio of about 3. The same ratio has been obtained in studies of the partition of alcohol between micelles and surrounding solution (14, 18). The observed changes of α thus appear to be caused by the dissolution of the alcohol into the micelle and not by some modification of the water structure due to the added alcohol.

The results shown in Figure 2 indicate an increase of α, and cmc with temperature. The change of α, however, is much smaller than that obtained upon addition of alcohol. The increase of α indicates that the reaction by which a bound Br^- dissociates from a micelle is endothermal. This result is unexpected, as the reaction of dissociation of an ion-pair made of oppositely charged ions is usually exothermal when only electrostatic interactions are involved, and when the ion-pair is of the outer-outer sphere type (19).

Figure 3 shows the changes of cmc and α near the cmc for alkyltrimethylammonium bromides of increasing chain length in water; water + 1.18$\underline{M}$ propanol; water + 0.5$\underline{M}$ butanol and water + 0.15$\underline{M}$ pentanol. The alcohol contents have been selected in order to have about the same value of α for TTAB in the three water-alcohol mix-

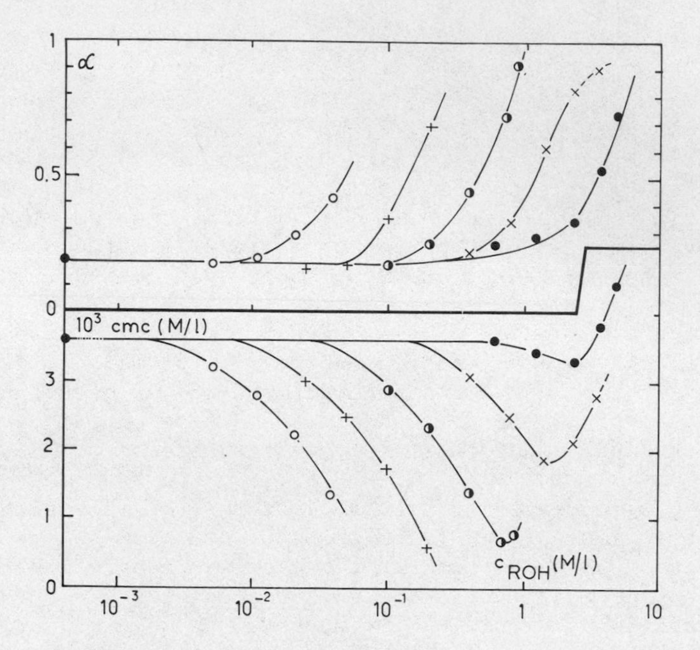

Fig. 1. Tetradecyltrimethylammonium bromide at 25°. Variation of cmc (lower curves) and ionization degree (upper curves) with the concentration c_{ROH} of 1-alcohols: (●) ethanol; (x) propanol; (◑) butanol; (+) pentanol; and (o) hexanol.

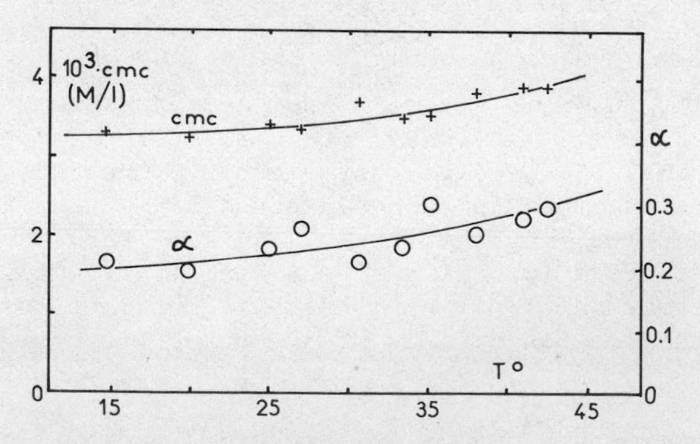

Fig. 2. Tetradecyltrimethylammonium bromide in water. Variation of cmc (+) and α (o) with temperature.

tures. From the results of Figure 1 these compositions closely correspond to systems where the number of alcohol molecules per micelle of a given detergent is practically constant, independently of the alcohol chain length. Straight lines have been found for the four solvent systems. However, while the lines for water and the water-propanol mixtures run parallel, their slope is clearly smaller than those for the water-butanol and water-pentanol mixtures. This may indicate that the alkyl chains of butanol and pentanol are long enough to reach the micellar core, thereby affecting both the electrostatic and hydrophobic contributions to micelle formation. The propanol chain would be too short to reach this core and this alcohol would only act on the electrostatic contribution to micelle formation.

Figure 3 also shows that the α vs. m curves for the three alcohols are almost coincident, within the experimental error, as to be expected from the choice of compositions of the alcohol-water mixtures. These curves, however, differ considerably from that in pure water and in particular show a minimum at around m = 10-12. The α vs. m curve for water shows a continuous decrease as m is increased. Similar, though less complete, results have been reported by other workers for alkyltrimethylammonium bromides (20, 21) and sodium alkylsulfates (22). This decrease is usually attributed to an increase of the micelle surface charge density with m, which results in a decreased ionization. In alcohol-water mixtures the α vs. m curves also show a decrease at low values of m, then another effect appears to be involved which changes the variation of α. Most likely this effect is related with the amount of alcohol solubilized by the micelles. Indeed, the α values are obtained near the cmc. As the cmc decreases exponentially upon increasing m, the amount of micelles present in a given alcohol-water mixture also decreases exponentially. Therefore the relative amount of micelle solubilized alcohol will increase rapidly with m. This, we believe, causes the upturn in the change of α with m (See Discussion Section).

Figure 4 shows the TTAB micelle molecular weight M_W at the cmc, in various water-alcohol mixtures, as obtained from light scattering. The changes are very important. In fact, micelle formation appears to be prevented in water-0.2M pentanol, where the intensity of scattered light showed no change up to detergent concentrations of 10^{-2}M. These large decreases of M_W upon increasing alcohol concentration explain the values of α close to 1 found at high alcohol content. It should be noted that the numerical values of M_W in Figure 4 have not been corrected to account for the binary nature of the solvent (23). This however will not qualitatively affect the observed variations of M_W.

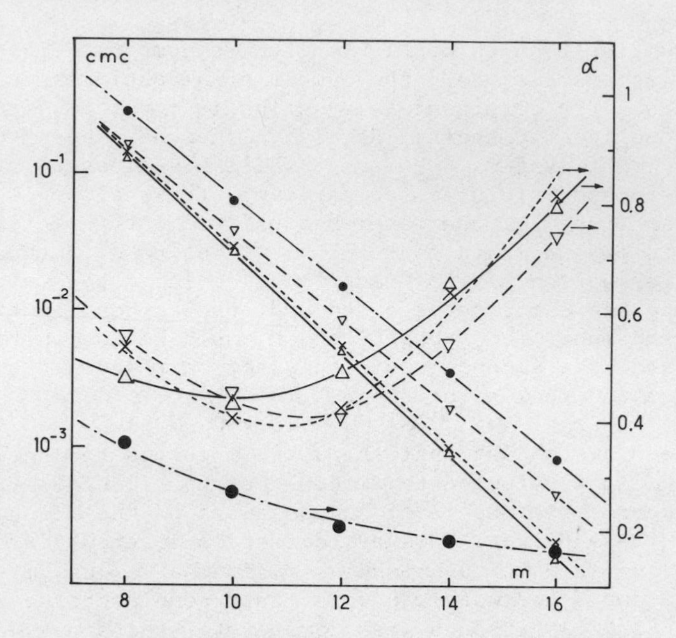

Fig. 3. Alkyltrimethylammonium bromides at 25° in H_2O (•,●); H_2O-1.18$\underline{M}$ propanol (▽,▽); H_2O-0.5$\underline{M}$ butanol (△,△); and H_2O-0.15$\underline{M}$ pentanol (x,X). Variation of cmc (•,▽,△,x) and α (●,▽,△,X) as a function of the number m of carbon atoms in the alkyl chain.

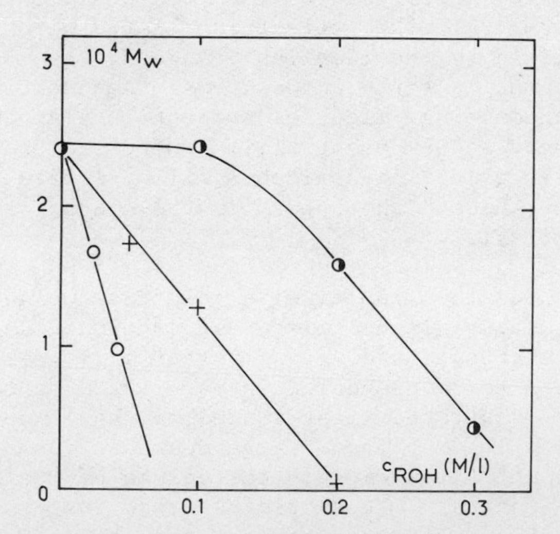

Fig. 4. Tetradecyltrimethylammonium bromide at 25°. Effect of alcohols on the micelle molecular weight, as determined from light scattering: (◑) butanol; (+) pentanol; and (○) hexanol.

Molecular weights determinations have also been performed by means of membrane osmometry in the presence of added salt: 0.025M KBr for hexadecyltrimethylammonium bromide (CTAB) and 0.1M KBr for TTAB (24). Butanol has been found to decrease the molecular weight of CTAB and TTAB micelles from 50,000 in the absence of alcohol to less than 10,000 at about 0.5M butanol. Thus both in the absence and in the presence of salt, butanol brings about a reduction of micelle molecular weight. With pentanol the changes of molecular weight are more complex. Whereas Figure 4 shows a decrease of M_W upon addition of pentanol, the presence of 0.1M KBr results in a M_W vs. c_{ROH} curve going through a maximum as high as 150,000 (24). These surprising changes of M_W have been confirmed by measurements of micelle hydrodynamic radius by means of quasi-elastic light scattering (24).

We finally show in Figures 5 and 6 two sets of typical results concerning the effect of alcohol on the dynamics of micellar systems. The change of the relaxation time τ_1 for the exchange of monomer between micelles and surrounding solution (8,25) for CTAB in presence of various alcohols is shown in Figure 5. In all instances, τ_1 decreases, that is the exchange occurs faster, upon addition of alcohol. The curves are very similar to those relative to the changes of α with c_{ROH}. The changes of α and τ_1 have obviously the same origin, that is the dissolution of alcohol into micelles. Figure 6 shows a dramatic decrease of the relaxation time τ_2 associated with the micelle formation-dissolution equilibrium (8,25) upon addition of pentanol. This decrease indicates an increased lability of the micellar structure as more and more alcohol is incorporated.

DISCUSSION

The main points to be discussed are the decrease of cmc, the increase of α and the decrease of τ_1 and τ_2 upon addition of alcohol, and the α vs. m curves at constant alcohol concentration.

For this purpose we first examine the effect of micellar dissolution of alcohols on micelle stability. This can be conveniently done by means of the schematic representation of part of an alcohol + detergent mixed micelle shown in Figure 7. It has been assumed that alcohol molecules dissolved into micelles have their hydroxyl group at the micelle surface or in the palissade layer (this may not be always the case, as is discussed below). The first effect of the dissolved alcohol molecules is a steric effect. Indeed alcohol molecules intercalated between ionic head groups may push them farther apart than in the absence of alcohol. This would result in a decrease of the surface charge density and a release of counterions, that is an increase of α. This steric effect may not be too important as long as not too many alcohol molecules are

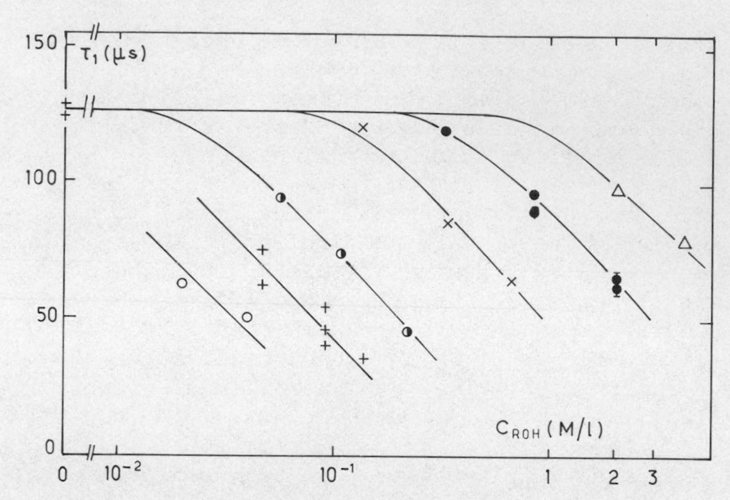

Fig. 5. Cetyltrimethylammonium bromide 2.2×10^{-3} at 24°. Variation of the relaxation time τ_1 for the exchange process upon addition of alcohol: (Δ) methanol; (●) ethanol; (x) propanol; (◐) butanol; (+) pentanol and (O) hexanol. Note the similarity between these curves and those relative to the change of α with c_{ROH}. The τ_1 values were measured by means of the shock-tube device at the laboratory of Prof. Hoffman (Bayreuth, W. Germany).

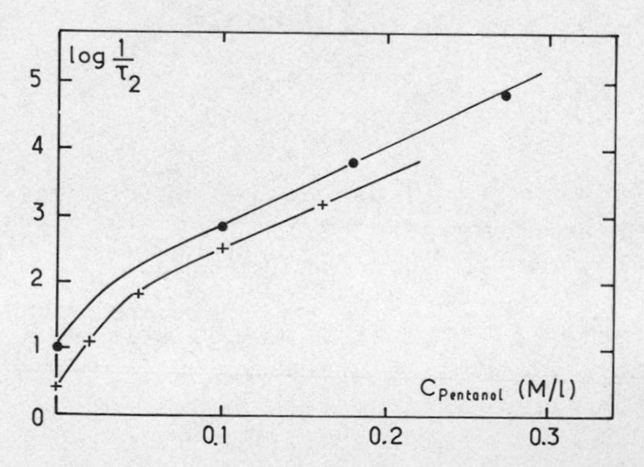

Fig. 6. Variation of the reciprocal of the relaxation time τ_2 for the micelle formation-dissolution process upon addition of pentanol to 0.1M TTAB (+) and 0.1M tetradecylpyridinium bromide (●) solutions in water at 25°. (The measurements were performed by means of T-jump; Eosine was used as a probe for the relaxation process in TTAB solutions (see C. Tondre, J. Lang and R. Zana, J. Colloid Interface Sci., 52, 372 (1975).)

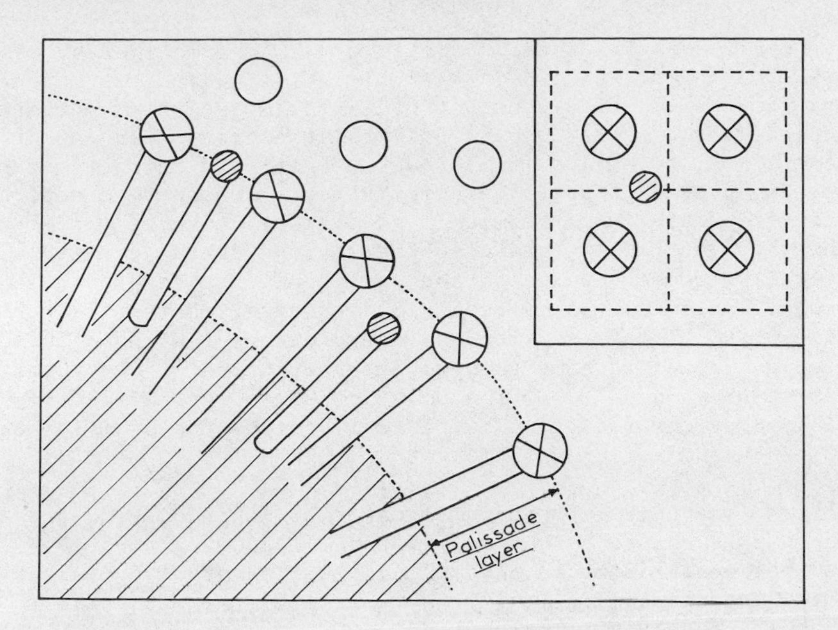

Fig. 7. Highly schematic representation of an exploded view of a part of the mixed micelle (detergent + alcohol) close to the micellar surface (⊗: detergent head group; ⊘: hydroxylic group; ○: counterion). The shaded area corresponds to the micelle hydrophobic core. For the sake of clarity the alkyl chains have been represented as linear. The two dotted lines delineate the palissade layer which contains the detergent head groups, the hydroxylic groups and water. The insert shows the projection of the micellar surface on a plane. For the sake of clarity, the detergent head groups have been arranged in a planar square lattice, instead of an hexagonal one. The head groups cover a certain surface area, but there may still be enough space between adjacent groups for hydroxylic groups.

incorporated into micelles. Indeed, the surface area per trimethylammonium head group at the micellar surface is of about 80Å^2, while the area covered by the head group is about 40Å^2. On the other hand, the surface area of the hydroxylic group is less than 10Å^2, therefore for the alcohol one has to consider the surface of the cross section of the alkyl chain, i.e., 20Å^2 rather than that of the OH group. In any case, there is ample surface area not covered by the head group for the intercalation of one alcohol molecule per detergent ion.

The second effect of solubilized alcohol molecules is the change of local dielectric constant ε_ℓ in the palissade layer, due to the replacement of water molecules by alcohol molecules. The values reported for ε_ℓ range from 40 to 50 (26), that is larger than the bulk dielectric constant of alcohols including methanol. The replacement of water by alcohol will therefore bring about a decrease of ε_ℓ, resulting in increased electrostatic repulsions between ionic head groups. This in turn will cause some detergent ions to dissociate from the micelles, thereby reducing the detergent aggregation number, and also the surface charge density since the overall process amounts to the replacement of detergent ions by alcohol molecules. As a result α increases. Note that the decrease of ε_ℓ associated with the dissolution of alcohol into micelles has received some support in experiments using pyrene as a fluorescence probe (27). The polarity sensed by micelle-solubilized pyrene was found to be decreased by the addition of alcohol to the micellar solution.

There is a third effect which may also contribute to the observed changes caused by the addition of alcohol. It is the increased molecular disorder in the palissade layer of the mixed alcohol + detergent micelles, with respect to the situation in the absence of alcohol (28).

The first two effects that have been described provide us with a simple explanation for the decrease of molecular weight and increase of α upon alcohol addition. They also explain the decrease of cmc, which now appears to be the result of the lower charge density of the mixed alcohol + detergent micelle, with respect to the pure detergent micelle (the repulsions being less strong, the micelles start forming at lower concentration). These effects also permit us to explain the presence of a minimum on the α vs. m curves of Figure 3. For the reason given above, in the absence of alcohol, α decreases as m increases. In presence of alcohol this effect is still operative but there is also an increase of ionization caused by the micellar dissolution of alcohol. This last effect becomes rapidly predominant because at detergent concentration close to the cmc, the number of micelles in the system decreases exponentially, just like the cmc, as m is increased. There are therefore less and less micelles to solubilize alcohol, and at constant c_{ROH}, the amount of solubilized alcohol per micelle increases, and the micelle becomes more and more ionized, as as m is increased.

The complex changes of molecular weight of TTAB micelles in H_2O-0.1$\underline{M}$ KBr upon addition of pentanol may be explained in terms of a distribution of alcohol between different micelle solubilization sites. In H_2O, the alcohol may essentially be dissolved in the palissade layer, thereby decreasing the molecular weight by the effects discussed above. In H_2O-0.1$\underline{M}$ KBr, pentanol is salted out of water and may preferentially dissolve into the micelle hydrophobic core, thereby increasing the micelle molecular weight.

When the core is saturated the pentanol may start dissolving into the palissade layer, thereby causing a decrease of molecular weight by the effects discussed above.

We now turn to the kinetic results. Aniansson (29) has recently reported an extension to mixed micelles, made out of two detergents, of his theoretical treatment of the kinetics of simple micellar systems (8,25). The expressions given for τ_1 and τ_2 can be used to qualitatively assess the effect of alcohol on these relaxation times, since the alcohol molecules can be considered to form mixed micelles with the detergent. The ratio C_m/cmc, where C_m is the micelle concentration, appears in the predominant term in Aniansson's equation for $1/\tau_1$. At constant detergent concentration, close to the cmc, an addition of alcohol brings about an increase of C_m and a decrease of cmc. The ratio C_m/cmc, and thus $1/\tau_1$ should increase with the alcohol concentration, as is indeed observed.

The expression of τ_2 is much more complex than τ_1 and its dependence on alcohol concentration cannot be simply assessed. However for a qualitative discussion, we recall that $1/\tau_2$ is proportional to $(\text{cmc})^{r-1}$ (8), where r is the number of amphiphilic ions constituting the associated species at the minimum of the micelle size distribution curve, and which may be considered as micelle nuclei (8). The value of r for normal detergents, in the absence of alcohol, is of about 10 (8,30). In previous studies rapid changes of τ_2^{-1} have been explained in terms of changes of cmc and r (31). A similar explanation is likely to hold for the effect of addition of alcohol on τ_2, but the situation may somewhat be more complex as alcohol may also be incorporated to micelle nuclei. If we neglect this effect, take r = 10 and consider the effect of 0.1M pentanol on the τ_2 of a 0.1M TTAB solution, one is led to assume a decrease of r from 10 to 6 to explain the observed decrease of τ_2, as the cmc is decreased by a factor of 2 upon addition of 0.1M pentanol.

It should be noted that the exchange of alcohol between the mixed micelles and the surrounding solution also gives rise to a very fast relaxation process (in the 10-100 ns range) which has been investigated separately (32). Aniansson's theory for mixed micelles predicts this very fast process and gives the expression of the corresponding relaxation time.

A quantitative analysis of the above kinetic results on the basis of Aniansson's theory of mixed micelles is not possible at the present time because it requires the knowledge of such quantities as the total micelle molecular weight at every concentration of alcohol and detergent, and the micelle composition. Work is now in progress in our laboratory in order to determine these quantities.

CONCLUSIONS

The effect of alcohols on dilute micellar solutions is to decrease the cmc and micelle molecular weight and to increase the micelle ionization. Also, the exchange of detergent ions between micelles and surrounding solution and the micelle formation-dissolution are strongly accelerated upon addition of alcohol (micelles become more labile in presence of alcohol). All these variations can be explained on the basis of the effects associated with the dissolution of alcohol into micelles, leading to mixed alcohol + detergent micelles.

The extreme labilization of micellar structure caused by the addition of alcohol may reflect the molecular disorder at the micelle-solution interface, due to the presence of chains of different lengths (28). It may be responsible for the much faster equilibration of detergent + oil systems in the presence of alcohol, with respect to what is observed in the absence of oil. This faster equilibration is one of the reasons of the addition of alcohol to detergent + oil systems in tertiary oil recovery.

ACKNOWLEDGMENTS

The author is pleased to thank Prof. Hoffmann (Bayreuth, W. Germany) for the use of the shock tube and Drs. Yiv, Candau and Strazielle for the permission to mention unpublished results (quasi-elastic and classical light scattering, osmometry). The financial support of the DGRST under Grant No. 77 7 1456 is gratefully acknowledged.

REFERENCES

1. K. L. Mittal, ed., "Micellization, Solubilization and Microemulsions," Vols. I and II, Plenum, New York, 1977.
2. K. Shinoda and S. Friberg, Adv. Colloid Interface Sci., 4, 281 (1975).
3. E. Sjöblom and S. Friberg, J. Colloid Interface Sci., 67, 16 (1978).
4. L. E. Scriven, in "Micellization, Solubilization and Microemulsions," K. L. Mittal, ed., Vol. II, Plenum, New York, p. 877, 1977; Nature, 263, 123 (1976).
5. Y. Talmon and S. Prager, J. Chem. Phys., 69, 2984 (1978).
6. E. Ruckenstein, J. Colloid Interface Sci., 66, 369 (1978); E. Ruckenstein and J. Chi, J. Chem. Soc. Faraday Trans. II, 71, 1690 (1975).
7. A. Skoulios and D. Guillon, J. Phys. Lett., 38, L-137 (1977).
8. E.A.G. Aniansson, S. Wall, M. Almgren, H. Hoffmann, I. Kielman, W. Ulbricht, R. Zana, J. Lang, and C. Tondre, J. Phys. Chem., 80, 905 (1976).

9. R. Zana, J. Colloid Interface Sci., in press.
10. H. Coll, J. Phys. Chem., 74, 520 (1970).
11. K. S. Birdi, S. Backlund, K. Sorensen, T. Krag, and S. Dalsager, J. Colloid Interface Sci., 66, 118 (1978).
12. J. Lang, C. Tondre, R. Zana, R. Bauer, H. Hoffmann, and W. Ulbricht, J. Phys. Chem., 79, 276 (1975).
13. C. Tondre, J. Lang, and R. Zana, J. Colloid Interface Sci., 52, 372 (1975).
14. K. Hayase and S. Hayano, Bull. Chem. Soc. (Japan), 50, 83 (1977).
15. K. Shirahama and T. Kashiwabara, J. Colloid Interface Sci., 36, 65 (1971).
16. M. Manabe and M. Koda, Bull. Chem. Soc. (Japan), 51, 1599 (1978).
17. H. Singh and S. Swarup, Bull. Chem. Soc. (Japan), 51, 1534 (1978).
18. M. Manabe, K. Shirahama, and M. Koda, Bull. Chem. Soc. (Japan), 49, 2904 (1976).
19. This can be simply demonstrated by deriving with respect to T the expression of the free energy of formation of an ion-pair. See for instance M. Emerson and F. Holtzer, J. Phys. Chem., 71, 3320 (1967).
20. J. T. Pearson, J. Colloid Interface Sci., 37, 509 (1971).
21. K. J. Mysels, J. Colloid Interface Sci., 10, 507 (1955).
22. A. Yamauchi, T. Kunisaki, T. Minematsu, Y. Tomokiyo, T. Yamaguchi, and H. Kimikuza, Bull. Chem. Soc. (Japan), 51, 2791 (1978).
23. G. Parfitt and J. Wood, Kolloid Z.Z. Polymere, 229, 55 (1969).
24. R. Zana, J. Candau, C. Strazielle, and S. Yiv (unpublished results).
25. E.A.G. Aniansson and S. Wall, J. Phys. Chem., 78, 1024 (1974) and 79, 857 (1975).
26. J. Cardinal and P. Mukerjee, J. Phys. Chem., 82, 1614 (1978); P. Mukerjee and J. Cardinal, ibid., 82, 1620 (1978), and references therein.
27. P. Lianos and R. Zana, Chem. Phys. Lett., in press.
28. P. Maelstaf and P. Bothorel, Compt. Rend. Acad. Sci. (Paris), Ser. C, 288, 13 (1979); D. O. Shah, private communication.
29. E.A.G. Aniansson, in "Techniques and Applications of Fast Reactions in Solutions," D. Reidel Publ. Co., Dordrecht, Holland, p. 249, 1979.
30. H. Hoffmann, Ber. Bunsenges. Phys. Chem., 82, 988 (1978), and references therein.
31. W. Baumuller, H. Hoffmann, W. Ulbricht, C. Tondre, and R. Zana, J. Colloid Interface Sci., 64, 418 (1978).
32. S. Yiv and R. Zana, J. Colloid Interface Sci., 65, 286 (1978).

THE EFFECT OF ALCOHOL ON SURFACTANT MASS TRANSFER ACROSS THE OIL/BRINE INTERFACE AND RELATED PHENOMENA

M.Y. Chiang* and D.O. Shah

Departments of Chemical Engineering and Anesthesiology
University of Florida
Gainesville, Florida 32611, U.S.A.

The effect of alcohol on surfactant mass transfer from bulk solution to the oil/dilute micellar solution interface was studied. Various interfacial properties of the surfactant solutions and their ability for displacing oil were determined. For the surfactant-oil-brine systems studied, the interfacial tension (IFT) and surfactant partition coefficient did not change when isobutanol was added to the following systems: 0.1% TRS 10-410 in 1.5% NaCl vs. n-dodecane and 0.05% TRS 10-80 in 1.0% NaCl vs. n-octane. On the other hand, the interfacial viscosity, oil drop flattening time (i.e. the time required for an oil droplet to flatten out after being deposited on the underside of a polished quartz plate submerged in the micellar solution) and oil displacement efficiency were influenced markedly by the addition of alcohol.

In the presence of isobutanol, the oil/dilute micellar solution interface became more fluid and the flattening time decreased from 90 seconds to less than a second or 420 seconds to less than a second, and the final oil saturation decreased from 30% to 5.36% and 11.73% to 1.28% respectively for the two systems mentioned above. Furthermore, it was observed that after the arrival of the oil bank, the ΔP leveled off for the isobutanol containing systems, whereas it continuously increased for the systems without isobutanol. This observation is consistent with the proposed role of alcohol in lowering the interfacial viscosity and promoting coalescence of oil ganglia in porous media.

---

*Present Address: Dr. M. Y. Chiang, American Can Company, Post Office Box 50, Princeton, New Jersey 08540, U.S.A.

The flattening time was strikingly lower for the surfactant + alcohol system as compared to the flattening times in the presence of the surfactant or alcohol alone in the brine, suggesting that the rate of achieving ultralow IFT at the oil/micellar solution interface is strikingly enhanced by the presence of isobutanol resulting in greater oil recovery.

In order to delineate the effect of surfactant mass transfer on in situ behavior of oil ganglia, we carried out several oil displacement experiments using equilibrated and nonequilibrated oil/micellar solution systems. For equilibrated systems, the oil displacement efficiency showed an excellent correlation with IFT and capillary number. However, for unequilibrated systems, the oil displacement efficiency depended on salinity. Below optimal salinity, the oil displacement efficiency almost remained the same for both equilibrated and nonequilibrated systems, whereas at and above optimal salinity the oil displacement efficiency was higher for nonequilibrated systems as compared to equilibrated systems. This was attributed to mass transfer rate effects in these systems. Both sandpacks and Berea cores gave similar results. The results of this study demonstrate the importance of transient phenomena at oil/dilute micellar solution interface for oil displacement process with emphasis on the effect of alcohol and salinity.

INTRODUCTION

Laboratory studies on oil displacement efficiency by surfactant-polymer flooding process have been reported by a number of investigators (1-10). In general, the process is such that after being conditioned by field brine or preflush, a sandstone core or a sandpack is oil-saturated to the irreducible water content. It is then waterflooded to the residual oil level. Finally, a slug of surfactant solution followed by a mobility buffer is injected. The slug of surfactant solution can either be aqueous or oleic with a surfactant plus alcohol concentration of 5-15%.

Because of the cost and the time factors involved, oil displacement studies are always preceded by certain test tube screening procedures. Specifically, the interfacial tension (IFT) of less than 0.01 dyne/cm is recognized to be the necessary but not the sufficient criterion for selection of a surfactant system. Many investigators (10-15) have shown that ultralow IFT of less than 0.001 dyne/cm can be achieved with less than 0.1 wt. % surfactant solution. Since this low surfactant concentration system is several hundred times more dilute than the ones used in a typical surfactant-polymer flooding process, the economics dictates that the oil displacement by such low surfactant concentration solution should be explored. Moreover, it should be established that the

IFT obtained in the oil/brine/surfactant system after a rigorous equilibration procedure in a test tube can be indeed achieved in situ when the surfactant solution passes by the entrapped oil ganglia in porous media. In other words, is the rate of mass transfer of surfactant to and across the interface a major limiting factor in achieving ultralow IFT in porous media? The present paper attempts to answer some of these questions by comparing the behavior of equilibrated and nonequilibrated oil/surfactant solution systems in porous media.

In general, the surfactant formulations used for enhanced oil recovery contain a short chain alcohol. The addition of alcohol can influence the viscosity, IFT and birefringent structures of micellar solutions as well as coalescence rate of oil ganglia. The present paper reports the effect of addition of isobutanol to a dilute petroleum sulfonate (< 0.1% conc) solution on IFT, surface shear viscosity, surfactant partitioning, the rate of change of IFT (or flattening time) of oil drops in surfactant solutions and oil displacement efficiency. The two surfactant systems chosen for this study indeed exhibited ultralow IFT under appropriate conditions of salinity, surfactant concentration and oil chain length (11,15,19).

EXPERIMENTAL

Surfactant solutions: Commercial petroleum sulfonate TRS 10-80 (80% active) or TRS 10-410 (61.2% active) obtained from Witco Co. and Fisher A.C.S. certified grade NaCl crystals 1% NaCl) were dissolved in distilled, deionized water to make the surfactant stock solutions by weight. Then, they were diluted by brine (1% NaCl) to the desired concentration just before the start of each run, so that the surfactant aging effect was minimized. The purity of n-octane or n-dodecane (Chemical Samples Co.) was > 99% and was used as the oil to equilibrate the surfactant solution at the volume ratio of 1:2 in a glass-stoppered 1-liter separatory funnel. After vigorous shaking, the surfactant and oil mixture was left standing for 10 days at room temperature until a clear mirror-like interface was reached. The equilibrated aqueous and oleic solutions were then drained into separate storage bottles. The effect of alcohol was studied by adding 99% pure isobutanol (IBA) (Chemical Samples Co.) to the surfactant solution at 1:1 weight ratio for the active component in TRS 10-80 or TRS 10-410.

Interfacial tension measurements: Interfacial tension between various oleic and aqueous phases was measured using the Spinning Drop Tensiometer at 25°C. The spinning time and rate were kept constant so that comparative results could be obtained.

Interfacial viscosity measurements: Interfacial viscosity (IFV) was measured using a viscous-traction interfacial viscometer constructed according to Wasan (16). Teflon particles were used to measure the centerline velocity of oil/water interface.

Contact angle in quartz/brine/oil systems: The wettability of the quartz surface used to simulate the surface of sandstones, was studied by a contact-angle goniometer. Using a microsyringe, an oil drop was deposited on the underside of a smooth, polished quartz surface submerged in the aqueous solution at 25°C. The angle through the oil phase was measured and Polaroid pictures of the oil drop were taken at different time intervals.

Surfactant concentration measurements: The surfactant concentration in the effluent stream was measured by the two-phase titration method according to Reid et al. (17).

Oil displacement in porous media: Horizontally mounted sandpacks and Berea cores encased in an air-circulating constant temperature box were used for oil displacement efficiency tests. The sandpacks, 1.06" diameter by 7.0" long, had an average porosity of 38% and permeability of 3.0 darcy. The Berea cores were 1" square by 12" long cast in epoxy resin within 1.5" diameter by 14" long PVC pipes. They had an average porosity of 18% and permeability of 220 millidarcy.

Having been dry-filled with sands under vibration and tapping, the pack was flushed vertically with carbon dioxide to displace interstitial air. Deionized water was then pumped through and the pore volume (PV) was measured. Since carbon dioxide easily dissolves in water, trapped gas in the pack can be greatly reduced or eliminated. New sandpacks and fresh Berea cores were used for each run. The brine salinity in porous media was the same as the salt concentration of surfactant solution. The injected oil and aqueous solutions were either pre-equilibrated or nonequilibrated. Constant fluid velocity of 10 ft/day was maintained during the oil saturation and 2.3 ft/day was maintained during aqueous solution or brine flooding. Because the viscosity of n-octane was 0.5 cp, a favorable mobility was assumed for aqueous surfactant solution flooding. Therefore, no polymer was added in the dilute surfactant solution.

Displacement tests were conducted for equilibrated and nonequilibrated systems. For the effect of surfactant concentration on oil recovery, the total amount of surfactant injected was the same, i.e., the slug size times the concentration was equal (70% PV x 0.5% = 35) for each run.

RESULTS AND DISCUSSION

Effect of isobutanol on oil displacement efficiency: The following systems, 0.1% TRS 10-410 with/without 0.06% IBA in 1.5% NaCl vs. n-dodecane and 0.05% TRS 10-80 with/without 0.04% IBA in 1.0% NaCl vs. n-octane were examined.

Figure 1 is the cumulative oil recovery profile of the systems studied. It shows that with the addition of 0.06% IBA into the TRS 10-410/n-dodecane system, the oil recovery by direct surfactant solution flooding (i.e., without waterflooding) is improved from 84.37% to 98.32% after 3.5 PV surfactant solution injection. The TRS 10-80/n-octane system showed an increase in oil recovery from 60% to 91% by the addition of isobutanol (Figure 1). It should be noted that the increase in oil recovery occurs only after the major oil bank comes out (i.e., after 1 PV of produced fluid). We propose that the presence of isobutanol promotes the coalescence of oil droplets in porous media leading to a better oil recovery efficiency. A much more drastic difference is seen in the TRS 10-80/n-octane system, where the tertiary oil recovery increased from 0% without IBA to 76.84% with IBA (Table 1) after 2.7 PV surfactant solution injection. Thus, for both secondary and tertiary oil recovery processes (i.e., with or without brine flooding stage) carried out in these laboratory scale experiments, the addition of isobutanol enhances the oil recovery efficiency presumably by promoting the coalescence in porous media.

Table 1 shows the effect of the addition of isobutanol on various properties of oil/brine/surfactant systems for TRS 10-410 and TRS 10-80. Because the same IFT values were obtained for the systems with and without IBA (Table 1), the observed differences in oil recovery cannot be explained in terms of any change in IFT. The presence of alcohol did not significantly influence the partition coefficient of surfactant in n-dodecane or n-octane. It is important to emphasize that the partition coefficient changes sharply near the ultralow IFT region (19). Thus, the partition coefficient does not appear to correlate with the oil displacement efficiency. However, the presence of isobutanol decreases the interfacial viscosity and markedly influences the flattening time of the oil droplets. It has been suggested (18) that a rigid potassium oleate film at the oil/water interface can be liquefied by the penetration of the hexanol molecules in order to produce spherical microemulsion droplets. It has been shown (14) also that for a commercial petroleum sulfonate-crude oil system, the oil droplets with the alcohol coalesce much faster than the ones without alcohol. For the systems studied here, IBA is believed to have penetrated the petroleum sulfonate film as seen by the decrease in IFV. The decrease in interfacial viscosity would presumably promote the coalescence in porous media.

Table 1. The Effect of IBA on Flattening Time, IFT, IFV, Partition Coefficient, and Oil Displacement Efficiency

System	0.1% TRS 10-410 in 1.5% NaCl vs. n-Dodecane	0.1% TRS 10-410 + 0.06% IBA in 1.5% NaCl vs. n-Dodecane	0.05% TRS 10-80 in 1% NaCl vs. n-Octane	0.05% TRS 10-80 + 0.04% IBA in 1% NaCl vs. n-Octane
Run	S100-48	S100-43	S100-02	S100-44
Flattening Time	90 sec	< 1 sec	420 sec	< 1 sec
IFT (dynes/cm)	0.086	0.088	0.025	0.024
Interfacial Viscosity (s.p.)	0.096	0.086	0.023	0.018
Partition Coefficient	0.010	0.009	0.3	1.36
Secondary Recovery				
By Brine Flooding	--	--	61.2%	60.08%
By Surfactant Soln. Flooding	84.37%	98.32%	60%	91%
Tertiary Recovery	--	--	0	76.84%
Final Oil Saturation	11.73%	1.28%	30%	5.36%

*All displacement experiments are carried out with nonequilibrated systems in sandpacks at 25°C; dimensions and flow rates same as given in Table 2.

Secondary and tertiary oil recovery values are percent of oil-in-place, whereas final oil saturation is percent of total pore volume.

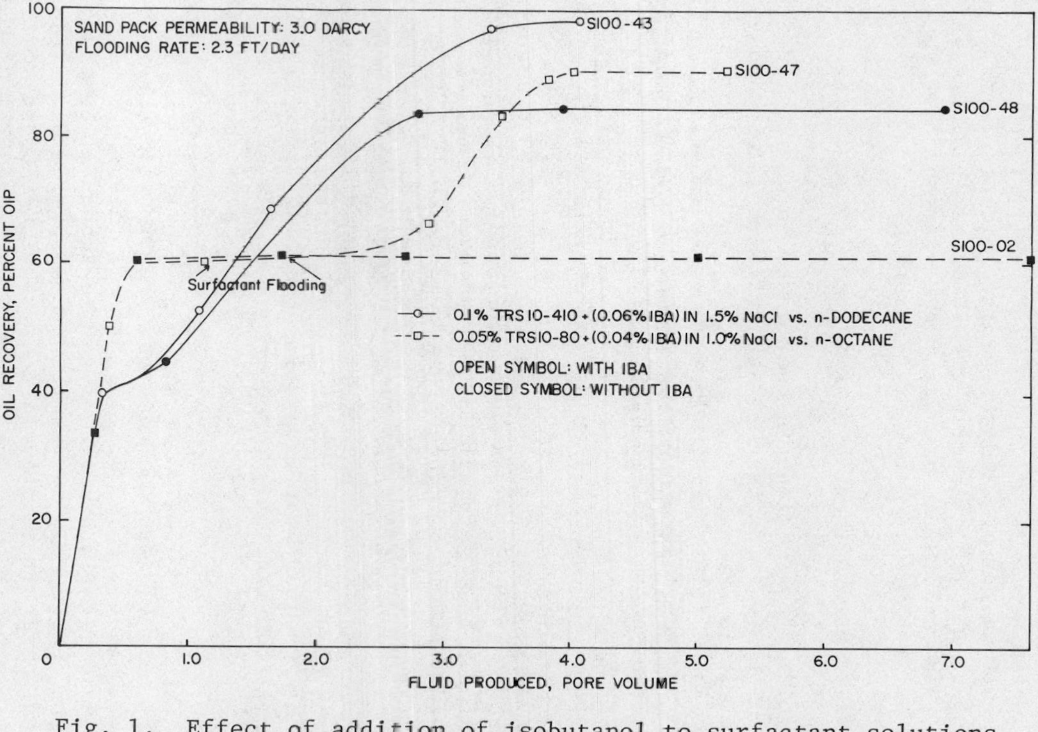

Fig. 1. Effect of addition of isobutanol to surfactant solutions on oil recovery for octane and dodecane in sandpacks by continuous injection of surfactant solutions, 25°C.

Since the shape of an oil droplet is an indication of IFT as measured by sessile drop method, the oil droplet flattening time reflects the rate of change in IFT. The results clearly show that the presence of alcohol increases the rate of achieving the final value of interfacial tension. This implies that the surfactant molecules come to the interface much faster in the presence of alcohol. Zana (20) has shown that the kinetics of micellization is more rapid in the presence of alcohol. This is presumably due to loose packing of mixed micelles containing surfactant and alcohol. Thus, it appears that the kinetics of micellization could influence the rate at which molecules saturate the surface by the breakdown of micelles to provide monomers for adsorption.

We have shown (19) that the interfacial concentration of surfactant depends on the partition coefficient of the surfactant. When the partition coefficient is near unity, a maximum surface concentration of the surfactant is achieved. In flow through porous media, it is expected that achieving the equilibrium condition may take much longer time. Therefore, we investigated the equilibrated and nonequilibrated systems in porous media to elucidate their effect on the oil recovery efficiency.

A comparison of equilibrated and nonequilibrated systems for oil displacement efficiency: Figure 2 shows the IFT and the percent oil recovery as a function of initial TRS 10-80 concentration in 1% NaCl for equilibrated and nonequilibrated systems. It was observed that for the pre-equilibrated system, 94% oil was recovered at 0.05% TRS 10-80 concentration corresponding to minimum IFT at this concentration. However, for nonequilibrated systems, the maximum oil recovery shifted from 0.05% to 0.1% TRS 10-80 concentration. The maximum oil recovery for nonequilibrated systems was much lower than that observed for equilibrated systems (Figure 2). Since the amount of surfactant injected was the same for each run (0.125 gm), the maximum oil recovery was interpreted as a result of the capillary number vs. final oil saturation correlation (21).

However, this correlation does not seem to hold under the typical (i.e., nonequilibrated) tertiary oil recovery conditions (Case A in Table 2). In order to find the amount of tertiary oil that can be recovered, the sandpacks were saturated with fresh (i.e., nonequilibrated) n-octane and were brine-flooded to the residual oil level. A fresh surfactant slug of 0.05% TRS 10-80 in 1% NaCl was then pumped through the sandpacks. It was interesting to note that in this case even after an injection of 10 PV surfactant slug, no or very little oil was recovered (Case A in Table 2). Because the effluent surfactant concentration approached that of the injected surfactant concentration, the poor oil recovery cannot be explained by the adsorption of the surfactant on sand particles. The observed excellent oil recovery for the equili-

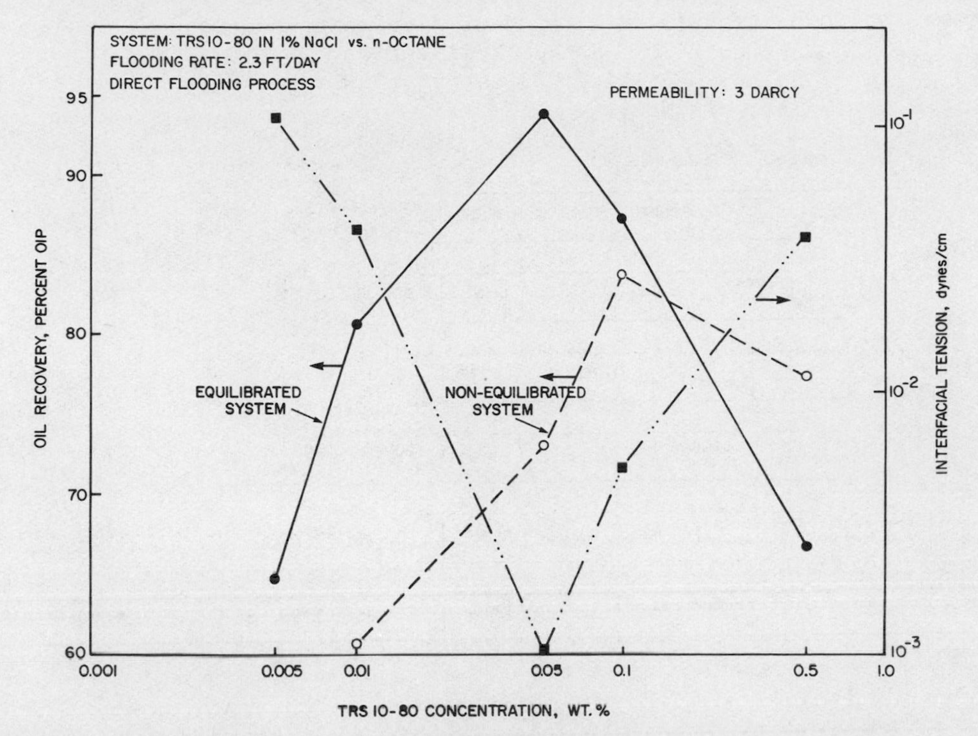

Fig. 2. The effect of surfactant concentration on oil displacement in sandpacks at 25°C. A correlation of IFT with oil recovery for equilibrated oil/brine/surfactant systems. The oil recovery is strikingly different for nonequilibrated systems (sandpacks, dimensions are given in Table 2).

brated system is then believed to be due to the effective surfactant partitioning during equilibration procedure. Thus, for equilibrated systems, the ultralow IFT is achieved quickly in porous media which results in an excellent correlation of oil displacement efficiency with IFT (Figure 2). In general, the oil recovery is better for equilibrated systems as compared to nonequilibrated systems except at 0.5% TRS 10-80 concentration (Figure 2). For this nonequilibrated system, due to slow mass transfer process the interfacial concentration might be similar to equilibrated low surfactant concentration systems. This will cause lower IFT and hence better oil recovery (Figure 2).

Systematic and comprehensive studies on oil displacement by various fluids were made and the results are listed in Table 2 and Figure 3. It is clear that oil recovery in all cases was nearly complete at the end of the fresh PV injection of the surfactant solution (Figure 3). Case A corresponds to the typical tertiary

Table 2. 0.05% TRS 10-80 in 1% NaCl displacing n-octane in sandpacks at 25°C. Sandpack dimension: 1.06" dia. x 7" long; permeability: 3 darcy; flow rate: 2.3 ft/day; brine: 1% NaCl.

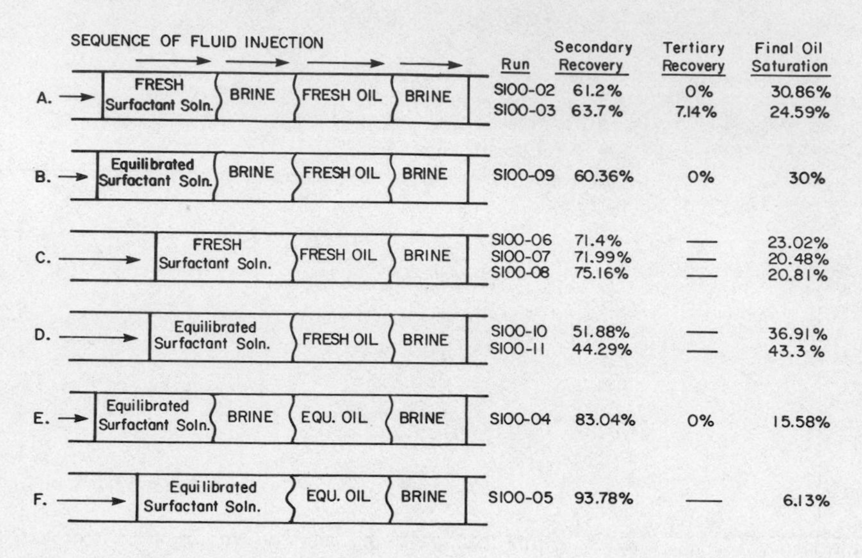

	SEQUENCE OF FLUID INJECTION	Run	Secondary Recovery	Tertiary Recovery	Final Oil Saturation
A.	FRESH Surfactant Soln. → BRINE → FRESH OIL → BRINE	SI00-02	61.2%	0%	30.86%
		SI00-03	63.7%	7.14%	24.59%
B.	Equilibrated Surfactant Soln. → BRINE → FRESH OIL → BRINE	SI00-09	60.36%	0%	30%
C.	FRESH Surfactant Soln. → FRESH OIL → BRINE	SI00-06	71.4%	—	23.02%
		SI00-07	71.99%	—	20.48%
		SI00-08	75.16%	—	20.81%
D.	Equilibrated Surfactant Soln. → FRESH OIL → BRINE	SI00-10	51.88%	—	36.91%
		SI00-11	44.29%	—	43.3 %
E.	Equilibrated Surfactant Soln. → BRINE → EQU. OIL → BRINE	SI00-04	83.04%	0%	15.58%
F.	Equilibrated Surfactant Soln. → EQU. OIL → BRINE	SI00-05	93.78%	—	6.13%

oil recovery process while Case F shows 94% recovery of the equilibrated system (Table 2). A fair comparison of the equilibrated with nonequilibrated systems is Case F vs. Case C, the direct oil displacement by surfactant solution without brine-flooding. The equilibrated system (Case F) is better by 22% (94% vs. 72%). This is a clear indication of the importance of surfactant partitioning during oil displacement.

As fresh n-octane in Case B and equilibrated n-octane in Case E were being displaced by both brine and equilibrated surfactant solutions, an oil recovery of 60% and 83% respectively, was observed. Again, the recovery of the equilibrated oil is better by 23%, a difference of the same magnitude as the equilibrated system in Case F being compared with the nonequilibrated system in Case C. Thus, the equilibration of oil appears to be important for the observed oil recovery differences between the equilibrated and nonequilibrated surfactant solutions (Cases A and B or Cases C and D) it is observed that there is either no difference in oil recovery or the equilibrated performs worse than the nonequilibrated. In order to interpret the results shown in Table 2, let us consider the mechanism shown in Figure 4.

The commercial petroleum sulfonate such as TRS 10-80 is known to be a mixture of various low and high equivalent weight sulfonates. The higher equivalent weight species tend to be more oil-

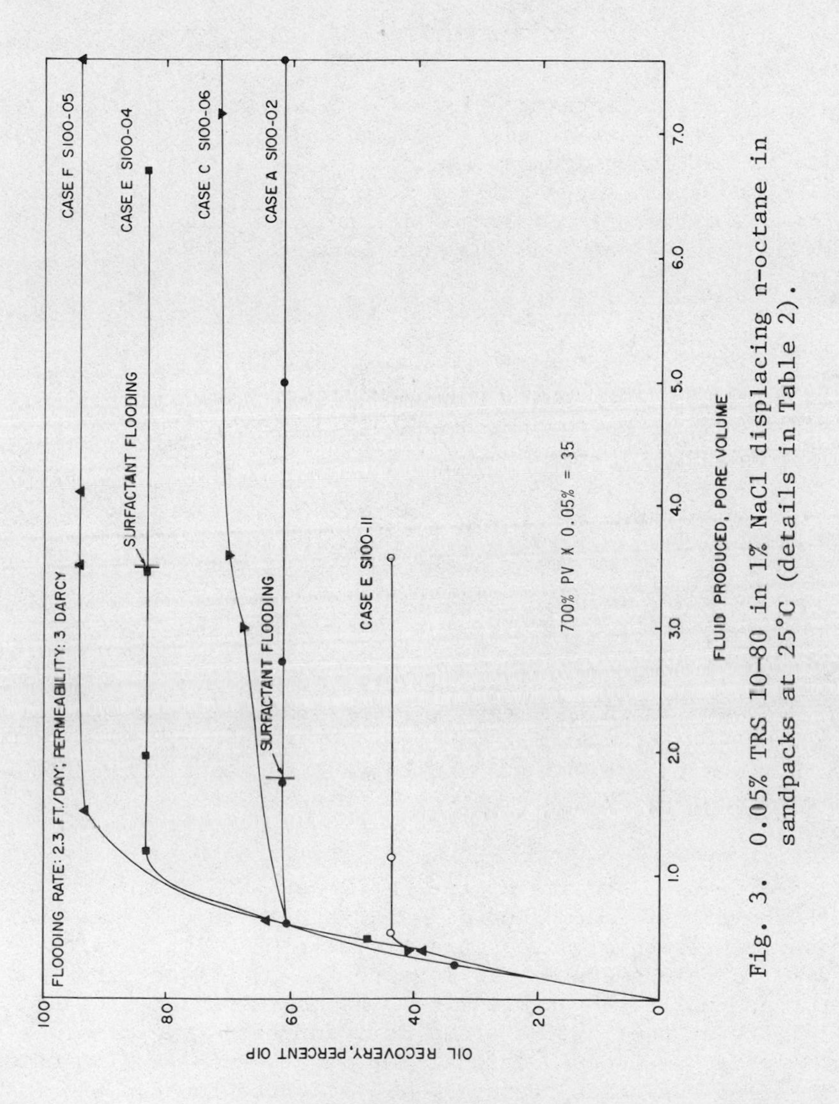

Fig. 3. 0.05% TRS 10-80 in 1% NaCl displacing n-octane in sandpacks at 25°C (details in Table 2).

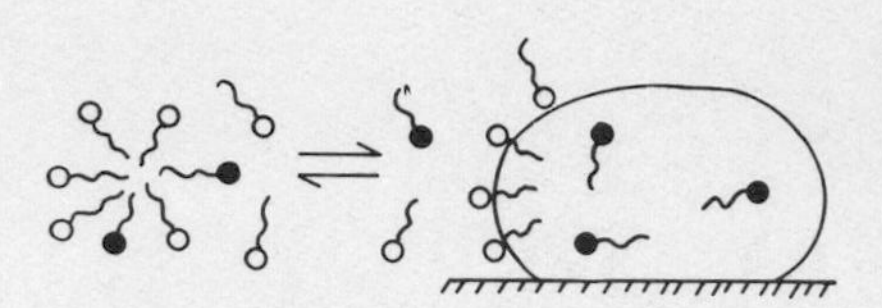

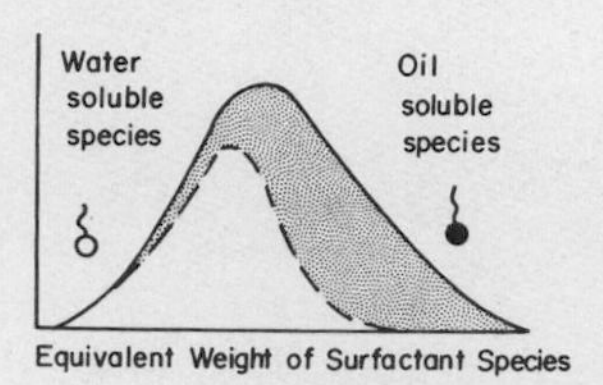

Fig. 4. Schematic representation of and various steps involved in mass transfer of petroleum sulfonate from aqueous solution to the interface and then to the oil phase. The right hand side of the diagram illustrates the role of preferentially water soluble and oil soluble surfactant species in partitioning of the petroleum sulfonate.

soluble or more hydrophobic, while the lower equivalent weight species tend to be more water-soluble or more hydrophilic. Schematically, it is depicted by the diagram on the right hand side of Figure 4. When such a surfactant is added to an oil/water mixture, each species partitions in the oil and brine according to its hydrophilic-lipophilic balance. The stipled region is proportional to the fraction partitioning in the oil, whereas, the clear region below is proportional to the fraction of water-soluble species.

Initially, the surfactant is dissolved in the aqueous solution. However, as this aqueous solution is equilibrated with an oil, the oil-soluble species partitions into the oil phase. From IFT data shown in Table 3 and the later discussion, it is evident that the oil/brine IFT is similar to that reported by Gale and Sandvik (7).

The molecular species at the interface are in equilibrium with those in the aqueous and oil phases. If we consider the addition of a fresh oil drop in a micellar solution (Figure 4), the surfactant monomers should move to the interface first and then to the inside of the oil drop. As monomers get depleted in the vicinity of the interface due to adsorption, the micelles break down and produce additional monomers. From the interface, the oil-soluble species preferentially migrate towards the inside of the oil droplet.

We propose that the interface is occupied with both water-soluble and oil-soluble species. For equilibrated systems, the surfactant species come from both sides of the interface and saturate the interface with surfactant molecules more quickly as

Table 3. IFT, Flattening Time and Oil Recovery Efficiency of 0.05% TRS 10-80 in 1% NaCl vs. n-Octane at 25°C.

	System	IFT (mN/m)	Flattening Time* (seconds)	Oil Recovery+ (% OIP)
I.	Fresh Oil/1% NaCl	≈50.8**	∞	61-63
II.	Fresh Oil/Equilibrated Surfactant Solution	0.731	6600	44-52
III.	Fresh Oil/Fresh Surfactant Solution	0.627	480	75-77
IV.	Equilibrated Oil/1% NaCl	0.121	900	83
V.	Equilibrated Oil/Equilibrated Surfactant Solution	0.0267	240	94
VI.	Equilibrated Oil/Fresh Surfactant Solution	0.00209	15	--

*Flattening time is defined as the time required for the n-octane drop to gradually flatten out.

**Octane/H_2O, 20°C, IFT = 50.8 mN/m, "Interfacial Phenomena," Davies and Rideal, Chapter 1, p. 17, Table 1, Academic Press, N.Y., 1963.

+Porous media dimensions, sequence and rate of fluid injection are given in Table 2.

compared to the nonequilibrated systems in which all surfactant species come only from one side (the aqueous phase) of the interface, containing more stable mixed micelles of water and oil-soluble surfactant species. Moreover, for the nonequilibrated surfactant slug, the water-soluble species may form a film at the oil/brine interface deterring the mass transfer from the aqueous phase to the oleic phase of the oil-soluble species.

A comparison of Cases B and D in Table 2 suggests that predominantly water-soluble species of the equilibrated aqueous phase of the surfactant solution worsen the oil displacement process as compared to brine flooding presumably due to the formation of stable emulsions or a decrease in coalescence rate in porous media. It is hypothesized that a rigid surfactant film forms on the oil droplet when displaced by the equilibrated aqueous phase of the surfactant solution. This film prevents the coalescence of oil droplet in the narrow channels of the sandpack. It was observed that the differential pressure (ΔP) across the sandpack increases continuously beyond the water breakthrough peak when flooded by the equilibrated surfactant solution, but ΔP decreases or levels off after the water breakthrough when flooded by 1% NaCl. Hence, the apparent paradox in capillary number-oil recovery correlation (systems I and II in Table 3) can be resolved if the interfacial viscosity (16) is considered in addition to the IFT. Indeed, as alcohol was incorporated into the system, IFV decreased and oil recovery increased (Table 1).

The results of Cases B, C, and D suggest the beneficial effect of the presence of oil-soluble species in improving oil recovery in Case C. The reason that equilibrated surfactant solution displaces less oil than the fresh surfactant solution as in Cases D and C in Table 2 is partially due to the fact that there is less surfactant in the equilibrated solution as compared to the fresh solution. During the equilibration process, some of the surfactant species must have migrated from the aqueous phase to the oleic phase resulting in a reduction in surfactant concentration in brine. This is substantiated by the measurement of surfactant concentration of 0.01% for the original 0.05% surfactant solution after equilibration.

Also, the results of Cases A and C as well as Cases E and F indicate that a lower final oil saturation, S_{of}, was obtained, if the sandpack was flooded directly by the surfactant solution without a secondary flooding by brine.

To explain the effect of equilibration on oil recovery, the liquid-liquid and liquid-rock interfaces (i.e., the IFT's and contact angles) were studied for these systems and the results are listed in Table 3. Except for the system of fresh oil/1% NaCl, the contact angle measurements followed the pattern shown in Figures 5 and 6. The oil drop formed nearly a sphere on the quartz surface initially. It then flattened out and finally, in some cases, disintegrated or emulsified into many small droplets. The time between the formation of the initial spherical droplet and the final emulsification is defined as the oil droplet flattening time. Except for system III, there is a good correlation between the flattening time, the IFT value and the oil displacement efficiency.

Among systems I through V (Table 3), the lowest IFT existed for the interface between equilibrated oil and equilibrated surfactant solution. A drastic increase in IFT occurred as either equilibrated oil or equilibrated surfactant solution was replaced

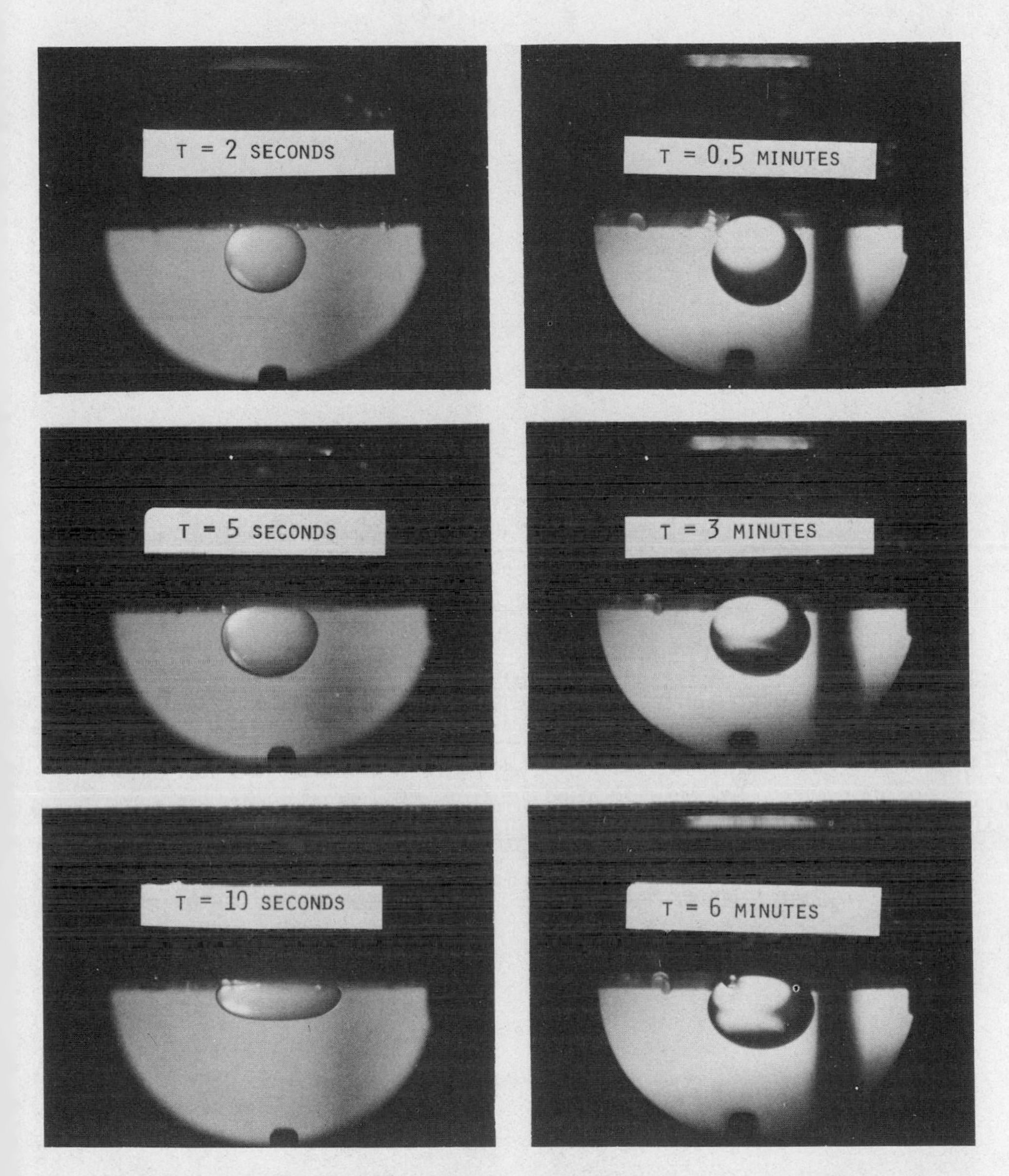

Fig. 5. The left hand side column illustrates the spreading of a drop of equilibrated n-octane on a quartz surface submerged in fresh 0.05% TRS 10-80 in 1% NaCl solution (Case VI in Table 3). The right hand side column illustrates the spreading of equilibrated n-octane drop on quartz in 1% NaCl solution (Case IV in Table 3).

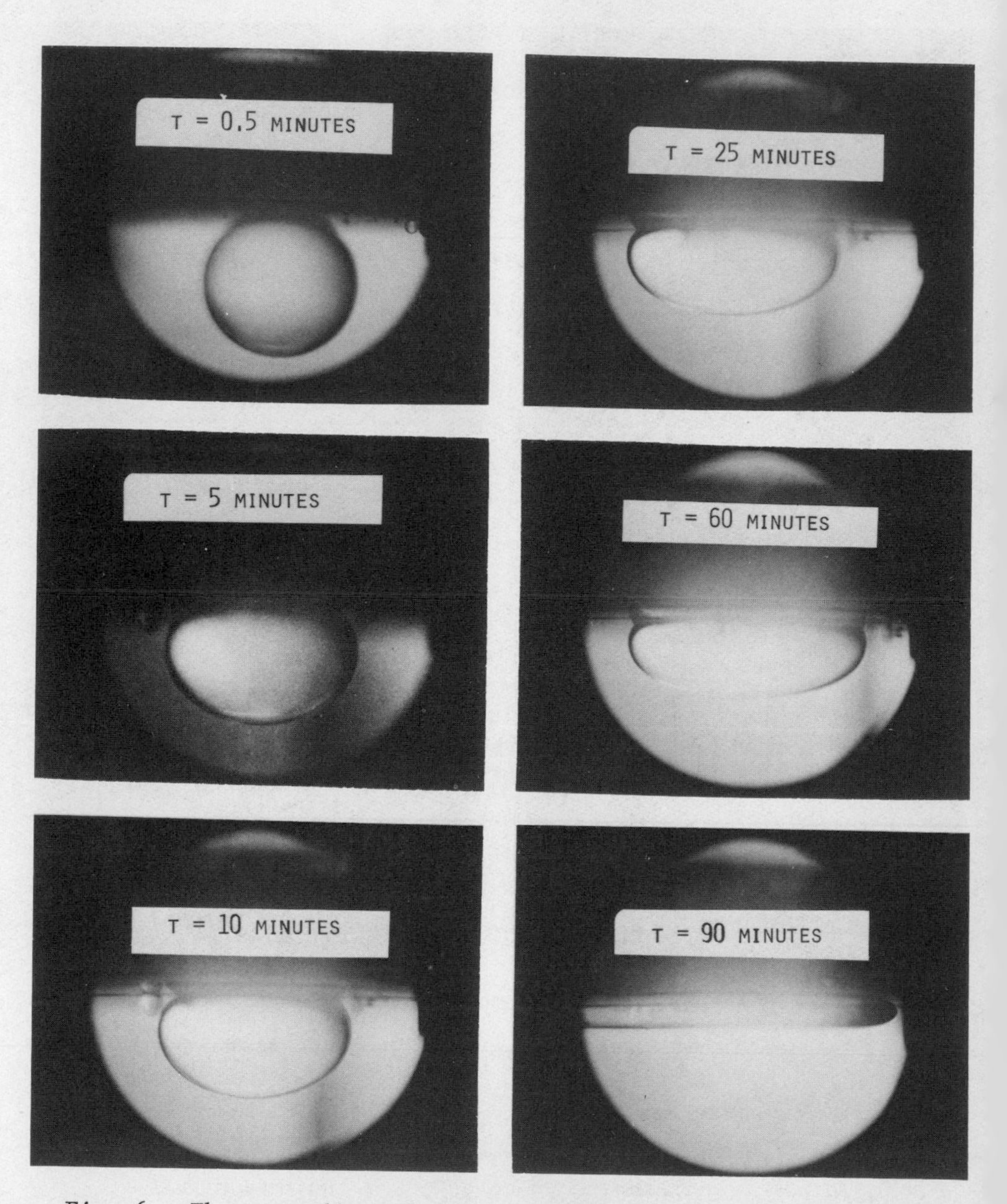

Fig. 6. The spreading of a drop of n-octane on quartz surface submerged in the equilibrated aqueous phase of 0.05% TRS 10-80 in 1% NaCl solution (Case II in Table 3).

by fresh oil or fresh surfactant solution. However, examining systems II and IV, it is evident that the equilibrated oil rather than the equilibrated surfactant solution is responsible for the lowering of IFT. This suggests that the oil-soluble species are the low tension producing sulfonates in this system.

These hydrophobic species are mainly responsible for the oil droplet flattening phenomenon. The flattening time of a single oil drop has a direct bearing on the oil displacement efficiency. Because there are large numbers of oil droplets within the porous media, the amount of oil recovered depends on how easily each of them can be mobilized. The faster they are flattened, the easier it would be to mobilize, interconnect and displace them. Cash et al. (12) demonstrated that oil displacement by the spontaneously emulsifying systems is better than the systems lacking spontaneous emulsification.

In Table 3, the longest flattening time corresponds to the system that has the least amount of oil-soluble species present and the worst oil recovery. The only exception is system III, although the oil drops flattened faster than system IV, it gave poorer recovery than system IV. The following explanation is suggested. While flattening time is being measured, the oil-soluble species from the fresh surfactant solution quickly adsorb onto the quartz surface, which facilitates the flattening of the oil drop. However, the IFT is much higher in case III as compared to that in case IV. In agreement with the capillary number concept, we observed a better oil recovery in case IV than in case III.

To sum up, the following mechanism is proposed to account for the observed effects in IFT and oil droplet flattening phenomenon. As shown in Figure 4, mixed micelles in equilibrium with surfactant monomers are formed by the water-soluble and oil-soluble species in the bulk aqueous solutions. During equilibration, the surfactant monomers transfer to the water/oil interface and then to the interior of the oil drop resulting in a reduction of IFT. The concentration of oil-soluble species in the surfactant solution dictates the absolute value of IFT and the rate of surfactant mass transfer, which, in turn, determines the flattening time of the oil drop.

Because different batches of TRS 10-80 were used in making the sets of surfactant solutions in Figure 2 and Table 3, small variation in values of IFT for the equilibrated oil and equilibrated 0.05% TRS 10-80 in 1% NaCl was observed. Nevertheless, the trend of high and low IFT within each set remained the same. Therefore, the interpretation of IFT based on these values is believed to be valid.

Effect of salinity on oil displacement efficiency of equilibrated and nonequilibrated systems: Figure 7 shows the effect of salinity on the oil recovery and IFT of 0.1% TRS 10-410 + 0.05% IBA vs. n-dodecane. It shows that at 0.5% and 1.0% NaCl concentrations, the oil recovery is the same for both systems. Only at and above the optimal salinity (i.e., 1.5% and 2.0% NaCl), the nonequilibrated systems produce better oil recovery than the equilibrated systems. A possible explanation of this effect is as follows. It has been shown that for salt concentrations higher than the optimal salinity, the tendency for the surfactant to migrate from the aqueous phase to the oil phase increases. Therefore, when one takes a nonequilibrated system at or above optimal salinity, there is a significant driving force for the surfactant to migrate from the aqueous to oil phase. Moreover, the presence of alcohol in such solutions enhances the mass transfer of surfactant across the interface. Therefore, as the nonequilibrated surfactant solution contacts the oil ganglia, presumably a rapid mass transfer occurs resulting in ultralow interfacial tension. The oil ganglia thereby flatten out or spontaneously disintegrate. A successful flattening and subsequent coalescence of the oil ganglia in the initial stages presumably lead to the formation of an oil-water bank which then successfully sweeps additional oil ganglia along the porous media by coalescence process. By maintaining the ultralow IFT at the oil bank/surfactant solution interface decreases entrapment of the oil from the oil-water bank. Therefore, the improved performance of nonequilibrated systems at and above optimal salinity is related to the effective mass transfer of surfactant from the aqueous phase to the oil phase and the concomitant generation of ultralow IFT and presumably low IFV and associated spontaneous flattening of oil ganglia. This explanation is consistent with the results of oil displacement in Berea cores by the same surfactant system as shown in Figure 8. It shows the effect of salinity on the amount of oil recovery as a percent of oil-in-place and a percent of final oil saturation. It indicates that more oil was displaced at and above optimal salinity and that close to 90% oil recovery was obtained.

Figure 9 is a production history of a typical run. The cumulative oil recovery, pressure difference (ΔP) across the porous bed, normalized effluent surfactant concentration and percent of oil cut have been plotted. The cumulative oil recovery curve and the ΔP curve rise sharply initially then change their slopes at 0.4% PV. The oil recovery curve further increases at a constant rate while ΔP decreases, then both change slopes again at 5 PV and, finally, the oil recovery graph reaches a constant value and ΔP keeps on rising continuously. Throughout the flooding process, the effluent surfactant concentration increases very slowly from 0% initially to 15% of the injected surfactant concentration at 6.5% PV. It jumps to 37% at 7 PV and eventually reaches 42% at the end of the run. The oil cut drops drastically from the 100% at the

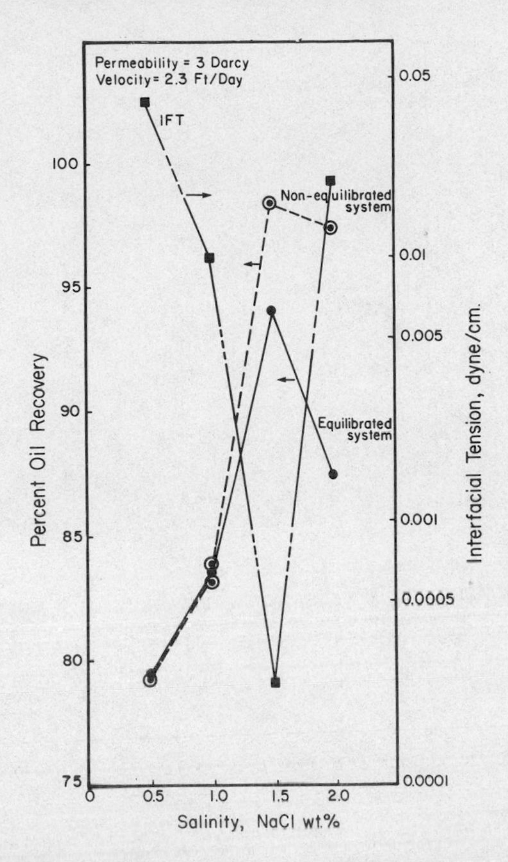

Fig. 7. Effect of salinity on oil recovery by continuous injection of 0.1% TRS 10-410 + 0.06% IBA on n-dodecane displacement in sandpacks at 25°C. (Dimensions of sandpacks are the same as given in Table 2).

beginning to 7% at 0.5% PV, then it maintains a 4% recovery for 4.5 PV fluid production.

The initial fast rise of the oil recovery curve and the ΔP curve correspond to the 100% oil recovery in the effluent stream for the fully oil saturated Berea core. This is evident from the oil cut curve. The slopes change when water breaks through at the exit. In the next stage, oil is then produced in the form of oil-water bank, which is composed of the coalesced oil droplets mobilized by the surfactant solution. As oil is recovered at a constant rate, ΔP decreased gradually.

Toward the end of this constant rate of oil production, oil comes out as the trailing end of the oil-water bank. At the same time, enough surfactant has been accumulated in the sandstone core to form emulsions with the oil droplets in situ. Consequently,

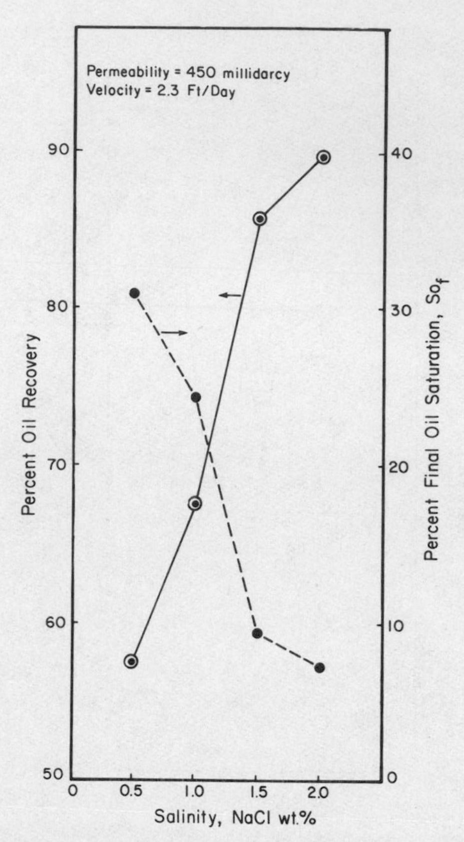

Fig. 8. Effect of salinity on oil recovery by continuous injection of 0.1% TRS 10-410 + 0.06% IBA on n-dodecane displacement in Berea cores at 25°C. (Details given in the text.)

ΔP increased due to the blockage of the small pores and narrow channels by these oil-swollen surfactant-rich emulsions. As the process progresses, the surfactant-rich emulsion breaks through as a white opaque solution and manifests itself as a step increase on the C/C_o curve at 7 PV. Finally, as the end of the flooding process is approached, oil recovery diminishes, ΔP keeps on increasing as before, and C/C_o levels off.

It is interesting to note that the shape of the cumulative oil recovery curves in the unconsolidated sandpack is similar to that in the consolidated Berea core (Figures 1 and 9), except that oil is produced at a much faster rate for the sandpacks. Therefore, the oil displacement mechanism is presumably the same in these two porous media for the continuous dilute surfactant solution flooding process. Chou and Shah (22) have shown that 1 or 4 ft sandpacks give identical results for oil recovery and the fluid production

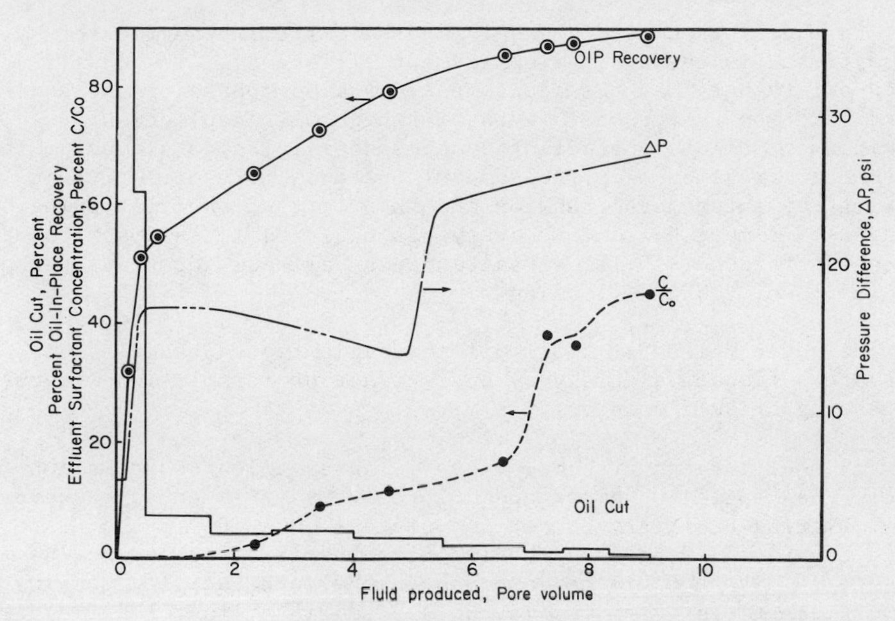

Fig. 9. Oil displacement history of produced fluids upon continuous injection of 0.1% TRS 10-410 + 0.06% IBA in 2% NaCl brine to displace n-dodecane in Berea cores, 25°C. (Berea core details given in the text.)

profile when plotted as a function of PV injected. Therefore, we believe that the use of small sandpacks is still meaningful for showing the phenomena in porous media.

It should be emphasized that the entire study reported in this paper relates to the low surfactant concentration (< 0.5%) and does not involve the formation of middle phase microemulsions (23), etc. in this oil displacement process. At all times, the oil/brine/surfactant systems were composed of only two phases, oil and brine, with surfactant distributed in both phases. Also, this study is carried out at low salinity (< 2% NaCl) although we have reported elsewhere on high salinity formulations (24-26) which can produce ultralow IFT in millidynes/cm range at salt concentrations as high as 32%.

CONCLUSIONS

(1) The study revealed that the addition of isobutanol to dilute TRS 10-80 or TRS 10-410 petroleum sulfonate solutions did not influence significantly IFT, or surfactant partitioning but decreased interfacial viscosity and markedly reduced flattening time of oil drops and increased oil displacement efficiency, presumably by promoting coalescence of oil ganglia in porous media.

(2) The equilibrated and nonequilibrated oil/brine/surfactant systems differed in their oil displacement efficiency. The equilibrated oil rather than the equilibrated aqueous phase of the surfactant solution is responsible for the high oil displacement efficiency of dilute surfactant systems containing no alcohol. The oil soluble fraction of petroleum sulfonate is more effective in lowering the interfacial tension and in promoting the flattening of oil drops. Almost 94% oil recovery was achieved in sandpacks by a low concentration ($\approx$ 0.1%) surfactant plus alcohol formulation when used in place of brine flooding.

(3) The lower values of final oil saturation were obtained for the systems flooded directly by the surfactant formulation without first being brine-flooded.

(4) For equilibrated systems, there is an excellent correlation between the capillary number and oil recovery efficiency. However, in calculating capillary number for nonequilibrated systems, care should be exercised because the IFT measured in vitro may not be achieved in situ and, in certain cases, the interfacial viscosity and not interfacial tension, may be a predominant factor influencing the oil displacement efficiency.

(5) The effect of salinity on oil displacement efficiency revealed that for the alcohol containing formulations, the nonequilibrated system was more efficient for oil recovery as compared to the equilibrated system at and above optimal salinity. It is proposed that not only the equilibrium values of the parameters such as interfacial tension and interfacial viscosity are important but the dynamic process of surfactant partitioning is also important in mobilization of oil ganglia. The conditions that promote the efficient mass transfer from the aqueous phase to the interface promote the deformation and mobilization of oil ganglia. This would facilitate an early formation of oil bank and displacement of oil from porous media.

ACKNOWLEDGMENTS

The authors wish to express their sincere thanks and appreciation to the National Science Foundation-RANN, ERDA and the Department of Energy (Grant No. DE-AC1979BC10075) and the consortium of the following Industrial Associates for their generous support of the University of Florida Enhanced Oil Recovery Research Program during the past five years: 1) Alberta Research Council, Canada, 2) American Cyanamid Co., 3) Amoco Production Co., 4) Atlantic Richfield Co., 5) BASF-Wyandotte Co., 6) British Petroleum Co., England, 7) Calgon Corp., 8) Cities Service Oil Co., 9) Continental Oil Co., 10) Ethyl Corp., 11) Exxon Production Research Co., 12)

Getty Oil Co., 13) Gulf Research and Development Co., 14) Marathon Oil Co., 15) Mobil Research and Development Co., 16) Nalco Chemical Co., 17) Phillips Petroleum Co., 18) Shell Development Co., 19) Standard Oil of Ohio Co., 20) Stepan Chemical Co., 21) Sun Oil Chemical Co., 22) Texaco Inc., 23) Union Carbide Corp., 24) Union Oil Co., 25) Westvaco Inc., 26) Witco Chemical Co., and the University of Florida.

The authors gratefully acknowledge the permission granted by SPE to publish this paper which was first presented at the SPE Fifth International Symposium on Oilfield and Geothermal Chemistry, held in Stanford, California, May 28-30, 1980, SPE copyright 1980.

REFERENCES

1. L. W. Holm, J. Pet. Tech., 1475 (Dec. 1971).
2. J. J. Taber, Soc. Pet. Eng. J., 3 (March, 1969).
3. R. N. Healy and R. L. Reed, in "Improved Oil Recovery by Surfactant and Polymer Flooding," Vol. I, D. O. Shah and R. S. Schechter, eds., Academic Press, N.Y., 383, 1977.
4. R. L. Cash, Jr., J. L. Cayias, M. Hayes, D. J. MacAllister, T. Schares, R. S. Schechter, and W. H. Wade, SPE 5562 presented at Fall SPE Meeting, Dallas, Texas, Sept. 28-Oct. 1, 1975.
5. S. O. Trushenski, D. L. Dauben, and D. R. Parrish, Soc. Pet. Eng. J., 633 (Dec. 1974).
6. G. L. Stegemeier, in "Improved Oil Recovery by Surfactant and Polymer Flooding," Vol. I, D. O. Shah and R. S. Schechter, eds., Academic Press, N.Y., 1977.
7. W. W. Gale and E. I. Sandvik, Soc. Pet. Eng. J., 191 (Aug. 1973).
8. W. B. Gogarty and W. C. Tosch, J. Pet. Tech., 1407 (Dec. 1968).
9. S. C. Jones and K. D. Dreher, SPE 5566 presented at Fall SPE Meeting, Dallas, Texas, Sept. 28-Oct. 1, 1975.
10. W. R. Foster, J. Pet. Tech., 205 (Feb. 1973).
11. W. C. Hsieh and D. O. Shah, SPE 6594 presented at SPE International Symposium on Oilfield and Geothermal Chemistry, La Jolla, CA, June 27-28, 1977.
12. R. L. Cash, J. L. Cayias, R. G. Fournier, J. K. Jacobson, T. Schares, R. S. Schechter, and W. H. Wade, SPE 5813 presented at SPE Improved Oil Recovery Symposium, Tulsa, OK, March 22-24, 1976.
13. D. R. Anderson, M. S. Binder, H. T. Davis, C. D. Manning, and L. E. Scriven, SPE 5811 presented at SPE Improved Oil Recovery Symposium, Tulsa, OK, March 22-24, 1976.
14. D. T. Wasan and V. Mohan, in "Improved Oil Recovery by Surfactant and Polymer Flooding," Vol. I, D. O. Shah and R. S. Schechter, eds., Academic Press, N.Y., 161, 1977.

15. K. S. Chan, Ph.D. Dissertation, University of Florida, Gainesville, FL, 1978.
16. D. T. Wasan, L. Gupta, and M. K. Vora, AIChE J., 17, 1287 (Nov. 1971).
17. V. W. Reid, G. F. Longman, and E. Heinerth, Tenside, 4, 292 (1967).
18. D. O. Shah, A. Tamjeedi, J. W. Falco, and R. D. Walker, AIChE J., 18, 1116 (Nov. 1972).
19. K. S. Chan and D. O. Shah, J. Disp. Sci. & Tech., 1(1), 55 (1980).
20. R. Zana, in "Surface Phenomena In Enhanced Oil Recovery," D. O. Shah, ed., Plenum Press, N.Y. (in press).
21. J. C. Melrose and C. F. Brandner, J. Canadian Petr. Tech., 58 (Oct.-Dec. 1974).
22. S. I. Chou and D. O. Shah, in "Surface Phenomena In Enhanced Oil Recovery," D. O. Shah, ed., Plenum Press, N.Y. (in press).
23. K. S. Chan and D. O. Shah, SPE 7896 presented at SPE International Symposium on Oilfield and Geothermal Chemistry, Houston, TX, Jan. 22-24, 1979.
24. V. K. Bansal and D. O. Shah, J. Coll. Interface Sci., 65, 451 (1978).
25. V. K. Bansal and D. O. Shah, Soc. Pet. Eng. J., 167 (June 1978).
26. V. K. Bansal and D. O. Shah, J. Am. Oil Chem. Soc., 55, No. 3, 367 (1978).

COMPLEX ACIDS AND THEIR INTERACTION WITH

CLAYS IN OIL SANDS SLIMES

M. A. Kessick

Chemistry Division
Alberta Research Council
Edmonton, Alberta, Canada T6G 2C2

Organic material closely associated with the clays in low grade oil sand and in oil sands clay slimes has been extracted and examined by IR and NMR spectroscopy, and shown to contain carbonyl, hydroxyl and possibly polyphenolic functionality. It is postulated that the closely bound organic material may consist of vestigial tannins or lignins in various stages of degradation. The importance of ferric iron in promoting adsorption of this organic material to clays has been demonstrated using tannic acid with illite and kaolinite as model systems. This type of adsorption appears to be important in determining the settling characteristics of oil sands clay slimes.

INTRODUCTION

Athabasca oil sand is a complex, variable mixture of sand particles, clays, bitumen and water (1). Currently, there exist two commercial operations to extract the bitumen from this matrix and upgrade it to synthetic crude oil, a partially refined product. Both operations use the hot water extraction process (2) on surface mined material.

In this extraction process the mined matrix is pulped with hot alkali solution. On further dilution with hot water the bitumen then detaches from the resulting slurry and is allowed to float to the surface in a primary separation tank, where it is collected for dewatering and upgrading. The yield is increased further by dissolved air flotation of the diluted slurry, after sand separation, in a subsequent process loop. During the separation very large volumes of tailings are produced, consisting mainly of sand and a dispersion of various clays containing residual organic matter. The sand settles rapidly from the tailings and presents no real disposal problem. It is in fact used to build containment dikes for the clay dispersions, which, although they will settle to provide some water for recycle, do so only to a sludge or slime that itself is still very high in water content, and that shows little tendency to dewater further, even when subjected to mechanical dewatering procedures (3). It is the buildup of these settled clay sludges, or slimes, necessarily contained behind the sand dikes, which presents not only an environmental problem but also a significant repository for non-recycleable water.

The reason for the intractability of the clay slimes has been a subject of considerable study. Attempts have been made to explain the phenomenon in terms of a composite equilibrium settling volume calculated on the basis of individual settling volumes measured for pure samples of the various clay and other minerals known to be present (4). In this approach, correlation between theoretically derived values and those measured experimentally depends upon the presence of significant quantities of amorphous iron oxide. The approach does not take into account the effect of residual organic material in the slimes. Organic substances, of the types present in bitumen and heavy oils, are known to adsorb strongly to clay minerals (5,6).

Work in this laboratory has indicated that the interaction of organics with the clay minerals is indeed important in determining the nature of the oil sands slimes (7), and that this interaction is specific, and involves a fairly well-defined class of organic compounds. In many ways, the interaction seems to be of the same type as that reported to be operative in the adsorption of fulvic acid to montmorillonite containing Cu^{2+}, as well as other multivalent cations, including Fe^{3+} (8). The interaction is thought to provide a hydrophobic character to the clay particle surfaces, allowing bridging through residual bitumen to set up a weak gel structure.

Solvent extraction of dried clay slimes indicates the presence of two major fractions of residual organic material. One fraction is readily extractable with methylene chloride, and occurs in amount approximately 0.1 to 0.2g per gram of dried slime. The infrared pattern of this fraction shows only absorption maxima that can be attributed to aliphatic stretching and deformation vibrations (Figure 1). The second fraction occurs in amount ca. 0.02g per gram of dried sludge, and is more strongly associated with the clays. It can only be dislodged by strongly polar solvents, such as methyl ethyl ketone (MEK)/water mixtures, under mildly acidic conditions (9).

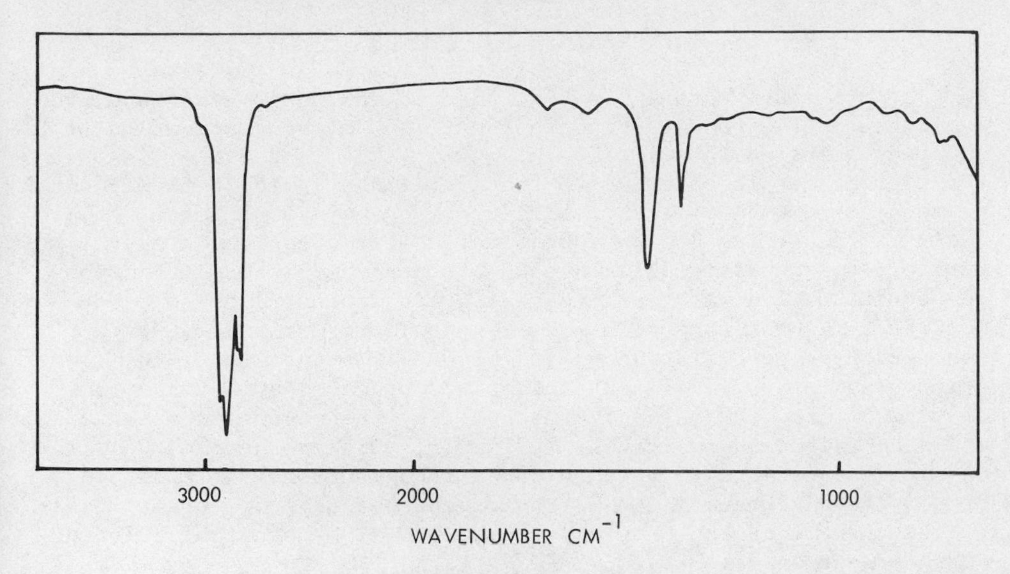

Fig. 1. IR spectrum of bitumen readily extractable from dried clay slimes.

It is believed that the Fe^{3+} ion provides a key link between this second fraction and the clay surface. The presence of iron was originally noticed as a buff-red discoloration on ignition of clay that still held the closely-bound material. Its presence was confirmed by examination under an electron microscope with an energy dispersive attachment. Also, the organic material removed by this second-stage extraction was found by atomic absorption analysis to contain about 9% by weight of iron, confirmed to be in the ferric form by the absence of color formation with o-phenanthroline. This implied that the organic substances in this fraction were strong chelators, and that they had probably interacted with ferric iron itself adsorbed to the clays as an exchangeable cation.

The organic material has been subjected to further separation by chromatography on an alumina column, using acid/MEK as an eluent. The chromatography fractions were associated with varying amounts of iron, and the infra red spectra of the fractions showed varying amounts of ketonic and conjugated ketonic functionality. It is now believed that these fractions may represent aliphatic degradation products of partially oxidized polyphenolic materials, the degradation most likely occurring on the alumina column itself.

The composition of these materials has been investigated further by IR and NMR spectroscopy, and the absorption of a tannin as a representative polyphenolic compound to iron-containing and iron-free clays investigated. A sample of low grade oil sand, containing a considerable percentage of clay, was also examined for the presence of clay-bound organic materials.

MATERIALS AND METHODS

Methyl ethyl ketone was distilled as the water azeotrope from a mixture of the reagent grade solvent (350 ml) and acidified ferric chloride solution (50 ml H_2O, 1 ml. conc. HCl, 1 g $FeCl_3$). After addition of ferric chloride to the distillate (0.1g in 50 ml) and subsequent removal of the distillate by evaporation, infra red analysis of the residue showed no detectable organic content. Purified hydrochloric acid (3.75N) was prepared by equilibration between distilled water and reagent grade concentrated acid both contained in beakers inside a desiccator. The distilled water absorbed hydrogen chloride gas to produce the purified acid. Acid/methyl ethyl ketone extractant was prepared by adding 0.1 ml of this acid to 100 ml of the methyl ethyl ketone/water azeotrope, which contained approximately 12% water. Fisher Chemical Co. tannin (tannic acid) was taken to be representative of this class of polyphenolic materials and no attempt was made at further purification. All other chemicals used, including methylene chloride, were reagent grade.

Buffer solutions: Buffer solutions were made up to be approximately 0.1M. pH 4 buffer was prepared from acetic acid and sodium hydroxide and pH 8.4 buffer from sodium bicarbonate and distilled water equilibrated with atmospheric CO_2.

Clays: A sample of kaolinite (Fisher Colloidal Kaolin) was washed several times with hydrochloric acid at pH 2, and then with distilled water and dried. Another sample was stirred for 24-48 hours with 10^{-2}M $FeCl_3$, also at pH 2, washed with distilled water and dried, to give a dry product with a pronounced yellowish tinge.

Illite (Silver Hills, Montana) was ground into a fine powder, and organic matter and free iron oxides removed by standard procedures (10). A portion was then rinsed with hydrochloric acid at pH 2, washed with distilled water, and dried. Iron treated illite was then prepared from another portion using the procedure described above for kaolinite.

Extraction procedures: Low grade oil sand and dried oil sand slime were initially extracted with methylene chloride in a Soxhlet apparatus until the eluate from the thimble became colorless. The residue from the thimble was then dried, and stirred with acid/methyl ethyl ketone extractant at room temperature. The resulting suspension was filtered, and the clear filtrate evaporated down and dried at room temperature over phosphorous pentoxide in a desiccator. The dried residue was then used for IR and NMR analysis.

Chromatography: Chromatography was carried out by redissolving the acid methyl ethyl ketone in methylene chloride and placing a portion on a basic alumina (Brockmann 5016-A) column and eluting with mildly acid methyl ketone. The fractions were collected, evaporated down and dried over P_2O_5 for further characterization.

Adsorption studies: Adsorption experiments were carried out by shaking for 24 hours the appropriate ground clay (0.5g) in 50 ml aliquots of a 10^{-2}M buffer solution containing from ca. 4 to 70 mg/l tannic acid. The suspensions were then centrifuged and the equilibrium concentrations of tannic acid solution were determined from the UV absorbance at 270 nm. It had previously been established that Beer's Law held for this material at this wavelength.

RESULTS AND DISCUSSION

The closely bound organic material extracted from the dried oil sands slimes by acid methyl ethyl ketone shows an infra red spectrum (Figure 2) indicating the presence of chelated and possibly polymeric hydroxyl compounds (2500-3200 cm^{-1} absorption) and carbonyl groups (1700-1725 cm^{-1} absorption) most likely ketonic or carboxylic in character. The absorption at ca 1600 cm^{-1} could indicate a conjugated or hydrogen-bonded carbonyl group or a carboxylate anion. Some aromatic absorption may also be occurring in the 1500-1600 cm^{-1} wavenumber range.

The ^{13}C - NMR spectrum of this material (Figure 3) shows significant presence of substituted aromatic material (125 ppm) as well as aliphatic carbon content. This spectrum shows a considerably reduced presence of polysaccharide decomposition products compared to fulvic acid extracted from wastewater (11). The ^1H - NMR spectrum (Figure 4) also indicates significant presence of substituted aromatic compounds (ca. 7 ppm).

The extract of closely bound organic material was redissolved in methyl ethyl ketone and reprecipitated and washed with 10% hydrochloric acid in order to liberate the organic material from chelation with iron, since it was known from previous data (9) that the iron content would be significant. The hydrolyzed extract, after drying and redissolving in methylene chloride, was placed on the basic alumina chromatography column packed in methylene chloride. Subsequent elution with acid/MEK yielded two major fractions. The residue from both fractions had very similar IR spectra (Fraction 1 is shown in Figure 2). The spectra show the appearance of a well-defined hydroxyl absorption at 3500 cm^{-1}, as well as the appearance of another well-defined absorption, possibly olefinic at 1620 cm^{-1}. The ^1H NMR spectra for the residue from both fractions were found also to be remarkably similar. That for Fraction 1 (Figure 5) shows a complete loss of aromatic resonance and the

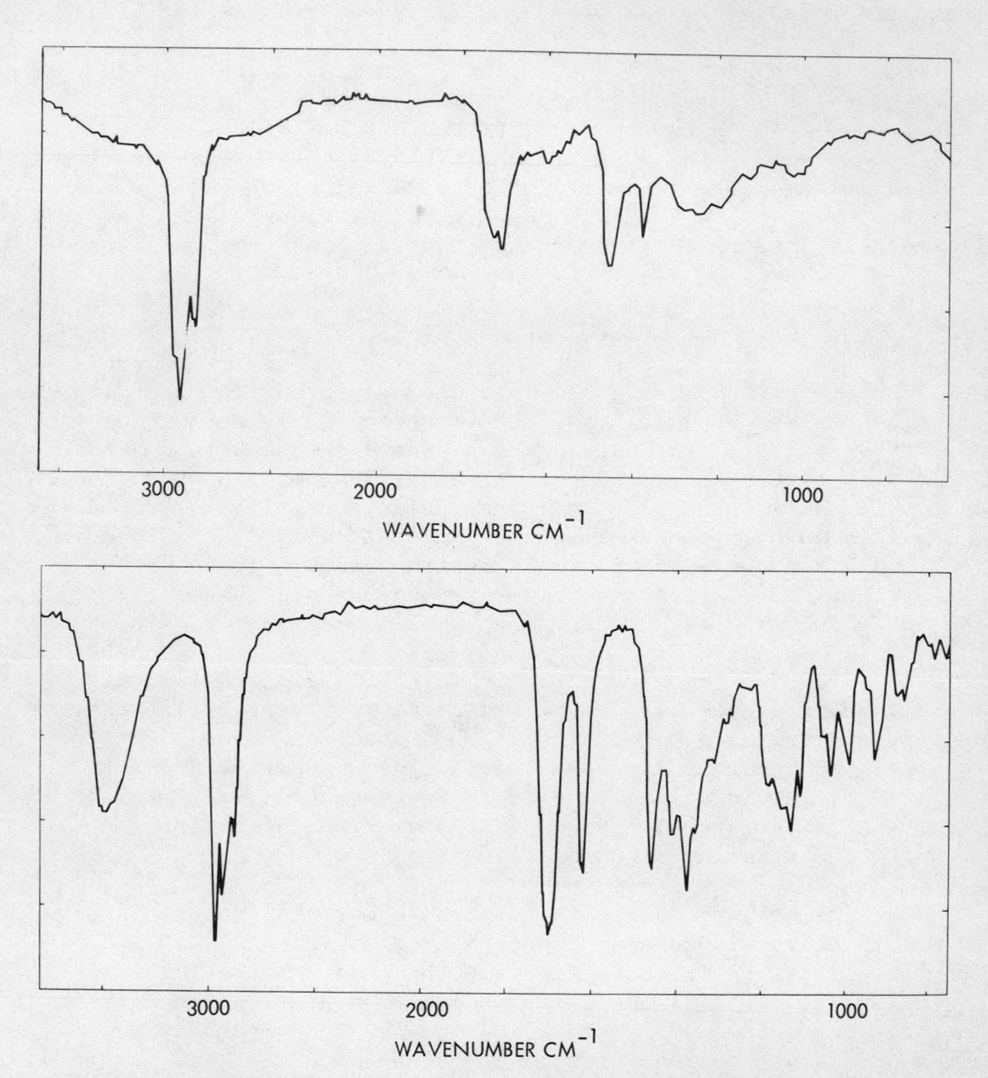

Fig. 2. IR spectra of acid/MEK extract of dried clay slime (A) and the first chromatography fraction (B).

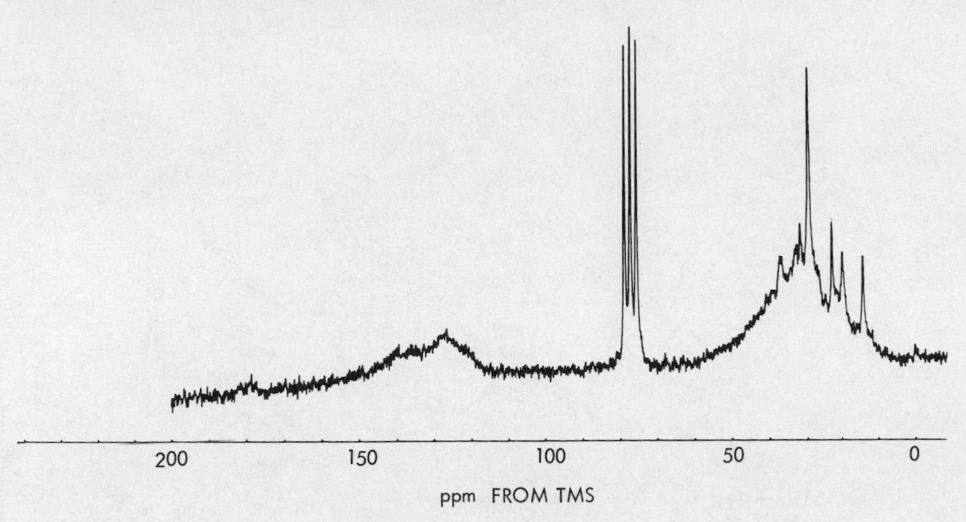

Fig. 3. ^{13}C - NMR spectrum of acid/MEK extract of dried clay slime

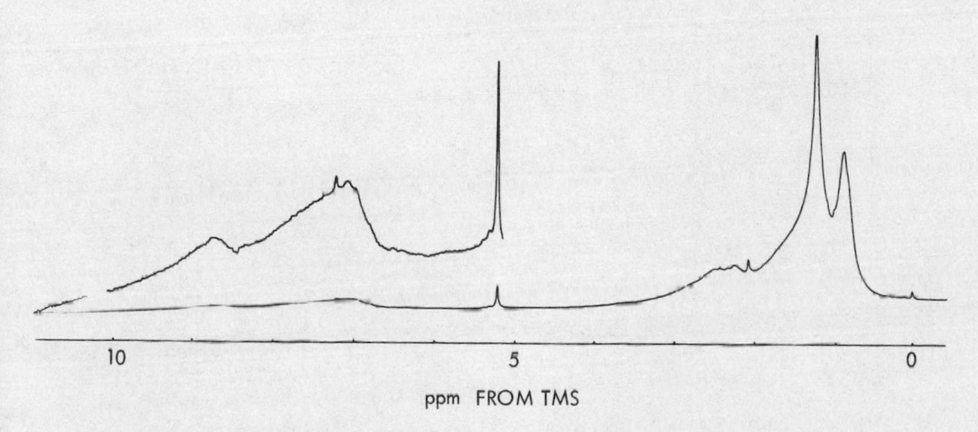

Fig. 4. ^{1}H - NMR spectrum of acid/MEK extract of dried clay slime

appearance of a well-defined olefinic proton resonance. It is believed these results show that the material originally extracted is strongly chelated and possibly polyphenolic in nature, and that it may break down when liberated from the chelate complex, and possibly degrade further during chromatography. These probable degradation products are very pungent and appear to contain aliphatic ketone, hydroxyl and olefinic groups, consistent perhaps with ring-opening degradation of polyphenolic materials. The IR (Figure 6), ^{13}C NMR (Figure 7) and ^{1}H NMR (Figure 8) spectra of closely bound organic material extracted from low grade oil sand (7.8% bitumen, 7.2% water, 83.8% solids) which contained a considerable amount of clay, ca. 7%, show that very similar compounds to those adsorbed onto the slimes clays were also adsorbed onto the clays present in the low grade oil sand matrix.

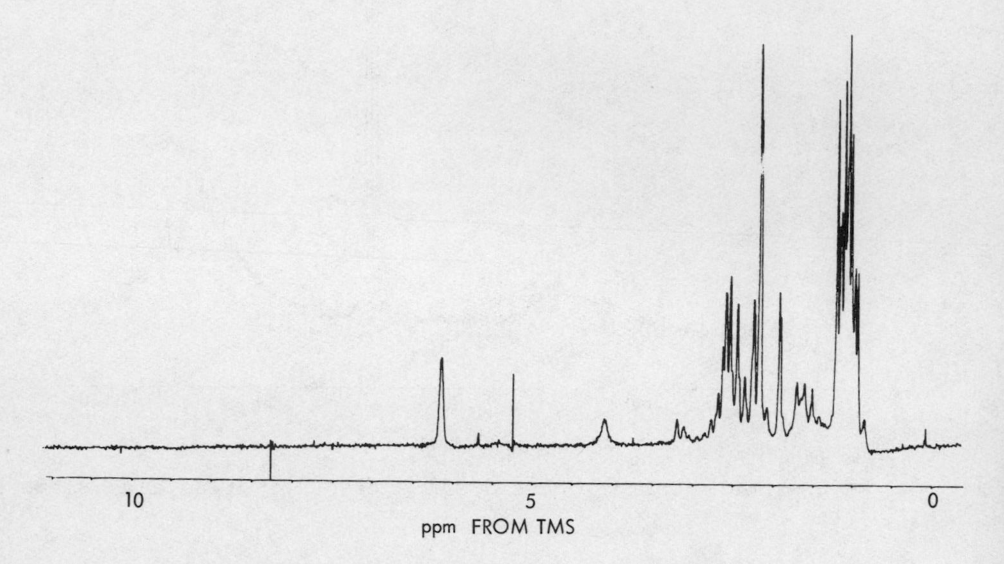

Fig. 5. ^{1}H - NMR spectrum of 1st chromatography fraction of acid/MEK extract of dried clay clime.

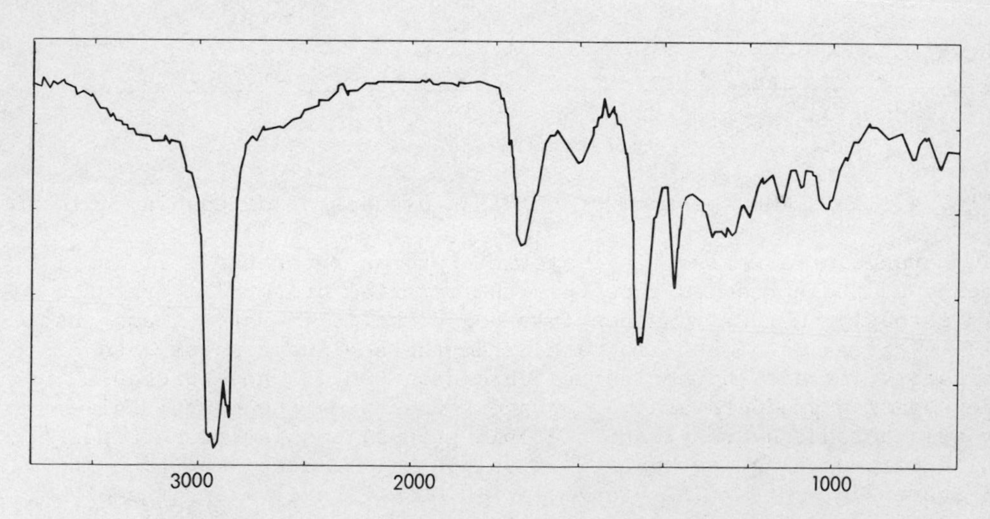

Fig. 6. IR spectrum of acid/MEK extract of low grade oil sand matrix.

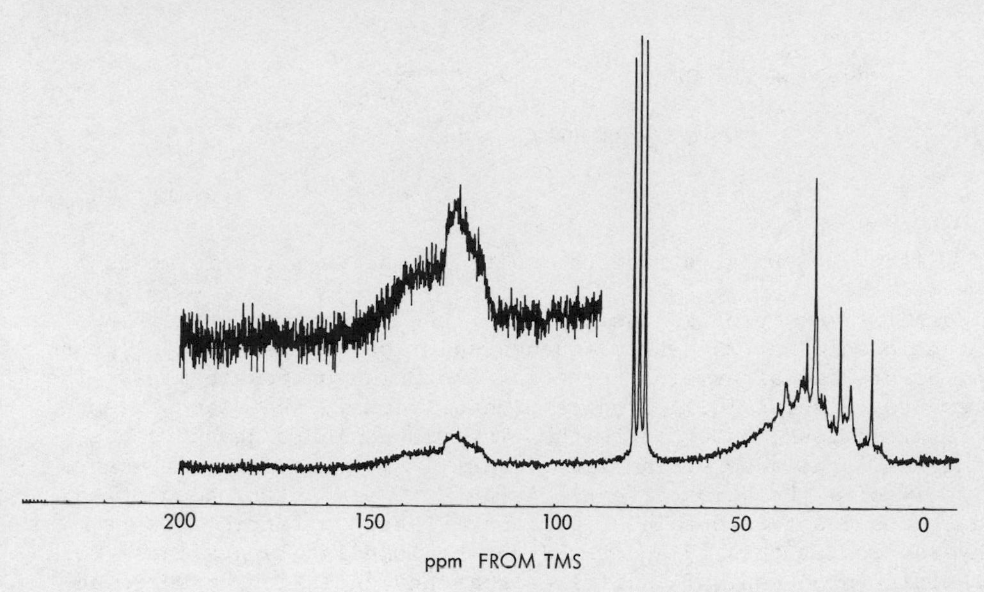

Fig. 7. ^{13}C - NMR spectrum of low grade oil sand acid/MEK extract.

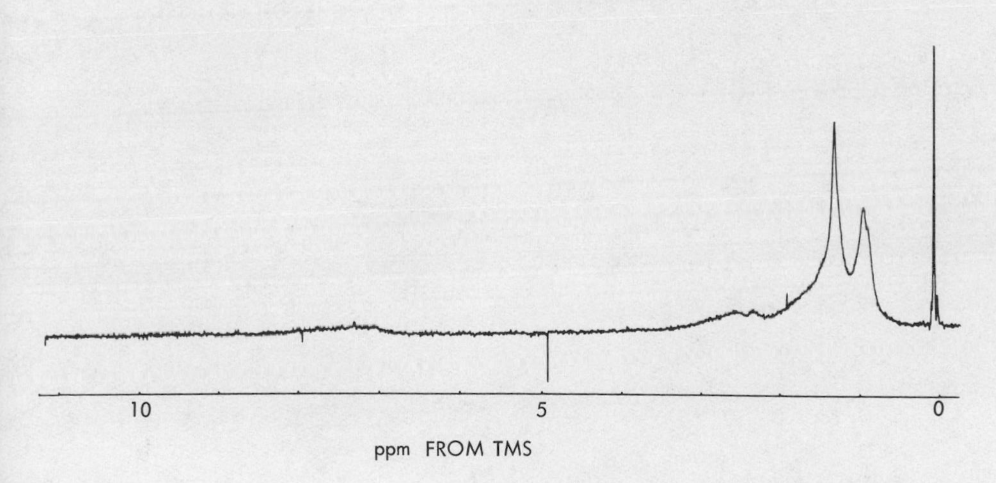

Fig. 8. ^{1}H - NMR spectrum of low grade oil sand acid/MEK extract.

It is inferred from these spectral data that the closely bound organic material may consist of vestigial tannins or lignins, or tannins or lignins which have undergone partial diagenesis as well as possible oxidation and ring-opening degradation, leaving intact, however, some polyphenolic or other iron-chelating functionality. Typical tannic acid structural units are shown in Figure 9. It has been hypothesized (2) that ferric iron is instrumental in binding this type of material to the predominant clays found in oil sands and oil sands slimes (kaolinite ca. 65%, illite ca. 35%).

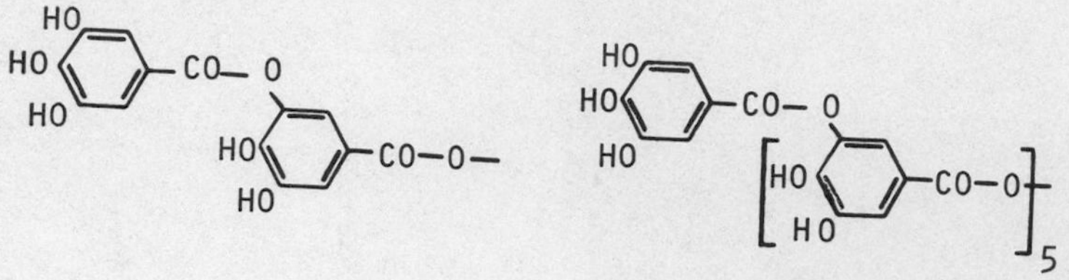

Fig. 9. Typical structural units in the tannic acid molecule

Isotherms were plotted therefore for the adsorption of tannin (tannic acid) at 25°C from aqueous solution to these clays, at pH 4, before and after electrostatic adsorption of ferric ions. The results (Figures 10,11) indicate unequivocally that the presence of ferric ions, adsorbed to the clays in an ion-exchange sense, increases the adsorption significantly. This is a result most likely of a chelation effect. At pH 8.5, meaningful adsorption data could not be obtained, presumably because ferric iron is hydrolyzed at this pH and would not be immediately available for chelate formation. This implies that any chelation-promoted adsorption present in the clay slimes must have taken place prior to the industrial hot water extraction process.

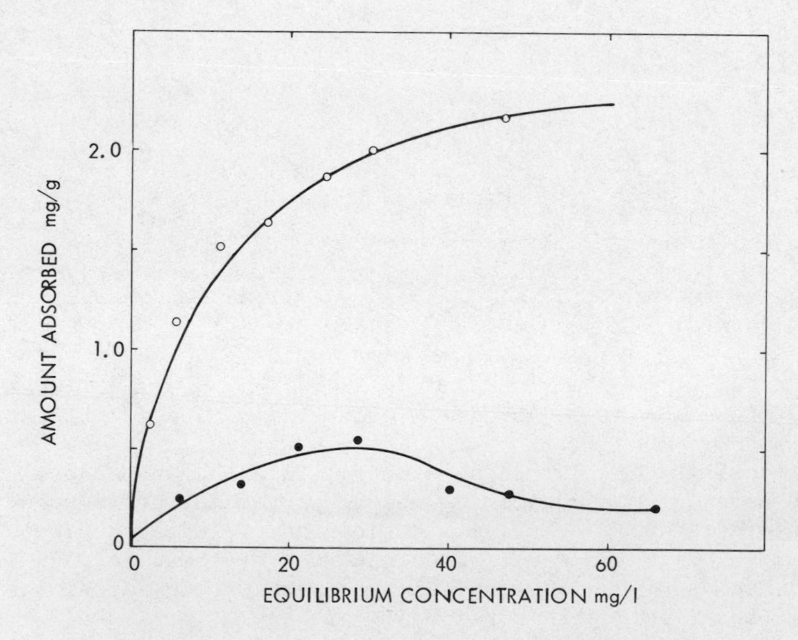

Fig. 10. The adsorption of tannic acid to illite, iron free ●, iron treated O.

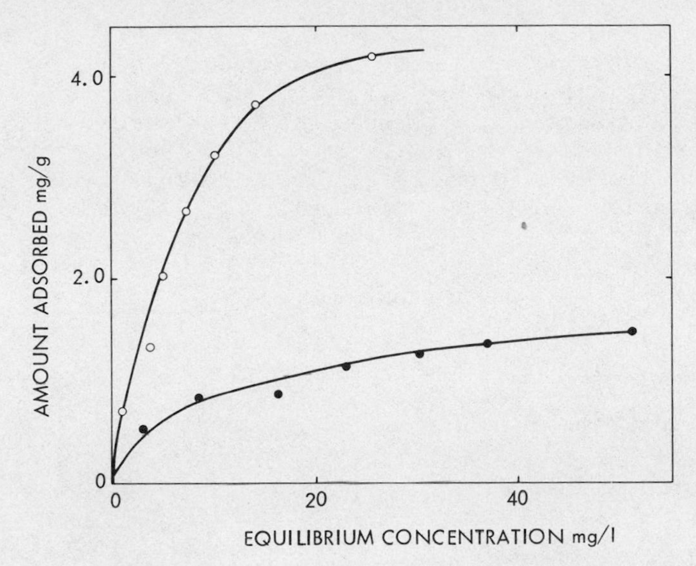

Fig. 11. The adsorption of tannic acid to kaolinite, iron free ●, iron treated O.

CONCLUSIONS

Vestigial tannin or lignin material is believed to be strongly adsorbed to clays found in oil sands matrix and is carried through in this form to oil sands slimes. This type of adsorption occurs readily under mildly acid conditions, probably through chelation with electrostatically bound ferric irons. At neutral or slightly alkaline pH the material remains adsorbed and promotes structure formation in the oil sands clays slimes, possibly by imparting hydrophobic character to clay surfaces which can then interact with residual bitumen setting up a weakly cross-linked gel.

REFERENCES

1. F. W. Camp, The Tar Sands of Alberta, Canada. Cameron Engineers Inc., Denver (1969).
2. A. R. Allan and E. D. Sanford, Alberta Research Council IS 65, 103 (1974).
3. F. W. Camp, Can. J. Chem. Eng., 55, 581 (1977).
4. R. N. Yong and A. J. Sethi, J. Can. Pet. Tech., 17(4), 76 (1978).
5. B.K.G. Theng, The Chemistry of Clay-Organic Reactions, Ahalsted Press (1974).
6. D. M. Clementz, Clays and Clay Min., 24, 312 (1976).
7. M. A. Kessick, J. Can. Pet. Tech., 25, 49 (1979).

8. M. Schnitzer and H. Kodama, Clays and Clay Min., 20, 359 (1972).
9. M. A. Kessick, Clays and Clay Min., in press.
10. M. L. Jackson and O. P. Mehra, Proc. 7th Natl. Conf. Clays Clay Minerals, Washington 1958, 317 (1960).
11. G. Sposito, G. D. Schaumberg, T. G. Perkins, and K. M. Holtzclaw, Envir. Sci. Technol., 12, 931 (1978).

ON THE COALESCENCE CHARACTERISTICS OF LOW TENSION OIL-WATER-SURFACTANT SYSTEMS

R. W. Flumerfelt, A. B. Catalano and C-H. Tong

Department of Chemical Engineering
University of Houston
Houston, Texas 77004, U.S.A.

The coalescence behavior of isolated oil drops in microemulsion systems and of microemulsion drops in brine systems is studied. A new experimental method is presented for conducting such controlled coalescence tests in an inclined spinning drop apparatus. Observations on five oil-water-surfactant systems are presented, three involving iso-octane as the oil phase and two involving crude oils. Coalescence times were measured in terms of the applied coalescence force and the droplet-droplet contact radius. The results show a wide range of behavior with coalescence rate per unit coalescence force varying over three orders of magnitude, with the specific magnitude for any given system being quite sensitive to the nature of the system and the associated phase changes with varying NaCl concentration. In nearly all cases, the most rapid coalescence occurred with the lowest tension systems. This apparently reflects the increased tendency of low tension systems to exhibit film rupture at relatively large values of the critical film thickness. The specific role of such effects and interfacial viscous effects are discussed.

INTRODUCTION

In the recovery of residual oil by low tension surfactant flooding, one can identify several key steps which are important to an efficient process. First the oil ganglia must be mobilized through the reduction of interfacial tension. These oil ganglia must then reconnect to form larger and larger ganglia, eventually leading to the formation of an oil bank. This oil bank must then sweep the production zone and effectively reconnect additional ganglia in the sweep path.

To date, much of the research on low tension surfactant flooding has been devoted to the identification of efficient surfactant systems and conditions for mobilizing oil ganglia. Although the mobilization step is certainly necessary to the process, it, in itself, is not sufficient to insure success. In particular, it has been shown by investigators in this laboratory (1,2) that without reconnection, isolated ganglia will eventually experience breakup and/or re-entrapment, and the ultimate efficiency of the process will be quite low.

The reconnection of oil ganglia is fundamentally a problem of coalescence. Coalescence occurs between droplets (or ganglia) when the thin film separating the drops is drained to such an extent that hydrodynamic disturbances and interfacial attractive forces eventually collapse the film (3-6). Over the years, droplet-droplet and droplet-interface coalescence phenomena have been studied by numerous investigators (see References 7-9 for summary reviews, and References 10-14 for the more recent and advanced works). Only recently, have such phenomena been recognized as important in surfactant flooding processes (1,2,15).

In the present work, we summarize recent observations on the coalescence behavior of several oil-water-surfactant systems. We are particularly interested in the behavior exhibited by low tension systems and the role of bulk phase viscous effects and dynamic interfacial properties such as interfacial shear and dilatational viscosities. Also, a new experimental approach for conducting coalescence studies is described which involves tests in an inclined spinning drop device. This approach allows for coalescence tests under controlled conditions and provides an efficient method for quickly screening and evaluating different surfactant systems. In addition to providing direct measures of coalescence kinetics, such tests can also be used in the indirect determination of interfacial viscous properties.

THE FILM DRAINAGE MODEL OF COALESCENCE

Jeffreys and Davies (7) identify several stages of a typical coalescence process between two droplets, or between a drop and an interface. First, the drops approach one another resulting in inertial induced deformation and possible oscillation. The oscillations are damped by viscous effects and eventually a thin film of the continuous phase is formed between the drops. As time proceeds, the film is drained thinner and thinner until eventually rupture occurs. The process is ended with the rapid disappearance of the film and the combination of the contents of the drop phases.

The film drainage step tends to be rate limiting. Most investigators have been concerned with describing the kinetics of this step through the use of film drainage models (7-14). Such analyses attempt to describe the relation between the coalescing force F, which promotes drainage, and the bulk phase and interfacial effects which resist drainage.

In Figure 1(a) we illustrate a flat plate interpretation of the film drainage process; in Figure 1(b) we show a typical time course of the film thickness δ. As illustrated, when the two drops are relatively far apart, δ changes very rapidly with time under the action of a coalescing force F. (In any given application, the coalescing force may be a buoyant force, an externally applied force, a force resulting from dynamic pressure gradients in the continuous phase, or any other force causing the drops to come together.) As the drop surfaces approach one another, the viscous and interfacial resistance forces increase, and $d\delta/dt$ decreases. This continues to slow the film drainage process as shown in 1(b) until a certain distance δ_c is reached where hydrodynamic disturbances and molecular attraction forces cause the film to rupture. If the molecular interactions are repelling, then $d\delta/dt \rightarrow 0$, and we have a stable emulsion.

The various hydrodynamic models presented for the film drainage process are too numerous to review here. The works of Reed et al. (10), Ivanov and coworkers (11,13), Barber and Hartland (12), and Jones and Wilson (14) represent some of the more recent contributions. The work of Barber and Hartland is of particular

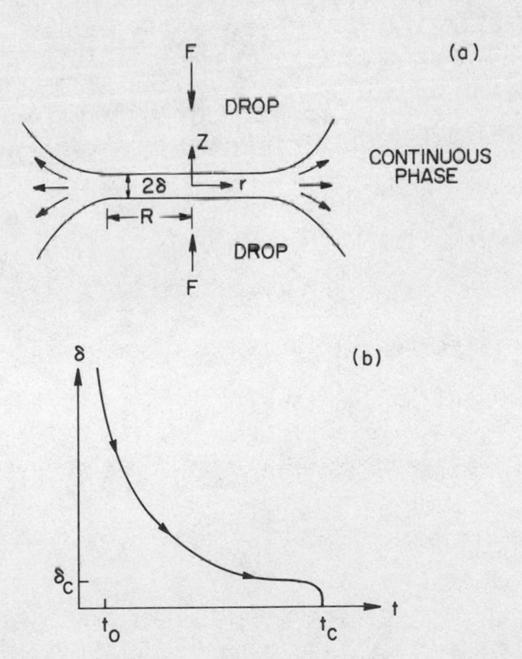

Fig. 1. Film drainage model of coalescence.

importance to the studies here since (1) dynamic interfacial effects such as interfacial tension gradients and interfacial viscosities are included in the analysis, and (2) analytical solutions are obtainable, hence providing a convenient basis for interpreting and evaluating our experimental observations. More accurate interpretations might be possible with more complete theories, however, this would be achieved only at the expense of considerable effort in numerical computations.

The important assumptions in the Barber and Hartland theory are:

1) The film drainage process is quasi-steady state.
2) The interfaces are flat as shown in Figure 1(a).
3) Drop phase effects, viscous and inertial, are negligible.
4) The interfacial material behavior is described by a Newtonian surface fluid model involving shear and dilational viscosities, ε and κ, respectively.
5) Interfacial tension gradient effects are represented as apparent dilational viscosity effects.

Barber and Hartland present results for several assumed boundary conditions at r=R. Although the quantitative results are different in each case, the qualitative results are consistent. The case used here is that where the shear stress is assumed to vanish at r=R. Although Barber and Hartland present their results in integral form, one can integrate their film drainage rate equation to obtain the following relation between the coalescence time t_c, the applied force F, the effective contact radius R, the bulk phase viscosity of the film μ, the critical collapse distance δ_c, and the combination $\eta = \kappa' + \varepsilon$ of the apparent interfacial dilational viscosity and the intrinsic interfacial shear viscosity:

$$t_c = \frac{\pi\eta^2}{3\mu F}\left[\chi^2 - 4\ \ln I_o(\chi)\right] \quad (1)$$

where

$$\chi \equiv (6\mu R^2/\delta_c\eta)^{1/2} \quad (2)$$

Here I_o is the modified Bessel function of zero order.

Two limiting cases of Equation (1) should be noted. When χ is sufficiently large, say greater than 80 (for 5% error or less), the interfacial resistance to drainage is small, and Equation (1) becomes

$$t_c = \frac{2\pi R^2\eta}{F\delta_c} \quad (3)$$

which we term the mobile interface limit. On the other hand, if χ is sufficiently small, say less than 0.5 (again for 5% error bound), the interfacial viscous effects render the interface essentially rigid and Equation (1) takes the form:

$$t_c = \frac{3\pi\mu R^4}{4F\delta_c^2} \tag{4}$$

which is the rigid interface limit.

It should be observed that the dependence of t_c on R varies between these limits, being second order in the mobile interface case, and fourth order in the rigid interface case. Also, the dependence on δ_c changes from δ_c^{-1} in Equation (3) to δ_c^{-2} in Equation (4).

In Figures 2 through 4 we illustrate the effects of η, δ, μ, and R on t_cF as predicted by Equation (1). In Figures 2 and 3, it can be seen that the effects of η and δ are quite significant over the entire domain, whereas, the effect of μ, shown in Figure 4, is important only in the rigid interface region. This is physically reasonable, since when the interfaces are rigid, the rate of film drainage will be largely governed by the viscosity of the film. On the other hand, when the interface is mobile, the flow in the film will be nearly uniform, i.e., v_r will be nearly constant across the film, and corresponding shear rates and bulk phase viscous effects will be small.

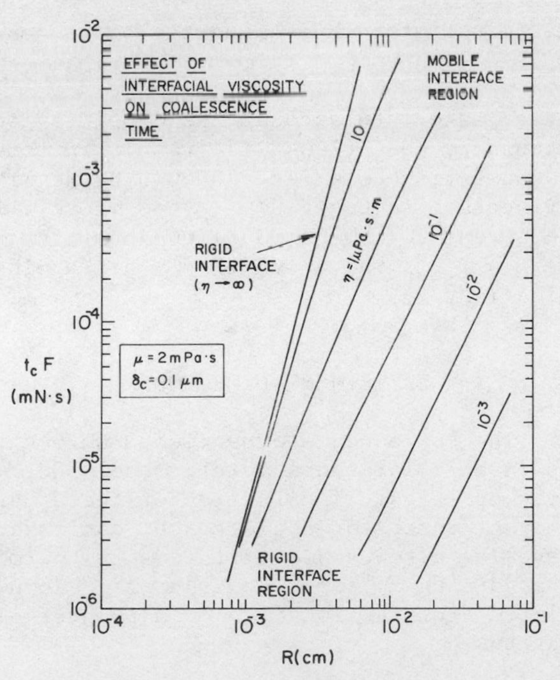

Fig. 2. Figure illustrates the dependence of t_cF on the interfacial viscosity coefficient η.

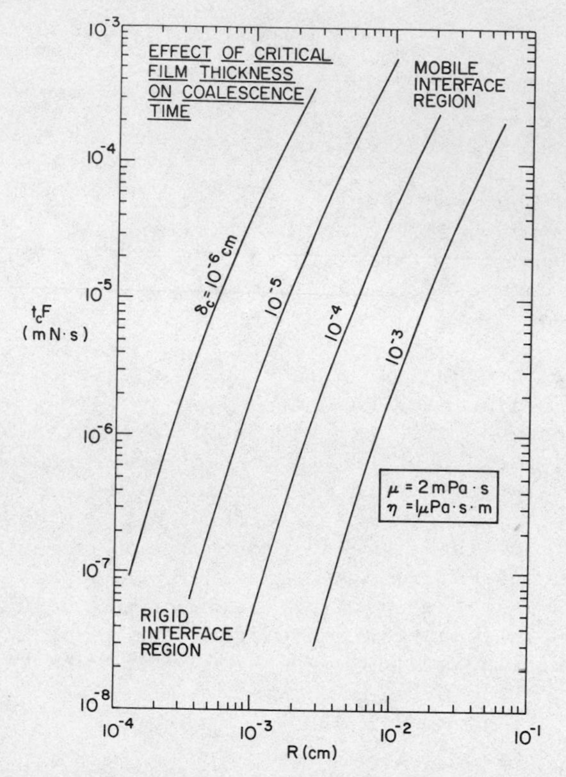

Fig. 3. Figure illustrates the dependence of t_cF on the critical collapse distance δ_c.

Theories that take into account the droplet viscosity $\hat{\mu}$ or use different boundary conditions will result in somewhat different results. Drop phase viscous effects should be negligible compared to film phase viscous effects if $\hat{\mu}/\mu << \hat{L}/\delta$ where $\hat{L}$ is some characteristic length measure of the drop dimensions (say the undeformed radius). Since $\hat{L}/\delta >> 1$, the above theory has a wide range of practical application.

EXPERIMENTAL METHODS

Apparatus: The coalescence experiments were carried out in a spinning drop interfacial tensiometer, Model 300, manufactured by the University of Texas. Inclination of the apparatus, which provides the buoyant force for coalescence, can be accomplished by turning a leveling screw. Use of this screw permits the tilting of the apparatus up to 7° upwards or 3° downwards. The angle is calculated from the location of a pointer placed on the end of the apparatus.

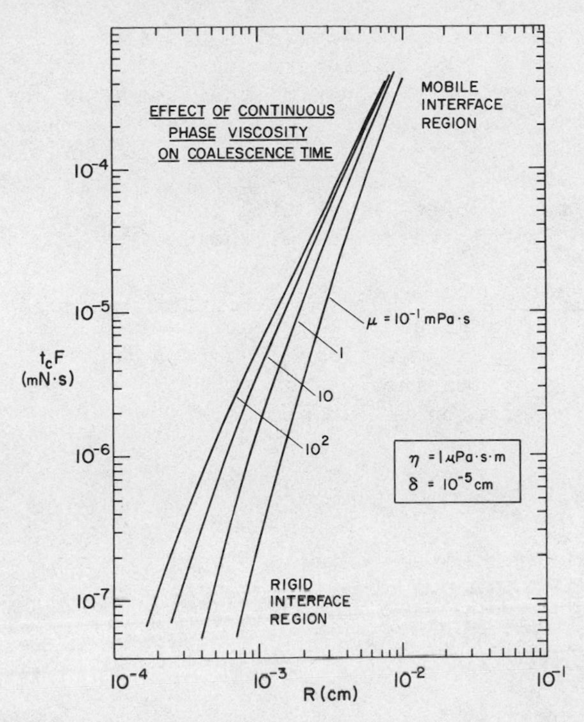

Fig. 4. Figure illustrates the dependence of t_cF on the film phase viscosity μ.

Pre-equilibrated solutions were introduced into the capillary tube with syringes. In most cases a number of oil drops existed after injection. Inclination of the apparatus forced the droplets to the higher end of the capillary tube where they coalesced to form one large drop. This large drop wetted the end of the tube, and careful manipulation of the tilt and rotational speed caused one or more drops to break off. If more than one free droplet was formed, the apparatus was again tilted to coalesce the extra drops. The size and number of drops broken off were influenced by the way the speed and inclination were varied.

When one free drop had been formed and was far from the stationary drop, the apparatus was leveled off, and the rotational speed that was desired for the experiment was chosen. The interface which had been altered during the previous process was then given the opportunity to come to equilibrium. A waiting period of at least one hour was chosen, since times longer than this seemed to have no effect on the coalescence behavior. The radius and length of the free droplet were then measured with the cathetometer, which was previously calibrated.

To induce coalescence the apparatus was inclined, moving the free drop back towards the stationary one. Since the two drops were far apart initially, by the time the free drop neared the stationary one the stationary drop had already reached its new equilibrium shape. The droplets appeared to contact each other (see Figure 5) at which time an extremely thin film of bulk phase fluid separated the droplets. The time interval between apparent contact and film rupture was taken as the coalescence time.

Equations for the determination of contact radius R and coalescence force F: In Figure 6 the geometry of the coalescence experiment is shown. Since the observed interface at $z=\delta$ is essentially flat, the pressure on the film side of the interface must equal the pressure on the drop side:

$$p(r,\delta) = \hat{p}(r,\delta) \tag{5}$$

Taking the reference pressure at $z=L$, and assuming quasi-static conditions, we obtain the following expression for $\hat{p}(r,\delta)$

$$\hat{p}(r,\delta) = \frac{2\sigma}{a} - \hat{\rho}g(L\sin\theta + r\sin\phi\cos\theta) + \frac{1}{2}\hat{\rho}\omega^2 r^2 \tag{6}$$

Here, a is the radius of curvature at $z=L$ and $r=0$, σ the interfacial tension, $\hat{\rho}$ the drop phase density, g the acceleration of gravity, L the length of the drop, θ the inclination angle, ϕ the angle measured from the horizontal line in the plane of the film ($z=\delta$ plane), and ω the angular rotation speed of the capillary. Similarly, we can write the pressure $p(r,\delta)$:

$$p(r,\delta) = -\rho g(L\sin\theta + r\sin\phi\cos\theta) + \frac{1}{2}\rho\omega^2 r^2 + p'(r,\delta) \tag{7}$$

where ρ is the density of the continuous phase and $p'(r,\delta)$ the dynamic pressure in the film, i.e., the pressure over and above that which would exist if there were no film flow.

Equating Equations (6) and (7) we obtain:

$$p'(r,\delta) = \frac{2\sigma}{a}\left[1 + \frac{\Delta\rho g a}{2\sigma}(L\sin\theta + r\sin\phi\cos\theta) - \frac{\Delta\rho\omega^2 r^2 a}{4\sigma}\right] \tag{8}$$

Now, for long free drops in spinning drop geometries, Princen et al. (16) have established that

$$a_f = \frac{2}{3} R_{fo} \tag{9}$$

and

$$\frac{\Delta\rho\omega^2 a_f^3}{2\sigma} = \frac{16}{27}\,; \tag{10}$$

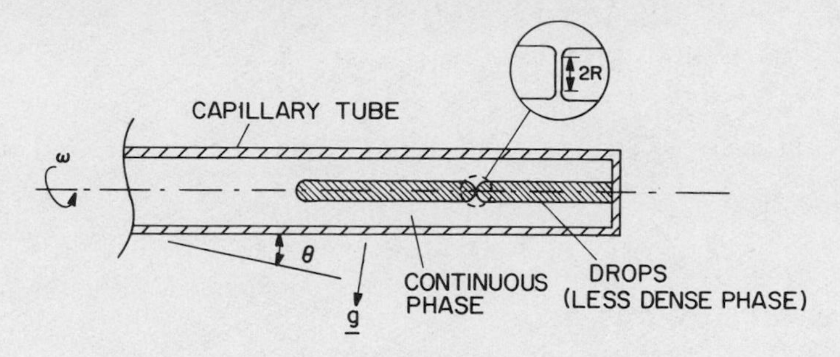

Fig. 5. Drop coalescence experiments in inclined spinning drop apparatus.

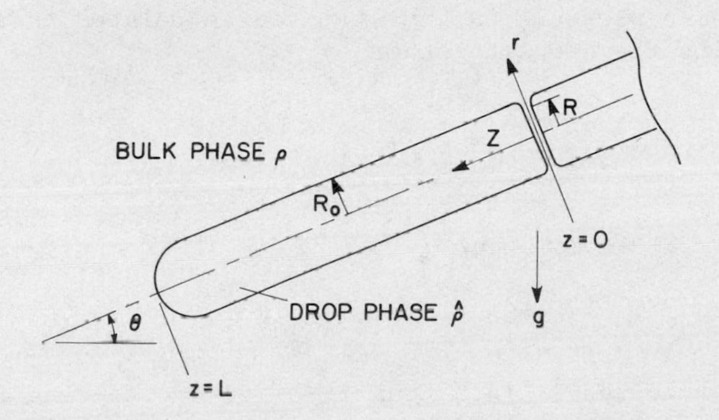

Fig. 6. Geometry of drop coalescence experiment.

these results being considered valid when the free drop length L_f is more than 8 times the free drop radius R_{fo}.

For the droplet-droplet coalescence experiment, we assume that the shape of the coalescing drop is similar to that of a free drop, except in the neighborhood the droplet-droplet contact (z=0), and, hence,

$$a \simeq a_f \tag{11}$$

$$R_o \simeq R_{fo} \tag{12}$$

It then follows that if R/R_o is small, the term

$$\frac{\Delta\rho\omega^2 r^2 a}{4\sigma} \leq \frac{\Delta\rho\omega^2 R^2}{4\sigma} \cdot \frac{2R_o}{3} = \frac{2}{3}\left(\frac{R}{R_o}\right)^2 \tag{13}$$

in Equation (8) will be small compared to 1. Also, if $g(L\sin\theta + R)/\omega^2R^2 << 1$, then the term

$$\frac{\Delta\rho g a}{2\sigma}\,(L\sin\theta + r\sin\phi\cos\theta) \leq \frac{4}{3}\,\frac{g(L\sin\theta + R)}{\omega^2 R^2} \tag{14}$$

will be negligible. Under these conditions, Equation (8) reduces to

$$p'(r,\delta) = \frac{2\sigma}{a} \tag{15}$$

or from Equations (9)-(12)

$$p'(r,\delta) = \frac{3}{4}\,\Delta\rho\omega^2\, R_{fo}^2 \tag{16}$$

When applicable, these results imply that the dynamic pressure is constant.

The dynamic pressure in the film can be related to the buoyant force F pushing the drops together by

$$F = 2\pi \int_0^R p'(r,\delta)\,r\,dr, \tag{17}$$

or, since $p'(r,\delta)$ is constant,

$$F = \pi R^2 p'(r,\delta) \tag{18}$$

Substituting this result into Equation (16) and solving for R, we obtain

$$R = \left[\frac{4F}{3\pi\Delta\rho\omega^2 R_{fo}^2}\right]^{1/2} \tag{19}$$

Finally, using

$$F = \Delta\rho V g \sin\theta\ , \tag{20}$$

we obtain:

$$R = \left[\frac{4Vg\sin\theta}{3\pi\omega^2 R_{fo}^2}\right]^{1/2} \tag{21}$$

The drop volume V is generally not measured; instead, the free drop measurements L_f and R_{fo} are obtained. From the analysis of Princen et al. (16),

$$V = \pi R_{fo}^2 L_f - \frac{4}{3}\,\pi R_{fo}^3 \tag{22}$$

In summary, the determinations of the coalescence force F and the contact radius R require measurements of $\Delta\rho$, R_{fo}, L_f, θ, and ω. In view of the assumptions in Equations (11) and (12), R_o and L could be measured instead of R_{fo} and L_f. In the experimental work here, we measured the free drop dimensions.

Finally, we should note that coalescence tests in the spinning drop device always involve coalescence of the less dense phase in the more dense phase. For an oil-microemulsion system (O/ME), this generally means coalescence of oil drops in the microemulsion solution; for a microemulsion-water system (ME/W) it means coalescence of ME drops in water. Coalescence of the reverse systems, ME drops in a less dense oil and water drops in a less dense ME system, are not possible.

FLUID SYSTEMS AND PROPERTIES

The fluid systems tested are shown in Table 1 along with the important fluid and interfacial properties. The constituents of Systems A and B are the same; they differ in that B has twice as much surfactant and alcohol cosolvent as A. System C has an iso-octane-CCl_4 mixture as the oil phase instead of iso-octane alone as in A and B. The CCl_4 was used to make the oil, water, and microemulsion phases of nearly equal density. This neutral buoyancy condition was needed in our measurements of interfacial viscosities with the drop deformation and orientation method (17,18).

The surfactant in Systems A and B was Witco TRS 10-80, while in System C, an Exxon C-12 orthoxylene sulfonate was used. The phase behavior exhibited by A and B was that observed with most oil-brine-petroleum sulfonate systems. At low salt concentrations, two phase systems are observed with the surfactant residing in the water phase. (Note that in Table 1 the phase that contains the highest concentration of surfactant is considered the microemulsion (ME) phase.) At intermediate salt concentrations, a three phase region is observed with the ME phase being the middle phase. The lower values of interfacial tension are observed with these three phase systems, with the lowest tensions, at least for A and B, being those associated with the ME-W interfaces. System C does not exhibit a three phase region, but still, a rather low tension is observed for the ME/W interface of the 1.3% NaCl case.

Systems D and E are crude oil-brine-surfactant systems, with System D involving Witco 10-80 surfactant and System E the Exxon C-12 orthoxylene sulfonate system. The large amount of surfactant and alcohol used in System D was to insure a large middle phase volume, the latter being required for interfacial viscosity measurements on a viscous traction instrument.

Table 1. Fluid Systems and Properties

System	Wt.% NaCl	Fluid Phases	Int. Tension (mN/m) O/ME[b]	Int. Tension (mN/m) ME/W	Viscosity (mPa·s) O	Viscosity (mPa·s) ME	Viscosity (mPa·s) W	Density (kg/m^3) O	Density (kg/m^3) ME	Density (kg/m^3) W
A. EQUAL VOLUMES - ISO-OCTANE & BRINE + 1 WT.% WITCO 10-80 + 5 WT.% TERT. BUTANOL	0.9	O/ME	0.099	--	0.470	1.51	--	687.1	988.4	--
	2.6	O/ME/W	0.034	0.014	0.480	6.46	1.27	687.5	948.6	1004
	6.1	ME/W	--	0.107	--	0.549	1.25	--	702.1	1031
B. EQUAL VOLUMES - ISO-OCTANE & BRINE + 2 WT.% WITCO 10-80 + 10 WT.% TERT. BUTANOL	0.4	O/ME	0.21	--	0.492	2.00		688.9	973.7	--
	0.8	O/ME	0.099	--	0.482	2.70		690.8	974.2	--
	1.5	O/ME/W	0.013	0.0105	0.508	5.16	1.67	695.0	873.2	987.6
	2.3	ME/W	--	0.068	--	0.692	1.65	--	709.5	996.0
	3.0	ME/W	--	0.108	--	0.850	1.64	--	720.7	999.3
C. BY VOLUMES - 32.5% ISO-OCTANE & 17.5% CCl_4 & 50% BRINE + 0.5 WT.% EXXON C-12 OXS[a] + 10.0 WT.% TERT. BUTANOL	0.5	O/ME	0.20	--	0.639	1.76	--	970.4	985.4	--
	1.3	ME/W	--	0.016	--	0.682	1.50	--	969.3	990.0
	2.0	ME/W	--	0.12	--	0.677	1.59	--	971.8	995.5
D. EQUAL VOLUMES - DELAWARE-CHILDERS CRUDE OIL & BRINE + 4 WT.% WITCO 10-80 + 20 WT.% TERT. BUTANOL	2.4	O/ME	0.51	--	8.58	4.21	--	854.0	969.4	
	9.1	O/ME/W	0.21	0.48	8.89	11.3	1.80	850.7	902.8	1058
	15.2	ME/W	--	0.55	--	10.9	1.68	861.6	--	1009

E. EQUAL VOLUMES -	1.0	O/ME	0.95	--	10.3	1.55	--	904.4	1058	--
DELAWARE-CHILDERS	2.0	O/ME	0.37	--	10.7	1.76	--	904.7	1065	--
CRUDE OIL & BRINE +	4.0	O/ME/W	0.041	0.18	10.3	15.2	1.09	907.6	970.0	1093
1 WT.% EXXON C-12 OXS[a]	6.0	O/ME/W	0.038	0.41	11.2	15.3	1.51	911.6	974.9	1118
5 WT.% TERT. BUTANOL	8.0	O/ME/W	0.014	0.49	10.8	14.9	1.40	921.3	963.3	1121
	11.0	ME/W	--	0.92	--	11.1	1.55	--	920.4	1138
	15.0	ME/W	--	1.91	--	11.4	1.61	--	920.3	1177

[a] OXS - orthoxylene sulfonate

[b] ME - microemulsion; defined here as the phase containing the most surfactant.

Both systems exhibit a three phase region at intermediate salt concentrations. Also, the interfacial tensions associated with the three phase region are somewhat lower than those associated with the two phase cases. The lowest tensions are observed with the O/ME interfaces of the three phase systems. This is different from the behavior for Systems A and B where the ME/W interface produced the lowest tension.

Finally, it should be noted that the viscosities associated with the crude oil systems are considerably larger than those of the iso-octane systems; the highest viscosities being produced by the ME phase.

The order of mixing and preparation of oil-water-surfactant systems is of great importance, as shown by Puig et al. (19), and can influence the time required for equilibration, the structure of the microemulsion, and the interfacial tension. All solutions were prepared in the same manner to avoid order of mixing effects. First, the brine solution was prepared, to which a solution of surfactant in alcohol was added. After mixing this solution, the oil was added, and the mixture was stirred vigorously with a magnetic stirrer for at least one day. (The crude oil was filtered before its addition.)

In order to eliminate the complications of mass transfer and changing droplet size in the coalescence experiments, the solutions were allowed to equilibrate for at least one month at 25.0°C ± 0.5°C. Interfacial tension was determined on the spinning drop apparatus in the course of the coalescence experiments. Viscosities were determined using Canon-Fenske viscometers. Density measurements were made on an analytical balance by measuring the buoyancy of a solid plummet immersed in the fluid.

EXPERIMENTAL OBSERVATIONS

A typical experiment involved placing drops in the apparatus and running a number of tests at different rotational speeds, drop volumes, and inclination angles. This allowed coalescence time measurements over a range of contact radius values. In each experiment, the coalescence time t_c, the coalescence force F, and the contact radius R were obtained.

In Figure 7 we show the results for System A. The ordinate of this plot, t_cF, represents the reciprocal of coalescence rate per unit coalescence force, and the abscissa is the contact radius R. For any given value of R, the system with the fastest coalescence rate would be the one with the smallest value of t_cF, and that corresponding to the slowest rate, the largest value.

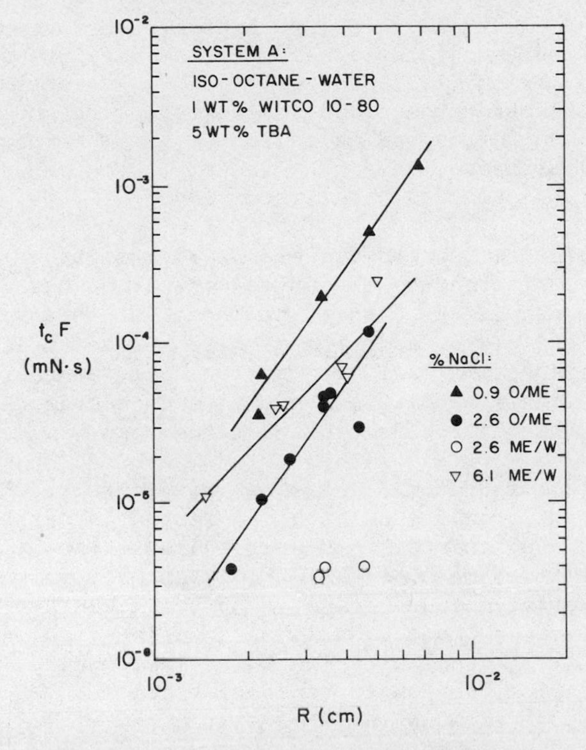

Fig. 7. Coalescence data, System A.

In Figure 7, we see that the fastest coalescence rates are associated with the 2.6 ME/W case. In fact, the coalescence times measured for this system were so small that only a few reliable data were obtainable. In Figure 7, these are shown in the vicinity of $t_c F \sim 3.5 \times 10^{-6}$ mN•s and are clearly an order of magnitude below the values associated with the 2.6 O/ME system, which represents the next fastest coalescing system. The slowest coalescence was observed with the 0.9 O/ME system, and the $t_c F$ values for this system were almost two orders of magnitude above the 2.6 ME/W system.

Based upon the dependence of $t_c F$ on R implied by the mobile and rigid interface limits of Equations (3) and (4), it follows that the 6.1 ME/W case exhibits more mobile interface behavior than do the 0.9 O/ME and 2.6 O/ME cases in Figure 7. This suggests faster drainage rates for the 6.1 ME/W system; however, since the observed coalescence times are higher for this case than the 2.6 O/ME case, it follows that the associated critical collapse distance must be smaller. Evidently, the lower interfacial tension associated with the 2.6 O/ME system must make the film less stable and δ_c much larger.

In Figure 8 the results for System B are shown. Here again, the fastest coalescence rates are associated with the systems in the three phase region (1.5% O/ME and 1.5 ME/W), where the lowest interfacial tensions are observed. Also, the more mobile interfaces (i.e., those corresponding to the smallest slopes on the t_cF vs. R plot) are associated with the higher concentration salt solutions, just as observed with System A.

The similarities between Systems A and B might be expected since only the surfactant-cosolvent concentration has been changed from one to the other. Evidently, at these surfactant-cosolvent concentration levels, the addition of more surfactant-cosolvent simply shifts the optimal salinity region to lower salt concentrations, but has little effect on the coalescence characteristics of the systems in the respective two and three phase regions.

The results for System C are shown in Figure 9. This system is quite different from Systems A and B in that a different surfactant has been used and CCl_4 has been added to the oil phase (iso-octane). First, we note that the lowest t_cF values in Figure 9 are an order of magnitude greater than the lowest values in Figures 7 and 8, even though the lowest tensions achieved in these cases are similar. Also, in System C the most mobile interface is that associated with the lowest tension system (1.3 ME/W), and the slowest coalescence rate is observed at the lowest salt concentration, rather than the highest as in Systems A and B. One important similarity between Systems A, B, and C, however, is the consistent inverse relationship between coalescence rate and interfacial tension, i.e., coalescence rate increases as the interfacial tension decreases.

In Figures 10, 11, and 12 we show the results for Systems D and E. These are crude oil-brine systems, involving the same oil (Delaware-Childers), but with different surfactants and different surfactant-cosolvent concentrations. System D, which has 4% Witco 10-80 and 20% TBA, exhibits a three phase O/ME/W region, but the tensions are not low (refer to Table 1). The corresponding coalescence times, even in the three phase region, are quite high, with the lowest t_cF values in Figure 10 being almost 100 times greater than those of Systems A or B. Also, with System D, the lowest tension system (9.1 O/ME) does not exhibit the fastest coalescence rates; both the 2.4 O/ME and 9.1 ME/W systems show slightly lower coalescence times and, correspondingly, faster coalescence rates. One possible explanation is that continuous phase viscosity of the 9.1 O/ME case, i.e., the viscosity of the microemulsion phase, is quite high compared to the continuous phase viscosities in the other cases.

The coalescence results for System E are shown in Figures 11 and 12, with the lower salt concentration results being presented

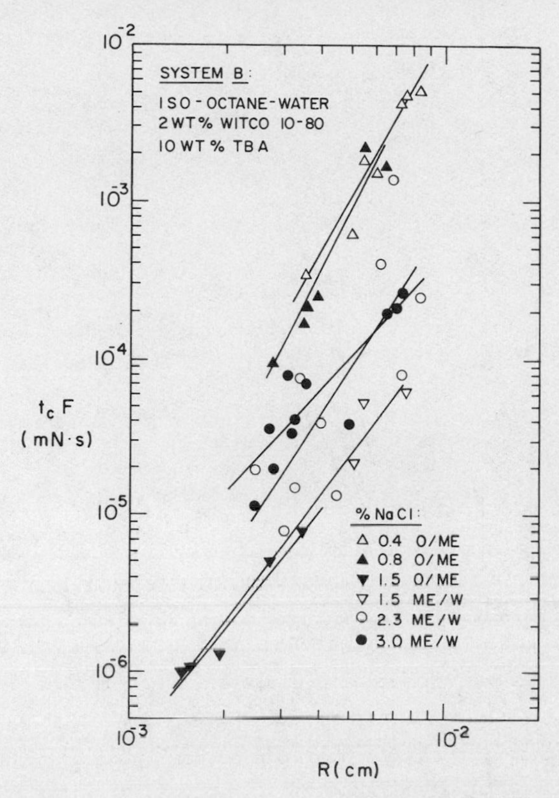

Fig. 8. Coalescence data, System B.

in Figure 11 and the higher concentration results in Figure 12. In considering these results, we first note that relatively low values of t_cF were obtained with this system. In fact, if one extrapolates the results of the fastest coalescing system (the 8.0% NaCl ME/W system) into the range of contact radii measured for the rapid coalescing 2.6 ME/W system of B, we find that the same magnitude of t_cF values are obtained. Clearly, System E, like B, possesses certain characteristics favorable to coalescence. First, low tensions are obtained at several salt concentrations in the three phase region, and the fastest coalescence rates are observed with these low tension systems. Also, the slopes of the curves in Figures 11 and 12 are, as a group, somewhat lower than those of the other systems indicating more mobile interfaces. Both the low tensions and the mobile interface characteristics would tend to result in less stable films, larger values of δ_c, and in the end, shorter coalescence times.

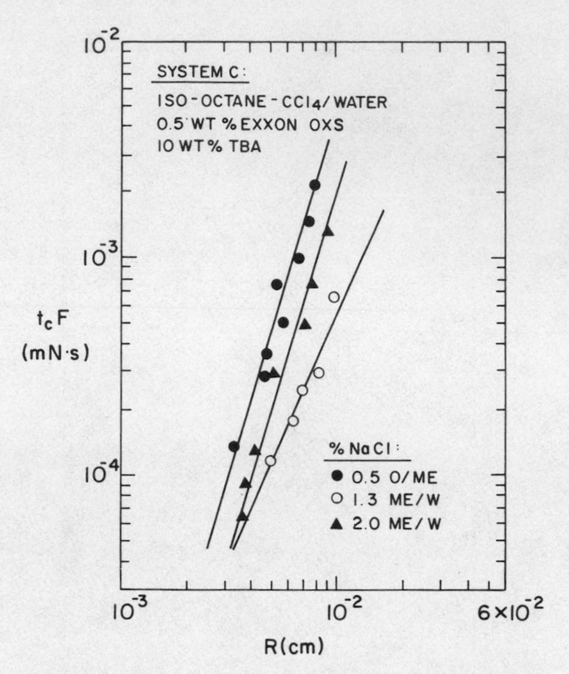

Fig. 9. Coalescence data, System C.

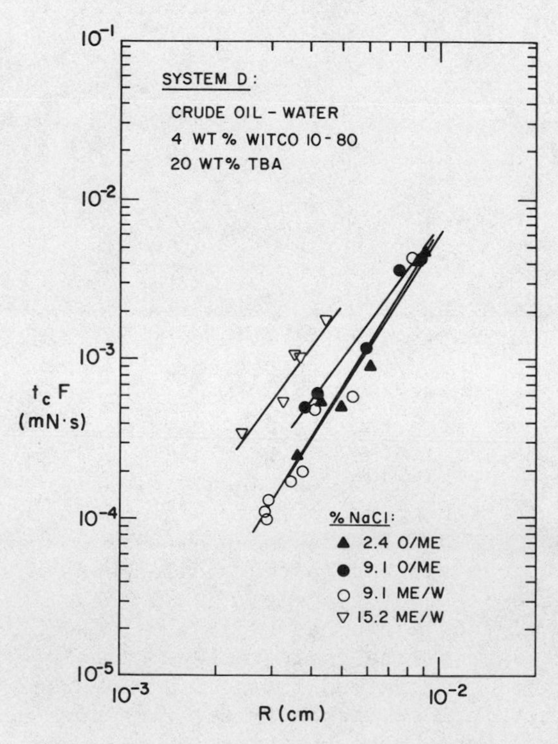

Fig. 10. Coalescence data, System D.

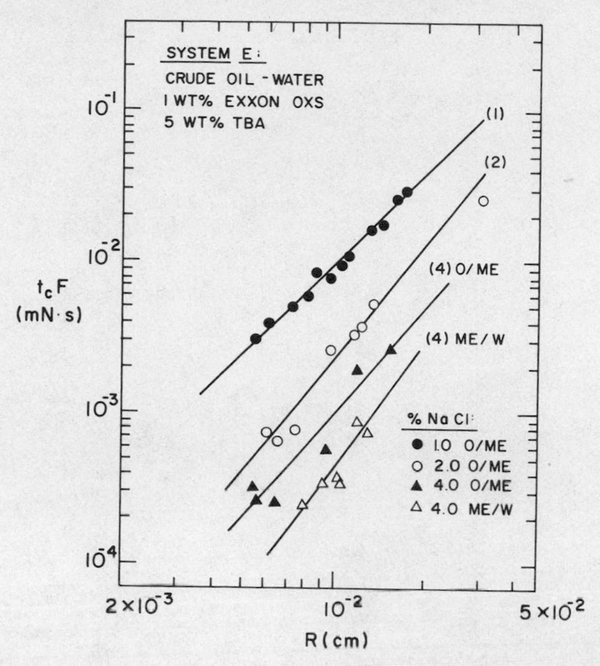

Fig. 11. Coalescence data, System E - lower salt concentrations.

ANALYSIS OF RESULTS RELATIVE TO BARBER-HARTLAND THEORY

In order to more completely assess the implications of the results in Figures 7 through 12, we were interested in analyzing the results in terms of the Barber-Hartland theory. Basically, this involved predictions of η and δ using the coalescence data and Equation (1). Although this does not provide a check of the theory, it does provide a framework for interpreting the results and analyzing the contributions of the various important factors (bulk viscous effects, interfacial viscous effects, critical collapse distance, etc.). Of course, the accuracy of these interpretations and analyses is strictly dependent on the accuracy of the theory. The evaluation of the latter requires independent determinations of η and δ_c, which, at the time of this study, were not available in this laboratory. Tests of this type are currently being set up, and the results will be reported at a later date.

Now, to find η and δ_c for each of the systems studied requires a non-linear, multidimensional regression analysis in which the best fit values of these parameters are obtained consistent with some minimization criterion. Here, we found η and δ_c which minimized

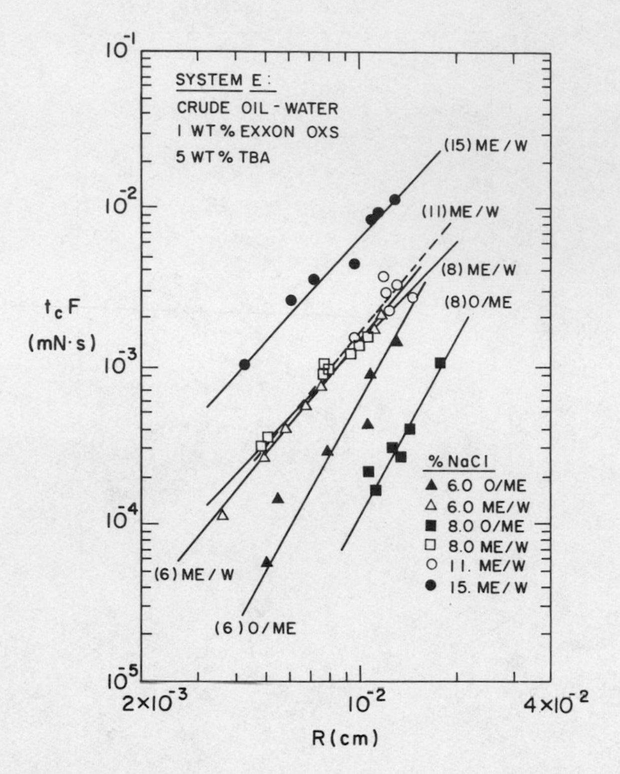

Fig. 12. Coalescence data, System E - higher salt concentrations.

$$S = \sum_{i=1}^{N} \left\{ \frac{(t_c F)_{pi} - (t_c F)_{mi}}{(t_c F)_{mi}} \right\}^2 \tag{23}$$

for each fluid system data set. Here $(t_c F)_{pi}$ and $(t_c F)_{mi}$ are the predicted and measured values of $t_c F$ corresponding to each measured value of the contact radius R_i. A multidimensional Newton-Raphson method was used to solve the minimization equations.

In Table 2 we provide the results of these computations, as well as summary of the experimental coalescence data. The first four columns in the table identify the particular fluid system and gives the important physical properties, i.e., the interfacial tension and the film phase (or continuous phase) viscosity. In the next two columns we summarize the ranges of the measured contact radii and the coalescence times, the latter being put in the form $t_c F/2\pi R^2$ which provides a convenient measure of reciprocal coalescence rate per unit coalescence stress. By coalescence stress we mean that stress which drives the interfaces together and promotes the film drainage. We use $t_c F/2\pi R^2$ instead of $t_c F$ here because the former is less sensitive to the magnitude of the

Table 2. Summary of Coalescence Results and Estimates of Dynamic Interfacial Properties

System	Wt.% NaCl	σ (mN/m)	μ (mPa•s)	R (μM)	$t_cF/2\pi R^2$ (Pa•s)	Int.Mob. $t \sim R^n$	η/δ_c (Pa•s)	δ_c (μm)	η (μN•s/m)	η_M (OR ε_M) (μN•s/m)
A	0.9 O/ME	0.099	1.51	22-68	125-450	n=3.0	810	0.11	8.9	(0.1)
	2.6 O/ME	0.034	6.46	18-47	20-87	3.1	58.	0.32	1.9	(5.2)
	2.6 ME/W	0.014	1.27	33-45	3-6	–	4.2	–	–	(12.)
	6.1 ME/W	0.107	1.25	15-52	84-154	2.0	120.	–	–	(0.4)
B	0.4 O/ME	0.21	2.00	35-82	380-1340	3.9	–	0.19	–	(<0.1)
	0.8 O/ME	0.099	2.70	27-64	200-1180	4.0	–	0.11	–	(<0.1)
	1.5 O/ME	0.013	5.16	14-34	5-11	2.7	15	0.54	0.81	(7.3)
	1.5 ME/W	0.0105	1.67	15-76	6-26	2.7	22	0.34	0.75	(33)
	2.3 ME/W	0.068	1.65	27-84	10-160	~2	~50	–	–	(0.2)
	3.0 ME/W	0.108	1.64	23-76	25-76	2.0	57	–	–	(0.5)
C	0.5 O/ME	0.20	1.76	33-80	199-543	3.1	900	0.18	16.	–
	1.3 ME/W	0.016	1.50	49-98	67-115	2.2	92	0.12	1.1	~7
	2.0 ME/W	0.12	1.59	36-93	79-242	3.0	460	0.24	11.	–
D	2.4 O/ME	0.51	4.21	35-91	298-996	3.3	2300	0.26	59	(58)
	9.1 O/ME	0.21	11.3	37-88	498-1040	2.5	1100	0.20	22	(4.0)
	9.1 ME/W	0.48	1.80	28-82	190-1070	3.0	1140	0.14	16	(30)
	15.2 ME/W	0.55	1.68	24-44	942-1430	2.8	2000	0.05	10	(17)
E	1.0 O/ME	0.95	1.55	54-173	12-19	2.0	15.5	–	–	–
	2.0 O/ME	0.37	1.76	60-129	2-5	~2	4.7	–	–	–
	4.0 O/ME	0.041	15.2	54-160	1.1-2.4	2.2	1.7	3.3	0.56	–
	4.0 ME/W	0.18	1.09	78-126	0.5-1.2	2.6	1.3	3.8	0.48	–
	6.0 O/ME	0.038	15.3	52-139	0.35-1.2	2.9	1.6	15.	2.4	–
	6.0 ME/W	0.41	1.51	36-112	1.7-2.2	2.4	2.8	1.2	0.34	–
	8.0 O/ME	0.014	14.9	104-178	0.25-0.55	3.1	0.95	44.	4.2	–
	8.0 ME/W	0.49	1.40	48-106	2.0-2.6	2.1	2.2	–	–	–
	11.0 ME/W	0.92	1.55	93-150	2-4	2.0	2.4	–	–	–
	15.0 ME/W	1.91	1.61	42-139	8-12	2.0	10.2	–	–	–

contact radius. Hence, by comparing the various $t_cF/2\pi R^2$ values for each system, one can immediately identify the hierarchy of coalescence rate behavior. The system with the fastest coalescence rate will have the lowest value of $t_cF/2\pi R^2$, and the one with the slowest rate, the highest value.

In the last five columns we provide the values of the interfacial mobility parameter n (here n corresponds to power in $t_cF \sim R^n$), the predicted ratio of the surface viscosity coefficient η and the critical collapse distance δ_c, the predicted individual values of η and δ_c, and, where available, values of $\eta = \kappa'+\varepsilon$ or ε_M obtained by direct methods. In the latter cases, the η data were obtained from a drop deformation method (17,18), and the ε data from tests in a viscous traction instrument (20).

The interface mobility parameters were obtained from the slopes of the log t_cF and log R data for each system. It is an important parameter since it indicates when the limiting case results apply. When n=2, we have the mobile interface limit, and separate determinations of η and δ_c are not possible; only η/δ_c information can be obtained from Equation (3). In the rigid interface limit, i.e., n=4, only δ_c can be obtained by fitting the data to Equation (4).

Considering the results for Systems A and B, we see that, as noted before with respect to Figures 7 and 8, the fastest coalescence rates (smallest values of $t_cF/2\pi R^2$) are associated with the lowest tension systems. Also, for each system, the interface mobility increases with salt concentration, and, at the highest concentrations, the mobile interface limit (n=2.0) is realized. The η/δ_c value found in each case corresponds closely to the average value of $t_cF/2\pi R^2$. Large values of η/δ_c infer large values of interfacial viscosity and/or small values of the critical collapse distance. Large values of η would promote more rigid interfaces and longer coalescence times. Small values of δ_c mean that the film would have to drain longer before film collapse. Obviously, the larger η/δ_c, the longer the coalescence time t_c. On the other hand, if η is sufficiently small and δ sufficiently large, a rapid coalescence event can occur.

In using η/δ_c as an indication of the coalescing characteristics of systems, it is implicitly assumed that the interface is not rigid, i.e., n≠4. When a rigid interface occurs, the important physical parameter is μ/δ^2 [see Equation (4)], and not η/δ. Coalescence rates under such conditions increase with decreasing values of μ/δ_c^2. This is illustrated in the 0.4 and 0.8 O/ME cases of System B.

The individual values of δ_c and η can be estimated in those cases where an intermediate interface mobility is realized, i.e.,

$n \neq 2$ or 4. Only a few cases for Systems A and B were of this type. In those cases where δ_c values were obtained, it ranged between 0.11 and 0.54 μm, with the smallest values (and, correspondingly, the largest coalescence times) associated with the lowest salt concentrations, and the largest values with the salt concentrations in the three phase optimal salinity region. Values of δ_c have been directly measured for pure and surfactant solutions (21), and these are generally found to be less than 0.1 μm, and more typically 0.02 to 0.05 μm. The high values observed here may reflect the striking differences between low tension systems like those tested here, and the higher tension systems studied previously by other investigators. In particular, various critical collapse theories (22,23) suggest that the critical film thickness would increase with decreasing σ. For very low values of σ, the critical film thickness could be quite large and the coalescence event quite rapid.

Values of the interfacial viscosity coefficient $\eta = \kappa' + \varepsilon$ were obtained only in a few cases for Systems A and B. The values obtained appear to be fairly low, and there is some indication that the highest values are associated with the lowest salt concentrations (note the values predicted for the 0.9 and 2.5% cases of System A, and also note that the 0.4 and 0.8 O/ME cases for System B are rigid interface cases suggesting relatively large values of η).

In addition to the importance of the critical collapse distance, the rate of film drainage is important. In the case of mobile interfaces the ratio η/δ_c represents the most important physical parameter. For rigid interfaces, the important parameter is μ/δ^2.

ACKNOWLEDGMENT

The support of the Department of Energy (Grant E(40-1)-5075) throughout this study is gratefully acknowledged.

REFERENCES

1. A. C. Payatakes, K. M. Ng, and R. W. Flumerfelt, AIChE J., 26(3), 430 (1980).
2. K. M. Ng and A. C. Payatakes, AIChE J., 26(3), 419 (1980).
3. G. D. MacKay and S. G. Mason, Can. J. Chem. Eng., 41, 203 (1963).
4. S. Hartland, Trans. Inst. Chem. Engrs. (London), 45, T102 (1967).
5. T. D. Hodgson and D. R. Woods, J. Colloid Interface Sci., 30, 429 (1969).

6. K. A. Burrill and D. R. Woods, J. Colloid Interface Sci., 42, 15 (1963).
7. G. V. Jeffreys and G. A. Davies, in "Recent Advances in Liquid/Liquid Extraction," C. Hanson, ed., Pergamon Press, 495 (1971).
8. D. R. Woods and K. A. Burrill, J. Electroanal. Chem., 37, 191 (1972).
9. A.J.S. Liem and D. R. Woods, AIChE Symp. Ser., 70 (144), 8 (1974).
10. X. B. Reed, E. Riolo, Jr., and S. Hartland, Int. J. Multiphase Flow, 1, 411 (1974); ibid., 437 (1974).
11. I. B. Ivanov and T. T. Traykov, Int. J. Multiphase Flow, 2, 397 (1976).
12. A. D. Barber and S. Hartland, Can. J. Chem. Eng., 54, 279 (1976).
13. T. T. Traykov, E. D. Manev, and I. B. Ivanov, Int. J. Multiphase Flow, 3, 485 (1977).
14. A. F. Jones and S.D.R. Wilson, J. Fluid Mech., 87, 263 (1978).
15. D. T. Wasan, S. M. Shah, M. Chan, and J. J. McNamara, Soc. Pet. Eng. J., 18, 409 (1978).
16. H. M. Princen, I.Y.Z. Zia, and S. G. Mason, J. Colloid Interface Sci., 23, 99 (1967).
17. R. W. Flumerfelt, J. Colloid Interface Sci., 76(2), 330 (1980).
18. W. J. Phillips, R. W. Graves, and R. W. Flumerfelt, J. Colloid Interface Sci., 76(2), 350 (1980).
19. J. E. Puig, E. I. Franses, H. T. Davis, W. G. Miller, and L. E. Scriven, Soc. Pet. Eng. J., 19(2), 71 (1979).
20. A. B. Catalano, M.S. Thesis, University of Houston, May 1979.
21. K. A. Burrill and D. R. Woods, J. Colloid Interface Sci., 42, 15 (1973); ibid., 35 (1973).
22. I. B. Ivanov, B. P. Radoev, E. D. Manev, and A. Sheludko, Trans. Faraday Soc., 66, 1262 (1970).
23. E. Ruckenstein and R. K. Jain, Trans. Faraday Soc., 70, 132 (1974).

THIN FILMS AND FLUID DISTRIBUTIONS IN POROUS MEDIA

K.K. Mohanty, H.T. Davis and L.E. Scriven

Department of Chemical Engineering & Materials Science
University of Minnesota
Minneapolis, Minnesota 55455, U.S.A.

Investigations of thin-film effects by means of complete solutions of the augmented Young-Laplace equation are summarized, as are the implications for observable contact angle, certain wettability syndromes, and breakup of nonwetting phase as it is replaced in a porous medium by wetting phase. Apparent contact angle, capillary pressure, and thin-film thickness are found to be interdependent at equilibrium: contact angle in porous media may depend significantly on pore size, fluid proportions, and filling history.

INTRODUCTION

Oil-bearing sedimentary rocks are frequently classified as either water-wet or oil-wet porous media, although categories of intermediate and mixed wettabilities are sometimes also set up. Wettability refers variously to spontaneous imbibition of water or oil, to the shapes of the curves of the relative permeabilities versus saturation (fraction of pore volume occupied by a given fluid) or to an apparent contact angle, i.e. the visually observed angle of intersection of a water-oil meniscus with a smooth surface of the rock or of an ostensibly equivalent solid material. In some circumstances a fluid can be totally wetting to a porous medium. In such cases a thin film of wetting fluid covers the solid. If bulk amounts of wetting fluid are present they connect to thin film through transition regions in which the film thickens; the wetting fluid is then distributed not only in continuous and disconnected pendular states of bulk material, but also in thin film states. When the surface area of the solid is great, as it can be, for example because of clay minerals in sandstone oil reservoirs, the thin films can contain appreciable inventories of wetting fluid which is difficult to displace. The intermediate and mixed wettability syndromes probably stem from thin film phenomena. All in all, it appears that films are important determinants of the distribution and recoverability of fluids in porous media.

In this paper we summarize investigations of the conditions under which thin films form; the effect these films have on equilibrium meniscus shapes and on apparent contact angles; and the role that film instabilities can play in isolation and entrapment of nonwetting fluid (1-4). We do not attempt to cover the highly important effects of the geometric and compositional heterogeneity of the surfaces of most solids in reality (4).

The shape and stability of menisci between bulk fluids is described classically by the Young-Laplace equation and its associated energy equation. When thin films are present these equations must be modified to take into account the effect of the interaction of the phases confining the film with each other and with the film material. For films no more than several molecules thick (adsorbed states), classical analysis is best abandoned entirely and a molecular model used in its place. However, for films at least an interfacial width thick--what can be called a thin thin-film--the departures of equilibrium chemical potentials and, in particular, stresses from those in bulk material of the same composition and temperature can be adequately accounted for by augmenting the Young-Laplace equation with the *disjoining pressure*, an entity defined some time ago by Deryagin (5) and now in fairly common use in thin-film analysis. The thin films treated in this paper are thick enough to be described by the disjoining pressure concept; molecularly thin films are the actors in adsorption but have little mobility and carry negligible inventory of fluid and so lie outside the scope of the present discussion.

AUGMENTED YOUNG-LAPLACE EQUATION

It has long been realized (5) and experimentally verified (6) that the normal stress (normal denoting the direction perpendicular to the interface) in a sufficiently thin, flat, fluid film at rest is different from that of the same fluid in bulk state. The normal stress in a layer or ordinary film of bulk fluid is of course the thermodynamic pressure of hydrostatic equilibrium. The difference is commonly known as disjoining pressure,

$$\Pi(h) \equiv P_N(\mu,h) - P_N(\mu,\infty) \qquad (1)$$

where h denotes the thickness of the film.

Related to this is the concept of disjoining potential, $\Delta\mu(h) \equiv \mu(P_N,h) - \mu(P = P_N,\infty)$, which is the difference between the chemical potentials of a bulk fluid and a thin film under the same normal stress or pressure. Π and $\Delta\mu$ are related, the ratio being the negative of the density of the film phase if it is incompressible.

Disjoining pressure and potential arise because molecules in a thin film reside in a different environment than those in a bulk phase of the same composition, temperature, and chemical potential. Several contributors to the disjoining pressure are known (see Table 1). The literature about molecular and ionic-electrostatic components is well reviewed by Sheludko (7). The structural contribution formally identified by Deryagin et al. (8) seems not to have been fully examined theoretically. All of the contributions have multicomponent versions (9) and are more or less additive. Their dependences on thickness are more complicated than the approximations in Table 1. It turns out that depending on the constitution of the three phases involved, the disjoining pressure (or potential) can vary with thickness in a variety of ways. Figure 1 shows a number of representative possibilities catalogued by Dzyaloshinskii, Lifshitz and Pitaevskii (10). The functional form of disjoining pressure governs equilibrium film states and their stability, as we now show.

Table 1. Sources of Disjoining Pressure

Component	Molecular Origin	Approximate Dependence on Thickness
Molecular, Π_m	London-Van der Waals' dispersion force including electromagnetic retardation	Ah^{-3} for small h Bh^{-4} for large h
Ionic-electrostatic, Π_e	Overlapping of double layers	Ch^{-2} for water on quartz (15)
Structural, Π_s	Short range forces Hydrogen bonding	(Unsettled)

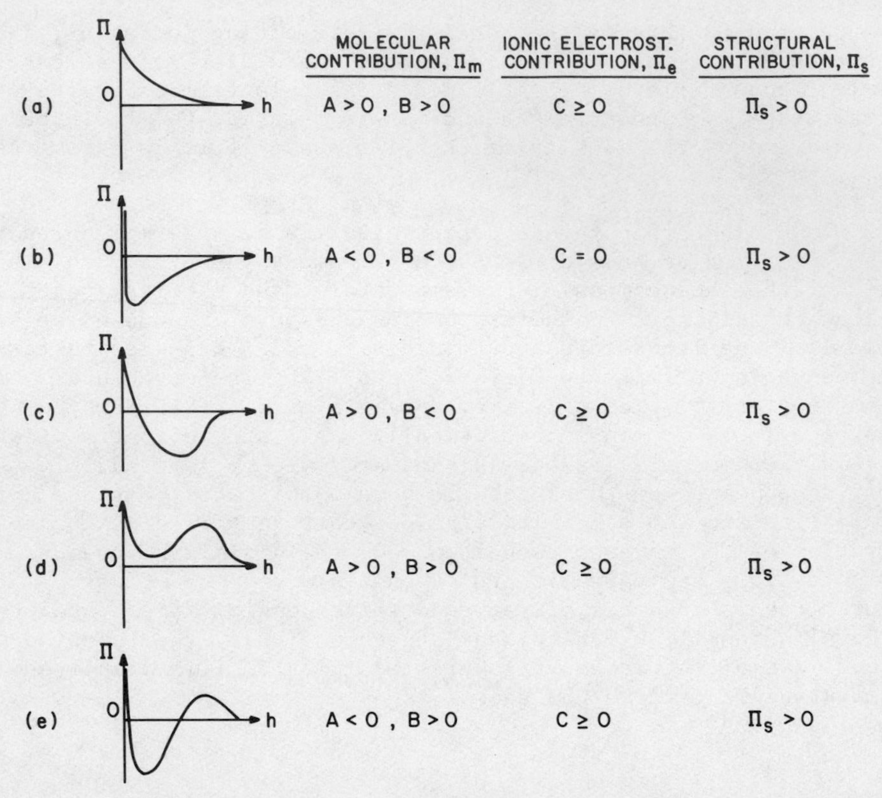

Fig. 1. Lifshitz et al.'s catalog of disjoining pressures (10) (and disjoining potentials).

In a fluid layer of nonuniform thickness the local disjoining pressure depends not only on thickness, but also on the gradient of thickness and even on higher-order gradient-like derivatives. The full story is not yet known, but if the thickness gradient is not large and the radii of film curvature are everywhere several times the interfacial width, then disjoining pressure can be regarded as a function of thickness alone (11).

If the film is not too thin--a "thick" thin-film rather than "thin" thin-film--the interfacial tension between it and contiguous fluid can be taken as independent of thickness (12).

The Young-Laplace equation of capillary (with interfacial tension a constant) can be brought to bear on equilibrium film shape in these circumstances. It is

$$P_{N,A} - P_{N,B} = 2H\sigma \tag{2}$$

where $P_{N,A}$ and $P_{N,B}$ denote normal stress in phases A and B, respectively, locally next to the interface. The mean curvature of the interface is $2H = r_1^{-1} + r_2^{-2}$, where r_1 and r_2 are the principal radii of curvature, which depend on spatial derivatives of film thickness. When gravity and other body forces are negligible or absent, the pressure throughout bulk phases at rest is uniform and we have $P_{N,A} = P_A$ and $P_{N,B} = P_B + \Pi$, whence

$$2H\sigma + \Pi = P_A - P_B \equiv \lambda \tag{3}$$

λ is a constant over the entire interface and is called the capillary pressure (though it is a pressure difference, a distinction worth remembering). This equation is the *augmented Young-Laplace Equation* that relates film curvature H and film thickness h. Because H is generated by a second order, non-linear differential operator on h and Π is a non-linear function of h, solving the equation for equilibrium film shape is often far from easy. If external body forces are present, the terms representing them are simply added to the right hand side of Equation (3). An essential feature of the equation is the additivity of disjoining and capillary pressure, which, it appears, was first recognized by Deryagin (13). Different forms of this equation have been used by Padday (14), Deryagin et al. (15), Philip (16) and Renk et al. (17). Berry (18) derived essentially the same equation from the statistical mechanics of a simplified molecular model.

FILMS ON SIMPLY SHAPED SOLIDS

If the film is supported by a solid of simple shape and the disjoining pressure is given by a polynomial in the thickness of the film, the augmented Young-Laplace equation becomes an *ordinary*, second-order differential equation, still nonlinear but solvable by straightforward quadrature, though in most cases the final integral has to be evaluated numerically (1,16). This useful fact was overlooked by previous investigators, who employed tedious asymptotic methods either informally (15) or more formally (17) for approximations to the (almost) uniform thin film and to the (almost) bulk film that exist in equilibrium; these approximations had to be matched together. Our solution by quadrature gives the entire film profile at once, including the region of transition between the asymptotic approximations. Sufficiently simple shapes of supporting solid include two-dimensional ones, e.g. two opposed parallel plates (slots and fissures), a bundle of parallel circular cylinders, intersections of plates (wedge-shaped pores), tubes of polygonal or axisymmetric cross-section, and rods and chains of polygonal or axisymmetric cross-section (e.g. a string of spheres).

When inside a slot or a tube there is a filled length behind a bulk meniscus that connects to a thin-film in a partly-filled length, the slot half-width or tube radius, R sets the thin-film thickness (Figure 2). The reason is that the slot or tube places a geometric lower bound on meniscus mean curvature H (i.e. an upper bound on radii of curvature); capillary pressure is proportional to meniscus curvature; chemical potential depends on capillary pressure; and thin film thickness is set by chemical potential, whether it is represented by disjoining potential or Deryagin's disjoining pressure. Thus

$$\Pi(h_o) \cdot (R - h_o) - \int_{h_o}^{R} \Pi(\eta)\, d\eta = \sigma \qquad \text{for slots}$$

$$\frac{\Pi(h_o) \cdot R + \sigma}{2} - \frac{R}{(R - h_o)} \int_{h_o}^{R} \Pi(\eta)\, d\eta = \sigma \qquad \text{for tubes} \tag{4}$$

Table 2 gives the thickness, h_o, of water and octane films connecting to bulk fluids in narrow quartz tubes that are partially invaded by vapor. The physical constants for water are taken from Deryagin et al. (15) and those for octane are from Churaev et al. (19). Because the disjoining pressure is a positive and decreasing function of film thickness, h_o increases as the tube radius increases.

If disjoining pressure changes sign and slope as in Figure 1c--e.g. $\Pi = Ah^{-3} - Bh^{-2}$, the first term being the molecular component and the latter the ionic-electrostatic component--then a thin film inside a tube can connect to an axisymmetric swelling containing bulk fluid under a meniscus that has a negative radius of curvature as shown in Figure 3a. Such a swelling is called a *collar* because of its similarity to the collars that can form around tight pore throats within porous media. The biconical pore shown in Figure 4a is an instructive, axially symmetric model of the converging-diverging nature of pore throats. Collars can form in both cylindrical and biconical pores when completely wetting fluid is displaced by a non-wetting fluid.

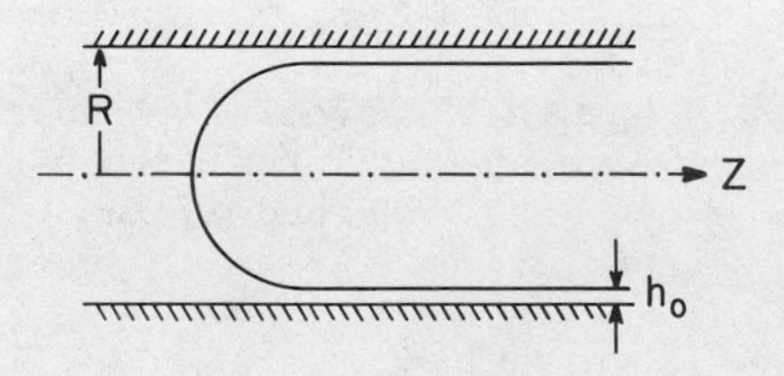

Fig. 2. Thin-films in a partially filled capillary or slot

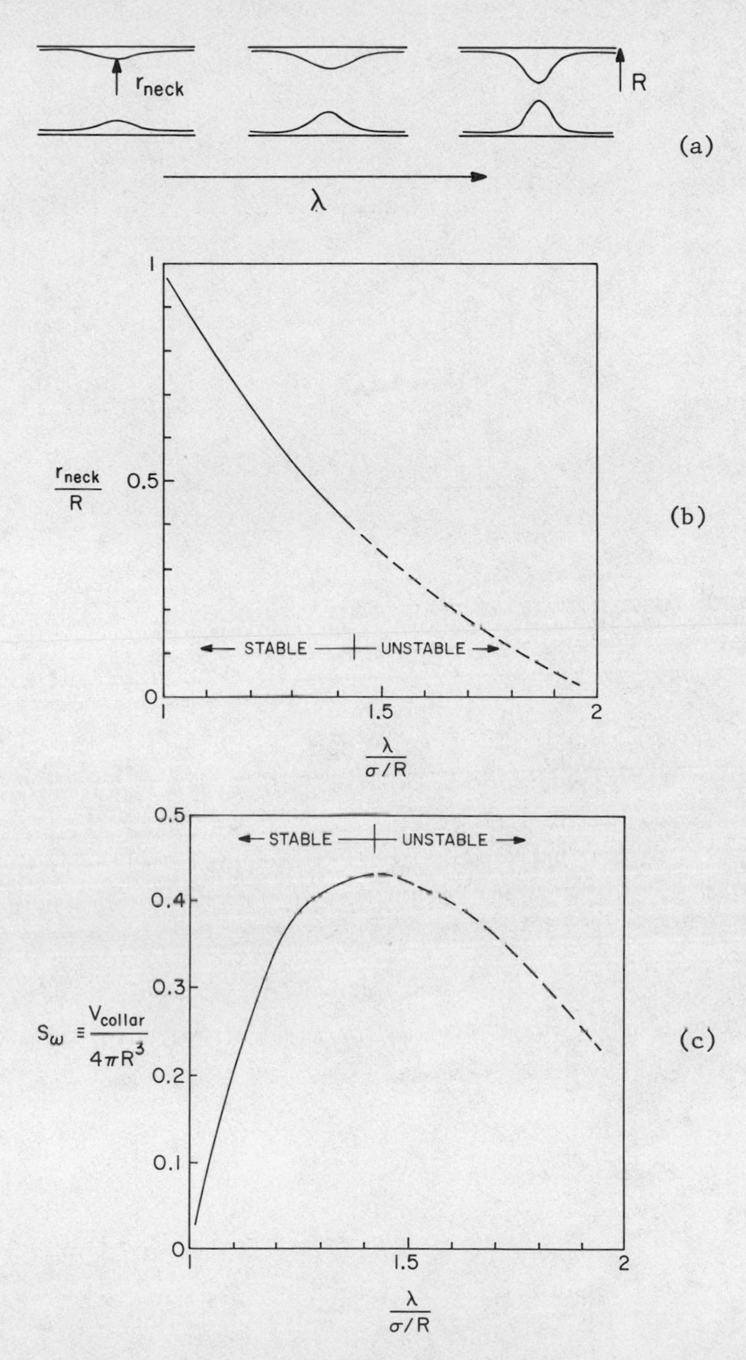

Fig. 3. Film with collar inside circular tube: (a) shapes having different neck radius; (b) neck radius vs. capillary pressure (dimensionless variables); (c) volume of the collar (local saturation) vs. capillary pressure.

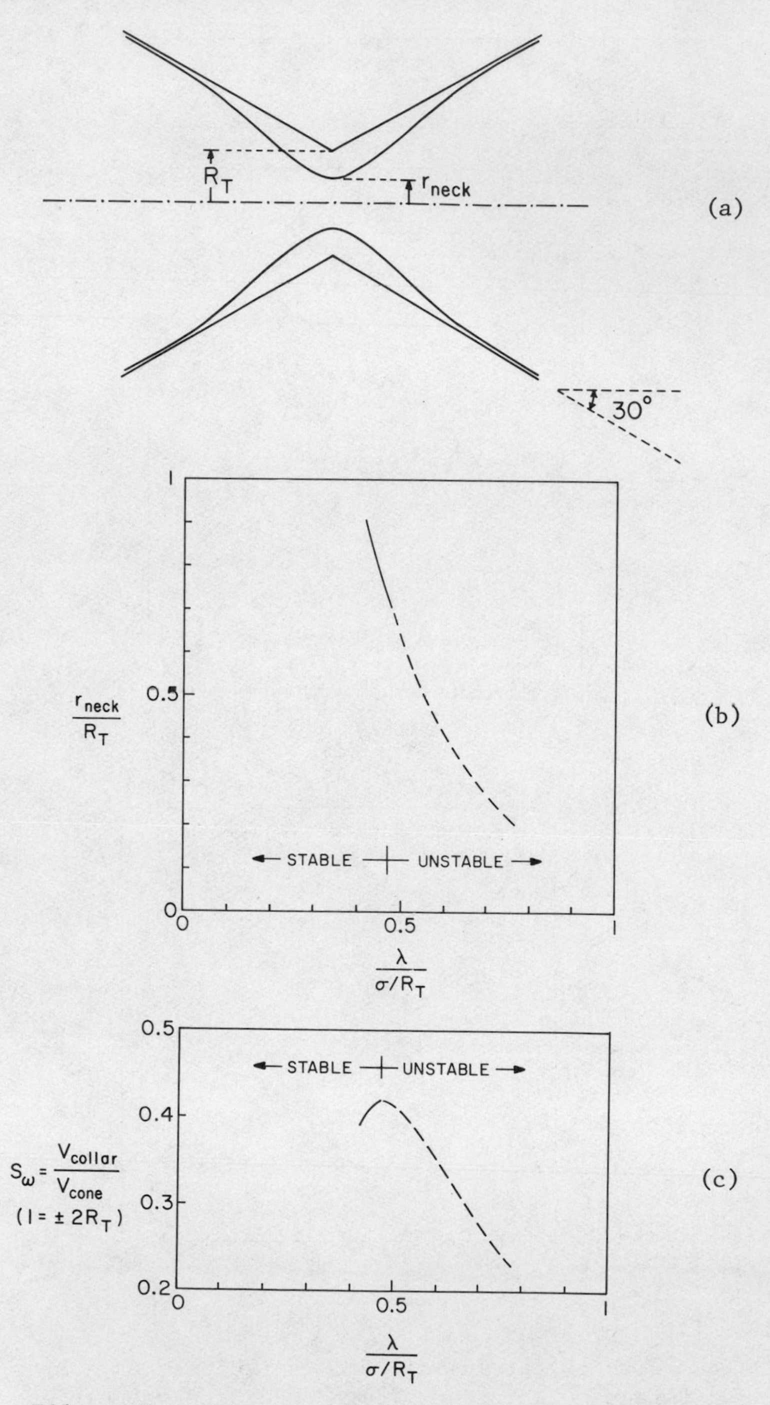

Fig. 4. Film with collar inside biconical pore: (a) representative shape; (b) neck radius vs. capillary pressure (dimensionless variables); (c) volume of the collar (local saturation) vs. capillary pressure.

Table 2. Effect of Tube Diameter on Limiting Film Thickness

	Film Thickness h_∞, Å	
	Quartz-Octane	Quartz-Water
Tube Radius, μm		
1	34.3	52.4
10	74.0	166.5
100	159.5	527.0
Parameters		
n	3	2
A	0.9×10^{-21} Joules	2×10^{-12} Newtons
σ	21.8 dynes/cm	72 dynes/cm

Such collars can become unstable and fill the pore locally under certain conditions, thereby breaking the continuity of non-wetting fluid in the pore. The non-wetting fluid is in effect "choked off," "pinched off," or "snapped off." Goren (20) and Everett and Haynes (21) have observed snap-off of non-wetting fluid in cylindrical pores. A given film configuration obtained as a solution of the augmented Young-Laplace equation is stable if the configuration minimizes an energy functional associated with the equation. Recently we investigated the stability question in detail (2). The results for the configurations considered here are summarized in the next section.

ENERGY STABILITY ANALYSIS

The appropriate energy functional, H to be minimized is the sum of surface energy, a pressure-related energy, and an interaction energy associated with disjoining pressure. For axisymmetric configurations r = f(z) it is

$$(f,f_z) \equiv \int [2\pi f\sqrt{1 + f_z^2} - \pi\lambda f^2 - 2\pi R\omega(h)] \, dz \qquad (5)$$

where $h \equiv R - f$, $\omega(h) \equiv -\int_h^\infty \Pi(\eta)\, d\eta$ and R is the radial distance of the pore wall from its axis of symmetry. It is convenient to employ the conjugate point condition for a local minimum (22). Figure 3b shows the capillary pressures λ and the neck radii r_n of collars that connect to uniform films in cylindrical pores, and Figure 3c shows the corresponding volumes of the collars. As the local wetting fluid saturation S_ω (defined for convenience here as the volume of the collar divided by $4\pi R^3$) increases the collars grow until at a critical neck radius, they pass into the unstable regime and sooner or later choke off the non-wetting phase. The critical neck radius is about 0.4 of the tube radius if the volume of wetting fluid in the collar is fixed, i.e. if the connecting thin-films do not conduct fluid appreciably. In the total absence of such a volume constraint all collars in cylindrical pores are unstable.

The collars at the throat of a biconical pore, as portrayed in Figure 4a, behave similarly. Again the maximum of volume corresponds to the stability limit provided the collar volume is left unchanged by shape perturbations. In the case illustrated the critical neck radius is about 0.7 of the throat radius of the biconical pore.

That a growing collar of wetting fluid in a converging-diverging pore throat does indeed become unstable and choke off or break the column of nonwetting fluid has now been demonstrated by S.R. Strand in our laboratory (3). In the experiments an aqueous solution was displaced by oil (Nujol) of the same density from a smooth, toroidally throated glass pore; the displacement was halted at various chosen stages; and the behavior of the aqueous collar that forms was observed and photographed. Small collars remain intact whereas larger ones are unstable and throttle the neck they surround, breaking the oil connection. The size of the collar that formed was controlled by the speed of displacement and the stage at which it was halted. Stability calculations have been completed for a photographed sequence of collar shapes. The predicted onset of instability agrees closely with the observations (4).

DISJOINING PRESSURE AND APPARENT CONTACT ANGLE

The length scale on which disjoining pressure is significant is typically less than the wavelength of light. Thus even when one of the fluids completely wets the solid in a thin-film state, the meniscus at observable distances from the solid may appear to intersect the solid at a nonzero angle. The optically observable contact angle, whether to the naked eye or through the usual microscope, is called the apparent contact angle and is, under conditions that are presumed to give equilibrium, what has long been called *the* equilibrium contact angle.

Knowing the entire interfacial profile one can find the angle at which the bulk interfacial profile appears to intersect the solid surface. The solid line in Figure 5 is the real interfacial profile described by Equation (3). The broken line is the extrapolated profile $h^*(z)$ resulting when the special properties of thin film are ignored, i.e., Π is equated to zero in Equation (3) to obtain

$$-\sigma \frac{d}{dh^*}\left[(1 + h_z^{*2})^{-1/2}\right] = \lambda \qquad (6)$$

for a two-dimensional meniscus. Integrating Equation (3) from far away from the solid to the ultimate thin-film thickness, h_o; similarly integrating Equation (6) down to the solid surface, $h = 0$; and subtracting the two results gives

$$(1 + h_z^{*2})^{-1/2}\Big|_{h^*=0} = \cos\theta = 1 + I(h_o)/\sigma \qquad (7)$$

where $I(h_o) \equiv \Pi(h_o) \cdot h_o + \int_{h_o}^{R} \Pi(\eta)\, d\eta$. R can be regarded as infinite if $\Pi \simeq 0$ for $h \geq R$. Because the augmented Young-Laplace equation is an equation of momentum balance at the fluid-fluid interface, its integral Equation (7) amounts to an overall force balance across the length of the transition zone. At equilibrium the system is conservative and so the same equation can be derived by a virtual work argument, as Ivanov et al. did (23).

Certain implications of Equation (7) can be illustrated with the help of Figure 1. According to Equation (7), apparent contact angle θ is real and nonzero only if $I(h_o)$ is negative. For positive $I(h_o)$ the extrapolated interface $h = h^*(z)$ does not intersect the solid surface $h = 0$, but an eye sees a zero contact angle. In Figure 1c, $I(h_o) < 0$ for h_o more than some thickness h_a, where $h_a < h_c$. Hence, the apparent contact angle is nonzero for $h_o > h_a$. Moreover, in this case as the limiting film thickness h_o increases so does the apparent contact angle. That h_o depends on pore diameter, R is already established. Hence even the apparent contact angle is influenced by the diameter of fine-bored tubes and pores (and by the width of narrow slots and fissures).

There are exceptions to this rule. In Figure 1a $\Pi(h) \geq 0$ and so $I(h_o) \geq 0$. Hence the apparent contact angle is zero for all menisci, and thus for all pore diameters. On the other hand in Figure 1b $\Pi(h) \leq 0$ for all films except adsorbed layers; hence the apparent contact angle is nonzero but independent of capillary pressure. In this case there may exist a non-zero submicroscopic contact angle θ_m at the adsorbed layer, but the macroscopically observed contact angle, or apparent contact angle, θ is different from θ_m, the difference originating in the disjoining pressure.

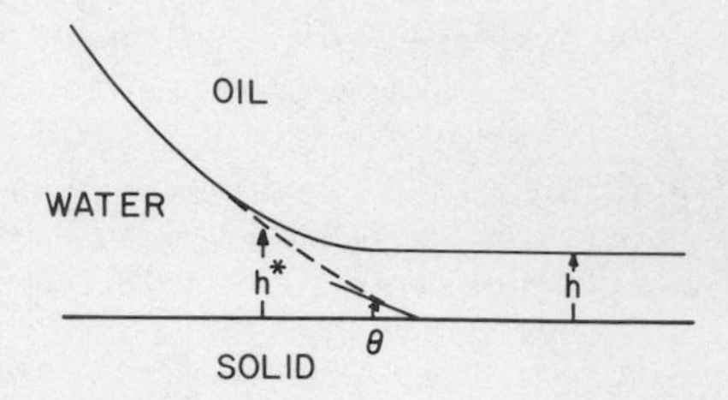

Fig. 5. Apparent contact angle is the angle of intersection of the eye's extrapolation on the bulk meniscus with the solid surface.

IMPLICATION FOR OIL-WATER DISTRIBUTION IN POROUS MEDIA

Altogether our investigations reveal that wettability phenomena in tight places need to be understood in terms of a closed loop of physically definable quantities, as diagrammed in Figure 6. The curvature of a meniscus in a pore is set by the pore radius and the apparent contact angle the meniscus makes with the pore wall. The curvature then sets the capillary pressure. The capillary pressure λ and the disjoining pressure $\Pi(h)$ determine the film thickness and this with the disjoining pressure $\Pi(h)$ determines the apparent contact angle, which closes the loop. At equilibrium the thin-film thickness, the apparent contact angle and the capillary pressure are all interrelated. For a given fluid/porous solid combination there may be no unique contact angle. In disequilibrium they continue to be interrelated but are affected by dynamic effects such as viscous stresses (11). The interplay of geometric quantities and physical properties evident in Figure 6 is the hallmark of fluid behavior in porous media.

In a water-wet, oil-bearing porous rock at a high enough oil saturation that only connate water is present, the water resides in films on mineral grains and the usually present clay structures, and in pendular accumulations in and around wall regions of high curvature such as necks of the solid matrix, pore throats, and grooves and pits on rough solid surfaces. From Table 1 it appears that the thickness of these wetting films can reach the order of 100Å. On this basis, in a 35% porosity pack of clean quartz spheres of 10 μm diameter the wetting liquid in thin films amounts to only about 1% of the pore space. This is very little compared to the typical connate water saturation in oil-bearing rocks. However, if the model sandstone contains clay minerals (clay content = 2% by weight, specific area of clay = 100 m^2/gm, and specific gravity of both clay and rock = 2.5), then the wetting fluid in thin films would account for 20% of the pore space (provided there is enough spacing between clay platelets for wetting films to reside). This gives an indication that the amount of wetting

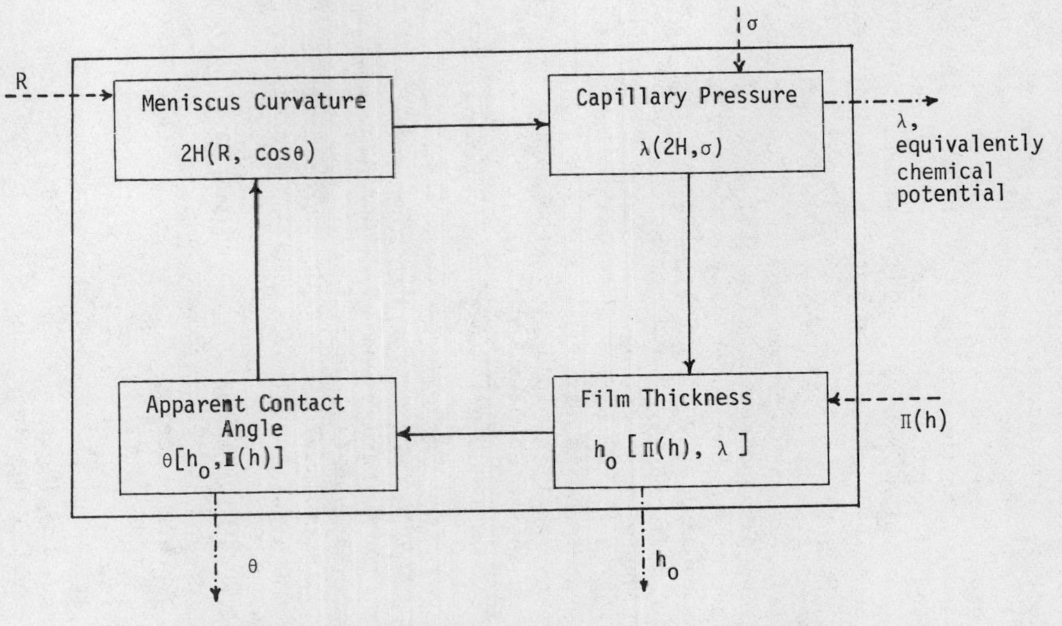

Fig. 6. Thin-films in equilibrium with bulk fluid behind menisci in slots, tubes, and porous media.

fluid in thin-films can be substantial in some circumstances, at least. If the wetting phase is oil, the thin film inventory could be of crucial importance to oil recovery. The fluid conductivity of such films is yet to be determined.

As the wetting fluid saturation grows--and it necessarily does during oil production from water-wet rock--it accumulates in the pendular states. The growing collars around the pore throats here and there turn unstable and choke off oil junctions. The results of our analysis show that the smaller the throat radius compared to the size of the adjacent pore bodies the lower the local wetting phase saturation at which the instability sets in. Thus in any irregular porous medium in which non-wetting fluid is being replaced with wetting fluid, the connections of the non-wetting fluid at each stage are more likely to be broken in the tighter throats in which they still exist. This process is aided by another mechanism of non-wetting phase disconnection called the "by-passing" or "pore doublet" mechanism (24) which is governed by the topology (i.e. the connectedness) as well as the geometry of the pore space. As displacement progresses, more and more breaks occur with the result that portions of non-wetting phase become completely disconnected from the connected pathways in which that phase is flowing. This process of isolation goes on until ultimately all of the remaining non-wetting fluid is in the form of isolated blobs that are scattered about the pore space (25,26).

REFERENCES

1. K. K. Mohanty, H. T. Davis, and L. E. Scriven, Bull. Amer. Phys. Soc., 23, 966 (1978).
2. K. K. Mohanty, H. T. Davis, and L. E. Scriven, Paper No. 121a presented at AIChE Annual Meeting, San Francisco, November 25-29 (1979).
3. C. J. Hurmence, S. R. Strand, H. T. Davis, L. E. Scriven, and K. K. Mohanty, Bull. Amer. Phys. Soc., 25, 1073 (1980).
4. K. K. Mohanty, Ph.D. Thesis, University of Minnesota, Minneapolis (1980).
5. B. V. Deryagin, Zh. Fiz. Khim., 5, 379 (1934).
6. B. V. Deryagin and M. Kusakov, Bull. Acad. Sci., USSR, 5, 471 (1936); Acta Physicochim, USSR, 10, 25 (1939).
7. A. Sheludko, Adv. Coll. Interface Sci., 1, 391 (1967).
8. B. V. Deryagin and N. V. Churaev, J. Colloid Interface Sci., 49, 249 (1974).
9. I. B. Ivanov and Chr. St. Vassilieff, Colloid Polymer Sci., 256, 1142 (1978).
10. I. E. Dzyaloshinskii, E. M. Lifshitz, and L. P. Pitaevskii, Sov. Physics JETP, 37, 161 (1960).
11. G. Teletzke, H. T. Davis, and L. E. Scriven, private communication.

12. H. T. Davis and L. E. Scriven, J. Stat. Phys. (in press).
13. B. V. Deryagin, Proc. 2nd Int. Congr. Surface Activity, London, 2, 153 (1957).
14. J. F. Padday, Special Discussions Faraday Soc., 1, 64 (1970).
15. B. V. Deryagin, V. M. Starov, and N. V. Churaev, Coll. J., 38, 875 (1976).
16. J. R. Philip, J. Chem. Phys., 66, 5069 (1977).
17. F. Renk, P. C. Wayner, and G. M. Homsy, J. Colloid Interface Sci., 67, 408 (1979).
18. M. V. Berry, J. Phys. A: Math. Nucl. Gen., 1, 231 (1974).
19. N. V. Churaev, Coll. J., 36, 287 (1974).
20. S. L. Goren, J. Fluid Mech., 12, 309 (1962).
21. D. H. Everett and J. M. Haynes, J. Colloid Interface Sci., 38, 125 (1972).
22. O. Bolza, "Lectures on the Calculus of Variation," Chelsea, N.Y. (1904/1973).
23. I. B. Ivanov, B. V. Toshev, and P. B. Radoev, in Proc. Soc. Chem. Ind. Meeting on Wetting, Spreading and Adhesion (Louguborough), J. F. Padday, ed., Academic Press, London (1978).
24. G. L. Stegemeier, in "Improved Oil Recovery by Surfactant and Polymer Flooding," D. O. Shah and R. S. Schechter, eds., Academic Press Inc., New York (1977).
25. K. M. Ng, H. T. Davis, and L. E. Scriven, Chem. Eng. Sci., 33, 1009 (1978).
26. R. G. Larson, L. E. Scriven, and H. T. Davis, Nature, 268, 409 (1977).

ON THE FATE OF OIL GANGLIA DURING IMMISCIBLE DISPLACEMENT IN WATER WET GRANULAR POROUS MEDIA

A.C. Payatakes, G. Woodham and K.M. Ng

Department of Chemical Engineering
University of Houston
Houston, Texas 77004, U.S.A.

In a series of previous publications a model was formulated for the study of the dynamics of oil ganglia populations during immiscible displacement in oil recovery processes. The model is composed of four components: a suitable model for granular porous media, a mobilization/breakup criterion, a Monte Carlo simulation method capable of predicting the fate (mobilization, breakup, stranding) of solitary oil ganglia, and two coupled ganglia-population balance equations--one applying to moving ganglia and the other to stranded ones. Central roles in this model are played by the probability of mobilization, the probability of breakup per rheon, the probability of stranding per rheon, the breakup coefficient and the stranding coefficient. These parameters have already been calculated for randomly shaped oil ganglia. However, stochastic simulations show that mobilized oil ganglia tend to get elongated and slender. The effect of slenderization on the aforementioned parameters is investigated here based on Monte Carlo simulation results.

INTRODUCTION

Primary and secondary oil recovery methods succeed in recovering, on the average, only one-third of the amount of oil originally in place. In view of the vast amounts of residual (non-recoverable with secondary processes) oil, and of the rapidly deteriorating petroleum demand-and-supply situation, development of enhanced oil recovery methods becomes an utter necessity, at least for the next few decades until more permanent energy sources can be developed.

At the end of secondary oil recovery processes, the residual oil is dispersed throughout the reservoir rock in the form of small oil ganglia (nodular blobs) each of which occupies one to, say, fifteen contiguous microchambers of the porous medium. The rest of the porous space is taken by formation water. It is the object of enhanced oil recovery methods to mobilize as much as possible of this residual oil by miscible and/or immiscible displacement.

Optimally designed micellar flooding begins as a (quasi) miscible displacement process, since the injected chemical slug has the capacity to solubilize the oil ganglia it encounters. However, this process inevitably deteriorates to a much less efficient immiscible mode (1-3). During the immiscible displacement phase of the process, the advancing chemical slug fails to mobilize all the oil ganglia it engulfs. To make matters worse, the mobilized oil ganglia frequently fission, producing large and small daughter ganglia (4,5). The smaller daughter ganglia become rapidly restranded, unless the capillary number of the flood is sufficiently large (say $N_{Ca} > 10^{-2}$) to keep mobilized even small ganglia occupying only a few microchambers. Unfortunately, capillary numbers of this magnitude are hard to achieve and much harder to maintain. Consequently, a micellar flood fails to mobilize all the oil it encounters along its path; furthermore, it only succeeds in carrying to the production wells just a fraction of the initially mobilized oil. Obviously, this less than ideal performance is due (apart from sweep-efficiency effects) to the fact that a large part of the flood operates under immiscible displacement conditions.

In view of the above, the mechanics of immiscible displacement of an oleic phase by an aqueous phase constitute a central aspect of micellar flooding, and warrant careful analysis. Much effort has been made to model the immiscible displacement of a single oil ganglion. The idea that mobilization hinges on the interplay between viscous forces, on one hand, and capillary forces on the other evolved gradually. The capillary number, N_{Ca}, which is a ratio of viscous to capillary forces has emerged as the most important parameter in immiscible flooding. Culminating a series of studies on the stability of liquid-liquid interfaces, Melrose and Brandner (6) proposed a criterion for the mobilization of isolated oil ganglia, which seems to be in satisfactory agreement with experimental data. More recent studies in the same direction were reported by Stegemeier (7), and Oh and Slattery (8). These works shed valuable light into the mechanics of mobilization and entrapment, but leave some important questions unanswered. For instance, what happens to an oil ganglion once it is mobilized? More importantly, what is the collective fate of a large population of oil ganglia engulfed by an immiscible flood? Of course, previous authors have been aware of these questions all along. Melrose (9) pointed out that in order to predict the set of experimental observations usually reported, knowledge of the cooperative behavior of an ensemble of a very large number of interconnected cavities is required. This statement contains the key to further progress. It pinpoints the main difficulty and at the same time sows the seeds for its resolution. Clearly, in order to develop a good understanding of oil ganglia dynamics, formulation of a realistic porous media model which can be used for the study of oil ganglia motion, coalescence, breakup and stranding is necessary.

Such a model is the point of departure of a recent series of studies: Payatakes, Flumerfelt and Ng (10-12), Payatakes, Ng and Flumerfelt (4) and Ng and Payatakes (5). The strategy of this effort is composed of the following steps.

- Development of a suitable porous media model.
- Development of a mobilization/breakup criterion.
- Development of a Monte Carlo simulation method to study the fate of solitary oil ganglia and to determine probabilities of various events as functions of capillary number, ganglion size and porous medium geometry.
- Development of a system of ganglia number balance equations governing the dynamics of oil ganglia populations.
- Development of methods of solution of the integrodifferential population balance equations to study the dynamics of oil ganglia in immiscible floods.

This effort is already at an advanced stage; steps one to four are, in essence, complete for the case of sandpacks and can be extended with little extra effort to apply to consolidated porous media. A collocation method for the integration of the population balance equations is also nearing completion.

The present study is concerned with an important question which pertains to the Monte Carlo simulation, and which arose during the study of the fate of solitary oil ganglia (5). Simply stated, this question is: what, if any, is the effect of the oil ganglion shape on the fate of the ganglion? We will attempt to answer this question here. To make the treatment comprehensible and self-contained we will start by reviewing briefly some background material, without proof or unnecessary detail. The interested reader is referred to the original references (see above) for particulars. Then, we proceed to treat the problem at hand.

POROUS MEDIA MODEL

The porous media model is a cubic network of unit cells, Figure 1, each one of which is a constricted segment of a sinusoidal tube, Figure 2. The dimensions of a unit cell, namely the

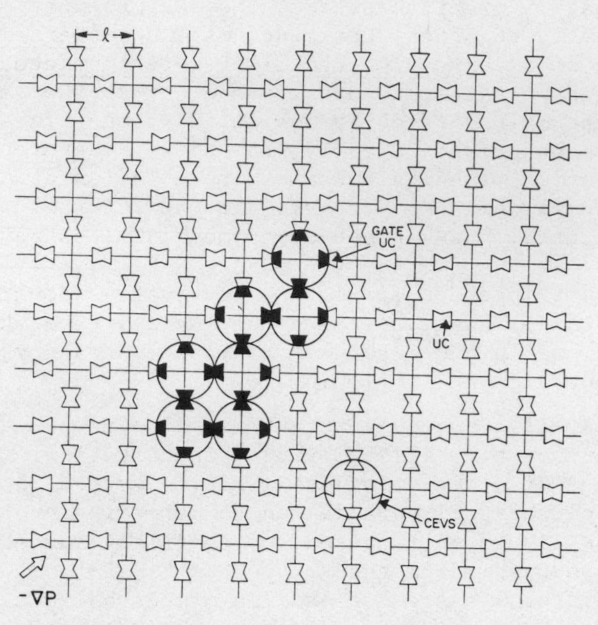

Fig. 1. Two-dimensional depiction of idealized cubic network of unit cells. A conceptual elemental void space (CEVS) and a unit cell (UC) are identified. An idealized oil ganglion occupying seven adjoining pores is also shown.

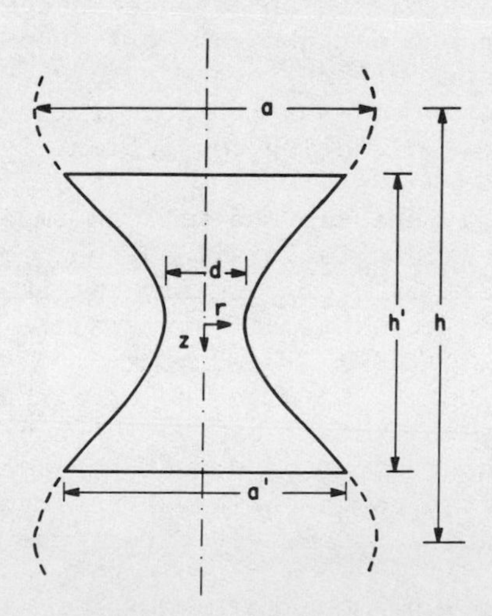

Fig. 2. Typical unit cell of the porous media model, ($-h'/2 \le z \le h'/2$). The wall profile is a sinusoidal function in z. The segment ($-h/2 \le z \le h/2$) is the "extended" unit cell.

throat diameter d, the maximum diameter a', the length h', the maximum diameter of the extended unit cell a, and the length of the extended unit cell h, are random variables. In the case of a pack composed of nearly uniform spheres or grains, these variables are highly correlated. Then, for simplicity, all the dimensions of a unit cell are assumed to be proportional to its throat diameter d; the resulting network consists of unequal but geometrically similar unit cells. The throat diameter distribution and the constants of proportionality required to calculate a, a', h and h' from d are derived readily from the grain size distribution, the porosity of the packing and the initial drainage curve (4,10,12).

Each node of the network is connected to six unit cells, Figure 1 (the pair normal to the plane of paper is not shown). One node and the adjacent six half-unit cells represent an elemental void space, namely a cavity of the porous medium including the halves of satellite throats leading to neighboring cavities. One node and the adjacent six half-unit cells comprise a conceptual elemental void space (CEVS).

If a given elemental void space is occupied by oil, the corresponding CEVS is assumed to be occupied by oil, too. According to the model, when a CEVS is occupied by oil, oil fills its six half-unit cells. The representation of an oil ganglion occupying several pores is made in an analogous way by filling the appropriate CEVS's, Figure 1. Unit cells containing an oil-water interface are called gate unit cells. As will be seen below, the sizes and relative positions of the gate unit cells determine the behavior of the oil ganglion.

QUASI-STATIC CRITERION FOR THE MOBILIZATION AND BREAKUP OF OIL GANGLIA

In studying the motion of oil ganglia, their shape, size, orientation relative to the macroscopic pressure gradient and topology of occupied space are important factors. Let i be the index identifying a gate unit cell, and let d_i be the diameter of its throat. The drainage curvature through this cell is given approximately by

$$J_{dr,i} = \frac{4}{d_i} \cos\theta \qquad (1)$$

where θ is the apparent contact angle. Generalizing the Melrose-Brandner (6) criterion, Ng and Payatakes (5) developed the following algorithm for the determination of the conditions for mobilization and of the locations of the xeron and hygron.

Step 1. Assign indices $i = 1,2,\dots N_{guc}$ to all gate unit cells (cells containing oil-brine interfaces) starting with the one furthest downstream.

Step 2. Determine a first approximation to the value of the critical pressure gradient needed for mobilization, $G_{cr} = |\nabla P|_{cr}$, by setting

$$G_{cr} \cong G_{cr}^{(0)} = \frac{\gamma_{ow} J_{dr,1}(\theta)}{L_p}$$

where L_p is the maximum ganglion length projected on the main flow direction.

Step 3. Let K be the index of the gate unit cell through which the hygron will take place, if the ganglion gets mobilized. Determine K as the index value for which

$$J_1^{(K)}(G_{cr};\theta) = \max\{J_1^{(j)}(G_{cr};\theta);\ j = 2,3,\dots N_{guc}\} \tag{2}$$

where, by definition,

$$J_1^{(j)}(G;\theta) \equiv J_{\ell b,j}(\theta) + \frac{\Delta L_{1j}\cos\theta_{1j}}{\gamma_{ow}} G \tag{3}$$

with

$$J_{\ell b,j} \equiv \frac{4}{a_j}\cos\theta \tag{4}$$

Step 4. Let I be the index of the gate unit cell through which the xeron will take place, if the ganglion gets mobilized. Let the quantities

$$\beta_{Ki} \equiv \frac{\Delta L_{Ki}\cos\theta_{Ki}}{[J_{dr,i}(\theta) - J_{\ell b,K}(\theta)]} \tag{5}$$

be defined as appendix mobility factors. Determine I as the index value for which

$$\beta_{KI} = \max\{\beta_{Ki};\ i = 1,2,\dots N_{guc}\} \tag{6}$$

Step 5. Determine a better approximation to G_{cr} by setting

$$G_{cr} = \frac{\gamma_{ow}}{\beta_{KI}} \tag{7}$$

Step 6. Use the updated value of G_{cr} and iterate Steps 3 to 5 until no changes in K, I and G_{cr} are observed (usually one to two iterations are sufficient). Then proceed to Step 7.

Step 7. Mobilization will take place if

$$|\nabla P| > G_{cr} \tag{8}$$

or, equivalently, if the appendix mobilization number

$$N_{am} \equiv \beta_{KI} \frac{|\nabla P|}{\gamma_{ow}} > 1 \tag{9}$$

or, equivalently, if

$$N_{Ca} = \frac{\mu_w V_f}{\gamma_{ow}} > \frac{k_{rw} k}{\beta_{KI}} \tag{10}$$

If the hygron (index K) takes place somewhere in the middle of the ganglion and the ganglion "thickness" at that point consisted, prior to the rheon, of a single CEVS (the one about to be invaded by the aqueous phase), fission of the oil-ganglion into two (very rarely three) smaller daughter ganglia results. From such point on, the two daughter ganglia are considered separately as distinct units.

MONTE CARLO SIMULATION OF THE FATE OF SOLITARY OIL GANGLIA

Such studies provide information pertaining to the behavior of individual ganglia and form the basis for the analysis of the collective behavior of large populations of interacting ganglia (4). The latter problem, in turn, is central to the understanding of oil bank formation and/or attrition.

The details of the computer-aided Monte Carlo simulation are given in Ng and Payatakes (5). Basically, an individual stochastic realization involves four steps:

- Specification of proper random values for the dimensions of the unit cells of the porous media model.
- Specification of the size, shape and position of the oil ganglion.
- Specification of the capillary number value.
- Application of the mobilization/breakup criterion.

The expected behavior of oil ganglia can be expressed in terms of the following probabilities (per rheon)

$$M = \Pr\{\text{The ganglion gets mobilized (without fissioning) from its present position}\} \tag{11}$$

$$B = \Pr\{\text{The ganglion fissions during the current rheon}\} \tag{12}$$

$$S = \Pr\{\text{The ganglion gets stranded at its present position}\} \tag{13}$$

which satisfy the obvious relation

$$M + B + S = 1 \tag{14}$$

Values of M, B and S for given porous medium and ganglion size can be calculated as functions of the capillary number based on hundreds of one-step Monte Carlo realizations, starting with random initial ganglia shapes. Examples of calculated results for a 100×200 sandpack and ganglia sizes of 2-CEVS, 3-CEVS, 5-CEVS and 15-CEVS are given in Figures 3, 4, 5 and 6, respectively (5). It is worth noting that these curves retain more or less constant shapes for oil ganglia volumes exceeding 3- or 4-CEVS. The main effect of increasing ganglion size is to transfer all curves to the left, towards smaller N_{Ca} values.

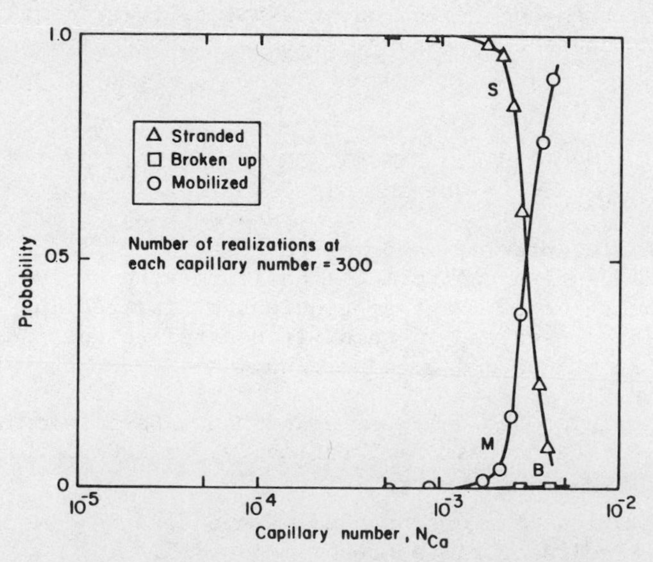

Fig. 3. Plot of the probability that a 2-CEVS blob introduced randomly into a 100×200 sandpack will find itself stranded (S), broken up (B), or mobilized (M) at that position versus capillary number.

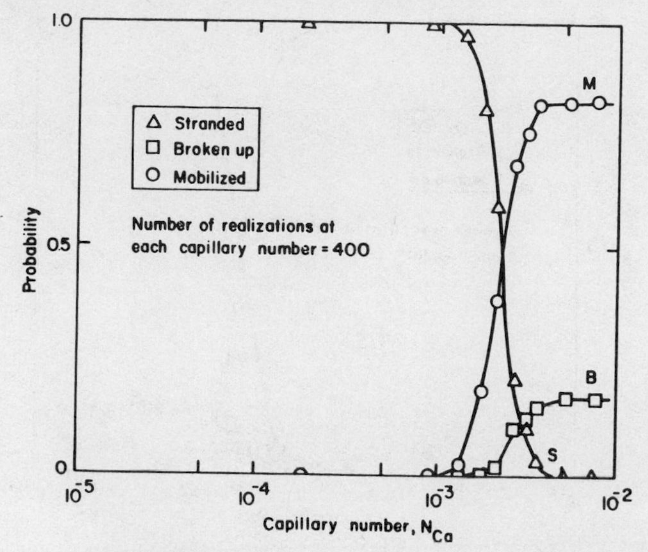

Fig. 4. Plot of the probability that a 3-CEVS blob introduced randomly into a 100×200 sandpack will find itself stranded (S), broken up (B), or mobilized (M) at that position versus capillary number.

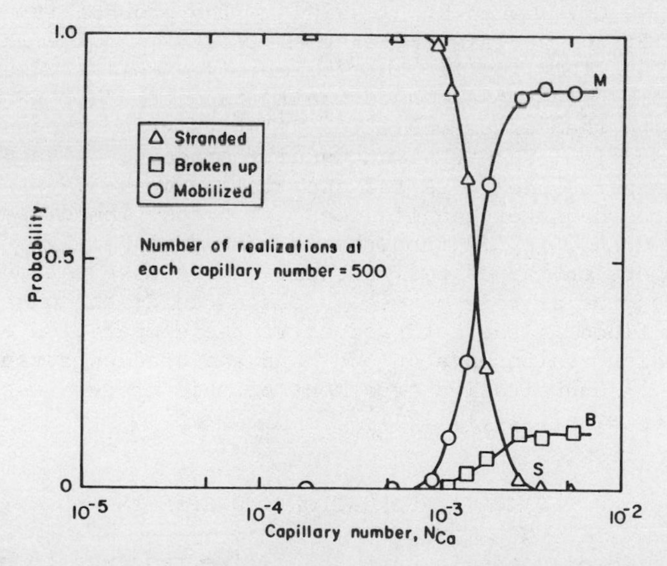

Fig. 5. Plot of the probability that a 5-CEVS blob introduced randomly into a 100×200 sandpack will find itself stranded (S), broken up (B), or mobilized (M) at that position versus capillary number.

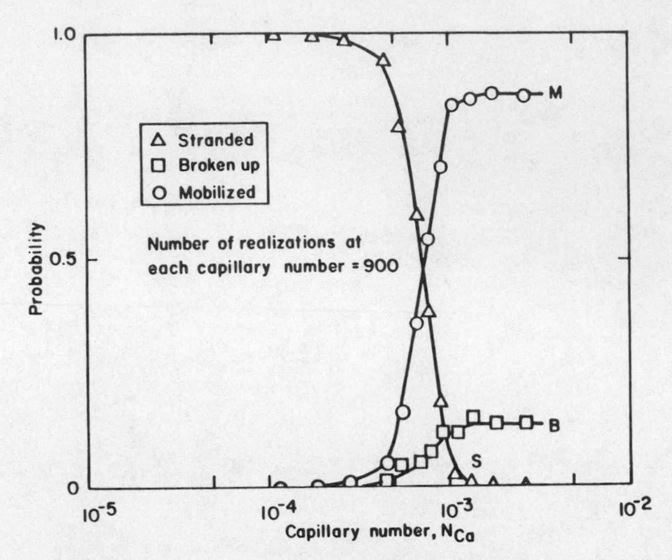

Fig. 6. Plot of probability that a 15-CEVS blob introduced randomly into a 100x200 sandpack will find itself stranded (S), broken up (B), or mobilized (M) at that position versus capillary number.

The probability of stranding per rheon, S, decreases rapidly with increasing N_{Ca} and ganglion size; it is negligible for large ganglia (> 10-CEVS) when $N_{Ca} > 5x10^{-3}$. The probability of breakup per rheon, B, for ganglia larger than 2-CEVS is quite large ranging up to ~ 0.15. This implies that a mobilized solitary ganglion (or one that does not have the opportunity to meet other ganglia and coalesce with them) fissions after relatively few rheons into two daughter ganglia. This fission usually produces quite unequal daughter ganglia, one large and one small, Ng and Payatakes (5). If any of the daughter ganglia does not become immediately stranded, but keeps moving, it, in turn, has a high probability of fissioning, unless it gets entrapped before that. These observations indicate clearly the crucial role played by coalescence; the outcome of an immiscible flood is the net product of the competition of mobilization/collision/coalescence on one hand and breakup/stranding on the other. A quantitative treatment of this process was developed by Payatakes et al. (4,11,12).

GANGLIA ELONGATION AND EFFECTS

Mobilized oil ganglia have a definite tendency to become long and slender as they undergo successive rheons (5). They also tend to align themselves with the direction of the macroscopic flow. A consequence of this is that, as the ganglion advances through a homogeneous porous medium, its per rheon probability of getting stranded, S, decreases. Its per rheon probability of fissioning, B, may also change. The extent to which these changes are significant is not clear from the outset.

To study this phenomenon we consider a thin "slug" of thickness δ composed of non-interacting oil ganglia undergoing immiscible displacement. One could create such a situation in the lab in either one of two ways. One could start with a very sparse population of ganglia arranged in a thin slug in a packing of, say, glass spheres; thanks to the low concentration, collisions between ganglia during immiscible flooding become highly unlikely. Or, one could displace one ganglion at a time, repeat the experiment sufficiently many times, and consider the results collectively. Here we perform an analogous theoretical analysis, instead.

Let $n(z,t;v)\Delta v$ be the number of moving v-ganglia* per unit porous medium volume, where z is the spatial coordinate in the direction of the flood (z=0 is the injection plane), and t is time measured from the beginning of the flooding process. A stranding coefficient λ and a breakup coefficient ϕ are defined by the following two expressions:

$$\left.\frac{\partial n}{\partial z}\right)_{\substack{\text{due only}\\ \text{to stranding}}} = -\lambda n \tag{15}$$

$$\left.\frac{\partial n}{\partial z}\right)_{\substack{\text{due only}\\ \text{to fission}}} = -\phi n \tag{16}$$

Hence, if we start with a monosized population of non-interacting v-ganglia, the population balance equation for v-ganglia becomes simply

$$\frac{\partial n(z,t;v)}{\partial t} + \bar{u}_z(v;\underline{a}_1)\,\frac{\partial n(z,t;v)}{\partial z} = -\,\bar{u}_z(v;\underline{a}_1)[\lambda(v;\underline{a}_2) + \phi(v;\underline{a}_2)] \times n(z,t;v) \tag{17}$$

In the problem at hand, we are not concerned with daughter ganglia generated by fissioning of v-ganglia.

Let a new time variable τ be defined as

$$\tau = t - \frac{z}{\bar{u}_z(v;\underline{a}_1)} \tag{18}$$

---

*v-ganglion is defined as one having volume between v and v + Δv.

Clearly, τ is constant for an observer moving with the slug of v-ganglia. It can be shown easily that Equation (17) is transformed into

$$\left[\frac{\partial n(z,\tau;v)}{\partial z}\right]_\tau = -[\lambda(v;\underline{a}_2) + \phi(v;\underline{a}_2)]n(z,\tau;v) \tag{19}$$

Assuming that λ and ϕ are not functions of z, Equation (19) can be integrated to get

$$n(z,\tau;v) = \frac{y_o}{\delta}\exp[-(\lambda+\phi)z] \; ; \quad \text{(moving v-ganglia)} \tag{20}$$

where y_o is the number of v-ganglia per unit cross-sectional area in the initial slug.

The coefficients λ and ϕ are related to the per rheon probabilities M, B, and S through the relations

$$\lambda = -\frac{S}{(1-M)}\frac{\ln M}{s_z} \tag{21}$$

$$\phi = -\frac{B}{(1-M)}\frac{\ln M}{s_z} \tag{22}$$

where s_z is the expected length of axial migration of the centroid of the ganglion per rheon, and is easily determined from stochastic simulations (5).

Let, now, $\sigma(z,\tau;v)\Delta v$ be the number of stranded v-ganglia per unit porous media volume, and let $\beta(z,\tau;v)\Delta v$ be the number of v-ganglia that fissioned per unit porous media volume (at z). We can show easily that, in the case under consideration,

$$\sigma(z,\tau;v) = \lambda\, y_o \exp[-(\lambda+\phi)z] \tag{23}$$

$$\beta(z,\tau;v) = \phi\, y_o \exp[-(\lambda+\phi)z] \tag{24}$$

In deriving Equations (20), (23) and (24) we neglected axial dispersion of v-ganglia. This simplification is justified, if we restrict our attention to small slug migration distances.

Equation (20) gives the number concentration of v-ganglia in the slug still moving at position z, Equation (23) gives the concentration of stranded v-ganglia left behind by the passing slug, and Equation (24) gives the corresponding number of v-ganglia fissions. If we plot $\ln(n\delta/y_o)$ versus z/ℓ we get a straight line with intercept unity and slope $-(\lambda+\phi)\ell$, where ℓ is the length of periodicity of the porous medium, (Figure 1). If we plot $\ln(\sigma\ell/y_o)$

versus z/ℓ we get a straight line with intercept $\ell n(\lambda\ell$ and slope $-(\lambda+\phi)\ell$. Similarly, if we plot $\ell n(\beta\ell/y_o)$ versus z/ℓ we get a straight line with intercept $\ell n(\phi\ell)$ and slope, again, $-(\lambda+\phi)\ell$. Of course, this linearity is a consequence of our assumption that λ and ϕ are not functions of z and have values which are averages over the ensemble of completely random shapes. Now, if we recall that a migrating solitary ganglion tends to elongate, and that M, B and S depend on the projected ganglion length L_p, we conclude that λ and ϕ may very well be functions of z, tending to values corresponding to the most elongated shape(s). It follows that $\ell n(n\delta/y_o)$ vs. z/ℓ, $\ell n(\sigma\ell/y_o)$ vs. z/ℓ and $\ell n(\beta\ell/y_o)$ vs. z/ℓ curves should actually deviate from linearity. This provides a basis for testing the magnitude of effects caused by elongation.

With this goal in mind, the stochastic simulation method was used to make numerical experiments, taking as basis the porous media model for a 100×200 sandpack (4). For fixed ganglion volume and random initial shape, one-thousand realizations were made with the same capillary number value ($N_{Ca} = 1.5\text{x}10^{-3}$) allowing in each realization the ganglion to migrate until it either fissioned or it got stranded. Superimposing the starting planes of all realizations and considering the entire ensemble of results, "experimental" values of $(n\delta/y_o)$, $(\sigma\ell/y_o)$ and of $(\beta\ell/y_o)$ can be determined. To this end, we consider several positions along z, and at each position z we count the number of moving v-ganglia, the number of stranded v-ganglia left in the interval $(z \pm \frac{1}{2}\,\Delta z)$ and the number of fissions of v-ganglia in the interval $(z \pm \frac{1}{2}\,\Delta z)$. Then we calculate readily the following variables

$$\chi = \frac{\text{number of v-ganglia still moving at position z}}{\text{initial number of v-ganglia in the slug}} \tag{25}$$

$$\psi = \frac{\text{number of v-ganglia stranded in the interval } (z \pm \frac{1}{2}\,\Delta z)}{\text{initial number of v-ganglia in the slug}} \tag{26}$$

$$\omega = \frac{\text{number of v-ganglia fissioned in the interval } (z \pm \frac{1}{2}\,\Delta z)}{\text{initial number of v-ganglia in the slug}} \tag{27}$$

Of course, ψ and ω depend on Δz, and in order for our results to be meaningful, Δz must be sufficiently small, say, $\Delta z << \ell$. The ratios χ, $\psi/\Delta z$ and $\omega/\Delta z$ correspond to $(n\delta/y_o)$, σ/y_o and β/y_o, and therefore the "experimental" values of $(n\delta/y_o)$, $(\sigma\ell/y_o)$ and $(\beta\ell/y_o)$ are χ, $(\psi\ell/\Delta z)$ and $(\omega\ell/\Delta z)$, respectively. The values of χ, $(\psi\ell/\Delta z)$ and of $(\omega\ell/\Delta z)$ thus obtained are based on one-thousand Monte Carlo realizations. To obtain a measure of the errors involved, we repeated the procedure five times, and used the χ, $(\psi\ell/\Delta z)$ and $(\omega\ell/\Delta z)$ values (at each z) as single observations. Using the Student t-distribution we calculated error bounds at the 95% confidence level. The results of this study for 5-CEVS, 7-CEVS, and 15-CEVS ganglia are plotted in Figures 7 to 14. A plot of $(\psi\ell/\Delta z)$ vs. (z/ℓ) for 15-CEVS is not given, because, as it turns out, too few large ganglia get stranded at $N_{Ca} = 1.5\text{x}10^{-3}$. The results in each of these figures are based on five-thousand realizations. In addition to these data, we plot the straight lines obtained from Equations (20), (23) and (24) using constant $\lambda\ell$ and $\phi\ell$ values, obtained by Ng and Payatakes (5), based on one-step realizations with entirely random-shaped ganglia.

Mere inspection of these results shows that deviation from linearity actually occurs, at least in certain cases. As discussed earlier, such deviations are due to elongation. They can be treated quantitatively as follows. Consider the conceptual slug experiment. Simple v-ganglia number balances give

$$\sigma = -\delta \left[\frac{\partial n(z,\tau;v)}{\partial z}\right]_{\substack{\text{due only}\\ \text{to stranding}}} \tag{28}$$

$$\beta = -\delta \left[\frac{\partial n(z,\tau;v)}{\partial z}\right]_{\substack{\text{due only}\\ \text{to fission}}} \tag{29}$$

Equations (15), (16), (28) and (29) give

$$\lambda = \frac{\sigma(z,\tau;v)}{\delta n(z,\tau;v)} \tag{30}$$

$$\phi = \frac{\beta(z,\tau;v)}{\delta n(z,\tau;v)} \tag{31}$$

Recalling that the "experimental" values of $(n\delta/y_o)$, σ/y_o and β/y_o are χ, $\psi/\Delta z$ and $\omega/\Delta z$, we get

$$\lambda\ell = \frac{(\psi\ell/\Delta z)}{\chi} \qquad \text{(at fixed } z/\ell) \tag{32}$$

$$\phi\ell = \frac{(\omega\ell/\Delta z)}{\chi} \qquad \text{(at fixed } z/\ell) \tag{33}$$

The last two expressions can be used to determine values of $\lambda\ell$ and $\phi\ell$ at various positions z/ℓ, since the quantities appearing in the right hand sides are already available in Figures 7 to 14. Values of $\lambda\ell$ vs. z/ℓ for 5-CEVS and 7-CEVS are plotted in Figure 15. Values of $\phi\ell$ vs. z/ℓ for 5-CEVS, 7-CEVS and 15-CEVS are plotted in Figure 16. A number of observations can be made:

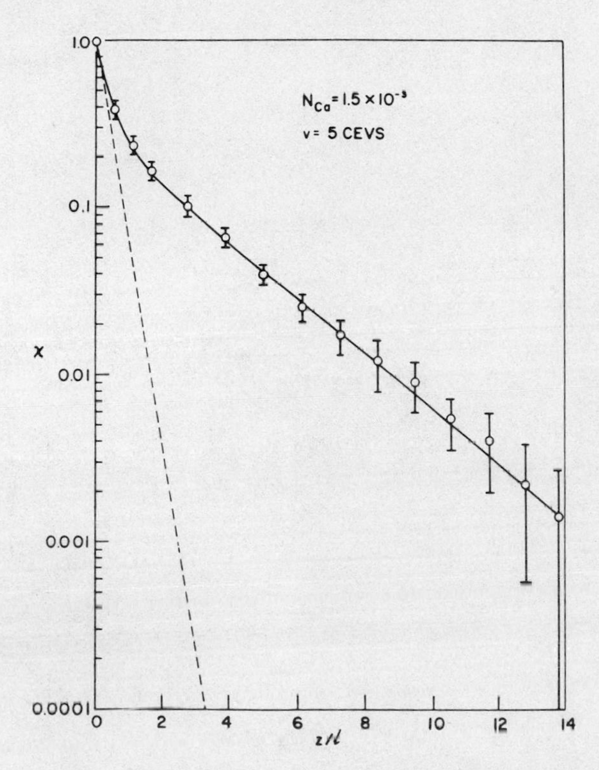

Fig. 7. Values of χ determined from Monte Carlo simulations versus normalized length of migration z/ℓ for 5-CEVS ganglia. The dotted line is based on λ and ϕ values corresponding to entirely random ganglia shapes.

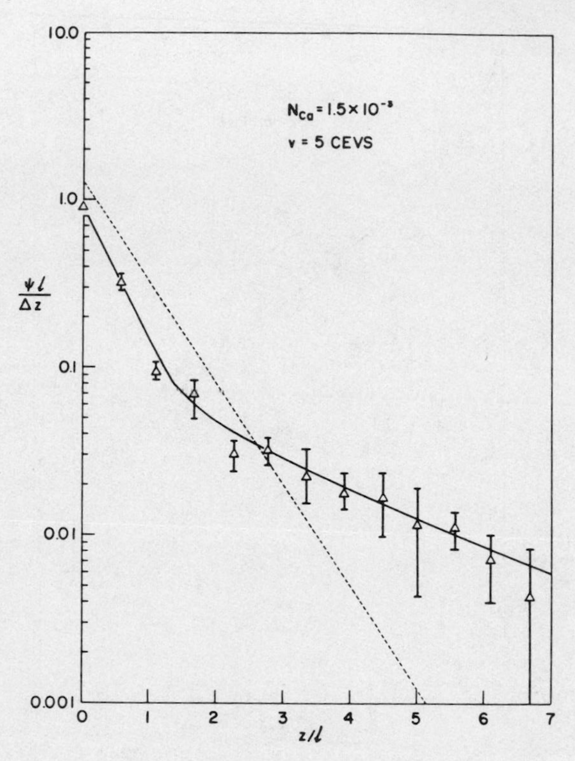

Fig. 8. Values of ($\psi\ell/\Delta z$) determined from Monte Carlo simulations versus normalized length of migration z/ℓ for 5-CEVS ganglia. The dotted line is based on λ and ϕ values corresponding to entirely random ganglia shapes.

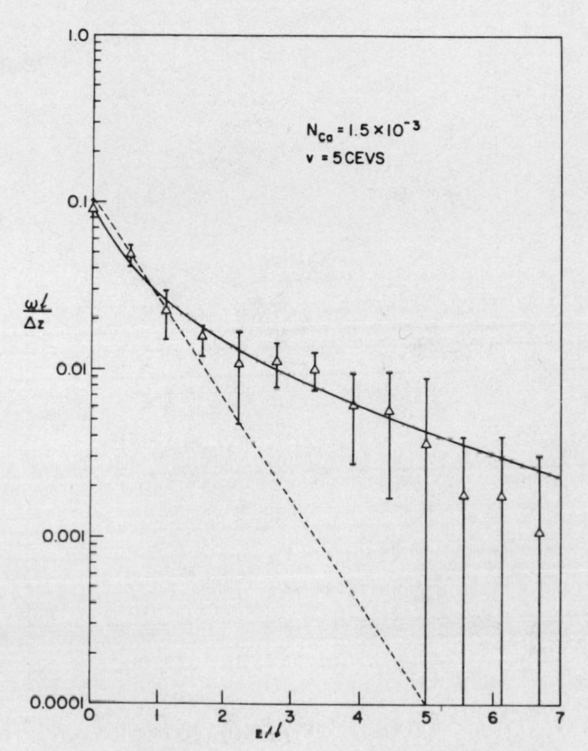

Fig. 9. Values of $(\omega\ell/\Delta z)$ determined from Monte Carlo simulations versus normalized length of migration z/ℓ for 5-CEVS ganglia. The dotted line is based on λ and ϕ values corresponding to entirely random ganglia shapes.

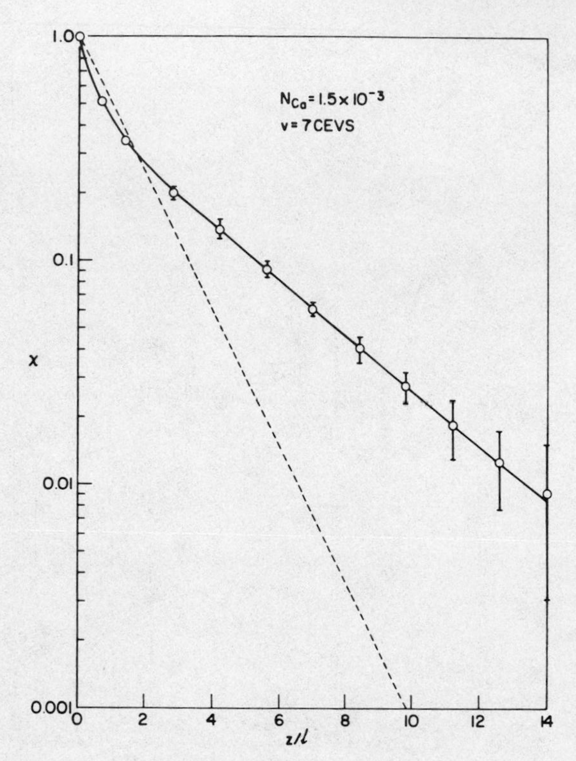

Fig. 10. Values of χ determined from Monte Carlo simulations versus normalized length of migration z/ℓ for 7-CEVS ganglia. The dotted line is based on λ and ϕ values corresponding to entirely random ganglia shapes.

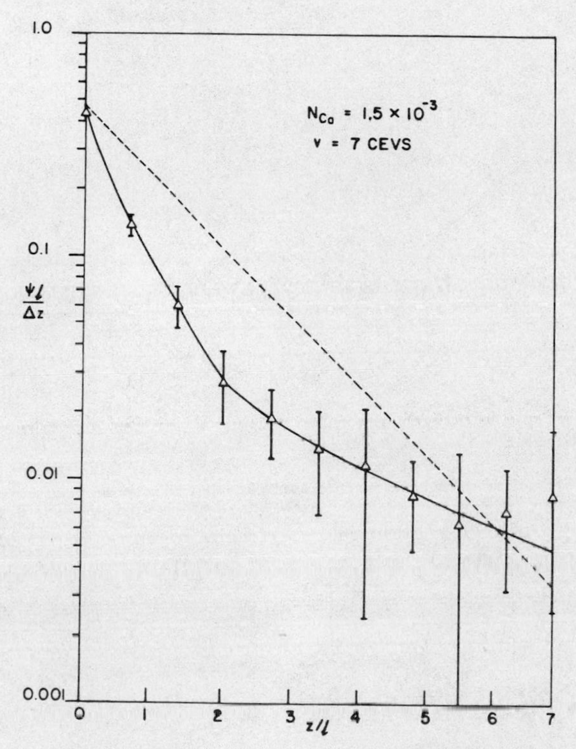

Fig. 11. Values of $(\psi\ell/\Delta z)$ determined from Monte Carlo simulations versus normalized length of migration z/ℓ for 7-CEVS ganglia. The dotted line is based on λ and ψ values corresponding to entirely random ganglia shapes.

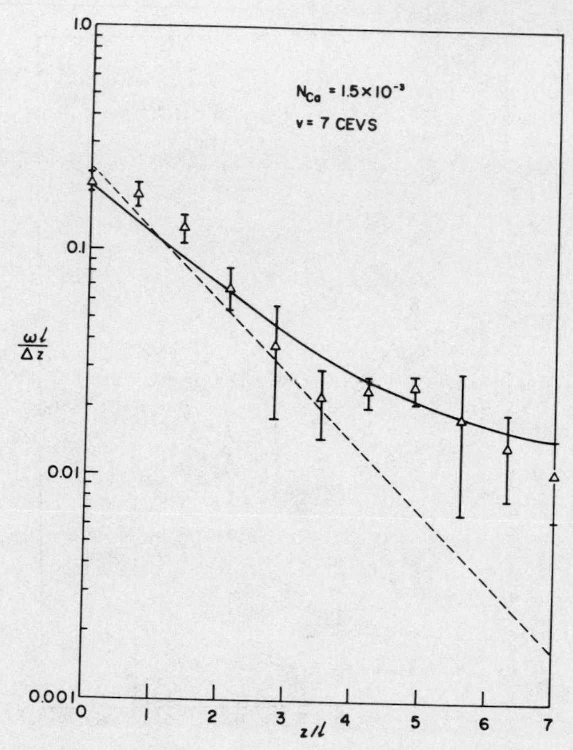

Fig. 12. Values of $(\omega\ell/\Delta z)$ determined from Monte Carlo simulations versus normalized length of migration z/ℓ for 7-CEVS ganglia. The dotted line is based on λ and ϕ values corresponding to entirely random ganglia shapes.

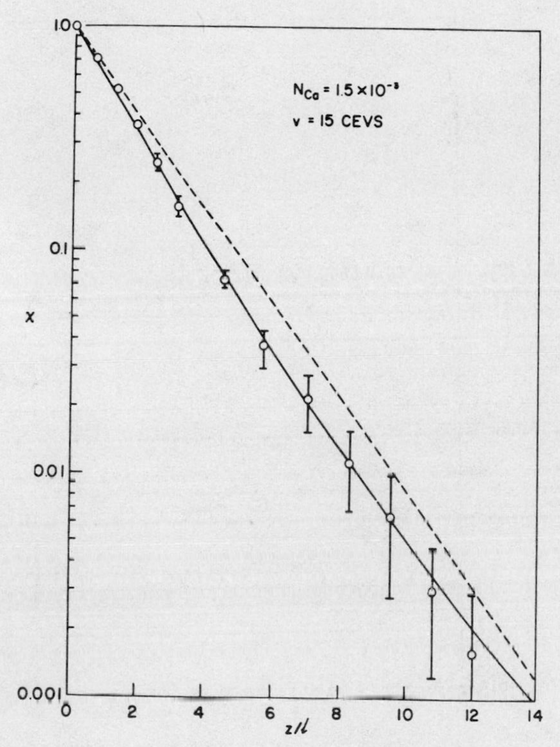

Fig. 13. Values of χ determined from Monte Carlo simulations versus normalized length of migration z/ℓ for 15-CEVS ganglia. The dotted line is based on λ and ϕ values corresponding to entirely random ganglia shapes.

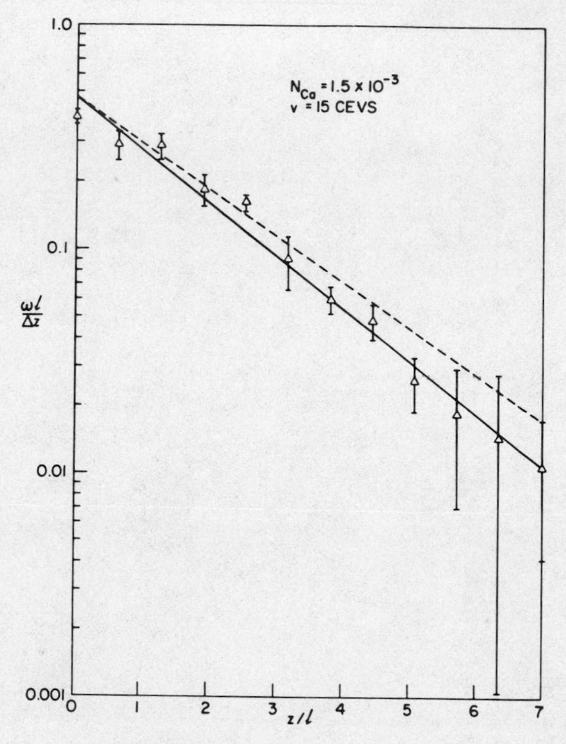

Fig. 14. Values of $(\omega\ell/\Delta z)$ determined from Monte Carlo simulations versus normalized length of migration z/ℓ for 15-CEVS ganglia. The dotted line is based on λ and ϕ values corresponding to entirely random ganglia shapes.

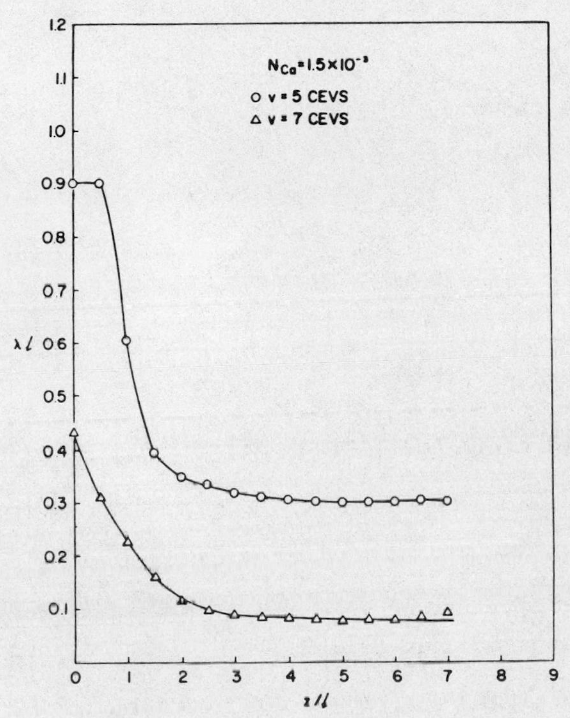

Fig. 15. Values of $\lambda\ell$ for random-but-evolving ganglia shapes versus normalized length of migration z/ℓ in a 100×200 sandpack. The value at $z/\ell = 0$ is the average over the ensemble of all shapes. Values of $\lambda\ell$ for 15-CEVS are not shown because they are very small for $N_{Ca} = 1.5\text{x}10^{-3}$.

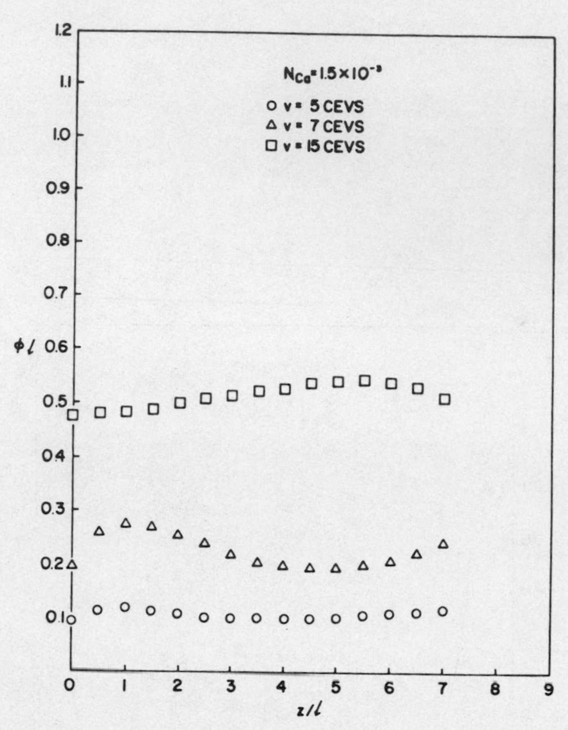

Fig. 16. Values of $\phi\ell$ for random-but-evolving ganglia shapes versus normalized length of migration z/ℓ in a 100×200 sandpack. The value at $z/\ell = 0$ is the average over the ensemble of all shapes.

- For all ganglia sizes, the values of $\lambda\ell$ and $\phi\ell$ for entirely random-shapes and for random-but-evolving shapes are in agreement at $z = 0$ (that is, prior to any shape evolution). The observed discrepancies can be easily attributed to random error associated with the Monte Carlo simulation.
- In the case of relatively small ganglia (3 to 10-CEVS), elongation produces a systematic and marked decrease of $\lambda\ell$, Figure 15. This deviation is much smaller for larger ganglia. The effect of elongation on ϕ, on the other hand, is neither marked nor systematic, Figure 16.
- In the case of relatively large ganglia (say > 15-CEVS), effects of elongation on λ and ϕ are not manifested during at least the first thirty, or so, rheons (Figures 13 and 14). It should be borne in mind that for ganglia of this size, thirty rheons correspond to a migration of only 8ℓ to 10ℓ.

These observations are explained as follows. In the case of a randomly shaped small ganglion, a few rheons, say three or more, produce significant elongation, Figure 17, resulting in sizeable decrease of λ. In the case of random-shaped large ganglia, even

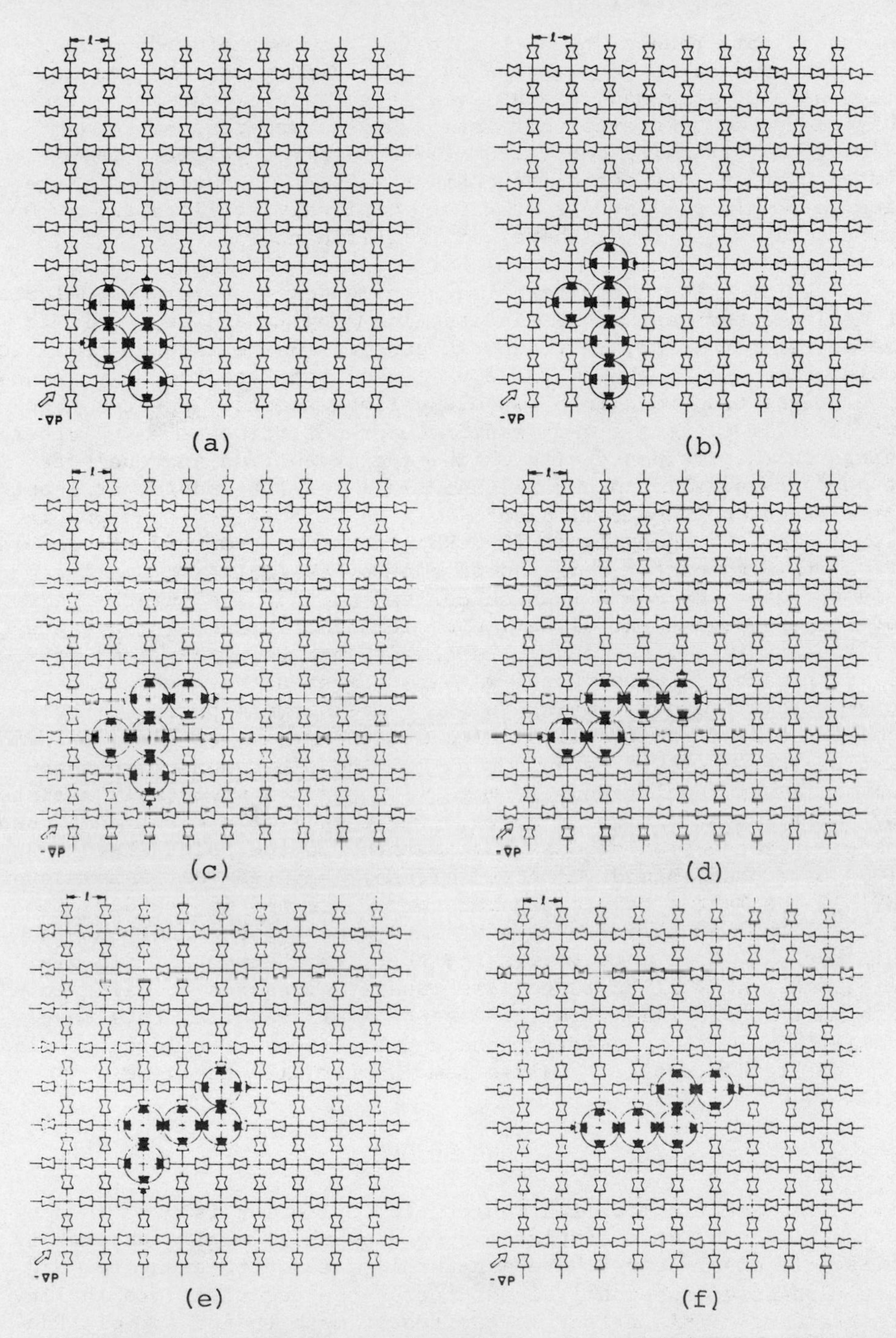

Fig. 17. Typical example of elongation of a mobilized 5-CEVS ganglion. (a) Starting position and shape; (b) after 1 rheon; (c) after 2 rheons; (d) after 3 rheons; (e) after 4 rheons; (f) after 5 rheons.

thirty or more rheons do not result in an amount of elongation, which is sufficient to reduce λ by much, Figure 18. Of course, after sufficiently many rheons even large ganglia can get noticeably elongated. However, for large ganglia and reasonable N_{Ca} values, the stranding coefficient λ is nearly negligible anyway, and elongation does not affect things much. Elongation of large ganglia should play a role only for relatively small capillary number, say $N_{Ca} < 10^{-4}$, where λ is significant (5).

The foregoing considerations become somewhat more complicated if we leave the case of non-interacting ganglia and we return to the actual situation, in which oil ganglia change through collision/coalescence and breakup. These processes generate randomly shaped ganglia and tend to dilute the elongation effects. Whether elongation still plays a significant role or not, depends among other things on the frequency with which ganglia collide and coalesce. It seems reasonable to assume that small ganglia undergo at least a few rheons between collisions, even in relatively dense populations. For such ganglia (< 10-CEVS) the elongation effects should be taken in account. This can be done by assigning to small ganglia values of M, B and S which correspond to the elongated forms. For larger ganglia the situation is not as clearcut, because these may repeatedly collide and coalesce with others within, say, thirty to forth rheons; furthermore, within this span they certainly undergo a few fissions, losing in the process small parts of their bodies. Side-to-side collision/coalescence tends to slow down or reverse the process of elongation. On the other hand, front-to-back collision/coalescence is more frequent and vigorously assists the formation of elongated forms. There is little doubt that when the oil reconnection process has advanced to the point where huge blobs have been formed (composed of, say, hundreds or thousands of CEV's), the latter have elongated forms, aligned in the direction of the flow. Advancing arrays of such super-blobs constitute an "oil bank." It is more realistic, then, to assume that all ganglia, small as well as large, are usually elongated and aligned with macroscopic flow. In the case of small ganglia, this means smaller λ values and nearly unchanged ϕ values. For large ganglia, λ is negligibly small; ϕ values remain virtually the same.

CONCLUSIONS

A new ganglia mobilization/breakup criterion is used in a series of Monte Carlo simulations to show that ganglia undergoing immiscible displacement have a natural tendency to elongate, and that elongation substantially decreases the per rheon probability of stranding. This effect is manifested much faster for oil ganglia in the range 4-CEVS to 10-CEVS than for ganglia with sizes larger than, say, 20-CEVS. Nevertheless, ganglia of all sizes

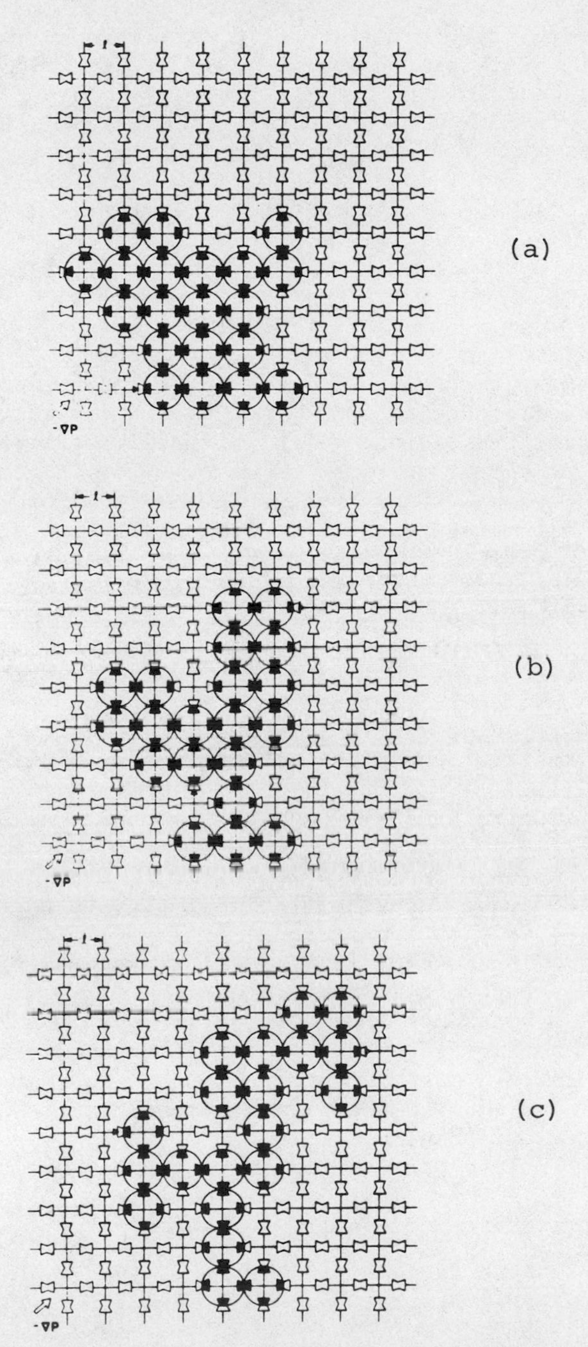

Fig. 18. Typical example of elongation of a mobilized 15-CEVS ganglion. (a) Starting position and shape; (b) after 5 rheons; (c) after 10 rheons.

become elongated, eventually. The effect of elongation on the per rheon probability of breakup does not seem to be very important.

ACKNOWLEDGMENT

This work was performed under U.S. Department of Energy Grant No. E(40-1)-5075.

NOTATION

a	maximum diameter of the periodically constricted tube corresponding to a given unit cell, Figure 2
a'	maximum diameter of a unit cell, Figure 2
$\underline{a}_1$, $\underline{a}_2$	parameter vectors
$B(v;\underline{a}_2)$	probability of breakup of a v-ganglion per rheon
d	minimum diameter of a unit cell, Figure 2
d_i	constriction diameter of the i-th gate unit cell
G_{cr}	critical value of $\lvert\nabla P\rvert$ at which a given ganglion gets mobilized
h	wavelength of the periodically constricted tube corresponding to a given unit cell, (length of extended unit cell), Figure 2
h'	length of unit cell, Figure 2
I	index of the gate unit cell through which the xeron takes place
i,j	indices used to identify gate unit cells (or constrictions)
J_{dr}	drainage curvature
$J_{dr,i}$	drainage curvature in the i-th gate unit cell
J_{imb}	imbibition curvature
$J_{\ell b,j}$	lower bound of interfacial curvature in the (extended) j-th gate unit cell, Equation (17)
$J_1^{(j)}$	curvature defined by Equation (3)
K	index of the gate unit cell through which the hygron takes place
k	absolute (single phase) permeability
k_e	effective permeability
k_o, k_w	absolute permeabilities to oil and water, respectively
$k_{rw} = \frac{k_e}{k_w}$	relative permeability to water
L_p	maximum ganglion length projected <u>on</u> the main flow direction
ℓ	length of periodicity of the porous medium, Figure 1
$M(v;\underline{a}_2)$	probability that a v-ganglion will undergo at least one rheon from its present position
N_{am}	appendix mobilization number, Equation (9)

$N_{Ca} = \frac{\mu_w V_f}{\gamma_{ow}}$ capillary number

N_{guc} number of gate unit cells of a given ganglion

$n(z,\tau;v)\Delta v$ number of moving v-ganglia per unit reservoir volume

P pressure

r radial coordinate in a unit cell, Figure 2

$S(v;\underline{a}_2)$ probability that a v-ganglion introduced randomly in the porous medium will find itself stranted before a single rheon takes place. Also, probability of stranding per rheon

s_z expected length of travel of the centroid of a v-ganglion in the z direction per rheon

t time measured from the initiation of the flood

$\bar{u}_z(v;\underline{a}_1)$ mean velocity of the centroids of v-ganglia in the axial direction

V_f superficial velocity of the aqueous phase

y_o number of v-ganglia initially in the slug, per unit cross-sectional area

Greek Letters

$\beta(z,\tau;v)\Delta v$ number of fissions of v-ganglia per unit volume

β_{Ki} appendix mobility factor (for a pair of appendices denoted by K and i); Equation (5)

β_{KI} maximum value among β_{Ki} for a given ganglion

γ_{ow} interfacial tension at the oil-water interface

ΔL_{ij} distance between two constrictions, denoted by i and j, for a given ganglion

Δz small increment of z

δ thickness of slug of v-ganglia

θ apparent contact angle as measured from the wetting phase

θ_{ij} angle between the line connecting gate constriction i to gate constriction j and the macroscopic flow direction

$\lambda(v;\underline{a}_2)$ stranding coefficient

μ dynamic viscosity

$\sigma(z,\tau;v)\Delta v$ number of stranded v-ganglia per unit reservoir volume

τ transformed time, Equation (18)

$\phi(v;\underline{a}_2)$ breakup coefficient

χ number of v-ganglia in the slug still moving at z, divided by the initial number of v-ganglia in the slug

ψ number of stranded v-ganglia in the interval $(z - \frac{\Delta z}{2}, z + \frac{\Delta z}{2})$ divided by the initial number of v-ganglia in the slug

ω number of v-ganglia fissions in the interval $(z - \frac{\Delta z}{2}, z + \frac{\Delta z}{2})$ divided by the initial number of v-ganglia in the slug

REFERENCES

1. R. N. Healy and R. L. Reed, Soc. Petrol. Eng. J., 14, 491 (1974).
2. R. N. Healy, R. L. Reed, and D. G. Stenmark, Soc. Petrol. Eng. J., 16, 147 (1976).
3. R. C. Nelson and G. A. Pope, SPE 6773 presented at 52nd Annual Fall Technol. Conf. and Exhib. SPE, Denver, Colorado, Oct. 9-12, 1977.
4. A. C. Payatakes, K. M. Ng, and R. W. Flumerfelt, AIChE J., 26, 430 (1980).
5. K. M. Ng and A. C. Payatakes, AIChE J., 26, 419 (1980).
6. J. C. Melrose and C. F. Brandner, Can. J. Petrol. Technol., 13 (4), 54 (1974).
7. G. L. Stegemeier, 81st National AIChE Meeting, Kansas City, MO, April 1976.
8. S. G. Oh and J. C. Slattery, Proceedings of 2nd ERDA Symposium on Enhanced Oil and Gas Recovery, Tulsa, Oklahoma, Sept. 1976.
9. J. C. Melrose, Can. J. Chem. Eng., 48, 638 (1970).
10. A. C. Payatakes, R. W. Flumerfelt, and K. M. Ng, 70th Annual AIChE Meeting, New York, Nov. 13-17, 1977.
11. A. C. Payatakes, R. W. Flumerfelt, and K. M. Ng, 84th National AIChE Meeting, Atlanta, Georgia, Feb. 26-March 1, 1978.
12. A. C. Payatakes, R. W. Flumerfelt, and K. M. Ng, Proceedings of 4th DOE Symposium on Enhanced Oil and Gas Recovery, Tulsa, Oklahoma, Aug. 29-31, 1978.

V: ADSORPTION, CLAYS AND CHEMICAL LOSS MECHANISMS

PRECIPITATION AND REDISSOLUTION OF SULFONATES AND THEIR ROLE IN ADSORPTION ON MINERALS

P. Somasundaran, M. Celik and A. Goyal

Henry Krumb School of Mines
Columbia University
New York, New York 10027, U.S.A.

The interaction of inorganic species, such as those of calcium and aluminum that are normally present in reservoir fluids, with surfactants is found to produce precipitation of the surfactants followed by their redissolution above the critical micelle concentration. A maximum is often observed in the adsorption isotherm of surfactants on reservoir rocks. The contribution of the surfactant precipitation/dissolution phenomenon to the occurrence of adsorption maximum has been investigated in this study using the kaolinite/sulfonate system. The magnitude of the adsorption maximum is found to be minimized when the precipitation/redissolution of the surfactant is taken into account, suggesting the important role of the latter phenomenon in determining the apparent adsorption.

INTRODUCTION

Adsorption isotherms from surfactant solutions have been reported to often exhibit maximum and sometimes even minimum in the region around critical micelle concentration (1-4). The phenomenon of maximum and minimum is of such theoretical interest as well as practical importance in such areas as enhanced oil recovery using surfactant flooding. The presence of maximum has been attributed in the past to mechanisms involving micellar exclusion from interfacial region due to electrostatic repulsion or structural incompatibility, presence of impurities, surfactant composition, adsorbent morphology, etc. (1,2). None of these mechanisms is, however, fully substantiated to be considered as a confirmed mechanism for surfactant adsorption from concentrated solutions particularly due to serious possibilities for experimental artifacts arising from such processes as precipitation and entrapment.

Since precipitation has often been observed during adsorption studies when surfactants are contacted with certain mineral suspensions, the role of it in determining the nature of isotherms for the important reservoir system made up of kaolinite and dodecylbenzenesulfonate was investigated in this study. Results showed a complex and interesting behavior of the system involving precipitation followed by redissolution at higher sulfonate concentrations that was markedly dependent upon the salt concentration and the type of inorganic ions involved (5). Also, the phenomenon of precipitation and redissolution whenever it occurred proved to have a definite influence on the shape of the isotherm.

PRECIPITATION AND REDISSOLUTION

Results obtained for light transmission of $Ca(NO_3)_2$, NaCl, NH_4NO_3 and $AlCl_3$ solutions to which different amounts of sodium dodecylbenzenesulfonate were added are given in Figures 1 and 2. It can be seen that in the salt concentration ranges studied there was precipitation occurring as the sulfonate is increased, but interestingly, in the cases of calcium and aluminum solutions the precipitate was found to redissolve. In the case of NaCl and NH_4NO_3 solutions, the precipitate did not dissolve upon increasing the sulfonate concentration. It is however to be noted that in the latter cases, unlike for $Ca(NO_3)_2$ solution, the sulfonate added did not ever exceed that equivalent to the Na^+ in solution. Also, the precipitate obtained in NaCl and NH_4NO_3 solutions did redissolve when water was added to the solutions. Addition of water did not cause immediate redissolution of the calcium salt precipitate. The observed increase in turbidity could have resulted from either precipitation of calcium sulfonates or coagulation of sulfonate micelles into large aggregates upon charge neutralization (6). Redissolution occurs when sulfonate added exceeds the value required for complete neutralization of calcium ions possibly due to solubilization of the precipitate by micelles of the excess sulfonate or due to incorporation of the excess sulfonate in the precipitated micelles. The latter results in the recharging of the coagulated micelles and consequent redispersion. These mechanisms suggested earlier (6) will be discussed in detail elsewhere.

It is to be noted that precipitation and redissolution of the above type can lead to an abstraction (abstraction of surfactant from solution) maximum and, if precipitation is not totally isolated from adsorption, it can consequently lead to an <u>apparent</u> adsorption maximum. In the case of clays, a number of monovalent and multivalent cations such as those of Na, Ca, Al, etc. can be expected to be present in a supernatant of it and to produce precipitations.

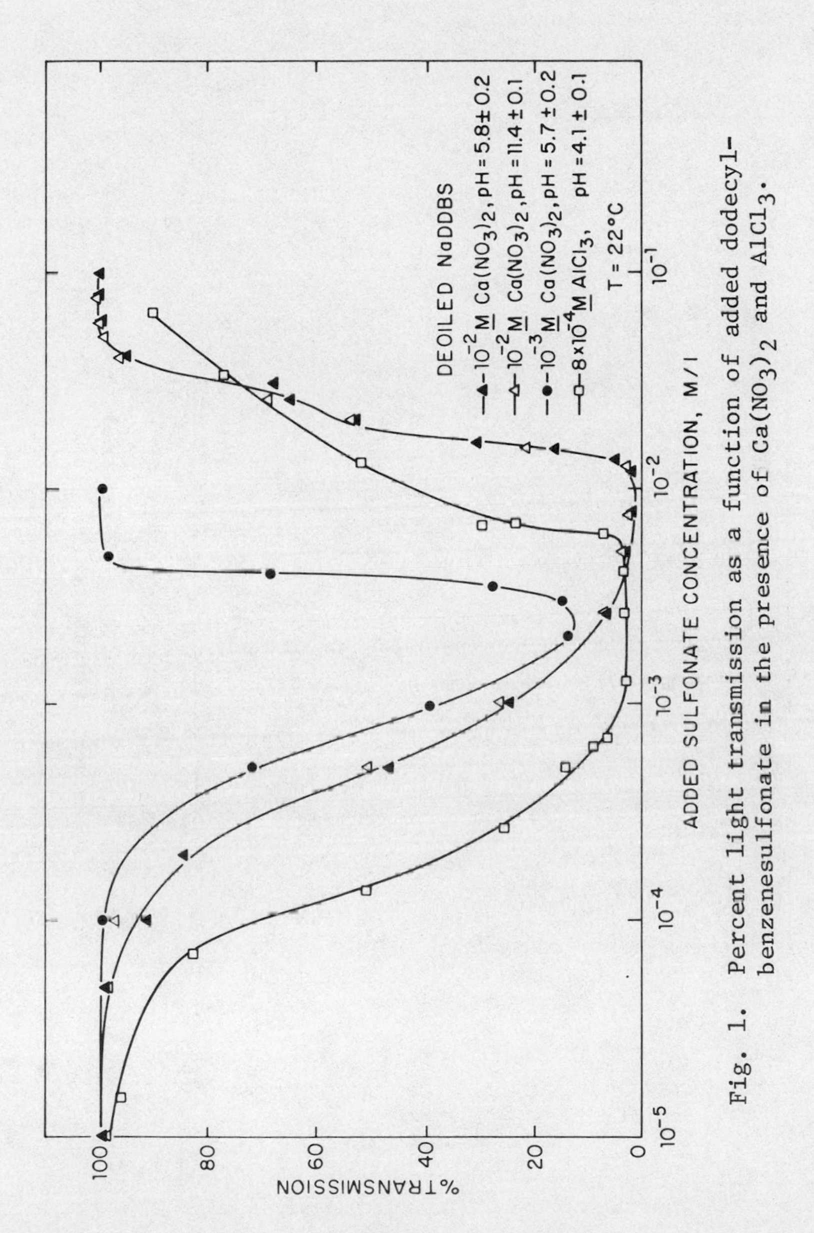

Fig. 1. Percent light transmission as a function of added dodecylbenzenesulfonate in the presence of $Ca(NO_3)_2$ and $AlCl_3$.

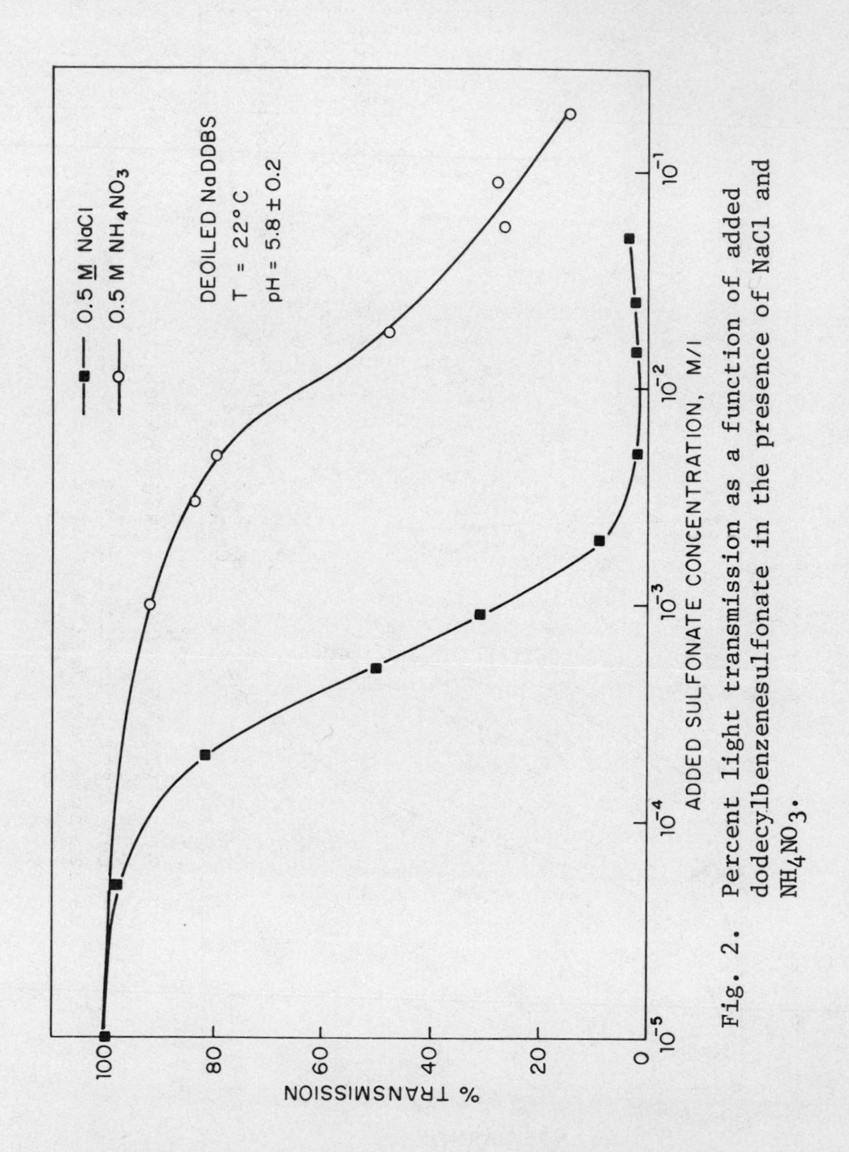

Fig. 2. Percent light transmission as a function of added dodecylbenzenesulfonate in the presence of NaCl and NH_4NO_3.

PRECIPITATION, ABSTRACTION AND ADSORPTION

Results obtained for abstraction of isomerically pure sodium dodecylbenzenesulfonate from a solution upon equilibrating it with homoionic Na-kaolinite are given in Figure 3. It can be seen that this isotherm is characterized by the presence of a maximum. The data for sulfonate abstraction upon contacting it with the supernatant from a kaolinite suspension is also given in this figure; these results should correspond to the sulfonate that is removed by precipitation due to its interaction with the dissolved mineral species. The supernatant was under conditions similar to those used in adsorption tests by equilibrating the Na-kaolinite in a 10^{-1}M NaCl solution at a solid to liquid ratio of 0.2 for seventy-two hours and then by removing the kaolinite by centrifugation. Depletion by precipitation was determined by contacting the surfactant with the supernatant for twelve hours and then by determining the residual concentration after centrifugation. The natural pH values obtained in this case were slightly higher than those obtained in the adsorption tests. Data for light transmission of the sulfonate-supernatant mixtures before centrifugation, also given in Figure 3, correlate with the data for sulfonate precipitation. Also, both the precipitation and transmission curves exhibit maxima in the concentration region of abstraction maximum.

Results obtained for the abstraction of sulfonate in the presence of kaolinite should include both the amount that is adsorbed on the mineral particles and that removed by precipitation. The amount of sulfonate that will precipitate due to interaction of it with dissolved mineral species in the presence of kaolinite need not be equal to that which will precipitate in a supernatant from which kaolinite has been removed. Assuming that any such discrepancy is minimal, adsorption isotherm for the present case is calculated from the difference between the two isotherms and is given in Figure 3. It can be seen that the major peak giving rise to the maximum is absent in the case of the adsorption isotherm suggesting that precipitation of sulfonate does indeed contribute towards determining the shape of the abstraction isotherm for the system considered here. It is to be noted that the resultant adsorption isotherm does still exhibit a maximum, even though much less prominent. The presence of such maximum can be attributed to the mechanisms that were mentioned earlier.

It is clear from these results that surfactants of the sulfonate type can undergo precipitation and redissolution upon interacting with solutions containing dissolved inorganic ions and that such phenomenon can play a major role in producing a maximum in abstraction isotherms.

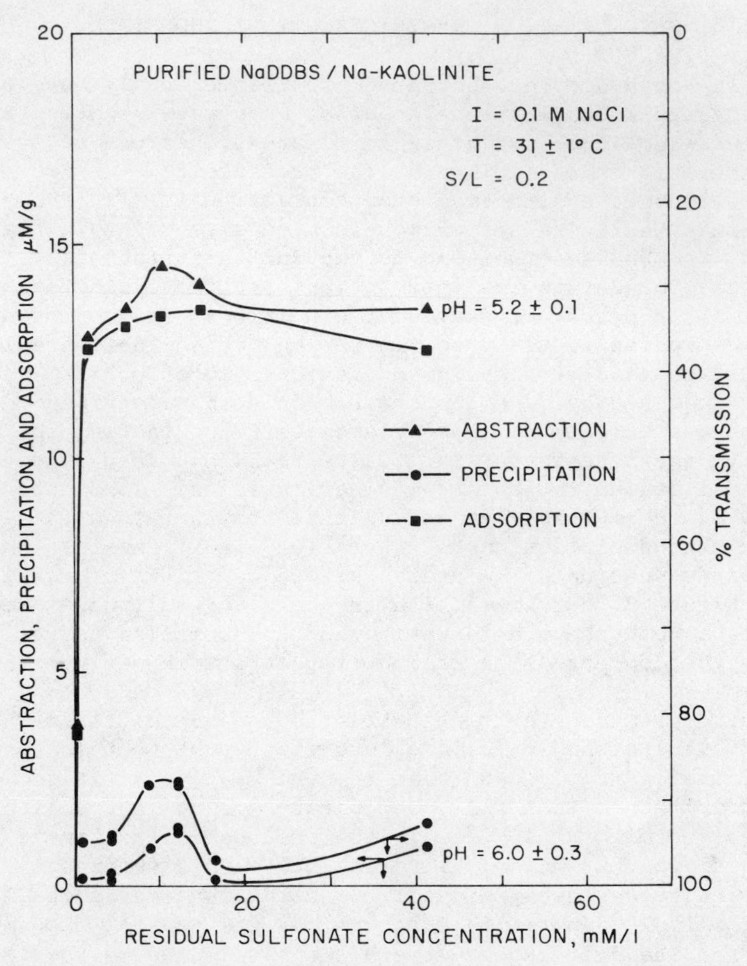

Fig. 3. Effect of precipitation on the shape of adsorption isotherm of sodium dodecylbenzenesulfonate on Na-kaolinite.

ACKNOWLEDGMENTS

Support of the Department of Energy (DE-AC19-79BC 10082), National Science Foundation (NSF ENG-78-11776), Amoco Production, Chevron Oil Field Research, Exxon Research and Engineering, Gulf Research and Development, Marathon Oil, Mobil Research & Development, Shell Development, Texaco, and Union Oil is gratefully acknowledged. We also thank Dr. I.B. Ivanov and Mr. K.P. Ananthapadmanabhan for helpful discussions.

REFERENCES

1. H. S. Hanna and P. Somasundaran, in "Improved Oil Recovery by Surfactant and Polymer Flooding," D. O. Shah and R. S. Schechter, eds., Academic Press, New York, 205 and 253, 1977.
2. P. Somasundaran and H. S. Hanna, SPE Symposium on Improved Oil Recovery, 241, April 16-19, Tulsa, Oklahoma, 1978.
3. S. P. Trushenski, D. C. Dauben, and D. K. Parrish, SPE Paper 4582 presented at 48th Annual SPE Meeting, Las Vegas (1973).
4. J. H. Bae and C. B. Petrick, SPE Paper 5819 presented at Improved Oil Recovery Symposium of SPE, Tulsa (1976).
5. H. S. Hanna and P. Somasundaran, Paper presented at the 52nd Colloid and Surface Science Symposium, June 12-14, Knoxville, 1978.
6. Annual Report Submitted to the National Science Foundation and a Consortium of Supporting Industrial Organizations, July 1978, Columbia University, New York.

THE EFFECT OF MICROEMULSION COMPOSITION ON THE ADSORPTION OF PETROLEUM SULFONATES ON BEREA SAND/MONTMORILLONITE CLAY ADSORBENTS

E. J. Derderian, J. E. Glass and G. M. Bryant

Union Carbide Corporation
Chemicals and Plastics
Research and Development Department
South Charleston, West Virginia 25303, U.S.A.

The adsorption from microemulsion of two petroleum sulfonates, PDM-334 and TRS 10-410, on Berea sand/montmorillonite clay adsorbents has been studied to determine: 1) the effect of microemulsion composition, specifically its relative oil and brine content, on sulfonate adsorption; 2) the effect of adsorption on the microemulsion composition and interfacial tension behavior. Whereas the degree of sulfonate adsorption can be determined by conventional methods (e.g. UV spectroscopy), one must utilize a microemulsion property which is a sensitive function of the relative oil and brine content of the microemulsion in order to determine the adsorption-induced changes in the microemulsion composition. This can be accomplished by the use of the microemulsion specific refraction.

The complementary use of the microemulsion specific refraction and UV spectroscopy leads to the following conclusions: 1) At constant temperature, pressure, total surface area and other equivalent conditions such as salinity, sulfonate adsorption is determined primarily by the composition of the microemulsion from which the sulfonate is adsorbed. PDM-334 and TRS 10-410, which are preferentially oil-soluble, tend to be adsorbed to a lesser degree from oil-rich than from brine-rich microemulsions. The degree of sulfonate adsorption can be generally correlated with the microemulsion specific refraction. Since the cosurfactant component can be used effectively to obtain a desired microemulsion composition, sulfonate adsorption is affected significantly by the choice of cosurfactant; 2) The microemulsion specific refraction increases as a result of adsorption in all cases which indicates that the oil to brine ratio in the microemulsion has increased. The post-adsorption microemulsion-oil and microemulsion-brine interfacial tensions reflect this change in composition.

A qualitative model is proposed which accounts for the observed adsorption behavior in terms of the effect of microemulsion composition on the sulfonate-microemulsion, sulfonate-adsorbent and microemulsion-adsorbent interactions.

In a separate set of experiments an array of twenty microemulsions, resulting from five blends of TRS 10-410 and TRS 18 and four cosurfactants, is characterized in terms of specific refractions and selected interfacial tensions. The locus of the minimum values of the controlling interfacial tension γ_c is explored as a function of sulfonate blend composition and cosurfactant hydrophilicity. The lowest minimum value of γ_c is observed for the microemulsion containing the most hydrophilic cosurfactant and the sulfonate blend which has the largest fraction of TRS 18 allowed by the locus domain.

INTRODUCTION

Low interfacial tensions and minimal surfactant loss due to interactions with reservoir solids are two of the most important conditions for effective oil recovery by displacement fluids in chemical flooding. These two requirements are, of course, related; surfactant adsorption from a microemulsion whose properties have been carefully designed leads to changes in the composition and therefore in the interfacial behavior of the microemulsion.

The importance of minimizing adsorption has provided the impetus for a number of adsorption studies of both anionic and nonionic surfactants on representative reservoir solids; most of these deal with surfactant adsorption from aqueous solution. In general it has been found that the adsorption of petroleum sulfonates on mineral adsorbents increases with decreasing solubility in the solvent. Gale and Sandvik (1) have found that petroleum sulfonate adsorption from brine on clay minerals increased with molecular weight and therefore decreasing solubility in brine. Trogus et al. (2) have also found that the adsorption of alkyl aryl sulfonates from brine on Berea and kaolinite increased with surfactant molecular weight. Lawson and Dilgren (3) have noted that sulfonate adsorption increased with brine salinity and therefore with decreased solubility. These results are in agreement with the theory that adsorption from solution depends in part on the magnitude of the solute-solvent interactions in the solution (4); in general, weak interactions, as manifested by low solubility, lead to large solute adsorption.

It is also recognized (4) that in the case of ionic adsorbents, adsorption is determined in part by the surface charge of the adsorbent. In general, H^+ and OH^- are potential - determining in mineral oxide-water systems; therefore the surface charge, and the adsorption of sulfonate anions, would depend on the pH of the system (5). Somasundaran and Hanna (5) have found in fact that sulfonate adsorption on kaolinite decreases with increasing pH since the fraction of positive sites on the surface is expected to decrease.

Adsorption studies of petroleum sulfonates from aqueous solution are generally restricted, however, to dilute solutions because of the limited solubility of these surfactants in water, particularly in the presence of salt. With respect to oil-recovery systems, it is of interest to extend the study of petroleum sulfonate adsorption to microemulsions where the sulfonate concentrations are generally much higher. Glover et al. (6) have found that sulfonate adsorption from small-bank laboratory microemulsion displacement tests increased with salinity at low salt concentrations; at higher salt concentrations sulfonate loss was attributed to phase trapping as proposed by Trushenski (7).

The objective of the present work is to determine the static adsorption of petroleum sulfonates from microemulsions on representative reservoir solids and to define the effect of microemulsion composition, specifically its relative oil and brine content, on sulfonate adsorption. It is also of interest to determine the effect of adsorption on the microemulsion oil and brine content because of the relationship between microemulsion composition and interfacial behavior. Consequently, the adsorption of a given petroleum sulfonate was determined from a series of microemulsions where each microemulsion contained different volume fractions of the same oil and brine. The difference in microemulsion composition within such a series was effected either by using a different cosurfactant in each microemulsion or by changing the total surfactant/cosurfactant concentration. The adsorbent was carefully reproduced in each experiment in terms of sand/clay composition and total surface area. All experiments within a series were therefore carried out at constant temperature, pressure, adsorbent composition and total surface area.

GENERAL CONSIDERATIONS

In order to accomplish the objectives of determining the effect of microemulsion oil and brine content on sulfonate adsorption as well as defining the changes in microemulsion composition as a result of adsorption, it is necessary to utilize a microemulsion property which is a sensitive function of the relative oil and brine content of the microemulsion. The microemulsion specific refraction satisfies this requirement.

We have shown previously (8) that in an equilibrated multiphase system there exists a correlation between the Lorenz-Lorentz specific refraction of the microemulsion and its interfacial tensions versus the excess oil and brine phases. The Lorenz-Lorentz specific refraction r is defined by

$$r = \frac{1}{\rho} \frac{n_D^2 - 1}{n_D^2 + 2} \tag{1}$$

where ρ and n_D are the density and index of refraction of the microemulsion at a given temperature. Healy and Reed have shown (9) that oil recovery from immiscible microemulsion core floods is maximized when the following condition is satisfied:

$$\gamma_{mo} = \gamma_{mb} \tag{2}$$

where γ_{mo} and γ_{mb} are the interfacial tensions of the microemulsion versus equilibrated excess oil and brine phases, respectively. Furthermore, the controlling interfacial tension γ_c, which in effect determines the magnitude of the capillary number, is a minimum when $\gamma_{mo} = \gamma_{mb}$ (9).

We have demonstrated (8) that the condition defined by Equation (2) always occurs at a constant value of the microemulsion specific refraction $r = r^* = 0.263 \pm 0.004$ cm^3/g. The value of r^* is independent of surfactant composition, cosurfactant composition, surfactant/cosurfactant concentration and ratio, salinity of brine, temperature, and composition of the simulated crude oil.

The correlation is based on the fact that the microemulsion specific refraction is a sensitive function of the relative oil and brine content of the microemulsion. The condition $r = r^*$ occurs only for microemulsions which contain equal volumes of oil and brine. In general, a theoretical value r_{th} for the microemulsion specific refraction can be obtained to a very good approximation from

$$r_{th} = (V_o + V_b)^{-1} (V_o r_o + V_b r_b) \tag{3}$$

where V_o and V_b are the volumes and r_o and r_b the specific refractions of the oil and brine in the microemulsion. At $V_o = V_b$ which corresponds to $\gamma_{mo} = \gamma_{mb} = (\gamma_c)_{min}$, Equation (3) becomes

$$r_{th} = 0.5 (r_o + r_b) = r_{th}^* \tag{4}$$

In the adsorption experiments described below, the microemulsion specific refraction was determined before adsorption and correlated with the initial oil and brine volume fractions in the microemulsion. Following adsorption, the microemulsion specific refraction was determined again and the qualitative change in the microemulsion relative oil and brine content was deduced from the change in the specific refraction.

EXPERIMENTAL

(a) Materials and preparation of systems: Systems of two and three phases, of which one was always a microemulsion phase, were prepared from cosurfactants, primary surfactants, hydrocarbon, and brine.

The cosurfactants used were: n-butanol, Butyl CELLOSOLVE, Butyl CARBITOL, and butoxytriglycol; the glycol ethers are, respectively, one, two and three mole ethoxylates of n-butanol. All are products of Union Carbide Corporation.

The primary surfactants used were: PDM-334, a monoethanolamine dodecyl orthoxylene sulfonate (av. MW = 427, 79.7% sulfonate), Exxon Chemical; and two petroleum sulfonates, TRS 10-410 (eq. wt. = 422, 61.6% sulfonate) and TRS 18 (eq. wt. = 510, 61.3% sulfonate) from Witco Chemical. All of the sulfonates are considerably more soluble in oil than in water.

The hydrocarbon used was nonane (99 mole % minimum purity, Phillips Petroleum). Brine was prepared from triply distilled water and NaCl and $CaCl_2$ which were analytical reagent grade compounds.

The Berea sand was obtained from Cleveland Quarries in Amherst, Ohio; the montmorillonite was a clay mineral standard from Ward's Natural Science Establishment, Inc. Both solids were characterized well in terms of particle size distribution and surface area. The crushed clay and sand were sieved separately using US sieve series; the fractions of sand and clay which passed through sieve No. 170 but not No. 200 were retained for use as adsorbents. The particle size distribution of these fractions is fairly narrow with the majority of particles in the 74-88 μm range (10). The surface areas of these adsorbents as determined by nitrogen adsorption were 2.86 m^2/g for the sand and 84.7 m^2/g for the clay.

Upon preparation, the multicomponent fluid systems were allowed to stand undisturbed at constant temperature for 3 days or until the resulting multiphase systems exhibited clear, transparent microemulsions and sharp interfaces. In all of these systems, the microemulsion phase contained virtually all of the sulfonate in the system.

(b) Methods of analysis: The microemulsions used in the adsorption experiments were characterized prior to adsorption in terms of the following: 1) phase behavior; 2) sulfonate concentration; 3) volume fractions of oil and brine; 4) specific refraction; 5) interfacial tensions versus equilibrated excess phases.

The sulfonate concentration in the microemulsion was determined from the equilibrated microemulsion phase volume and the known weight of sulfonate in the system; the assumption that the microemulsion phase contained all of the sulfonate was justified for all microemulsions. The volume fractions of oil and brine in the microemulsion were determined from the excess volumes of oil and brine, respectively. The microemulsion density and index of refraction needed to calculate the specific refraction (Eq. (1)) were measured on a Mettler-Paar DMA 40 digital density meter with accuracy of ± 0.0001 g/cm^3 and a Zeiss Abbe refractometer (±0.0001), respectively; the temperature was controlled with an Exacal 100 and Endocal 150 constant temperature circulator-baths connected in series. Interfacial tensions between the microemulsion and equilibrated excess phases were measured on a University of Texas Spinning Drop Tensiometer or a Spinning Drop Tensiometer from S&S Instrument Mfg.; measurements were carried out until equilibrium values were obtained as indicated by constant readings over a period of at least 1 hour.

In addition, a calibration graph of sulfonate concentration versus ultraviolet absorbance was obtained for each microemulsion prior to adsorption. A small known volume of microemulsion was diluted with methanol ("spectro" grade) to obtain a clear sulfonate stock solution which was used to prepare four still more dilute methanol solutions of varying sulfonate concentration. The UV absorbance spectra of these solutions were measured on a Cary 118 spectrophotometer; absorbances were determined at 271 nm for PDM-334 microemulsions and at 259 nm for TRS 10-410 microemulsions. All calibration plots were linear and passed through the origin.

A known volume of microemulsion was then added to a bottle containing weighed amounts of the Berea sand and montmorillonite clay; the bottle was sealed, placed on a roller and rotated gently at constant temperature for 48 hours.

Following the adsorption experiment, the microemulsion was separated from the solids by decantation and successive centrifugations; after each centrifugation the microemulsion density was determined and the procedure was repeated until a constant density was obtained. The separated microemulsion was always clear and transparent. The index of refraction was then measured and the specific refraction of the microemulsion was calculated. The interfacial tensions of this microemulsion versus the excess phases from the original system were then measured. The pre-adsorption dilution procedure was then repeated, the UV absorbances were measured and the concentrations of sulfonate in the methanol solutions were determined from the appropriate calibration plot. The amount of sulfonate in the post-adsorption microemulsion was then calculated from the average of the values obtained from each diluted methanol solution; the internal agreement between the averaged values was always excellent.

RESULTS

(a) Adsorption of sulfonate from four TRS 10-410 microemulsions: The adsorption of TRS 10-410 from four microemulsions on 4.0g Berea sand and 0.2g montmorillonite clay was determined at 24°C. The four multicomponent systems which provided the microemulsions were prepared from: 5.0 wt % TRS 10-410, 3.0 wt % cosurfactant, 46.0 wt % nonane, and 46.0 wt %(2.0 wt % NaCl, 0.2 wt % $CaCl_2$) brine. The four systems were therefore identical in the sulfonate, oil and brine components but each contained a different cosurfactant. The four cosurfactants were: n-butanol (microemulsion A), butoxymonoglycol (microemulsion B), butoxydiglycol (microemulsion C), and butoxytriglycol (microemulsion D). Consequently, the resulting differences in the equilibrium phase behavior of these systems and the compositions of the four microemulsions could be attributed directly to the cosurfactant component. Table 1 shows the effect of the cosurfactant on the phase behavior, composition and specific refraction of the four microemulsions. As the ethylene oxide content and therefore the hydrophilicity of the cosurfactant increases from n-butanol to butoxytriglycol, the microemulsion changes from upper to lower phase; the concentration of TRS 10-410 is highest in the middle-phase microemulsion. Correspondingly, the volume fraction of oil ϕ_o in the microemulsion decreases and the volume fraction of brine ϕ_b increases; the microemulsion specific refraction decreases as the brine to oil ratio in the microemulsion increases. Figure 1 clearly shows the correlation between r and ϕ_o and ϕ_b for microemulsions A, B, C, D and a fifth microemulsion ($r = 0.288$ cm^3/g) whose cosurfactant was a 50:50 blend by weight of butoxymonoglycol and butoxydiglycol. As the hydrophilicity of the cosurfactant increases from microemulsion A to microemulsion D, ϕ_o decreases, ϕ_b increases and the microemulsion specific refraction decreases. We have shown previously (8) that the specific refractions of these microemulsions would assume only values between r_o and r_b, the specific refractions of the particular bulk oil and brine components of the microemulsions; in this case $r_o = 0.342$ cm^3/g and $r_b = 0.204$ cm^3/g. The specific refractions of microemulsions A and D begin to approach these limiting values (Figure 1); this is reasonable in view of the ϕ_b and ϕ_o values of these microemulsions. It is important to note that at $\phi_o = \phi_b = 0.39$ the value of r is 0.265 cm^3/g which is in

Table 1. Effect of the cosurfactant component on the equilibrium composition and specific refraction of TRS 10-410 microemulsions at 24°C.

Microemulsion/cosurfactant	Position relative to excess phases	TRS 10-410 conc. (g/1)	ϕ_o	ϕ_b	$r(cm^3/g)$
A/n-butanol	upper phase	67.6	0.87	0.02	0.327
B/Butoxymonoglycol	upper phase	63.2	0.81	0.05	0.321
C/Butoxydiglycol	middle phase	129.7	0.28	0.50	0.250
D/Butoxytriglycol	lower phase	81.4	0.12	0.74	0.226

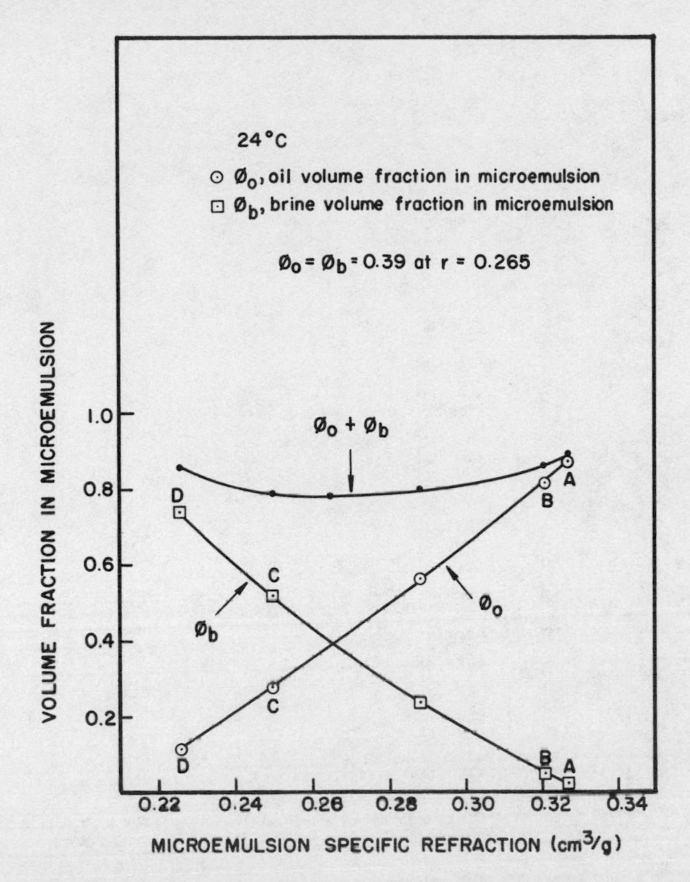

Fig. 1. Dependence of the microemulsion specific refraction on the oil and brine volume fractions in the microemulsion.

excellent agreement with the value of r^*. Interestingly, the sum $\phi_o + \phi_b$ exhibits a shallow minimum at about $r = 0.265$ cm^3/g which indicates that the microemulsion characterized by $\phi_o = \phi_b$ is the most concentrated in the sulfonate.

Table 2 presents the adsorption results for microemulsions A, B, C and D as well as their specific refractions before and after adsorption. The degree of TRS 10-410 adsorption from the microemulsions containing glycol ethers as cosurfactants (microemulsions B, C and D) clearly increases with decreasing initial r and therefore increasing brine content of the original microemulsion. For microemulsions B, C, and D the percent of TRS 10-410 adsorbed decreases linearly with increasing specific refraction of the microemulsion from which the sulfonate is adsorbed (Figure 2). The microemulsion with n-butanol as cosurfactant (microemulsion A) does not fit this behavior, however, and possible reasons for this deviation will be presented later.

Table 2. Sulfonate adsorption from TRS 10-410 microemulsions at 24°C and corresponding change in microemulsion specific refraction.

Microemulsion/cosurfactant	TRS 10-410 adsorbed, percent	$r(cm^3/g)$ before adsorption	$r(cm^3/g)$ after adsorption
A/n-butanol	13	0.327	0.329
B/Butoxymonoglycol	6	0.321	0.323
C/Butoxydiglycol	19	0.250	0.257
D/Butoxytriglycol	23	0.226	0.227

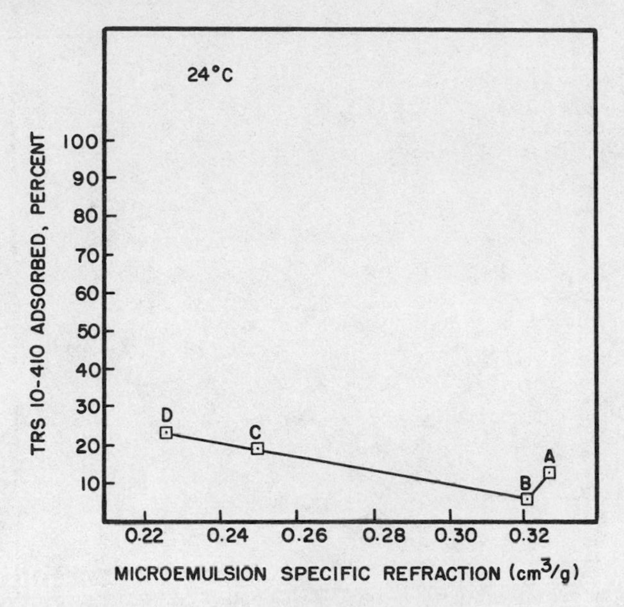

Fig. 2. Dependence of the fraction of adsorbed TRS 10-410 on the microemulsion specific refraction.

Table 2 also shows that the specific refraction of each microemulsion increases as a consequence of adsorption. This result indicates that all post-adsorption microemulsions have higher oil to brine ratios than the corresponding pre-adsorption microemulsions. The effect of these compositional changes is reflected in the interfacial tension behavior of the microemulsions (Figure 3). The microemulsion-oil and microemulsion-brine interfacial tensions both before and after adsorption exhibit the familiar correlation with r (8); however, all γ_{mo} values are decreased and all γ_{mb} values are increased as a consequence of adsorption. The minimum in the controlling interfacial tension $(\gamma_c)_{min}$ has also increased from 0.0070 dyne/cm to 0.0077 dyne/cm. It is interesting to note that the γ_{mb} values for the n-butanol microemulsion both before and after adsorption lie considerably above the γ_{mb} curves. The microemulsion at r = 0.288 cm^3/g was not part of the adsorption series and therefore only the initial values of γ_{mo} and γ_{mb} are shown.

(b) Adsorption of sulfonate from four PDM-334 microemulsions: The adsorption of PDM-334 from four microemulsions on 4.0g Berea sand and 0.2g montmorillonite clay was determined at 24°C. The four multicomponent systems which yielded the microemulsions were prepared from: 5.0 wt % PDM-334, 3.0 wt % cosurfactant, 46.0 wt % nonane, and 46.0 wt % (2.0 wt % NaCl, 0.2 wt % $CaCl_2$) brine. The same four cosurfactants were used as in the TRS 10-410 series; the microemulsion with n-butanol was designated E, the butoxymonoglycol microemulsion F, the butoxydiglycol microemulsion G, and the butoxy-

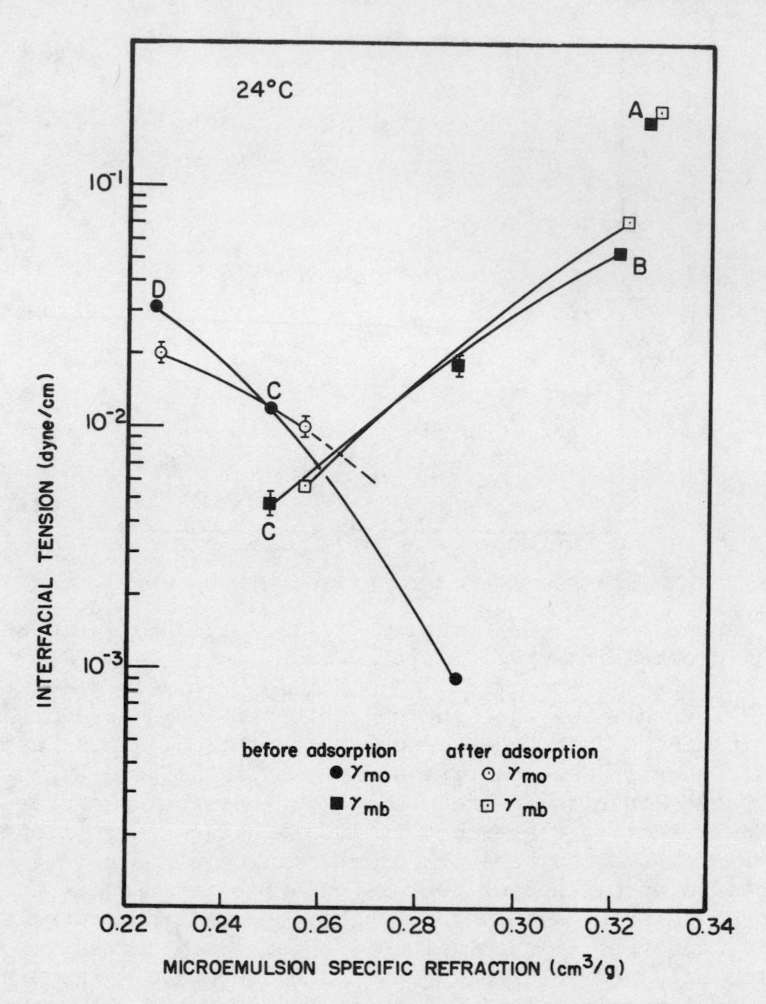

Fig. 3. Effect of sulfonate adsorption on the microemulsion specific refraction as well as the microemulsion-excess oil and microemulsion-excess brine interfacial tensions.

triglycol microemulsion H. Again, differences in the equilibrium phase behavior of these systems as well as in composition of the four resulting microemulsions could be attributed directly to the cosurfactant. Again as the hydrophilicity of the cosurfactant increases, ϕ_o in the microemulsion decreases and ϕ_b increases (Table 3); these compositional changes are again accurately reflected by the microemulsion specific refraction and a relationship between ϕ_o, ϕ_b and r similar to the one observed with the TRS 10-410 microemulsions (Figure 1) can be obtained.

The adsorption results as well as the microemulsion specific refractions before and after adsorption are given in Table 4. The

Table 3. Effect of the cosurfactant component on the equilibrium composition and specific refraction of PDM-334 microemulsions at 24°C.

Microemulsion/cosurfactant	Position relative to excess phases	PDM-334 conc. (g/l)	ϕ_o	ϕ_b	r(cm^3/g)
E/n-butanol	upper phase	64.0	0.82	0.02	0.325
F/Butoxymonoglycol	upper phase	60.0	0.77	0.10	0.312
G/Butoxydiglycol	lower phase	77.4	0.11	0.70	0.231
H/Butoxytriglycol	lower phase	81.4	0.10	0.74	0.226

Table 4. Sulfonate adsorption from PDM-334 microemulsions at 24°C and corresponding change in microemulsion specific refraction.

Microemulsion/cosurfactant	PDM-334 adsorbed, percent	$r(cm^3/g)$ before adsorption	$r(cm^3/g)$ after adsorption
E/n-butanol	14	0.325	0.326
F/Butoxymonoglycol	8	0.312	0.315
G/Butoxydiglycol	16	0.231	0.236
H/Butoxytriglycol	17	0.226	0.227

percent of adsorbed PDM-334 from the glycol ether microemulsions F, G and H increases with decreasing r and therefore with increasing brine content of the initial microemulsion. For these microemulsions the fraction of PDM-334 adsorbed decreases linearly with increasing specific refraction of the microemulsion from which it is adsorbed (Figure 4). The n-butanol microemulsion again does not fit this behavior.

Table 4 shows that, as in the TRS 10-410 series, the specific refraction of each PDM-334 microemulsion increases following adsorption. All post-adsorption microemulsions therefore again have higher oil to brine ratios than the initial microemulsions. The interfacial tension behavior is also similar to that observed with the TRS 10-410 microemulsions (Figure 3); all γ_{mo} values are decreased and all γ_{mb} values are increased as a result of adsorption. Again the γ_{mb} values for the n-butanol microemulsion both before and after adsorption lie above the γ_{mb} curves.

(c) <u>Adsorption of sulfonate from four PDM-334/butoxytriglycol microemulsions</u>: In view of the adsorption results obtained with the TRS 10-410 and PDM-334 microemulsions, it is of interest to determine the adsorption of one of these sulfonates from microemulsions which, while different in composition, have the same cosurfactant component. The adsorption of sulfonate from four PDM-334/butoxytriglycol microemulsions was therefore determined on 8.0g Berea sand and 2.0g montmorillonite clay. The multicomponent systems which yielded the microemulsions were prepared from: 0.63x vol % PDM-334, 0.37x vol % butoxytriglycol, 0.5(100-x) vol % nonane, and 0.5(100-x) vol % (3.0 wt % NaCl, 0.3 wt % $CaCl_2$) brine, where x, the combined concentration of PDM-334 and butoxytriglycol, assumed the values 3.3, 4.4, 6.5 and 8.8. The resulting differences in microemulsion phase behavior and composition are therefore due to the variations in the total amount of surfactant and cosurfactant in the system; increasing the overall PDM-334/butoxytriglycol concentration at constant ratio (63/37) resulted in a gradual decrease in microemulsion specific refraction (Table 5). The adsorption data are presented in Table 6 along with the microemulsion specific refractions before and after adsorption; the fraction of PDM-334 adsorbed from all microemulsions increased with decreasing specific refraction and therefore with increasing brine content of the microemulsion. In addition, all r values increased on adsorption indicating again that the post-adsorption microemulsions all had higher oil to brine ratios than the corresponding pre-adsorption microemulsions. The percent of PDM-334 adsorbed increases fairly linearly with decreasing specific refraction of the microemulsion from which it was adsorbed (Figure 5). The actual amounts of PDM-334 adsorbed per gram of adsorbent are also illustrated in Figure 5. The large amount of sulfonate adsorbed from the middle phase microemulsions is somewhat disconcerting since it indicates

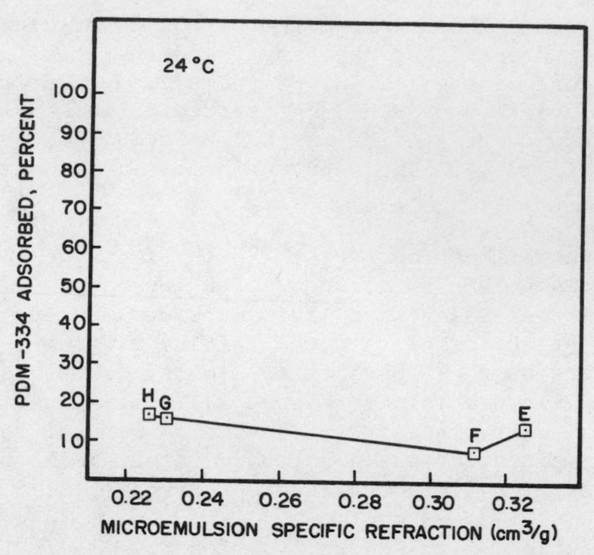

Fig. 4. Dependence of the fraction of adsorbed PDM-334 on the microemulsion specific refraction.

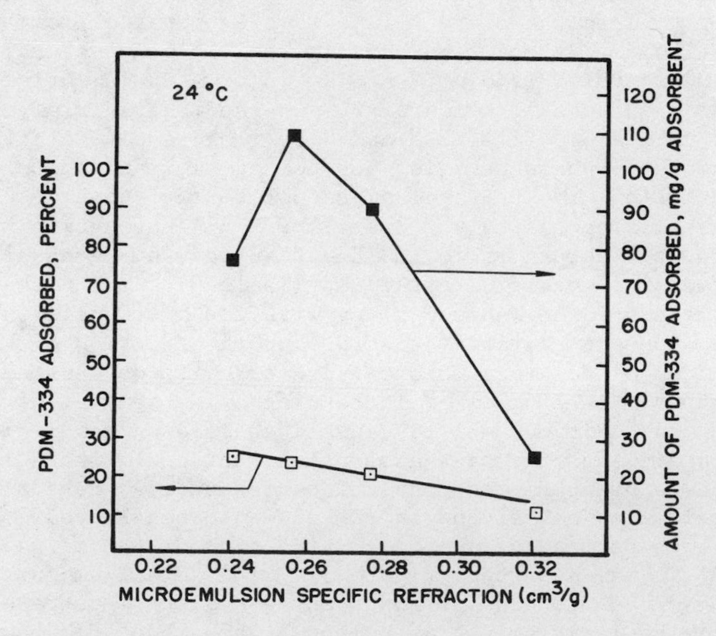

Fig. 5. Dependence of the fraction as well as amount of adsorbed PDM-334 on the microemulsion specific refraction.

Table 5. Effect of total PDM-334/Butoxytriglycol concentration on the microemulsion specific refraction at 24°C.

Microemulsion	Position relative to excess phases	PDM-334 conc. (g/l)	$r(cm^3/g)$
x = 3.3	upper phase	37.6	0.321
x = 4.4	middle phase	124.6	0.277
x = 6.5	middle phase	131.5	0.257
x = 8.8	lower phase	107.6	0.241

that within a series of microemulsions the greatest total sulfonate adsorption occurs from microemulsions which exhibit the condition $\phi_o \simeq \phi_b$ and therefore $\gamma_{mo} \simeq \gamma_{mb}$. This result, however, is not surprising; Figure 1 demonstrates that in general within any such series the microemulsions with $r \simeq 0.26$ have the highest sulfonate concentration. Consequently, even though there is a linear relationship between the fraction of sulfonate adsorbed and the microemulsion composition, there is a maximum in the total amount of sulfonate adsorbed at $r \simeq 0.26$.

(d) Adsorption of PDM-334 from aqueous and nonane solutions: In order to determine the effect on adsorption levels of the magnitude of the interactions between the sulfonate and the environment from which it is adsorbed, PDM-334 was adsorbed from aqueous and nonane solutions, which also contained n-butanol and butoxytriglycol; the adsorbent consisted of 8.0g Berea sand and 2.0g montmorillonite clay. The results in Table 7 show that the fractions of PDM-334 adsorbed from both the aqueous and nonane solutions are much larger than those adsorbed from any of the PDM-334/butoxytriglycol microemulsions. Furthermore, the fractions of sulfonate adsorbed from the aqueous solutions are significantly larger than those from the nonane solutions. It is interesting to note that in both the aqueous and nonane systems, the identity of the cosurfactant has no effect on the adsorption of the sulfonate.

(e) Effect of surfactant and cosurfactant on the magnitude of $(\gamma_c)_{min}$: The results of the static adsorption experiments described in the previous sections indicate that sulfonate adsorption, as expected, affects microemulsion-excess phase interfacial tensions significantly even within the controlled framework of these laboratory experiments. Nevertheless, it is important to identify microemulsion compositions which exhibit the condition $\gamma_{mo} = \gamma_{mb} = (\gamma_c)_{min}$ (9); in addition, the magnitude of $(\gamma_c)_{min}$ itself should be minimized.

It is of interest therefore to examine systematically the effect of the surfactant and cosurfactant components on microemul-

Table 6. Sulfonate adsorption from PDM-334/Butoxytriglycol microemulsions at 24°C and corresponding change in microemulsion specific refraction.

Microemulsion	PDM-334 adsorbed, percent	$r(cm^3/g)$ before adsorption	$r(cm^3/g)$ after adsorption
x = 3.3	11	0.321	0.331
x = 4.4	21	0.277	0.295
x = 6.5	24	0.257	0.271
x = 8.8	25	0.241	0.260

Table 7. PDM-334 adsorption from aqueous and nonane solutions at 24°C.

Solution	PDM-334 conc. (g/1)	Cosurfactant	PDM-334 adsorbed, percent
water	10.0	n-butanol	95
water	10.0	butoxytriglycol	96
nonane	10.0	n-butanol	84
nonane	10.0	butoxytriglycol	86

sion composition and therefore on the microemulsion-excess phase interfacial tensions. With this goal in mind, an array of twenty microemulsions was prepared which also included the TRS 10-410 microemulsions A, B, C and D. The general design of the array was: (5.0-y) wt % TRS 10-410, y wt % TRS 18, 3.0 wt % cosurfactant, 46.0 wt % nonane, and 46.0 wt % (2.0 wt % NaCl, 0.2 wt % $CaCl_2$) brine, where y assumed the values 0.0, 1.25, 2.50, 3.75, and 5.0 for each of the four cosurfactants n-butanol, butoxymonoglycol, butoxydiglycol, and butoxytriglycol. The microemulsions were characterized in terms of their specific refractions and utilizing the general property of the correlation between $(\gamma_c)_{min}$ and r^*, only selected interfacial tensions were measured. Figure 6 shows the twenty microemulsions in terms of the surfactant composition, cosurfactant component and microemulsion specific refraction; also shown are the corresponding interfacial tensions. Microemulsions A, B, C and D (filled circles) are described in Table 1; microemulsion X (3.75 wt % TRS 10-410, 1.25 wt % TRS 18, butoxydiglycol) and microemulsion Y (3.75 wt % TRS 10-410, 1.25 wt % TRS 18, butoxytriglycol) are middle phase; Z (2.50 wt % TRS 10-410, 2.50 wt % TRS 18, butoxytriglycol) is upper phase as are the remaining microemulsions in the array. The shape of the surface described by the r values of these microemulsions reflects the following compositional properties of the microemulsions. For any constant TRS 10-410/TRS 18 composition, the volume fraction of oil ϕ_o increases and the volume fraction of brine ϕ_b decreases as the cosurfactant becomes less hydrophilic; in this respect, the behavior of the adsorption microemulsions A, B, C and D whose r values form the y=0 edge of the surface is typical. Furthermore, for any particular cosurfactant ϕ_o increases and ϕ_b decreases as y, the fraction of the more oil-soluble TRS 18, increases. The rate of change of r, and therefore microemulsion composition, is very rapid for $y \leq 2.50$ when the cosurfactant is either butoxytriglycol or butoxydiglycol. Similarly, the rate of change of r as a function of the cosurfactant component from butoxytriglycol to butoxymonoglycol is very rapid for sulfonate compositions corresponding to y = 0.0 and y = 1.25.

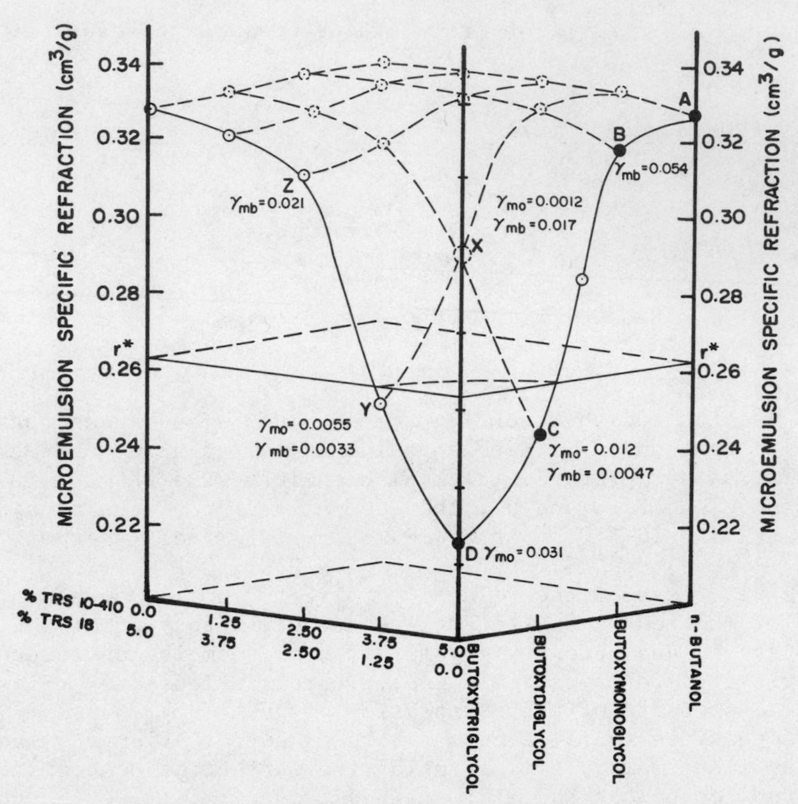

Fig. 6. Effect of surfactant composition and cosurfactant component on the microemulsion specific refraction; also shown are the interfacial tensions for selected microemulsions and the locus of γ_c minima formed by the intersection of the array surface and the r^* plane.

The interfacial tensions along the curve formed by the intersection of the array surface and the plane at $r = r^* = 0.263\ cm^3/g$ are of interest because of the correlation between r^* and the condition $\gamma_{mo} = \gamma_{mb} = (\gamma_c)_{min}$; in effect since the intersection curve lies in the r^* plane, it represents a locus of γ_c minima. It is important therefore to determine the actual magnitudes of these minima along the intersection curve as a function of the cosurfactant and sulfonate components.

A composite plot of the measured interfacial tensions and specific refractions of microemulsions B, C, D, X, Y and Z is given in Figure 7. Interpolated values of four minima of γ_c were then obtained from: B, C and D (0.0070 dyne/cm); C and X (0.0070 dyne/cm); X and Y (0.0045 dyne/cm); and D, Y and Z (0.0039 dyne/cm). Figure 8 shows a top view of the r^* plane with the values of the γ_c minima at the corresponding cosurfactant and surfactant compositions. The magnitudes of the minima decrease as the hydro-

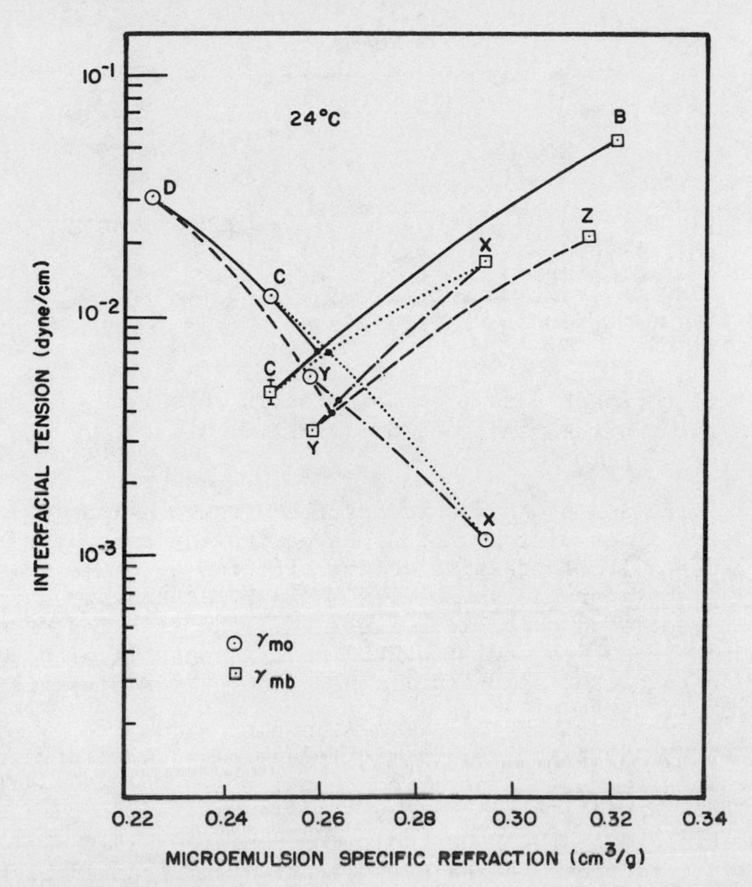

Fig. 7. Determination of the magnitudes of $(\gamma_c)_{min}$ from the dependence of the interfacial tensions on the microemulsion specific refraction.

philicity of the cosurfactant and the amount of TRS 18 in the sulfonate blend increase. This result is intuitively appealing since the right combination of a hydrophilic cosurfactant and a hydrophobic primary surfactant of a wide molecular weight distribution would lead, through enhanced interactions, to maximization of the absolute amounts of oil and brine in the microemulsion, a corresponding minimization of both γ_{mo} and γ_{mb} and therefore a very low value for the appropriate minimum in γ_c.

DISCUSSION

It is useful at this point to summarize briefly the results presented in the last section.

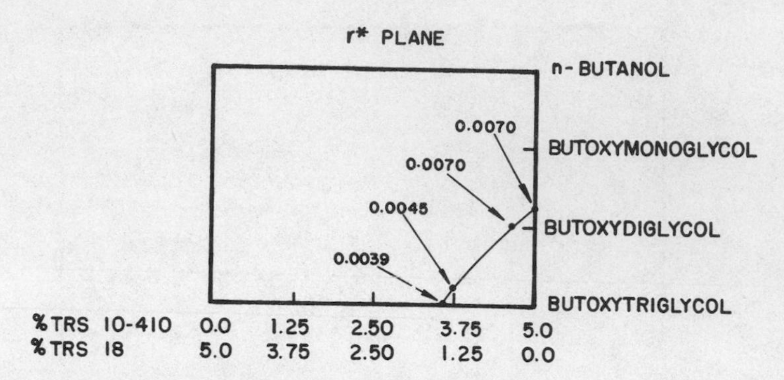

Fig. 8. Effect of surfactant composition and cosurfactant composition on the magnitude of $(\gamma_c)_{min}$.

- The fraction of sulfonate adsorbed from the TRS 10-410/glycol ether microemulsions on equivalent adsorbents increases with increasing volume fraction of brine in the microemulsion and in fact is linearly dependent on the microemulsion specific refraction. The fraction of sulfonate adsorbed from the TRS 10-410/n-butanol microemulsion under the same conditions, however, is significantly higher.
- The PDM-334/glycol ether microemulsions qualitatively exhibit the same adsorption behavior as the TRS 10-410/glycol ether microemulsions. The fraction of PDM-334 adsorbed from the n-butanol microemulsion under the same conditions is again, however, significantly higher.
- The fraction of sulfonate adsorbed from the four PDM-334/butoxytriglycol microemulsions increases with total surfactant/cosurfactant concentration and therefore with increasing volume fraction of brine in the microemulsion; the dependence on the microemulsion specific refraction is again linear. The absolute amount of PDM-334 adsorbed per gram of adsorbent exhibits a maximum for the middle phase microemulsions where the surfactant concentrations are largest.
- The microemulsion specific refraction increases as a result of adsorption in all cases. This result indicates that the oil to brine ratio in the microemulsion has increased.
- The fraction of PDM-334 adsorbed from aqueous and nonane solutions was much larger than from any of the PDM-334/butoxytriglycol microemulsions. Furthermore, adsorption was significantly higher from aqueous than from nonane solutions; the identity of the cosurfactant in these solutions had no effect on sulfonate adsorption.

The lowest value along the locus of (γ_c) minima, defined by the intersection of the array surface and the r^* plane, was obtained with butoxytriglycol, the most hydrophilic of the four cosurfactants in the array, and the most hydrophobic TRS 10-410/TRS 18 blend within the intersection.

In view of these results, a qualitative model is proposed for the isothermal, static adsorption of sulfonates on mineral solids from microemulsions. We assume that the adsorption behavior is primarily determined by the clay fraction of the adsorbent because of its high specific area. In general, the adsorption process can be regarded as a partitioning of the sulfonate anions between the microemulsion phase and the microemulsion-adsorbent interface. At equilibrium the amount of adsorbed sulfonate is a measure of the relative magnitudes of the sulfonate-microemulsion, sulfonate-adsorbent and microemulsion-adsorbent interactions. Since the sulfonate is charged, one must in principle consider electrostatic as well as chemical forces.

Let us consider qualitatively the effect of the microemulsion composition on the sulfonate-microemulsion, sulfonate-adsorbent and microemulsion-adsorbent interactions. We assume that the interaction potential between a single sulfonate anion and its nearest-neighbor molecules and ions in the oil-brine interfacial region in the microemulsion is nearly constant within the appropriate series of microemulsions (e.g. A, B, C and D). As a first approximation therefore we consider the sulfonate-microemulsion interactions to be unaffected by microemulsion composition within such a series.

The sulfonate-adsorbent interactions, however, are expected to be strongly influenced by the microemulsion-adsorbent interactions due to the effect of the microemulsion composition on the electrical properties of the adsorbent surface. It is well known (11) that when montmorillonite is brought into contact with water, the latter is adsorbed in the interlayer spacings to a thickness of several monolayers. A fully hydroxylated surface would then be expected with H^+ and OH^- as the potential-determining ions (12); the surface charge would then depend on the pH of the system. It is possible that when a brine-external (lower phase) microemulsion is brought into contact with the adsorbent, brine would be adsorbed in an analogous fashion. The adsorption of sulfonate anions from such a microemulsion would then depend in part on the number of positive sites on the clay particles. When montmorillonite is brought into contact with an oil-external (upper phase) microemulsion, the environment at the adsorbent would be predominantly oleic, although there may be some limited adsorption of brine from the dispersed phase. It is possible that in such a system the overall number of positive sites on the adsorbent would be less than in the brine-external microemulsion-montmorillonite system.

Such a difference in surface charge would account in part for the larger fraction of sulfonate adsorbed from the brine-external microemulsions.

In addition, if the sulfonate is adsorbed from a microemulsion where $\phi_o >> \phi_b$, then the adsorbed sulfonate would interact with a predominantly oleic environment; conversely, if the sulfonate is adsorbed from a microemulsion where $\phi_b >> \phi_o$, then it would interact with a predominantly aqueous environment. If the adsorbed sulfonate is inherently more soluble in oil than in water, as is the case with TRS 10-410 and PDM-334, then its tendency to leave the surface and re-solubilize would be much greater if the surrounding environment consisted mostly of oil. Therefore the relative adsorption of an oil-soluble sulfonate would be larger from a brine-external microemulsion than from an oil-external one.

This model accounts for the observation that the fraction of adsorbed sulfonate increases with microemulsion brine content within each of the TRS 10-410/glycol ether, PDM-334/glycol ether, and PDM-334/butoxytriglycol microemulsion series. Sulfonate adsorption increases with increasing brine content of the microemulsion for two reasons: the sulfonate-adsorbent interactions are enhanced due to the greater positive charge on the surface of the adsorbent; the adsorbed sulfonate tends not to re-solubilize because of its inherently low solubility in brine.

The large fractions of PDM-334 adsorbed from the aqueous and nonane solutions point out the effect on adsorption of the strength of the interaction between the sulfonate and the environment from which it is adsorbed. In general, the interactions between a sulfonate anion and its nearest-neighbor molecules and ions in the microemulsion are much stronger than the attractive forces responsible for the solubility of the sulfonate in either nonane or water even in the presence of a cosurfactant. Consequently, adsorption from both aqueous and nonane solutions is much larger than from microemulsions.

The larger adsorptions from the TRS 10-410/n-butanol and PDM-334/n-butanol microemulsions may be due to weaker interactions between the respective sulfonate anion and its nearest-neighbor molecules and ions as a result of the very low brine content in these microemulsions; these microemulsions exhibit inordinately high interfacial tensions versus brine. It appears that adsorption from the n-butanol microemulsions may begin to approach the adsorption behavior exhibited by the nonane sulfonate solutions. It may be that the assumption that sulfonate-microemulsion interactions are equivalent within each microemulsion series such as A, B, C and D holds adequately only for the glycol ether microemulsions.

We consider finally the significance of the experimental observation that the microemulsion specific refraction always increases as a result of adsorption. The direction of change in r indicates that the oil to brine ratio has increased. This observation is in agreement with the assumption that brine is adsorbed from all the microemulsions. There is, however, an additional reason for the observed increase in the microemulsion specific refraction. An increase in r can be qualitatively effected for example in any one of the PDM-334/butoxytriglycol microemulsion in the absence of an adsorbent by any one of the following compositional changes: a simultaneous decrease in the amount of both PDM-334 and butoxytriglycol in the microemulsion; a decrease in the amount of butoxytriglycol with no change in the amount of PDM-334; an increase in the amount of PDM-334 with no change in the amount of butoxytriglycol. The last two violate the experimentally observed decrease in PDM-334 due to adsorption and are therefore unrealistic. Consequently, the increase in microemulsion specific refraction as a result of the adsorption experiment appears to be due, at least in part, to the loss of both surfactant and cosurfactant from the microemulsion phase.

CONCLUSIONS

Static petroleum sulfonate adsorption on simulated reservoir solids is affected significantly by the composition of the microemulsion from which the sulfonate is adsorbed. Petroleum sulfonates which are preferentially oil-soluble tend to be adsorbed to a lesser degree from oil-external than from brine-external microemulsions under equivalent conditions. Since the cosurfactant component can be used effectively to obtain a desired microemulsion composition, sulfonate adsorption is affected significantly by the choice of cosurfactant. If an oil-external microemulsion is to be used in a flood, it is preferable to employ a glycol ether cosurfactant, rather than a simple alcohol, because of lower surfactant adsorption; in addition, the glycol ether microemulsion would exhibit enhanced salt tolerance which would affect favorably its interfacial tension properties.

Finally, the microemulsion specific refraction can be correlated with the fraction of sulfonate adsorbed because it is a sensitive function of the relative oil and brine content of the microemulsion. Its utility should also extend to dynamic adsorption experiments. The specific refraction is therefore an extremely useful microemulsion property in adsorption studies as well as in terms of its correlation with microemulsion interfacial tensions versus equilibrated excess phases.

ACKNOWLEDGMENTS

We would like to thank Dr. T.B. MacRury for a critical reading of the manuscript and Dr. J.C. Hatfield for helpful discussions.

REFERENCES

1. W. W. Gale and E. I. Sandvik, Soc. Pet. Eng. J., 191 (1973).
2. J. B. Lawson and R. E. Dilgren, Soc. Pet. Eng. J., 75 (1978).
3. F. J. Trogus, T. Sophany, R. S. Schechter, and W. H. Wade, Soc. Pet. Eng. J., 337 (1977).
4. A. W. Adamson, "Physical Chemistry of Surfaces," Interscience, New York, 1967.
5. P. Somasundaran and H. S. Hanna, SPE Paper 7059, SPE Symposium on Improved Oil Recovery, Tulsa (1978).
6. C. J. Glover, M. C. Puerto, J. M. Maerker, and E. I. Sandvik, SPE Paper 7053, SPE Symposium on Improved Oil Recovery, Tulsa (1978).
7. S. P. Trushenski, in "Improved Oil Recovery by Surfactant and Polymer Flooding," D. O. Shah and R. S. Schechter, eds., Academic Press, New York, 1977.
8. E. J. Derderian, J. E. Glass, and G. M. Bryant, J. Colloid Interface Sci., 68, 184 (1979).
9. R. N. Healy and R. L. Reed, Soc. Pet. Eng. J., 129 (1977).
10. CRC Handbook of Chemistry and Physics, 57th edition, 1976-1977.
11. B.K.G. Theng, "The Chemistry of Clay-Organic Reactions," John Wiley & Sons, New York, 1974.
12. P. Somasundaran and D. W. Fuerstenau, J. Phys. Chem., 70, 90 (1966).

ADSORPTION OF PURE SURFACTANT AND PETROLEUM SULFONATE AT THE SOLID-LIQUID INTERFACE

J. Novosad

Petroleum Recovery Institute
Research Division
Calgary, Alberta, Canada T2L 2A6

This paper is concerned with the retention of surfactants considered for tertiary oil recovery in reservoir rocks.

The first part includes a fundamental analysis of experimental variables which shows that a maximum in the surfactant adsorption isotherm should be expected. Advantages and disadvantages of using experimental variables such as surface excess, amount adsorbed and selectivity are discussed and the relationships among these are developed. For example, it is shown that selectivity is the most suitable variable for extrapolation of experimental data.

The second part discusses the adsorption of surfactant mixtures. Difficulties in analyzing experimental data are pointed out and a method for determining adsorption of surfactant mixtures from the data on adsorption of individual surfactants is suggested. Calculations showing that the overall isotherm may be quite different from the isotherms of individual surfactants are performed.

The third part is concerned with solubilities of surfactants in brines in the presence of alcohol cosurfactants. Experimental data presented indicate that quite large amounts of alcohol are needed to keep surfactants fully dissolved in brine solutions.

The fourth part includes experimental results on adsorption of pure surfactant and petroleum sulfonates on Berea sandstone. Retention of surfactants is related to their solubility limits in the brine.

INTRODUCTION

Success or failure of a surfactant flood may depend on the degree of retention of surfactants during the course of the flood and one of the possible mechanisms of surfactant retention is the adsorption at the solid-liquid interface. Several papers dealing with the adsorption of commercially available surfactants have been published (1,6,13,14) but a meaningful comparison of reported data is quite difficult since surfactants of various degrees of purity have been used and the concentration ranges, brine salinities, and temperatures have varied considerably. In addition, comparisons of the results are difficult because different experimental techniques have been employed.

It is the objective of this paper to discuss fundamental aspects of the thermodynamics of adsorption at the solid-liquid interface, with emphasis on providing proper definitions of experimental variables such as the surface excess, selectivity, amount adsorbed, and the relationships among them. Types of surfactant adsorption isotherms for binary systems are discussed, and it is shown that an extreme caution must be taken when interpreting isotherms for surfactant mixtures. It is hoped that this discussion will facilitate a better understanding and interpretation of experimental results reported in the literature.

Experimental results include adsorption data for systems containing surfactant, cosurfactant, brine, and Berea sandstone. Adsorption values of pure surfactant (Texas #1) and commercially available petroleum sulfonates (TRS 10-80) are reported and the retention of surfactants in the Berea rock is related to a surfactant solubility limit in the brine.

EXPERIMENTAL VARIABLES

Adsorption isotherm--binary systems: Surfactant adsorption from a liquid onto a solid surface is usually determined from a change of surfactant concentration in the bulk liquid phase after contact of the liquid with the solid adsorbent:

$$n_1^e = n^o \,(x_1^o - x_1) \qquad 0 \le x_1 \le 1 \qquad (1)$$

where n_1^e = surface excess of component 1 (g)
n^o = mass of the liquid phase (g)
x_1^o = original concentration of the component 1 (fraction)
x_1 = equilibrium concentration of the component 1 in the bulk phase (fraction)

Equation (1) defines the surface excess as a thermodynamic quantity which is determined from the difference in concentrations of the bulk phase before and after adsorption takes place. There is neither difficulty nor ambiguity in its experimental determination because no measurement is required in the adsorbed phase or in the interface region.

The surface excess should not be equated to the amount of surfactant adsorbed which is defined as the amount of surfactant present in the adsorbed phase (10). The distinction between these two variables is shown by noting the following material balances for the liquid phase:

$$n^o = n + n' \tag{2}$$

$$n_1^o = n_1 + n_1' \tag{3}$$

where the (') denotes the adsorbed phase
n = mass of the bulk phase (g)
n' = mass of the adsorbed phase (g)

Combination of Equations (1), (2) and (3) leads to a useful expression for the surface excess:

$$n_1^e = n' \, (x_1' - x_1) \tag{4}$$

Equation (4) shows that the surface excess is a relative measure of adsorption. It is the excess of component 1 in the adsorbed phase over the hypothetical amount of the same component which would be there if the concentration of the bulk phase was uniform all the way to the solid surface (e.g. for the case where there is no effect of the solid on the liquid phase). This is shown schematically in Figure 1.

Even though the surface excess is the fundamental variable describing adsorption, it is the amount adsorbed which is of practical interest and the relationship between these quantities is expressed in the rearranged equation (4).

$$n_1' = n_1^e + n' x_1 \tag{5}$$

It is important to note that in order to calculate the amount adsorbed n_1', the amount of liquid in the adsorbed phase, (n'), must be known. Since this quantity is not subject to a direct measurement it should be recognized that the amount adsorbed is not a unique variable and that it always should be related to a specific definition of the adsorbed phase or to a specific experimental technique used for separating the solid and the adsorbed phase from the bulk liquid phase. This point is quite obvious from Figure 1 as a shift in the dividing line between adsorbed and bulk phases would not change the value of the surface excess, but the amount adsorbed would be changed.

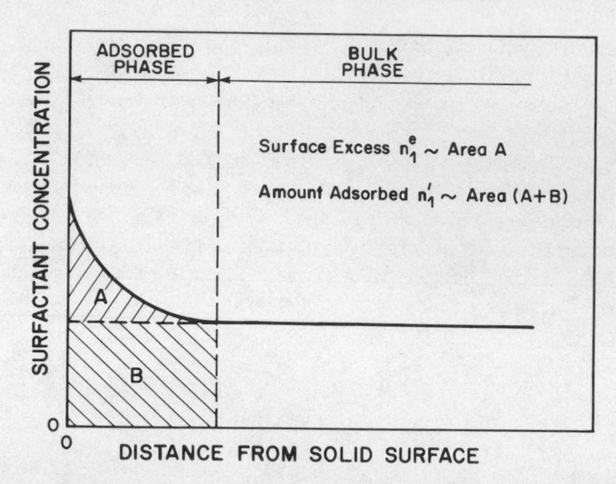

Fig. 1. Schematic presentation of the surface excess and the amount adsorbed.

The difference between the amount adsorbed, n_1', and the surface excess, n_1^e, is important in some cases but may be negligible in others. On one hand, for a low concentration of component 1:

$$n_1' \simeq n_1^e \quad \text{because} \quad x_1 \to 0 \tag{6}$$

On the other hand, the difference will be accentuated in cases where there is a weak specific uptake of one component into the adsorbed phase because the contribution of n_1^e to n_1' will be small and the second term, on the right-hand side of Equation (5), will contribute more to the amount adsorbed. Then approximation (6) cannot be used and Equation (5) must be utilized and therefore, a definition of the adsorbed phase, (n'), must be specified. This is commonly done by the acceptance of the monolayer concept of adsorption from liquid solutions. It is assumed that only one layer of molecules covering the solid surface is affected by the solid and hence only this monolayer differs from the bulk liquid phase. This concept allows the number of molecules present in the adsorbed phase to be calculated based on a known specific area of solids and known cross-sectional areas of molecules of the adsorbate and the solvent:

$$\frac{1}{n'} = \frac{x_1'}{m_1} + \frac{x_2}{m_2} \tag{7}$$

where m_1 and m_2 are amounts of components 1 and 2 needed for a full monolayer coverage of the surface of the adsorbent.

It should be noted that the differentiation between the surface excess and the amount adsorbed is a general phenomenon in adsorption as it is applicable even to adsorption at the solid-gas interface. However, in this latter case, the difference in density of the adsorbed phase and the bulk phase is so large that the dividing line between these two phases is always uniquely placed. Also, the contribution of the bulk phase to the amount adsorbed ($n'x_1$) is negligible due to the much lower density of the bulk phase, and thus the surface excess is, for all practical purposes, equal to the amount adsorbed. Since unique definition or separation of the adsorbed and the bulk phases at the solid-liquid interface is impossible, the values of the amount adsorbed depend on the experimental procedures used to separate the solid from the liquid phase.

A typical adsorption isotherm for two completely miscible components in terms of the surface excess and the amount adsorbed is shown in Figure 2. It can be seen from Figure 2 that the surface excess, even though it is a fundamental variable which is always measured, may not be the most convenient variable to work with. For example, since it is equal to zero at $x_1 = 0$ and $x_1 = 1$, the surface excess must go through a maximum, making it a difficult variable to extrapolate.

A more convenient variable to use is the selectivity S_{12} defined by (10):

$$S_{12} = \frac{x_1'/x_2'}{x_1/x_2} \qquad (8)$$

For positive adsorption of component 1 ($n_1^e > 0$), $S_{12} > 1$; for negative adsorption ($n_1^e < 0$), $S_{12} < 1$; for no adsorption ($n_1^e = 0$) (concentrations in the surface layer are equal to those in the bulk phase), $S = 1$. A relation between the surface excess and the selectivity can be obtained by combining Equations (4) and (8):

$$n_1^e = \frac{n' x_1 x_2 (S_{12} - 1)}{S_{12} x_1 + x_2} \qquad (9)$$

Equations (8) and (9) indicate that the selectivity is related to a specific model of adsorption in the same way as the amount adsorbed as the adsorbed phase must be defined in order to calculate concentrations in it.

The combination of Equations (7) and (9) leads to a relation between the surface excess and the selectivity for the monolayer model of adsorption:

$$n_1^e = \frac{m_1 x_1 x_2 (S_{12} - 1)}{S_{12} x_1 + x_2 \, m_1/m_2} \qquad (10)$$

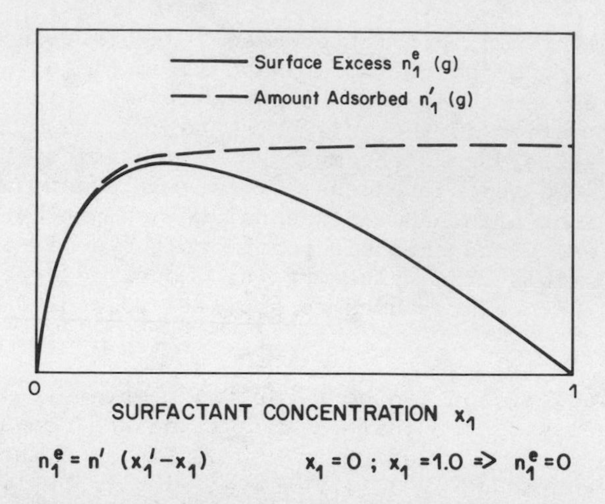

Fig. 2. Adsorption isotherm for fully miscible liquids.

Figure 3 shows three classes of adsorption isotherms and the related selectivities calculated from Equation (9). Advantages of using the selectivity for extrapolations are obvious.

A literature survey of surfactant adsorption data indicates that the most common isotherms are the ones in which the selectivity is declining with increasing surfactant concentration and that adsorption isotherms cover a very limited concentration range because the surfactants have low solubility in brines. Some of the results indicate the presence of a maximum in the isotherm, while some show only a monotonic increase in adsorption with increasing concentration. Extensive efforts have been made to explain the presence of the maxima in some of the isotherms.

It was previously shown that the surface excess must go through a maximum, so that the emphasis in discussions about the presence of maxima in surfactant adsorption isotherms should be directed to whether the maximum in the isotherm is present within the measured concentration range.

Adsorption of surfactant mixtures: Several authors independently reached a conclusion that it is necessary to consider commercially available surfactants as mixtures of surfactants in order to explain some of the results of coreflooding experiments. Suffridge (12) concluded from studies of elution of adsorbed petroleum sulfonates that there are substantial differences in adsorption behavior of monosulfonates and disulfonates. Trogus et al. (13) attempted to explain maxima in adsorption isotherms using theory based upon the assumption of two surfactants with different critical micelle concentrations in the solution. Gale and Sandvik (4) showed that higher molecular weight sulfonates were adsorbed preferentially on the Berea sandstone, while Somasundaran (11)

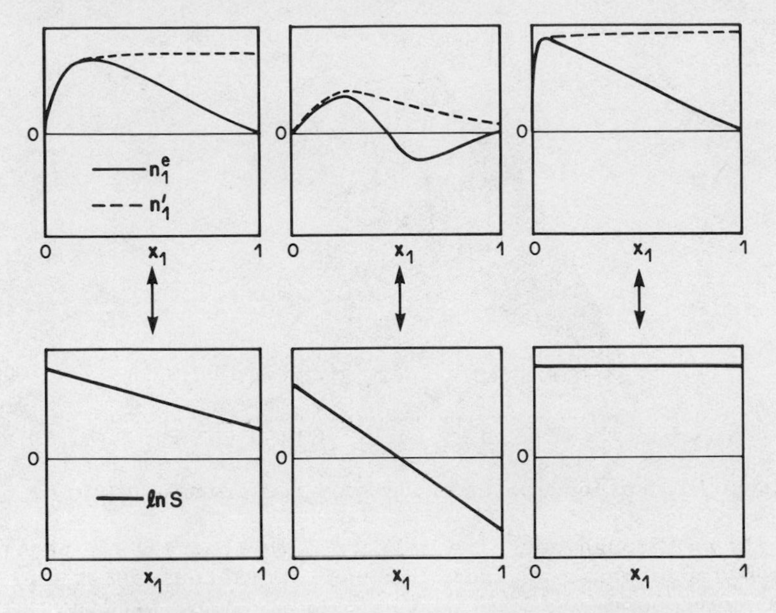

Fig. 3. Three classes of adsorption isotherms.

recently proposed that the so-called structure-making counterions are responsible for maxima in adsorption isotherms.

Consideration is now given to the experimental variables relating to the adsorption of surfactant mixtures. The adsorption isotherm of component 1 in a ternary system is again defined by Equation (1). However, the isotherm is now three-dimensional and for a consistent two-dimensional presentation, a constant ratio of x_2/x_3 seems to be a logical choice (components 1 and 2 are surfactants, component 3 is water or brine). Figure 4 represents a ternary adsorption isotherm surface.

Since concentrations x_1, x_2, x_3 represent equilibrium values (i.e. concentrations in the bulk phase after adsorption takes place) it is impossible to prepare the original samples of ternary solutions in such a way that the x_2/x_3 ratio stays constant without prior knowledge of the adsorption isotherm. This is the reason that adsorption isotherms seem to depend on the solid/liquid ratio in the system. An increase in the amount of the solid phase increases the total amount of surfactants adsorbed, which results in a change of x_2/x_3 ratio and a shift of the experimental point on the adsorption isotherm surface. Obviously, this effect is more pronounced in systems with large differences in individual surfactant adsorption characteristics.

Evaluation of flow-type experiments is even more difficult because the x_2/x_3 ratio changes continuously as the surfactant

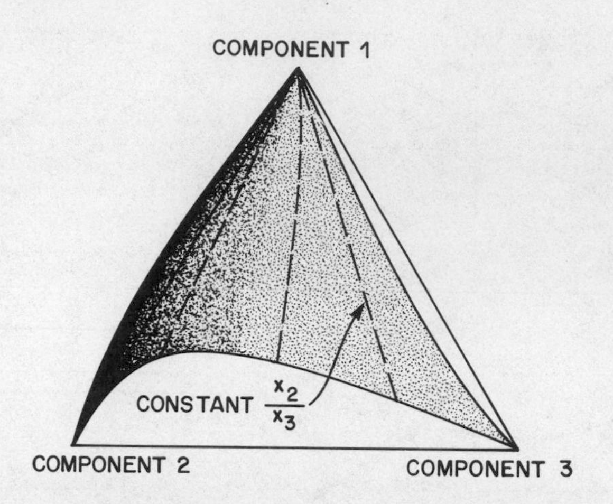

Fig. 4. Adsorption isotherm surface for ternary systems.

solution flows through the core. It follows that flow experiments (especially those where a finite slug of surfactant solution is injected) yield some averaged values of adsorption because the x_2/x_3 ratio may vary in a complicated way across the surface of the isotherm.

Minka and Myers (8) have extended the concept of surface excess and selectivity to multicomponent mixtures. They applied a theory of an ideal adsorbed phase to predict the adsorption behavior of ternary mixtures from adsorption measurements in binary systems. Having binary data in the form of Equation (10) a ternary isotherm is calculated as follows:

$$(n_1^e)_t = \frac{x_1 x_2 (S_{12} - 1) S_{13} + x_1 x_3 (S_{13} - 1) S_{12}}{S_{12} S_{13} \frac{x_1}{m_1} + S_{13} \frac{x_2}{m_2} + S_{12} \frac{x_3}{m_3}} \quad (11)$$

$$(n_2^e)_t = \frac{x_1 x_2 \left(\frac{1}{S_{12}} - 1\right) S_{13} + x_2 x_3 \left(\frac{S_{13}}{S_{12}} - 1\right)}{S_{13} \frac{x_1}{m_1} + \frac{S_{13}}{S_{12}} \frac{x_2}{m_2} + \frac{x_3}{m_3}} \quad (12)$$

Most literature data on adsorption of surfactants which are being considered for surfactant flooding report batch experiments in which the total surfactant adsorption $[(n_1^e)_t + (n_2^e)_t]$ is measured as a function of total surfactant concentration. This is in effect measuring the surfactant adsorption isotherm in such a way that the ratio of the original surfactant concentrations, x_1^o/x_2^o, is constant rather than the ratio of the equilibrium concentrations, x_2/x_3. Therefore, isotherms for different surfactant systems are measured along a different path on the surface of the ternary isotherm, making comparisons of adsorption values of various surfactants impossible.

Examples of calculated isotherms which could be obtained from such an experimental arrangement are shown next. Calculations involve a solution to the following six equations: Equations (11), (12) and

$$C = x_1^o / x_2^o \tag{13}$$

$$n_1^e = \frac{m_1 x_1 x_3 (S_{13} - 1)}{S_{13} x_1 + \frac{m_1}{m_3} x_3} \tag{14}$$

$$n_2^e = \frac{m_2 x_2 x_3 (S_{23} - 1)}{S_{23} x_2 + \frac{m_2}{m_3} x_3} \tag{15}$$

$$x_1 + x_2 + x_3 = 1 \tag{16}$$

Equations (14) and (15) are employed to calculate selectivities S_{13} and S_{23} from known binary adsorption data (n_1^e, n_2^e) and the constant C is determined from the original composition of the surfactant mixture. Some examples of the possible shapes of isotherms determined by the described procedure are given in Figures 5 and 6, and parameters used in the calculations are listed in Table 1.

Table 1. Parameters Used in Determining Isotherms Shown in Figures 5 and 6

Parameter (units)	Figure 5	Figure 6
$C = x_1^o/x_2^o$	1.0	1.0
n^o/m (g/g)	1.0	1.0
S_{12}^o	10.0	10.0
S_{13}^o	1.10^6	1.10^6
$m_1 = m_2$ (mg/g)	3.1	3.1
m_3 (mg/g)	0.36	0.36
A	-180.0	-180.0
B	-120.0	- 20.0
$\ln S_{ij} = Ax_1^2 + Bx_1 + S_{ij}^o$		

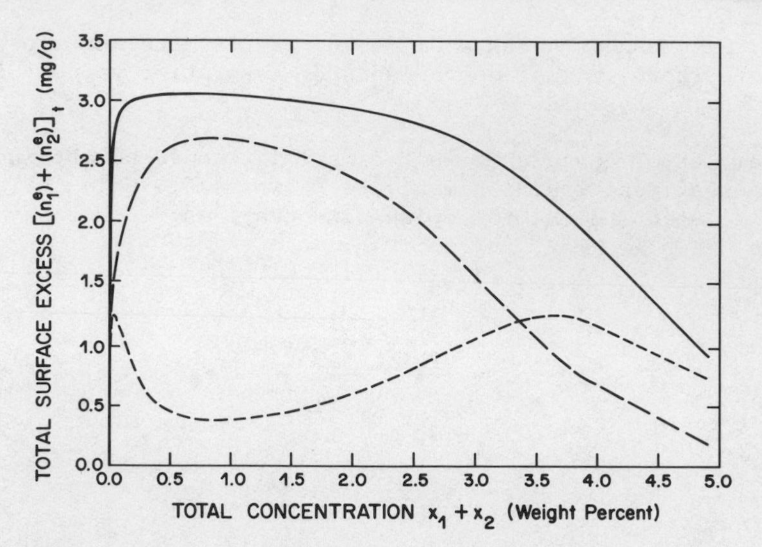

Fig. 5. Adsorption isotherm for surfactant mixture (— overall isotherm; --- individual surfactants).

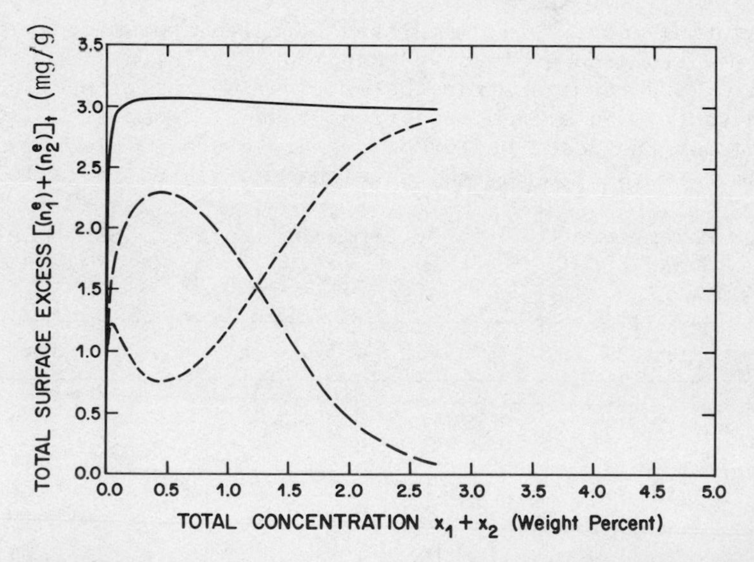

Fig. 6. Adsorption isotherm for surfactant mixture (— overall isotherm; --- individual surfactants).

It is interesting to observe that the overall isotherm in Figure 5 has a maximum around 0.5% total surfactant concentration but that one surfactant shows two maxima and a minimum. Even more interesting is the isotherm in Figure 6 which exhibits a rather sharp rise at very low concentrations and then levels off, although the adsorption isotherms of the individual surfactants do not even resemble the overall isotherm.

These examples indicate a need for very careful examination of surfactant adsorption data in terms of the adsorption of individual components of surfactant mixtures. Also, if only one surfactant in the mixture produces ultralow interfacial tensions, then its own isotherm may have a more important effect on the results of a surfactant flood than the overall isotherm.

In summary, just the acceptance of the fact that commercially available surfactants are mixtures of various molecular weight surfactants may explain the unusually shaped adsorption isotherms that have been observed.

EXPERIMENTAL

Adsorption isotherms have been measured for pure surfactants and commercially prepared mixtures of petroleum sulfonates. Batch and flow experiments have been performed.

Chemicals: Sodium 8-phenyl-n-hexadecyl-p-sulfonate (Texas #1) was obtained from Professor Wade of the University of Texas and has been used as received. According to Frances et al. (2), the purity of the sample exceeds 98%.

Witco's TRS 10-80 was used as an example of a commercially available petroleum sulfonate. Samples were desalted and deoiled according to the procedures described by Shah et al. (9).

Determination of adsorption isotherms--batch experiments: Berea sandstone was crushed into 2-3 mm pieces and placed in an oven at 110°C for more than 12 hours. The cooled rock was immersed in surfactant solutions at room temperature. The containers were 15 ml centrifuge tubes closed by Teflon-lined caps and the ratio of liquid to solid, (n^o/m), was kept approximately equal to unity for all samples. The samples were placed in a thermostatically controlled air box and shaken for 24 hours. They were then left standing until the liquid phase cleared. Small aliquots of the liquid phase were withdrawn with a syringe and diluted with distilled water to a concentration level suitable for a UV spectrophotometric analysis of the surfactant concentration (about 20 mg of surfactant/1).

Spectra scans were taken with a Varian Superscan III in the range of 380-200 nm and the peak heights at 208 nm and 236 nm were measured to determine surfactant concentrations. New calibration curves were prepared for each experiment and each analysis was repeated three times. It is estimated that the average error in the surfactant concentration measurements was between 1-2%.

The error in the surface excess is, of course, substantially larger because adsorption levels are quite low and concentration changes due to adsorption are therefore small. An example of uncertainties in the measurements of surface excess for surfactant systems caused by errors in analytical procedures is shown in Figure 7. This error analysis shows clearly that batch methods should not be used for measuring surfactant adsorption on Berea sandstone for surfactant concentrations above 1% unless extremely accurate analytical procedures are developed.

Flow-type measurement of adsorption isotherms--10, 20, and 30 cm long pieces of square Berea core (2.5 x 2.5 cm) were cut, dried, and cast in epoxy. Several pore volumes of NaCl brine were injected into the cores and the pore volumes were determined from differences in core weights. Surfactant solutions were then injected into the cores at linear velocities not exceeding 1.25 cm/hour (1 ft/day) and samples were collected in enclosed vials since the early experiments showed that a substantial error occurs due to the evaporation of the solution if the outlet of the core is left open to the air. Volumes of solutions injected into the cores were measured and the concentration of surfactant in each sample was determined.

Two types of experiments were performed. In the first, surfactant was injected into the core continuously at a constant concentration and the surface excess was determined from the delayed arrival of the surfactant at the core outlet. In the second one, a slug of surfactant was injected, followed by the injection of the brine and a loss of surfactant was calculated from a surfactant

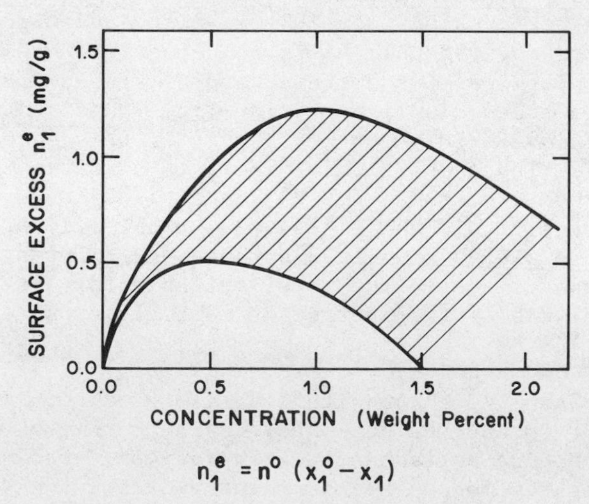

Fig. 7. Uncertainty in surface excess determination caused by 3% error in surfactant concentration measurements.

material balance. As mentioned earlier, the latter experiment yields some average value of adsorption because the surfactant concentration in the core is not kept constant. In addition, the cosurfactant and the surfactant chromatographically separate and this may result in the solubility limits of the surfactant being exceeded during the experiment. It will be shown that this affects the results of adsorption experiments substantially.

Surfactant solubility determination: Solubility limits of surfactants in brine solutions in the presence of alcohol cosurfactants were determined from visual observations of the surfactant solutions and by a spectroturbidimetric method similar to the one described by Frances et al. (2). The visual observation consisted of centrifuging surfactant solutions in a table-top centrifuge at room temperature and measuring the amount of the surfactant sediment at the bottom of the tube. Surfactant was not considered soluble in a given brine if any surfactant sediment was observed.

The spectroturbidimetric method confirmed the results obtained from visual observations as the turbidity of solutions measured at 436 nm disappeared for solutions producing no surfactant sediment when centrifuged.

DISCUSSION

All experiments have been performed in parallel for pure surfactants and the petroleum sulfonate mixture. No substantial differences between these two types of surfactants were observed in the experiments reported in this paper.

Adsorption isotherms for TRS 10-80 and Texas #1 surfactant are shown in Figures 8 and 9. Experimental points indicated by circles were obtained from batch experiments, and triangles indicate results of flow experiments. There is excellent agreement between these two methods of measuring adsorption.

A most noticeable feature is the maximum in both adsorption isotherms. Since the maximum is present even in the isotherm for the pure surfactant it can be explained only by accepting the idea of declining selectivity with increasing surfactant concentration. Selectivity values for TRS 10-80 surfactant, calculated from Equation (10) with monolayer values determined from cross-sectional areas of surfactant ($22Å^2$) and water ($8.3Å^2$) molecules, are shown in Figure 10. The specific area for Berea sandstone was assumed to be $1\ m^2/g$.

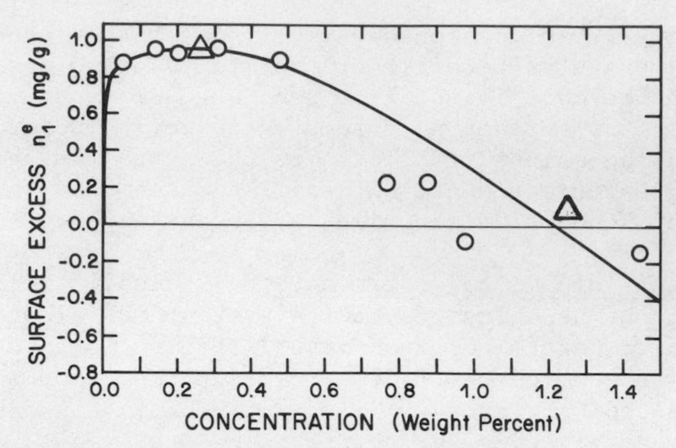

Fig. 8. Adsorption isotherm for 1/10 TRS 10-80/sec-butylalcohol in 1% NaCl brine on Berea sandstone at 22°C.

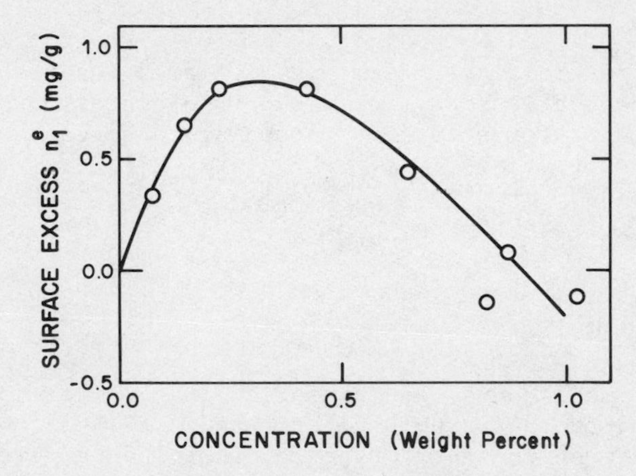

Fig. 9. Adsorption isotherm for 1/10 Texas 1/sec-butylalcohol in 1% NaCl brine on Berea sandstone at 22°C.

It should be noted that the adsorption experiment described above involves systems that only contain surfactant, cosurfactant, brine and reservoir rock. There is no oil present in the system, and this is the reason why the ratio of cosurfactant to surfactant is higher than usually reported by other researchers. It was found that the higher ratio was necessary for complete dissolution of surfactants in the brine since a lower alcohol/surfactant ratio caused the solutions to become cloudy. This latter condition resulted in substantial increases in surfactant retention in flow experiments. The observation of the relationship between the solution condition and retention lead to experiments which could better define the phenomenon.

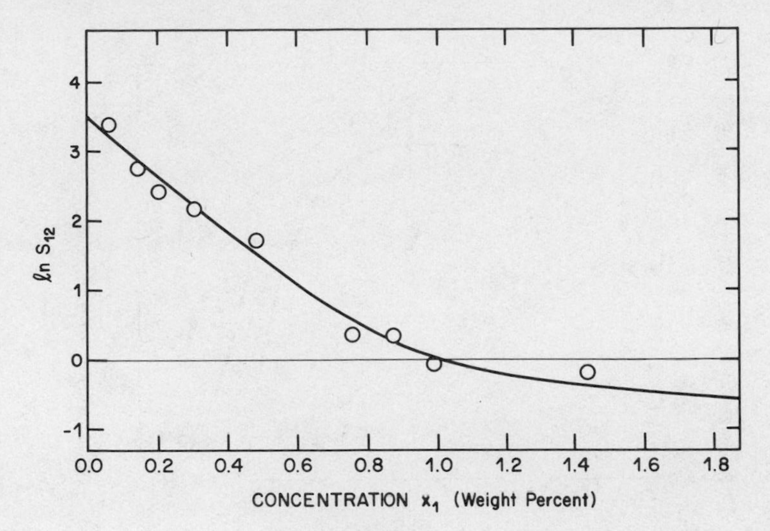

Fig. 10. Selectivity as a function of surfactant concentration for 1/10 TRS 10-80/sec-butylalcohol in 1% NaCl brine on Berea sandstone at 22°C.

The initial experiments involved preparing surfactant solutions in brine and centrifuging them for about 30 minutes in a table-top centrifuge at approximately 3000 rpm. If a surfactant sediment was observed at the bottom of the centrifuge tube, a small amount of secondary butylalcohol was added to the solution and the system was centrifuged again and this was repeated until no surfactant was being centrifuged out. The amount of alcohol in the system at this point was considered to be the minimum alcohol content needed for a complete surfactant dissolution. It was found that this point coincides with the limit of surfactant solubility as measured by spectroturbidimetry. It is assumed that surfactant solutions containing less than the minimum amount of alcohol needed for complete surfactant dissolution contain surfactant in a dispersed state. Figures 11 and 12 shows results of solubility measurements for TRS 10-80 and Texas #1 surfactants.

A comparison of adsorption characteristics of surfactant solutions above and below their solubility limits was then performed. Solutions of the same surfactant concentrations containing different amounts of alcohol were injected into Berea cores and surfactant breakthrough curves were determined. The results are shown in Figures 13 and 14.

The difference in surfactant retention is striking. Surfactant concentration at the core outlet has not reached the injected concentration even after 3 pore volumes of a dispersed surfactant solution were injected into the core. The retention of surfactants from the dispersed solution is one order of magnitude higher

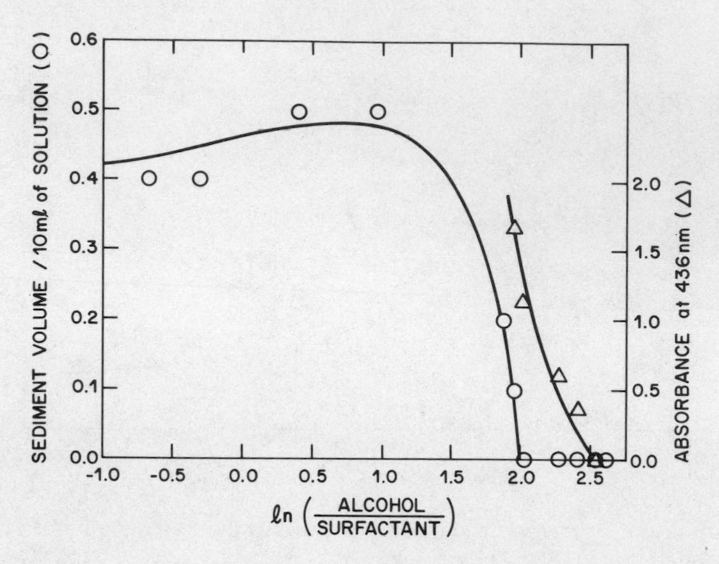

Fig. 11. Solubility of 1% TRS 10-80/sec-butylalcohol in 1% NaCl at 22°C.

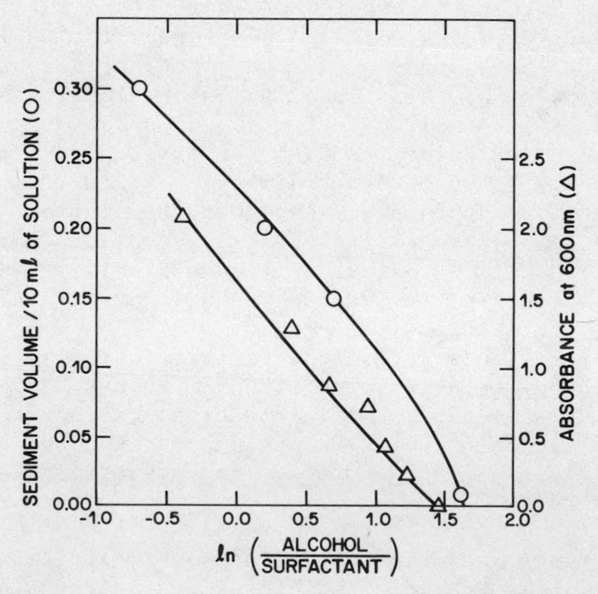

Fig. 12. Solubility of 1% Texas 1/sec-butylalcohol in 1% NaCl at 22°C.

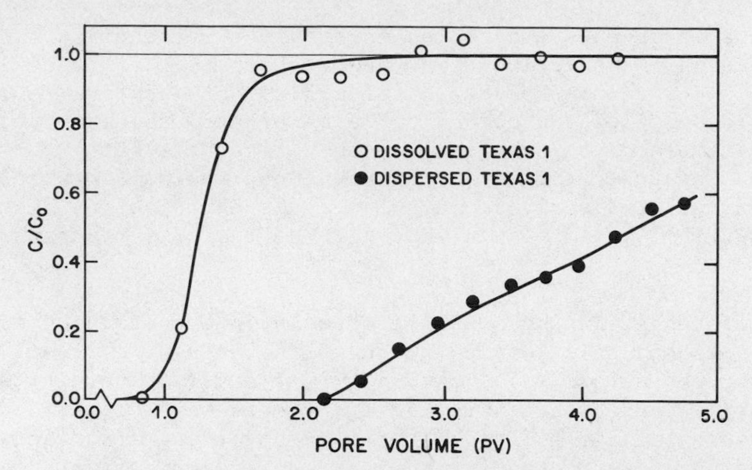

Fig. 13. Retention of 0.26% Texas 1 solution in 1% NaCl in Berea sandstone at 22°C (dissolved solution contains 10% sec-butylalcohol, dispersed solution contains 4% sec-butyl-alcohol).

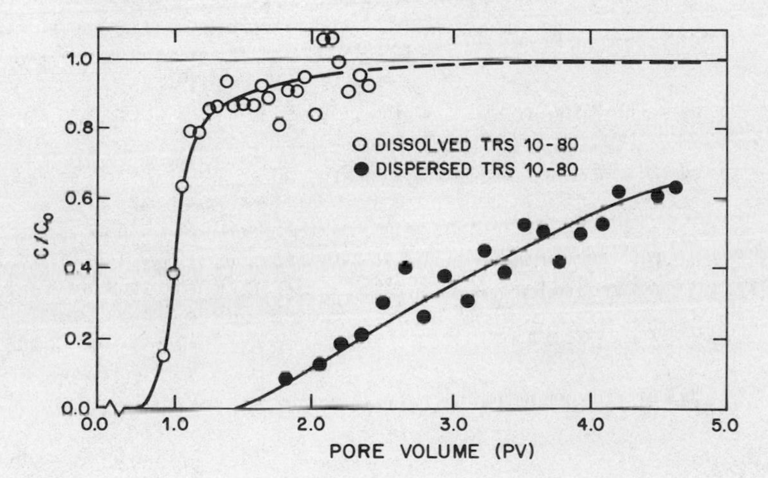

Fig. 14. Retention of 0.3% TRS 10-80 solution in 1% NaCl in Berea sandstone at 22°C (dissolved solution contains 7% sec-butylalcohol, dispersed solution contains 3% sec-butyl-alcohol).

than the retention of surfactant from a solution in which the surfactant is completely dissolved.

It has been shown by Glover et al. (5) that the phase behavior of microemulsion systems can affect the retention of surfactants in porous media more than their physical adsorption on the solid surface. Similarly it is suggested here that a loss of surfactants in porous media may be affected more by the state of surfactant molecules in the solution than their adsorption at the solid-liquid interface. Data on surfactant solubility show that quite large amounts of alcohol are needed to keep surfactants dissolved and since alcohols distribute themselves between the brine and the oil it could happen that in a surfactant flood the surfactants, which were originally dissolved, may become dispersed as the flood progresses. This would result in a high degree of surfactant retention.

Differences in the behavior of solutions containing dispersed surfactants and solutions in which surfactants were dissolved have been observed before. Batycky and McCaffery (15) have noted that long term interfacial tension aging effects were eliminated upon the addition of alcohol to the brines containing a surfactant. In the studies of surfactant systems in the absence of cosurfactants, Scriven and Davis (3) observed ultralow interfacial tensions only for surfactant systems containing a third "surfactant-rich" phase at the interface. They suggest that the dispersed and not the dissolved surfactant is responsible for the ultralow tensions. This would imply that two competing requirements must be satisfied for effective surfactant systems. On one hand, surfactant dispersion at the interface is needed for the ultralow interfacial tensions but, on the other, a complete solubility of surfactants in brine is required in order to keep their retention at low levels.

The results shown in Figure 15 indicate that relatively smaller amounts of alcohol are needed to keep surfactants dissolved in higher concentration solutions. This combined with the fact that surfactant surface excess is also lower at higher concentration may be the reason for a better performance of high surfactant concentration--low pore volume floods when compared to low surfactant concentration--large pore volume floods.

In order to confirm the ideas expressed in this paper, it will be necessary to perform surfactant retention experiments in cores containing residual oil. Such experiments are now in progress.

CONCLUSIONS

1. Fundamental analysis of experimental variables for the adsorption at the solid-liquid interface shows that the adsorption isotherm in terms of the surface excess can have a maximum.
2. Selectivity is proposed as a useful variable for extrapolation of adsorption data.
3. Examples of calculated adsorption isotherms for surfactant mixtures indicate that both maxima and minima in isotherms are possible.

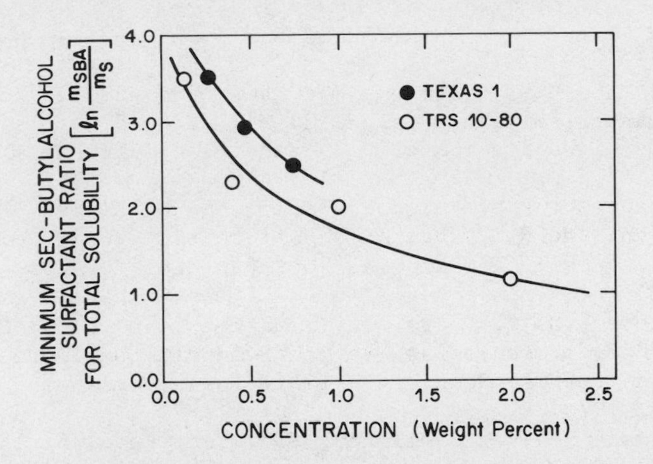

Fig. 15. Dependence of surfactant solubility on concentration.

4. Adsorption isotherms of individual components of a surfactant mixture may have completely different shapes than the overall isotherm.
5. Quite large amounts of alcohol cosurfactant are required to keep surfactants completely dissolved in brines, particularly for lower concentrations of the primary surfactants.
6. Losses of surfactants in porous media may be affected more by retention of dispersed surfactants than by its adsorption at the solid-liquid interface.

NOMENCLATURE

$C.$ = ratio of surfactant concentrations in the original surfactant mixture
m = mass of the solid adsorbent (g)
m_i = mass of component i needed for full monolayer coverage of the surface of the adsorbent (g)
n = mass of the bulk phase (g)
n' = mass of the adsorbed phase (g)
n^o = mass of the liquid phase (g)
n_i^e = surface excess of component i (g)
$(n_i^e)_t$ = surface excess of component i in ternary systems (g)
S_{ij} = selectivity
x_i = equilibrium concentration component i in the bulk phase (fraction)
x_i' = equilibrium concentration component i in the adsorbed phase (fraction)
x_i^o = original concentration of component i (fraction)

REFERENCES

1. J. H. Bae and C. B. Petrick, Soc. Pet. Eng. J., 17, 353 (1977).
2. E. I. Frances, H. T. Davis, W. G. Miller, and L. E. Scriven, 175th ACS National Meeting, Anaheim, California, March 13-17, 1978.
3. E. I. Frances, J. E. Puig, Y. Talmon, W. G. Miller, L. E. Scriven, and H. T. Davis, Report to DOE (EW-78-S-19-0002).
4. W. W. Gale and E. I. Sandvik, Soc. Pet. Eng. J., 13, 191 (1973).
5. C. J. Glover, M. C. Puero, J. M. Maerker, and E. I. Sandvik, SPE 7053, presented at the 5th Symposium on Improved Methods for Oil Recovery held in Tulsa, Oklahoma, April 16-19, 1978.
6. J. B. Lawson and R. E. Dilgren, Soc. Pet. Eng. J., 18, 75 (1978).
7. O. G. Larionov and A. L. Meyers, Chem. Eng. Sci., 26, 1025 (1971).
8. C. Minka and A. L. Meyers, AIChE J., 19, 453 (1973).
9. D. O. Shah and R. D. Walker, Jr., Semi-Annual Report, June 1977, University of Florida, Gainesville.
10. S. Sircar and A. L. Meyers, AIChE J., 17, 186 (1971).
11. P. Somasundaran and H. S. Hanna, SPE 7059, presented at the 5th Symposium on Improved Methods for Oil Recovery held in Tulsa, Oklahoma, April 16-19, 1978.
12. F. E. Suffridge and D. L. Taggart, SPE 6596, presented at the 1977 SPE Symposium on Oilfield and Geothermal Chemistry held in La Jolla, California, June 27-28, 1977.
13. F. Trogus, S. Thach, R. S. Schechter, and W. H. Wade, Soc. Pet. Eng. J., 17, 337 (1977).
14. S. P. Trushenski, D. L. Dauben, and D. R. Parrish, Soc. Pet. Eng. J., 14, 633 (1974).
15. J. P. Batycky and F. G. McCaffery, Research Note No. 6, Petroleum Recovery Institute, January 1978. Also Paper No. 78.29.26 presented at the 29th Annual Meeting of CIM in Calgary, June 13-16, 1978.

ION EXCHANGE ON MIXED IONIC FORMS OF MONTMORILLONITE AT HIGH IONIC STRENGTHS†

W. J. Rogers,* S. Y. Shiao, R. E. Meyer,
C. G. Westmoreland and M. H. Lietzke

Chemistry Division
Oak Ridge National Laboratory
Oak Ridge, Tennessee 37830, U.S.A.

Adsorption equilibria of inorganic ions between water and minerals in formations control brine composition and, through these, affect many aspects of the chemistry of surfactant solutions. For example, alkaline earth ions influence phase and interfacial properties of aqueous-hydrocarbon systems containing surfactants. Precipitation of surfactants may occur in the presence of alkaline earth ions. Thus information concerning the distribution of alkaline earth and alkali metals between solutions and minerals is basic to the prediction of behavior of surfactant solutions in reservoirs. Because many clays have large cation exchange capacity, they may dominate adsorption properties of those formations in which they are present. Sodium and calcium are very common and representative of the ions under consideration. This paper summarizes studies of the distribution of these ions between a common clay, montmorillonite, as well as several other clays, and a series of solutions of constant total ionic strength (I) with varying ionic strength fraction of sodium. Distribution coefficients D

---

*Ph.D. Candidate, University of Tennessee, Knoxville, Tennessee, supported in thesis work by an appointment to the Laboratory Graduate Participation Program, administered by Oak Ridge Associated Universities. Present address: TVA, River Oaks Building, Mussel Shoals, Alabama 35660.

†Research sponsored by Fossil Fuel Extraction, the U.S. Department of Energy, under contract W-7405-eng-26 with the Union Carbide Corporation.

for Na(I) and Ca(II) were determined by batch equilibrations using isotope dilutions with radioactive tracers. Equilibrium quotients (K/Γ) for the exchange of sodium and calcium were then calculated and the effects of solution composition, of solution phase activity coefficients, of ionic strength, of degree of purification, and of source of clay were investigated. Equilibrium quotients with adjustment for solution-phase activity coefficients did not vary greatly with I, except at low loading of sodium on the calcium form of montmorillonite, where D_{Na} became anomalously high. Values of K/Γ for illite and attapulgite were within an order of magnitude of those for montmorillonite.

INTRODUCTION

The importance of brine compositions to the performance of surfactant formulations in enhanced oil recovery is well known. Presence of multivalent ions can shift optimal salinities, change interfacial tensions, precipitate surfactants and polymers, and change the viscosity of polymer solutions. Divalent ions, particularly Ca(II) and Mg(II) are ubiquitous in brines, but in addition, the formation minerals normally constitute a reservoir of these and other multivalent ions, which may desorb under the influence of salinity changes or of the complexing properties of flooding chemicals. The ionic composition of solutions and the minerals in equilibrium with them can influence the adsorption of substances added to tag the flow of aqueous banks through formations and thus affect their utility as tracers. In order to anticipate possible effects, and to design control measures, preflushes for example, information is needed on the equilibria between solutions and formation minerals over the range of conditions likely to be encountered.

The range is indeed wide. Oil field brines vary from dilute brackish waters to virtually saturated salt solutions, and many different cations can be present. The variety of minerals which may be encountered is enormous. Variations of other parameters such as temperature and pH may also be important factors. Accumulation of information sufficient to predict in detail properties of any formation which becomes of interest appears to be a formidable task.

It may be, however, that by selecting the most important aspects, useful patterns can be delineated. It would appear reasonable to start with those substances having the highest ion-exchange capacity, to study solution-solid equilibria for the most common ions, to carry out measurements over a wide range of compositions, and to evaluate the extent to which results can be correlated by conventional adsorption equations, for example, those describing ion exchange.

We have attempted to follow this course. We have selected clays, particularly montmorillonite, for initial studies, and have concentrated on Na(I)/Ca(II) exchange, although we shall present some results for other ions for comparison. Most of the work reported here is for ionic strengths of 0.01 to 1 molal, but some measurements (1) at higher concentrations and with related pairs of ions, such as Na(I)/Sr(II), indicate that extrapolation to higher salinities can be done with reasonable confidence. We have in addition investigated the properties of montmorillonites from different sources and subjected to different degrees of purification. Along with comparison with literature data (2,3), these results allow us to infer indications of how much results may be distorted by contamination with other species and how much confidence one can have in applying results to a formation of interest.

EXPERIMENTAL

Materials: Our samples of clay minerals include the following:

A. Source clays, Department of Geology, University of Missouri.
   1. STx-1 Ca-Montmorillonite (white), Gonzales County, Texas.
   2. SWy-1 Na-Montmorillonite, Crook County, Wyoming.
   3. CMS-PF1-1 Attapulgite, Florida.
   4. Oklahoma Illite, Beaversbend.

B. Synthetic Montmorillonite, Industrial Chemicals Division, NL Industries, Inc., Houston, Texas.

The clays were sometimes used without purification. For other experiments, the sand fraction of the clay minerals was separated by slow-speed centrifugation of a suspension of the clay. The clay was then treated to remove insoluble carbonates by using 1 M acetate buffer (pH 5). For some samples, the complete Jackson procedure (4) was used in which, after the carbonate removal step, organic matter is removed by using 30% hydrogen peroxide and iron oxides by using sodium citrate and sodium dithionite. Finally the clay was converted to the appropriate form (calcium or sodium) by contacting the clay not less than three times with concentrated solutions of the appropriate chloride salt. The clay was then dried either in a vacuum desiccator or by freeze drying. Moisture content of the clay was determined from the loss in weight by heating a small portion of the clay at 110°C in an oven.

Solutions: Solutions were made by dilution of stock solutions of $CaCl_2$, NaCl, NaAc (Ac refers to acetate), and $CaAc_2$. The calcium solutions were allowed to stand overnight and were then filtered. Stock solutions of Ca(II) were analyzed by EDTA titration, the titrant being standardized against $CaCO_3$; sodium was analyzed by atomic absorption against standards made up from NaCl. Preparation of solutions from stocks was by weight. Adjustment of pH was by addition of acetic acid in most cases.

Most of the Na(I)/Ca(II) results reported here were obtained at pH 5, maintained by a 0.01 to 0.1 M acetate buffer. Our results indicate, in accord with the literature on complexing, that acetate had insignificant effects. An investigation of one composition as a function of pH indicated that Ca(II)/Na(I) equilibria are little affected by a change from pH 5 to 9. Consequently, although most of our solutions were more acidic than the usual oil-formation-brines, ordinarily neutral, the results should be applicable to cases of interest.

Methods: The measurements were carried out by batch equilibration, most by an isotope dilution technique. Samples of clay were pre-equilibrated several times with NaCl-$CaCl_2$ solutions of fixed compositions, until successive equilibrations showed no change in concentration. To separate samples (separate because of the difficulty of discriminating between the gamma emission of ^{22}Na and ^{47}Ca radioisotopes) were added known amounts of Na and Ca tracers, and the solutions were allowed to equilibrate for 2-5 days. The solid was centrifuged down, and aliquots of the supernatant solutions were counted. By material balance, the fractions of the activity adsorbed were computed, and, from these, the distribution coefficients, D

$$D \equiv \frac{(A^{a+})_{clay}}{(A^{a+})} = \frac{(n_A)_{clay}/w}{(n_A)/V} = \frac{[(n_A)_i - (n_A)]}{(n_A)} \frac{V}{w} \quad (1)$$

where (A^{a+}) is the concentration A^{a+} in solution, moles/liter.
$(A^{a+})_{clay}$ is the concentration on the solid, in moles/kg dry solid,
$(n_A)_{clay}$ and (n_A) are the counts of radioactive tracer on the solid and in the solution respectively at equilibrium (or the amounts of A, if analysis is done by other methods).
$(n_A)_i$ is the initial total counts in the aqueous phase.
w is the weight of dry solid in monoionic form in kg,
V is the volume of solution in liters.

Montmorillonite, especially the sodium form in low-ionic strength solutions, swells; consequently, centrifugation in moderately high fields is needed to separate the clay particles and the solutions. Clay suspensions were usually centrifuged with a DuPont Sorvall RC-5 refrigerated super-speed centrifuge with an angle-head rotor (FM-24) at 15,000 rpm (28,000g) for 15 minutes, and the supernatant was then withdrawn for analysis.

Erratic aspects in results to be reported led us to repeat some of the measurements by a slightly modified procedure. The isotope dilution sequence was essentially the same, except that no acetate was present (pH at equilibrium ranged from 5 to 6.5) and the supernatant solution after equilibration with the tracers was transferred to another tube and centrifuged a second time. Both centrifugations were for thirty minutes, and a barely perceptible deposit of solids was obtained in the second step. In addition to determination of D and capacity by isotope dilution, Na(I), Ca(II), and chloride were displaced from the clay packs by successive contact with 2 M NH_4NO_3 or KNO_3 solutions. These solutions were analyzed for the displaced ions by EDTA titration, by atomic absorption, and by a Büchler-Cotlove chloridometer, to allow computation of ion-exchange capacity. The clay (a Wyoming sample) had been treated with pH 5 buffer to remove carbonates. Sand had not been removed, but its quantity was known and its weight was subtracted from the weight of clay in computation of results. The sand was assumed not to adsorb Na(I) or Ca(II).

In experiments at trace level or very low loading of one ion, the isotope dilution procedure is not required. All that is necessary is that the exchangeable ions on the solid be those of the salt present in macro concentration. The computation of results, which followed that conventionally used in the determination of distribution coefficients, ignores possible effects of coion exclusion from water in the clay pack; it is implicitly assumed that the water in the pack has the same coion concentration as the supernatant solution. It is possible that coion exclusion has an effect which may be significant on distribution coefficients (particularly when they are low) and on capacities computed from the results. We shall discuss this interference later in connection with the capacities measured by displacement of both cations and anions from the clay pack and associated water. Coion exclusion has been further discussed in references 1 and 7 (see also 8, 9, and 10).

Equations: The distribution of electrolyte components AX_a and BX_b, X^- being a monovalent anion, between solution and a clay or other ion exchange phase, are governed by the conditions

$$a_{AX_a} = m_A m_X^a \gamma_{\pm AX_a}^{a+1} = k_A (a_{AX_a})_{clay} = k_A \left((A^{a+})(X^-)^a \right)_{clay} (\gamma_{\pm AX_a})_{clay}^{a+1} \tag{2}$$

$$a_{BX_b} = m_B m_X^b \gamma_{\pm BX_b}^{b+1} = k_B (a_{BX_b})_{clay} = k_B((B^{b+})(X^-)^b)_{clay}(\gamma_{\pm BX_b})^{b+1}_{clay} \tag{3}$$

where a is activity; m is concentration in the aqueous phase in terms of moles/kg of water; parentheses, ( ), are concentrations in the clay phase which could also be in terms of moles/kg water in the exchanger phase, but which in this paper are expressed as moles/kg of dry clay; $\gamma_{\pm}$ is the mean ionic activity coefficient of the electrolyte; and k adjusts for reference states in the two phases, if selected to be different. From the quotient obtained by dividing equation (2) raised to the bth power by equation (3) to the ath, the expression

$$K = \frac{(A^{a+})^b_{clay} m_B^a}{m_A^b (B^{b+})^a_{clay}} \cdot \Gamma_{AB} = D_A^b \frac{m_B^a}{(B^{b+})^a_{clay}} \cdot \Gamma_{AB} \tag{4}$$

is obtained, where the activity coefficient quotient is defined by

$$\Gamma_{AB} = \frac{(\gamma_{\pm AX_a})^{b(a+1)}_{clay} (\gamma_{\pm BX_b})^{a(b+1)}}{(\gamma_{\pm AX_a})^{b(a+1)} (\gamma_{\pm BX_b})^{a(b+1)}_{clay}} = \Gamma_{clay}\Gamma_{aq} = \Gamma_{clay} \frac{(\gamma_{\pm BX_b})^{a(b+1)}}{(\gamma_{\pm AX_a})^{b(a+1)}} \tag{5}$$

This is equivalent to the equation obtained on the basis of the more usual way of expressing the equilibrium

$$bA^{a+} + aB^{b+}_{clay} \rightleftarrows bA^{a+}_{clay} + aB^{b+} \quad . \tag{6}$$

We have followed the alternative route through equations (2) and (3) to emphasize that at equilibrium, there will also be coions (with cation exchangers such as clays, anions) "invading" the solid phase.

For the case of primary interest here, equilibria between clays and aqueous $NaCl$-$CaCl_2$ solutions,

$$K/\Gamma = \frac{(Ca^{++})_{clay}}{m_{Ca}} \cdot \frac{m_{Na}^2}{(Na^+)^2_{clay}} = D_{Ca}/D_{Na}^2 \ , \tag{7}$$

$$\Gamma = \frac{(\gamma_{\pm CaCl_2})^3_{clay}}{(\gamma_{\pm CaCl_2})^3} \cdot \frac{(\gamma_{\pm NaCl})^4}{(\gamma_{\pm NaCl})^4_{clay}} \ , \tag{8}$$

$$\Gamma = \Gamma_{clay} \cdot \frac{(\gamma_{\pm NaCl})^4}{(\gamma_{\pm CaCl_2})^3} = \Gamma_{clay}\,\Gamma_{aq} \tag{9}$$

and

$$K/\Gamma_{clay} = \frac{\mathrm{D}_{Ca}}{\mathrm{D}_{Na}^2}\,\frac{(\gamma_{\pm NaCl})^4}{(\gamma_{\pm CaCl_2})^3} \;. \tag{10}$$

Results in many cases are presented in terms of the ionic strength I

$$I = \frac{1}{2} \sum m_i z_i^2 \;, \tag{11}$$

the summation being over i ions of charge z_i, of the solutions in equilibrium, or the fraction of the total ionic strength contributed by the salt of one ion. For NaCl-$CaCl_2$,

$$I_{total} = I_{NaCl} + I_{CaCl_2} = m_{NaCl} + 3\, m_{CaCl_2} \;. \tag{12}$$

In the figures the cation is used in the subscript to designate the salt, i.e., I_{Ca} denotes I_{CaCl_2}. In evaluating solution-phase activity coefficients from equations for NaCl-$CaCl_2$ aqueous solutions, acetate concentration is added to chloride.

Customarily, in the solution phase, the symbol γ is used in conjunction with concentrations in molality, or moles per kg of solvent. We express solution-phase concentration in the equations in molality m; D is the distribution coefficient symbol corresponding to this convention (moles per kg of (dry) clay/moles per kg of water). In most of the results here, which are for 1 m electrolyte concentration or less, there is little difference between molality and molality (or between D and D). The symbols in some cases are used here interchangeably; however, differences in the range of 10% are incurred near saturated NaCl.

The separation of Γ_{AB} into ratios for the solution and solid phases is useful when solution phase activity coefficients are available (5) because it allows evaluation of variations from ideality of the clay phase alone over a range of experimental conditions. When measurements for the aqueous mixed electrolyte systems in question are not available, adequate estimates can frequently be made by Debye-Hückel equations or by various methods from measurements on two-component systems (6).

Certain special cases are of interest here. When one ion (A) is present at concentrations low enough for its loading of the solid phase to be only a small fraction of ion exchange capacity,

C, expressed in equivalents/kg of solid, and if exclusion of coion X^- is essentially complete, $(B^{b+})_{clay} \simeq C/b$.

$$K_{AB}/\Gamma_{AB} \simeq \mathrm{D}_A^b \frac{m_B^a}{(C/b)^a} \simeq \frac{D_A^b (B^{b+})^a}{(C/b)^a} \quad (13)$$

If Γ_{AB} is constant over the range of conditions, Equation (13) can be differentiated to give

$$\frac{d \log D_A}{d \log (B^{b+})} = -a/b \quad , \quad (14)$$

i.e., a plot of log D_A vs. log (B^{b+}) will be linear, with slope $-(a/b)$, or, e.g., -2 if A^{a+} is Ca^{2+} and B^{b+} is Na^+.

RESULTS

The results presented here are selected from a considerably larger body to illustrate certain aspects of behavior, principally the effects of ionic strength and of fraction of ion-exchange capacity occupied by the two ions, properties of clays from different sources, and effect of purification procedures. A more complete account is available in reference (10).

Effect of pH: In order to carry out experiments at controlled acidity and to avoid possible difficulties from $CaCO_3$ precipitation, the convenient acetic acid-acetate buffer was used to hold the pH at 5 in most experiments. No effects of acetate complexing of Ca(II) were expected or observed. The pH is somewhat lower than in many groundwaters of interest. However, although pH variation from 5 to 7 has been shown to affect values of D of some transition-metal ions on montmorillonite substantially (11), it does not appear important with the alkaline earth ion, Sr(II) (12). It also does not seem of great concern with Ca(II). Figure 1 summarizes some values of D_{Ca} at about one percent Ca(II) loading from about 0.1 m NaCl solutions. Within scatter, values are constant up to pH 9, and the sharp increase above pH 9 likely indicates precipitation. Cycling pH up and down with HCl and NaOH did not significantly affect distribution coefficients. The clay samples in this case had been treated to remove carbonates and to convert them to the sodium form.

Distribution coefficients: Typical values of the distribution coefficients on one clay sample are presented in Figures 2 and 3. The clay was treated with a pH 5 buffer to remove carbonates and was pre-equilibrated with $NaCl$-$CaCl_2$ solutions as previously described. As in other results to be presented, acetate concentration was 0.1 m at I = 1, 0.01 m at I = 0.1, and was the only anion

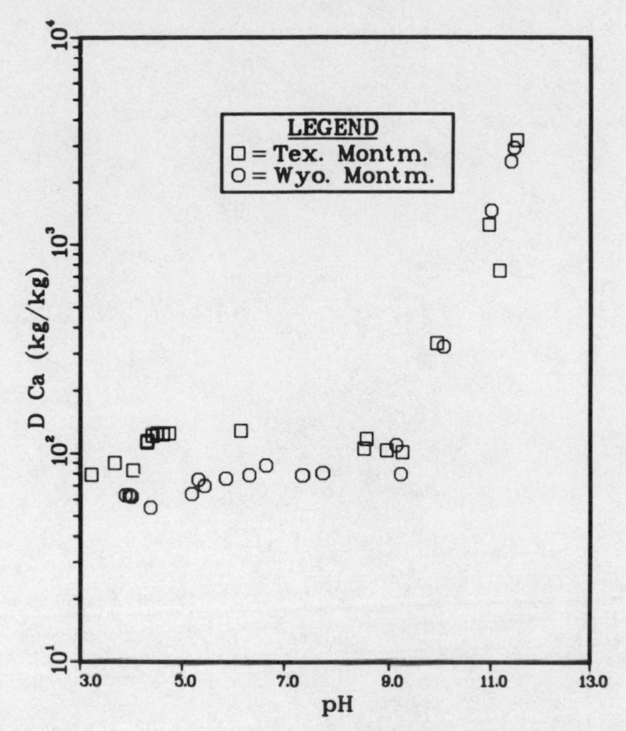

Fig. 1. $\underset{\sim}{D}_{Ca}$ as a function of pH in 0.0955m NaCl.

present at I = 0.01. Figure 2 shows D_{Ca} at three ionic strengths as a function of the fraction of total solution ionic strength contributed by sodium salts. Distribution coefficients decline with increasing ionic strength, and increase with increasing sodium ionic-strength fraction in the solution phase. The trend is fairly regular, except for the points at trace Ca(II). The clay is particularly dispersible in the sodium form, especially at low salt concentrations, and these points perhaps indicate the inaccuracy of measurements under these conditions. Figure 3 summarizes the corresponding results for D_{Ca}. At trace sodium level, the points seem incongruent with the others. The reason for the incongruency, as we shall discuss later, is that the adsorption isotherm appears to be highly non-linear at low fractional sodium loading of the ion-exchange capacity.

With the exception of the trace-level points, the behavior is fairly close to what would be expected of an ion exchanger, but this is not apparent from these plots of distribution coefficients. We shall consequently present the same results later in terms of equilibrium quotients. Besides allowing one to see at a glance the adherence to and departures from ideal behavior, K/Γ varies much less over a range of conditions than distribution coefficients and

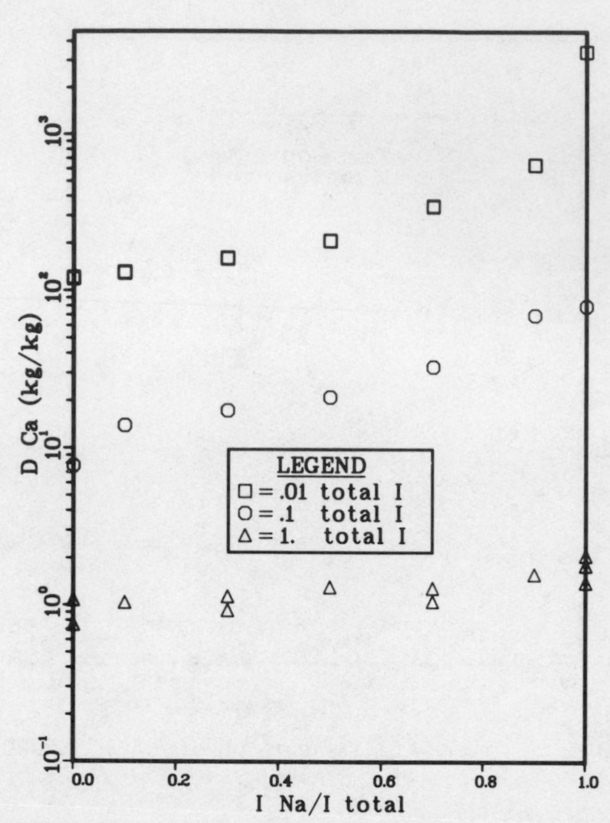

Fig. 2. Distribution coefficient of calcium on untreated Wyoming montmorillonite as a function of solution-phase ionic-strength fraction of sodium (molal basis) at three total ionic strengths, pH = 5.0, sodium-calcium exchange.

is consequently more convenient to use in a simulation, for example, of a preflush.

Effect of purification and clay source: The effect of various purifications of a sample of Wyoming montmorillonite is shown in Figure 4. Comparison of K/Γ_{clay} at ionic strength of one is made of a sample untreated except for the equilibration to establish the various Na(I)/Ca(II) ratios on the solid; a sample treated with pH 5 acetate buffers to remove calcium and other carbonates; and a sample subjected to the complete Jackson procedure, which involves additional steps of settling to remove sand, peroxide treatment to remove organic matter, and dithionite-citrate treatment to remove non-constituent iron compounds, especially oxides. In addition a sample of synthetic montmorillonite from NL Industries, Inc., is included. Within scatter, values of K/Γ_{clay} for all of these samples are in agreement. The low values for the Ca-form correspond to the high D_{Na} mentioned earlier.

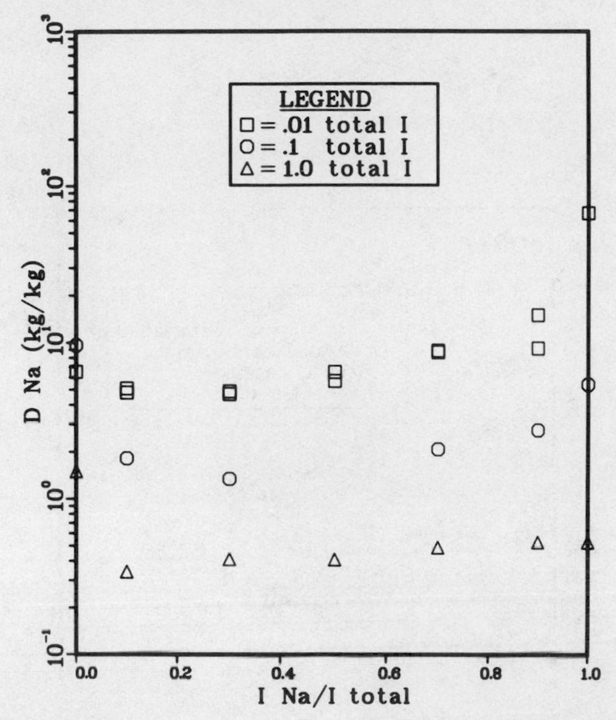

Fig. 3. Distribution coefficient of sodium on untreated Wyoming montmorillonite as a function of solution-phase ionic-strength fraction of sodium (molal basis) sodium-calcium exchange, three ionic strengths.

Another comparison of clays from different sources is given in Figure 5, in terms of K/Γ vs. ionic strength fraction of NaCl in the solution phase. Differences between the two materials are within scatter over most of the composition range. Insofar as one can conclude from the limited number of samples studied, it appears that these results should be useful in predicting behavior of montmorillonite in natural formations.

Effect of ionic strength: Figure 6 compares values of K/Γ_{clay} for one clay sample at three ionic strengths. Values at 0.1 and 1 molal ionic strength agree fairly well, but those at 0.01 appear lower, although their scatter casts some doubt on their reliability. Some of the possible difficulties previously mentioned with experiments at the lowest ionic strength may be causing trouble.

Comparison with other clays: Similar measurements were carried out on illite and attapulgite at 0.01 ionic strength. Distribution coefficients on both these clays are lower than on montmorillonite under comparable conditions. However, their ion-exchange capacities are also lower (up to a factor of ten lower;

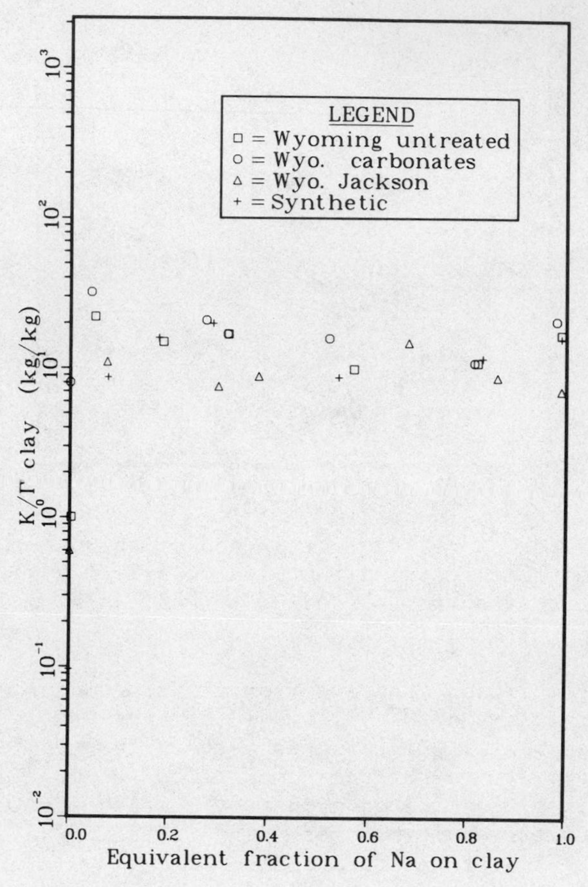

Fig. 4. K/Γ_{clay} as a function of equivalent fraction of sodium on the clay for four montmorillonite samples: untreated, carbonate-free, and Jackson-procedure-purified Wyoming montmorillonite, and synthetic montmorillonite, sodium-calcium exchange.

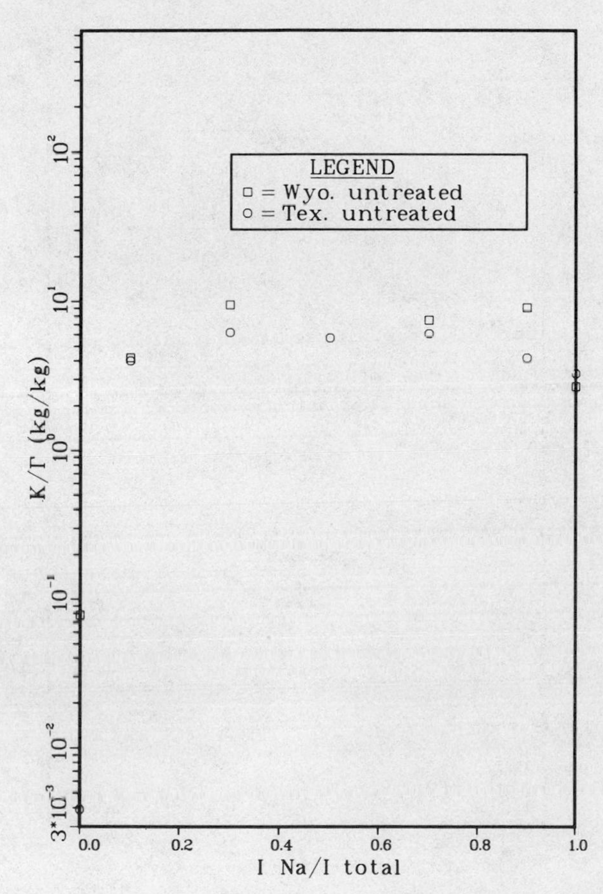

Fig. 5. K/Γ as a function of ionic-strength fraction (molal basis) of sodium in solution phase for untreated samples of Texas and Wyoming montmorillonites at a total ionic strength of .1m.

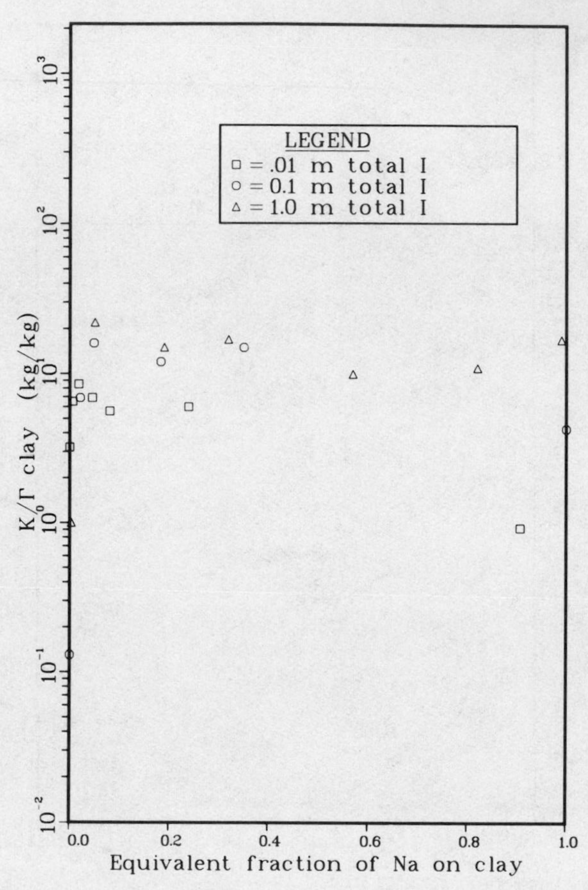

Fig. 6. K/Γ_{clay} versus equivalent fraction of sodium on untreated Wyoming montmorillonite at three ionic strengths. pH = 5.0. Total ionic strengths .01, .1, 1.0, sodium-calcium exchange.

about 0.1 equiv/kg for illite and 0.14 for attapulgite) and the values of K/Γ (Figure 7) are similar for attapulgite and a little higher for illite in comparison with montmorillonite.

Distribution coefficients at trace loading: For trace loadings of one ion, the concentration of the other ion on the solid is essentially the ion-exchange capacity. For constant Γ and complete exclusion of coions, a plot of log D of the trace ion vs. log concentration of the major ion should be linear with slope predictable from the valencies. This is illustrated with results in Figures 8 and 9, some of which were reported previously (1). Plots for trace Ca(II), Sr(II), Ba(II), and K(I) on the sodium form and for Sr(II) and Ba(II) on the Ca(II) form are shown in Figures 8 and 9. Linear plots are obtained but the slopes are only approximately those expected. The absolute values are appreciably less than 2 for alkaline earths/Na(I) and less than 1 for the others in Table 1. Adjustment for the solution phase activity coefficient changes would make disagreement slightly worse for Ca(II)/Na(I).

Equilibrium quotients at low sodium loading of montmorillonite: High distribution coefficients of Na, or low values of K/Γ at low fractional loading of ion-exchange capacity by sodium, have been pointed out in earlier figures. Figure 10 shows values, also previously reported (1), of equilibrium quotients in more detail in this region. The adsorption isotherm is seen to be quite nonlinear, even when only a thousandth of the capacity is occupied by sodium. Ideal behavior (K/Γ independent of loading) was not reached, because the sodium impurities in reagent grade $CaCl_2$ were too high. The results suggest that there is a relatively low fraction of capacity for which distribution coefficients of sodium are very high, but we have not identified the source of the behavior. Montmorillonites from all the sources investigated appear to have similar trends at low sodium loading, although not necessarily quantitatively the same.

Ion-exchange capacities: Measurements of distribution coefficients of both cations combined with knowledge of the concentration in the equilibrium aqueous solution should be sufficient information for computation of ion-exchange capacities. Values obtained in this way are expected to be less precise than those obtained by the more conventional method of converting the clay to a given ionic form, displacing these ions by ions of another type, and analyzing the displaced ions ("displacement").

Two measurements of distribution coefficients are required in the isotope dilution approach, and the errors in each are exacerbated by the fact that concentration on the solid phase is obtained by difference. Any errors from ion exclusion or from incomplete solid-liquid separation will therefore be magnified in the estimate of capacity. The values have however the advantage of providing

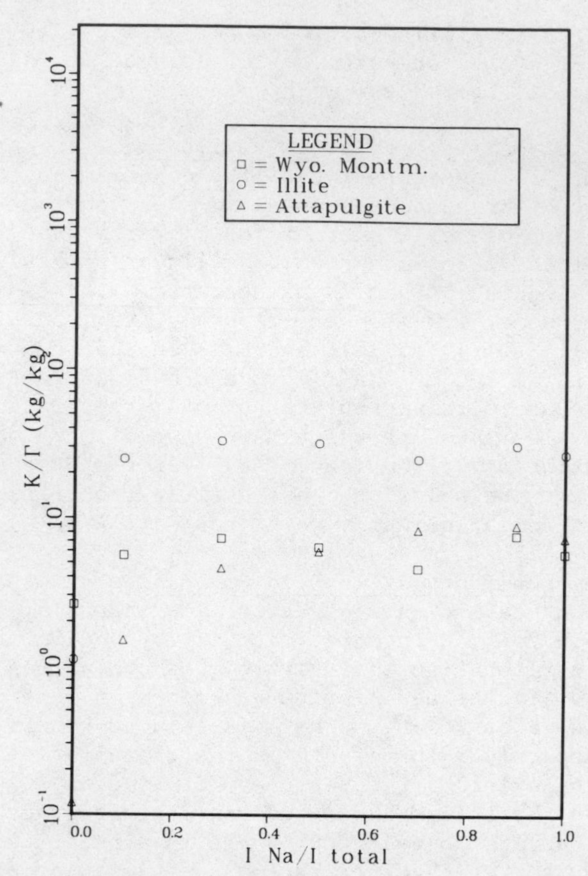

Fig. 7. K/Γ for Wyoming montmorillonite, illite, and attapulgite as a function of solution-phase ionic-strength fraction of sodium (molal basis) at a total ionic strength of .01m, pH = 5.0, sodium-calcium exchange.

information on capacity under the conditions of the equilibrations.

Ranges of capacity values are summarized in Table 2. There should be little difference in the samples labelled "untreated" and "carbonate removed," because alkaline earth carbonate impurities should be dissolved in the pre-equilibration to convert the clay to the specified ionic form. The results by isotope dilution are given only for those measurements in which the solution in equilibrium had a sodium chloride concentration contributing from 0 to 3/4 of the total ionic strength; values at low Ca(II) were erratic, perhaps in correspondence to the difficulties of measuring distributions for the sodium form, previously alluded to. There was a trend to lower apparent capacities with increasing Na(I) concentration. The sample which had been subjected to the complete Jackson purification was highly dispersible at I = 0.01, and results

Table 2. Ion-Exchange Capacity Measurements on Montmorillonite (equivalents/kg dry clay)

	Isotope Dilution (I_{Na}/I = 0 to 0.75)					Displacement[a] Wyoming			
			Wyoming			Untreated		Complete Jackson Purification	
I	Synthetic	Texas Untreated	Untreated	Carbonates Removed	Complete Jackson Purification	Na	Ca	Na	Ca
0.01	--	0.75–0.82	0.72–0.80		0.25–0.90	0.60		0.68	
0.03							0.85		0.94
0.1	--	0.75–0.78	0.71–0.85	0.68–0.80		0.79		0.95	
0.3							0.86		0.94
1	0.68–0.89		0.59–0.66	0.55–0.72	0.52–0.78				

[a] Reference 7. Ion exclusion correction applied.

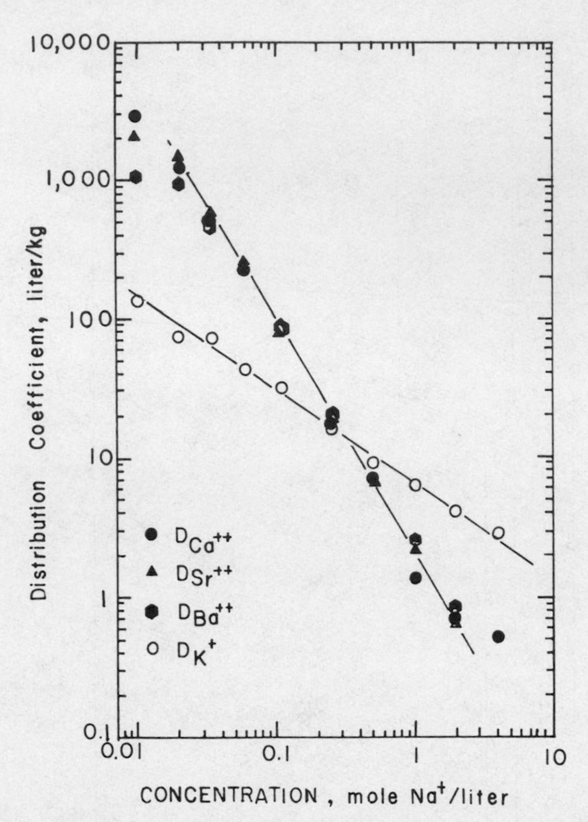

Fig. 8. Distribution coefficients for adsorption of Ca^{++}, Sr^{++}, Ba^{++}, and K^{+} on the sodium form of Wyoming montmorillonite, low or trace loading.

with it were very erratic. Some previously reported (7) values obtained by a displacement method are listed in the table. Values measured on the sodium form were also low at 0.01 ionic strength. At the other end of the range, measurements by isotope dilution at I = 1 are inherently less accurate because of low D values.

Even with all these sources of experimental difficulty, the scatter in the results is disquieting. The possibility that ion-exchange capacity varied with Na(I)/Ca(II) ratio on the clay seems improbable, but the indication of it appeared to be outside scatter on several samples, under different conditions. The values of cation-exchange capacity obtained by isotope dilution were generally less than most reported in the literature, about 1 equivalent/kg, and also less than the values in Table 2 we had obtained by displacement. Also, values of K/Γ, as we shall discuss later, were substantially different from values obtained from literature data.

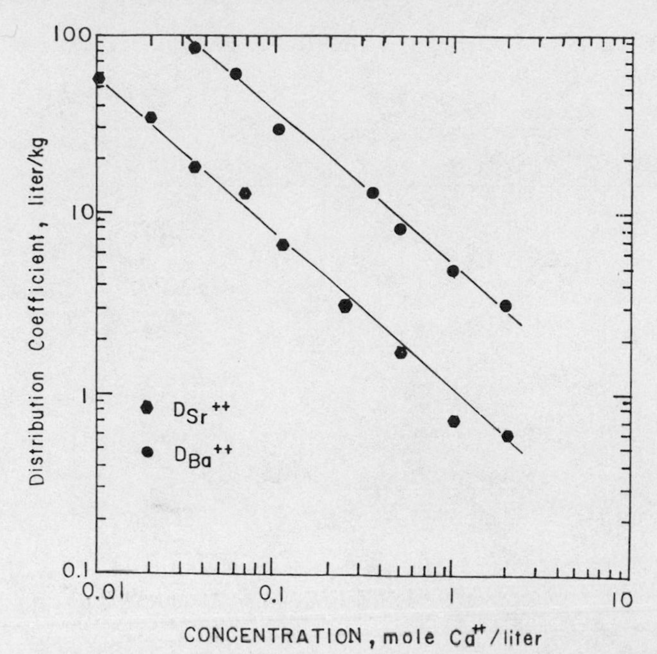

Fig. 9. Distribution coefficients for adsorption of Sr^{++} and Ba^{++} on the calcium form of Wyoming montmorillonite, loading of $Sr^{++} \sim 7\times10^{-5}$ to 4.5×10^{-3} mole/kg, loading of $Ba^{++} \sim 5\times10^{-4}$ to $\sim 7.6\times10^{-3}$ mole/kg, pH 5, acetate buffer (0.01M $Ca(Ac)_2$ + HAc), equilibration time 30 to 70 hours.

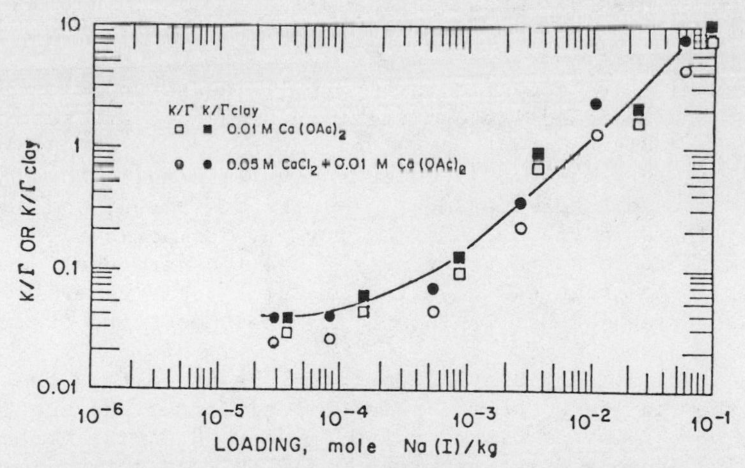

Fig. 10. Effect of loading on equilibrium quotients $[K/\Gamma = D_{Ca}/D_{Na}^2]$ of Wyoming montmorillonite predominantly in the calcium form (pH = 5, equilibration for 14 hours).

Table 1. Adsorption of Ions on the Na-Form of Wyoming Montmorillonite.[a] Log D = s log (Na^+) + b.

Adsorbing Ion	s	b
K^+	-0.67	+0.822
Ca^{++}	-1.667	+0.282
Sr^{++}	-1.682	+0.311
Ba^{++}	-1.56	+0.402
Adsorption on the Ca-Form		
Sr^{++}	-0.901	-0.030
Ba^{++}	-0.776	+0.734

[a] s and b by least-squares fit.

In view of these discrepancies, the isotope dilution measurements at I = 0.1 were repeated, with the procedure modified as previously described to attain a cleaner solid-liquid separation before analysis, and to obtain capacity values by ion displacement on the same samples on which isotope dilution determinations had been made.

The results are compared in Figure 11 (center plot). The results by the modified isotope dilution procedure are seen to imply a relatively constant capacity over the range of solution compositions, between 0.76 and 0.80 equivalents/kg. Those by the unmodified procedure are considerably more scattered, and substantially lower in the sodium-rich region. Comparison of the contributions to total capacity of the two ions are shown in the top sector of the figure. Results for Ca(II) seem generally higher, and those for Na(I), lower, for the unmodified procedure.

Values computed by displacement, on the assumption that the water in the clay pack had the same composition as the supernatant solution, were somewhat higher, by 0.04 to 0.08 equiv/kg, than by the modified isotope dilution procedure. As we described earlier, the solution displaced from the clay pack and associated water was also analyzed for chloride, and it was found the concentration of anion in the pack water was significantly less than in the supernatant solution, i.e., there appeared to be some salt exclusion. Computation of cation-exchange capacity by subtracting from total

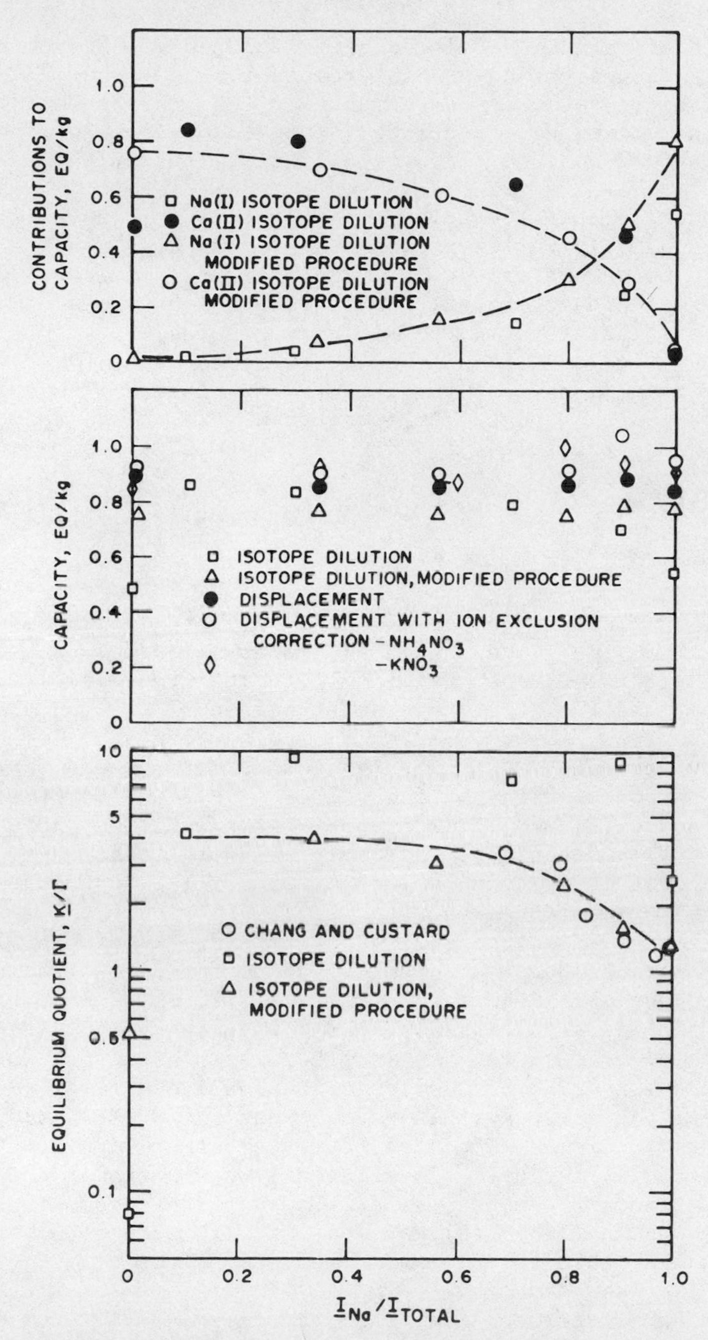

Fig. 11. Comparison of Ca(II)-Na(I) equilibrium quotients for exchange between aqueous $CaCl_2$-NaCl solutions, I_{total} = 0.1, and montmorillonite, measured by different isotope-dilution techniques. Comparison of ion-exchange capacities measured by isotope dilution and by displacement of ions.

equivalents of cations displaced the total equivalents of chloride displaced gave net cation-exchange capacities somewhat higher, 0.9-1 equiv/kg, and more scattered, than the displacement values without ion-exclusion correction. (The values from Reference 7 given in Table 2 were obtained by a similar procedure.)

We conclude that there is no reason to assume a variation of capacity with Na(I)/Ca(II) ratio. It also appears that there is appreciable coion exclusion. However, the present results do not allow determination of relative exclusion in water between clay particles and intralayer water within particles, if such a distinction is justified at all. Any anion-exchange capacity present would result in higher cation-exchange capacities. These unresolved questions introduce some uncertainty in the values of K/Γ inferred from our results, and from literature values measured by similar procedures.

DISCUSSION

Of a considerable number of literature studies known to us concerning Na(I)/Ca(II) exchange on montmorillonite, two, one by Chang and Custard (2) on a Wyoming bentonite from the American Colloid Co., and the other, by Van Bladel et al. (3) on a sample from Camp Berteau, Morocco, present results with which we can readily compare ours. Those of Chang and Custard were carried out at about 0.1 ionic strength, and values of K/Γ are compared in the bottom sector of Figure 11 with ours by the modified and unmodified isotope dilution technique. Agreement with the results by the modified procedure is good; values by unmodified procedure are about twice those of the other two sets. In Figure 12, results by the modified procedure, of Chang and Custard, and of Van Bladel et al. are compared. Because those of Van Bladel are for a lower ionic strength, the presentation in Figure 12 is of K/Γ_{clay} to eliminate the effect of Γ_{aq} differences between $I = 0.01$ and $I = 0.1$ on the comparison. As in other cases here, Γ_{aq} was computed from the isopiestic results of Robinson and Bower (13). Agreement of the three sets is good. Maes and Cremers (14) also carried out measurements on Camp Berteau montmorillonite, but we could not readily extract values for comparison from the form in which their results are presented. However, they compare theirs with those of Van Bladel (3), as well as with some of J. Dufey in a report unavailable to us. Agreement seems good.

It appears, therefore, that all of these studies, carried out by several different techniques, are in reasonable agreement on the equilibrium quotients for Na(I)/Ca(II) exchange on samples of montmorillonite from several different sources. Ion-exchange equations give a good correlation of the adsorption behavior on montmorillonite, in agreement with prior studies (see, e.g., References

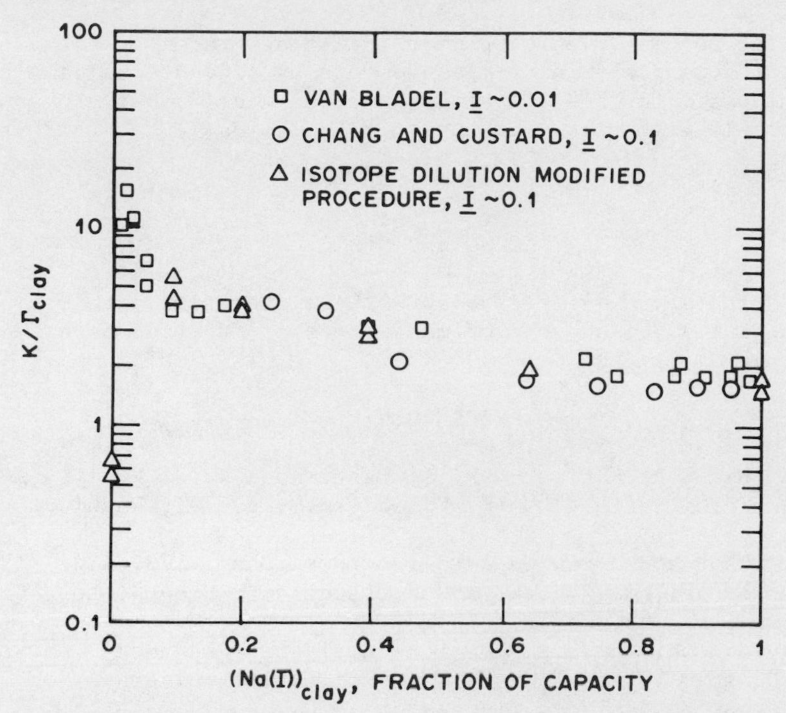

Fig. 12. Comparison of equilibrium quotients, adjusted for solution-phase activity coefficient ratios, for Na(I)-Ca(II) exchange on montmorillonite, with literature values.

15-20). The best present values for K/Γ_{clay} would appear to be those summarized in Figure 12. There appears to be a downward trend from values of about 5 for K/Γ_{clay} at 10% of capacity loaded with sodium to about 2 for the sodium form. The solution phases were 0.1 ionic strength or lower in Figure 12, but our results by the unmodified isotope dilution procedures, in spite of scatter, as well as for trace Ca(II) distribution coefficients, indicate that these quotients are reasonably valid up to $I = 1$ and higher (Figures 8 and 9).

High values of D_{Na} at low loading of sodium on the Ca(II) form were observed with all samples we investigated, with all the degrees of clay purification tried. The values are not well established, because of the extreme sensitivity to loading. None of the literature studies seems to have reached low enough loading for this non-ideality to become apparent. We do not know the explanation for the behavior.

From the results in Figure 8, there do not appear to be great differences between the equilibria between the alkaline earth ions Ca(II), Sr(II), and Ba(II) with sodium on montmorillonite. Mg(II) is of obvious practical interest, but is more difficult to study, owing to lack of a convenient, gamma-emitting tracer. We have carried out some preliminary measurements by conventional analytical techniques. The results of K/Γ scatter widely, but appear to be not greatly different than those of Ca(II)/Na(I): 1 to 3 at I = 1, and 1 to 10 at I = 0.1.

ACKNOWLEDGMENTS

The authors would like to express their gratitude to Drs. K.A. Kraus and J.S. Johnson for many hours of fruitful discussions.

REFERENCES

1. S.-Y. Shiao, P. Rafferty, R. E. Meyer, and W. J. Rogers, ACS Symposium Series 100 (S. Fried, ed.), p. 297, Washington, D.C., 1979.
2. H. L. Chang and H. C. Custard, J. Phys. Chem., 72, 4340 (1968).
3. R. Van Bladel, G. Gavira, and H. Laudelout, Proc. Inter. Clay Conf., p. 385, Madrid, Spain, 1972.
4. M. L. Jackson, in "Soil Chemical Analysis - Advanced Course," published by the author, Dept. of Soils, University of Wisconsin, Madison, Wisconsin, 1956.
5. R. M. Rush, ORNL-4402, Oak Ridge National Laboratory, Oak Ridge, Tennessee, April 1969.
6. G. Scatchard, R. M. Rush, and J. S. Johnson, J. Phys. Chem., 74, 3786 (1970).
7. R. E. Meyer,"Sesquiannual Report - April 1977-September 1978, Fossil Energy Division/DOE," DOE/W-7405-eng-26-1 (1979).
8. R. Triolo, N. E. Harrison, and K. A. Kraus, J. Chromatography, 179, 19 (1979).
9. J. E. Dufey and H. G. Laudelout, Soil Science, 121 (2), 72 (1976).
10. W. J. Rogers, Ph.D. Thesis, University of Tennessee, Knoxville, Tennessee, 1979.
11. Y. Egozy, Clay and Clay Minerals, 28, 311 (1980).
12. P. Rafferty, S.-Y. Shiao, C. M. Binz, and R. E. Meyer, J. Inorg. and Nuclear Chem., in press.
13. R. A. Robinson and V. E. Bower, J. Res. Nat. Bur. Stand., 70A, 313 (1966).
14. A. Maes and A. Cremers, J. Chem. Soc. Far. Trans. I, 73, 1807 (1977).
15. G. L. Gaines, Jr. and H. C. Thomas, J. Chem. Phys., 21(4), 714 (1953).
16. R. J. Lewis and H. C. Thomas, J. Phys. Chem., 67, 1781 (1963).
17. J. S. Wahlberg and M. J. Fishman, U.S. Geol. Surv. Bull., 1140-A, 1962.
18. J. S. Wahlberg, J. H. Baker, R. W. Vernon, and R. S. Dewar, U.S. Geol. Surv. Bull., 1140-C, 1964.

19. J. S. Wahlberg and R. S. Dewar, U.S. Geol. Surv. Bull., 1140-D, 1964.
20. T. Tamura, ORNL-P-438, Oak Ridge National Laboratory, Oak Ridge, Tennessee, 1964.

PERMEABILITY REDUCTION IN WATER SENSITIVITY OF SANDSTONES

K. C. Khilar and H. S. Fogler

Department of Chemical Engineering
The University of Michigan
Ann Arbor, Michigan 48109, U.S.A.

A model was developed to describe the rapid decline of permeability of Berea sandstone resulting from its water sensitivity. The model is based on the release and capture of clay particles. The release of clay particles from the pore wall is approximated as a first order decay process and the rate at which particles leave the suspension to plug the pores is assumed to be directly proportional to the particle concentration in the suspension. The resulting equations are solved in conjunction with a semi-empirical relationship relating the local permeability and the amount of clay particles captured to determine the overall permeability of the sandstone core as a function of time. Model predictions are in agreement with experimental observations.

INTRODUCTION

The water sensitivity of sandstone is a problem in colloid chemistry that has been known to exist for the past thirty years. The water sensitivity phenomenon is observed to occur when fresh water replaces salt water in sandstones. This phenomenon has been well documented by past investigators (1-5) and a consensus exists that the phenomenon occurs due to clay swelling or clay particle migration, or a combination of these effects. It has also been pointed out that clay particle migration may be, in fact, the more prevalent damage mechanism (2). This article focuses on some theoretical and experimental aspects of the damage due to clay particle migration.

The results of a typical experiment to demonstrate this phenomenon are shown in Figure 1. In this experiment, a one inch diameter sandstone core two inches in length is vacuum saturated with 6 wt % solution of NaCl and then placed in a core holder. Next, a 6 wt % NaCl solution is passed through the core in the axial direction for a period of time well beyond that after the flow is stabilized. Then flow through the core is abruptly switched from salt water solution to fresh water and pressure drop across the core is monitored as a function of time. One observes from Figure 1 that the overall permeability of the core drops drastically after only one or two pore volumes of fresh water have entered the system.

While water sensitivity was initially realized during the water flooding of petroleum reservoirs, it is now of serious concern in many other areas. It has been found that attempts to acidize wells with aqueous solutions of HCl and HF have resulted in failure due to a drastic permeability decrease near the well bore. Water sensitivity is also believed to be a cause of long term permeability decreases found in gas wells especially in the spring, due to the presence and movement of low salinity water. Although this phenomenon was discovered as early as the year 1945 (1), up until now mathematical models describing the water sensitivity phenomenon have not been put forth. Understandably, such a model is required to predict the phenomenon quantitatively and can be helpful in the design of preventive measures.

The overall mechanism of the water sensitivity phenomenon of sandstone containing little or no swelling clays involves release of clay particles from the pore wall, followed by migration and capture at the pore constrictions (2,6). However, the fundamental mechanisms of basic processes of release and capture of clay particles are not well understood. The lack of knowledge, particularly relating to the release process, can partly be attributed

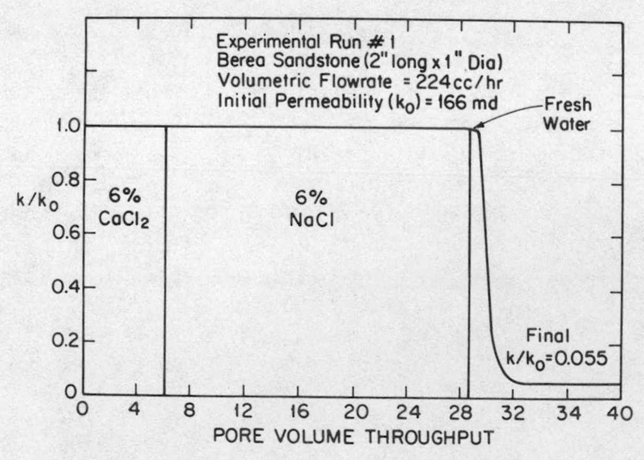

Fig. 1. Permeability decrease in a typical shock experiment.

to the difficulty in carrying out precise experiments to investigate the mechanisms involved in the processes. In this article, results of two new experiments are reported: 1) flow reversal with salt water, 2) pressure measurement along the axial distance of the sandstone core. The experimental observations are useful in further understanding the mechanisms of release and capture as well as developing the model presented in the next section.

EXPERIMENTAL

Apparatus: A schematic diagram of the experimental apparatus used in this work is shown in Figure 2.

The fluid is pumped to the sandstone core by a twin cylinder Ruska Pump. This is a constant volumetric positive displacement pump which will deliver fluid at a constant flow rate (0-240 cc/hr) against a pressure of up to 2000 psi. The cylinders are stainless steel, to avoid potential corrosion problems. Immediately following the pump are two Millipore filters which will trap any suspended particles larger than 0.45μ that might be in the fluid.

Two separate sets of pressure transducers were acquired (0-8, 0-20, 0-100, and 0-500 psi), to allow the accurate measurement of a wide range of pressure drops. An electronic interface converts the signal from the transducers to a voltage signal which is fed to the recorder. A dual channel recorder is used to monitor the pressure drop across the core and the electrical resistance of the fluid flowing through the core.

The fluid flows through the switching valve (V) and enters the core holder (shown in Figure 2). The core holder is placed in a constant temperature bath. The valve (V) is a six port, high pressure micro-volume valve which allows the fluid being pumped to the core to be switched quickly and sharply. Once the fluid leaves the core holder, it is either collected for analysis or discarded.

A detailed diagram of the core holder is shown in Figure 3. The metal parts of the core holder are stainless steel and it is designed to withstand 2000 psi. The core sample, 1, is held in place by the top, 3, and bottom, 4, core sample end plugs and a sleeve of tygon rubber tubing, 2. The tubing is held tightly against the core sample by an overburden pressure of 500 psi or more, obtained by filling up the core holder with viscous oil and then pressurizing with nitrogen gas which enters through a port in the vessel, 6. This overburden pressure prevents the fluid from flowing along the sides of the core. This type of system allows the pressure drop across the core to reach nearly 500 psi. The nitrogen is contained by the end plates 4, 5, which are in turn

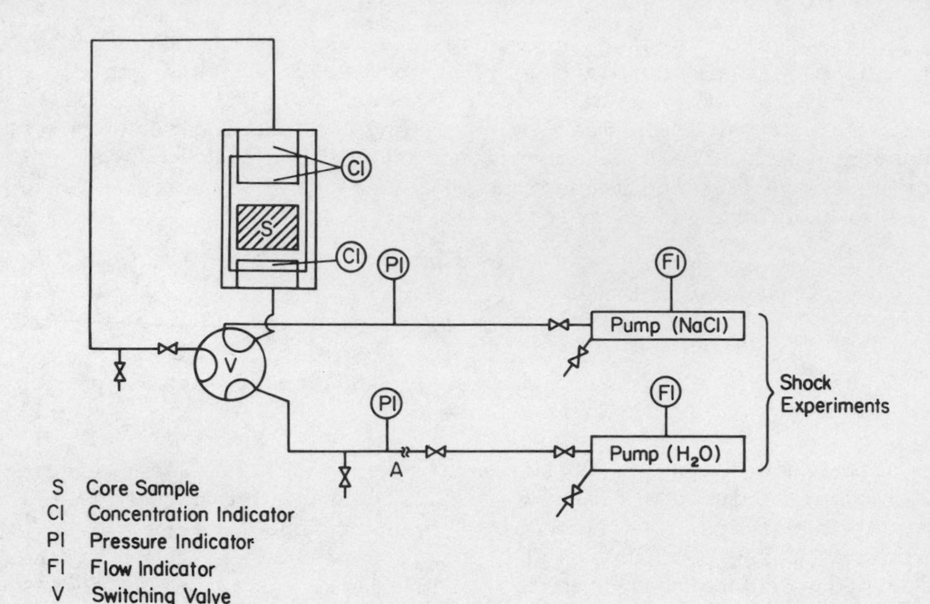

Fig. 2. Schematic diagram of experimental apparatus.

held in place by the screw-on caps. Rubber O-rings provide the actual seals between the metal parts. The fluid flows into the holder through the inlet port, 8, and out through the outlet port, 7. A switching valve is used to reverse the flow, during some experiments, in which case the fluid will enter through port 7.

A plexiglass fluid distributor and collector are located between the end plugs, 3, 4, and the core sample, 1. In addition to collecting and distributing the fluid at the core faces, they provide support for the platinum electrodes. Two pairs of electrodes are located both in the distributor and in the collector. These electrodes are used to measure the resistivity of the fluid entering and leaving the core. The salt concentration can be determined from these resistivity measurements. The electrodes have been plated with platinum black to minimize the effect of polarization. The resistance across the electrodes is measured by a Barnstead water purity meter. This meter, like the pressure transducers, also requires an electronic interface to convert its output to the appropriate range of voltage. The resistivity of the core effluent, and thus its concentration, is then recorded by the second channel of the recorder.

In summary, core flood types of experiments can be conveniently carried out at constant temperature using the apparatus described above. The pressure drop across the core can be measured by adaptor-transducer unit for the range of 0-500 psi. The salt concentration can also be measured both in inlet and outlet streams

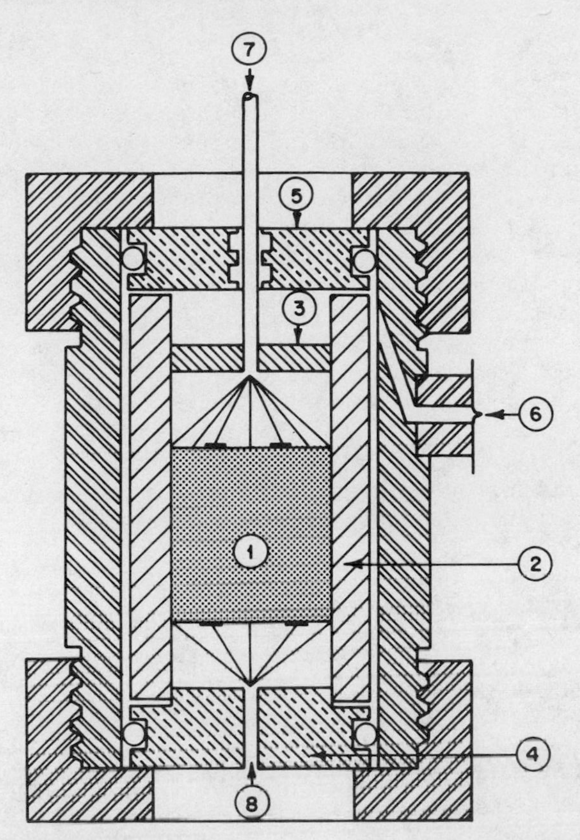

Fig. 3. Sandstone core holder: 1) sandstone core; 2) tygon rubber sleeve; 3) top core end plug and electrodes; 4) bottom core end plate, end plug, and electrodes; 5) top core holder end plate; 6) pressure inlet for N_2 overburden pressure; 7) outlet stem; 8) water inlet.

by means of platinum electrodes and a resistivity meter and with a sodium specific ion electrode.

Experimental procedure: Two types of experiments were carried out in this work: 1) Shock experiments, 2) shock followed by flow reversal with salt water.

A shock experiment begins with vacuum saturation of a sandstone core for three days. Berea sandstone cores of diameter 1" and lengths 1" and 2" were used. The reason for using Berea is that it contains no swelling clays and it resembles many formation sandstones in its clay mineralogy (2).

The saturated core is placed in the core holder and its initial permeability (K_o) of the sandstone core is then measured by pumping a 30,000 ppm of sodium chloride solution at a constant flow rate through the core by means of a Ruska positive displacement pump. Routinely, around 50 pore volumes of salt solution are passed through the core during the initial permeability measurements. The core is then suddenly contacted with fresh water using a six port switching valve (shown in Figure 2). The pressure drop across the core and the salt concentration of both the entering and exiting stream are respectively measured by means of a Unimeasure adaptor-transducer unit and platinum electrodes. After the shock is completed, around 50 pore volumes of salt solution are again passed through the damaged core. The direction of flow of salt water is then reversed by means of the six port switching valve. The circuit for reverse direction is shown in Figure 2. The flow in the reverse direction is continued until flow is switched to fresh water to produce the next shock or until the termination of the experiment. The permeability (K) is calculated using Darcy's law for one dimensional laminar flow of a homogeneous fluid which is given as below.

$$q/A_c = \frac{K}{\mu}\left(\frac{\Delta P}{L}\right) \tag{1}$$

Flow reversal with salt water: Experiments were carried out where the procedure of having the core shocked, the flow reversed, the core shocked again, was repeated three times. Salt water was passed through the core before and after each shock as described in the procedure above. The results of this flow reversal are shown in Figure 4. As one can see in Figure 4, the permeability of damaged core consistently increased up to 80 and 90 percent of the original permeability (K_o) with the reversal with salt water. More important, this gain in permeability with salt water reversal was found to be permanent, unlike the temporary gain in permeability for reversal with fresh water (2,6). As a matter of fact, routinely, the flow direction was reversed with salt water after each time the core was damaged in a shock experiment. The results obtained are similar to those given in Figure 4.

The observed permeability restoration of a damaged core with salt water reversal can be explained as follows. A reversal in flow of fresh water should release the clay particles blocking the constrictions in pores and these particles then flow with the fluid until they block a constriction in the opposite direction. Thus a temporary increase in permeability should be expected. However, reversing the flow using salt water, a flocculating type solution, allows the particles to coagulate and adhere to the pore wall before they can block constrictions in the opposite direction. Therefore, one observes a permanent increase in permeability. This finding signifies the importance of the phenomena of peptization and flocculation in water sensitivity and also indirectly supports the accepted overall mechanism of release and capture of clay particles.

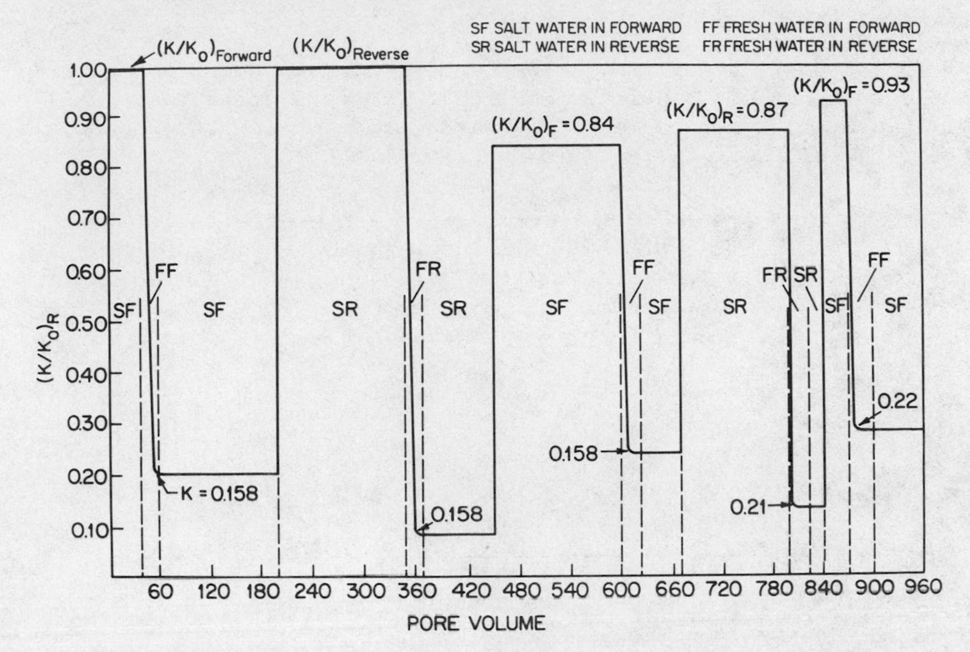

Fig. 4. Permeability restoration in salt water reversal.

The movement of the permeability front: A shock experiment was carried out using a 2" Berea core in which the pressure was measured at two points--at the inlet of the core and at the mid-point of the core. The pressure drop $(\Delta P)_1$ across the entry section was calculated from the algebraic difference between the two measured pressure drops $[(\Delta P)_{overall}$ and $(\Delta P)_2]$. From these measurements, permeabilities of two sections of the core entry section (K_1) and exit section (K_2) were computed.

The variations in K_1 and K_2 with cumulative pore volume of fresh water sent through the core are illustrated in Figure 5. As one observes in Figure 5, the permeability of entry section (K_1) begins to decrease at around 0.25 pore volume whereas the permeability of exit section (K_2) remains unaffected until 0.8 cumulative pore volumes. This shows that the sandstone core is damaged successively along its length. In other words, a permeability front moves along the core as fresh water replaces salt water in the sandstone core.

DEVELOPMENT OF A MATHEMATICAL MODEL

In modeling sandstone core, the core is considered to consist of a number of unit cells which are connected in series in the axial direction. A schematic representation for a core conceptualized to contain 10 unit cells is shown below (Figure 6). Particles are released from the walls of these cells and are captured at their constriction (outlet) as schematically shown in Figure 7. Fluid contents of each individual cell are considered to be well mixed. It is also assumed that there is no transport of particles between cells when the core is shocked.

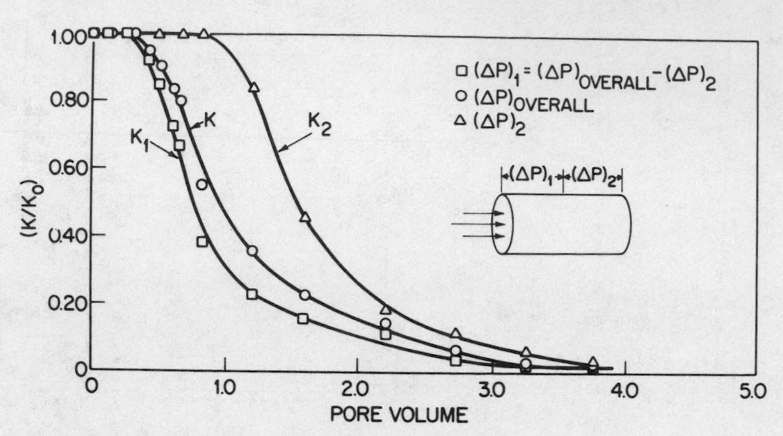

Fig. 5. Movement of the permeability front. Berea, 1" diameter, 2" length; flow rate = 16 cc/hr.

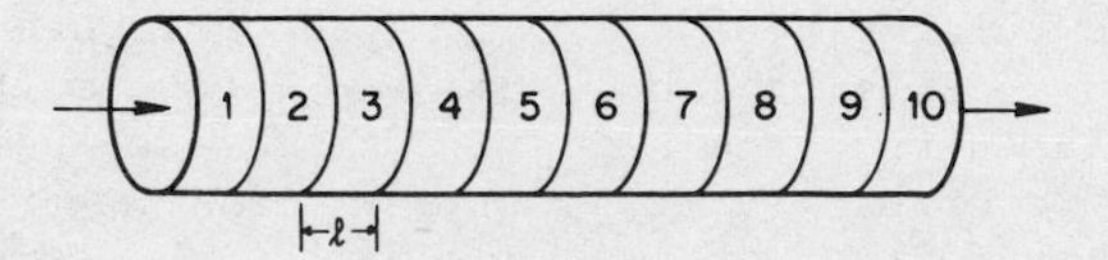

Fig. 6. Schematic diagram for a core containing 10 unit cells.

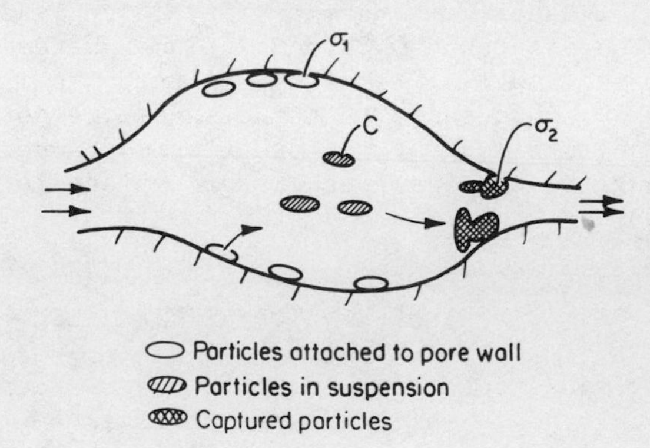

Fig. 7. Schematic diagram of a unit cell.

With these assumptions, a material balance on the suspended clay particles contained in a unit cell (as shown in Figure 7) yields:

$$\frac{dC}{dt} = r_r - r_c \tag{2}$$

r_r = rate of release of clay particles from pore wall (gm of clay/(cc of suspension) (min)

r_c = rate at which particles leave the suspension to be captured (gm of clay/(cc of suspension) (min))

C = concentration of clay particles in suspension (gm/cc of suspension)

t = time (min)

Release rate equation: The process of release of clay particles is assumed to be a first order decay process and can be written mathematically as

$$r_r = -\frac{\partial \sigma_1}{\partial t} = \alpha \sigma_1 \tag{3}$$

Here, σ_1 is the concentration of clay particles on the pore walls, which can be released due to ionic shock, expressed in gm/(cc of suspension), and α is the release coefficient expressed in min^{-1}. The intensity of the release process is believed to be dependent on many system variables. A long, but by no means complete, list of variables includes the concentration of particles and the physical and chemical nature of their attachment to the pore wall, the chemical nature of permeating fluid, the ionic strength and the rate of change of ionic strength of the permeating fluid, and the amount of shear force acting on the particles due to flow of fluid. Therefore, given a particular sandstone, the release coefficient depends on the chemical nature, ionic strength, rate of change in ionic strength (effects of all these variables are termed as chemical effects) and on the flow rate of the permeating fluid. Experimentation to determine the chemical effects on the release coefficient (α) is currently underway.

Capture rate equation: The rate (r_c) at which particles leave the suspension is assumed to be proportional to the particle concentration in the suspension. Therefore, with β as the capture coefficient and σ_2 as the concentration (gm/cc of suspension) of captured particles, one has the equation

$$r_c = \frac{\partial \sigma_2}{\partial t} = \beta C \tag{4}$$

The capture rate equation is similar to the widely accepted rate equation which defines the filter coefficient (λ) in deep bed filtration. The rate of removal of particles from a suspension with respect to depth is proportional to the concentration of particles in the suspension (7)

$$-\frac{\partial C}{\partial z} = \lambda C \tag{5}$$

Noting that $z = Ut$, and the rate of removal of particles is equivalent to the rate of capture in the present case, it can be readily shown that:

$$\beta = U\lambda \tag{6}$$

The filter coefficient (λ) is, in general, a function of specific deposit (σ) and initial porosity (ε_o). In deep bed filtration, λ varies significantly due to large variation of σ during the filtration process. However, during the rapid damage in the water sensitivity phenomenon, the volume of deposited particles per unit volume of core (σ) is believed to be small. One experimental observation supporting the above view is that the difference in porosity between damaged and fresh sandstone core is virtually immeasurable. Therefore, unlike deep bed filtration, the capture coefficient (β), as defined in the model, remains approximately constant for a given flow rate. The variation in the filter coefficient (λ) is usually given by an empirical relationship (7,8). Theoretical work leading to approximate relationships between the filter coefficient (λ) and other variables has been reported (9, 10). The filter coefficient (λ) is related to the collection efficiency (η) and collection efficiency is computed for a model collector using fundamental deposit mechanisms.

<u>Axially non-uniform release of clay particles</u>: The release of clay particles due to ionic shock in various cells will begin at different times depending on their relative location from the front of the core. This occurrence can be understood when one considers the front-like movement of ionic shock along the core (dispersion with high Peclet number). Therefore, one should account for this axial variation of release of clay particles. In fact, accounting for the axial variation of the release process becomes necessary in this model. Otherwise, the model predictions of the transient permeability (K) of the damaging sandstone core will be axially independent. Such a result is in direct contradiction with the experimental observation that the permeability (K) varies with axial distance (graphically shown in Figure 5).

Therefore, the variation in release rate in different cells will now be accounted for with the following approximation.

Cells are assumed to be equally long and the length is ℓ. The average time for the permeating fluid to traverse through a unit cell is then,

$$\tau = \frac{\ell}{q/A_c\varepsilon} \tag{7}$$

where q is the flow rate in cc/min and $A_c\varepsilon$ is the open cross sectional area of the cylindrical sandstone core in cm^2.

It is furthermore assumed that the release coefficient, (α), is a non-zero constant for the cell which has been completely traversed by the fluid (fresh water) and is zero for the remaining cells. Expressing it mathematically, for

$$0 < t < i\tau$$

$$\alpha_j = \alpha \qquad j=1,2,\ldots(i-1) \tag{8}$$
$$\alpha_j = 0 \qquad j=i,i+1,\ldots n$$

where n is the total number of cells, i ($i<n$) is the number of a cell, and t measures the time after the fresh water is sent to the core. Note, cells are numbered successively from the entrance to the exit of a sandstone core (as shown in Figure 6).

Calculation of the concentration of captured particles: Equation (2) is now combined with Equations (3) and (4) to yield:

$$\frac{dC}{dt} = \alpha\sigma_1 - \beta C \tag{9}$$

with $\sigma_1 = \sigma_{10}$ at $t = 0$, Equation (3) is readily integrated to obtain the concentration of particles on the pore wall (σ_1).

$$\sigma_1 = \sigma_{10}\exp(-\alpha t) \tag{10}$$

Substituting for α_1 in Equation (9) and rearranging:

$$\frac{dC}{dt} + \beta C = \alpha\sigma_{10}\exp(-\alpha t) \tag{11}$$

The above differential equation is solved easily with an initial condition, $C = 0$ at $t = 0$, and the resulting expression for concentration C of particles in the fluid, is given by

$$C = \frac{\alpha\sigma_{10}}{\beta-\alpha}\left[\exp(-\alpha t) - \exp(-\beta t)\right] \tag{12}$$

Now, the concentration of captured particles as a function of time for a given cell can be calculated from Equation (4)

$$\sigma_2 = \int_0^t \beta C dt$$

Integrating after substituting for C from Equation (12)

$$(\sigma_2/\sigma_{10}) = 1 - \left[\frac{\alpha \exp(-\beta t) - \beta \exp(-\alpha t)}{\alpha - \beta}\right] \quad (13)$$

Prediction of the overall reduction in permeability: In order to compute the permeability of a cell, a relationship between local permeability and the local concentration of captured particles is needed. A square relationship based on the buildup of particles on the wall of a cylindrical pore or on the constricted space was developed. However, the authors realize that the buildup or deposit of particles in a real situation would be different from the assumed symmetrical buildup. Nevertheless, the square relationship was found to be satisfactory in the model calculations, as will be seen in a later section. The relationship relating local permeability to concentration of captured particles in the ith cell is given by:

$$K_i/K_o = \left(1 - A\frac{\sigma_{2i}}{\sigma_{10}}\right)^2 \quad (14)$$

where $A = 1 - \left(\frac{K_f}{K_o}\right)^{1/2}$ is a constant for a particular sandstone. At this stage, the model is sufficiently developed to compute the permeability of various cells of the core. Equations (8), (13), (14) and the values of parameters are used to calculate the transient permeability, K_i of each cell. The overall permeability, K, of the sandstone core can now be easily calculated using Darcy's law for each cell which is placed in series.

$$\frac{L}{K} = \sum_{i=1}^{n} \frac{\ell_i}{K_i} \quad (15)$$

where K and L are overall permeability and length of the core respectively. K_i and ℓ_i are permeabilities and lengths of the ith cell.

For n equally long cells:

$$L = \sum_1^n \ell_i = n\ell \quad (16)$$

$$K = n\left[\sum \frac{1}{K_i}\right]^{-1} \qquad i=1,\ldots,n \tag{17}$$

Equation (17) can further be simplified by the use of Equation (8). Noting in Equation (13) that $\sigma_2 = 0$ when $\alpha = 0$, it follows from Equation (14) that $K_i = K_o$ for a cell having $\alpha = 0$. That is, cells having no release of clay particles do not encounter damage.

Therefore, for the ith cell,

$$K_i = K_o \qquad \text{for } t < i\tau$$

$$K_i = K_i(t - i\tau) \qquad \text{for } t > \tau$$

Letting $t' = t - i\tau$

$$K(t) = n\left[\sum \frac{1}{K_i(t')}\right]^{-1} \tag{18a}$$

where

$$K_i = K_o\left[1 - A\,\frac{\sigma_{2i}(t')}{\sigma_{10}}\right]^2 \tag{18b}$$

and

$$\sigma_{2i}(t') = 1 - \left[\frac{\alpha\exp(-\beta(t-i\tau)) - \beta\exp(-\alpha(t-i\tau))}{\alpha - \beta}\right] \tag{18c}$$

With the knowledge of A, α, β and n, Equations (18a-c) can be used to calculate the overall permeability of the core as a function of time.

Comparison of model predictions with experimental observations: Eight shock experiments were carried out using the procedures described earlier for the flow rate range of 2.0 to 224.0 cc/hr. The predictions of the model described above were compared with transient permeability data obtained for various flow rates. Information necessary for model calculation includes flow rate, length and diameter of the core, porosity, initial and final permeability, parameters α and β and the total number of cells. Since α and β, at this stage, cannot be predicted from first principles, the strategy was to choose α and β to agree with the experimental data. The number of cells (n) were chosen equal to 5 for cores of length 1" and 10 for cores of length 2". Predictions with higher number of cells were found to be virtually the same as those with 5 or 10 cells. The constant A in Equation (14) was calculated from initial and final permeabilities for each shock experiment.

A comparison of model predictions with the experimental data for five different flow rates is shown in Figures 8-12. One observes the agreement is very good, especially when one considers the simplicity of the model as compared to the complexity of the phenomenon it describes. The slight disparity between model predictions and the data at very low flow rates can partly be attributed to the inaccuracy in the measurement of very low pressure drops (order of 10^{-1} psi).

Recent experiments show a critical salt concentration exists below which the clay particles are released from the pore wall. While this finding can be incorporated in the present model, preliminary results show that for most cases, minor corrections occur and agreement between model predictions and experimental data is quite good.

The numerical values of the release and capture coefficient at eight different flow rates are summarized in Table 1.

Variation of parameters (α and β) with flow rate: The parameters α and β were found to be a function of the flow rate. The chemical effects on the release coefficient (α) were presumably the same since the shock in ionic strength was produced by a sudden replacement of 3% (by weight) sodium chloride solution by fresh water in all shock experiments. Hence, only flow rate effects are considered here.

Table 1. Values of Capture and Release Coefficient

Flow Rate (q) (cc/hr)	Initial Permeability (K_o) (md)	Release Coefficient (min^{-1})	Capture Coefficient (min^{-1})
2.0	10.2	0.011	0.03
6.0	6.8	0.033	0.01
10.0	33.3	0.030	0.01
16.0	13.20	0.08	0.10
40.0	6.30	0.22	0.30
48.0	4.10	0.35	0.45
56.0	144.0	0.31	0.45
224.0	166.0	1.20	1.40

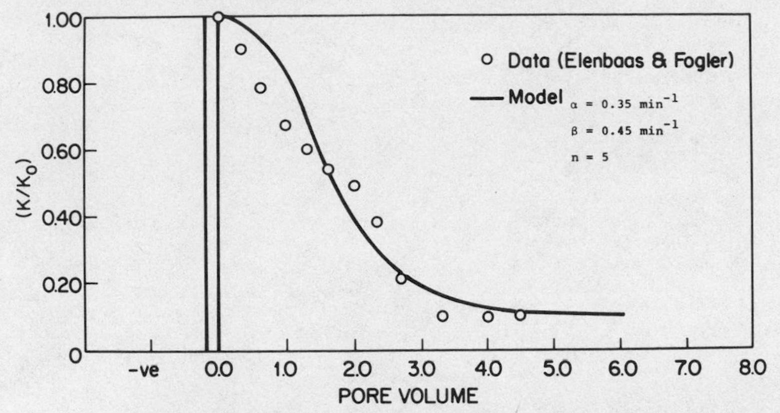

Fig. 8. Comparison of model with data. Berea, 1" diameter, 2" length; flow rate = 48 cc/hr.

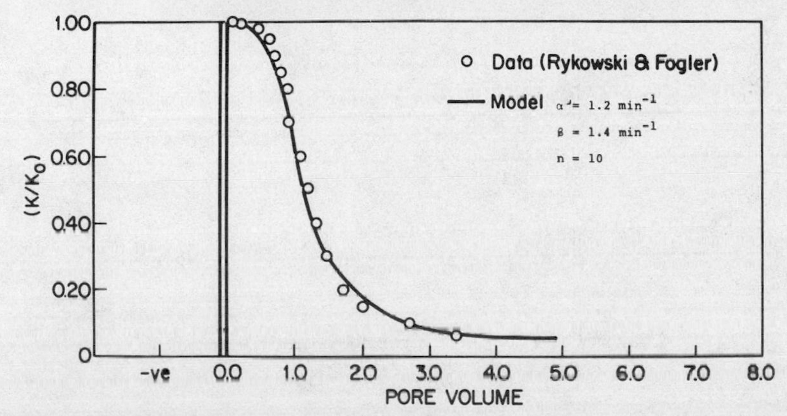

Fig. 9. Comparison of model with data. Berea, 1" diameter, 2" length; flow rate = 224 cc/hr.

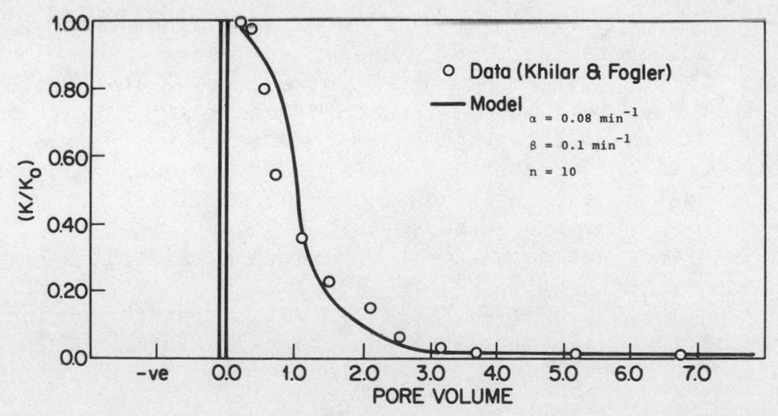

Fig. 10. Comparison of model with data. Berea, 1" diameter, 2" length; flow rate = 16 cc/hr.

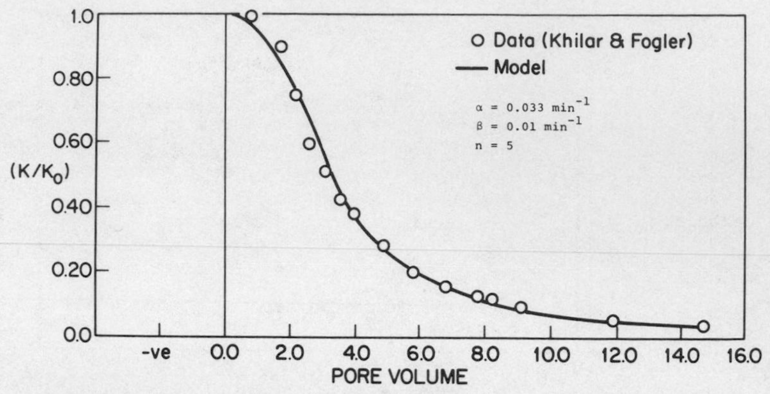

Fig. 11. Comparison of model with data. Berea, 1" diameter, 1" length; flow rate = 6 cc/hr.

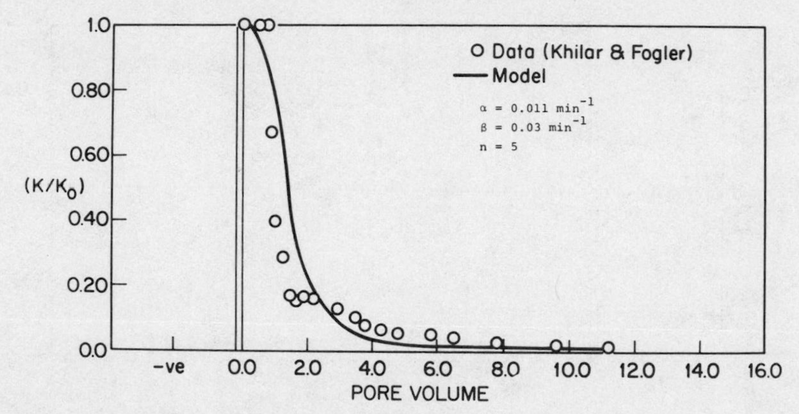

Fig. 12. Comparison of model with data. Berea, 1" diameter, 1" length; flow rate = 2 cc/hr.

The release coefficient (α) is shown as a function of flow rate (q) as in Figure 13. The release coefficient (α) was found to be directly proportional to the flow rate (q). This observed proportionality between the release coefficient (α) and the flow rate (q) is unexplained at this point in our study. However, on a speculative basis, the nature of the above relationship is comprehensible, since at higher flow rates the particles attached to the pore wall will be subjected to higher shear force. It should be emphasized that the shear force of moderate flow rates acting alone cannot release particles attached to the pore wall. However, shear force even at slow flow rates can detach particles from the pore walls if they have already been loosened by other means; for example, electric double·layer repulsion.

Capture coefficients (β), shown as the function of flow rate, are in Figure 14. Interestingly, excluding the values at very low

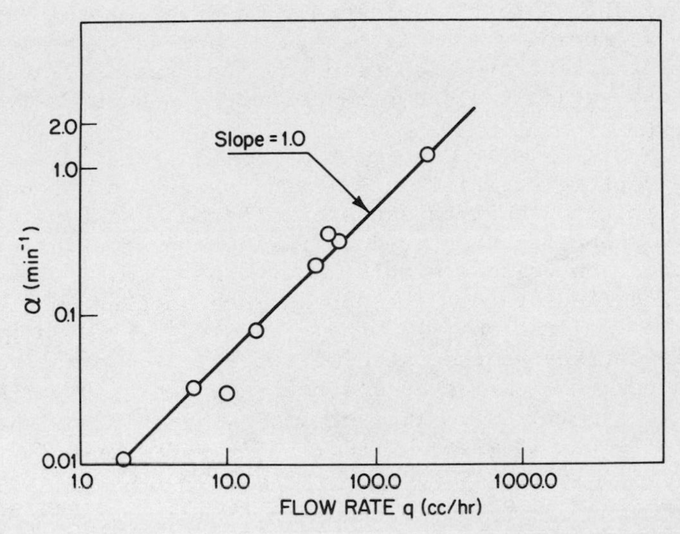

Fig. 13. Variation of release coefficient with flow rate.

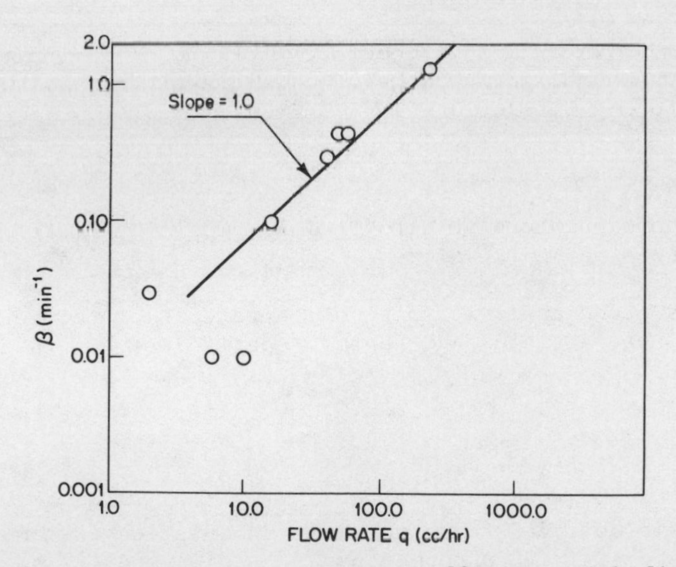

Fig. 14. Variation of capture coefficient with flow rate.

flow rates, the capture coefficient (β) is roughly proportional to the flow rate. The dispersed data points at low flow rates can partly be attributed to the inaccuracy in experimental determination of (K/K_o) during the early period of damage. The computation of (K/K_o) under the above situation involves ratios of very low pressure drops, which could not be measured as accurately as high pressure drops.

As the capture coefficient (β) was found to be proportional to the flow rate for a given sandstone (Berea sandstone was used in this work), one is encouraged to further substantiate the capture rate equation using a simple concept of deposit of particles on spherical collectors. This approach has been successfully used in waste water filtration (9). The efficiency (η) of a single collector is defined as the ratio of two rates--the rate at which particles strike the collector and the rate at which particles flow toward the collector. Rate of flow of particles toward a spherical collector is given by $(UC)\ \pi d^2/4$ where U is the undisturbed flow velocity, C is the concentration of particles and $\pi d^2/4$ is the projected area of a spherical collector of diameter (d). Therefore the rate of capture of particles for a single collector is equal to $u(e\eta\ \pi d^2/4)C$, where e is the retention efficiency.

Pore constriction is geometrically modeled as the space between two spherical collectors of diameter (d), placed in the wall of a pore of radius (r^1), where $r^1 > d$. Based on the above geometrical model of pore constrictions, it is fairly conceivable to define a number (N) as the number of collectors of diameter (d) per unit pore volume which is necessary for the total capture of particles on each collector to be equal to the actual rate of capture at all pore constrictions in a unit pore volume.

Then,

$$\left[\begin{array}{l}\text{Rate of capture of par-}\\\text{ticles per unit volume}\end{array}\right] = \frac{d\sigma_2}{dt} = \left(Ne\eta\ \frac{\pi d^2}{4}\ U\right)C \qquad (19)$$

Based on this crude analysis, rate of capture is proportional to concentration of particles and capture coefficient (β) is given by

$$\beta = \frac{\pi d^2}{4}\ e\eta NU \qquad (20)$$

The single collector efficiency (η) is approximately given by η_I where η_I is the efficiency due to direct interception only. Direct interception is believed to be the most dominant capture mechanism in the water sensitivity phenomenon. A special and extreme case of this mechanism is the so-called sieve filtration. Considering the particle sizes involved in this phenomenon (few microns in size) and the range of flow velocities used in this work

(5.4×10^{-3} cm/sec to 1.2×10^{-2} cm/sec), the contributions of other mechanisms like Brownian diffusion and particle inertia to total capture can be neglected in comparison to that of the mechanism of direct interception.

For a single spherical collector, η_I is given by (9)

$$\eta_I = \frac{3}{2}\left(\frac{d_p}{d}\right)^2 \tag{21}$$

where d_p is the particle diameter. Substituting the expression for η_I in Equation (20), one has

$$\beta = \frac{3}{8}\pi e N d_p^2 U \tag{22}$$

The values of β were plotted against interstitial velocity U on a rectangular coordinate for a flow rate range of 16 cc/hr to 224 cc/hr. The plot was approximated by a straight line with a slope equal to 0.37 cm^{-1}. Using this numerical value of slope, the order of magnitude estimate of N was found equal to 3×10^7/pore volume for an average particle size of 1μ.

CONCLUSIONS

Experiments using Berea Sandstone were carried out to elucidate the water sensitivity phenomenon. It has been shown that the permeability loss of Berea Sandstone can partially be restored by reversing the flow with salt water. Flow reversal releases the clay particles blocking pore constrictions and unlike in fresh water, these clay particles coagulate in the presence of salt water and readhere to the pore wall. The mathematical model developed to describe the rapid decline of permeability of Berea Sandstone due to the phenomenon of water sensitivity gives good agreement with experimental observations.

REFERENCES

1. N. Johnson and C. M. Beeson, Trans. AIME, 160, 43 (1945).
2. D. H. Gray and R. W. Rex, 14th National Conference on Clays and Clay Minerals, 355, 1966.
3. N. Mungan, J. Pet. Tech., 17, 1449 (1965).
4. N. Mungan, J. Can. Pet. Tech., 113 (1968).
5. F. O. Jones, Jr., J. Pet. Tech., 16, 441 (1964).
6. C. H. Hewitt, J. Pet. Tech., 15, 813 (1963).
7. K. J. Ives , "Filtration and Separation," Uplands Press, U.K., March/April, 125, 1967.
8. J. P. Herzig, D. M. Leclere, and R. LeGoff, Indust. Engr. Chem., 62, 9 (1970).

9. K. Yao, M. T. Habibbian, and C. R. O'Melia, Envir. Sci. Tech., 5, 1105 (1971).
10. R. Rajagopalan, Chi Tien, AIChE J., 23, 523 (1976).

VI: POLYMER RHEOLOGY AND SURFACTANT-POLYMER INTERACTIONS

SYNTHETIC RANDOM AND GRAFT COPOLYMERS FOR UTILIZATION IN ENHANCED OIL RECOVERY--SYNTHESIS AND RHEOLOGY

C. L. McCormick, R. D. Hester, H. H. Neidlinger
and G. C. Wildman

Department of Polymer Science
University of Southern Mississippi
Hattiesburg, Mississippi 39401, U.S.A.

Polyelectrolyte and non-polyelectrolyte acrylamide random copolymers and graft copolymers of dextran with acrylamide were synthesized to evaluate the effects of polymer composition and structure on viscosity modification and solution behavior in water and brine. Polymer solution rheological behavior and precipitation properties were measured.

The polyelectrolyte acrylamide copolymers poly(acrylamide-co-sodium acrylate), poly(acrylamide-co-acrylic acid), and hydrolyzed polyacrylamide were used to study the effect of charge density and charge distribution. The non-polyelectrolyte acrylamide copolymers N,N-dimethylacrylamide and N,N-diethylacrylamide, were prepared to study the effect of N-substituted monomers.

Dextran-g-acrylamides were synthesized with two initiation systems--a highly controllable Ceric ion/HNO_3 system and the Ferrous ion/H_2O_2 system. Mechanistic and kinetic studies were conducted with macromolecular characterization techniques including use of a unique gel permeation chromatographic method. Uncharged random- and graft-copolymers showed Newtonian behavior at low shear rates while the charged polymers showed pseudoplastic behavior.

INTRODUCTION

In recent years an increasingly large number of commercial and laboratory water-soluble polymers have been prepared for utilization as displacement fluids in enhanced oil recovery (EOR).

Unfortunately, the potential usefulness of these polymers is not easily assessed. Variations in macromolecular structure, molecular weight, branching, ionic character, and hydrogen-bonding capacity of each polymer type will, of course, cause drastic behavioral differences under a given set of field conditions.

In our continuing investigations into water-soluble polymers for EOR, we shall in this report describe the synthesis, characterization, and solution behavior of a series of random- and graft-copolymers. In this investigation our efforts have been centered on homopolyacrylamides (PAM) and their partially hydrolyzed derivatives (HPAM); random copolymers of acrylamide with sodium acrylate (PAMNaA), with acrylic acid (PAMAA), with N,N-dimethylacrylamide (PAMDMAM), or with N,N-diethylacrylamide (PAMDEAM); and acrylamide-grafted copolymers of dextram (D-g-AM). Model polymers of the above types are of great interest since they represent a wide range of molecular structure, shape, charge density, and hydrogen-bonding capability.

MACROMOLECULAR SYNTHESES OF MODEL POLYMERS

Homopolymer synthesis: Homopolymers of acrylamide, acrylic acid, and sodium acrylate were prepared in aqueous solution utilizing ammonium persulfate initiators (reactions 1, 2 and 3) (1). After polymerization, the polymers were precipitated into acetone and dried under vacuum at 50°C for 48 hours.

The nitrogen content of the homopolyacrylamide as determined by the Kjeldahl method was 19.46% as compared to a predicted value of 19.71%. Viscosity average molecular weights were calculated for two synthesized PAM samples to be 1.38×10^6 and 2.96×10^6 respectively.

Hydrolysis of acrylamide homopolymer: The synthetic PAM samples and a commercial PAM sample from Polysciences, Inc. with a reported $\bar{M}_w$ of 5 to 6 x 10^6 were hydrolyzed in aqueous solutions of 0.250 m sodium hydroxide. Hydrolysis reactions (reaction 4) were conducted under conditions shown in Table 1 utilizing 1000 ml, 3-necked flasks equipped with reflux condensers and overhead stirrers. Nitrogen was bubbled through the solutions during the reaction. At specified intervals, the reaction solutions were precipitated into acetone. The polymers were dissolved and precipitated three times and were further purified by extraction with a 20/80 water/methanol solution to remove residual sodium hydroxide. Six hydrolyzed polyacrylamide samples (Table 1) were prepared from the Polysciences standard having degrees of hydrolysis ranging from 19.2% to 65.7%.

Table 1. Reaction Parameters for Hydrolysis of Homopolyacrylamide of $\bar{M}_w$ 5 to 6 x 10^6

Sample #	PAM Conc. [m]	NaOH [m]	T (°C)	Time (min.)	% of Nitrogen	Degree of Hydrolysis
HPAM-21	0.250 ± 0.005	0.250 ± 0.005	40	10	14.99	19.2
HPAM-22	0.250 ± 0.005	0.250 ± 0.005	40	30	13.85	24.2
HPAM-23	0.250 ± 0.005	0.250 ± 0.005	40	60	13.60	25.2
HPAM-24	0.250 ± 0.005	0.250 ± 0.005	40	180	11.96	32.9
HPAM-25	0.250 ± 0.005	0.250 ± 0.005	60	300	6.66	59.7
HPAM-26	0.250 ± 0.005	0.250 ± 0.005	90	720	5.58	65.7

$$n\ CH_2{=}CH(C{=}O)NH_2 \longrightarrow {+}CH_2{-}CH(C{=}O)NH_2{+}_n \quad (1)$$

$$n\ CH_2{=}CH(C{=}O)OH \longrightarrow {+}CH_2{-}CH(C{=}O)OH{+}_n \quad (2)$$

$$n\ CH_2{=}CH(C{=}O)O^{\ominus}Na^{\oplus} \longrightarrow {+}CH_2{-}CH(C{=}O)O^{\ominus}Na^{\oplus}{+}_n \quad (3)$$

$${+}CH_2{-}CH(C{=}O)NH_2{+}_n \xrightarrow[H_2O]{NaOH} {+}CH_2{-}CH(C{=}O)O^{\ominus}Na^{\oplus}{+}_n + NH_4OH \quad (4)$$

$$n\ CH_2{=}CH(C{=}O)NH_2 + m\ CH_2{=}CH(C{=}O)O^{\ominus}Na^{\oplus} \longrightarrow \{{-}[CH_2{-}CH(C{=}O)NH_2]_n{-}[CH_2{-}CH(C{=}O)O^{\ominus}Na^{\oplus}]_m{-}\}\ \text{random} \quad (5)$$

$$n\ CH_2{=}CH(C{=}O)NH_2 + m\ CH_2{=}CH(C{=}O)OH \longrightarrow \{{-}[CH_2{-}CH(C{=}O)NH_2]_n{-}[CH_2{-}CH(C{=}O)OH]_m{-}\}\ \text{random} \quad (6)$$

$$n\ CH_2{=}CH(C{=}O)NH_2 + m\ CH_2{=}CH(C{=}O)N(CH_3)_2 \longrightarrow \{{-}[CH_2{-}CH(C{=}O)NH_2]_n{-}[CH_2{-}CH(C{=}O)N(CH_3)_2]_m{-}\}\ \text{random} \quad (7)$$

$$n\ CH_2{=}CH(C{=}O)NH_2 + m\ CH_2{=}CH(C{=}O)NEt_2 \longrightarrow \{{-}[CH_2{-}CH(C{=}O)NH_2]_n{-}[CH_2{-}CH(C{=}O)NEt_2]_m{-}\} \quad (8)$$

Random copolymer synthesis: Poly(acrylamide-co-sodium acrylates) (1) were prepared (reaction 5) by ammonium persulfate initiation in distilled water at a 10% monomers concentration at 25°C. The copolymers were isolated by precipitation into methanol, followed by freeze drying. The reactivity ratios were determined and the predicted copolymer composition was in excellent agreement with that experimentally determined by the micro Kjeldahl nitrogen analysis. Molar feed ratios of acrylamide to sodium acrylate varied from 96/4 to 55/45.

Poly(acrylamide-co-acrylic acids) (2) were prepared in feed ratios of Am/AA varying from 96/4 to 55/45. Potassium persulfate initiator was used in water at 30°C. Reactivity ratios were determined to be $r_1 = 0.39 \pm 0.05$ and $r_2 = 1.40 \pm 0.11$ for acrylamide and acrylic acid respectively. Copolymer compositions agreed well with those theoretically projected values.

Random copolymers of acrylamide with N,N-dimethylacrylamide and with N,N-diethylacrylamide were prepared as illustrated in reactions (7) and (8). The respective copolymers, PAMDMAM and PAMDEAM, have been synthesized to elucidate more fully the role of hydrogen-bonding and N-substitution on viscosity modification. These polymers were synthesized using potassium persulfate initiators in water or 30% methanol/water solutions. Reaction conditions and monomer feed ratios are given in Table 2. The resulting copolymers were purified by precipitation into acetone followed by vacuum drying at 50°C for 60 hours.

GRAFT COPOLYMERIZATION

Poly(dextran-g-acrylamides) from ceric ion initiation (2): Dextran-g-acrylamide copolymers were prepared by ceric ion initiation in water at 25°C using 0.05 M nitric acid. The reactions were conducted for three hours with an initial dextran concentration of 0.0617 M (moles of glucose units) and an acrylamide concentration of 1.0M. The molecular weight of the dextran obtained from ICN Pharmaceuticals, Inc. was reported to be $\bar{M}_v$ = 200,000 to 300,000. Ceric ion concentrations were varied from 2.5×10^{-3} moles/liter to 10.0×10^{-3} moles/liter.

The crude reaction products, which contained unreacted dextran, homopolyacrylamide, and graft copolymer were weighed and then analyzed for nitrogen using the Kjeldahl method. These polymers were then extracted using a soxhlet apparatus with a solvent mixture of N,N-dimethyl formamide and acetic acid. The intrinsic viscosities were determined at 30°C using a Cannon-Fenske viscometer. The nitrogen content was again determined. The lack of efficiency of the solvents tested for selectively separating homopolymer from graft copolymer for our dextran-g-acrylamide system led us to extend our efforts in polymer refinement by fractionation techniques. Thus, after the polymerization reaction, 250 ml of the product solution was diluted with distilled water to a total volume of 2500 ml. This solution was equally divided into ten parts and fractionally precipitated with increasing amounts of acetone. The concentration of dextran which was present in the precipitate was determined by the anthrone test. The relatively smooth fractional precipitation curves obtained are consistent with rather uniform grafting of the polysaccharide substrates with the ceric ion. The percentage of acrylamide grafting onto the

Table 2a. Reaction Parameters for the Synthesis of Poly N,N-dimethylacrylamide-co-acrylamide in 30% MeOH/H_2O Solution

Sample #	Acrylamide (AM) Conc. [m]	N,N-Dimethyl-acrylamide (DMAM) Conc. [m]	$K_2S_2O_8$ Conc. [m]	Mole Fraction of DMAM in Feed	Rxn Temp. (°C)	Rxn Time (hr)	Conversion %
PAM-5	2.111	--	4.245×10^{-3}	0.00	40	3	98.8
PAMDMAM-1	1.688	0.2982	3.971×10^{-3}	0.15	40 60	6 3	98.4
PAMDMAN-2	1.322	0.5664	3.774×10^{-3}	0.30	40 60	8 2	97.7
PAMDMAM-3	0.8886	0.8870	3.578×10^{-3}	0.50	40 60	8 2	99.8
PDMAM-1	--	1.514	3.089×10^{-3}	1.00	40 60	8 2	96.3

Table 2b. Reaction Parameters for Preparation of Polyacrylamide-CO-N,N-diethylacrylamides in water

Sample #	Acrylamide (AM) Conc. [m]	Diethyl-acrylamide (DEAM) Conc. [m]	$K_2S_2O_8$ Conc. [m]	Mole Fraction of DEAM in Feed	Rxn Temp. (°C)	Rxn Time (hr)	Conversion %
PAMDEAM-1	1.603	0.2832	3.773×10^{-3}	0.15	40	12	92.3
PAMDEAM-3	3.188	1.3630	9.114×10^{-3}	0.30	60	5	94.6

dextran as determined by nitrogen using the Kjeldahl method and the intrinsic viscosity are given in Table 3. Gel permeation chromatography studies indicated very little homopolyacrylamide or unreacted dextran in the fractionated samples.

Poly(dextran-g-acrylamides) from ferrous ion--hydrogen peroxide initiation: Graft copolymerization of acrylamide onto dextran with ferrous ion-hydrogen peroxide initiation is being studied in detail. The effect of variation of each reaction parameter on graft copolymerization behavior may be measured by direct gel permeation chromatographic analysis of the final product polymer solution.

The graft copolymerizations were conducted in a three-neck flask fitted with a nitrogen bubbler, stirrer, and dropping funnel. In a typical reaction, dextran ($\bar{M}_v$ 100,000-200,000) was dissolved in distilled water under a nitrogen atmosphere with stirring for 10 minutes. The ferrous ion solution was added and 30 minutes were allowed for adsorption. Acrylamide monomer was added and stirred for 10 minutes and then 50 ml of hydrogen peroxide solution were added dropwise for 20 minutes. The variations of reaction parameters are shown in Table 4. Reaction series PA, PB, PC, PC, and PE correspond to variations of ferrous ion concentration, hydrogen peroxide concentration, reaction time, monomer concentration, and dextran substrate concentration, respectively.

After each reaction, the product polymer solution was diluted to about 0.5 g/dl., and 50 ml of this solution was precipitated into 1000 ml of methanol. The precipitated polymer was filtered and dried in a desiccator under reduced pressure overnight and weighted to determine conversion.

The distribution of the polymer product was followed by gel permeation chromatography utilizing a Waters Associates Liquid Chromatograph. The instrument conditions were: elution solvent, distilled and irradiated water; temperature, 37°C; column, stainless steel tubing of inner diameter 0.462 cm, length 60 cm; flow rate, 1.042 ml per minute; pressure, 750 psi. The column packing material was Electro Nucleonics Inc. glass of nominal size 3000Å and 1400Å. Three of 3000Å and one of 1400Å pore size columns were connected in series and placed in an oven which was maintained at 37°C. Sample concentration was 0.3 g/dl and 50 μl of each sample were injected through the injection loop. Detection was performed with a Waters Associates differential refractometer Model R401. A Waters Associates UV detector (Model 440) was used at a wavelength at 254 nm.

The results of extensive mechanistic studies (3) indicate that optimum grafting conditions (reduced homopolymerization) appear to be 0.64 x 10^{-5} of Fe^{+2} ion per gram of dextran, 30-40

Table 3. Reaction Parameters for the Graft-Copolymerization of Dextran with Acrylamide Using Ce(IV).

Sample Number	Dextran* [m]	Acrylamide [m]	Ce(IV) [m]	T(°C)	Time (hrs)	% AM in Copolymer	[η]
AD 17-1	0.062	1.00	0.25×10^{-2}	25	3	71	5.1
AD 17-2	0.062	1.00	0.50×10^{-2}	25	3	57	4.5
AD 17-3	0.062	1.00	0.75×10^{-2}	25	3	59	3.6
AD 17-4	0.062	1.00	1.00×10^{-2}	25	3	57	3.1

*Dextran $\bar{M}_v$ = 200,000 - 300,000

Table 4. Reaction Parameters for Synthesis of Dextran-g-acrylamide copolymers

Sample	Reaction Conditions					Conv.
	Dextran	Acrylamide	Fe^{+2}	H_2O_2	Rxn Time	%
PA-1	1% (w/w)	0.5 M	½ EA*	20 X**	3 hr.	61.7
PA-2	1%	0.5 M	1 EA	20 X	3 hr.	71.4
PA-3	1%	0.5 M	2 EA	20 X	3 hr.	81.8
PA-4	1%	0.5 M	4 EA	20 X	3 hr.	82.2
PB-1	1%	0.5 M	1 EA	10 X	3 hr.	63.1
PB-2	1%	0.5 M	1 EA	30 X	3 hr.	71.6
PB-3	1%	0.5 M	1 EA	40 X	3 hr.	76.0
PB-4	1%	0.5 M	1 EA	50 X	3 hr.	74.7
PB-5	1%	0.5 M	1 EA	60 X	3 hr.	70.3
PC-1	1%	0.5 M	1 EA	20 X	½ hr.	34.5
PC-2	1%	0.5 M	1 EA	20 X	1 hr.	41.7
PC-3	1%	0.5 M	1 EA	20 X	2 hr.	58.5
PC-4	1%	0.5 M	1 EA	20 X	3 hr.	71.4
PC-5	1%	0.5 M	1 EA	20 X	4 hr.	86.2
PD-2	1%	1 M	1 EA	20 X	3 hr.	82.0
PD-3	1%	1.5 M	1 EA	20 X	3 hr.	70.3
PE-2	2%	0.5 M	1 EA	20 X	3 hr.	68.9
PE-3	4%	0.5 M	1 EA	20 X	3 hr.	69.2

*EA refers to equivalent amount of adsorption; 0.64×10^{-5} moles per 1 g dextran.

**X refers to times the amount of Fe^{+2} concentration.

times this concentration of H_2O_2, a reaction time of 4 hours, 1% dextran ($\bar{M}_v$ for this experiment 200,000-300,000), and an acrylamide monomer concentration of 1 to 2 moles/liter.

The dextran-g-acrylamide sample PA-2 (Table 4) exhibited an intrinsic viscosity of 3.4 in water as compared to 0.3 for the unreacted dextran. Micro Kjeldahl analysis indicated 65% grafting (defined as weight of chain/total weight of graft copolymer x 100).

MACROMOLECULAR CHARACTERIZATION

Gel Permeation Chromatography: During the initial phases of our studies of high molecular weight, water-soluble polymers, a number of formidable problems were discovered. Useful, pressure-stable packings with pore sizes large enough for efficient separation of high molecular weight, highly extended macromolecules in water solvent alone are not readily available. Secondly, fractionated, water-soluble polymeric standards of very high molecular weight are not available. Finally, ionic strength (salt and pH) are extremely important, especially for charged polymers.

The limitation of not having large enough pores in the packing material (stationary phase) for efficient separation can be overcome by changing polymer-solvent interactions to effectively "shrink" the macromolecules so that they properly fit into lower pore size packings. This "shrinkage" may be accomplished by adding specific quantities of salt or nonsolvent to a "good" solvent in the mobile phase. If one assumes that the "Universal Calibration" technique, widely used in gel permeation chromatography, is valid and that a linear relationship exists between shrinkage of standards and unknowns under the same conditions, molecular weights based on size may be assigned.

Porous glass packings exhibit the desired long term stationary phase stability under high pressures and were chosen for our present work. In water solution, the choice of salt concentration and pH are critical for efficient separation with the porous glass packing. The effects of salt concentration, pH, and the stationary phase have been recently reviewed in an excellent report by Cooper and Van Derveer (4). Buytenhuys and Van Der Maeden (5) reported a mixed solvent of tetramethylammonium hydroxide in water buffered to a pH of 5.0 to be efficient with porous glass column packing in separation of water-soluble polymers. We have modified the above solvent with methanol in our analysis of a variety of water-soluble random copolymers and graft copolymers.

An experiment was conducted on a Waters Liquid Chromatograph to evaluate the potential molecular "shrinking" technique for determining molecular weights for high molecular weight polymers in controlled porous glass. Figure 1 illustrates the desired effect of adding a nonsolvent (methanol) to the water solutions of a selected high molecular weight polymer (*) and standard polymers represented by o and x on the ln[η]M vs. V_e plot where [η] is intrinsic viscosity, M is molecular weight (based on hydrodynamic volume) and V_e is the elution volume. In water solutions this high molecular weight polymer (*) would elute at the void volume, V_o, of the column if the molecules were too large to pass into the available pores. On the other hand, very small molecules would pass freely through the pores and should theoretically elute at the total permeation volume, V_p. Obviously, for efficient separation of molecular species, pore sizes should be carefully chosen such that large changes in elution volume should be obtained for relatively small changes in hydrodynamic volume. The hydrodynamic volume is directly related to molecular weight and intrinsic viscosity.

As methanol is added to the aqueous mobile phase for a given set of columns, the value of ln[η]M for a specific sample should be reduced (consistent with hydrodynamic volume shrinkage for highly water-soluble polymers). Additionally, assuming that pH, ionic strength, etc. are constant for the studied system, elution volume should also vary in a linear fashion when changing from water to a methanol-water solution. A high molecular weight polymer (∞) which elutes at the void volume in water may be brought on scale by the addition of nonsolvent (Figure 1) and thus the relative molecular weight can be determined by comparison to the standards. It should be stressed that in order for studies of this type to be useful with polyelectrolytes, salt concentrations and pH values must be adjusted to minimize both the charge-charge interaction between the polymers and the stationary phase and also to suppress the Donnan effects.

Excellent results were obtained utilizing poly(styrene sulfonate) standards in 20% methanol/water solution with 0.2 M tetramethylammonium hydroxide buffered to a pH of 4.5 with acetic acid. At this pH and ionic strength the polystyrene sulfonates may be largely in the acid form and thus would be expected to behave as uncharged macromolecules. Figure 2 shows a linear response of ln[η]M vs. elution volume for a four column set (three with 3000Å glass packing and one with 1400Å glass packing). Columns were stainless steel, 60 cm in length and 7.62 mm in diameter.

Molecular weight was calculated for one polyacrylamide sample obtained from Polysciences using the above calibration curve. A value of 5 to 6 x 10^6 was reported by the manufacturer; our experimentally determined value was 8.4 x 10^6.

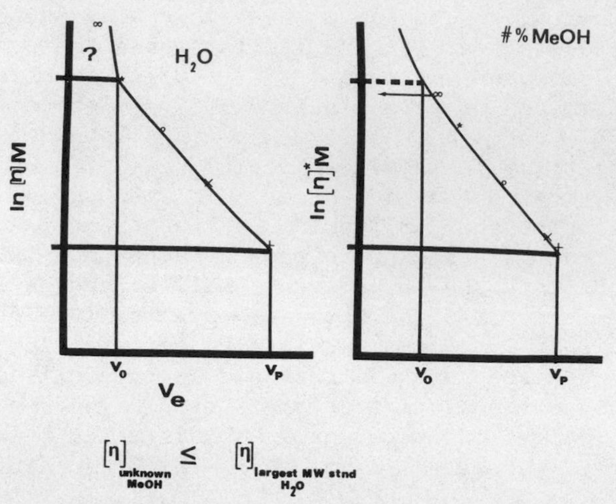

Fig. 1. Effect of charges in solvent on the elution volume vs. ln[η]M calibration curve.

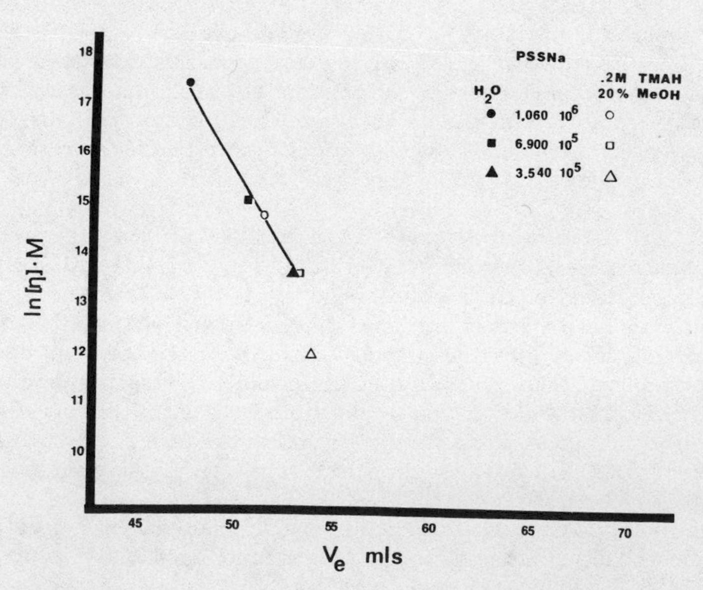

Fig. 2. Elution volume vs. ln[η]M for a series of poly(sodium sulfonate styrenes).

DILUTE SOLUTION STUDIES

Dilute solution behavior of the synthesized copolymers and the corresponding homopolymers has been investigated by intrinsic viscosity studies at different temperatures in deionized water and salt solutions of various ionic strengths. Intrinsic viscosities, $[\eta]$, were obtained in the usual manner (1) utilizing the Huggins equation, Kraemer equation, Schulz-Blaschke equation and Fuoss equation. Viscosity-average-molecular weights for homo-polyacrylamide samples were determined from a $[\eta]$-M relationship given by Scholtan (6):

$$[\eta] = 6.31 \times 10^{-3} M^{0.8} \text{ in water at } 25°\text{C} .$$

$[\eta]$-M relations for PAMNaA copolymers have been recently reported in literature (7).

Since aqueous homo-polyacrylamide solutions show a solution viscosity instability or aging effect, viscosities were also determined as a function of time. The rate of change of reduced viscosity with time is greatest for high concentrations of polyacrylamide, but decreases for more dilute samples. The aging effect is only observed above a critical molecular weight ($\bar{M}_v \simeq 10^6$) and increases with increasing molecular weight (8). Figure 3 shows the effect of ionic strength on the aging behavior of a PAM of $\bar{M}_v$ = 1.36×10^6. After approximately 100h storage viscosity stability was achieved. Reduced as well as intrinsic viscosities pass obviously through a maximum with increased salt concentration. This trend is also observed in the presence of LiCl and urea. Such observations may be related to a change in intramolecular H-bonding that would be destroyed in the presence of salt or urea and to the existence of an osmotic effect at higher salt concentrations that would cause a decrease in the viscosity.

Figure 4 shows the change in reduced viscosity for the above PAM (polymer concentration = 0.0318 g/dl) in 1.5% NaCl solution as a function of time at different pH-values. Stabilization can be observed after approximately 100 hours for pH-values 7 and 2. At pH = 12 we observe an increase in the reduced viscosity due to saponification of the PAM. The synthesized dextran-g-acrylamide copolymers do not show any aging effects indicating small PAM-graft lengths. We believe that the effect of the solution instability may be explained not by degradation but by disentanglement of the polymer molecules and/or a change of conformation of the solution structure due to changes in intramolecular hydrogen bonding. It thus appears that suitable standard conditions should be found for preparation of PAM solutions in order to obtain reliable viscosity values.

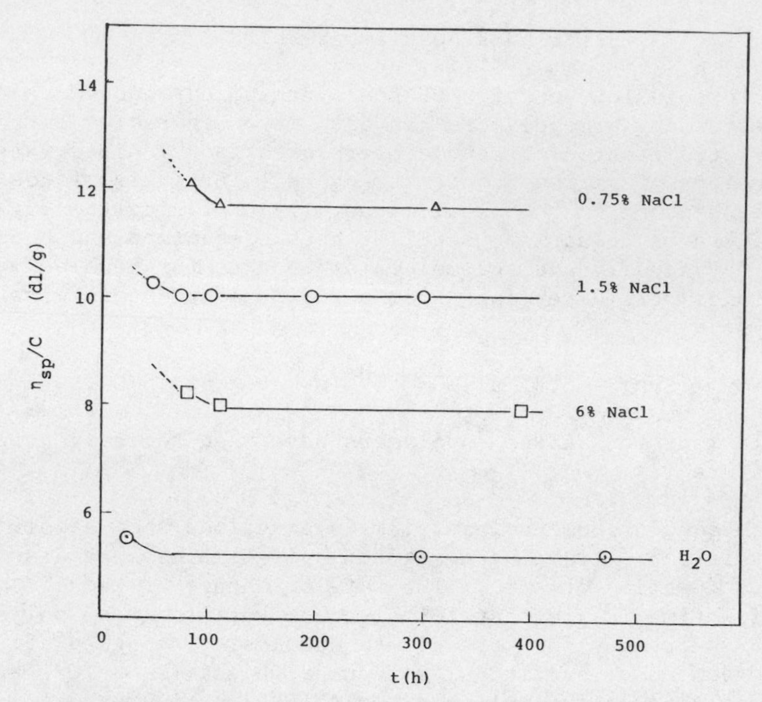

Fig. 3. Reduced viscosity of PAM C-44 (c = 0.063 g/dl) as a function of time for various NaCl concentrations.

Table 5 and Table 6 give some representative viscosity data for PAMAA and D-g-AM respectively in 1.5% NaCl solution after allowing for viscosity stability. The last column in the table displays the ratio of the intrinsic viscosity in deionized water, $[\eta]H_2O$, and in 1.5% NaCl solution, $[\eta]NaCl$. The ratio for PAMAA is a function of copolymer composition and structure and shows a maximum at about 50 mole % acrylamide; we observe here an almost 50-fold decrease in $[\eta]$ for a 1.5% NaCl solution, whereas $[\eta]$ for homo-PAM is relatively insensitive to salt effects. The intrinsic viscosity for D-g-AM in salt solution shows no significant change from the intrinsic viscosity in water. The ratio $[\eta]H_2O/[\eta]NaCl$ is comparable to that of dextran.

Figure 5 gives some representative viscosity data for partially hydrolyzed PAM of different molecular weights in 1.5% NaCl solution. $[\eta]$ passes through a maximum at about 50 mole % Na-acrylate content. It is obvious, that even in the presence of high amounts of salt, the copolymer has a higher value of $[\eta]$ than the corresponding homopolymer. The $[\eta]$-maximum may be interpreted on the basis of an increase in the molecular hydrodynamic volume due to hindered rotation caused by cyclization between amide and carboxylic-anion group (9) which might be formulated as follows:

Table 5. Viscosity Data for Poly(acrylamide-co-acrylic Acid) Samples in 1.5% NaCl Solution at 30°C

Polymer	Copolymer Composition [mole % acrylamide]	[η] (dl/g)	$[\eta]H_2O/[\eta]NaCl$
PAMAA-29	71.5	3.6	35.3
PAMAA-30	62.2	4.7	42.6
PAMAA-31	51.2	5.8	49.5
PAMAA-32	42.7	8.4	34.9

Table 6. Viscosity Data for Poly(dextran-g-acrylamide) Samples in 1.5% NaCl Solution at 30°C

Polymer	Ce^{+4} [m]	Grafting Efficiency (%)	[η] (dl/g)	$[\eta]H_2O/[\eta]NaCl$
D-g-AM 10-1	0.0025	28.1	1.64	1.16
D-g-AM 10-2	0.0050	31.6	2.68	1.08
D-g-AM 10-3	0.0075	32.6	2.25	1.20
D-g-AM 10-4	0.0100	33.7	2.05	1.17
Dextran	--	--	0.29	1.08

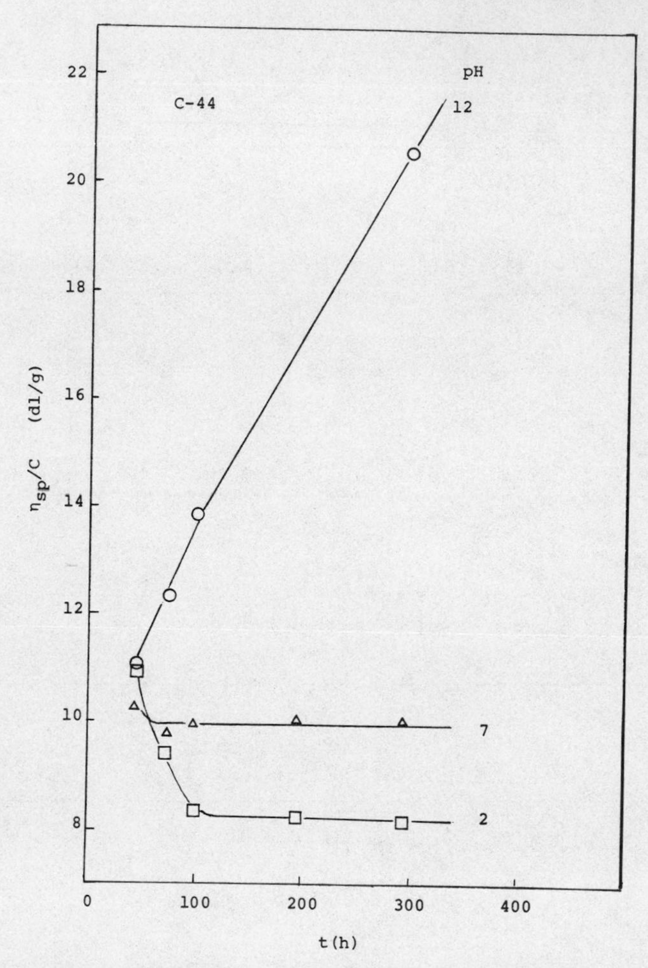

Fig. 4. Reduced viscosity of 1.5% NaCl solutions of PAM C-44 (C = 0.0318 g/dl) as a function of time for different pH values.

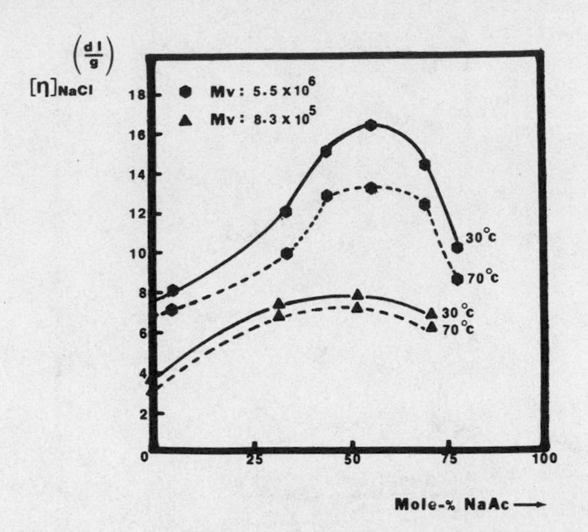

Fig. 5. Intrinsic viscosity of partially hydrolyzed polyacrylamides in 1.5% NaCl solution.

```
                 H2
        H        C        H
~~ H2C—C    /        \    C—CH2 ~~
          /                \
       O=C                  C=O
          \                /
           O⊖             N
     Na⊕      ·· H  /       \ H
```

Figure 6 shows the ratio $[\eta]H_2O/[\eta]NaCl$ as a function of copolymer composition. The maximum at about 30 mole % Na-acrylate can be attributed to increased cyclization in pure water as compared in NaCl solutions. The increase in the ratio at higher extents of hydrolysis results from an increase in the hydrodynamic volume in pure water due to increased electrostatic interactions along the molecule.

The effect of ionic charge density and added salts on the hydrodynamic volume of the copolymers affects not only the rheological performance but also influences the thermodynamic stability of these polymer solutions. Investigated were HPAM samples of varying extent of hydrolysis (from 0 to 66 mole % Na-acrylate) in salt solutions of different valences. For monovalent salts, no precipitation has been observed up to saturated salt concentrations. The addition of multivalent gegenions, on the other hand, leads to precipitation that depends on the concentration of polymer and added salt and on the degree of hydrolysis. Ca^{2+} salts precipitate highly hydrolyzed samples more rapidly than Mg^{2+} salts, but are

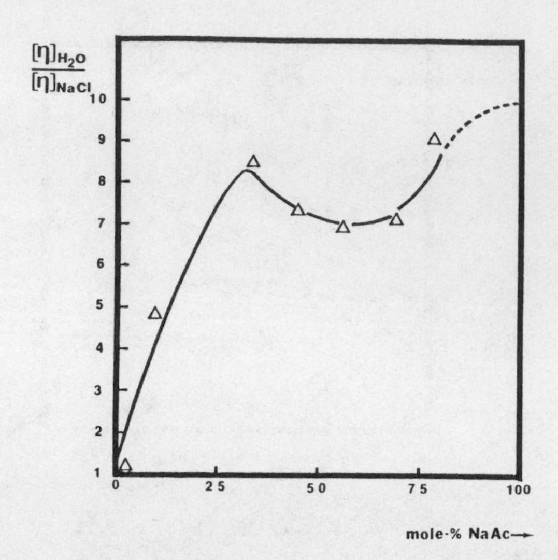

Fig. 6. Relative viscosity change of partially hydrolyzed polyacrylamides in water and brine at 30°C as a function of copolymer composition, ($\bar{M}_v = 5.5 \times 10^6$).

less effective than Zn^{2+} and Al^{3+} salts. In some cases the precipitation is observed at an intermediate polymer concentration and the solutions become clear again as polymer concentration increases (Figure 7). The value and position of the cloud point depends for a given polymer concentration on the nature of the multivalent salt, on the degree of hydrolysis, and on the salt concentration.

In conclusion, the influence of solvents on the dilute solution behavior of the copolymers can be ascribed primarily to changes in the molecular expansion brought about by solvent action, and changes in the chain flexibility. From a practical point of view, there is no advantage in using too highly hydrolyzed polyacrylamides in enhanced oil recovery, especially when the water contains large amounts of salts. For highly hydrolyzed samples the viscosity sharply decreases when mono- and especially multivalent salts are added; precipitation then occurs as more multivalent salt is added. Effectiveness will depend on exact field conditions, especially on the nature and the concentration of added salts, and the temperature and pressure.

POLYMER SOLUTION PSEUDOPLASTICITY AT LOW SHEAR RATE CONDITIONS

Many water soluble polymers show a solution behavior which is non-Newtonian. When flow curves are experimentally obtained that plot the fluid shear stress versus shear rate, the curves show that in most cases the solution rheology is pseudoplastic. For pseudo-

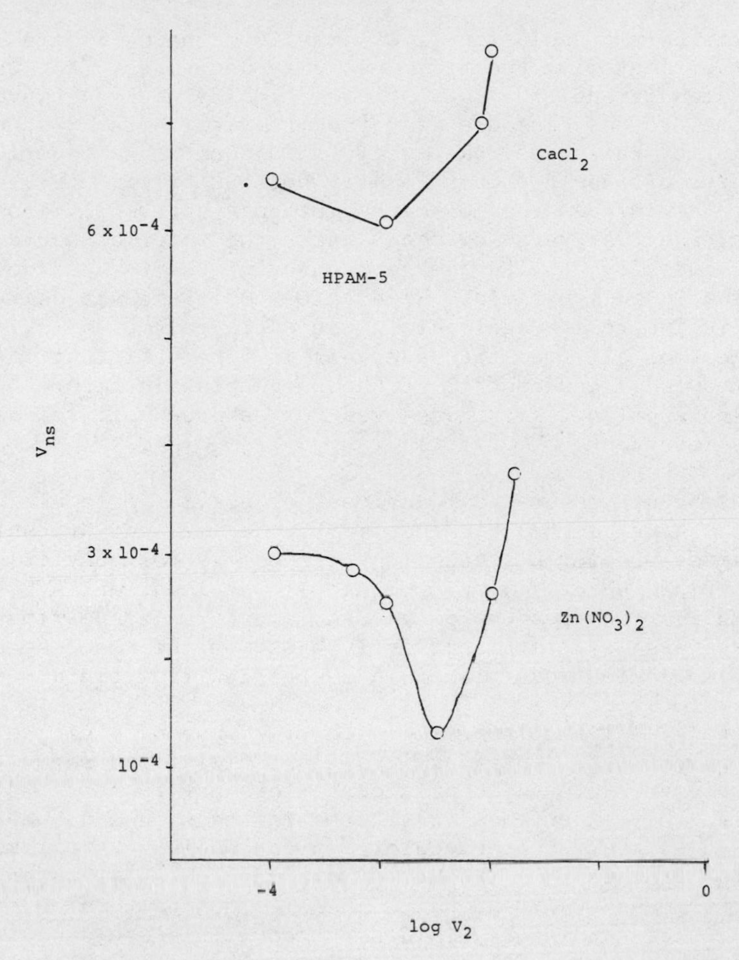

Fig. 7. Dependence of the volume fraction of the precipitant V_{ns} on the logarithm of the volume fraction V_2 of the polymer at the cloud point at T = 30°C.

plastic fluids, the apparent viscosity, which is the ratio of fluid shear stress to shear rate, decreases as the shear rate increases. A pseudoplastic or shear thinning property is desirable for polymer solutions which are used in enhanced oil recovery.

High shear rate conditions exist near the well bore when polymer solutions are injected into an oil reservoir. At high shear rates the solution viscosity is low because of its pseudoplastic nature. This lower solution viscosity results in less fluid flow resistance and thus less power is required for injection of pseudoplastic polymer solutions into a reservoir.

As the polymer solution flows away from the well bore into the reservoir interior the fluid velocity decreases. For perfect plug flow conditions, the local fluid velocity would be inversely proportional to the distance away from the well bore. As the fluid velocity decreases, the local fluid shear rate experienced in the reservoir would proportionally decrease (10). As the fluid shear rate diminishes the polymer solution's apparent viscosity would increase. At very low shear rates the solution would have a constant viscosity, called the zero shear viscosity, and would behave as a Newtonian fluid. High values of fluid viscosity away from the well bore are desirable because the polymer solution's displacement of oil from the reservoir is more efficient (11). Thus, a knowledge and control of the pseudoplastic nature of the water soluble polymer solutions used for enhanced oil recovery is extremely important.

Two separate theoretical analyses of polymer molecular behavior in solution have been carried out by Bueche (12) and Graessley (13). Both analyses predict a pseudoplastic rheology for polymer solutions in simple shear. However, Bueche's explanation is based on the response of independent polymer molecules to a shearing flow field and Graessley's explanation is based on the response of entangled molecules when placed in a shearing flow field.

Bueche's analysis is based on the response of a single free draining polymer coil rotating in a shear field. As segments of the molecule are rotated about a center of gravity by the shearing flow field, they experience an alternating tension and compression. Bueche assumed that each molecular segment would behave like a small mass attached to a linear spring. At low shear rates the molecular segments can respond to the alternating tension and compression forces, and energy is dissipated. This would be detected by a higher apparent solution viscosity. The viscous behavior of such a system was treated quantitatively and the result obtained by Bueche for dilute polymer solutions is:

$$\frac{\eta-\eta_o}{\eta_o-\eta_s} = 1 - \frac{6}{\pi^2} \sum_{n=1}^{N} \frac{(\dot{\gamma}\lambda)^2}{\eta^2(\eta^4+\dot{\gamma}^2\lambda^2)} \left\{2 - \frac{(\dot{\gamma}\lambda)^2}{\eta^4+\dot{\gamma}^2\lambda^2}\right\} \qquad (1)$$

In equation (1), γ is the shear rate, η_s is the solvent viscosity, η_o is the solution zero shear viscosity, γ is a characteristic molecular response time, η is the solution viscosity, and N is the total number of molecular segments in a macromolecular coil.

Bueche also developed a simplification of equation (1) by dropping terms which are not important except at high shear rates. Thus at low shear rates:

$$\frac{\eta-\eta_o}{\eta_o-\eta_s} = 1 - (\lambda\dot{\gamma})^{1/2} \tag{2}$$

The characteristic molecular response time in equation (2) was defined for a linear molecule as:

$$\lambda = \frac{12(\eta_o-\eta_s)M\alpha}{\pi^2 RTc} \tag{3}$$

In equation (3), c is the concentration of polymer in solution, T is the absolute temperature, M is the polymer molecular weight, R is the gas constant, and α is a shielding factor.

The shielding factor compensates for the fact that polymer molecules are not freely draining. As the polymer molecule restricts the flow of solvent within the polymer coil, more energy is required to rotate the coil. For free draining polymer coils of linear molecules the shielding factor is one; for a nondraining coil the shielding factor is less than one. In general, for long molecules the shielding factor is a function of the ratio of segment weight to polymer molecular weight.

$$\alpha = f\left(\frac{M_o}{M}\right) \tag{4}$$

In equation (4), M_o is the effective molecular weight of the molecular segments forming the polymer coil. Polyelectrolytes and other stiff molecules have a large polymer segment weight and thus solutions of these polymers will show a larger degree of pseudoplasticity than equivalent molecular weight nonionic polymers. Also molecular response times of branched polymers are usually greater than the response times for equivalent molecular weight linear polymers (14).

The shear stress, τ, versus shear rate, $\dot{\gamma}$, data obtained on polymer solutions by a shear viscometer can be analyzed by using a linear form of equation (2).

$$\eta = \frac{\tau}{\dot{\gamma}} = \eta_o - (\eta_o-\eta_s)\lambda^{1/2}\,\dot{\gamma}^{1/2} \tag{5}$$

As seen by equation (5), when the apparent viscosity is plotted versus the square root of the shear rate, a linear function will be obtained. From a linear regression analysis of viscometer flow curve data, an intercept, a, and slope, b, can be determined. The zero shear viscosity is equal to the intercept, and the molecular response time is a function of the slope and the intercept as shown below:

$$\eta_o = a \tag{6}$$

$$\lambda = \frac{b^2}{(\eta_o - \eta_s)^2} \tag{7a}$$

$$\lambda = \frac{b^2}{(a - \eta_s)^2} \tag{7b}$$

In concentrated solutions, Bueche's idea of single polymer coils undergoing rotation without interaction with other coils is not reasonable. In contrast to Bueche, Graessley explained polymer solution pseudoplasticity by the dynamics of polymer molecules entangling and disentangling. The degree of entanglement that exists between molecules approaching one another in a shear field is dependent upon the average response time required for molecular segments to move and intermesh, λ, compared to the time the molecules remain sufficiently close to contact one another and become entangled. In simple shear, the contact time is inversely proportional to the shear rate. If the shear rate is low, the contact time will exceed the molecular response time and a high degree of molecular entanglement will exist. As entanglement increases more energy is dissipated. This results in a higher apparent solution viscosity at lower shear rates. Graessley reduced his theoretical analysis to the following function:

$$\frac{\eta}{\eta_o} = \frac{2}{\pi}\left[\cot^{-1}\theta + \frac{\theta(1-\theta^2)}{(1+\theta^2)^2}\right] \tag{8a}$$

where

$$\theta = \frac{\eta\dot{\gamma}\lambda}{4\eta_o} \tag{8b}$$

In equation (8), η is the apparent solution viscosity, η_o is the zero shear viscosity, and $\dot{\gamma}$ is the shear rate. The characteristic molecular response time, λ, is defined by equation (3) when the shielding factor is one. In equation (8), the terms in the bracket can be approximated by two series expansions.

$$\cot^{-1}\theta = \frac{\pi}{2} - \theta + \frac{\theta^3}{3} - \frac{\theta^5}{5} + \ldots \tag{9}$$

$$\frac{\theta(1-\theta^2)}{(1+\theta^2)^2} = \theta - 3\theta^2 + 5\theta^5 - 7\theta^7 + \ldots \tag{10}$$

Substitution of these series expansions into equation (9) gives:

$$\frac{\eta}{\eta_o} = \frac{2}{\pi}\left[\frac{\pi}{2} - \frac{8\theta^3}{3} + \frac{24\theta^5}{5} - \frac{48\theta^7}{7} + \ldots\right] \tag{11}$$

If θ is less than unity, then the higher degree terms of θ can be neglected and equation (11) can be approximated by:

$$\frac{\eta}{\eta_o} \cong 1 - \frac{16\theta^3}{3\pi} \tag{12}$$

when

$$\theta < 1$$

Substitution of the definition of θ, equation (8b), into equation (12) gives:

$$\frac{\eta}{\eta_o} = 1 - \frac{1}{12\pi}\left(\frac{\eta\dot{\gamma}\lambda}{\eta_o}\right)^3 \tag{13}$$

Equation (13) can be used to analyze viscometer flow curve data only when the θ parameter is less than one. The θ parameter can be made small by collecting solution viscosity data at the inception of pseudoplastic behavior (low shear rate). Equation (13) can be rearranged into a linear form.

$$\eta = \frac{\tau}{\dot{\gamma}} = \eta_o - \frac{\lambda^3}{12\pi\eta_o^2}\tau^3 \tag{14}$$

Equation (14) can be used to determine the molecular response time and the zero shear viscosity by plotting the apparent solution viscosity versus the cube of the shear stress. From a linear regression analysis of viscometer data, an intercept, a, and slope, b, can be determined. The intercept is equal to the zero shear viscosity, and the molecular response time is a function of the slope and intercept.

$$\eta_o = a \tag{15}$$

$$\lambda = (-12\pi b\eta_o^2)^{1/3} \tag{16a}$$

$$\lambda = (-12\pi ba^2)^{1/3} \tag{16b}$$

From a knowledge of the molecular response time, polymer concentration, temperature, and the zero shear viscosity, a polymer molecular weight can be calculated from equation (3).

Although both the Bueche and the Graessley analyses were developed for monodisperse polymers in solution, experimental evidence suggests that both analyses can be used for polydisperse systems if the viscosity average molecular weight is used in the definition of the average molecular response time (15). More complicated relationships for dealing with polydispersity can be found in the literature (16,17).

The molecular theories of Bueche and Graessley are similar in that both theories relate the pseudoplastic nature of polymer solutions as a function of a dimensionless Deborah number which represents a ratio of the response time of the polymer molecules in solution, λ, to the time scale of the flow process (18). In simple shearing the time scale of the flow process is inversely proportional to the shear rate. Thus both equation (5), developed from the Bueche theory, and equation (14), developed from the Graessley theory, can be expressed in terms of the Deborah number, N_{De}.

$$\frac{\eta}{\eta_o} = 1 - (N_{De})^{1/2} \tag{17}$$

$$\frac{\eta}{\eta_o} = 1 - \left(\frac{\eta^3}{12\pi\eta_o^3}\right) N_{De}^3 \tag{18}$$

where:

$$N_{De} \propto \lambda\dot{\gamma} \tag{19}$$

$$\eta_o >> \eta_s \tag{20}$$

In conclusion, both the Graessley and Bueche theories confirm the general experimental observation that the reduced solution viscosity, η/η_o, is a function of $\lambda\dot{\gamma}$. Therefore, solution rheological behavior can be described by two material parameters, the zero shear viscosity and polymer response time. In turn, these parameters are functions of macromolecular structure and solvent properties.

A Haake® coaxial cylinder viscometer (model RV100 with sensor system ME30) was used to study the rheological properties of several water soluble polymer solutions. Aqueous solution sets of five polyacrylamides having various concentrations were prepared using deionized water. The techniques for producing the hydrolyzed polyacrylamides have been previously discussed in the synthesis section of this paper. Chemical analysis of the polyacrylamides showed that the degrees of hydrolysis for the five polymers were 0, 19, 33, 60 and 66%. The zero % hydrolysis polymer was analyzed by GPC, and had a molecular weight of approximately 8 million. Flow curves on all solutions were obtained by the viscometer at 20°C by shearing the solutions from a shear rate condition of 30

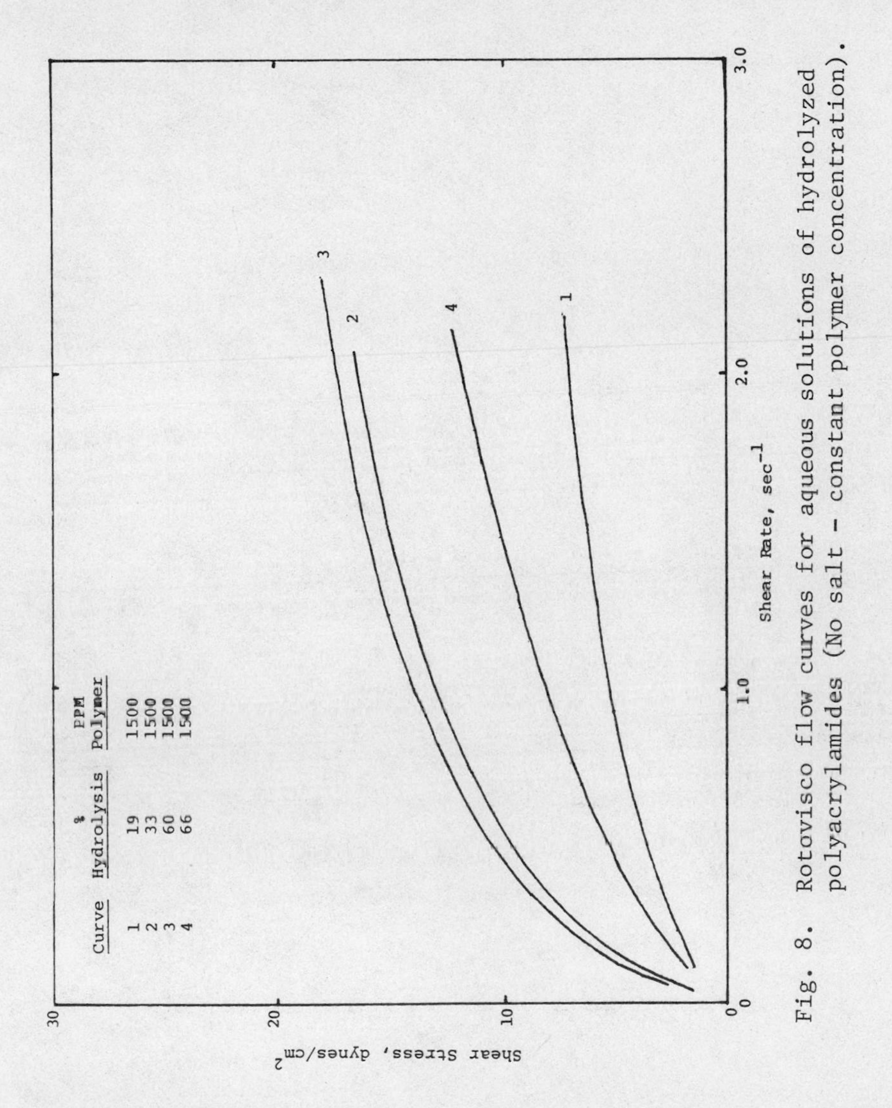

Fig. 8. Rotovisco flow curves for aqueous solutions of hydrolyzed polyacrylamides (No salt - constant polymer concentration).

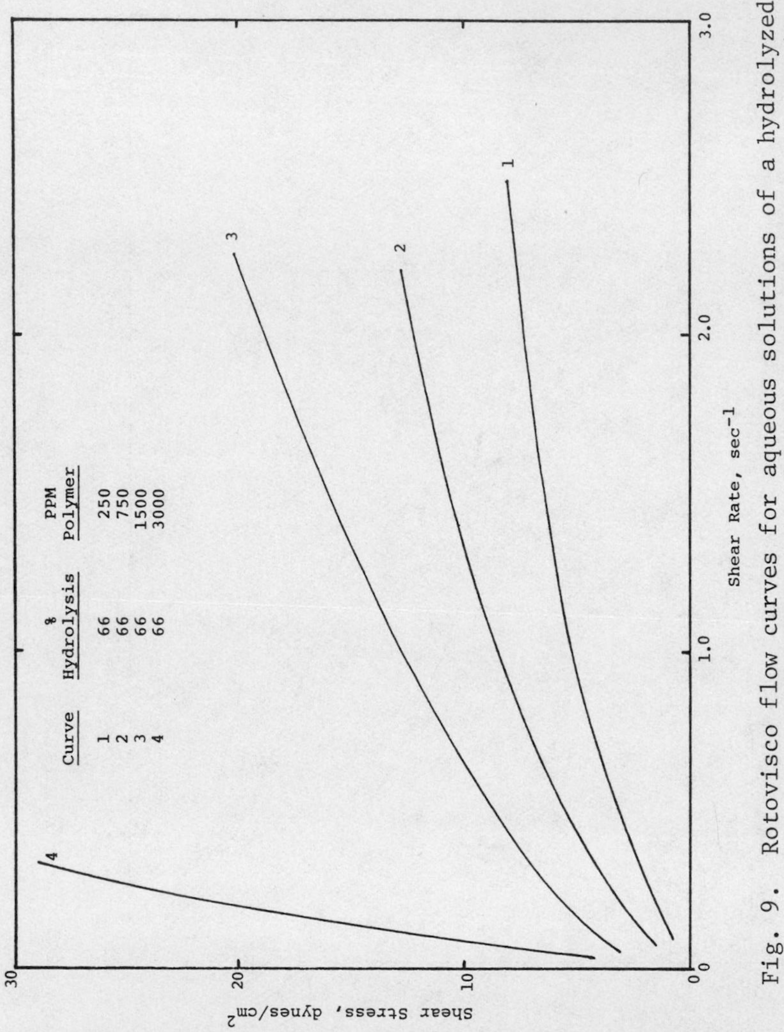

Fig. 9. Rotovisco flow curves for aqueous solutions of a hydrolyzed polyacrylamide (No salt - various polymer concentrations).

to zero sec^{-1} at a program scan rate of $-0.5\ sec^{-1}/min$. At this low scan rate, the deceleration forces of the rotating cylinder could be neglected and therefore no corrections on the shear stress outputs were made.

A set of flow curves for the HPAM polymer solutions at 1500 ppm concentration are shown in Figure 8. This figure is typical of the rheological behavior at other concentrations. Because the molecular weights of all the hydrolyzed polymers are approximately equal and the polymer concentration for all solutions was 1500 ppm, the difference in solution behavior shown by the flow curves must be due to the degree of hydrolysis. At any given shear rate, an increase in the degree of polymer hydrolysis up to 60% causes an increase in the shear stress or has increased the apparent solution viscosity. At a 66% degree of hydrolysis, the trend is reversed and the apparent solution viscosity decreases. Also it should be noted that solution pseudoplastic behavior is enhanced by increasing polymer hydrolysis up to 60%. Increased pseudoplastic behavior can be detected by a more rapid decrease in the flow curve slope as the shear rate increases.

As expected, increasing the polymer concentration in pure deionized water increases the fluid viscosity and the solution's pseudoplastic behavior; see Figure 9. Previous results have shown that the presence of salt ions significantly decreases both the fluid viscosity and solution's pseudoplastic behavior (6). The presence of salt ions neutralizes the charged nature of the hydrolyzed polymers and this reduces the volume of solution occupied by the ball-shaped polymer coils. With less solution volume occupied by single macromolecules, less intermolecular entanglement would exist because the molecules are separated by greater distances. Therefore, the solution viscosity of charged polymers in salt containing solvents would be less than in pure aqueous solvent. Similar results were found for poly(acrylic acid) solutions (19).

Each of the flow curves obtained by the Haake® viscometer were analyzed in the low shear rate range by using the theoretical models of both Bueche and Graessley. The shear stress, shear rate data taken from each flow curve were fitted to both equation (5) and equation (12) by linear regression. Best values for molecular response time and zero shear viscosity were then calculated using equations (6), (7), (15), and (16). These results are presented in Table 7.

Results from the Bueche analysis give higher values for the zero shear viscosity when compared to results from the Graessley analysis, but lower values for the molecular response time. However, the Bueche parameters are linearly correlated with the Graessley parameters. The Graessley molecular response times are linearly proportional to the Bueche response times. The zero shear

Table 7. Rheological Properties of Pseudoplastic Polymer Solutions

Aqueous Solution Composition			Bueche Analysis		Graessley Analysis		
Polymer Type	Polymer Conc. ppm	% Hydrolysis	Zero Shear Viscosity poise	Response Time sec.	Zero Shear Viscosity poise	Response Time sec.	Molecular Weight* x 10^{-6}
PAM	3000	0	0.09	--	0.09	--	--
HPAM	1500	19	14.6	0.52	11.2	7.8	21.0
HPAM	750	19	6.0	0.32	4.9	6.5	20.0
HPAM	3000	33	107.0	0.71	79.0	9.6	7.3
HPAM	1500	33	55.0	0.91	39.0	11.0	8.5
HPAM	750	33	22.2	0.79	16.2	10.3	9.6
HPAM	250	33	5.66	0.64	4.2	8.9	10.6
HPAM	3000	60	112.0	0.68	84.0	9.5	6.8
HPAM	1500	60	57.7	0.87	41.0	10.9	8.0
HPAM	750	60	21.0	0.59	15.9	8.6	8.1
HPAM	250	60	4.47	0.45	3.5	7.4	11.0
HPAM	3000	66	60.5	0.73	42.9	8.8	12.0
HPAM	1500	66	23.0	0.64	16.4	7.8	14.0
HPAM	750	66	14.4	0.80	10.0	9.0	14.0
HPAM	250	66	4.22	0.44	3.25	6.9	11.0

*Calculated from equation (3), with the shielding factor unity.

solution viscosities calculated from the Bueche analysis are approximately one and one-third times the zero shear solution viscosities determined from a Graessley analysis. This correlation of both sets of parameters infers that Bueche's and Graessley's models of polymer solution rheological behavior are different forms of the same relationship. Possibly, a relationship in which the Deborah number is the controlling factor as suggested by equations (17) and (18).

Because of the concentrated nature of the hydrolyzed polyacrylamide solutions, equation (3) was used to estimate the polymer viscosity average molecular weights from the zero shear and response time calculated from a Graessley's analysis of flow curve data. These values are shown in Table 7. Although the molecular weights ranged from 7 to 21 million, the average value for all solutions was 10 million, which was close to the expected 8 million value obtained by GPC analysis. The large variation in molecular weight is probably due to inadequate experimental technique in maintaining a constant viscometer zero adjustment.

CONCLUSIONS

- A series of random- and graft-copolymers of acrylamide with varying molecular weights, shapes, ionic densities, and hydrogen-bonding capabilities have been synthesized and characterized.
- Aqueous solutions of the above polymers show widely different behavior under carefully controlled conditions of concentration pH, temperature, and mono- multivalent electrolyte concentration.
- High molecular weight, ionic copolymers yield higher viscosities in water than do uncharged graft-copolymers viscosities of the former, however, are drastically affected by the addition of mono- or multivalent electrolytes.
- Ionic acrylamide copolymers show a definite time dependency of apparent viscosity after shearing at moderate rates. This "aging" effect, repeatedly demonstrated in a number of solution studies, does not appear to be due to irreversible breakage of primary bonds but is apparently the result of reversible molecular entangling or hydrogen-bonding interactions.
- Gel permeation chromatography experiments with aqueous polymer solutions under carefully controlled conditions may be used to characterize a number of widely different polymer types. Solvents may be chosen which "shrink" molecules to dimensions which are resolvable in commercially available porous glass packings. For a series

of standards, rather accurate calculations have been made of molecular weight.

- Practical implications of the GPC studies are that solvent character of the aqueous solutions may be varied by addition of organic solvents, buffers, or salt solutions to reduce molecular dimensions. Such changes may be necessary in reservoirs with small pore sizes to reduce excluded pore volume.
- The Graessley model of pseudoplasticity appears promising for use in describing the rheological behavior of dilute polymer solutions at the low shear rate conditions experienced in a reservoir.
- Use of pseudoplastic solution flow curve data and the Graessley model may enable prediction of polymer viscosity average molecular weight. Confirmation of this method for accurate molecular weight determination is presently underway.

ACKNOWLEDGMENTS

Research support from the U.S. Department of Energy under Contract No. EF-77-S-05-5603 is gratefully acknowledged.

NOMENCLATURE

a = intercept of a linear function
b = slope of a linear function
c = polymer concentration, g/cm^3
f = function of
M = molecular weight, g/mole
$\bar{M}_v$ = polymer viscosity average molecular weight, g/mole
M_o = polymer segment molecular weight, g/mole
m = concentration in moles/liter
N = total number of polymer molecular segments
N_{De} = Deborah number, dimensionless
R = gas constant, 8.314×10^7 dyne cm/°K mole
T = absolute temperature, °K
V_e = sample elution volume, cm^3
V_o = GPC column total void volume, cm^3
V_p = GPC column total permeation volume, cm^3
$\dot{\gamma}$ = fluid shear rate, sec^{-1}
η = apparent solution viscosity, dyne sec/cm^2
η_o = apparent solution viscosity at very low shear rates, "zero shear viscosity," dyne sec/cm^2
η_s = solvent viscosity, dyne sec/cm^2
$[\eta]$ = intrinsic viscosity

θ = parameter defined by equation (8)
α = solvent shielding factor, see equation (4), dimensionless
λ = average molecular response time, defined by equation (3), sec.
τ = fluid shear stress, dynes/cm^2
D-g-AM = acrylamide-grafted copolymers of dextran
HPAM = partially hydrolyzed polyacrylamides
PAM = homopolyacrylamides
PAMAA = random copolymers of acrylamide with acrylic acid
PDMAM = homopolydimethylacrylamide
PAMDEAM = random copolymers of acrylamide with N,N-diethylacrylamide
PAMDMAM = random copolymers of acrylamide with N,N-dimethylacrylamide
PAMNaA = random copolymers of acrylamide with sodium acrylate

REFERENCES

1. C. L. McCormick, H. H. Neidlinger, R. D. Hester, and G. C. Wildman, First Annual Report, BETC-5603-5, U.S. Department of Energy, April, 1979.
2. C. L. McCormick, H. H. Neidlinger, R. D. Hester, and G. C. Wildman, BETC-5603-7, U.S. Department of Energy, March, 1979.
3. C. L. McCormick, H. H. Neidlinger, R. D. Hester, and G. C. Wildman, BETC-5603-6, U.S. Department of Energy, December, 1979.
4. A. R. Cooper and D. S. Van Derveer, J. Liq. Chromat., 1, 693 (1978).
5. F. A. Buytenhuys and F.P.B. Van Der Maeden, J. Chromat., 149, 489 (1978).
6. G. C. Wildman, H. H. Neidlinger, C. L. McCormick, and R. D. Hester, DOE Quarterly Report, October 1 - December 31, 1978.
7. Elias, Macromol. Chem., 50, 1 (1961).
8. G. Smets, Macromol. Chem., 29, 190 (1959); J. Klein and R. Heitzmann, Makromol. Chem., 179, 1895 (1978).
9. B. H. Zimm, J. Chem. Phys., 16, 1099 (1948).
10. C. L. McCormick, R. D. Hester, G. C. Wildman, and H. H. Neidlinger, Proceedings 4th DOE Symposium on Enhanced Oil and Gas Recovery and Improved Drilling Methods, Tulsa, OK, August 29-31, 1A, p. B 1/1 - B 1/25, 1978.
11. A. E. Scheidegger, in "The Physics of Flow Through Porous Media," 3rd edition, University of Toronto Press, p. 219, 1974.
12. F. Bueche, J. Chem. Phys., 22, 1570 (1954).
13. W. W. Graessley, J. Chem. Phys., 43, 2696 (1965).

14. W. W. Graessley, Accounts of Chem. Res., 10, 332 (1977).
15. J. E. Dunleavy and S. Middleman, Trans. Soc. Rheol., 10, 157 (1966).
16. W. W. Graessley, J. Chem. Phys., 47, 1942 (1967).
17. S. Middleman, J. Appl. Polymer Sci., 11, 417 (1967).
18. A. B. Metzner, J. L. White, and M. M. Denn, Chem. Eng. Prog., 62, 81 (1966).
19. J. C. Brodnyan and E. L. Kelley, Trans. Soc. Rheol., 5, 205 (1961).

RESPONSE OF MOBILITY CONTROL AGENTS TO SHEAR, ELECTROCHEMICAL, AND BIOLOGICAL STRESS

T. P. Castor, J. B. Edwards and F. J. Passman

Energy Resources Co. Inc.

Cambridge, Massachusetts 02138, U.S.A.

Bench-top studies were carried out in the laboratory to evaluate the degradation of polyacrylamides and polysaccharides under simulated conditions of shear, electrochemical, and microbial stress. Effects were evaluated in solutions with varying concentrations of univalent and polyvalent electrolytes (0-100,000 ppm) at different values of temperature. The effect of microbial stress was evaluated by inoculating pure cultures of anaerobes and aerobes into solutions of the biopolymers and copolymers and measuring the bulk properties as a function of time to determine kinetics of biological and chemical degradation. The viscoelastic responses of the polyacrylamides as indicated by their screen factor and the filterability of the polysaccharides were monitored at the above conditions. Core flooding tests were carried out to further evaluate the chemical integrity of the mobility control agents under simulated field conditions. Average pressure drop and breakthrough concentrations as a function of pore volume throughput were used to evaluate the rheological and retentive behavior of the copolymers and biopolymers in brine-saturated and partially oil-saturated sandstone cores. The rheological behavior of fluids in the reservoir cores is correlated with the degradation of polymers from bulk phase measurements of shear viscosities, screen factors and millipore filter ratios.

INTRODUCTION

Most chemical enhanced oil recovery processes utilize mobility control agents to improve the displacement efficiency of the mobilized oil. In situ gelling and cross-linking of mobility control agents are being used increasingly to correct for gross heterogeneities such as high permeability streaks and underlying aquifers. The utility of these mobility control agents in the reservoir depends on their ability to sustain chemical integrity throughout the lifetime of a chemical flood at reservoir conditions and field rates. Bench-top and core flooding experiments can be carried out to evaluate the chemical integrity of the mobility control agents under different states and levels of stress. The degradation of mobility control agents in the bulk and in the reservoir cores, as a result of different levels of shear, electrochemical, and microbial stress can then be used to develop parametric curves. Parametric curves allow cross-interpretation for effective evaluation of the polymer rheology under typical reservoir and flood conditions.

Effects of electrochemical and shear stresses on rheologic and retentive behavior: The compatibility of polymers with the river and formation waters was evaluated in terms of sensitivity to salt, to shear. Most of the compatibility tests were performed in an 80/20 mixture of synthetic river and synthetic formation waters since injected polymer solutions are expected to contact significant concentrations of formation brine (> S_{wir}) even after a very efficient freshwater or alkaline preflush. Rheological and retentive behavior of polymers in reservoir cores were tested with an 80/20 mixture of river and formation waters as well as with a 100-percent river and a 100-percent formation waters. The analyses of the formation water and the river water are displayed in Table 1. Twelve polymers from nine manufacturers were evaluated; eight of these polymers are polyacrylamides and four are polysaccharides. The polymers are identified by an alphabetical index in terms of manufacturer, product, state and activity in terms of weight percent in Table 2; different polymers from the same manufacturer are numbered, e.g., C1 and C2 from American Cyanamid.

Viscosity as a function of electrolyte concentration: The extreme sensitivity of these copolymers to electrolytes is evident in the electrochemical degradation factors listed in Table 3. The electrochemical degradation factor is defined as the ratio of shear viscosities (measured in a Brookfield viscosometer with a UL adaptor) of polymer in river water and 80/20 mix of river/formation waters at 7.3 sec^{-1} and 95°F. The polyacrylamide copolymers are much more viscous than the polysaccharide biopolymer at equivalent concentrations in fresh water but are much less viscous than the biopolymer in the saline water (10,000 ppm TDS). The high molecular weight of the polyacrylamides, which provides the high viscosity of these copolymers in fresh water, results from the reactive nature of the individual polymer molecules. The diffuse double layer, generated by the high charge density of the reactive units, is compressed in the presence of high concentrations of uni- and polyelectrolytes. The collapsed double layer lowers the electrostatic forces of repulsion and causes the polymer chain to

Table 1. Composition of Formation and River Water in Storms Pool Project.

Constituent	Waltersburg Formation Water (ppm)	Little Wabash River Water (ppm)
Sodium	13,715	18
Calcium	732	23
Magnesium	358	8
Chloride	23,250	25
Sulfate	7	5
Iron	1	0
Bicarbonate	403	110
Carbonate	0	0
Total	38,466	189
Total Iron	1	9
Barium	18	0
H_2S	2.4	0
pH at 24°C	6.8	7.1
Specific gravity at 24°C	1.03	1.0
Resistivity at 24°C (ohm-m)	0.2	43.0
Turbidity (Nephelometric Turbidity Units)	54	77

Table 2. Identification of Polymers by Manufacturers, Products, Type, State and Activity in Terms of Weight Percent.

	Polymer	Manu-facturer	Product	State	Active Product (Wt %)
POLYACRYLAMIDES	A	Amoco	Sweepaid 103	W/O Emulsion	25
	B	NALCO	NAL-FLO F HMWS	W/O Emulsion	30
	C1	American Cyanamid	Cyanatrol 950S	W/O Emulsion	30
	C2		Cyanatrol 960S	W/O Emulsion	30
	D	Allied Colloid	1100-L	W/O Emulsion	50
	E1	CORT*	N-Hance 335	Aqueous Log	30
	E2	CORT	N-Hance 330	Aqueous Log	27
	F	Dow	XD-30226.0	W/O Emulsion	28
POLYSACCHARIDES	G1	Pfizer	Flocon 1035	Broth	2.7
	G2	Pfizer	Flocon 4880	Broth	5.7
	H	Abbott		Broth	2.75
	J	Rhone-Poulenc	Rhodopol	Dry	~88

*Joint venture company of Phillips Petroleum and Hercules.

Table 3. Electrochemical Degradation of 500 ppm Active Polymer at a Shear Rate of 7.3 sec^{-1} and 95°F.

Polymer	Viscosity in River Water (cp)	Viscosity in 80/20 Mix of River/Formation Waters (cp)	Electrochemical Degradation Factor
Copolymers			
A	36.3	2.6	14.0
B	49.8	4.3	11.6
C	43	2.5	17.2
D1	64	3.6	17.8
D2		2.8[a]	
D3		2.5	
E	42.4	3.8	11.2
F	62.3[b]	4.3[b]	14.5
Biopolymers			
G1	8.9	8.3	1.1
G2	13	12.5	1.04
H	13.4	8.4	1.6
J[c]	12.6	10.5	1.2

[a]This measurement was made on an unfiltered solution; all other solutions were filtered with a depth filter.

[b]The electrochemical degradation factor can be used for comparative purposes although the results of these tests are at 75°F and the other tests are at 95°F.

[c]Dry polymer (all others are liquids or gels).

ball up as a result of Van der Waal's forces of attraction. The effect of the electrochemical environment on the bulk rheological behavior of the polyacrylamide molecules could be an entropic rather than an energetic phenomenon, i.e., the internal energy of the polymer decreases in the presence of salt and the polymer chain mechanically ruptures during elongation deformation as the entropy or degree of randomness increases. Thus, the viscosity of the high molecular weight polyacrylamides drops, as illustrated in Figure 1, by an order of magnitude in the presence of uni- and polyelectrolytes at approximately 10,000 ppm TDS. The viscosity of the biopolymers, labelled G1 and G2 in Figure 1, are not affected by the presence of the electrolytes because of the lower charge density on the polysaccharide monomers. It should be noted that the viscosities of polyacrylamide B in the 80/20 mix of formation and river waters are not that much lower than the viscosities of the biopolymers.

Screen factor and millipore filter ratios of dilute polymeric solutions: Screen factor is defined as the ratio of the flow time for the polymer solution to the flow time of the appropriate solvent (water or brine) through a random stack of five 100-mesh SS screens. The screen factor is a measure of the viscoelastic response of the polymer, i.e., the ability, or lack thereof, of the polymer solution to sustain sudden elongational deformation and the resulting normal stresses (1). Elongational deformation occurs as the solution accelerates through the screen openings of the screen viscosometer. Similar deformation will occur in porous media during the flow of fluids through the constrictions of the convergent-divergent channels. The induced normal stresses can be large enough to break the polymer bonds and mechanically degrade the polymer. Thus, the screen factor is a better measure of mechanical degradation than is viscosity which responds to shear rather than normal stresses; this measurement is, however, polymer-specific and cannot be used to compare different polymers. The screen factors of the copolymers in 100 percent river water and an 80/20 mix of river/formation waters at room temperature are listed in Table 4.

Polyacrylamide E1, with the lowest electrochemical degradation factor of 11.2 in Table 3, experiences the smallest reduction of resistance factor in the presence of univalent and divalent electrolytes, from 55.9 in river water to 49.5 in an 80/20 mixture of river and formation waters. These unusually large resistance factors probably resulted from the hydrodynamic resistance of the long linear polymer chain which is a unique characteristic of its gamma radiation manufacturing process. There appears to be some correspondence between the effect of electrolytes on viscosity and screen factor since polymers C and D1 with the lowest electrochemical degradation exhibit the greatest reduction in screen factor on

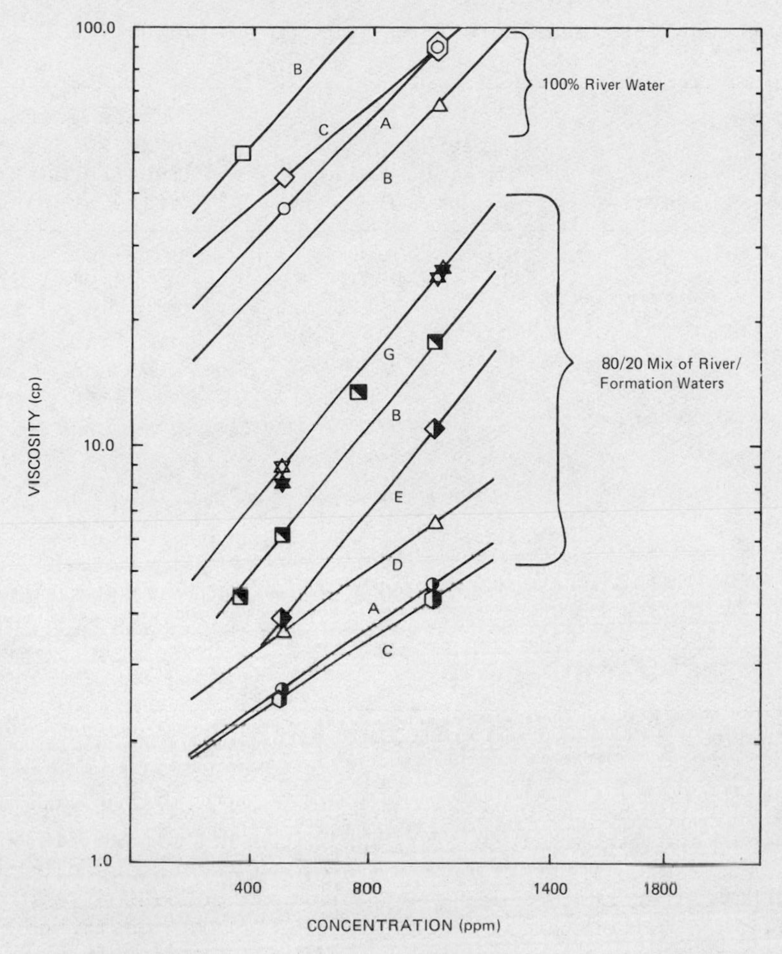

Fig. 1. Viscosity as a function of polymer concentration in 100% river water and in 80/20 mix of river and formation waters.

increase of electrolyte concentration. The correspondence exists because the electrochemical environment affects the orientation of the polymer molecules and their response to normal and shear stresses. Screen factor has been correlated with mobility reduction or resistance factor of dilute polymeric solutions in reservoir cores (2). The screen factors listed in Table 4 directly reflect the active and residual resistance factors of the reservoir core floods described in a later section.

Tests for filterability of polysaccharides through millipore filters are similar in procedure to screen factor tests for polyacrylamides but yield different interpretations. The millipore filter ratio is defined as the ratio of the time required for the last 250 cc to the first 250 cc of 1000 cc of 500 ppm polymer

Table 4. Screen Factors of 500 ppm Active Polymer

	Polymer	Screen Factor in River Water at 75°F	Screen Factor in 80/20 Mix of River/Formation Waters at 75°F
POLYACRYLAMIDES	A	11.66	7.01
	B	25.7	18.1
	C	6.62	2.84
	D1	31.35	15.8[a]
	D2		16
	D3		14.46
	E1	55.9	49.5
	F		23

[a]Unfiltered.

solution to flow through a presaturated 1.2 micron millipore filter under a constant gas pressure of 40 psi. The millipore filter ratio is a measure of the propensity of polysaccharide molecules to plug the filter and, by inference, the oil formation. Solutions with millipore filter ratios greater than 1.5 are considered unacceptable; recent specifications of polysaccharide solutions by oil companies require that the filter ratio be no greater than 1.3. The flow time at 1000 seconds is a parameter which should be considered in conjunction with the filter ratio. Filter ratios alone could be misleading in that a high degree of constant plugging could produce a low filter ratio. Pfizer has recommended that runs with flow times greater than 1000 sec/liter be considered unacceptable.

Table 5 lists the millipore filter ratios and the cumulative flow times of Pfizer's polysaccharide G. The diluents used for these runs were all filtered through a 0.65-micron filter under pressure. The increase in millipore filter ratio between the 100 percent river water (run 10), 80/20 mixture (runs 11, 12, and 13), and a 100 percent formation water (run 14) is consistent with the high retention of polysaccharide G2 in reservoir cores in the 80/20 mixture (321 lb/acre-ft) and in 100 percent formation water (542 lb/acre-ft). The results of filterability tests are plotted

Table 5. Polysaccharide G[a] Millipore Filter Ratios of 500 ppm Active Polysaccharide Product Under a Pressure of 40 psi.

Run No.	Broth Conc. (wt %)	%	Diluent[d] (% River[b] Formation[c] Waters)	Viscosity (cp @ 6 rpm)	Flow Time (@ 1000 cc)	Millipore Filter Ratio
5	2.7		80/20	9.9	40	1.14
6	2.7	g1	80/20	9.9	43	1.17
7	2.7		80/20	9.3	35	1.32
8	2.7		80/20	9.3	31	1.30
10	7.4		100/0	10.3	61	1.16
11	7.4	g3	80/20	10.1	83	1.37
12	7.4		80/20	9.9	80	1.30
13	7.4		80/20	9.9	75	1.30
14	7.4		0/100	10.3	285	2.9
15	8.4		100/0	10.0	51	1.60
16	8.4	g4	100/0	10.0		1.57
17	8.4		80/20	10.6	63	1.76
18	8.4		80/20	10.6		1.60

[a]Polysaccharide from manufacturer G is made in four different concentrations (G1 = 2.7%, G2 = 5.7%, G3 = 7.4%, G4 = 8.4%).

[b]250 ppm TDS with 2:1:: $Na^{+}:Ca^{++}$

[c]38,500 ppm TDS with 19:1:: $Na^{+}:Ca^{++}$

[d]500 ppm TDS with 10:1:: $Na^{+}:Ca^{++}$

in terms of filter rate (cc/sec) versus cumulative throughput (cc) in Figure 2. These representations combine the effects of incremental plugging (slope of curve) and initial plugging tendencies (initial flow rate) of dilute polysaccharide solutions. These curves indicate that biopolymer G1 has a better filterability than biopolymer H but that biopolymer G1 is much more affected by the presence of divalent and univalent electrolytes than biopolymer H. These results support the assertion that high retention of polysaccharide G1 in reservoir cores is caused by microgel formation cross-linked with polyelectrolytes. The filterability of other candidate polysaccharides are plotted in Figure 3 and summarized in Table 6. The millipore filter ratios indicate that with respect to filtration time and rate the Pfizer polysaccharides performed best. Filtration of the polysaccharides in distilled water indicated that the Kelco, Rhone-Poulenc and Abbott broths would require diatomaceous earth filtration before injection.

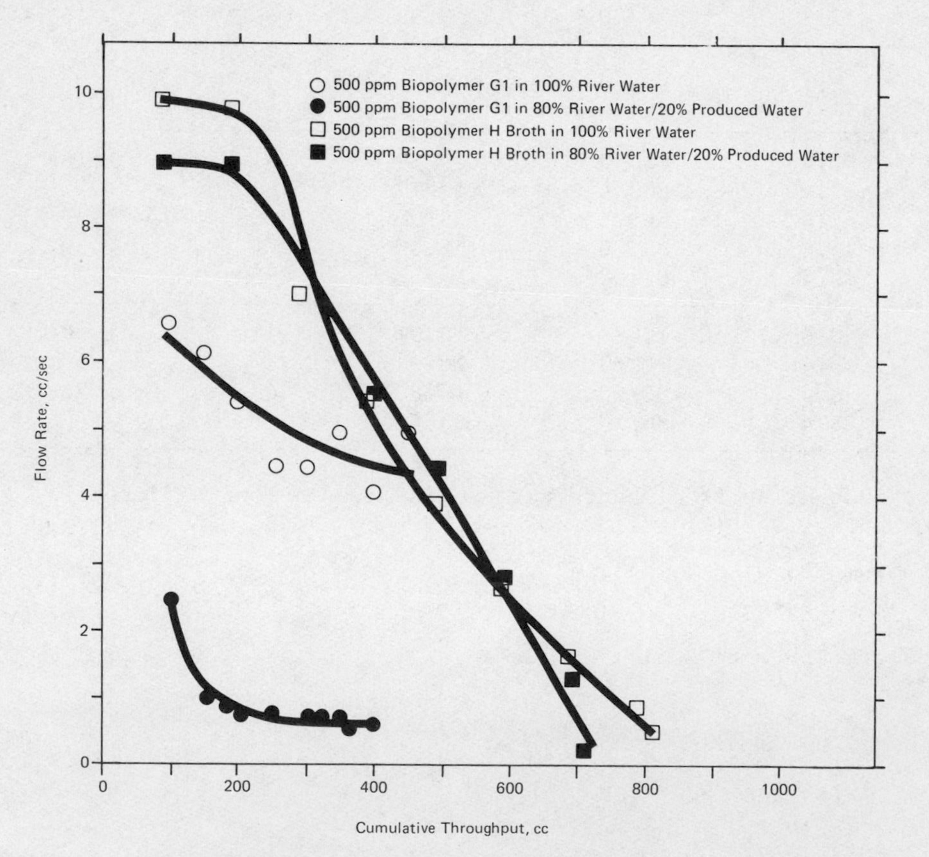

Fig. 2. Filterability of polysaccharide solutions as a function of electrolyte concentration.

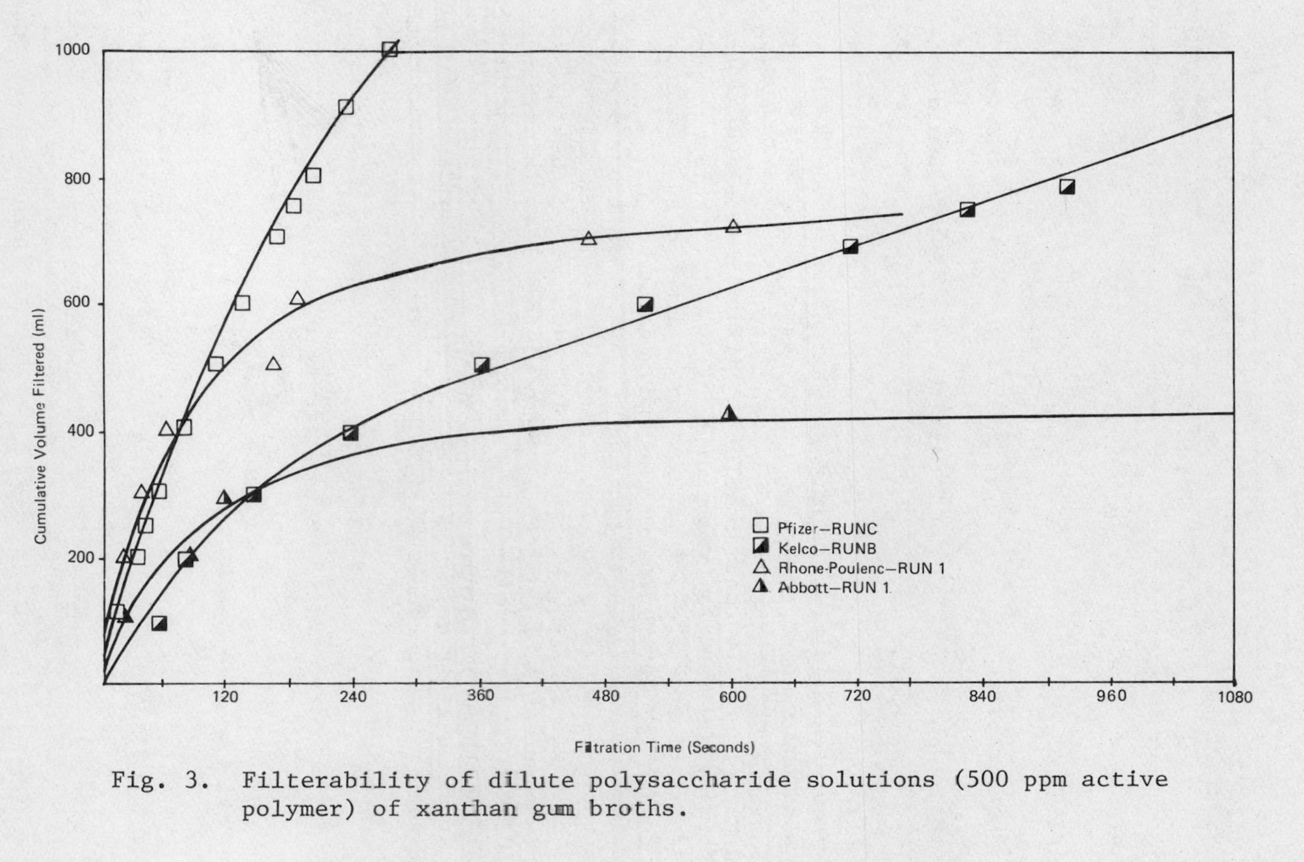

Fig. 3. Filterability of dilute polysaccharide solutions (500 ppm active polymer) of xanthan gum broths.

Table 6. Millipore Filter Ratios of 500 ppm Active Polysaccharide Product under a Pressure of 40 psi.

Manufacturer	Millipore Filter Ratio	
	First Pass	Second Pass
Pfizer (gl)	1.98	1.41
Kelco	7.43	1.68
Rhone-Poulenc	Terminated	2.66
Abbott	Terminated	Terminated

Shear sensitivity of dilute polymer solutions: Figure 4 illustrates the viscous response of 500 ppm of the test polymers to shear at reservoir temperatures in 100 percent river water and an 80/20 mix of river/formation waters. All the data points were measured by a rotational viscosometer, Brookfield with a UL adaptor. The sensitivity of the polymers to shear appears independent of the electrochemical environment. The independence of the shear-viscosity relationships in electrolyte concentrations is further illustrated in Figure 5 for a dilute biopolymer solution. Over the range of shear studied, the viscosity reduction of the test polymers is about 40 to 50 percent. Some of this loss is permanent for the polyacrylamide copolymers. Most of this loss will occur at the high velocities experienced at the wellbore if the surface facilities are adequately designed to minimize mechanical shear. The shear rate, τ, in porous medium can be calculated from the following equation (3):

$$\tau = \frac{1.42 \times 10^4 \times Q}{A \sqrt{k\phi}}$$

where: $Q \equiv$ cc/sec; $A \equiv cm^2$; $k \equiv$ darcy; $\phi \equiv$ porosity.

The shear rate at typical open-hole completions is approximately 2,000 sec^{-1} at an injection rate of 600 bbl/day; the shear rate in the reservoir is approximately 20 sec^{-1} at a reservoir rate of 1 foot/day. The viscosity of copolymer B at the wellbore condition is approximately 44 percent of the viscosity at a typical laboratory shear condition of 7.3 sec^{-1}; this value was obtained from extrapolation along curve B for the 80/20 mix in Figure 4. Mechanical degradation of the copolymers under shear, as measured by viscosity loss, in field (Wilmington) and laboratory (4) inves-

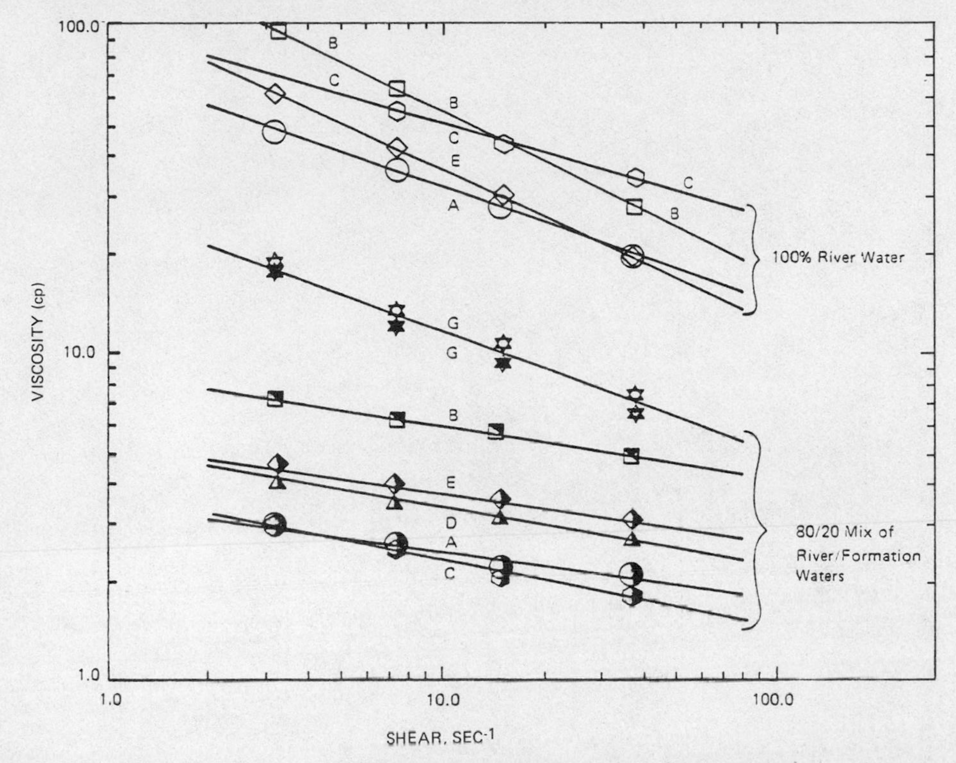

Fig. 4. Viscosity at 500 ppm as a function of shear with river water and with 80/20 mix of river and formation waters.

tigations suggests that some proportion of the viscosity reduction under increased shear will be irreversible on removal of the high shear field. The viscous response of the biopolymer to shear does not indicate significant hysteresis effects since the polysaccharide polymer molecules are relatively inelastic in nature.

Rheological and retentive behavior of polymers in reservoir cores: The rheological and retentive behavior of polymer in reservoir cores at 95°F and in an 80/20 mix of river/formation waters is summarized in Tables 7 and 8. One weight percent polymer solutions in river water were made up according to each manufacturer's suggested mixing procedure. The 1 weight percent solution was then diluted to the requisite concentration in an 80/20 mix of synthetic river and formation waters. The dilute polymer is filtered in a Seals No. 10 Candle Filter (a microporous porcelain filter) and readied for injection. Some of the flow experiments were conducted in the vertical upflow mode with the reservoir rock held in a Hasseler sleeve. A confining pressure of 300 psig was applied radially on the core and the back pressure was held to 40 psig. The pressure drop across the inlet and outlet lines of the core was recorded continuously during the run. The pore volumes of the

Table 7. Rheological Behavior of Polymers in Reservoir Cores at 95°F and ≃ 2 ft/day.

Test Number	Polymer	RF to 80/20 River/ Formation (TDS ≃ 10,000 ppm)	RRF to 80/20 River/ Formation (TDS ≃ 10,000 ppm)	RRF to 0/100 River/ Formation (TDS ≃ 40,000 ppm)
1	B	18	4.5	
2	B[a]	17.9	7.1	2.3
3	B[a]	16	6.3	1.9
4	C	14	7.5	
5	D	39 @ τ = 10 PV 19.5 @ τ = 5 PV	13.5	
6	E1	45.6	32.2	
7	E2	41.81	9.3	
8	E2[a]	23	18.1	3.6
9	E2[a]	19.2	11.7	1.5
10	E1[a] + Aluminum citrate	19.1 to 150	35	31
11	G2	15	9.3	
12	H	6	2	

[a]Carried out with residual oil.

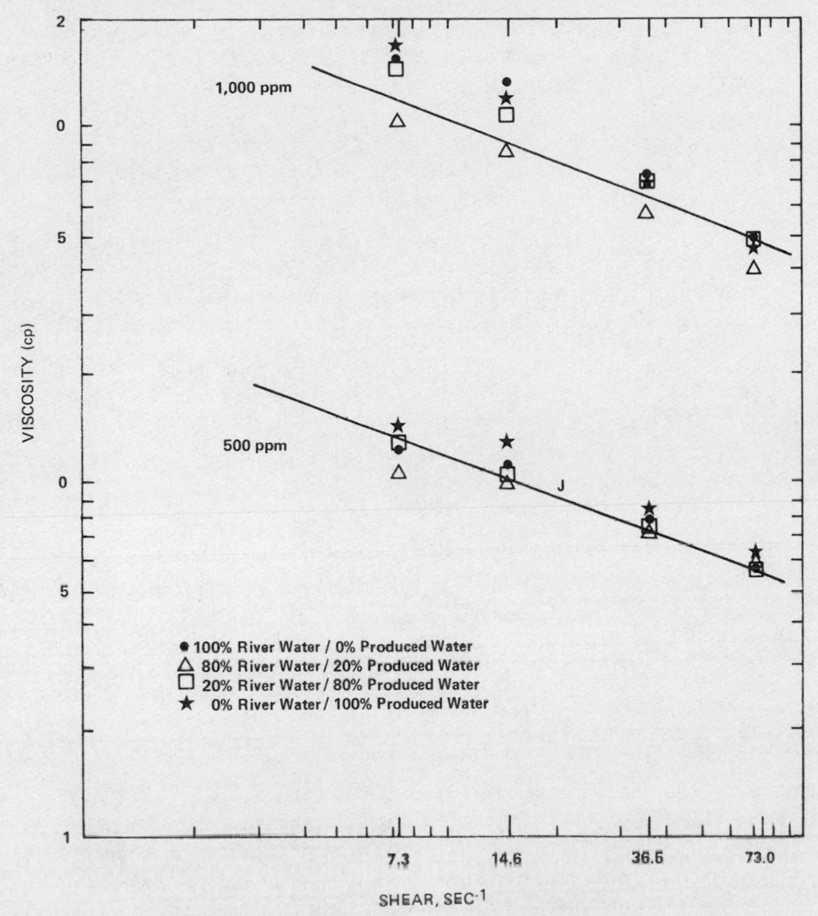

Fig. 5. Viscosity as a function of shear for biopolymer J at 95°F and 500 and 1,000 ppm of active polymer concentration with salt concentration as a parameter.

cores tested were between 13 cc and 15 cc. The polymer concentrations in the effluent stream were measured by precipitation of the polymer in the 1.0 cc to 1.5 cc samples and turbidimetric evaluation of the precipitated amounts. The polymer retentions were evaluated from the area under the retention/adsorption curve and the elution/desorption curve. These values are then the "irreversible" retention/adsorption loss of polymer.

Polyacrylamide E1 exhibits the highest active and residual resistance factors (45.6 and 32.2 respectively) of the polymers tested. Polyacrylamide E1 also exhibited the highest resistance factors in river water and in a mixture of river and formation waters. The transient resistance factor and normalized concentration curves as a function of pore volumes injected are illustrated

Table 8. Retentive Behavior of Polymers in Reservoir Cores at 95°F and ≃ 2 ft/day with an 80/20 Mix of River and Formation Waters.

Test Number	Polymer	Polymer Retention lb/acre-ft	Polymer Retention μg/g	Polymer Retention Fractional Retention of Polymer Injected
1	B	309	55.1	0.131
2	B[a]	385	≃68.6	≃0.163
4	C	236	40.5	0.094
6	E1	299	53	0.106
7	E2	227	39.8	0.084
8,9	E2[a]	330	≃58.9	≃0.140
11	G2	321	≃55.3	0.136

[a]Carried out with residual oil.

in Figure 6. The residual resistance factor (RRF) of polymer E1 dropped from 32.2 to 3.7 after the upstream end of the sample was trimmed. This is indicative of sandface plugging but not filter-cake buildup since the transient resistance factor (RF) curve flattens out at large pore-volume throughput. Sandface plugging is a probable cause for the high value of polymer retention (299 lbs/acre-foot). This core flooding experiment was repeated with a lower molecular weight polymer, E2. The RRF was more favorably reduced (to 9) and the retention for this run was 227 lbs/acre-foot. The RRF of polyacrylamide B in Figure 7 does not give any indication of plugging although the transient RF curve shows a continual increase with pore-volume throughput. The slight upward trend in the transient RF is suggestive of some plugging behavior at high values of pore volume throughput.

The transient RF and concentration curves of polysaccharide G2 are illustrated in Figure 8. The transient RF curve reflects some plugging behavior but not to the degree indicated by the high retention factor (321 lb/acre-foot). The high RRF of 9.3 reflects some polymer retention/adsorption phenomena. The large retention could have resulted from association of polysaccharide molecules with divalent ions such as Mg^{++} and Ca^{++} to form reversible complexes in the 80/20 connate water at the flood front. Such a

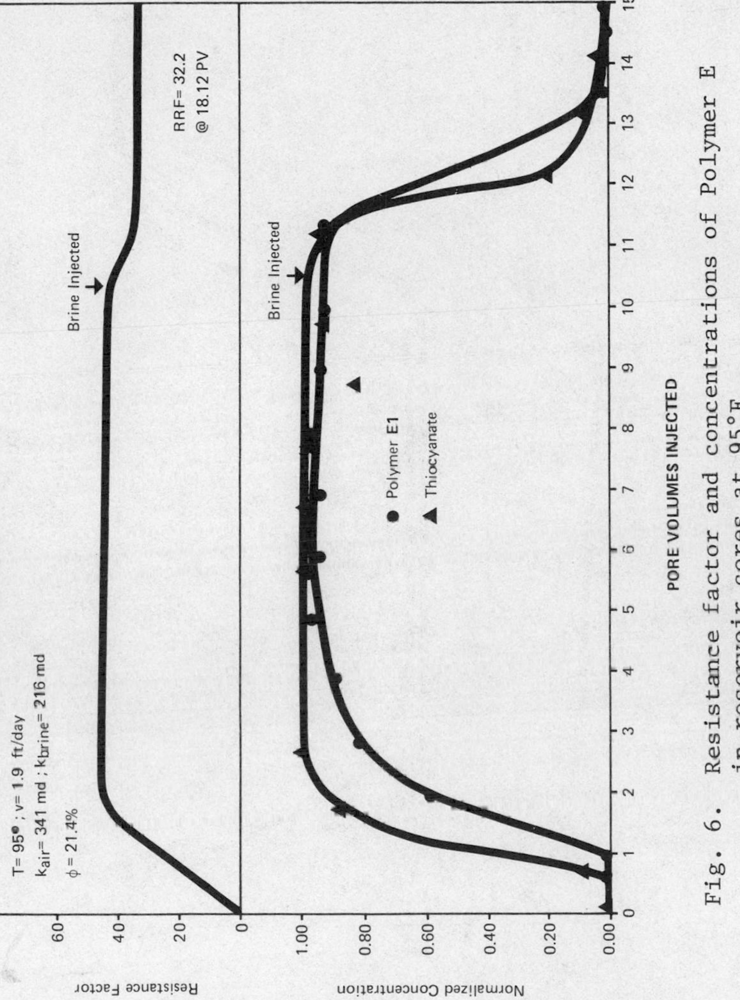

Fig. 6. Resistance factor and concentrations of Polymer E in reservoir cores at 95°F.

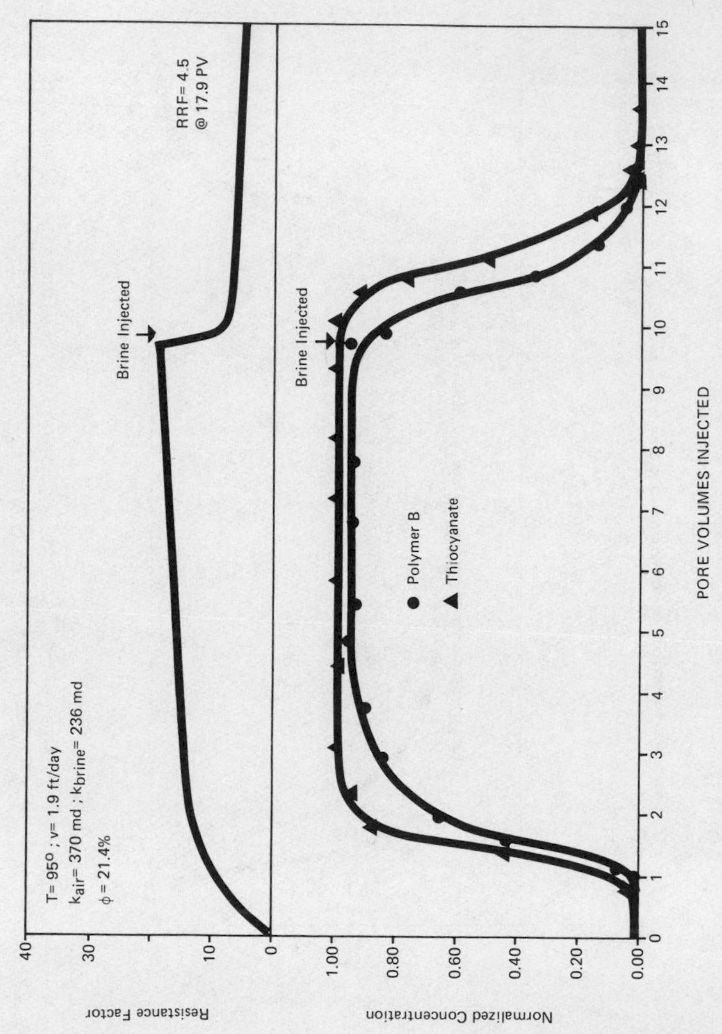

Fig. 7. Resistance factor and concentrations of Polymer B in reservoir cores at 95°F.

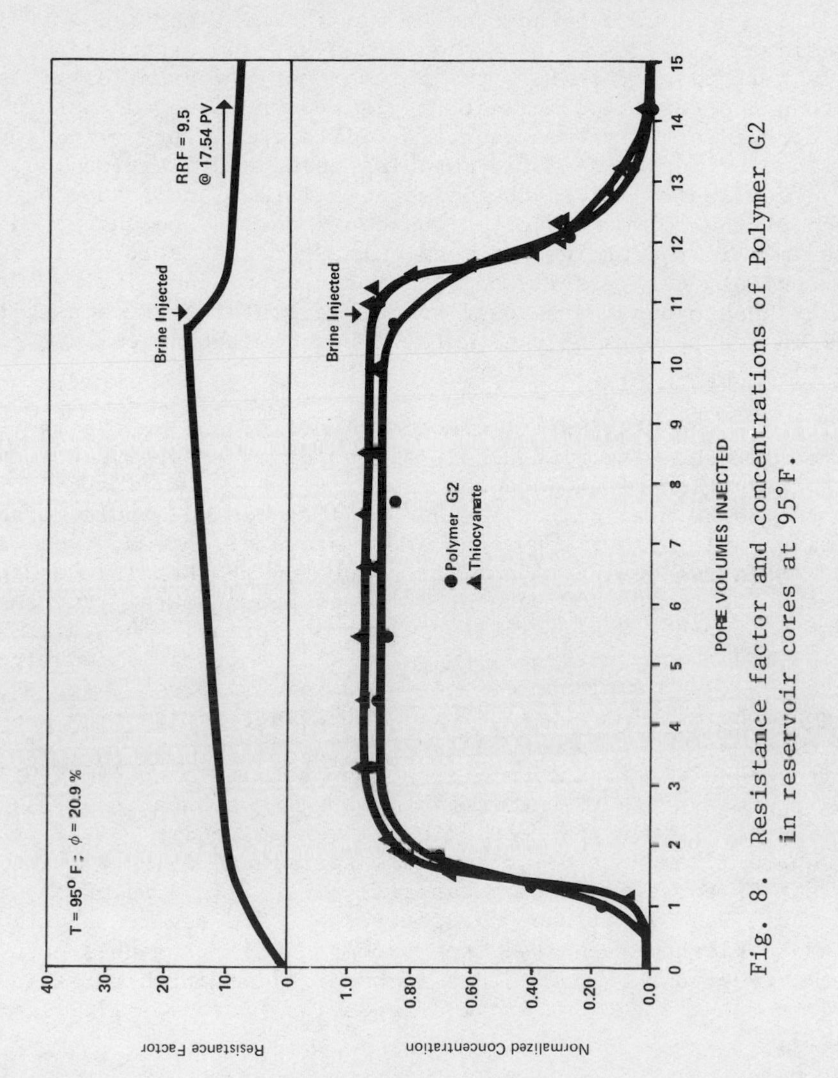

Fig. 8. Resistance factor and concentrations of Polymer G2 in reservoir cores at 95°F.

distribution of reversible complexes would explain the filter cake plugging of the polymeric agglomerates and the increase in millipore filter ratios listed in Table 5. High pH could have been responsible for some of the plugging behavior (5).

Polymer rheology tests on copolymer E2 (Tests 8 and 9) indicate some difference between these tests and the test carried out without residual oil (Test 7). Although the RF from Test 8 is approximately half that of Test 7, the polymer retention in Test 8 is more than the retention in Test 7. It should be noted that in the tests carried out with residual oil, differential pressure drops were measured across a slowly rotating core and retention was measured after recirculating the injected solution until the concentration of the polymer injected was equal to the concentration of the polymer in the effluent stream. The care taken to avoid capillary end effects, gravity segregation, and nonequilibrium adsorption in these experiments could be the reasons for the apparent seesaw in the results of the tests with and without residual oil saturation.

Test 10 was carried out with copolymer E1 and an aluminum citrate mixture. The high RF of 145 indicates effective permeability reduction of the aqueous phase in the core. The RRF of 35 to the 80/20 mix indicates that mobility control would be very effective during the fresh-water drive stage of the improved waterflood. This residual permeability reduction appears irreversible since the RRF to the 100 percent brine is approximately the same as that of the 80/20 mix of fresh water and brine. This high RRF to brine indicates that mobility control by a copolymer/aluminum citrate treatment would be effective during the fresh water and subsequent brine floods which follow a polyacrylamide slug.

The transient RF and concentration profiles of polymer C2 in Figure 9 suggests that some in-depth plugging of the reservoir plug occurs. A previous test with a lower molecular weight polymer (C1) showed little active or residual resistance effects. The transient RF curve indicates that filter cake plugging of the reservoir core was significant for polymer D. The severe injectivity problems could have resulted from incompatibility of this copolymer with the reservoir brines. This incompatibility problem is inferred from the electrochemical degradation factor of 17.2 listed in Table 3.

A polyacrylamide and a polysaccharide (which performed best in the compatibility studies and in the core experiments) were tested for rheological and retentive behavior in the cores in a 100 percent brine environment. These tests were performed in order to examine the interaction of the polymers with brines of unswept regions in the reservoir. The polymers were dissolved in distilled water and then mixed with 100 percent formation brine to the

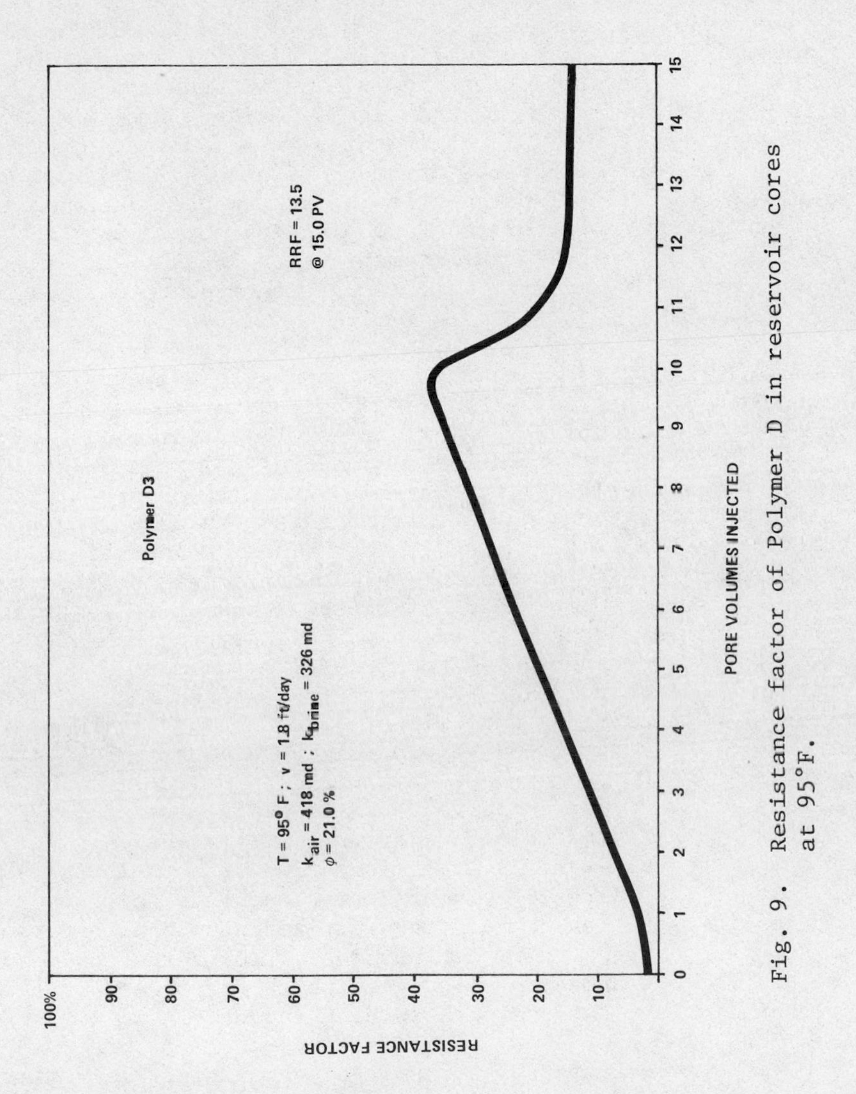

Fig. 9. Resistance factor of Polymer D in reservoir cores at 95°F.

desired concentration of 500 ppm. The dilution was such that the mixture concentration was greater than 99 percent formation brine. The results of the compatibility studies are listed in Table 9. The viscosity of polysaccharide G2 is relatively insensitive to the increasing degrees of salinity (0 to 10,000 to 40,000 ppm TDS). The viscosity of polyacrylamide E2 collapses by a factor of 11.2 after increase of brine salinity from 0 to 10,000 ppm TDS and then by a factor of 2.9 on increase of brine salinity from 10,000 to 40,000 ppm TDS. The rheological and retentive behavior of these tests is graphed in Figures 10 and 11. As evident from the transient RF curve in Figure 10, the injectivity of polysaccharide G2 is not severely impaired by the presence of high salinity brines. The transient RF curve of polyacrylamide E2, Figure 11, implies that a significant amount of sandface plugging occurs as the direct result of the high salinity environment. The RF, RRF and retention of these polymers in different electrolyte environments are tabulated in Table 10.

The high salinity environment of the dilute polyacrylamide solutions affects the viscosity of the bulk solution more than the hydrodynamic behavior of the polymeric solutions in reservoir cores. It appears that the situation is reversed for polysaccharides (at least for polysaccharide G) in that high salinity environments affect the active and residual resistance factors more than the viscosity of dilute polymeric solutions. The rheological and retentive properties of filtered and unfiltered polysaccharide solutions were measured to determine if microgel formation was responsible for high RFs and RRFs in the high salinity environment. The measurements were made with 500 ppm of active polymer dissolved in filtered and unfiltered river water and in filtered and unfiltered formation brine; solutions were filtered through a 7 to 10-micron filter under a constant pressure head of up to 40 psig. The retentions and resistance factors are summarized in Table 11. The retention of the unfiltered polysaccharide G1 in river water is double that of the filtered polymer/river water solution. The retention of the unfiltered polymer/formation brine solution is triple that of the filtered polymer/formation brine solution. The differences in retention between the filtered and unfiltered solutions indicate that a significant portion of the retention results from mechanical entrapment of microgels which are formed in the presence of divalent cations such as Mg^{++} and Ca^{++}.

The resistance factor of 500 ppm polysaccharide G1 in filtered river water is specific to the flooding velocity of 16.5 ft/day. The resistance factor at field rates (e.g., 1 ft/day) can be calculated since the shear rate is directly proportional to the flooding velocity and the logarithm of viscosity is directly related to the logarithm of shear rate. The shape of the resistance factor curves for the xanthan gum in formation brine indicates

Table 9. Electrochemical Degradation of 500 ppm Active Polymer at a Shear Rate of 7.3 sec^{-1} and 95°F.

Polymer	μ in 100% River Water (cp)	μ in 80/20 Mix of River/ Formation Water (cp)	Electrochemical Degradation Factor	μ in 100% Formation Water (cp)	Electrochemical Degradation Factor
Polyacrylamide E2	42.4	3.8	11.2	1.3	32.6
Polysaccharide G2	13	12.5	1.04	12.5	1.0

Table 10. Rheological and Retentive Behavior of a Polysaccharide and a Polyacrylamide in Reservoir Cores at 95°F and ≃ 2 ft/day.

	RF	RRF	Retention	
			μg/g	lb/acre-foot
Polysaccharide G2 in 80/20[a]	15	9.3	55.3	321
Polysaccharide G2 in 0/100	9	2	93.3	542
Polyacrylamide E2 in 100/0	49	10		
Polyacrylamide E2 in 80/20	41.8	9.3	39.8	227
Polyacrylamide E2 in 0/100	27	27	153.9	889

[a]80/20 stands for a mixture of 80 percent river water and 20 percent formation water.

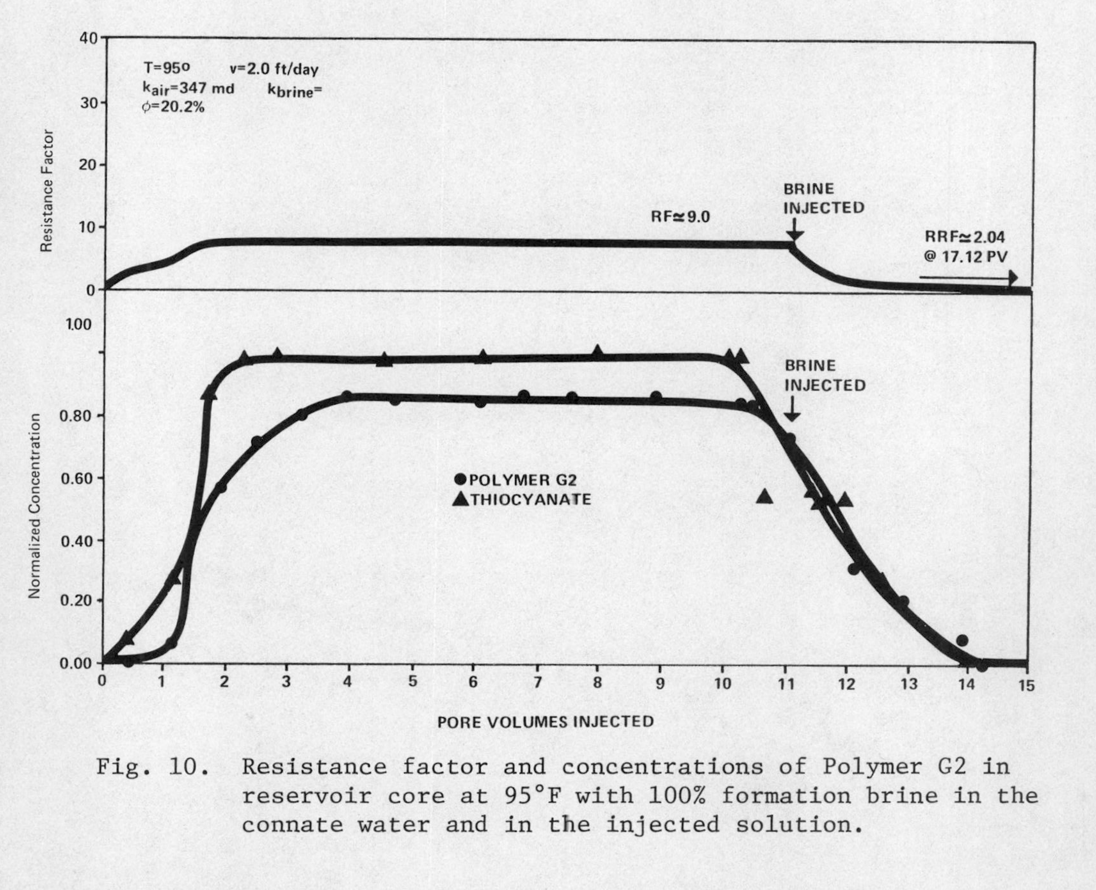

Fig. 10. Resistance factor and concentrations of Polymer G2 in reservoir core at 95°F with 100% formation brine in the connate water and in the injected solution.

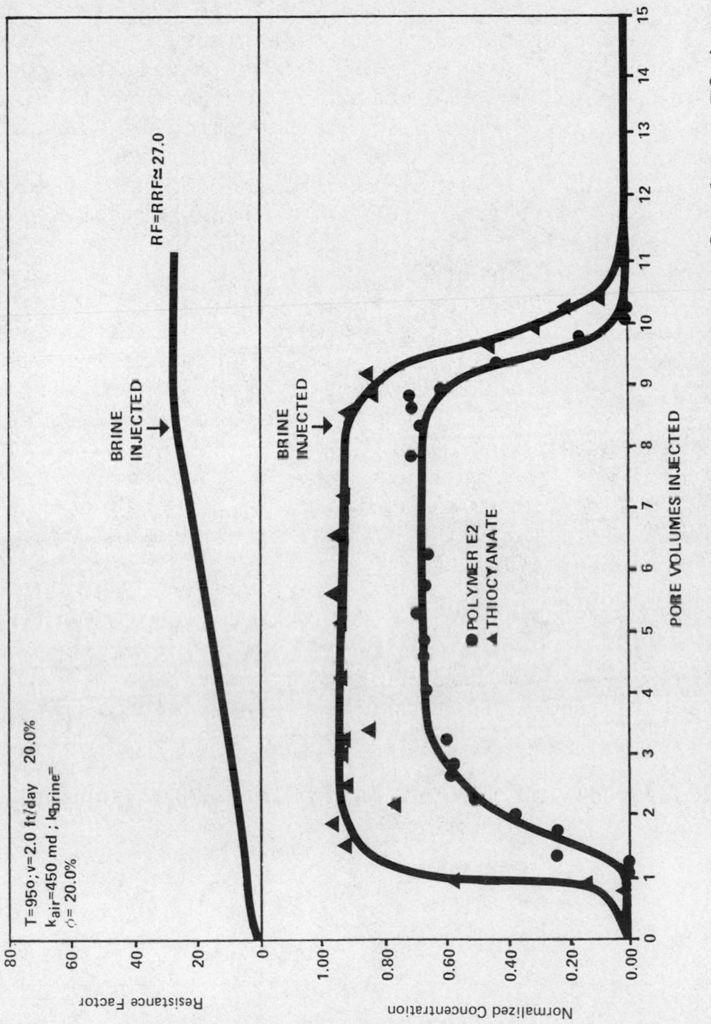

Fig. 11. Resistance factors and concentrations of Polymer E2 in reservoir core at 95°F with 100% formation brine in the connate water and in the injected solution.

that some plugging occurs for the filtered case (Figure 12) but that severe plugging occurs for the unfiltered case (Figure 13). Figure 14 plots the normalized concentration versus pore volumes injected curve for the unfiltered polysaccharide/river water solution; this figure illustrates the mass balance method by which the retentions in Table 11 were calculated.

The effect of microbial stress on rheologic behavior: Bacteria in surface and formation water can pose several potential problems in the design of a polymer or polymer-micellar flood. The predominance of aerobic and/or facultatively anaerobic (bacteria which can subsist either aerobically or anaerobically depending on the oxygen tension level) bacteria in the surface waters can cause (i) a catastrophic reduction in viscosity, and (ii) a decrease in filterability due to plugging of the oil bearing formation's sandface with bacterial cells. The activity of anaerobic sulfate reducing bacteria in the formation produces H_2S which can dissolve in the formation fluids to cause severe corrosion problems of the production strings and which can react with trace metals, such as iron, to form colloidal suspensions, complex suspended particulates, polymer and bacteria, and subsequently, plug the formation. Synergistic effects, as the result of cometabolism of the nutrient base by the aerobic and anaerobic bacteria, can rapidly reduce the integrity of a polymer slug in the reservoir. The addition of biocides to the wellbore, after the proliferation of bacterial colonies, often fail to inhibit microbial growth completely because effective concentrations are lower in situ due to the adsorptive or stripping characteristics of the reservoir rock. Proper design of a polymer-based flood could circumvent these potential problems. This is especially true for floods using extracellular polysaccharides since these fermentation products are prime nutrient sources for bacteria and are very susceptible to microbial degradation.

Table 11. Rheological and Retentive Behavior of Polysaccharide G1 in Reservoir Core at 75°F and 16 ft/day

			Retention	
	RF	RRF	μg/g	lb/acre-foot
River water (filtered)	3.2	1.2	16.3	95
Formation water (filtered)	2.3	1.02	20.7	120
River water (unfiltered)	3.0	1.32	31.0	180
Formation water (unfiltered)	plugged (RF=8.0 @ τ = 10PV)	1.45	63.8	370

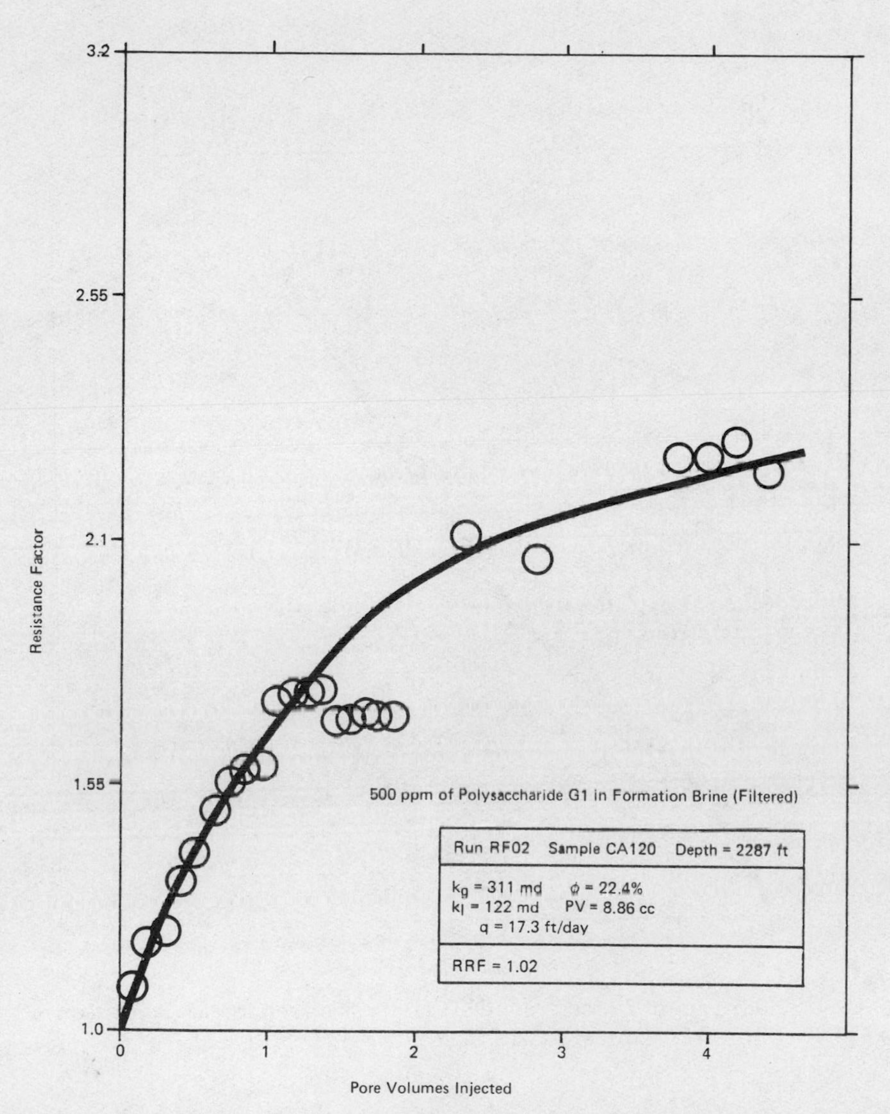

Fig. 12. Resistance factor of dilute polysaccharide solution from reservoir cores (Waltersburg Formation) at 75 ± 2°F.

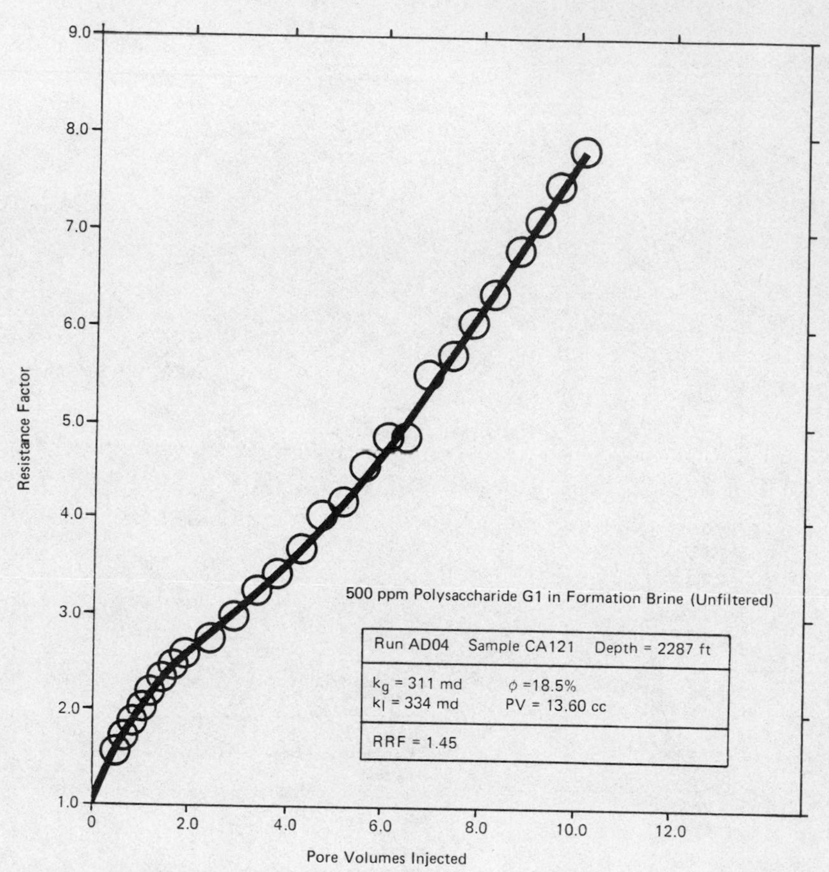

Fig. 13. Resistance factor of dilute polysaccharide solution from reservoir cores at 75 ± 2°F.

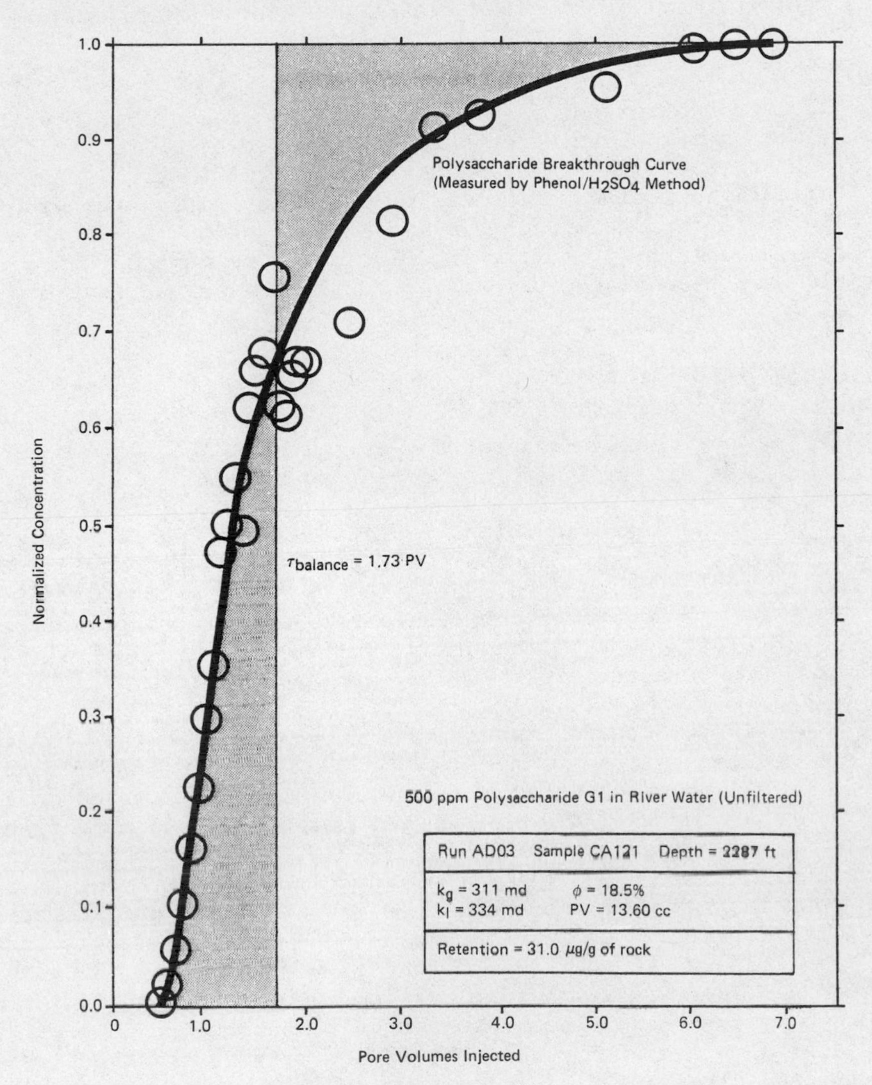

Fig. 14. Breakthrough concentration profile of dilute polysaccharide solution from reservoir cores (Waltersburg Formation) at 75 ± 2°F.

Potential microbial stress problems can be alleviated with a three phase evaluation program. In the first phase, the predominant microbial species in the surface and formation waters are isolated. The second phase of the microbiological study is geared to determine the deleterious effects of the isolated species. A third phase can focus on preventative measures, e.g., use of biocides, for ameliorating these effects without degrading the polymer solution itself.

Isolation of predominant species: Water samples from each source were membrane-filtered through 0.45-micron Gelman nucleopore membranes to count the colony-forming units (CFU) per ml of sample. The following dilution series was filtered in triplicate: 100 ml of sample, 10 ml of sample, 10 ml of 1:10 dilution, 10 ml of 1:100 dilution, and 10 ml of a 1:1,000 dilution. One membrane of each dilution was then placed on each of the following nutrient media:

1. Nutrient agar on 10-mm plate.
2. API agar plate prepared according to Allred (6). Sulfate reducing colonies become black on this medium.
3. Media-free 500-mm plates overlaid with API agar to exclude air and therefore select anaerobes.

Nutrient agar and aerobic API plates were incubated aerobically at 20°C for 7 days. Anaerobic API plates were placed in a BBL Gas Pak system and incubated anaerobically in an atmosphere of H_2 and CO_2 for 6 weeks, being observed periodically during that period. At the end of the incubation period, sulfate reducing CFU/ml (black colonies) and total anaerobic CFU/ml were counted.

A similar procedure was followed to quantify the anaerobic sulfate reducers in the formation. Two core samples 2,272 feet subsea were cracked open and the interior sections removed aseptically into a sterile mortar. They were ground with a mortar and pestle to a fine sandy consistency. The following procedure was performed in quadruplicate: 1 g of sample was transferred to a tube containing 9 ml API broth. This was stoppered and flowed back and forth four times to mix the inoculum. One ml from this tube was then transferred to a second tube of 9 ml API agar and mixed as before. The serial transfer was continued until a dilution of 1:10,000 was reached. The tubes were then capped and incubated at 20°C over a period of 5 weeks. The tubes were checked each week for intense black colonies indicating sulfate reducing bacteria.

To isolate a representative bacterial population from the water samples, 10 representative colonies from each master plate were restreaked onto nutrient agar and API agar plates, then incubated for 3 days at 20°C. They were then transferred to 5°C for storage. Single colonies from each of the nutrient and API agar plates were then streaked onto nutrient and API agar slants, respectively, and incubated for 3 more days at 20°C. A second set of nutrient agar and API plates was restreaked again from these slants to obtain pure cultures. During the experimental period, the isolates were restreaked onto fresh slants three times at 2-week intervals to maintain the freshness of the cultures. All cultures were maintained at 20°C and all microscopy and physiological tests were made with 48- to 72-hour-old cultures.

The isolates were screened taxonomically according to the procedures delineated in Energy Resources' taxonomy manual. The following tests were performed on each isolate: Gram stain, cell motility and morphology, colony morphology, physiological tests for catalase, oxidase and urease activity, carbohydrate utilization, gelatin hydrolysis, cellulose degradation, nitrate reduction, 0/129 sensitivity, and citrate reduction. All liquid and solid media and stains were Difco-Bacto materials. Phase microscopy of all organisms was performed using a Zeiss universal-phase epifluorescence scope.

Quantitative data for aerobic and anaerobic heterotrophs are given in Table 12. Aerobic heterotrophs in the sampled formation water, river water, and sediment were numerically significant in all three sources. Titers ranged from 60 CFU/ml in the sediment and formation water samples to 450 CFU/ml in the river water sample. Aerobes are considered to be significant at titers greater than or equal to 9 CFU/ml. The anaerobic population was significant (at least 1,500 CFU/ml) in the sediment sample only.

The taxonomy of the formation and river water samples (Table 13) showed a typical cross section of aerobic bacteria associated with soil and water. Three genera predominated: *Bacillus*, *Pseudomonas*, and *Arthrobacter*. Anaerobic sulfate reducers were found only in the formation water sample but constituted 40 percent of the isolates from this source. Four of the five sulfate reducing isolates were *Desulfovibrio* species; one was *Desulfotomaculum* species.

Deleterious effects of isolated species: The second phase of the biological investigation was divided into two stages. In the first stage, pure cultures were grown on substrates of polyacrylamide B and polysaccharide G2 to determine if these polymers could serve as a nutritional base for the microorganisms of the river water and of the sediments and water of the formation. In the second stage a time study on the effect of bacterial isolates from the river water and the formation on the viscosities of dilute polyacrylamide (B) and polysaccharide (G2) solutions were carried out. Taxonomic screening produced 40 isolates from the Little Wabash River and the Waltersburg formation. Microscopic observations were performed to verify that these isolates were pure cultures rather than combinations of organisms. The organisms, in a pure culture state, were grown on API agar substrate with and without polyacrylamide B and polysaccharide G2.

To assess the capability of each of the isolates to grow on polyacrylamide B, biopolymer G2, and a model hydrocarbon mixture

Table 12. Colony-forming Units of Bacteria in Formation Water, River Water and Sediment

	River Water CFU/ml	Sediment CFU/ml	Formation Water CFU/ml
Aerobes	450	100-150	70
Facultative anaerobes	310	13,000-26,000	93
Strict anaerobes	0	1,000-19,000	7
Sulfate reducers	0	100- 1,000	5

Note: The aerobic population is significant if more than 9 CFU/ml are found; the anaerobic population is significant if more than 1,500 CFU/ml are found.

Table 13. Storms Pool Unit Bacterial Taxonomy: Predominance and Distribution of Bacterial Taxa from Sampled River and Formation Waters

A. Aerobic Heterotrophs in River and Formation Water

Taxonomic Group	Number of Isolates from Total of 39	Number of River Water Isolates (18)	Number of Formation Water Isolates (21)
Arthrobacter	3	2	1
Bacillus	7	5	2
Pseudomonas	7	3	4
Micrococcus	5	1	4
Sporosarcina	4	3	1
Aerococcus	2	1	1
Cornebacterium	3	1	2
Flavobacter	1	1	-
Mycobacterium	1	-	1
Leuconostoc	1	-	1
Hyphomicrobium	1	1	-
Citrobacter	1	-	1
Lactobacillus	1	-	1
Actinobacillus	1	-	1
Paracoccus	1	-	1

B. Anaerobic Heterotrophs in River and Formation Water

Taxonomic Group	Number of Isolates from Total of 13	Number of River Water Isolates (3)	Number of Formation Water Isolates (10)	Sulfate Reduction
Vibrio (facultative)	5	3	2	0
Bacillus (facultative)	3	0	5	+a
Desulfovibrio	4	0	3	+++b
Desulfotomaculum	1	0	2	+++b

+a = < 10 percent sulfate reduction activity.
+++b = > 90 percent sulfate reduction activity

(11.1 percent each of C_{12}-C_{22} n-alkanes) the following procedure was used: the test polymer or model hydrocarbon mixture absorbed onto 10 percent silica gel was mixed 15 percent weight-to-volume with sterilized Makula and Finnerty's (7) river media, plus 1.5 percent agar. Thorough mixing was accomplished by sonicating the mixture 20 minutes in sterile 1-liter polypropylene bottles and vortexing for 5 minutes at high speed. This procedure was required for homogenizing the medium plus the nutrient source mixture. The media were then poured into 100-mm petri plates. Each isolate was streaked in duplicate onto each of the three media. The isolates were also streaked onto Makula and Finnerty's river media and 15 percent agar without the supplemental polymer or hydrocarbon source to check for the utilization of agar as a sole carbon source. All plates were incubated at 20°C for 10 days. Isolates growing on test media were transferred from the initial test plate and re-streaked onto a second set of plates to ensure that growth on the primary plate was not due to nutrient carry-over. Isolates growing on test nutrients but not on unsupplemented agar were considered confirmed utilizers of the test carbon sources. Positive growth of the isolates on the streaked plates was identified by the sequence illustrated in Figure 15.

Data from the nutrient screen are summarized in Table 14. Biopolymer G2 supported extensive growth but polyacrylamide B did not. Identification of each of the isolates showed a predominance of three taxa that are commonly associated with soil and water. Of the isolated from each of the three groups, *Arthrobacter*, *Bacillus*, and *Pseudomonas*, one-third to one-half were able to grow on biopolymer G2 with 30 ppm formaldehyde, and 60-100 percent were supported by the hydrocarbon mixture. Two other isolates, identified as *Micrococcus* and *Citrobacter* species, were able to utilize the hydrocarbon mixture only. Only one isolate was able to utilize the polyacrylamide B. The isolate has tentatively been identified as a *Micrococcus* species, but such nutritional flexibility is unusual with this species and the isolate's identity should be re-examined.

The testing procedure used here is subject to some variation due to the difficulty in handling the polymers in emulsion/broth form. Polyacrylamide B appeared able to degrade the agar support matrix into which it had been mixed, causing the media to turn into a sticky slime which would not adhere to the test plate after 3 or 4 days' incubation. This condition occurred erratically among the test plates and made it difficult and in some cases impossible to determine if growth occurred. Apparently this was not due to bacterial degradation, because the condition was originally discovered in uninoculated plates. After 4 weeks of incubation, fungal growth was observed on several of the polyacrylamide B plates. The fungus may have been introduced to the plates as a contaminant or may represent part of the microflora. Further investigation of this fungus is under way.

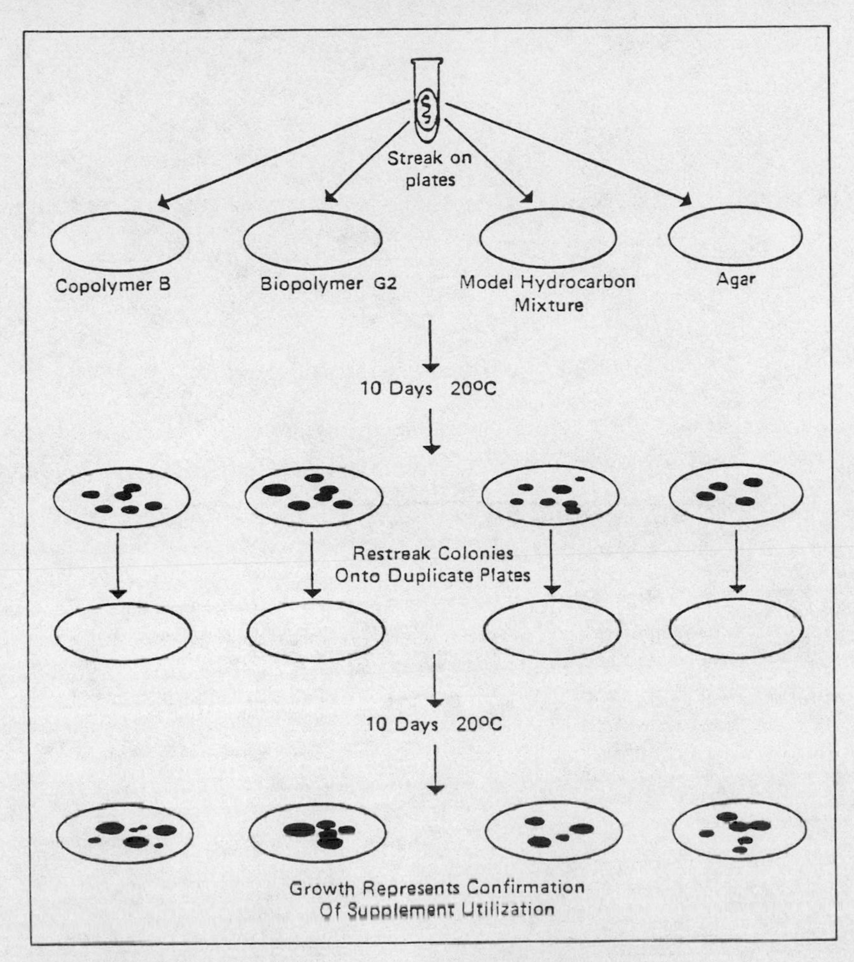

Fig. 15. Screen for nutritional capabilities of isolates from formation and river waters.

In the second stage of the preliminary survey of the biodegradation of biopolymers and copolymers, viscosity measurements were used as the criterion for polymer degradation. Large molecular weight carbon compounds, such as the polymers in this study, may be degraded by direct bacterial metabolization and by cometabolism, in which the presence of a supplemental nutrient source enables the bacteria to proliferate and degrade the polymer. In this way, compounds which initially are refractory to biodegradation may be degraded.

A test organism was inoculated into the biopolymer and the copolymer with four nutrient supplement combinations. Viscosity readings were taken at 2-week intervals over a period of 6 weeks. The organism chosen was originally isolated from the formation

Table 14. Nutritional Capabilities: Growth on Polyacrylamide B, Polysaccharide G2, and Model Hydrocarbon Mixture

Taxonomic Group	Number of Colonies Growing		
	Polysaccharide G2[a]	Polyacrylamide B	Hydrocarbon Mixture
Arthrobacter	1	0	2
Bacillus	4	0	7
Pseudomonas	2	0	6
Micrococcus	0	1	1
Citrobacter	0	0	1

[a]Contains 300 ppm formaldehyde.

water, and has been identified as a Pseudomonas sp. or isolate 91. In addition a negative control organism also isolated from the surface waters was tested. This species is identified as isolate 1. Each organism was grown on a fresh slant culture and then suspended in a phosphate buffer. This cell suspension was inoculated into the following nutrient-biopolymer combinations with 500 ppm active polymer product:

1. Unsupplemented polymers.
2. Polymers + 5 percent hydrocarbon.
3. Polymers + 5 percent (cas-amino acids and glucose).
4. Polymers + 5 percent (hydrocarbon + cas-amino acids and glucose).

The inoculated mixtures were placed in a shaker bath at 200 rpm at room temperature; 15-ml samples were aseptically removed and placed in sample vials each week for viscosity readings. A set of uninoculated mixtures was also incubated as controls. This test series was the same for each organism and polymer. The inoculated mixtures were placed in a shaker bath at 200 rpm at room temperature; 15-ml samples were aseptically removed and placed in sample vials each week for viscosity readings. A set of uninoculated mixtures was also incubated as controls. This test series was the same for each organism and polymer.

The viscosity-time relationships are plotted in Figures 16 through Figure 23. The viscosity-time relationships for the biopolymer and copolymer controls and the inoculated polymer substrate are plotted in Figures 16 and 17. The extreme degradation of the copolymer in Figure 16 at the first 2-week interval is undoubtedly

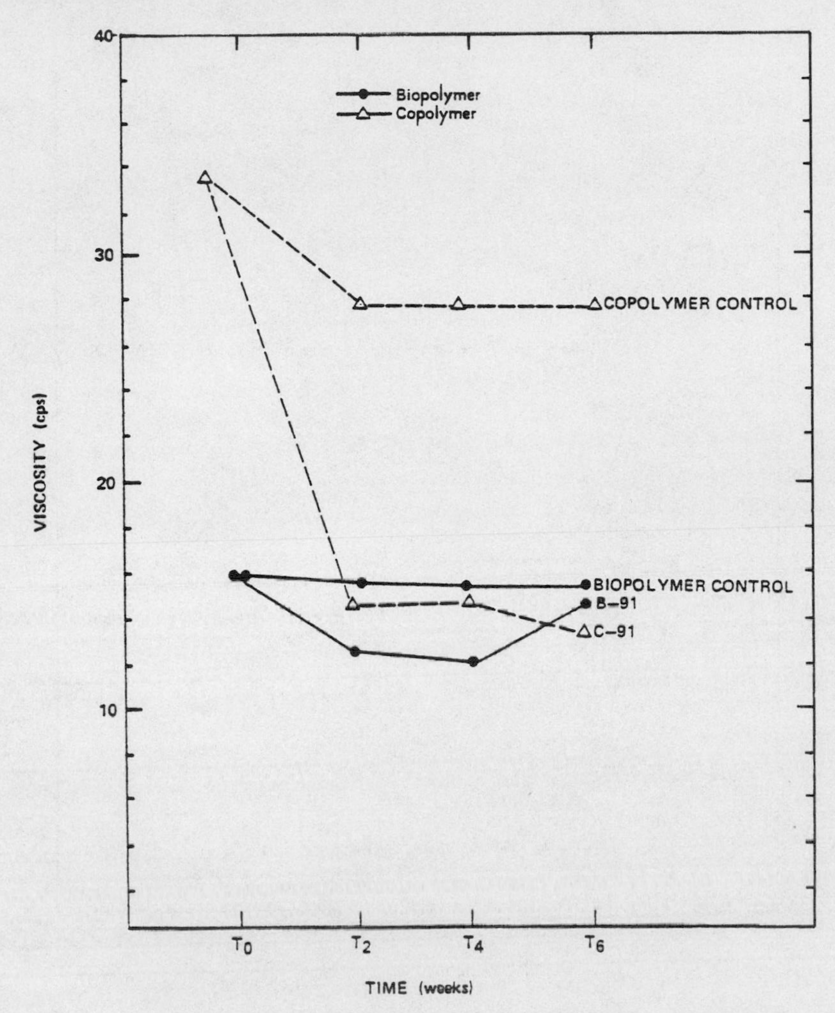

Fig. 16. Viscosity-time relationships for unsupplemented polymers inoculated with isolate 91.

caused by hydrolysis which is accelerated in the presence of high concentrations of oxygen. The high concentrations of oxygen result from the constant aeration which was required to remove fermentation gases such as CO_2, H_2, etc. Further reduction in viscosity is experienced on inoculating with isolates 91 and 1. These isolates act as biological catalysts which accelerate the parting of unsaturated and double bonds by free or combined oxygen. The reduction of viscosity is severe (~ 50%) and constant for the polyacrylamides. In the case of the biopolymers, cometabolism occurs and the viscosity increases as secondary polymers are formed. There is a lag time in the viscosity increase because of the finite residence time for secondary enzymatic action, which tags broken

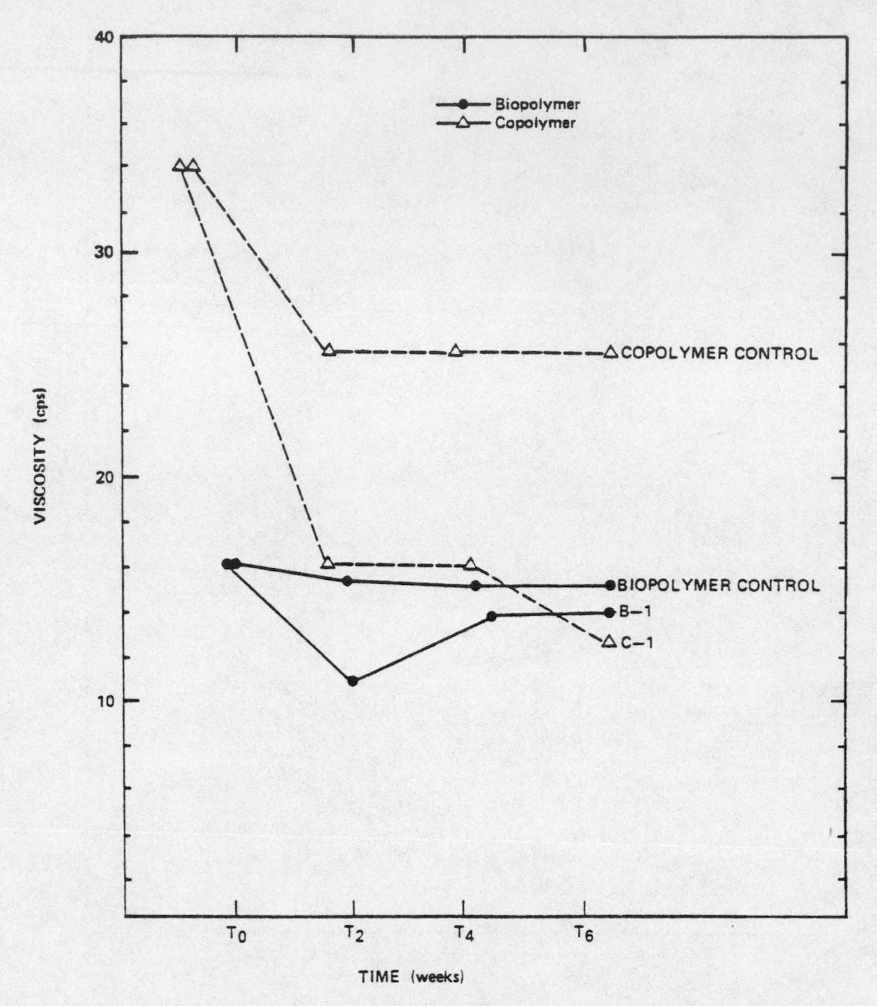

Fig. 17. Viscosity-time relationships for unsupplemented polymers inoculated with isolate 1.

ends of the hydrocarbon chains, to take place. The hydrocarbon nutrient supplement is used to simulate the hydrocarbon food source that are naturally available in the reservoir. This supplement depresses the viscosity control of the biopolymer and the copolymer as illustrated in Figures 16 and 17. The extreme depression of the viscosity of the copolymers with cas-amino acids and glucose in Figures 20 to 23 most probably results from acid hydrolysis. Note that viscosity of the biopolymer is not as susceptible to acid hydrolysis and consistently exhibit a minimum with an increase in viscosity approaching, and sometimes exceeding, the control viscosity after a two week period.

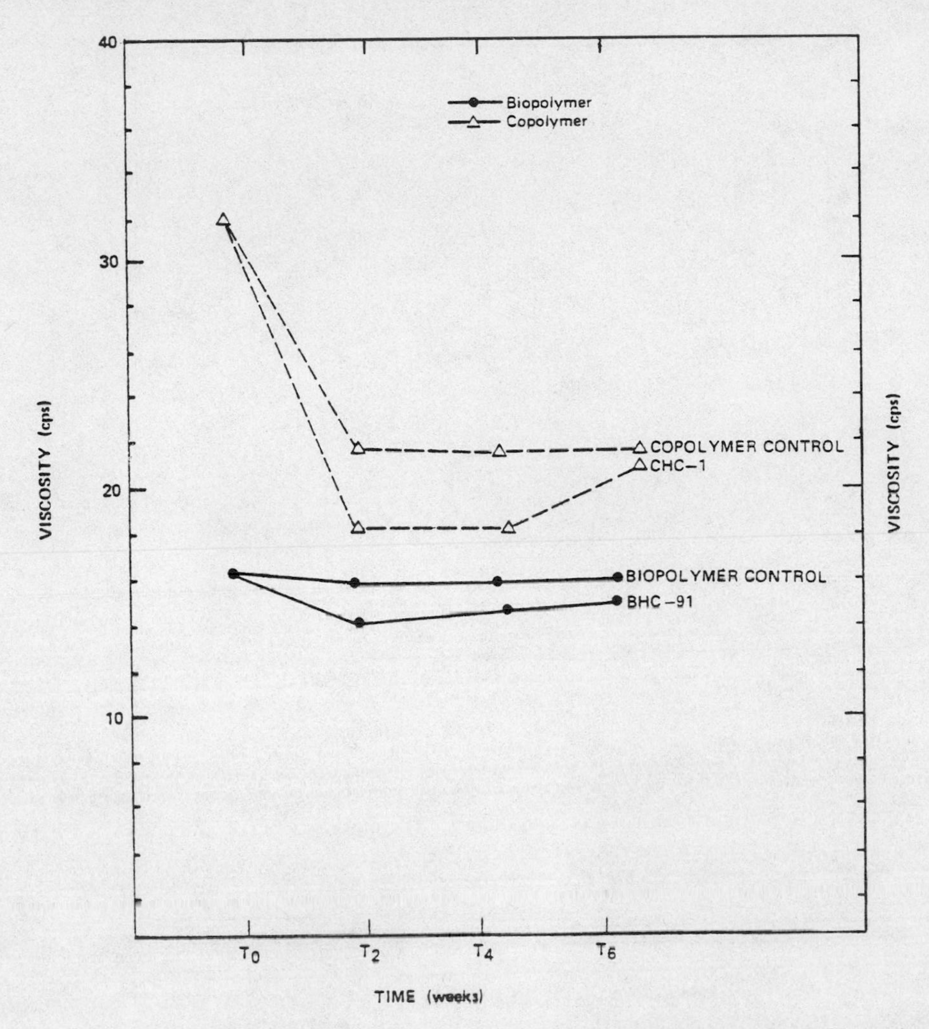

Fig. 18. Viscosity-time relationships for polymers plus 5% hydrocarbon inoculated with isolate 91.

Preventative measures: Three biocides are being evaluated to ameliorate microbial effects without degrading the polymer solution itself. The three biocides are acrolein, formaldehyde, and Visco 3991 (2,2-dibromo-3-nitrilopropionamide). The biocides are evaluated in terms of compatibility with the xanthan gum polymer and ability to restrain bacterial growth.

The polymer compatibility was evaluated in terms of viscosity reduction and filterability of dilute polymeric (500 ppm) solutions. The viscosity reduction curves are plotted in Figure 24 for different concentrations (10, 50, and 100 ppm) of formaldehyde. The curves indicate that viscosity is reasonably insensitive to con-

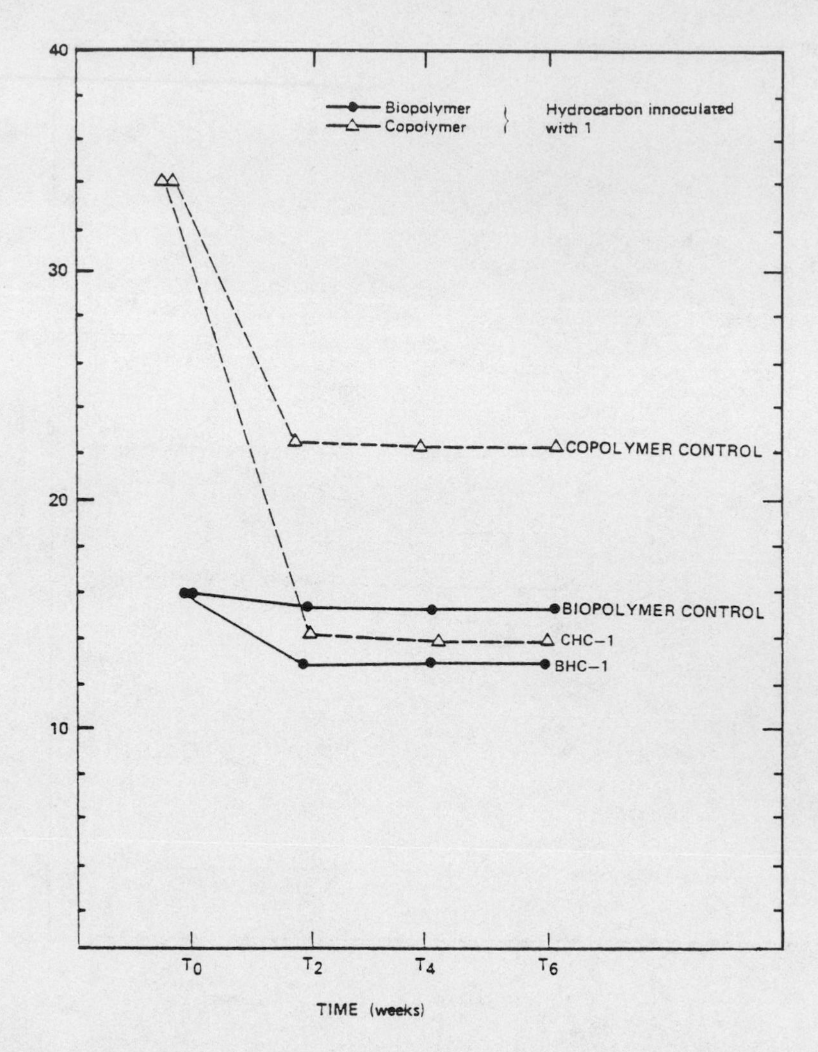

Fig. 19. Viscosity-time relationships for polymers plus 5% hydrocarbons inoculated with isolate 91.

centration of biocide below the 50 ppm level of formaldehyde and very sensitive (as much as 30-70 percent increase) to concentration above the 100 ppm level; similar curves for acrolein and Visco 3991 indicate that viscosity is insensitive (less than 2-10 percent increase) to concentrations of acrolein below 100 ppm, and that viscosity is only fairly sensitive (10-30 percent increase) to Visco 3991 above 10 ppm and below 100 ppm. The filterabilities of the solutions appear to decrease with increasing concentrations of biocide, as illustrated in Figures 25 and 26 for formaldehyde, but there was no consistent pattern to the decrease in filterability for any of the biocides examined.

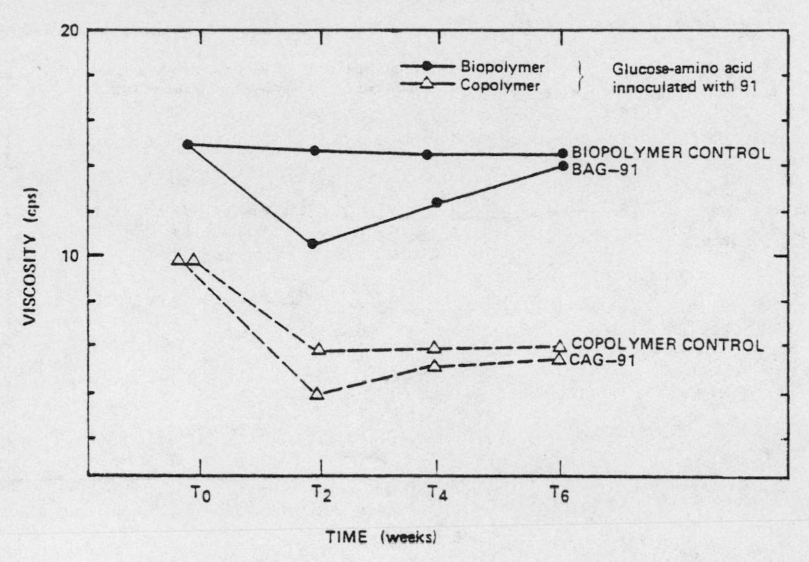

Fig. 20. Viscosity-time relationships for polymers plus 5% glucose-amino acids inoculated with isolate 91.

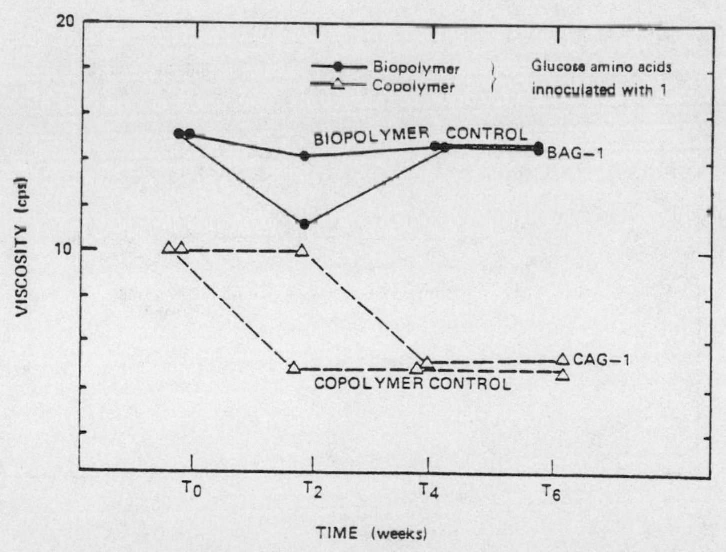

Fig. 21. Viscosity-time relationships for polymers plus 5% glucose-amino acids inoculated with isolate 1.

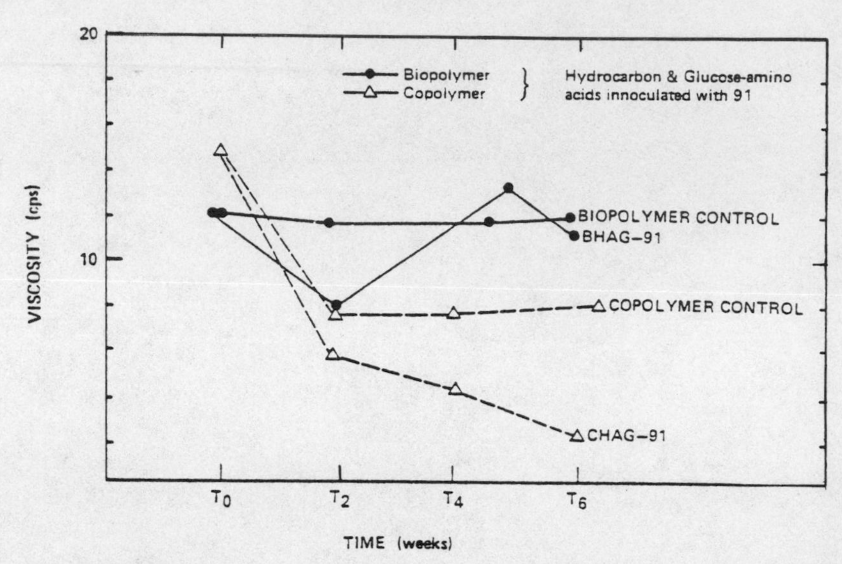

Fig. 22. Viscosity-time relationships for polymers plus 5% hydrocarbons and glucose-amino acids inoculated with isolate 91.

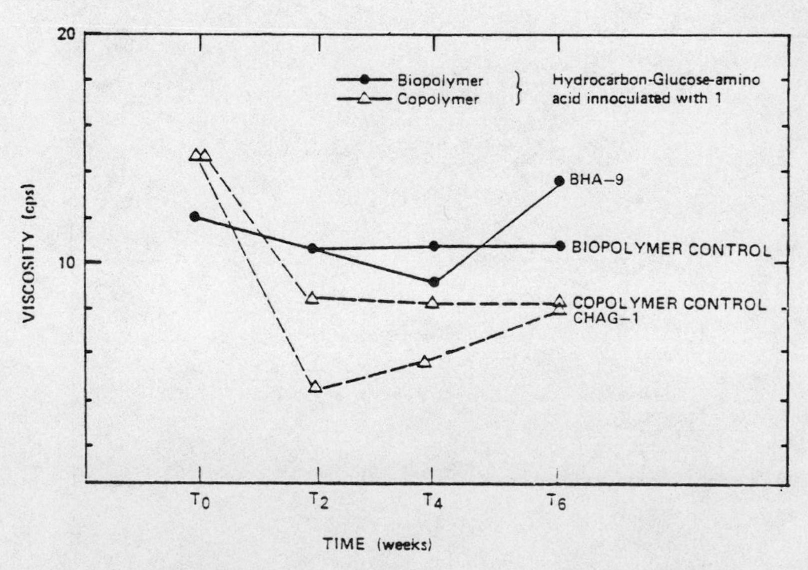

Fig. 23. Viscosity-time relationships for polymers plus 5% hydrocarbons and glucose-amino acids inoculated with isolate 1.

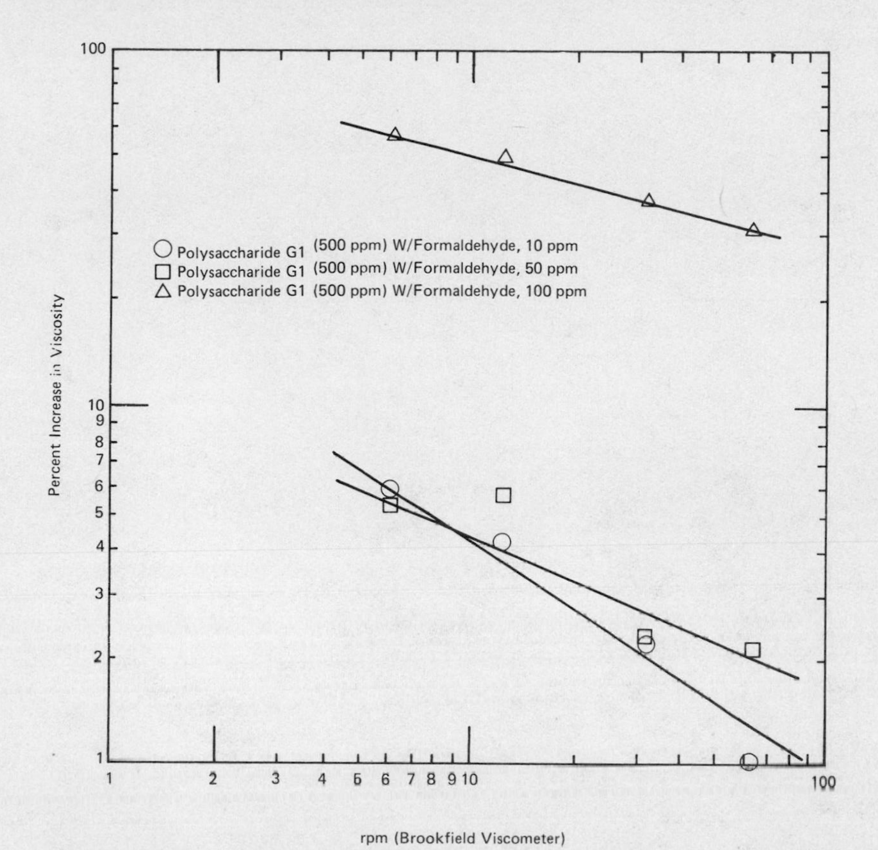

Fig. 24. Percent change in viscosity as a function of rpm for polysaccharide G1 with formaldehyde.

The impact of each biocide on microbial populations, together with the compatibility studies, can be used to select the most effective biocidal treatment. The biocides are first evaluated for their ability to inhibit metabolic activity at the time of maximum growth activity; this maximum time is determined from growth curves of the microbial populations of the formation and river wastes. Growth curves are then measured for the selected biocide(s) at different concentrations to determine the optimum level of biocide required and the period necessary for alternate slugging of the injection waters. The biocidal treatment is then determined from the growth-concentration curves and the compatibility studies.

CONCLUSIONS AND SIGNIFICANCE

- The polyacrylamide copolymers are much more viscous than the polysaccharide biopolymers at equivalent concentrations of fresh water, but the viscosities of these copolymers are

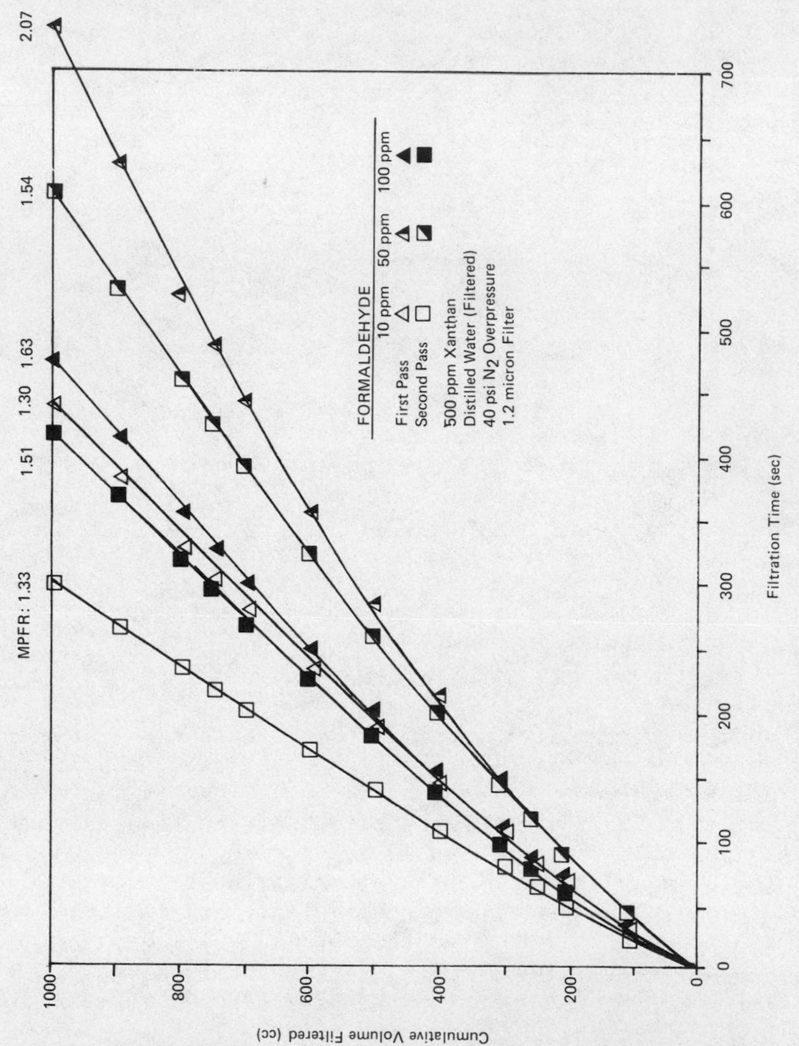

Fig. 25. Filtration time curves for polysaccharide G1 with formaldehyde.

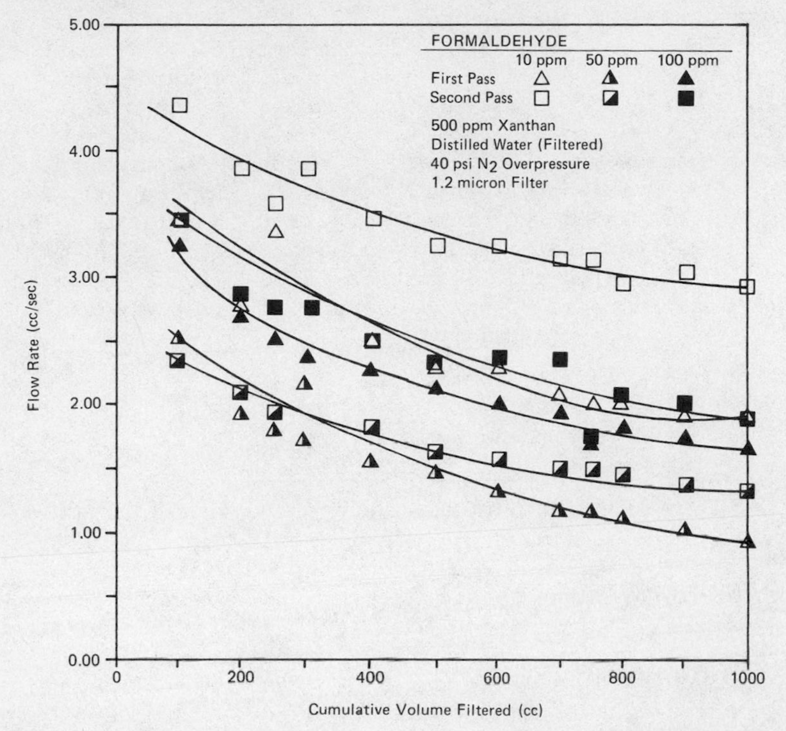

Fig. 26. Filtration time curves for polysaccharide G1 with formaldehyde.

much more sensitive to saline water than the biopolymers. The viscosity of the copolymers degrades by an order of magnitude on increase of the salt concentration to 10,000 ppm. Analysis of the sensitivity of mobility control agents to rotational shear suggests that some permanent loss of viscosity would occur for the polyacrylamide, but not the polysaccharide, at the well-bore.

- The RF and RRF of copolymers can be correlated with screen factors. The high values of resistance factors of copolymer E and the relative ease of injection could have resulted from the extended length and linearity of these polymeric molecules. The long, linear chains of polymer E are the direct result of the gamma radiation manufacturing process for gelling polymer E into aqueous logs without leaving residual traces of cross-linking agents as is done in most polymer emulsion manufacturing processes.

- Polyacrylamides such as C and D which had high electrochemical degradation factors, exhibited plugging behavior and low resistance factors in the reservoir cores which were presaturated with formation brine.

- Polysaccharide G2 had an unusually high residual resistance factor (9.3) and retention (321 lbs/acre-foot) in an 80/20 mix of river/formation waters. The high retention and resistance factor resulted from mechanical entrapment of microgels (reversible complexes of divalent cations with polysaccharide molecules) formed in the 80/20 mixture of river and formation waters. Further testing in 100% formation water indicated higher retention (542 lbs/acre-foot) in the reservoir cores. Correspondent increases in millipore filter ratios and filtration times were observed to occur with increase of univalent and divalent electrolytes.

- The surface and formation waters contained a large cross section of taxonomic groups. Three genera predominated: _Bacillus_, _Pseudomonas_, and _Arthrobacter_. Anaerobic sulfate reducers were found in the formation water. They were of the _Desulfovibrio_ and _Desulfotomaculum_ species.

- All sulfate-reducing bacteria were found in the formation water and sediments. This indicated a close association of the sulfate reducers with the oil-bearing formation. This substantiates previous evidence that the subsurface water microflora associated with oil deposits contain extremely high concentrations of sulfate reducers (8) even though there was a noted absence of sulfate reducers in the surface waters of the southern Illinois oil fields (9). The sulfate reducers found in the sediments were all strict anaerobes and thus capable of living in the reducing environment of the reservoir. The sulfate reducers found in the sediments were orders of magnitude greater than the sulfate reducers found in the formation waters. These sulfate reducers could be responsible for the 2.4 ppm of H_2S found in the formation waters. The quantity of H_2S in the formation water and the degree of corrosion could be accelerated if the polymers injected serve as nutrient bases for these sulfate reducers.

- It was determined that the polysaccharide molecules supported growth of a large percentage of isolates from the formation whereas the polyacrylamide molecules supported growth of only 2 percent of the isolates. Viscosity degradation occurred for both copolymers and biopolymers. Degradation of copolymers occurred mainly by oxygen and acid hydrolysis which was catalyzed by microbial action. The biopolymers experienced some degradation with subsequent enhancement of viscosity due to secondary enzymatic action which causes the cometabolic production of secondary polymers.

- The viscosity of polysaccharides increases with increasing concentration of the biocides. This increase is attributed to cross-linking of the polysaccharide molecules. There were correspondent increases in millipore filtration ratios and filtration times on the addition of biocides or decrease in filterability. A methodology was established to select the most effective biocide on the basis of compatibility with the polymer and the ability to retard metabolic activity in the microflora of the surface and formation waters.

In addition to the above, the potential of a bacterial problem in the reservoir appears acute because of the species which were isolated in API shaker tubes and which were microphotographed in the SEM work on random samples of reservoir rock. This SEM microphotograph is illustrated in Figure 27. The rods shown in the microphotograph are very similar to the photomicrographs of bacteria [0.3 to 0.5 micron diameter by 0.7 to 2.0 micron length (10)] associated with Kelzan M.F. biopolymer. Such bacteria can cause wellbore impairment and reduced injectivity (10).

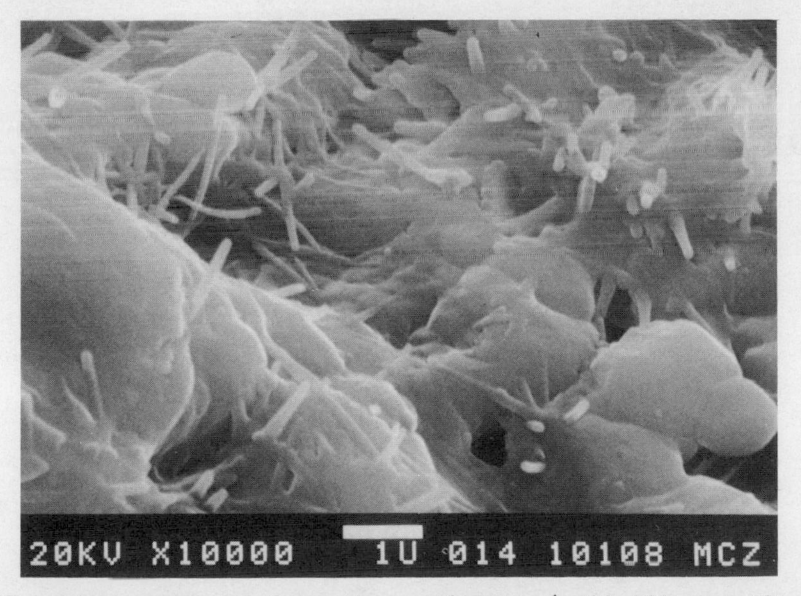

Fig. 27. Photograph of formation sediment (x 10,000 magnification)

REFERENCES

1. J. M. Maerker, Soc. Pet. Eng. J., 311 (August 1975).
2. R. R. Jennings, J. H. Rogers, and J. J. West, J. Pet. Tech., 391 (March 1971).
3. B. B. Sandiford, personal communication (October 1978).
4. J. M. Maerker, Soc. Pet. Eng. J., 172 (August 1975).
5. F. G. McCaffery and F. L. Edillson, Petroleum Recovery Institute, Research Note RN-5 (January 1, 1977).
6. R. C. Allred, in "API recommended practice for biological analysis of subsurface injection waters," Amer. Petrol. Institute, New York, p. 4, 1965.
7. R. Makula and W. R. Finnerty, J. Bacteriol., 95, 2102 (1969).
8. E. S. Bastin, Science, 63, 21 (1926).
9. E. S. Bastin and Greer, Bull. Am. Assoc. Petrol. Geologists, 14, 143 (1930).
10. D. Lipton, SPE 5099 presented at the Rocky Mountain Regional Meeting of the SPE-AIME, Denver, Colorado (April 1975).

FACTORS INFLUENCING POLYACRYLAMIDE ADSORPTION IN POROUS MEDIA AND THEIR EFFECT ON FLOW BEHAVIOR

I. Lakatos, J. Lakatos-Szabó and J. Tóth

Petroleum Engineering Research Laboratory
of the Hungarian Academy of Sciences
Miskolc-Egyetemváros, P.O. Box 2, Hungary

The dynamic adsorption of different polyacrylamides in unconsolidated porous media is discussed. The effect of average molecular mass, degree of hydrolysis, and concentration of polymers, quality and quantity of foreign electrolytes, wettability, chemical composition, and pore structure of porous media on adsorbed amount was studied. The influence of alcohol, surface active agents and alkaline materials is given for practical application of micellar-polymer, caustic-polymer methods.

The phenomena observed are explained by the actual properties and structure of random coils and solution and by molecular and molecule-wall interactions. On account of simultaneity of the adsorption and the mechanical retention, an exact mathematical formulation of the adsorption phenomena in consolidated porous media is difficult.

INTRODUCTION

At the present time the improvement of areal and vertical (volumetric) sweep efficiency takes a great deal of room in secondary and tertiary oil recovery. One of the widely used and perspective methods is mobility control by diluted aqueous solutions of different polyacrylamides (1,2). In the middle of the sixties some authors (3,4) proposed that the viscosity enhancement and the non-Newtonian flow behavior of the solutions were responsible for the reduction of phase mobility. Mungan (5,6), Gogarty (7), Dauben and Menzie (8) have pointed out, however, that the sorption phenomenon plays a decisive role in the flow characteristics of the polymer solutions and carrier phases. In the papers devoted to clarification of displacement mechanism (9-14) the polymer adsorption is considered to be one of the most important factors which determines the flow properties of fluids in reservoirs (15).

Although many valuable statements concerning the adsorption of the polymers in porous media exist until now a detailed and comprehensive paper has not appeared. In this paper investigations of adsorption phenomena carried out during the past decade in the Petroleum Engineering Research Laboratory of the Hungarian Academy of Sciences will be summarized (16,17).

EXPERIMENTAL

The studies were focused on hydrolyzed and unhydrolyzed polyacrylamides produced specifically for polymer flooding. The average molecular mass and degree of hydrolysis were between 0.7×10^6 - 4.8×10^6 and 0-40%, respectively. To measure the dynamic adsorption an unconsolidated porous model was made of silica sand having a particle size between 100-200μm. Other properties of the model were as follows:

length	21.3	cm
diameter	4.9	cm
porosity	38-42	%
permeability	1.7-2.0μm^2	

The specific surface area of the adsorbent was 0.18 m^2g^{-1} by krypton adsorption and 0.1 m^2g^{-1} measured by mercury permeometry.

The polymers were dissolved by the conventional technique and the filtered solutions were injected through porous model with a linear flow rate of 150 cm per day. The polymer concentration of the samples was determined by turbidimetry. The adsorbed amount was calculated on the basis of the saturation curves.

GENERAL FEATURES OF STATIC ADSORPTION OF POLYACRYLAMIDES

The static adsorption of different polyacrylamides was studied in detail by Mungan (6), Smith (9), Schamp and Huylebroeck (18), Szabó (12), Dawson and Lantz (19). Their conclusions can be summarized as follows:

1. The adsorption of unhydrolyzed and partially hydrolyzed polyacrylamides at different adsorbent can be characterized by Langmuir I isotherms.
2. The sorption phenomena show high irreversibility which is attributed to the hydrogen bridge bonds and the chemical linkages between carboxyl groups and multivalent cations of the surface (20).

3. A great difference exists in the adsorbed amount measured under static and dynamic conditions.

The latter statement is explained by the facts, among others, that

a) the specific surface area of consolidated and unconsolidated porous materials having the same particle size distribution is different, and

b) the distribution curves characterizing the pore size and the actual hydrodynamic diameter of the molecules usually overlap each other and, therefore, the surface area is accessible for the polymer molecules only in part (19,21).

As a result of point 2, the adsorption is accompanied by dynamic and irreversible entrapment of the molecules when a natural porous core or model is flooded by polymer solutions. This makes the application of a compatible polymer-model system imperative in adsorption studies. In the course of our investigations, this requirement was fulfilled as far as possible. Despite all our efforts, however, it should be emphasized that the mechanical retention can have a share in the adsorbed amount if the polymer concentration is over the critical value.

RESULTS AND DISCUSSION

Effect of average molecular mass and degree of hydrolysis: Martin and Sherwood (22) published data with regard to the effect of hydrolysis. According to the experimental results, the amount adsorbed shows a declining tendency as the degree of hydrolysis increases. No other quantitative data can be found in the literature relating to the effect of the molecular mass and the degree of hydrolysis.

In our experiments the polymer concentration was 0.5 gdm^{-3} and the solutions also contained sodium chloride to simulate the ion strength of conventional connate waters. Figures 1 and 2 show how the adsorbed polymer amount depends on the average molecular mass and the degree of hydrolysis, respectively. On the basis of the curves shown in Figure 1 the adsorbed amount slightly decreases with the molecular mass. The relative position and tendency of the curves predict what Figure 2 evidently show, namely the degree of hydrolysis to a large extent influences polymer adsorption.

The interpretation of the experimental results can be accomplished similarly to Willhite and Dominguez (15,23) in terms of structure of the solutions and random molecular coils. Results from dynamic tests for a number of polymers a relation was found between the adsorbed amount and the viscosity, specifically, the

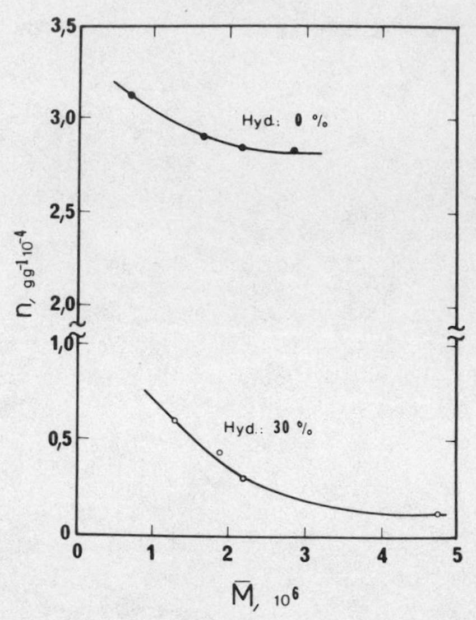

Fig. 1. Dependence of amount adsorbed on average molecular mass.

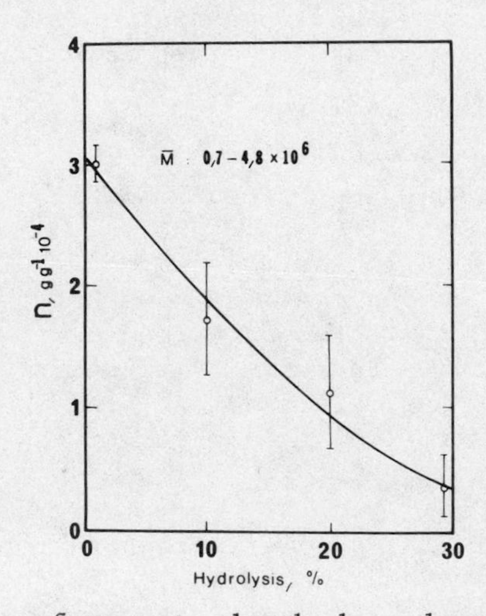

Fig. 2. Dependence of amount adsorbed on degree of hydrolysis.

intrinsic viscosity of the polymer solutions. Because macroscopic properties are connected with the properties of the random coils, correlations between the adsorbed amount, the equivalent diameter, and the density of the molecules prove to be logical (Figures 3 and 4).

The equivalent diameter of the random coil in a "good" solvent is proportional to cube-root of the average molecular mass and the intrinsic viscosity. In contrast, the density of the coils changes

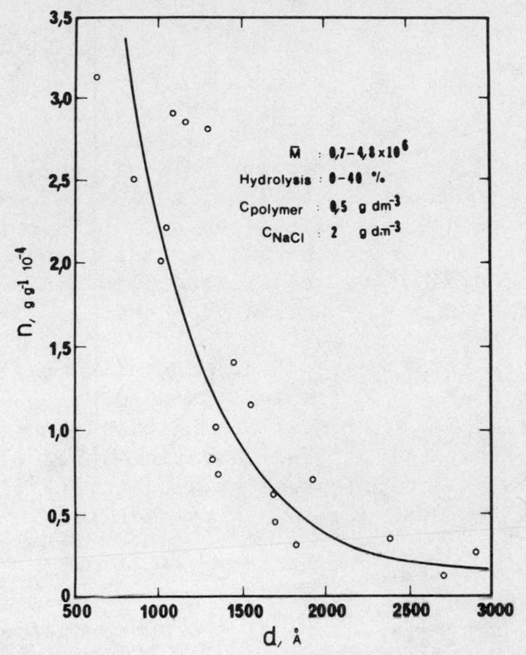

Fig. 3. Dependence of amount adsorbed on equivalent diameter of random coils.

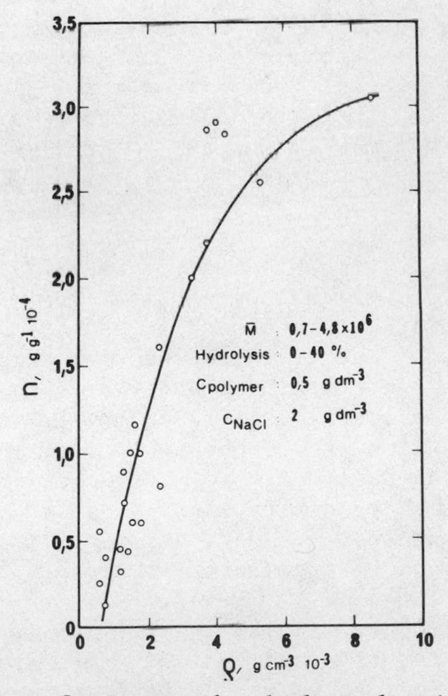

Fig. 4. Dependence of amount adsorbed on density of random coils.

very little. At least one order of magnitude difference exists in the density of unhydrolyzed and partially (30%) hydrolyzed polyacrylamides. Our final conclusion is that as a result of the increase in the average molecular mass and the degree of hydrolysis, the polymer retention in the porous media and hence the adsorbed amount decreases because the area per molecule increases; this is also accompanied by a significant lowering in the coil density. Considering that the coil density depends mainly on the number of carboxyl groups on the chain, the polymer adsorption is determined primarily by the degree of hydrolysis and only secondarily by the average molecular mass of the polyacrylamides.

Effect of polymer concentration: The polymer flooding is usually carried out by polymer solutions of 0.2-1.5 gdm^{-3}. Taking into consideration that under static conditions the adsorbed amount does not depend on the polymer concentration over 0.1-0.2 gdm^{-3} (12,15), the dynamic adsorption was studied in distilled water and in 2 gdm^{-3} NaCl solutions while their polymer content changed between 0.1-2.0 gdm^{-3}.

Dependence of the adsorbed amount on the polymer concentration is given for a typical unhydrolyzed and a partially hydrolyzed polyacrylamide in Figures 5 and 6. Apart from the fact that the absolute values indicate considerable differences, it is characteristic that n increases with the polymer concentration. Within this general tendency, as it appears from these curves, the steep increase occurs after a definite polymer concentration in all cases. Before this critical concentration--similar to the equilibrium section of static adsorption isotherms--the adsorbed amount does not depend significantly on polymer concentration. Incidentally, the critical polymer concentration in 2 gdm^{-3} NaCl is increased twofold over values measured in distilled water for both polymers.

These characteristic changes cannot be explained by the change of the equivalent diameter and the density of the random coils, because these molecular parameters shift only slightly with the polymer concentration. Rather the structure of the solutions changes with polymer concentration, and the solutions become gradually weaker in free solvent. According to the curves shown in Figure 7, the critical polymer concentration is 0.57 gdm^{-3} for unhydrolyzed polymer and 0.14 gdm^{-3} for partially hydrolyzed polyacrylamide. Over these concentrations the solutions can be regarded to be oversaturated. In NaCl solutions the critical values modified to 2 gdm^{-3} and 0.95 gdm^{-3}, respectively.

Willhite and Dominguez (15,23), referring to unpublished data, reported that molecular interactions and aggregation can take place long before the critical polymer concentration. Consequently a real solution does not exist in the vicinity of the critical con-

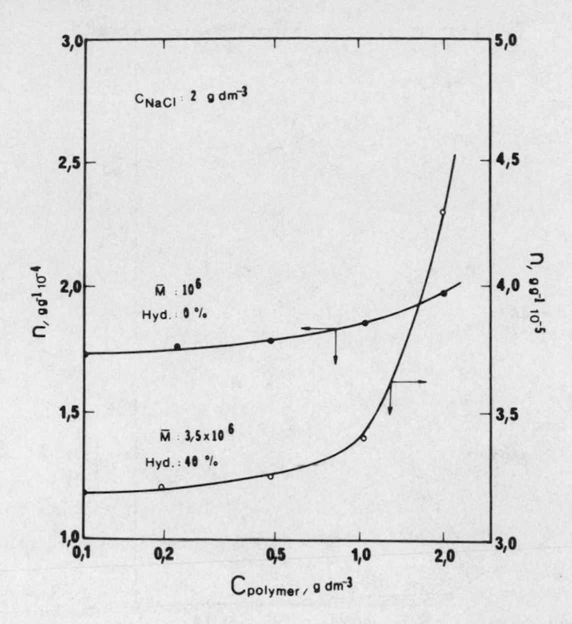

Fig. 5. Dependence of amount adsorbed on polymer concentration in 2 gdm^{-3} NaCl solution.

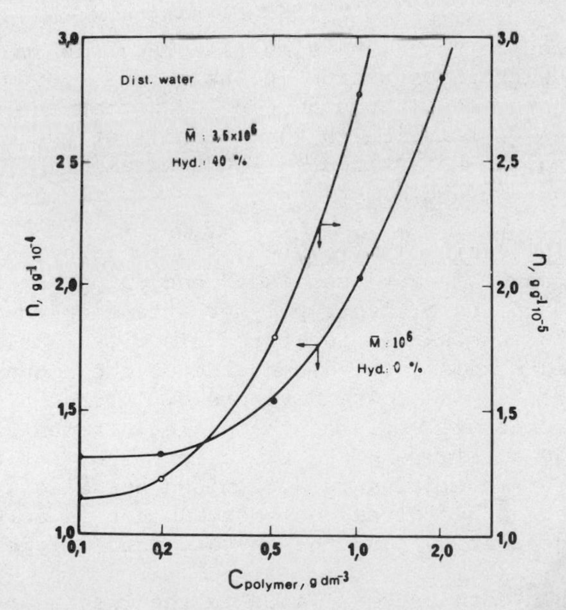

Fig. 6. Dependence of amount adsorbed on polymer concentration in distilled water.

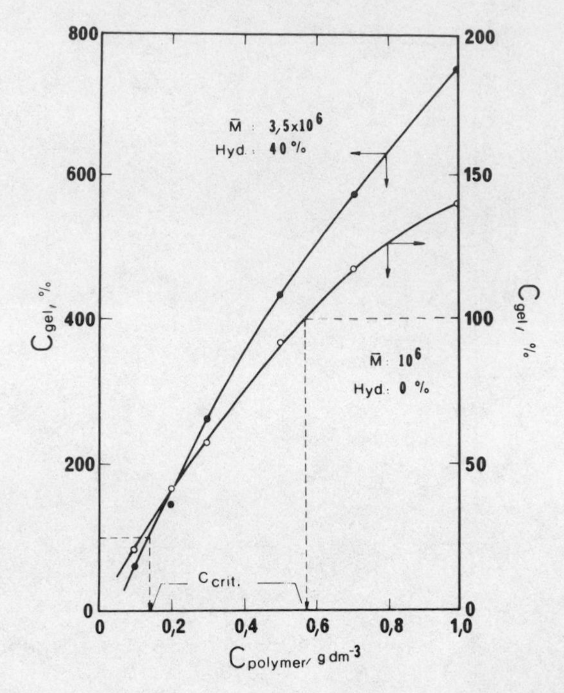

Fig. 7. Dependence of gel concentration on polymer concentration.

centration, but the whole solution is in a gel state. In this way aggregates rather than single molecules adsorb on the surface. As a result of particle size expansion, however, the accessible pore volume and surface decrease and simultaneously the mechanical entrapment of the aggregates starts in the porous system. Summarizing the experimental results it can be concluded that the steep tendency of the n-$c_{polymer}$ relation in the vicinity of the critical polymer concentration can be attributed fundamentally not to adsorption but to filtration phenomena.

Effect of inorganic electrolytes: In the previous section some statements have already been made concerning the salt effect. Comparing Figures 5 and 6 higher polymer retention was found in 2 gdm^{-3} NaCl solution than in ion-free, distilled water. On the basis of detailed studies, the dependence of the amount adsorbed on NaCl concentration is shown in Figure 8. Although one order of magnitude difference exists in n, the scale increments at the right and the left side of Figure 8 are the same. Thus not only the trend of the n-c_{NaCl} relation is determined but also its measurement. Adsorption of hydrolyzed polyacrylamides is more sensitive to the salt concentration than the unhydrolyzed polymers do.

Some valuable data are available in the literature concerning the effect of different inorganic salts present in the natural

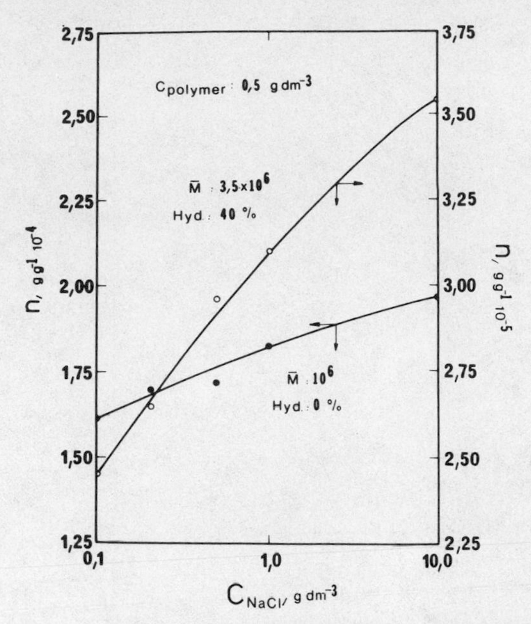

Fig. 8. Dependence of amount adsorbed on sodium chloride concentration.

connate waters on polymer adsorption (9-11,13,24). According to unequivocal opinions, the adsorbed amount increases with the salt concentration. The cause of this trend, however, is rarely discussed. In our earlier publication it was shown that the mono- and bivalent cations repress the dissociation of the carboxyl groups, and hence the contraction of the random coils is responsible for the unfavorable deterioration in the rheological properties of the polymer solutions (25). To present additional data the equivalent diameter and the density of the molecular coils are plotted against the NaCl concentration (Figures 9 and 10).

The coil size and density first change quickly with the salt content, then nearly reach equilibrium, and probably the structure of the solution is not modified over 2 gdm^{-3} NaCl concentration. The structure of the polymer solutions is influenced by cations to a higher extent in case of partially hydrolyzed polyacrylamides than in case of unhydrolyzed polymers, while the role of the average molecular mass is negligible (26). This answers the question why the adsorbed amount shows a relatively high sensitivity for salts in solutions of hydrolyzed polyacrylamides.

In connection with the experimental findings the question should be raised of mechanical entrapment which can also contribute to polymer retention in porous media if high polymer concentration is used. In our experiment the polymer concentration was 0.5 gdm^{-3} and that corresponds to an oversaturated state in case of partially

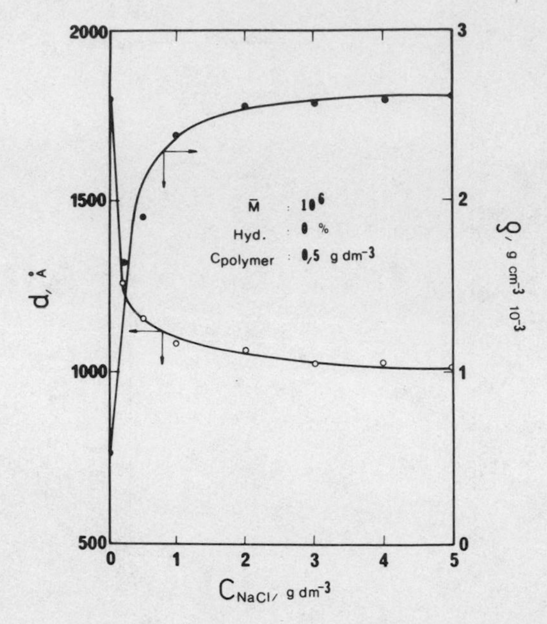

Fig. 9. Dependence of equivalent diameter and density of random coils on sodium chloride concentration (unhydrolyzed polyacrylamide).

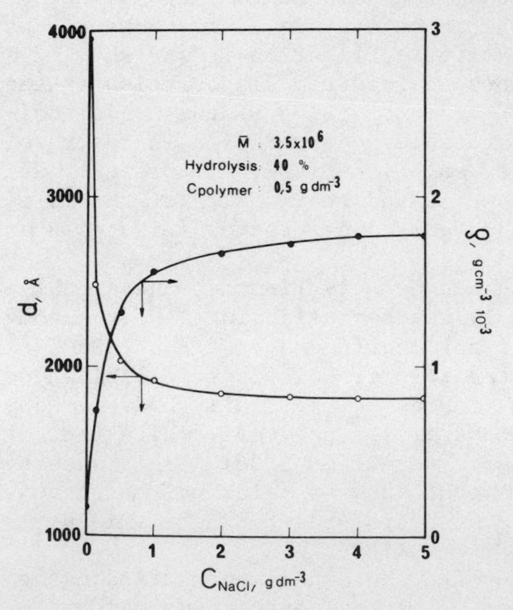

Fig. 10. Dependence of equivalent diameter and density of random coils on sodium chloride concentration (partially hydrolyzed polyacrylamide).

hydrolyzed polyacrylamide. The excess retention cannot be estimated, but Szabó (13) concluded that the mechanical retention depends on the salt concentration in the same way as the adsorption does.

The quality of inorganic electrolytes is not indifferent to polymer adsorption. Bi- and multivalent cations produce 10-30% higher coil contraction than the alkaline metals (25). Therefore, in the presence of multivalent metal salts, the amount adsorbed gradually increases in accordance with the dissociation state of the ionizable groups and the valence state of the cations. In 2 gdm^{-3} $CaCl_2$ solution, for example, the adsorbed amount was 18% higher than in 2 gdm^{-3} NaCl solution under otherwise identical conditions. After all, the salt effect is contradictory from the point of view of polymer flooding. On one hand, the inorganic salts increase the polymer adsorption in the porous system, and on the other they stabilize the structure of the polymer solution. The salt content of the connate water is usually above 1-2 gdm^{-3}, and that is sufficiently high to decrease the gel concentration below 100%. The conventional polymer flooding can be carried out in compatible polymer-rock system, where the permeability reduction is determined primarily by the interactions between the single molecules and the wall.

Effect of wettability: Until now little has been said about the dependence of the adsorption of polyacrylamides on the wettability. Smith (9) has found that the wetting character and the presence of oil do not effect the chemisorption of the polymers. On the other hand, a decrease in the quantity of polymer retarded in the presence of oil was reported by Mungan (5,6). Jennings et al. (27) agree with this, although they added that the mobility reduction of the fluids is independent of the polymer losses. Also worthy of attention are the works of Sarem (10) and Dominguez and Willhite (15,23), who have suggested for example that on a low energy Teflon surface the specific adsorption of acrylamides is much less than in reservoir rocks. The cause of the contradictions may be attributed to the fact that no differentiation is made by the authors as regards the type of retention (physical or chemical sorption). On the other hand, they are disposed to talk about a mechanical entrapment when, in fact, a sorption is concerned. In order to elucidate the question, we have made detailed investigations to determine the role of the wetting character (2,16,17). For the study unconsolidated silica sands having different surface character were used. The curves characterizing the saturation of the porous models are given in Figure 11, while the measured adsorbed amounts are given in Table 1.

From the data listed in the liquid taken off the oil-wet model, the polymer appears directly in the initial (injection) concentration; thus, no adsorption can be observed. From this it

Table 1. Adsorption of an Unhydrolyzed, Low Molecular Mass (10^6) Polyacrylamide on Silica having Different Surface Characteristics.

Surface Character	Adsorbed Amount n, g/g
Oil wet (siliconized)	0
Intermediate (siliconized to 20%)	$0.65 \cdot 10^{-5}$
Natural	$1.90 \cdot 10^{-5}$
Water wet (extracted)	$6.87 \cdot 10^{-5}$

follows that in an oil-wet reservoir, or on an oil film, the sorption of polyacrylamides is zero. On the other hand, the greatest retention occurs on the absolute water-wet model, and the value obtained is about three times as great as that measured on a natural silica surface. On the silica model, which contains the mixed adsorbent (20% of siliconized silica), the amount adsorbed is situated, according to the expectation, between the specific loss obtained on the natural model and that obtained with the model having hydrophobic character.

As a final result, it has been proven that the adsorption of polyacrylamides on reservoir rocks depends to a great extent on the wettability. In all probability no difference occurs whether the intermediate or oil-wet character of the rock surface is a result of adsorbed monolayer or thick oil film. A certain parallelism can be observed between the adsorption of polyacrylamides and methylene blue on a rock surface. By this far-reaching analogy a new methodological possiblity for semi-quantitative determination of the wetting character in reservoir rock is afforded, using the polymer adsorption (29) similar to the dye adsorption method of Holbrook and Bernard (28).

From the point of view of flow characteristics of reservoir fluids, the wettability in the reservoir is of primary importance. It seems to be established, on the basis of the fundamental works of Brown and Fatt (30), Johansen and Dunning (31), Owens and Archer (32), Mungan (33), and Salathiel (34) that the wettability in a reservoir is of heterogeneous character, that is, the small pores and consolidation points are water wet, while the inner surface of the large pores are oil wet. As a result of this, it may be expected that the sorption distribution of the polymer injected into the reservoir will likewise be characterized by heterogeneity.

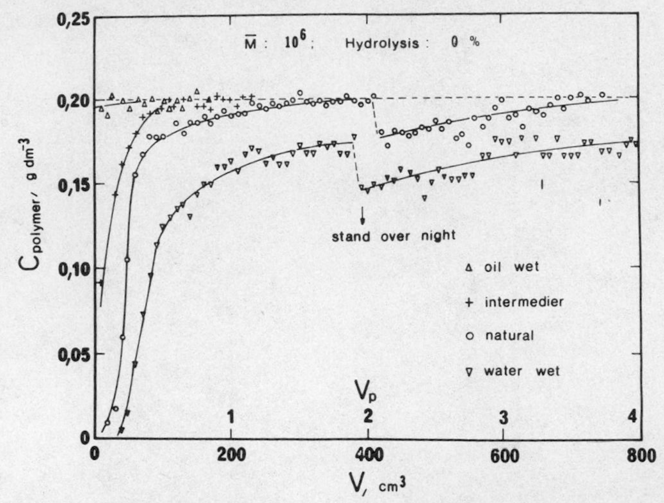

Fig. 11. Saturation curves for silica models having different surface character.

Thus, the flow behavior of the polymer solutions have to change considerably between wide limits: in the large pores only the viscosity enhancement will manifest itself, while the viscosity enhancement and the decrease in the apparent permeability will operate jointly in the small pores.

Effect of molecular interactions and additives: In the reservoirs the polymer solutions mix with different chemicals injected before and after the polymer slug. In this way the interaction between surface active agents and polymers or between caustic materials and polymers is unavoidable at the disintegrated fronts if combined flooding technology is used in practice. These chemicals, which can also be in the polymer solutions, have an influence on the wettability and the solution structure.

It is well known that the polyacrylamides are prone to form hydrogen bridges with very different compounds. As a characteristic example, the effect of isopropyl alcohol (IPA) will be shown here. This alcohol is used widely as a solubilizing agent in micellar solutions. The dependence of the amount adsorbed on alcohol content of the solutions can be seen in Figure 12 for hydrolyzed and unhydrolyzed polymers. For the sake of consistent explanation, the equivalent diameter and the density of the random coils are plotted against the IPA concentration in Figure 13. The trend of the curves in Figure 12 can be explained by different facts. First, the alcohols can influence the wettability of the adsorbent. Water-wet silica was used here, therefore this possibility must be rejected. (In case of intermediate or oil-wet porous media, the situation is not so simple.) Second, alcohols are not as good

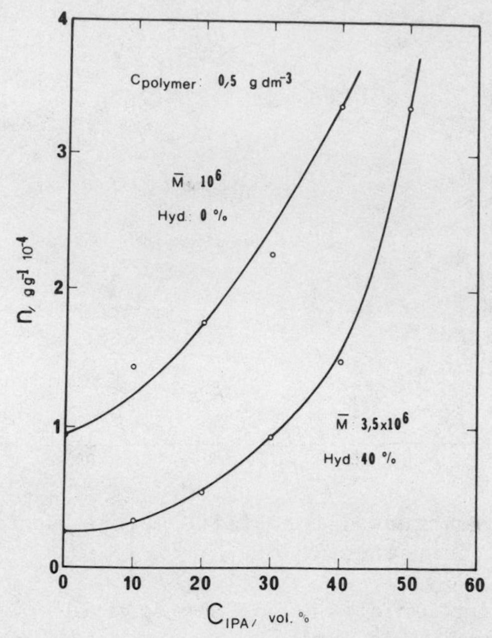

Fig. 12. Dependence of amount adsorbed on isopropyl alcohol concentration.

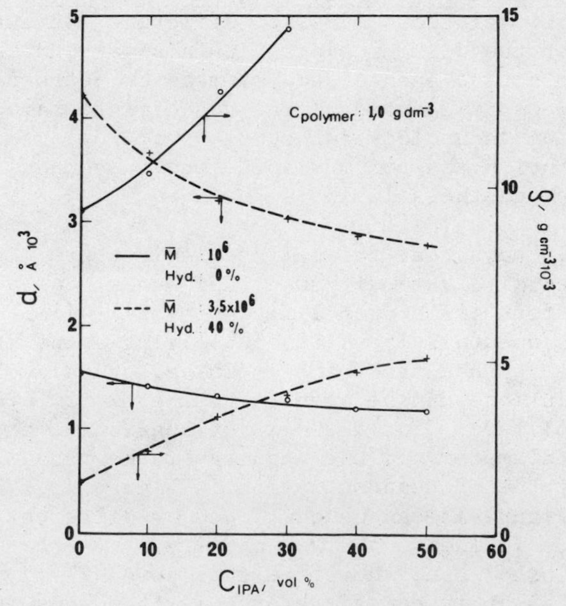

Fig. 13. Dependence of equivalent diameter and density of random coils on isopropyl alcohol concentration.

solvents as water is, and this appears in contraction of the molecules (Figure 13). And at last, not only water but alcohol and water-alcohol (4:2) aggregates also take part in solvation of the polymer molecules. This leads to enlargement of the hydrodynamic diameter of the polymer molecules, which can promote mechanical retention. In our example the latter two facts result in an increasing tendency of the amount adsorbed with alcohol concentration.

The effect of surface active agents and caustic materials is also complex. Such investigations are under way, and therefore our preliminary results will be summarized qualitatively here.

The effect of alkaline materials can cause wettability alteration if the rock surface is intermediate or oil-wet. The enhanced cation concentration influences the size and density of the random coils in a known way. Both facts increase the amount adsorbed. The effect of pH on chain structure (post-hydrolysis) should be taken into consideration if an unhydrolyzed polyacrylamide is studied. But very small difference was found between unhydrolyzed and partially hydrolyzed polymers in this respect.

Nearly similar statements can be made for surface active agents. Petroleum sulfonates and alkyl aryl sulfonates were investigated in detail. In a water-wet porous system the amount of adsorbed polymer decreased considerably. In the oil-wet model, however, the polymer retention increased to a certain degree. This calls our attention to the importance of the orientation of surface active agent in the adsorbed layer.

If the polymer solution also contains surface active agents an additional factor can be the interaction of the polymer and tensid molecules and the mutual adsorption of the molecules and micelles. To determine how these factors influence the dynamic adsorption of polymers in porous media, more work is required, and we face the problem that all special natural system must be treated individually.

Effect of chemical composition and structure of porous media: During the past few years the effect of chemical composition of rocks has been studied intensively. Sandstones, carbonates, silica and different clay minerals were used as adsorbents (6,9,12,18,19, 35). The conclusions do not show any divergence. To give a comprehensive picture of the adsorption of polyacrylamides, some data are given in Table 2 for the effect of adsorbent type on amount of different polymers adsorbed.

The measured amount adsorbed on carbonates and clay minerals is always higher (sometimes by orders of magnitude) than on sand-

Table 2. Effect of Rock Type on Adsorption at Different Polyacrylamides

Polymer $\bar{M}$	Hyd. %	Adsorbed Amount n, g/g Carbonate	Silica
10^6	0	$3.56 \cdot 10^{-5}$	$1.90 \cdot 10^{-5}$
3.5×10^6	40	$1.87 \cdot 10^{-5}$	$0.98 \cdot 10^{-5}$
medium	low	$1.79 \cdot 10^{-5}$	$1.63 \cdot 10^{-5}$
high	high	$1.43 \cdot 10^{-5}$	$0.99 \cdot 10^{-5}$

stones. The difference is explained by two obvious facts. In the interaction of the polymers with the multivalent cations on the rock surface, a compound is formed which dissociates very slowly (9). As mentioned earlier, the bi- and multivalent cation decrease the equivalent diameter of the molecules which results in an increase of the adsorbed amount. Thus, we have to agree with the statement of Jewett and Schurz (36) and many others, that the chemical heterogeneity of the reservoirs considerably influences the adsorption of the polyacrylamides.

The structure of the adsorbed polymer film and that which is inseparable from this, the structure of the porous media, is debated in the literature. Some authors (12,37,38) regard the adsorbed polymer film to be a monolayer remarking, however, that the coverage is usually higher than unity. Deviations are attributed to different effects (39,40). In our case, coverage was calculated for the equilibrium sections of the curves in Figure 5. In contrast to the data in the literature cited here, 2.21 was found for the unhydrolyzed and 1.98 for the partially hydrolyzed polymers. These values increase with polymer concentration. The phenomenon is more sensitive if consolidated porous system is used instead of unconsolidated silica bed. The structure of a natural sandstone core (Csongrád-Dél formation, South-Hungary) provides a good example, as seen in Figure 14.

The permeability of the core is 3.10^{-2} μm^2 determined by mercury porosimetry. This value is relatively low but not too rare. From the pore size distribution, 80% of the pore volume consists of pores having a diameter greater than 10^4Å, while 80% of the surface area (7.10^3 cm^2g^{-1}) is made up of pores having a diameter smaller than 10^3Å. This surface area is not accessible even to the low molecular mass unhydrolyzed polymer. Although the polymer solution flows only in the large pores, which comprise 20% of the

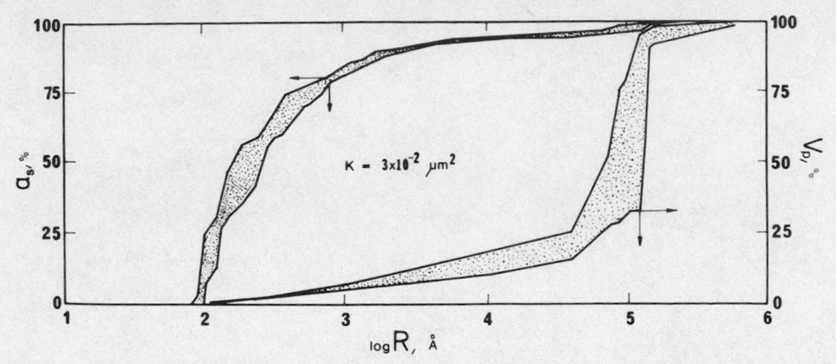

Fig. 14. Dependence of surface area and pore volume on pore radius.

surface area, the mechanical entrapment is high, so that the calculated coverage far exceeds unity. As a result of the simultaneous phenomena (sorption and mechanical entrapment), description of the dynamic adsorption of polyacrylamides in porous media is not possible by exact mathematical formulas (41).

THE ROLE OF ADSORPTION IN FLOW BEHAVIOR OF POLYMER SOLUTIONS

The papers devoted to analysis of adsorption phenomena also usually gave a general scope for flow behavior of the polymer solutions. The important role of adsorption phenomena on flow of mobility controlled phases will be presented in an oil and a water-wet system.

The requirements for the presentation of the general flow characteristics of polymer solutions were best met by high molecular mass ($3.5 \cdot 10^6$) hydrolyzed (40%) polymer. Concentration of the polymer was 100 ppm, viscosity of the solution under laboratory conditions (24°C) was 3.05 mPas. Taking into consideration the rock's permeability ($59 \cdot 10^{-3}$ and $62 \cdot 10^{-3}$ μm^2) and the equivalent diameter of the random coils (2600Å), the polymer-rock system may be regarded as compatible. In the course of the experiments, comparison is made between the flow phenomena observed in the porous cores considered to be water wet and oil wet, respectively. When the permeability was determined by "connate water" (2% sodium chloride solution), the injection sequence of the fluids was as follows: 1) polymer solution, 2) connate water, and 3) distilled water.

The difference between the two systems has been determined for each flow parameter, but Figures 15, 16 and 17 show data for resistance factors, residual resistance factor, apparent permeability and viscosity ratio only. In our figures the curves characterizing the water-wet cores are represented by solid lines, and those characterizing the oil-wet cores, by dotted ones.

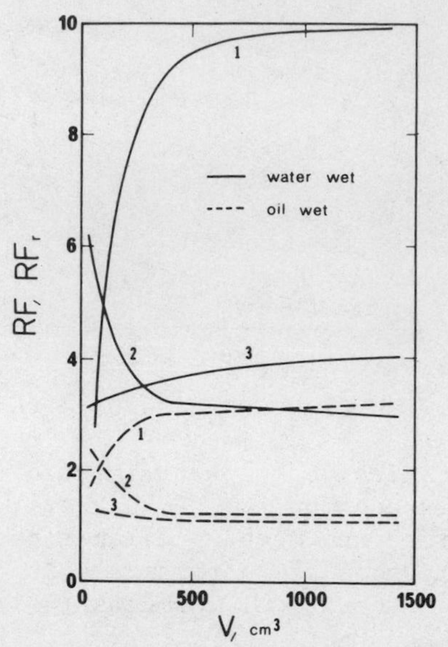

Fig. 15. Dependence of resistance factor and residual resistance factor on the fluid volume injected into porous media. (1, polymer solution (0.5 gdm^{-3}); 2, connate water; 3, distilled water.)

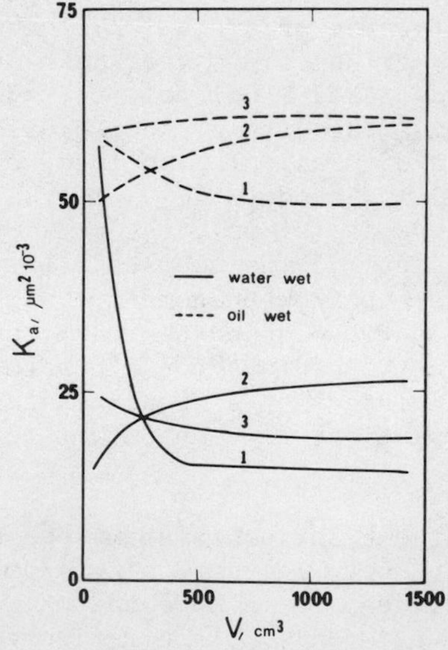

Fig. 16. Dependence of apparent permeability on the fluid volume injected into porous media. (1, polymer solution (0.5 gdm^{-3}); 2, connate water; 3, distilled water.)

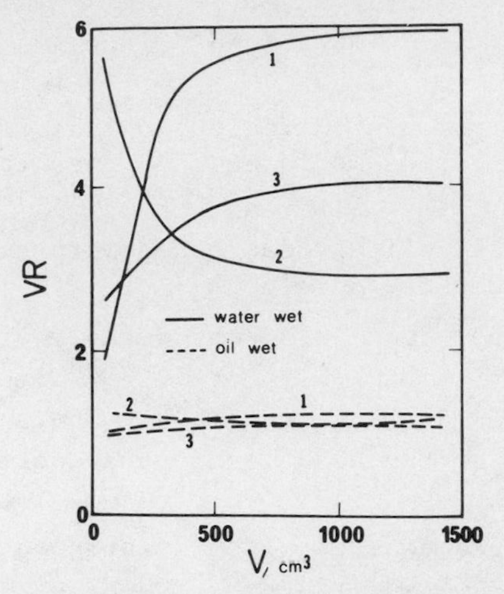

Fig. 17. Dependence of viscosity ratio on fluid volume injected into porous media. (1, polymer solution (0.5 gdm^{-3}); 2, connate water; 3, distilled water.)

In the water-wet model well-known trends may be observed. On the other hand, in the oil-wet system a substantial change takes place only in the case of polymer solution while the mobility of either the connate water or the distilled water does not exhibit any substantial difference as compared to the initial state.

On the basis of Figures 15-17, it may be stated that in an oil-wet system the flow of the polymer solution advancing connate water and distilled water takes place with its initial viscosity and mobility; further, that in the porous media no apparent permeability reduction occurs. With knowledge of the close relationship between the polymer adsorbed on the rock surface and the flow characteristics, this phenomenon may be traced to the absence of an irreversible gel layer over the pore surface. On an oil-wet surface the quantity of polymer bound by chemisorption, regardless of the type, is zero, and thus the "sweep" effect of the carrier phase is also perfect.

As regards industrial application of the method, the phenomenon entails undesirable consequences. In an oil-wet reservoir, an identical hydrodynamical effect can be approximated only if the apparent viscosity exhibited in the water-wet medium is compensated by an increase in the laboratory viscosity of the solution employed. This actually requires the use of 3-6 times as much of polymer. Even then, the permanent decrease in the apparent permeability and the permanent improvement of the flow characteristics of the carrier phase cannot be guaranteed. In practical application of the polymer flooding, the trend of the compromises, namely the technology and its parameters therefore should be decided on the basis of economic considerations.

CONCLUSIONS

The dynamic adsorption of different polyacrylamides was studied in unconsolidated porous media and the results can be summarized as follows:

1. The amount adsorbed slighly decreases with the molecular mass, but the change is negligible.

2. The amount adsorbed decreases considerably with the degree of hydrolysis.

3. The effect of the molecular mass and the degree of hydrolysis can be explained well by the actual size and density of the random coils.

4. The amount adsorbed depends significantly on the polymer concentration: until a critical value no change is observed, but over this limit the amount adsorbed increases exponentially.

5. The steep tendency of the n-$c_{polymer}$ relation in the vicinity of the critical concentration can be attributed to mechanical entrapment.

6. The amount adsorbed increases with the concentration of inorganic electrolytes, and the quality of the salts play an important role in the dynamic adsorption of the polyacrylamides.

7. The effect of the inorganic salts can also be explained by the changes of coil structure.

8. The adsorption of the different polyacrylamides depends to a great extent on the wettability. On an oil-wet surface the sorption of the polymers is zero.

9. The effect of surface active agents, caustic materials, alcohols, etc., can be explained through wettability alteration, molecular interaction, mutual adsorption, and changes in the coil and the solution structure. No general explanation exists and all special systems must be treated individually.

10. The chemical composition of the porous media fundamentally influences the adsorption. The rocks abundant in bi- and multivalent cations (carbonates and clay minerals) adsorb more polymer than the sandstones and silica type minerals do.

11. In natural porous systems only a part of the pores and the surface can take part in the adsorption, and the calculated coverage usually far exceeds unity.

12. In consolidated porous system the adsorption is always accompanied by mechanical entrapment, therefore the description of the dynamic adsorption by exact mathematical formulas and terms is difficult.

13. In general, the unhydrolyzed and partially hydrolyzed polyacrylamides behave differently. The dynamic adsorption of hydrolyzed polyacrylamides is more sensitive to all factors studied than the unhydrolyzed ones.

The adsorption or the lack of adsorption of the polyacrylamides influences strongly the flow behavior of the polymer solution and carrier phases in the porous systems. If no adsorption takes place in the porous media, an identical hydrodynamic effect with the system containing adsorbed film can be maintained by use of 3-6 times as much of the polymer. This fact unfavorably decreases in practice the effectiveness of the polymer flooding.

SYMBOLS

a_s surface area, %
d equivalent diameter of random coil, Å
n amount of adsorbed polymer, g/g
C concentration, gdm^{-3}
K permeability, μm^2
K_a apparent permeability, μm^2
$\bar{M}$ average molecular mass
R pore radius, Å
RF resistance factor
RF_r residual resistance factor
V volume, cm^3
V_p pore volume
VR viscosity ratio
δ density of random coil, gcm^{-3}

REFERENCES

1. D. O. Shah and R. S. Schechter, in "Improved Oil Recovery by Surfactant and Polymer Flooding," Academic Press, Inc., New York, 1977.
2. I. Lakatos and M. Munka, in "A polimeres kiszoritás elmélete és gyakorlata, I. rész Bányászati Szakirodalmi Tájékoztató, NIMDOK Budapest, 1978.
3. D. J. Pye, J. Pet. Tech., 8, 911 (1964).
4. B. B. Sandiford, J. Pet. Tech., 8, 917 (1964).
5. N. Mungan, F. W. Smith, and J. L. Thompson, J. Pet. Tech., 9, 1143 (1966).
6. N. Mungan, Can. J. Pet. Tech., 2, 45 (1969).
7. W. B. Gogarty, Soc. Pet. Eng. J., 6, 161 (1967).
8. D. L. Dauben and D. E. Menzie, J. Pet. Tech., 8, 1065 (1967).
9. F. W. Smith, J. Pet. Tech., 2, 148 (1970).
10. A. M. Sarem, SPE preprint No. 3002 (1970).

11. M. T. Szabó, SPE preprint No. 4668 (1973).
12. M. T. Szabó, Soc. Pet. Eng. J., 8, 333 (1975).
13. M. T. Szabó, J. Pet. Tech., 5, 561 (1979).
14. G. Chauveteau and N. Kohler, SPE preprint No. 4745 (1974).
15. G. P. Willhite and J. G. Dominguez, in "Improved Oil Recovery by Surfactant and Polymer Flooding," D. O. Shah and R. S. Schechter, eds., Academic Press, New York, 1977.
16. I. Lakatos and J. L. Szabó, Köolaj és Földgáz, 106, 336 (1973).
17. I. Lakatos, Köolaj és Földgáz, 108, 215 (1975). Acta Geodaet. Geophys. Montanist. (Budapest), 13, 325 (1978).
18. N. Schamp and J. Huylebroeck, J. Polym. Sci., 42, 553 (1973).
19. R. Dawson and R. B. Lantz, Soc. Pet. Eng. J., 12, 448 (1972).
20. A. S. Michael and O. Morelos, Ind. Eng. Chem., 47, 1801 (1955).
21. C. P. Thomas, SPE preprint No. 5556 (1975).
22. F. D. Martin and N. S. Sherwood, SPE preprint No. 6339 (1975).
23. J. G. Dominguez and G. P. Willhite, SPE preprint No. 5835 (1976).
24. N. Mungan, SPE preprint No. 3521 (1971).
25. I. Lakatos and J. L. Szabó, Köolaj és Földgáz, 106, 309 (1973).
26. J. W. Herr and W. G. Routson, SPE preprint No. 5098 (1974).
27. R. R. Jennings, J. H. Rogers, and T. J. West, J. Pet. Tech., 4, 391 (1971).
28. O. C. Holbrook and G. G. Bernard, Trans. AIME, 213, 261 (1958).
29. I. Lakatos, Hung. Patent No. 170. 460 (1977).
30. R.J.S. Brown and I. Fatt, J. Pet. Tech., 11, 262 (1956).
31. R. T. Johansen and H. N. Dunning, Prod. Monthly, 9, 20 (1959).
32. W. W. Owens and D. L. Archer, J. Pet. Tech., 7, 873 (1971).
33. N. Mungan, Soc. Pet. Eng. J., 5, 398 (1972).
34. R. A. Salathiel, SPE preprint No. 4104 (1972).
35. L. Desremaux and G. Chauveteau, Coll. Association Rech. Techn. Exploitation Petr., Com. No. 28, Paris (1971).
36. R. L. Jewett and G. F. Schurz, J. Pet. Tech., 6, 675 (1970).
37. G. C. Thakur, SPE preprint No. 4956 (1974).
38. C. P. Thomas, SPE preprint No. 5556 (1975).
39. B. W. Greene, J. Colloid Interface Sci., 37, 144 (1971).
40. A. Silberberg, J. Colloid Interface Sci., 38, 217 (1972).
41. G. J. Hirasaki and G. A. Pope, Soc. Pet. Eng. J., 14, 337 (1974).

OPTIMAL SALINITY OF POLYMER SOLUTION IN SURFACTANT-POLYMER FLOODING PROCESS

S. I. Chou and D. O. Shah

Departments of Chemical Engineering and Anesthesiology
University of Florida
Gainesville, Florida 32611, U.S.A.

A systematic study of the effects of the salinity of connate water and polymer solution on oil-displacement by aqueous and oleic surfactant formulations was carried out. For both types of surfactant formulations, the maximum oil recovery was observed when the salinity of the polymer solution was at the optimal salinity of the preceding surfactant formulation. It is shown that the processes occurring at the surfactant slug-polymer solution mixing zone dominate the oil recovery efficiency. Polymer solution having the optimal salinity of the preceding surfactant formulation assures that the interfacial tension of oil ganglia or middle phase microemulsions with polymer solution will remain low. The low interfacial tension together with adequate mobility control displaces the surfactant slug in a piston-like manner.

For porous media containing substantial amount of clays and divalent ions, surfactant loss is significant and oil recovery efficiency is greatly decreased. Using the salinity shock design for polymer solution (i.e. an abrupt change in salinity), lower surfactant loss, adequate mobility control, equally favorable interfacial tension and less polymer requirement can be achieved as compared to that of constant salinity design for polymer solutions. High oil recovery in Berea cores was obtained using this design even in the presence of high concentrations of $CaCl_2$ in connate water. The role of the salinity of polymer solution in affecting the *in situ* surfactant partitioning, surfactant retention and interfacial tension is discussed.

INTRODUCTION

The salinity of polymer solution can influence four major parameters of surfactant-polymer flooding process, namely, interfacial tension, mobility control, surfactant loss and phase behavior. When polymer solution of various salinities are equilibrated with surfactant solution in oil, the formation of lower, middle and upper phase microemulsion has been observed (1) similar to the effect of increasing connate water salinity (2,3). In general, there is an optimal salinity (2) which produces minimum interfacial tension and maximal oil recovery (1,4). On the basis of interfacial tension alone, the salinity of polymer solution should then be designed at or near the optimal salinity of the preceding surfactant formulation.

Another factor in designing the polymer solution is the viscosity. It is well known that the viscosity of polyacrylamide[1] solution decreases sharply with salinity. This factor then favors the use of fresh (very low salinity) polymer solution. More recently it has been shown that fresh polymer solution also gives lowest surfactant loss (1,5,6). This is because the dispersed and entrapped surfactant slug can be redissolved into low salinity polymer solution. A compromise between these opposing factors, interfacial tension vs. viscosity and surfactant loss, may lead one to design the salinity of polymer solution at a lower value than the optimal salinity of surfactant formulation (6) or a contrast salinity design (6,7) of the preflush-micellar-polymer solution process.

It should be noted, however, that higher surfactant concentration does not necessarily produce lower interfacial tension (8). A low salinity polymer solution capable of dissolving high concentration of surfactant may not have sufficiently low interfacial tension to mobilize and displace the oil ganglia left behind the surfactant slug.

In this paper, the effect of the salinity of polymer solution on tertiary oil recovery by aqueous surfactant formulation and soluble oils in both sand packs and Berea cores are investigated. Tertiary oil recovery efficiency is discussed in light of the interfacial tension, surfactant loss and mobility control. A salinity shock design for the mobility buffer polymer solution for surfactant-polymer flooding has been proposed. The role of surfactant partitioning with this design, particularly in the presence of high concentrations of $CaCl_2$ in connate water, is delineated.

---

[1]Partially hydrolyzed.

EXPERIMENTAL

A commercial petroleum sulfonate, TRS 10-410 (Witco Chemical Company) was used as received. This surfactant has an average molecular weight of 420 and is 62% active. The cosurfactant (isobutanol of 99.9% purity) and the oil (dodecane or hexadecane of 99% purity) were obtained from Chemical Samples Company, Columbus, Ohio. The composition of surfactant formulation will be specified in each section. Pusher-700^{TM} (Dow Chemical Company) was used to prepare the polymer solutions. Brine was prepared by dissolving NaCl or $CaCl_2$ in distilled water. Short sand packs with dimensions of 13 in. length x 1 in. diameter were used as the porous media for most of the studies. Four-foot long sand packs were used in section 2. The sand packs had a porosity of 38% and a permeability of about 4.2 Darcy. Rectangular Berea cores of dimension 12 in x 1 in. x 1 in. were also employed in oil displacement tests, which had a porosity of 22% and permeabilities of 200 to 400 milli-Darcy. Interfacial tensions between effluent oil, brine and surfactant phase (microemulsion) were measured immediately after collecting from porous media by a spinning drop tensiometer. The viscosity of soluble oils and polymer solutions were measured by a Brookfield viscometer with UL adapter or a cone and plate viscometer at a shear rate of 6.6 sec^{-1} and reading taken 30 minutes after starting the rotation. Surfactant concentrations in effluent oil and brine were measured by a two-phase, two-dye titration method (9).

Porous media were prepared by saturating with CO_2 at 50 psi to displace air, then were flooded with distilled water (for sand packs) or with 1.0% NaCl (for Berea cores) up to 10 pore volumes (PV). Subsequently, the porous medium was flooded with oil which reduced the brine approximately to 18% PV for sand packs and to 33% PV for Berea cores and was subsequently brine flooded to irreducible oil content (approximately 25% PV for sand packs and 34% PV for Berea cores). A surfactant slug was then injected and displaced by polymer solution. The linear displacement velocity in the water flooding and subsequent surfactant slug and polymer solution flooding was 2.78 ft/day for sand pack tests and 2.5 ft/day for Berea core tests (except for those shown in section 2). The size of polymer slug injected was 1 PV which was followed by brine of the same salinity as that of the polymer solution. Oil displacement experiments were performed at room temperature (23 ± 1°C). The salinities of the connate brine, brine for water flooding and polymer solution will be specified in each section. Oil displacement results were expressed as percent oil recovery according to

$$\text{Tertiary Oil Recovery, \%} = \frac{\text{total oil produced}}{\text{residual oil + oil injected}} \times 100\%$$

RESULTS AND DISCUSSION

1. The Importance of the Salinity of Polymer Solution in Oil Displacement Process

In this section we will show the effect of the salinity of polymer solution and connate brine on tertiary oil recovery by aqueous surfactant formulations (in 1.5% NaCl) and soluble oils. The optimal salinity of both types of formulations used in the present study was 1.5% NaCl (1,8).

Oil displacement experiments were performed under different salinity conditions: (1) constant salinity of polymer solution at 1.5% NaCl (i.e. optimal salinity of the soluble oil formulation), and variable connate water salinity; (2) constant salinity of connate water at 1.5% NaCl and variable salinity of polymer solution and (3) the salinity of polymer solution equals the salinity of connate water, both varied simultaneously. Sand packs were chosen as the model porous media in order to avoid the effects of porous media heterogeneity, clays and surfactant adsorption loss. The compositions of aqueous formulation and soluble oils are specified in Figures 1 and 2. The difference between their compositions reflects the density difference between water and dodecane whereas the surfactant and alcohol concentrations (w/v) are the same in both types of formulations. The polymer solution used was 1000 ppm PUSHER-700 in brine. For polymer solution in distilled water, the polymer concentration was reduced to 250 ppm to avoid excessive viscosity. Several experiments were repeated and the reproducibility was established to be within $\pm 2\%$ in tertiary oil recovery.

Both Figures 1 and 2 show the following results: (1) Maximal oil recovery was obtained when the salinity of both polymer solution and connate water was 1.5% NaCl. (2) When the salinity of polymer solution was 1.5% NaCl, oil recovery was favorable over a wide range of connate water salinities (0 to 4% NaCl). Oil recovery was 77% for aqueous formulation and 62% for soluble oils when the salinity of connate water was as high as 10% NaCl. (3) On the other hand, when the salinity of polymer solution was shifted from 1.5% NaCl (the optimal salinity of the surfactant slug), oil recovery drastically decreased irrespective of connate water salinities.

These results suggest that, for the present system, polymer solution salinity is far more critical than connate water salinity in tertiary oil recovery. In other words, the processes occurring at surfactant slug-polymer solution interface determines the final oil saturation and oil recovery efficiency in the systems reported here. Gupta and Trushenski (6) also suggested that oil recovery is controlled by the composition developing in the micellar-polymer

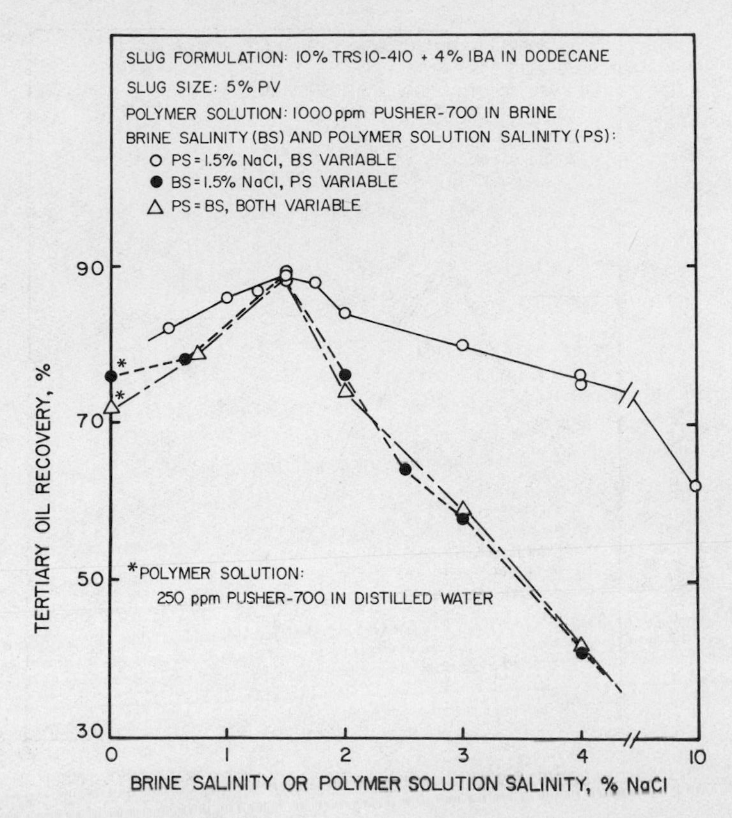

Fig. 1. The effect of salinity of connate water and polymer solution on tertiary oil recovery by soluble oil slug in sand packs.

mixing zone for aqueous micellar systems with crude oil in Berea core experiments.

The mechanism of the above results which has been described in detail elsewhere (1) is summarized as follows. When the salinity of polymer solution is at the optimal salinity of the preceding surfactant formulation, middle phase microemulsion is produced *in situ*, which has ultra-low interfacial tension with resident brine, residual oil and polymer solution. This ultra-low interfacial tension together with adequate mobility control will allow the displacement of the surfactant slug in a piston-like manner.[2] Consequently, favorable oil recovery can be obtained over a wide range

[2]The propagation of the surfactant slug in sand packs can be visually observed. When the salinity of polymer solution was at 1.5% NaCl, the slug front velocity was very close to the linear displacement velocity. The surfactant slug breakthrough, as

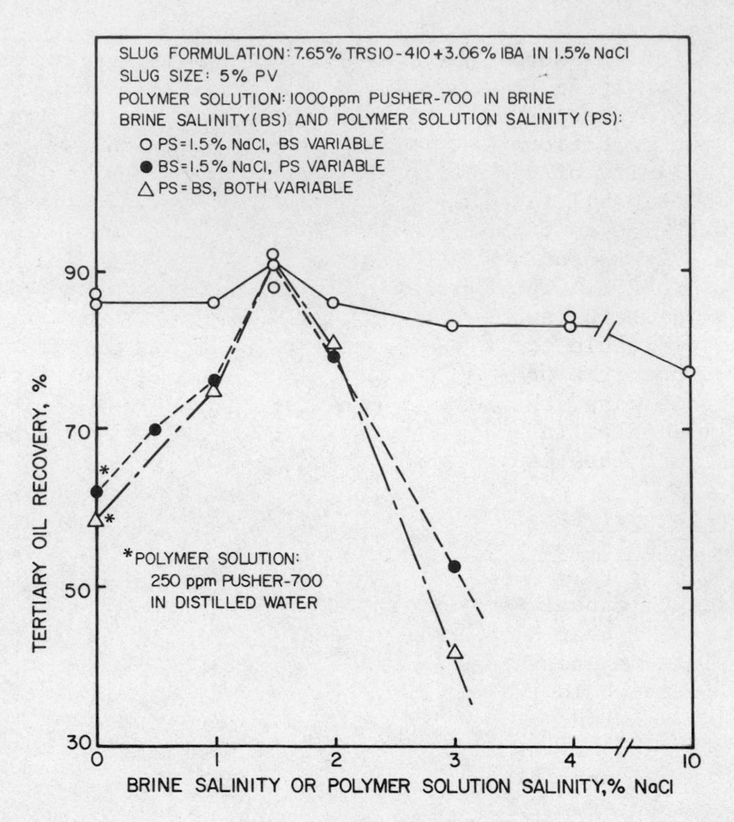

Fig. 2. The effect of salinity of connate water and polymer solution on tertiary oil recovery by aqueous surfactant slug in sand packs.

of connate water salinities when polymer solution is maintained at the optimal salinity of the preceding surfactant slug.

It should be mentioned that the necessary conditions for the validity of the above stated conclusions are that (1) the optimal salinity of the surfactant formulation remains approximately constant upon dilution, (2) there is a sharp minimum of IFT occurring at the optimal salinity, (3) mobility control is adequately

---

indicated by the first appearance of thick emulsions or an abrupt decrease of interfacial tension between produced oil and brine (or emulsion), was in the range from 0.9 to 1.0 PV irrespective of the salinity of connate water, although severe tailing of the surfactant slug was observed when the salinity of connate water was very high. On the other hand, when the salinity of polymer solution was high, no surfactant slug breakthrough was observed; and for fresh polymer solution, the surfactant slug breakthrough was smaller than 0.8 PV.

maintained and (4) surfactant loss is not too severe at optimal salinity. The first two conditions are generally recognized for the validity of the optimal salinity concept. The importance of the last two conditions can be realized from the following example. When the salinity of connate water and polymer solution was at 1.5% NaCl in soluble oil flooding, decreasing the polymer concentration from 1000 to 250 to 0 ppm, tertiary oil recovery decreased from 89% to 69% to 53% and surfactant recovery (at 1.3 PV) decreased from 67% to 42% to 25%. Thus in cases where mobility control is difficult to maintain (such as Berea core experiments with crude oils), it may be desirable to design the salinity of polymer solution somewhat lower than the optimal salinity of the surfactant formulation, since in this case, the lower phase microemulsions formed in porous media is miscible (in a broader sense) with the polymer solution and surfactant loss can be greatly reduced.

2. The Effect of Length of Sand Packs and Displacement Velocity on Oil Displacement Efficiency

The above experiments were performed in 13-inch long sand packs and the linear displacement velocity was 2.78 ft/day. It was desirable to understand how the length of the sand packs and fluid velocity would influence the major conclusions we have drawn in the last section. The effect of the salinity of polymer solution in oil recovery by soluble oils in 4 ft long sand packs was studied. The linear displacement velocity during tertiary flooding was reduced from 2.8 to 1.0 ft/day. The results together with some of the previous results in 1.1 ft long sand packs are shown in Table 1. It is shown that tertiary oil recovery was slightly higher (1 to 5%) for oil displacement in longer sand packs. Maximal oil recovery still occurred at 1.5% NaCl polymer solution. The interfacial tension between effluent oil and brine (or microemulsion) remains approximately the same (for both sand packs at a given salinity) as shown in Figure 3. The pressure drop history during tertiary flooding also shows nearly identical behavior for long and short sand packs. These observations indicate that the transport process in porous media is nearly the same for the 1.1 and 4 ft long sand packs and the conclusions drawn in section 1 should also be valid for oil displacement in 4 ft long sand packs.

3. The Effect of Salinity of Polymer Solution on Oil Displacement in Berea Cores

Tertiary oil recovery in Berea cores by soluble oils was studied. Hexadecane instead of dodecane was used in this study. Two connate water salinities were used. One was 2.1% NaCl (the optimal salinity of the surfactant formulation) and the other was 3% NaCl + 1% $CaCl_2$. The results are shown in Table 2. For both connate water salinities, maximal oil recovery was obtained when the salinity of polymer solution was 2.1% NaCl. Surfactant

Table 1. The Effect of the Salinity of Polymer Solution on Soluble Oil Flooding in 4 ft Long Sand Packs[a,b,c]

Salinity of Polymer Solution	Tertiary Oil Recovery	Surfactant Recovery @ 1.3 PV
0% NaCl[d]	80.7% (76%)[e]	95.3% (99%)[e]
1.1% NaCl	90.5% (85%)	89.0% (85%)
1.5% NaCl	92.2% (89%)	74.3% (67%)
4.0% NaCl	42.3% (41%)	~ 0 (~0)

a. Soluble oil formulation: 10% TRS 10-410 + 4% IBA in dodecane, 5% PV slug
b. Connate water salinity = 1.5% NaCl
c. Linear displacement velocity in 4 ft long sand packs was 1.0 ft/day whereas that in 1.1 ft long sand packs was 2.78 ft/day
d. Polymer concentration used in this case was 250 ppm, while in other cases it was 1000 ppm
e. The numbers in parentheses are the results in 1.1 ft long sand packs (Figure 1) while the rest are for 4 ft long sand packs

recovery is higher when the salinity of polymer solution is lower, however.

A comparison of these results with the results in sand packs (Table 1) shows that surfactant recovery is reduced by at least 50% for all cases while oil recovery is reduced by only 10% to 17%. Part of the decrease of oil recovery could be due to the lower permeability of Berea cores (200 to 400 milli-Darcy) than sand packs (4 Darcy). It seems that the surfactant recovered in the low salinity polymer solution is not very effective in displacing the residual oil ganglia left behind the surfactant slug due to high interfacial tension.

The presence of 1% $CaCl_2$ in connate water decreases the oil recovery substantially for all cases. Maximal oil recovery was 51% when the salinity of polymer solution was 2.1% NaCl (i.e. optimal salinity). It is interesting to note that this value is still much higher than the oil recovery of 25% when the salinity of connate water is 2.1% NaCl and the salinity of polymer solution is 4% NaCl (Table 2). This demonstrates that the salinity of polymer solution is more critical than the salinity of connate water in Berea cores similar to the results in sand packs (Figures 1 and 2).

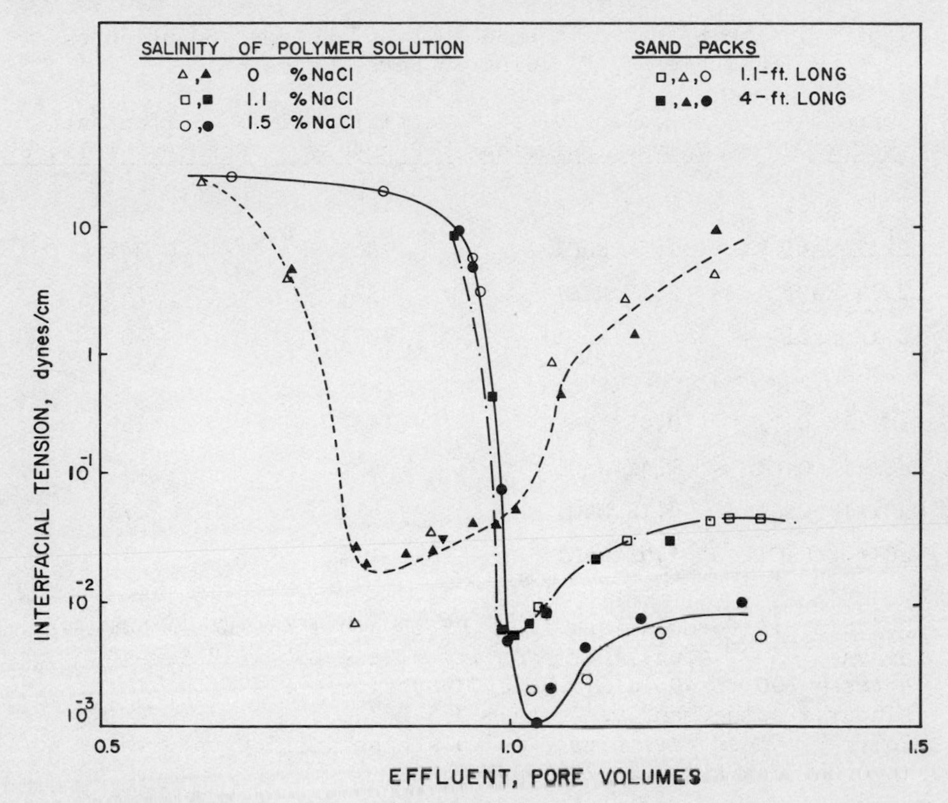

Fig. 3. The effect of the salinity of polymer solution on the interfacial tension between effluent oil and brine (or microemulsion) for soluble oil flooding in 1.1 ft and 4 ft long sand packs.

4. Salinity Shock Design for the Polymer Solution

It has been shown that there are two opposing factors, interfacial tension vs. surfactant loss and viscosity, in designing the salinity of polymer solution. Instead of making a compromise between these opposing factors, we propose here a method which can take advantage of all of these factors. The salinity of polymer solution need not be uniform and can be designed in any manner as long as ultra low interfacial tension is maintained near the surfactant slug/polymer solution interface. The salinity shock design (i.e. an abrupt change in salinity) of polymer solution employs two slugs of polymer solution in which the first slug of the polymer solution is at the optimal salinity of the preceding surfactant formulation and the second polymer slug in fresh water containing lower concentration of polymer. The design criterion for varying salinity and concentration is such that each successive slug of polymer solution has a higher viscosity. This is possible because

Table 2. The Effect of the Salinity of Polymer Solution on Soluble Oil Flooding in Berea Cores[a,b]

Connate Water	Salinity of Polymer Solution	Tertiary Oil Recovery	Surfactant Recovery @ 1.5 PV
2.1% NaCl	0.05% NaCl[c]	68%	44%
2.1% NaCl	1.5% NaCl	81%	34%
2.1% NaCl	2.1% NaCl	82%	14%
2.1% NaCl	4.0% NaCl	25%	~0
3% NaCl+1% $CaCl_2$	0.05% NaCl[c]	42%	25%
3% NaCl+1% $CaCl_2$	1.4% NaCl	40%	12%
3% NaCl+1% $CaCl_2$	2.1% NaCl	51%	7%
3% NaCl+1% $CaCl_2$	4.0% NaCl	11%	~0

a. Soluble oil formulation: 10% TRS 10-410 + 4% IBA in hexadecane; 5% PV slug. Permeability of the 13" x 1" Berea cores = 200 to 400 milli-Darcy, porosity = 22%
b. Linear displacement velocity = 2.5 ft/day
c. Polymer concentration used in this case was 500 ppm, while in other cases it was 2000 ppm

the viscosity of polyacrylamide solution decreases sharply with salinity (in the range 0 to 0.5% NaCl).

The results of tertiary oil recovery in Berea cores by both aqueous formulations and soluble oils with this design of polymer solution are shown in Table 3. For both soluble oils and aqueous surfactant formulations, tertiary oil recovery was substantially improved with this design. Surfactant recovery was comparable or higher than those by polymer solution in fresh water (Table 2). This design worked particularly well with connate water containing 1% $CaCl_2$ (Figures 4 and 5). As compared to previous results shown in Table 2, tertiary oil recovery increased from 51% to 88% or 89% (duplicate experiments) for soluble oil formulation and surfactant recovery increased from 7% to 48% (Tables 2 and 3 and Figure 5).

Figure 6 shows the cumulative oil recovery history. For both aqueous and soluble oil formulations, oil recovery was considerably delayed when the connate water contained 3% NaCl and 1% $CaCl_2$. Tertiary oil recovery was not complete until 1.5 PV effluents were

Table 3. Salinity Shock Design of Polymer Solutions for Oil Displacement in Berea Cores[a,b,c,d]

Slug Formulation	Connate Water Salinity	Tertiary Oil Recovery	Surfactant Recovery @ 1.5 PV
Soluble Oil	2.1% NaCl	95%	40%
Soluble Oil	2.1% NaCl	90%	45%
Soluble Oil	3% NaCl + 1% $CaCl_2$	88%	48%
Soluble Oil	3% NaCl + 1% $CaCl_2$	89%	--
Aqueous	2.1% NaCl	87%	43%
Aqueous	3% NaCl + 1% $CaCl_2$	86%	39%

a. Polymer solution was 2000 ppm PUSHER-700 in 2.1% NaCl, 0.4 PV slug; followed by 500 ppm in 0.05% NaCl, 0.6 PV slug
b. Soluble oil formulation = 10% TRS 10-410 + 4% IBA in hexadecane; 5% PV slug. Aqueous formulation = 7.65% TRS 10-410 + 3.06% IBA in 2.1% NaCl; 5% PV slug. Different concentrations (w/w) of oleic and aqueous formulations were employed to take into account their density difference so that the amount of surfactant and alcohol injected was the same
c. Permeability of the 13" x 1" Berea cores = 200 to 400 milliDarcy; porosity = 22%
d. Linear displacement velocity = 2.5 ft/day

produced. The delayed oil recovery can be easily understood from Figures 7 and 8. Figures 7 and 8 show the surfactant concentrations, partition coefficient and interfacial tension of effluent fluids for soluble oil and aqueous surfactant formulations respectively. The partition coefficient passes through unity near 1.35 PV effluents producing a minimum interfacial tension (0.002 dynes/cm). This low interfacial tension together with the high viscosity of fresh polymer solution can remobilize and displace the residual oil ganglia left behind the surfactant slug.[3] This caused the delayed high oil recovery.

[3]Judging from the surfactant concentration profile of the effluent fluids, the surfactant slug can be hardly classified as a slug in the latter stage of the displacement process. However, the dispersive mixing in Berea cores was not overly drastic as otherwise the minimum IFT would not occur near 1.35 PV effluents.

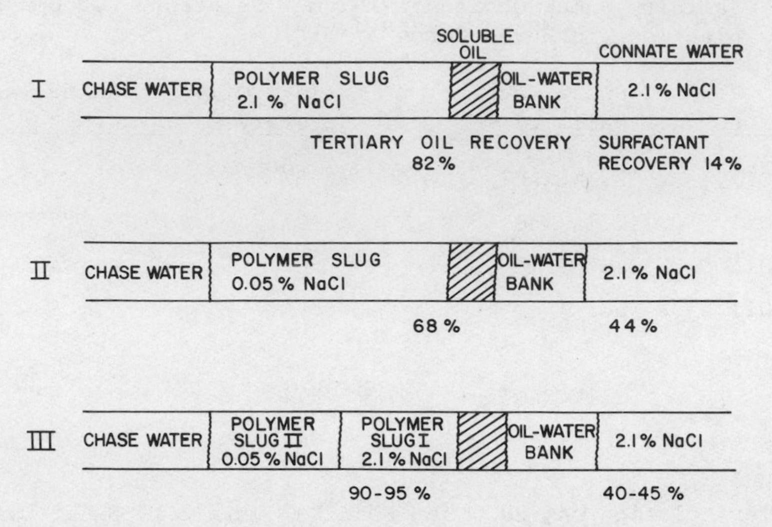

Fig. 4. The effect of salinity change of polymer solution on oil displacement efficiency and surfactant recovery for connate water of 2.1% NaCl.

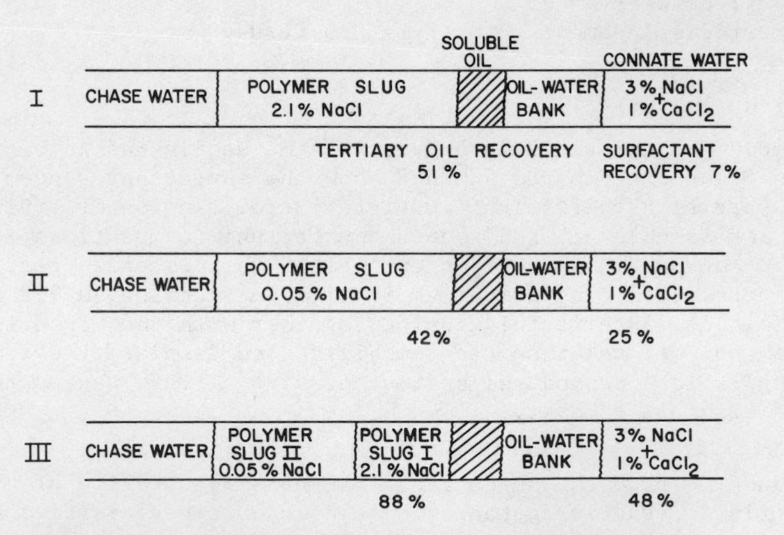

Fig. 5. The effect of salinity change of polymer solution on oil displacement efficiency and surfactant recovery for connate water of 3% NaCl + 1% $CaCl_2$.

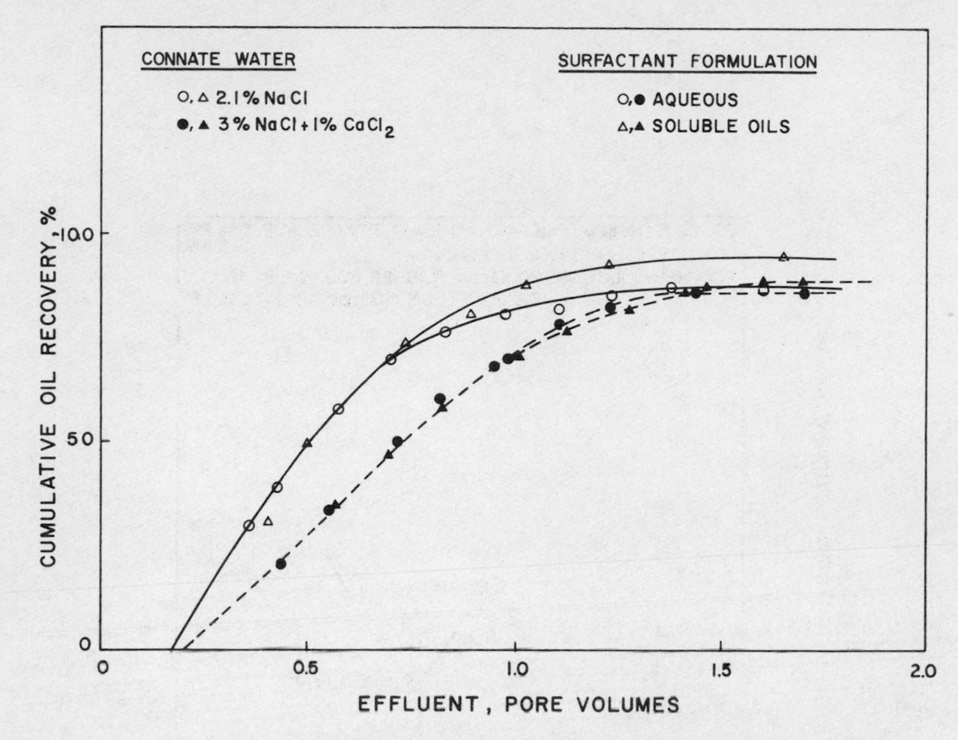

Fig. 6. Cumulative tertiary oil recovery of a salinity shock design of polymer solution (i.e. two polymer slugs) for soluble oil and aqueous surfactant formulation. Note the delayed oil recovery when connate water contains 1% $CaCl_2$.

The behavior of the partition coefficients shown in Figures 7 and 8 can be explained as follows. Near the leading edge of surfactant slug, the *in situ* phase behavior is of an upper phase type microemulsion due to the high salinity of connate water. Part of this microemulsion is trapped in porous media much the same as residual oil is trapped. The entrapped surfactant can be subsequently redissolved into the upcoming fresh polymer solution which lags 0.4 PV behind the tailing edge of surfactant slug. As a result, the surfactant partition coefficient between produced oil and brine passes through unity near 1.35 PV effluents. The detailed mechanism can be much more complicated than that mentioned because the salinity of polymer solution changes along the process.

In summary, as shown in Figure 9, the salinity shock design of mobility polymer solution can provide ultra low interfacial tension at microemulsion/polymer solution interface, reduce surfactant loss, and achieve high oil recovery efficiency. The poly-

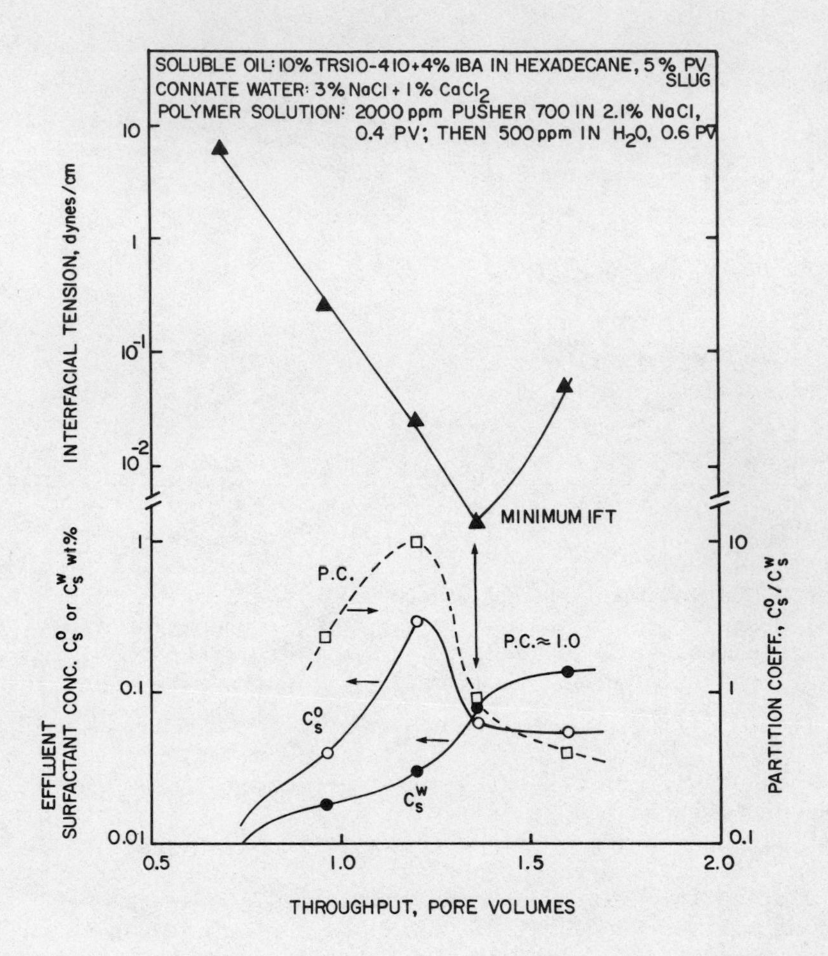

Fig. 7. Surfactant concentration in oil and brine, partition coefficient and interfacial tension of effluent fluids of soluble oil flooding in Berea cores containing high concentration of $CaCl_2$. The salinity shock design of polymer solution (i.e. two polymer slugs) was used.

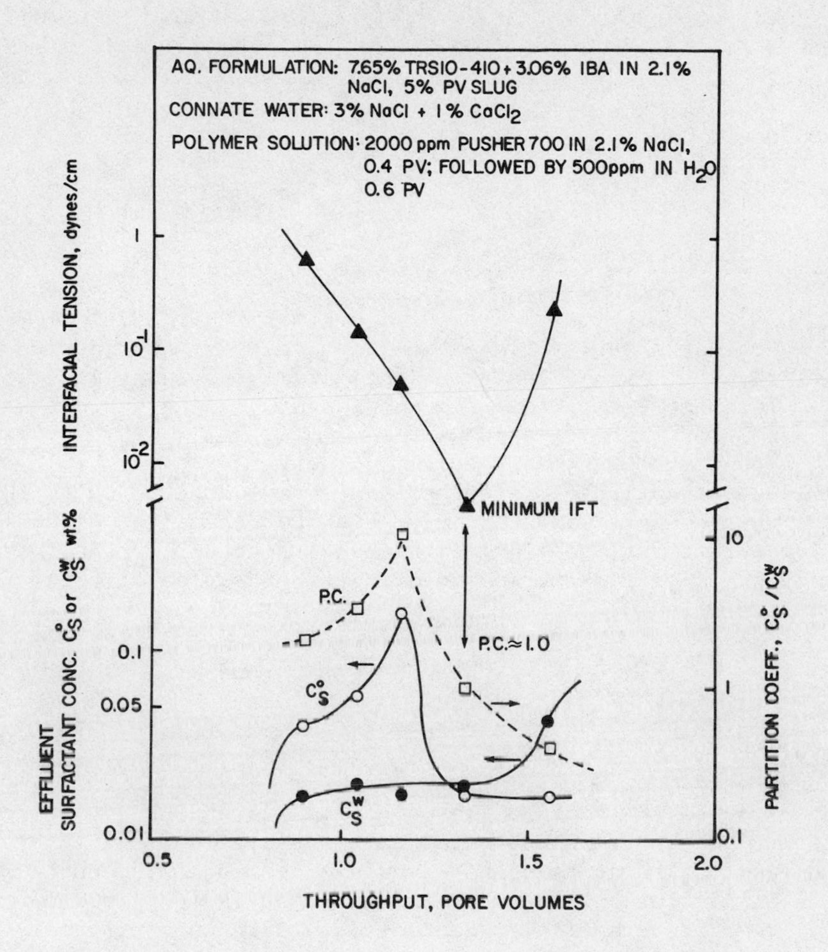

Fig. 8. Surfactant concentration in oil and brine, partition coefficient and interfacial tension of effluent fluids of aqueous surfactant slug flooding in Berea cores containing high concentration of $CaCl_2$. The salinity shock design of polymer solution (i.e. two polymer slugs) was used.

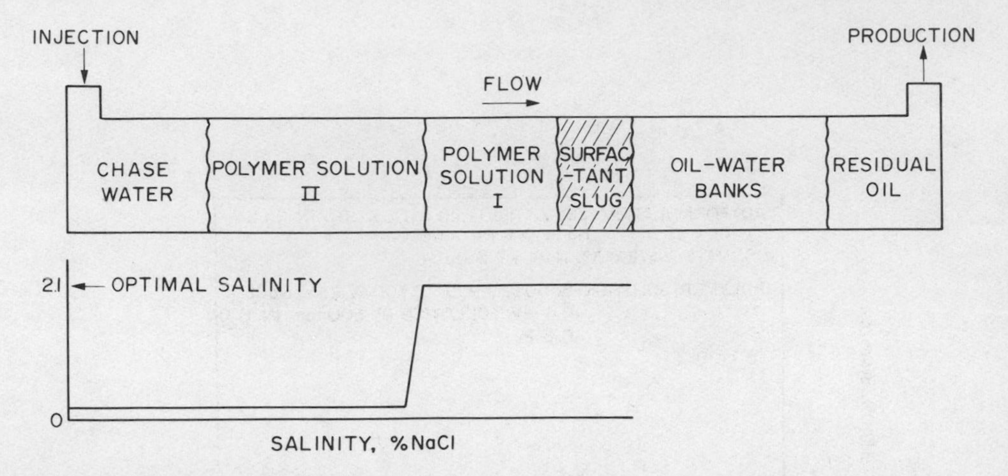

Fig. 9. Schematic representation of the salinity shock design (i.e. an abrupt change of salinity) of polymer solution for a more efficient oil displacement using surfactant-polymer flooding process.

acrylamide consumption can also be reduced by about 50% with this design. When connate water contains high concentration of $CaCl_2$, ultra low interfacial tension and high oil recovery can be obtained by achieving *in situ* partition coefficient of surfactant near unity.

CONCLUSIONS

The following major conclusions are drawn from the results of this study.

(1) When the salinity of polymer solution was at the optimal salinity of the preceding surfactant formulation, oil recovery in sand packs was favorable over a wide range of connate water salinities for both aqueous and oleic surfactant formulations. Oil recovery drastically decreased when the salinity of polymer solution was shifted from the optimal salinity even when the connate water was at the optimal salinity. These results indicate that the processes occurring at the surfactant slug-polymer solution mixing zone dominate the oil recovery efficiency.

(2) The increase of the length of sand packs from 1.1 to 4 feet and the decrease of displacement velocity from 2.8 to 1.0 ft/day during tertiary flooding did not change the oil recovery appreciably. The surfactant recovery, interfacial tension between effluent oil and brine as well as the pressure drop history remained approximately the same indicating that the transport process in porous media was nearly identical for these two cases.

(3) For oil displacement in Berea cores especially in the presence of 1% $CaCl_2$ in connate water, surfactant loss was quite significant and the oil recovery was greatly reduced. Maximal oil recovery was obtained when the salinity of polymer solution was at or slightly below the optimal salinity of surfactant formulation.

(4) The salinity shock design of polymer solution employs two slugs of polymer solution in which the first polymer slug is at the optimal salinity which provides ultra low IFT and maintains mobility control, while the second polymer slug at a much lower salinity is capable of reducing the surfactant loss. Oil recovery in Berea cores was as high as 86% even in the presence of 3% NaCl + 1% $CaCl_2$ in connate water.

(5) When the salinity of connate water was much higher than the optimal salinity, the mechanism of the high oil recovery by this polymer solution design was due to the achievement of *in situ* surfactant partition coefficient near unity which produces ultra low IFT behind the tailing edge of surfactant slug.

ACKNOWLEDGMENTS

The authors wish to express their sincere thanks and appreciation to the National Science Foundation-RANN, ERDA and the Department of Energy (Grant No. DE-AC1979BC10075) and the consortium of the following Industrial Associates for their generous support of the University of Florida Enhanced Oil Recovery Research Program during the past five years: 1) Alberta Research Council, Canada, 2) American Cyanamid Co., 3) Amoco Production Co., 4) Atlantic Richfield Co., 5) BASF-Wyandotte Co., 6) British Petroleum Co., England, 7) Calgon Corp., 8) Cities Service Oil Co., 9) Continental Oil Co., 10) Ethyl Corp., 11) Exxon Production Research Co., 12) Getty Oil Co., 13) Gulf Research and Development Co., 14) Marathon Oil Co., 15) Mobil Research and Development Co., 16) Nalco Chemical Co., 17) Phillips Petroleum Co., 18) Shell Development Co., 19) Standard Oil of Ohio Co., 20) Stepan Chemical Co., 21) Sun Oil Chemical Co., 22) Texaco Inc., 23) Union Carbide Corp., 24) Union Oil Co., 25) Westvaco Inc., 26) Witco Chemical Co., and the University of Florida.

REFERENCES

1. S. I. Chou, Dissertation, University of Florida, 1980.
2. R. N. Healy and R. L. Reed, Soc. Pet. Eng. J., 14, 491 (1974).
3. P. F. Boneau and R. L. Clampit, J. Pet. Tech., 29, 501 (1977).
4. R. N. Healy and R. L. Reed, Soc. Pet. Eng. J., 17, 129 (1977).

5. C. J. Glover, M. C. Puerto, J. M. Maerker, and E. L. Sandvik, Soc. Pet. Eng. J., 19, 183 (1979).
6. S. P. Gupta and S. P. Trushenski, Soc. Pet. Eng. J., 19, 116 (1979).
7. G. W. Paul and H. R. Froning, J. Pet. Tech., 25, 957 (1973).
8. K. S. Chan and D. O. Shah, SPE 7869, Paper presented at SPE International Symposium on Oilfield and Geothermal Chemistry, Houston, TX, January 22-24, 1979.
9. V. W. Reid, F. G. Longman and E. Heinerth, Tenside, 4, 212 (1967).
10. J. J. Rathmell, F. W. Smith, S. J. Salter and T. R. Fink, SPE 7067, Paper presented at SPE Improved Oil Recovery Symposium, Tulsa, OK, April 16-19, 1978.

POLYMER-SURFACTANT INTERACTION AND ITS EFFECT ON THE MOBILIZATION OF CAPILLARY-TRAPPED OIL

F. Th. Hesselink and M. J. Faber

Koninklijke/Shell Exploratie en Produktie
Laboratorium
P.O. Box 60, Rijswijk, Netherlands

Surfactant slug/polymer drive systems which have proved to be effective in recovering waterdrive residual oil in 30 cm long Bentheim sandstone cores have been found to be much less effective in longer (90 cm) cores. This is attributed to polymer/surfactant phase separation, the phenomenon which has little time to develop in experiments with short cores.

The effect of polymer on the surfactant/brine/oil phase behavior has been investigated, indicating that polymer is usually present in an aqueous phase, which can be highly concentrated. Furthermore, polymer will extract water from a microemulsion phase, thus increasing the surfactant concentration in the microemulsion and shifting the invariant point M away from the brine corner. This causes an increase in the microemulsion/brine interfacial tension. In a core flood this may lead to trapping of a microemulsion phase and thus to high surfactant losses. Polymer may also cause a drastic increase in the viscosity of the microemulsion (perhaps a non-equilibrium effect), which is an additional factor hampering the displacement efficiency and thus increasing surfactant loss.

These effects are discussed in terms of the well-known incompatibility of two different polymers in a single solvent, considering the microemulsion/swollen micelles as a pseudo-polymer system.

Reduction of the brine salinity reduces the size of the microemulsion 'particles' and therefore should suppress polymer/surfactant incompatibility. Indeed, core floods with a reduced salinity in the polymer drive have shown little indication of polymer/surfactant incompatibility, a very good recovery of residual oil and a drastic reduction of surfactant losses.

INTRODUCTION

The surfactant-brine-oil phase behavior relevant to the microemulsion flooding process of enhanced oil recovery has been described in a number of papers in the literature (1). In these papers it has been indicated that under suitable conditions a microemulsion phase can be found to be thermodynamically stable in contact with an aqueous solution and with oil. Furthermore, it has been described how this phase behavior depends on the type of surfactant and cosurfactant, ionic composition of the brine, type of oil and on temperature.

Middle-phase microemulsions, which can exist in equilibrium with almost pure brine and oil, have been shown by many authors (1) to be most suitable for mobilization of capillary-trapped waterdrive residual oil. This is explained on the basis of the extremely low interfacial tension between this type of microemulsion and oil ($\sim 10^{-3}$ mN/m), causing almost complete mobilization of capillary-trapped oil, and on the basis of the equally low interfacial tension between such a microemulsion and brine, which prevents the trapping of a surfactant-rich microemulsion phase.

We have previously described (2) how a surfactant system producing such an optimal microemulsion in situ upon contact with residual oil was able to recover 90% of the waterdrive residual oil from short (30 cm) Bentheim sandstone cores (Figure 1).

In this paper we report on surfactant floods in longer (0.9-1.5m) Bentheim sandstone cores. The purpose of these experiments has been to investigate the possible occurrence of incompatibilities between the surfactant slug and the polymer drive in cores. In test tubes we had observed that, upon contacting an aqueous surfactant and polymer solution, a liquid-liquid phase separation can become visible after some time (2-40 hours). Such a phase separation could be harmful in field application, whereas it may either not show up or have only little effect in 0.3m core experiments, since these take only 24 hours.

In connection with these experiments, data are presented on how the surfactant-brine-oil phase behavior is influenced by small amounts of a polysaccharide polymer dissolved in the brine. This polymer apparently complicates phase behavior considerably; in principle a fourth axis is required in the phase diagram. However, in view of the approximate nature of the ternary representation of the surfactant-brine-oil system, we shall not attempt to draw sys-

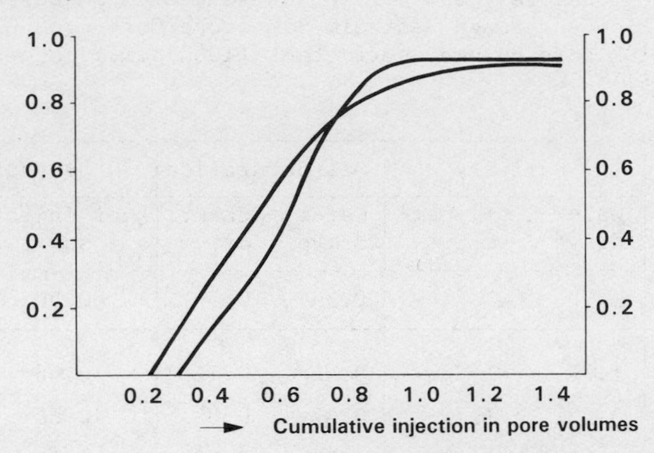

Fig. 1. Surfactant flooding experiments in Bentheim sandstone cores (30 cm long, T = 62°). Oil production versus cumulative injection.

tematic 3D phase diagrams, but only present the data required for the interpretation of the core flooding experiments.

EXPERIMENTAL

In the core flooding experiments the waterflood residual oil saturation was established in the usual manner, i.e., the core was evacuated, flooded with brine, then with oil and again with brine until no more oil was produced. The core was then flooded with a finite (0.13 pore volume) surfactant slug, followed by a biopolymer drive (Table 1). The most important variable in these experiments was the salinity of the polymer drive. The flooding rate was 1 ft/day. The surfactant mixture contained Witco's TRS 12B (62% a.i.) (a petroleum sulfonate) and Shell's Neodol 25-3S (60% a.i., an alkyl ethoxy sulfate) in a fixed ratio of 9:1 equivalents. The mixture further contained 12% isobutanol. The oil was a 1.9 mPa.s crude from North West Borneo. The 1.7% brine contained 5800 ppm NaCl, 10700 ppm $NaHCO_3$, 68 ppm $CaCl_2$, and 47 ppm $MgCl_2$. The polymer was Kelco's polysaccharide polymer Xanflood. In the phase behavior studies the data were obtained at least 30 days after preparation of the solutions. All data refer to 62°C.

RESULTS AND DISCUSSION

The results of the core flooding experiments are summarized in Table 1. In the 30 cm core and in experiment 1, all aqueous solutions (connate water, waterflood, surfactant slug, polymer

Table 1. Oil Recovery and Surfactant Retention in Surfactant Floods through Bentheim Sandstone Cores (62°C, Slug-size 13% pore volume, surfactant (100% a.i.) concentration 4%)

Experiment	Salinity		Oil Saturation		Surfactant Retention	
	water-drive	polymer-drive	water-drive	chem. drive	% of injected surf.	meq/100g
30 cm	1.7%	1.7%	0.39	0.04	0.50	0.040
1 (90 cm)	1.7%	1.7%	0.42	0.15	0.86	0.081
2 (90 cm)	0.32%	1.7%	0.37	0.09	0.68	0.063
3 (90 cm)	0.32%	1.7%	0.39	0.10	0.71	0.062
4 (90 cm)	1.7%	0.32%	0.37	0.04	0.41	0.040
5 (90 cm)	1.7%	0.32%	0.36	0.05	0.29	0.033
6 (150 cm)	1.7%	0.32%	0.39	0.05	0.46	0.042

drive) had the same (1.7%) salinity. In the long core experiment only 63% of the waterflood residual oil was recovered (Figure 2), whereas in a 30 cm core we had recovered about 90% of the residual oil. Cutting open the long core showed that the first part of the core had been swept almost completely clean of oil, whereas in the last part only about 50% of the core had been flushed. This effect of core length (residence time, flooding rate?) is a distinct warning against experiments in cores of too short length where subtle incompatibilities between injected fluids have no time to show up and may remain unnoticed until later.

In experiments 2 and 3, we brought the core to residual oil saturation with a 0.32% brine, while surfactant slug and polymer drive were made up in 1.7% brine. The results are somewhat better than in experiment 1, but still unsatisfactory.

Since we expected the problems to occur in the transition zone between surfactant slug and polymer drive, we investigated the phase behavior (phase volumes, viscosities) and interfacial activity of the surfactant/oil/brine system in the absence and presence of polymer (Figure 3). In the absence of polymer the design criteria for an active surfactant system are met; i.e., the oil/microemulsion and microemulsion/brine interfacial tensions are of the order of a μN/m (10^{-3} dyne/cm) or less, the microemulsion at the invariant point M contains about equal amounts of oil and brine and the microemulsion viscosities are about three times that of the oil

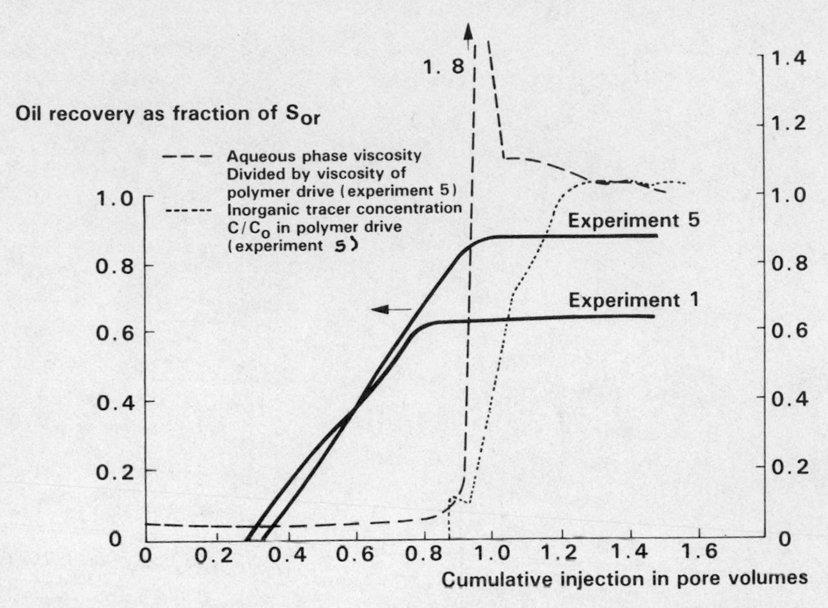

Fig. 2. Oil production versus cumulative injection.

over a range of compositions (Figure 3). Therefore, the mobilization of residual oil and the displacement of the oil bank should have proceeded satisfactorily.

With 1200 ppm polymer present in the brine, the position of the invariant point M does not shift much, but we observe further important deviations from phase behavior in the absence of polymer: the microemulsion/brine interfacial tensions are about 3-10 times higher than in the absence of polymer (Figure 3). This may lead to incomplete displacement of the microemulsion by the polymer drive, i.e., a certain amount of microemulsion will remain trapped by capillary forces. This contributes to the high surfactant retention observed in experiments 1-3.

Furthermore, the viscosities of both the microemulsion and brine phases are very high, about twice as high as the viscosity of the 1200 ppm polymer drive itself (Figure 3). This polymer drive may therefore be expected to bypass these viscous microemulsion phases, causing an unstable displacement and contributing to the high values for surfactant retention in the core that were observed in experiments 1-3. This phase separation problem is aggravated by the fact that polymer molecules tend to travel at a somewhat higher average velocity through the reservoir than the brine they are dissolved in, the so-called 'inaccessible pore volume' phenomenon. This phenomenon can also be observed in Figure 2, where the viscosity increase in the produced brine precedes the arrival of the inorganic tracer dissolved in the polymer drive.

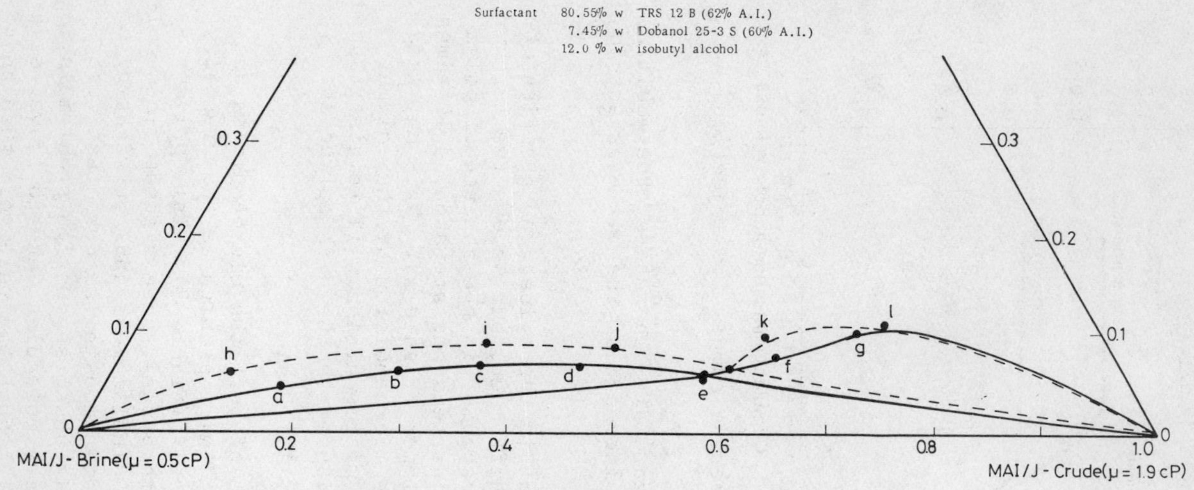

Fig. 3. Phase diagram, viscosities (μ) and interfacial tensions (γ) for the system: brine-surfactant-crude at 62°C. —— drawn curves: no polymer in brine; ---- dashed curves: brine contains 1200 ppm polymer (μ = 11.2 cP).

In the absence of polymer:

	γ(μ N/m)	$\mu_{microemulsion}$ (cP)
a	0.2	6
b	1.0	7
c	0.8	6
d	1.3	6
e	0.5/0.6	6
f	0.6	7
g	0.8	9

In the presence of polymer:

	γ(μ N/m)	$\mu_{microem.}$	μ_{brine} (cP)
h	2.7	23	32
i	3.3	18	21
j	3.5	14	21
k	0.4	7	49
l	1.0	9	--

Note: μ N/m = 0.001 dyne/cm.

In the right-hand part of the ternary diagram, where the microemulsion is in equilibrium with oil, the polymer is found as a hydrated mass on the bottom of the test tube, in fact as a third phase. The effects of increasing amounts of polymer on the surfactant-brine-oil phase behavior have been systematically studied with polymer concentrations up to 4000 ppm. Figure 4[1] shows the continuous shifting of the invariant point M to the right with increasing polymer concentration.

In general, it is observed that polymer will predominantly, if not totally, remain in an aqueous phase, if necessary extracting water from a microemulsion phase. Thus, the M point shifts to compositions with a lower water content and the left (two-phase) lobe in the ternary diagram is raised to higher surfactant concentrations, the surfactant being expelled from the polymer-containing phase and water being transferred from the microemulsion into the polymer-containing phase. Therefore, the aqueous phase and the microemulsion become more dissimilar in composition, resulting in higher interfacial tensions. This incompatibility of a polymer solution and an active surfactant system in which presumably very large micellar associates are present is a special case of the general phenomenon that two chemically different polymers dissolved in a single solvent usually show phase separation.

Several examples of polymer-polymer incompatibility in aqueous solution are given in Reference 3, whereas de Hek and Vrij (4) recently described a phase separation in a solvent in which both polymer and colloidal spheres were dissolved. Phase separation can be suppressed by reducing the molecular weight of the polymers. Reduction of the salinity reduces the size of the surfactant micelles and indeed also the polymer-surfactant incompatibilities (5,6). Actually, reduction of the salinity of the polymer drive, even without direct reference to polymer/surfactant incompatibilities, has recently become a favorable recipe for successful micellar floods (7-9).

A number of surfactant floods (experiments 4-6) were therefore run in cores of 0.9-1.5 m length in which the salinity of the polymer drive was only 20% of the salinity of the preceding water drive

---

[1] The phase behavior studies reported in Figure 4 were performed in a brine of similar salinity as in Figure 3, but containing additives to ensure long-term chemical stability of the polymer; among these additives was 0.4% isopropanol. These additives have a marginal effect on phase behavior and the same increase in microemulsion/aqueous phase interfacial tension has been observed. However, we did not find the drastic increases in viscosity reported in Figure 3.

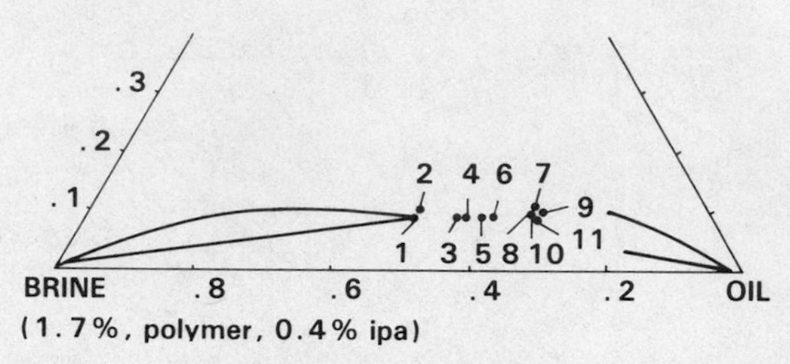

Fig. 4. Shift of 'invariant' point M with increasing polymer concentration.

and surfactant slug. At this salinity no polymer/surfactant phase separation was observed. This low salinity of the polymer drive caused a reduction of the salinity in the transition zone between the surfactant slug and polymer drive and therefore of the detrimental polymer/surfactant incompatibilities in this zone. Recovery of residual oil in these experiments (Table 1) was very good (85-90% of S_{or}), leaving about 4-5% of residual oil in the core after the chemical flood (S_{orc}). In addition, the surfactant retention in the core was considerably lower (Table 1). Apparently, the salinity reduction of the polymer drive effectively[2] suppressed the undesirable incompatibilities between surfactant slug and polymer drive.

---

[2] The peak in the viscosity of the aqueous phase following the oil bank shown in Figure 1 could be caused by either

a. remnants of polymer/surfactant incompatibilities not strong enough to influence oil recovery or surfactant retention and/or
b. inaccessible pore volume effects; polymer molecules proceeding faster through the core than the brine they are dissolved in cannot easily enter the microemulsion zone and therefore concentrate just behind this zone.

Similar but smaller peaks have been observed in experiments 4 and 6. These peaks are especially sensitive to alcohol and salinity.

ACKNOWLEDGMENTS

The authors wish to thank Messrs. G.A.E. Nuis, J.G. Meyers and P. Binda for performing the experiments described in this paper.

REFERENCES

1. R. N. Reed and R. N. Healy, in "Improved Oil Recovery by Surfactant and Polymer Flooding," D. O. Shah and R. S. Schechter, eds., Academic Press, New York, p. 383, 1977.
2. F. Th. Hesselink, Symposium on Enhanced Oil Recovery by Displacement with Saline Solutions, London, May 1977.
3. P. A. Albertsson, in "Partition of Cell Particles and Macromolecules," Almquist & Wiksell, Stockholm, 1971.
4. H. de Hek and A. Vrij, J. Colloid Interf. Sci., 70, 592 (1979).
5. S. P. Trushenski, in "Improved Oil Recovery by Surfactant and Polymer Flooding," D. O. Shah and R. S. Schechter, eds., Academic Press, New York, p. 555, 1977.
6. M. T. Szabo, Soc. Pet. Eng. J., 19, 4 (1979).
7. R. C. Nelson and G. A. Pope, Soc. Pet. Eng. J., 18, 325 (1978).
8. S. P. Gupta and S. P. Trushenski, Soc. Pet. Eng. J., 19, 116 (1979).
9. C. J. Glover, M. C. Puerto, J. M. Maerker, and E. L. Sandvik, Soc. Pet. Eng. J., 19, 183 (1979).

INDEX